P9-CKX-763

PROGRESS IN BRAIN RESEARCH

VOLUME 66

PEPTIDES AND NEUROLOGICAL DISEASE

Recent volumes in PROGRESS IN BRAIN RESEARCH

PROGRESS IN BRAIN RESEARCH

VOLUME 66

PEPTIDES AND NEUROLOGICAL DISEASE

EDITED BY

P. C. EMSON

MRC Group, Institute of Animal Physiology, Babraham, Cambridge CB2 4AT, UK

M. ROSSOR

University Department of Clinical Neurology, The National Hospital, Queen Square, London WC1N 3BG, UK

and

M. TOHYAMA

Department of Neuroanatomy, Institute of Higher Nervous Activity, Osaka University Medical School, 4-3-57 Nakanoshima, Kitaku, Osaka 530, Japan

ELSEVIER
AMSTERDAM – NEW YORK – OXFORD
1986

ISBN 0-444-80733-0 (volume)
ISBN 0-444-80104-9 (series)

Published by:
Elsevier Science Publishers B.V. (Biomedical Division)
P.O. Box 211
1000 AE Amsterdam
The Netherlands

Sole distributors for the USA and Canada:
Elsevier Science Publishing Company, Inc.
52 Vanderbilt Avenue
New York, NY 10017
USA

Library of Congress Cataloging in Publication Data

Main entry under title:

Peptides and neurological disease.

(Progress in brain research ; v. 66)
Includes bibliographies and index.
1. Neuropeptides--Testing. 2. Neuropeptides--Physiological effect. 3. Central nervous system--Diseases. I. Emson, P. C. II. Rossor, M. (Martin) III. Tohyama, M. (Masaya) IV. Series. [DNLM: 1. Central Nervous System--physiology. 2. Central Nervous System Diseases--pathology. 3. Nerve Tissue Proteins--metabolism. 4. Peptides--metabolism. 5. Peripheral Nerves--physiology. W1 PR667J v.66 / QU 68 P4242]
QP376..P7 vol. 66 [RC349.8] 612′.82 s 86-2165
ISBN 0-444-80733-0 (U.S.) [616.8]

Printed in The Netherlands

List of Contributors

Y. Agid, Laboratoire de Médecine Expérimentale - Unite 289 de l'INSERM, 91, Bd de l'Hôpital, 75634 Paris Cedex 13, France
L. F. Agnati, Department of Human Physiology, University of Modena, Modena, Italy
K. Andersson, Department of Histology, Karolinska Institutet, Stockholm, Sweden
N. Battistini, Department of Human Physiology, University of Modena, Modena, Italy
F. Benfenati, Department of Human Physiology, University of Modena, Modena, Italy
G. J. Bennett, Neurobiology and Anesthesiology Branch, National Institute of Dental Research, National Institutes of Health, Bethesda, MD 20892, USA
G. Bissette, Department of Psychiatry and the Center of Aging and Human Development, Duke University Medical Center, Durham, NC 27710, USA
S. J. Capper, Pain Relief Foundation, Walton Hospital, Rice Lane, Liverpool L9 1AE, England, UK
F. Cesselin, Laboratoire de Biochimie Médicale, CHU Pitié-Salpêtrière, 91, Bd de l'Hôpital, 75634 Paris Cedex 13, France
D. Dawbarn, MRC Neurochemical Pharmacology Unit, Hills Road, Cambridge, England, UK
R. Dubner, Neurobiology and Anesthesiology Branch, National Institute of Dental Research, National Institutes of Health, Bethesda, MD 20892, USA
P. C. Emson, MRC Group, Institute of Animal Physiology, Babraham, Cambridge CB2 4AT, England, UK
J. Epelbaum, Unité 159 de l'INSERM, 2 ter, rue d'Alésia, 75014 Paris, France
K. Fuxe, Department of Histology, Karolinska Institutet, Stockholm, Sweden
M. Goldstein, Department of Psychiatry, New York University Medical Center, New York, NY, USA
A. Härfstrand, Department of Histology, Karolinska Institutet, Stockholm, Sweden
T. Hökfelt, Department of Histology, Karolinska Institutet, Stockholm, Sweden
S. Hosokawa, Department of Neurophysiology, Neurological Institute, Faculty of Medicine, Kyushu University, Fukuoka 812, Japan
S. Inagaki, Third Department of Internal Medicine, Hiroshima University Medical School, 1-2-3 Kasumi, Minamiku, Hiroshima, 734 Japan
F. Javoy-Agid, Laboratoire de Médecine Expérimentale – Unite 289 de l'INSERM, 91, Bd de l'Hôpital, 75634 Paris Cedex 13, France
M. Kalia, Department of Neurosurgery, Tomas Jeffersson University, Philadelphia, PA, USA
M. Kato, Department of Neurophysiology, Neurological Institute, Faculty of Medicine, Kyushu University, Fukuoka 812, Japan
S. Kito, Third Department of Internal Medicine, Hiroshima University Medical School, 1-2-3 Kasumi, Minamiku, Hiroshima, 734 Japan
A. V. P. MacKay, Department of Psychological Medicine, University of Glasgow, Glasgow, Scotland, UK
C. D. Marsden, University Department of Neurology, Institute of Psychiatry and King's College Hospital Medical School, De Crespigny Park, London SE5 8AF, England, UK
F. Mascagni, Department of Histology, Karolinska Institutet, Stockholm, Sweden
J. S. Morley, Pain Relief Foundation, Walton Hospital, Rice Lane, Liverpool L9 1AE, England, UK
C. Q. Mountjoy, Department of Psychiatry, Addenbrookes Hospital, Cambridge, England, UK
C. B. Nemeroff, Departments of Psychiatry and Pharmacology and the Center of Aging and Human Development, Duke University Medical Center, Durham, NC 27710, USA
J. G. Parnavelas, Department of Anatomy and Embryology, University College London, London WC1E 6BT, England, UK
F. Plum, Department of Neurology, New York Hospital–Cornell Medical Center, New York, NY 10021, USA
M. N. Rossor, National Hospital, Queen Square, London WC1N 3BG, England, UK
M. Roth, Department of Psychiatry, Addenbrookes Hospital, Cambridge, England, UK

M. A. Ruda, Neurobiology and Anesthesiology Branch, National Institute of Dental Research, National Institutes of Health, Bethesda, MD 20892, USA

S. Shiosaka, Department of Neuroanatomy, Institute of Higher Nervous Activity, Osaka University Medical School, 4-3-57 Nakanoshima, Kitaku, Osaka 530, Japan

Y. Shiotani, Department of Neuroanatomy, Institute of Higher Nervous Activity, Osaka University Medical School, 4-3-57 Nakanoshima, Kitaku, Osaka 530, Japan

H. Takagi, 2nd Department of Anatomy, Kinki University School of Medicine, Sayama-cho, Minami-Kawachi-gun, Osaka 589, Japan

H. Taquet, Laboratoire de Biochimie Médicale, CHU Pitié-Salpêtrière, 91, Bd de l'Hôpital, 75634 Paris Cedex 13, France

M. Tohyama, Department of Neuroanatomy, Institute of Higher Nervous Activity, Osaka University Medical School, 4-3-57 Nakanoshima, Kitaku, Osaka 530, Japan

M. Zoli, Department of Human Physiology, University of Modena, Modena, Italy

Preface

During the last two decades there has been a dramatic increase in studies concerned with the measurement of neurotransmitters in neurological and psychiatric illness. Much of this interest stems from the pioneering studies of Horynkiewicz and colleagues of the dopamine deficit in Parkinson's disease and its amelioration with large doses of L-Dopa, which rapidly led to the investigation of the chemical pathology of Alzheimer's disease, Huntington's disease and schizophrenia. Outstanding in these studies were the observations of Bird and Iversen and the Perry's of the GABA deficit in Huntington's chorea, and the discovery by several groups of the cholinergic deficit in Alzheimer's disease.

In parallel with these findings of classical neurotransmitter deficits in neurological illness the 1970s also saw the peptide explosion with the discovery of a variety of "neuropeptides" localised to specific cells with a variety of hormonal, local or neurotransmitter actions. Surprisingly, but fortuitously, these studies soon revealed that many neuropeptides were remarkably stable post mortem. This post mortem stability of small peptides is still unexplained but does allow the neuroscientist to use the sensitive techniques of immunoassay or immunohistochemistry to investigate the organisation of the human brain in a way that was not previously possible.

A symposium was held in Cambridge in 1983 to review the possibilities of using neuropeptides as markers in neurological and psychiatric disease. To ensure a breadth of coverage an entire session was devoted to the organisation of the spinal cord since, although little was known about cord disease, this may provide clues to the treatment of chronic pain and the design of novel analgesics. For this reason the theoretical design of peptide antagonists was also reviewed.

Following this meeting it was considered that a book reviewing this area would be timely and the format of reviewing cerebral cortex, basal ganglia and spinal cord should be followed.

A number of chapters in this book are from the original contributors to the meeting. However, we took the opportunity in assembling the book to include some chapters which complement and extend the original contributions. These include a review of modern immunocytochemical techniques and a chapter on the physiological role of peptides, in this case in the basal ganglia.

Our knowledge of the physiological role of most neuropeptides is still "primitive" as emphasised by Hosokawa and Kito but it is difficult to believe that the specific and selective distribution of neuropeptides is not important. Hopefully studies of the longer term effects of peptides on transmitter turnover and the significance of the ubiquitous co-existence of peptide and short acting or classical transmitters, e.g. GABA and acetylcholine will provide important clues that may lead to the design of peptide-directed drugs for therapy of neurological illness. It was also our hope that in broadening the scope of this volume we might attract a wider audience of neuroscientists and interest them in studies of neurological and psychiatric illness.

This volume and the symposium it arose from was generously supported by: Fidia Research Laboratories, Sandoz Ltd., The Wellcome Foundation Ltd., Du Pont de Nemours and Co., Ciba-Geigy, Astra Pharmaceuticals Ltd., Imperial Chemical Industries Ltd., Smith Kline and French Laboratories Ltd., Sterling-Winthrop Group Ltd., American Cyanamid Company, The Boots Company Ltd., Merck, Sharp and Dohme Ltd., Roche Products Ltd., Pfizer Central Research, Johnson and Johnson, Beecham Pharmaceuticals, Parke Davis Research Group, The Wellcome Trust, The Parkinson's Disease Society.

We also gratefully acknowledge the help of many members of the former MRC Neurochemical Pharmacology Unit in the organisation of the meeting, especially Mrs. M. Wynn, Mrs. S. West, Ms. K. Pittaway and our co-organiser Dr. G. Reynolds. Finally neither the meeting nor this volume would have been possible without the help and encouragement of our wives and families.

P. C. Emson
M. N. Rossor
M. Tohyama

Frequently Occurring Abbreviations

-LI, -like immunoreactivity
-LIr, -like immunoreactive

ACE, angiotensin converting enzyme
ACH, acetylcholine
ACTH, adrenocorticotropin
APP, avian pancreatic polypeptide
AII, angiotensin II

CA, catecholamine
CCK, cholecystokinin
CCK-8, cholecystokinin octapeptide
CGRP, calcitonin gene-related peptide
ChAT, choline acetyltransferase
CNS, central nervous system
CRF, corticotropin releasing factor
CSF, cerebrospinal fluid

DA, dopamine
DYN, dynorphin

-END, -endorphin
-ENK, -enkephalin

FITC, fluorescein isothiocyanate

GABA, gamma aminobutyric acid

5-HIAA, 5-hydroxyindoleacetic acid
HPLC, high pressure liquid chromatography
HRP, horseradish peroxidase
5-HT, serotonin

LEU-ENK, leucine-enkephalin

MET-ENK, methionine-enkephalin
MSH, melanocyte stimulating hormone

NA, noradrenaline
NPY, neuropeptide Y
NT, neurotensin

5-OHDA, 5-hydroxydopamine
OT, oxytocin

PBS, phosphate-buffered saline

RIA, radioimmunoassay

SP, substance P
SRIF, somatostatin
TH, tyrosine hydroxylase
TRH, thyrotropin releasing hormone

VIP, vasoactive intestinal polypeptide
VP, vasopressin

WGA, wheat germ agglutinin

Contents

SECTION I

Introduction

P. C. Emson, M. N. Rossor and M. Tohyama (Eds.),
Progress in Brain Research, Vol. 66.

CHAPTER 1

Immunohistochemical techniques

S. Shiosaka and M. Tohyama

Department of Neuroanatomy, Institute of Higher Nervous Activity, Osaka University Medical School, Osaka, Japan

Introduction

The introduction of immunological techniques to morphology has enabled the identification of specific antigens such as enzymes or structural proteins in various tissues (Coons and Kaplan, 1950; Coons, 1958; Nakane and Pierce, 1966). Antisera against a number of neuromodulators or neurotransmitter substances such as neuropeptides, amines, acetylcholine and amino acids or their marker enzymes have been produced in the past two or three decades (for review see Emson, 1983). Using light microscopic immunocytochemical techniques, the distribution in the nervous system of these "neurotransmitters" has been explored. More recently, immuno-electronmicroscopic techniques have been improved and the morphology of the structures containing neuromodulators or neurotransmitters has been studied in more detail (Sternberger, 1967; Pickel et al., 1975; DeMay et al., 1981; Varndell et al., 1982). Much recent work has focused on the elucidation of the fibre pathways of neurotransmitter-defined neurons (Emson, 1983). Most of these studies have been based upon lesion experiments, but, in order to identify the fibre pathways precisely, more direct proof using a combination of immunocytochemistry and retrograde tracer technique is required. Two approaches have been used (Van der Kooy and Steinbusch, 1980; Van der Kooy and Wise, 1980; Bowker et al., 1981, 1982; Priestley et al., 1981; Sawchenko and Swanson, 1981). The first is to employ horseradish peroxidase (HRP) as a tracer and the second is to use fluorescent dyes. Both methodologies have disadvantages which will be described in detail below. We have recently succeeded in developing a highly sensitive combination method, using biotinized wheat-germ agglutinin (WGA) and HRP as a retrograde tracer marker (Shiosaka and Tohyama, 1984; Shiosaka et al., 1984, 1985). The success of this combined technique for electronmicroscopy has enabled us to identify the neurons to which the neuromodulator or neurotransmitter system projects and to determine the origin of the terminals that end on the immunoreactive cells. In this chapter, we will first describe these techniques in detail.

Another important problem concerns the now ubiquitous coexistence of multiple neuromodulators or neurotransmitter substances in single neurons (Hökfelt et al., 1977, 1984). This "coexistence" has been demonstrated by detecting different antigens in consecutive sections. However, direct analysis of different antigens on the same section would make it easier to demonstrate such coexistence. In addition, if such a technique is applied at the ultrastructural level, it will enable the identification of possible sites of interactions between two different neurotransmitter systems. Several groups, including ours, have developed double-staining methods for this purpose (Vandesande and Dierickx, 1975; Tramu et al., 1978; Lechago et al., 1979; Larsson, 1983; Hisano et al., 1984; Katayama-Kumoi et al., 1985; Lee et al., 1985). This chapter will therefore be devoted mainly to these aspects and the conventional techniques of immunocytochemistry, which have been described in detail by many reviewers, will be summarized only briefly.

Fundamental techniques of immunocytochemistry

Light microscopy

Pretreatment of animals

To enhance neuropeptide immunoreactivity in the cell soma, animals are injected with 5 μl/100 g body weight of colchicine solution (10 mg/ml) intraventricularly. Colchicine is known to inhibit fast axonal transport and has been used to increase the cell body content of a number of neurotransmitter or neuromodulator substances (Dahlström, 1968; Kreutzberg, 1969; Hökfelt and Dahlström, 1971; Ljungdahl et al., 1978). Usually, the agent is injected 12–24 h before the animal is anaesthetized and killed.

Fixatives

To detect a whole range of neuropeptides, we have routinely used the fixative Zamboni's fluid (Zamboni and De Martino, 1967) which contains 200 ml of 10% aqueous solution of paraformaldehyde (final conc. 2%), 2.0 g picric acid (final conc. 0.2%), and 500 ml of 0.2 M sodium phosphate buffer (pH 7.4; final conc. 0.1 M) per 1000 ml. A paraformaldehyde (4%)–phosphate buffer (0.1 M) solution is also commonly used when performing immunocytochemistry of neuropeptides and enzymes (Pearse, 1962; Hökfelt et al., 1975). A few immunocytochemical studies have also used a parabenzoquinone and formalin mixture (Pearse and Polak, 1975; Larrson, 1977), acrolein (King et al., 1983), Bouin's fixative, and periodic acid–lysine paraformaldehyde fixative (McLean and Nakane, 1974). These fixatives should be carefully selected at the first stage of the experiment, because immunoreactivity is significantly influenced by the fixative. In our experience, Zamboni's fixative is generally satisfactory for staining most neuropeptides and other antigens in the brain.

Tissue preparation

Animals are perfused via the heart or ascending aorta and postfixed for 1–3 days with an appropriate fixative (Fig. 1). Perfusion is usually carried out using a roller pump with 300–400 ml fixative/100 g body weight during a 30 min period. Then, the tissues are removed and cut into blocks of approximately 3–5 mm thickness. Tissues obtained by biopsy or tissues that cannot be fixed by perfusion are cut into small pieces and immersed in a fixative a little longer (approximately a week in the Zamboni's fixative for 3 × 3 × 3 mm blocks) than in the case of perfused tissue. Then the tissues are immersed in sodium phosphate buffer containing 30% sucrose and kept for 1–2 days in a refrigerator until the tissue is completely permeated. The tissue can be immersed up to one week before use, but a longer immersion in the sucrose buffer may cause a non-specific background staining. For long-term storage in sucrose, sodium azide (0.01%) should be used as a preservative.

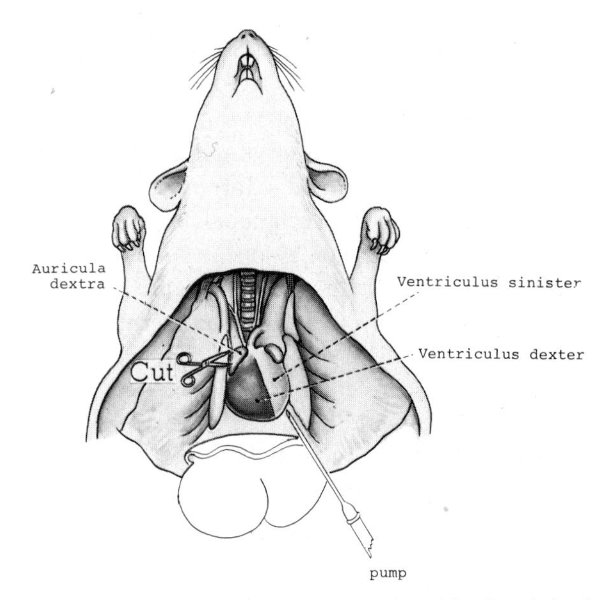

Fig. 1. Schematic illustration shows practice of fixation. Animal is anaesthetised with pentobarbital (Nembutal, 60 mg/kg body weight), and perfused from heart or ascending aorta with 50 ml of saline and followed by fixative. To restrict the flow of fixative to the upper body, the descending aorta is occasionally clamped.

Cryostat sectioning

The tissue blocks are rapidly frozen by dipping in solid CO_2 or immersing them in solid acetone. If the tissue is not frozen quickly, the sensitivity of immunostaining may be decreased. The frozen blocks are then sliced using a microtome in a cryostat set at −15 to −23°C. Sections can then be gently cut on a microtome. For subsequent processing of sections by the floating method a section thickness of 20–50 μm is recommended; for staining on the slide thinner sections of 5–10 μm are used. The gelatin-coated slide can be prepared as described in Appendix 1.

Flat-mount

Because it is difficult to determine the overall distribution of the immunoreactive cells and three-dimensional profiles from the frozen sections, the "flat mount" technique is often used.

Gastrointestinal tract and bladder (Costa et al., 1980). Animals are perfused with ice-cold saline. After perfusion, the organs are removed, opened, then pinned and spread out on a strip of balsa. Following fixation in Zamboni's fixative at 4°C for three days, the tissues are rinsed in 80% ethanol to remove picric acid. The tissues are dehydrated in 96% and 100% ethanol successively, and then in xylol (30 min each). They are then rehydrated successively in 100%, 80%, and 50% ethanol (30 min each) down to phosphate-buffered saline (PBS). In the PBS, the mucosal layer is separated from the muscle layer. The muscle layer can then be subjected to immunocytochemistry to demonstrate the relevant antigen. To enhance the penetration of the antiserum, the specimens are often frozen in liquid nitrogen for a few seconds and thawed at the stage either before dehydration or before subjecting the tissue to immunocytochemistry.

Retina (Ishimoto et al., 1982). The anaesthetized animals are perfused via the ascending aorta with 30 ml of ice-cold saline and fixed in Zamboni's fixative. After perfusion, the eyes are removed and put in the same fixative and the pigment epithelium is separated from the retina under an operating microscope. The isolated retina is postfixed in the same fixative for 1–2 days and then rinsed in 0.1 M phosphate buffer containing 30% sucrose. The isolated retina is then subjected to immunocytochemistry.

Blood vessels (Yamamoto et al., 1983). After perfusion of Zamboni's solution, the blood vessels are dissected out rapidly, postfixed in the same fixative overnight at 4°C and rinsed in 0.1 M phosphate buffer containing 30% sucrose for 1–2 days, and subjected to immunocytochemistry.

Indirect immunofluorescent method (Coons, 1958)

The method was devised by Coons in 1951, and a number of investigators are now using this technique because of its high sensitivity and stability. The principle of this technique is illustrated in Fig. 2a. The desired antigen for immunostaining is shown by the solid squares. The primary antiserum recognizes and binds to the antigen. If an antibody is used which is labeled by a marker, such as fluorescein isothiocyanate (FITC), rhodamine isothiocyanate (Riggs et al., 1958), Texas red (Titus et al., 1982) or HRP (Nakane and Pierce, 1966) then the method is termed the "direct antibody" method. On the other hand, if the primary antibody is not directly labeled but is labeled via a second antibody, which is made against the IgG of the same species as the primary antibody was raised in, the system is called the "indirect antibody" method. Because of its high sensitivity, the indirect immunofluorescent method using FITC as a fluorescent marker has been used most widely for the detection of neuroactive substances in the brain.

Procedure (Fig. 3). The fixative in the tissue sections is washed out by rinsing the sections in ice-cold PBS for at least 10 min. The sections are then incubated in PBS with the first antiserum which has been produced by injecting the appropriate antigen-conjugate into, for example, a rabbit (antiserum dilutions vary with the antiserum used, a good antiserum should be useable at 1:1000–5000)

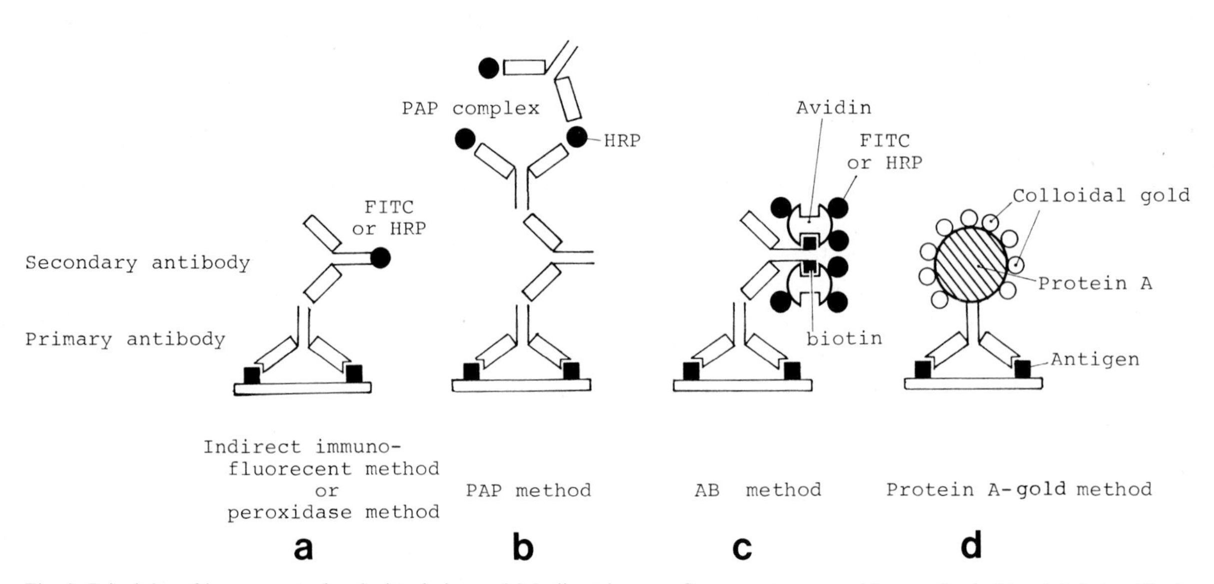

Fig. 2. Principles of immunocytochemical techniques. (a) Indirect immunofluorescent or peroxidase method; (b) unlabeled antibody peroxidase anti-peroxidase (PAP) method; (c) avidin– or streptavidin–biotin method; (d) protein A–gold method. Antigens are shown by the solid squares. Primary antibody reacts with antigens and secondary antibody which was made by injecting the same sort of immunoglobulin with primary antibody reacts with primary antibody. In the PAP method, secondary antibody bridges the primary antibody and PAP complex.

for 24 h at room temperature (or longer at 4°C). Addition of 0.1–0.3% Triton X-100 to the PBS facilitates permeation of antibodies through the cell membranes of the tissues. The tissue sections are then rinsed three times in PBS for 10–20 min each. They are then incubated with PBS solution containing the appropriate labeled secondary antiserum for 24 h at room temperature. After rinsing the sections three times in PBS for 10–20 min each, they are then mounted onto slideglass with one drop of glycerin/PBS (1:1, v/v) solution and covered by a coverslip.

The sections can then be examined under a fluorescent microscope equipped with the relevant filters, in this case a 495 μm filter for excitation and a 520 μm absorption filter for FITC.

The unlabeled antibody peroxidase anti-peroxidase (PAP) method (Sternberger, 1970)

The method was developed by Sternberger in 1970. Nakane and a colleague, however, first used the technique by which enzymes such as peroxidase can be used as markers of immunoreaction (immunoperoxidase method) (Nakane and Pierce, 1966, 1967). The merits of using HRP as a marker are: (1) the oxidised product of 3,3′-diaminobenzidine is stable and does not fade over long periods; (2) the visualization procedure is simple; (3) the product of 3,3′-diaminobenzidine is visible under an electronmicroscope; (4) it diffuses less from its site of formation than other substrates because the polymerized product is highly insoluble; and (5) its sensitivity is high. However, the immunoperoxidase method has been reported to have some weaknesses; these include possible impairment of antibodies by chemical alterations at conjugating antibodies with HRP, non-reacted unlabeled antibodies may interfere with immunocytochemical localization, and a dense background may originate from nonspecific labeling of tissue sections by the primary antiserum. When the incubation time is increased the background often becomes darker with specifically localized material. Moreover, many enzyme–antibody conjugates may be composed of large marker and immunoglobulin aggregates resulting in a lower tissue penetrability of the aggre-

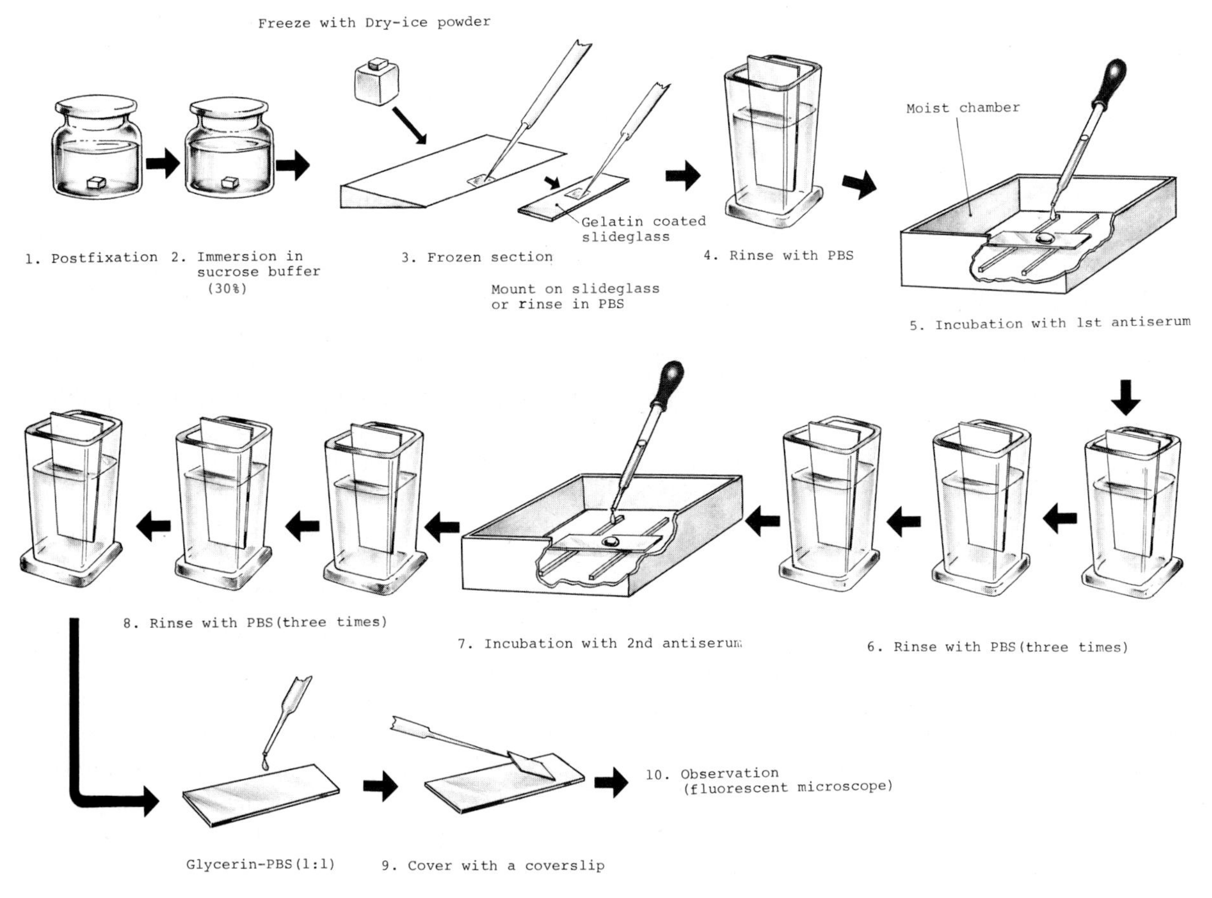

Fig. 3. Schematic illustrations showing the practice of the immunofluorescent technique.

gate (Sternberger, 1979). In order to overcome these weaknesses, the unlabeled antibody method was devised (Petrutz et al., 1975; Sternberger et al., 1970). Fig. 2b shows a schematic drawing of the principle of the unlabeled PAP method which is most popularly used. In this method, an unlabeled antibody is used at the second step. One of the combining sites of the second antibody reacts with the first antibody and another combining site remains free. The last step is accomplished with the PAP complex. An anti-peroxidase antibody is raised in the same animal as the first antibody. Since these (first antibody and anti-peroxidase) are the same immunoglobulins, the second antibody can couple them with its two binding sites. The HRP previously combined with anti-peroxidase is visualized with oxidative polymerization of 3,3′-diaminobenzidine, which results in a brown precipitate. The brown precipitate can be made into a purple to black precipitate of nickel (see Appendix 2 for details).

Procedure. Tissue sections prepared by cryostat sectioning are first incubated with 20% normal goat serum for 30 min to eliminate the nonspecific binding of antibodies to the surface of the sections, and then washed with PBS for 20 min at 4°C and incubated with a primary antibody (usually raised in the rabbit). The first antibody is diluted 1:2000–10 000 with PBS and the incubation period commonly used is 1–3 days at room temperature or 4°C. Sections are then washed three times with PBS for 20 min at 4°C, and incubated overnight at room temperature with an unlabeled second antibody (goat anti-rabbit IgG serum), diluted 1:100 with PBS. They are washed three times with PBS for 20 min each and incubated for 3–4 h at room temperature with a PAP complex (rabbit PAP complex) diluted 1:200 with PBS. They are again washed three times with PBS for 20 min each, and incubated first for 20 min in a solution of 0.05% 3,3′-diaminobenzidine–tetrahydrochloride (Dotite Chemical, Japan or Sigma, St. Louis, MO) dissolved in 0.05 M Tris-HCl buffer (pH 7.4), and then for 5–20 min in the same solution containing 0.01% hydrogen peroxide. Extreme care and the appropriate precautions should be exercised with the use of 3,3′-diaminobenzidene tetrahydrochloride as this is a potent carcinogen. Finally, they are rinsed in a Tris-HCl buffer (and mounted onto gelatin-coated slides if they have been prepared by the free-floating method). Light counterstaining with cresyl violet facilitates exact identification of brain regions. Sections are then dehydrated and mounted in balsam. Slides can be examined under a light microscope (Fig. 4). Differential interference microscopy can also be used for observation.

Streptavidin–biotin techniques or avidin–biotin technique (Hsu and Reine, 1981; Hsu et al., 1981)

The unique feature of the biotin–avidin system is the very high affinity interaction ($K_d = 10^{-15}$ M^{-1}) between biotin and the egg white protein, avidin (Green, 1963). The biotin–avidin bond is stronger than the antigen–antibody linkage ($K_d = 10^{-5}$–10^{-9} M^{-1}), and the tight coupling of the system enables sensitive detection with only a short incubation time whilst providing a very high sensitivity. Unfortunately, however, egg protein avidin is a glycoprotein and will react with other biological molecules, resulting in non-specific background (Green and Toms, 1970). Furthermore, it is highly positively charged at a neutral pH so that its high isoelectric point may cause non-specific binding to the negatively charged molecule (Hofmann, 1980). Streptavidin (a bacterial avidin) does not have the undesirable characteristics of avidin, and is easier to use. The typical application to immunocytochemistry of streptavidin or avidin is to use a biotinylated second antibody coupled with the first antibody which is reacted with the appropriate antigen. The free biotin molecule is linked with a streptavidin-conjugated marker substance, such as HRP, at the next step. The next reaction is then a diaminobenzidine reaction. The last step may be replaced by a pre-formed streptavidin–biotinylated peroxidase complex, providing higher sensitivity (Hsu et al., 1981). Fig. 2c shows the principle of this technique.

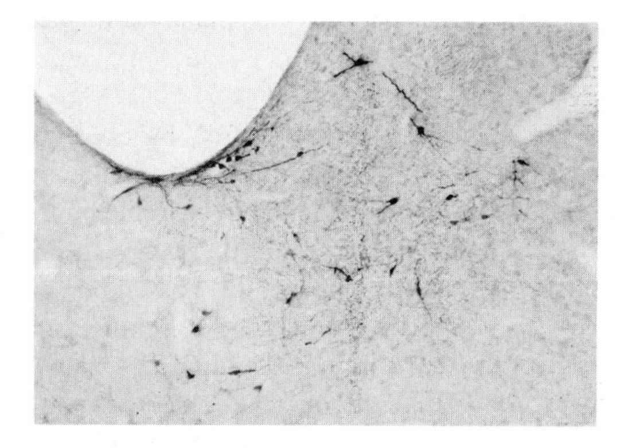

Fig. 4. Bright-field micrograph showing an example of immunostaining by PAP technique. Neuropeptide Y-like immunoreactive neurons in medial subnucleus of nucleus tractus solitarii were found. (From Yamazoe et al., 1985.)

Electronmicroscopic observation of immunoreactive structures in nervous tissue

Electronmicroscopic markers of immunoreaction

Observation of the immunoreactive structures under the electronmicroscope has predominantly used 3,3′-diaminobenzidine as the substrate for HRP. This is the commonest substrate used for detection of immunoreactivity by both the light microscope and electronmicroscope, and is widely used with the immunoperoxidase and PAP methods (Graham and Karnovsky, 1966; Nakane and Pierce, 1966). Electronmicroscopic observations of the oxidized 3,3′-diaminobenzidine precipitate on tissue sections were carried out in the early stages of immunocytochemistry (Nakane and Pierce, 1966; Sternberger et al., 1967). At present, a number of investigators use this material in spite of its carcinogenicity (Mesulam, 1978). The alternative non-carcinogenic marker, ferritin, has also been widely used (for review see Sternberger, 1979). However, since this molecule is very small and the electron density observed with an electronmicroscope is not so good, observations are made more difficult although ferritin is excellent for cell surface antigens (Bauer et al., 1975; Roth and Bender, 1978) (see also Fig. 14). Recently, colloidal gold, which has a high electron density and easily links with protein molecules such as immunoglobulins, has been reported to be a useful reagent for electronmicroscopic detection of immunoreactive substances in tissue (Faulk and Taylor, 1971; Romano et al., 1974; Roth et al., 1978). The material is easily combined with immunoglobulins (immunogold technique), Protein A, which has the ability to bind tightly to Fc fragments of immunoglobulins (Forsgreen and Sjöquist, 1966, 1967) (protein A–gold technique), and avidin and streptavidin (avidin biotin–gold technique).

Preembedding method

Originally, the PAP method was restricted to postembedding methods in electronmicroscopy, with the immunohistochemical manipulations being carried out on the plastic-embedded and ultrathin-sectioned tissue. As the PAP complex has a large molecular weight (420 000), the permeability of the complex into a tissue is considered to be low (Sternberger, 1979), although Pickel et al. (1975) proved that immunoreactivity of the antigen could be detected in rather thick sections by the preembedding reaction. This method is most useful because of its high sensitivity arising from omission of resin before staining, as antigenicity is often lost during embedding in the postembedding method. Recently this method has been widely used in neuroscience research. In our laboratory, the correlated light and electronmicroscopic technique of Somogyi and Takagi (1982), which is a modification of Pickel's postembedding technique, is frequently employed. This procedure is described below.

Procedures. Animals are perfused with saline and 300–400 ml of fixative (Eldred et al., 1983). One litre of fixative is made up by mixing 500 ml of 0.2 M sodium phosphate buffer (pH 7.4), 2.0 g of picric acid, 400 ml of 10% paraformaldehyde solution, 2 ml of 25% glutaraldehyde and distilled water. The final pH is 7.2–7.4. Areas of interest are cut into blocks (approximately 2 × 4 × 5 mm) and postfixed with the same fixative for 1 h at 4°C, followed by washing in 0.1 M phosphate buffer (pH 7.4). Then, the tissues are successively immersed in phosphate buffer with 10% sucrose for 1 h, and phosphate buffer with 20% sucrose overnight at 4°C. The blocks are snap frozen in liquid nitrogen and thawed in 0.1 M phosphate buffer at room temperature. This step is most important for permeation of unlabeled immunoglobulins and PAP complex into the tissue with concomitant preservation of fine structural details (Somogyi and Takagi, 1982). Sections are cut on a Vibratome (Oxford) with a section thickness of 50–70 μm, and washed overnight at 4°C in 0.1 M phosphate buffer. Incubation with antisera is carried out as follows: 20% normal goat serum, 20 min at 4°C; first antiserum diluted 1:2000–10 000 with PBS for 1–3 days at 4°C; washed with PBS, 3 × 20 min at 4°C; second antibody (antirabbit IgG when first antiserum was raised in a rabbit) diluted 1:100–200 with PBS, left overnight at 4°C; washed with PBS, 3 × 20 min at 4°C; PAP complex diluted 1:100 with PBS (rabbit), 3 h at room temperature; washed with PBS, 3 × 20 min; 0.05% 3,3′-diaminobenzidine–tetrahydrochloride–Tris–HCl (0.05 M, pH 7.6) solution, 20 min at 4°C; same solution added 0.01% hydrogen peroxide, 5–10 min at room temperature; and finally washed with 0.02 M PBS. The sections are then treated with 1% osmium tetroxide in 0.1 M phosphate buffer for 1 h at room temperature and dehydrated and uranyl-stained in the following order: 50% ethanol, 10 min; 70% ethanol, 10 min; 1% uranyl acetate in 70% ethanol, 40 min; 80% ethanol, 10 min; 90% ethanol, 2 × 5 min; 95% ethanol, 2 × 5 min; 99% ethanol, 2 × 5 min; absolute ethanol, 2 × 5 min; propylene oxide, 2 × 5 min; Epon and propylene oxide (1:1), overnight; and Epon, 1 h. These steps are all carried out at

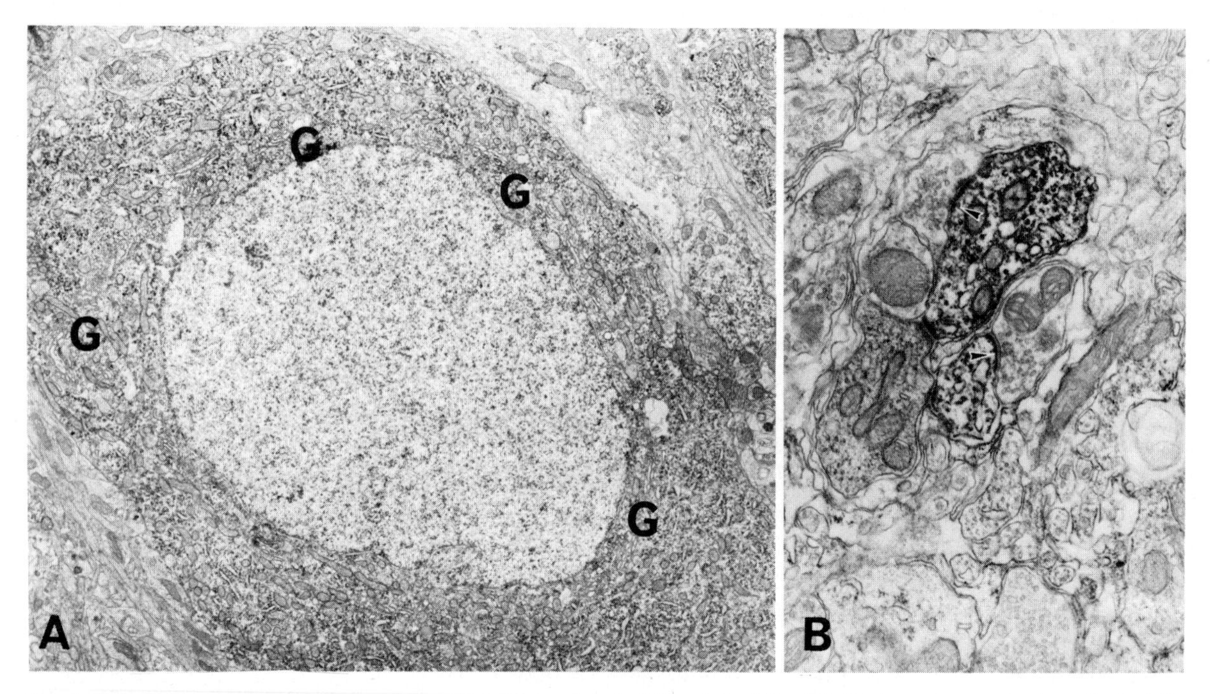

Fig. 5. Electronmicroscopic observation of immunoreactive cell body (A) and dendrites (B). Electronmicrographs show the histidine decarboxylase-like immunoreactive structures in the caudal magnocellular nucleus of rat hypothalamus. Endproducts were found diffusely throughout the perikarya (A) and dendrites (B). Arrowheads show axo-dendritic synaptic contacts between immunoreactive dendrites and nonimmunoreactive axon terminals. (From Hayashi et al., 1984.)

room temperature. Then, the sections are placed on a silicon-coated slide with one drop of Epon under a silicon-coated coverslip and cured for two days at 60°C. The resulting slides are photographed and/or sketched under a light microscope, and the coverslip is removed. A small piece of section is cut out with a surgical knife under an operation microscope and attached to the previously prepared cylinder of Epon. The subsequent steps are the same as ordinary ultramicroscopic analysis. Observation is carried out using an electronmicroscope (Fig. 5A, B). The protocol of this technique is given in Appendix 3.

Postembedding method

In the initial immunoelectronmicroscopic techniques, the postembedding method using the PAP-DAB reaction was employed (Sternberger, 1979). However, it is quite difficult to stain the many antigens by this procedure, because much of the antigenicity is destroyed at the embedding step. Recently, (1) a colloidal gold labeling method which enables easier detection under an ultramicroscope, and (2) a better embedding resin for retainment of antigenicity have been developed. For example, Lowicryl K4M (Polyscience) was recently introduced as an embedding medium for immunocytochemistry for electronmicroscopy (Roth et al., 1981a; Armbruster et al., 1982; Carlemalm et al., 1982) and its polymerization procedure was modified by Altman et al. (1984). The resin can polymerize at a very low temperature (−30 to −50°C), and for this reason, the antigenicity is maintained fairly well (Epon is polymerized at 60°C). These technical developments have enabled the use of the postembedding technique for immunocytochemistry in electronmicroscopy. The advantages of the post-embedding technique for the immunoelectronmicroscope analysis are that: (1) the permeability of immunoglobulins and markers into the tissue section can be ignored, (2) serial ultrathin sections enable easy identification of co-localized antigen, even in small structures such as a nerve terminal, and (3) the fine structure is preserved well. Staining procedures are given below.

Protein A–gold technique (Roth et al., 1978, 1981b)
First of all, the tissue is fixed with an appropriate fixative (see p. 4). Then, the tissues are sectioned into 50–100 μm thickness with a Vibratome, post-fixed, washed in phosphate buffer, dehydrated in graded ethanol series, and finally embedded in Lowicryl K4M. The osmification process is *omitted,* because it seriously destroys the antigenicity. Ultrathin sections are cut and mounted on nickel grids and then processed by the protein A–gold technique (Fig. 2d) (Romano and Romano, 1977; Roth et al., 1978, 1981b).

Procedure. Incubation of sections is carried out in the following order at room temperature unless otherwise stated. The grids are floated, and a droplet of 10% hydrogen peroxide (etching procedure) is dropped onto the sections. Ten to 20 min later, they are washed in PBS with 2% bovine serum albumin for 10 min. The grids are covered with a droplet of a specific antiserum at an appropriate concentration in a humid atmosphere overnight at 4°C, and washed three times for 5 min each change in PBS with bovine serum albumin. The grids are covered with a droplet of protein A–gold complex for 1 h, washed with PBS for 5 min, and stained with uranyl acetate and lead citrate. Protein A reacts and binds strongly with the Fc fragment of IgG molecules from several species (Forsgreen and Sjöquist, 1966, 1967; Kronvall et al., 1970). The procedure for the preparation of protein A–colloidal gold complex is described in Appendix 4 and 5.

Immunogold technique (Varndell et al., 1982, 1983)
The nickel grid-mounted sections of unosmicated tissue are used.

Procedure. Incubations are carried out as follows. The grids are etched in a droplet of 10% hydrogen peroxide for 10–20 min, washed with PBS with normal goat serum for 30 min, incubated with primary antiserum (rabbit) diluted with PBS containing 2% bovine serum albumin for 24 h at 4°C, washed with PBS–bovine serum albumin, incubated in the gold-labeled anti-rabbit IgG (goat) for 12 h at 4°C, washed with PBS–bovine serum albumin, and finally washed with double distilled water. The immunostained sections on grids are stained with uranyl acetate and lead citrate and then looked at by electronmicroscopy.

Combination technique of immunohistochemistry and retrograde tracing

Light microscopy

Investigations of fibre connections provide basic information for resolving the physiological functions of the central and peripheral nervous system. For this purpose, there are several methodological approaches. One is immunohistochemistry combined with experimental lesions. After destruction of the presumed parent cell bodies, or following the transection of their axons, the distribution of immunoreactive fibres in various brain regions is examined. In an area where the immunoreactive nerve terminals are decreased it is assumed that there are connections between the parent cells and that area (Figs. 6A, B and 7). However, the approach cannot deny the possibility of lesions of fibres-of-passage. In addition if there is a commissural connection, it may be impossible to detect any reduction in staining.

Retrograde tracing using HRP, WGA or dyes is a potent technique to identify fibre projections (LaVail and LaVail, 1972; Nauta et al., 1974; Kristenson, 1975; Kuypers et al., 1977). However, retrograde tracing used alone cannot show what bioactive substances the labeled neuron contains. For this purpose, it is necessary to use immunohistochemistry. Up to now, two kinds of methods have been used. One is the combination of the use of a fluorescent dye and immunohistochemistry (Van der Kooy and Steinbusch, 1980; Sawchenko et al.,

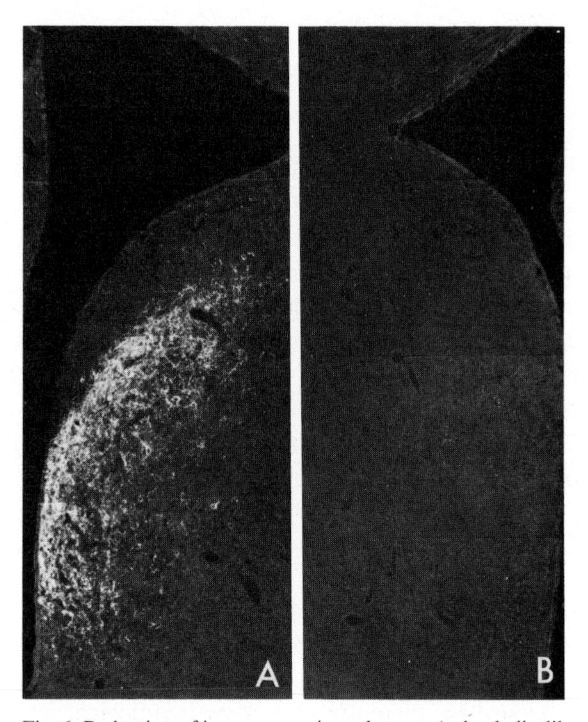

Fig. 6. Reduction of immunoreactive substance (enkephalin-like immunoreactivity) in the lateral septal area resulted from the destruction of the rat hypothalamus. (A) Control side, (B) operated side. Frontal section.

1981; and for review Hökfelt et al., 1983; Skirboll and Hökfelt, 1983; Skirboll et al., 1984), and the other is the combination of the HRP method and immunohistochemistry (Bowker et al., 1981; Priestley et al., 1981). Though these techniques are useful, they have some disadvantages. In the former combination, retrograde tracer dyes such as 4,6-diamidino-2-phenylindole (DAPI), true blue, fast blue or primulin can be used in the immunofluorescent technique of Coons, but the dyes injected into the brain diffuse widely from the injection site, and it is difficult to fix the dyes in the retrograde-labeled soma during the immunoreaction that follows. The detailed procedures concerning this combination technique are described in a recent review by Skirboll et al. (1984). The other combination involves the use of HRP as a tracer and the unlabeled PAP technique for immunohistochemistry. Bowker et al. (1981) used the HRP–diaminobenzidine reaction for a retrograde tracing, in which the diaminobenzidine reaction was blackened by cobalt chloride, and the PAP–diaminobenzidine reaction was visualized as a brown precipitate. These reactions can be observed under the same field of a light microscope, but the black reaction and brown precipitate are difficult to distinguish under a microscope. Moreover, the fixative appropriate for immunohistochemistry tends to inhibit the HRP enzyme activity which is used as an index of the tracer (Smolen et al., 1979), and so the sensitivity of the retrograde HRP technique in this method is less than that of the usual HRP technique. Recently, we have succeeded in developing a new and powerful combination technique involving biotinized protein as the tracer and immunohistochemistry (Fig. 8). This method has the following merits. (1) Biotinized tracer visualized by linkage with streptavidin–Texas red has properties similar to those of HRP or WGA alone (Schwab et al., 1978; Swanson et al., 1978; Gerfen and Sawchenko, 1984; Shiosaka et al., 1984; Shiosaka and Tohyama, 1984; Wilchek and Bayerm, 1984; Shimada et al., 1985). (2) Such linked products have strong red fluorescence and a granular appearance in the soma, and thus they can be readily distinguished from non-labeled background. (3) Since tracers accumulated retrogradely are visualized by the streptavidin–biotin reaction, not by enzyme activity, the immunofluorescence reaction is not interfered with. (4) The high affinity

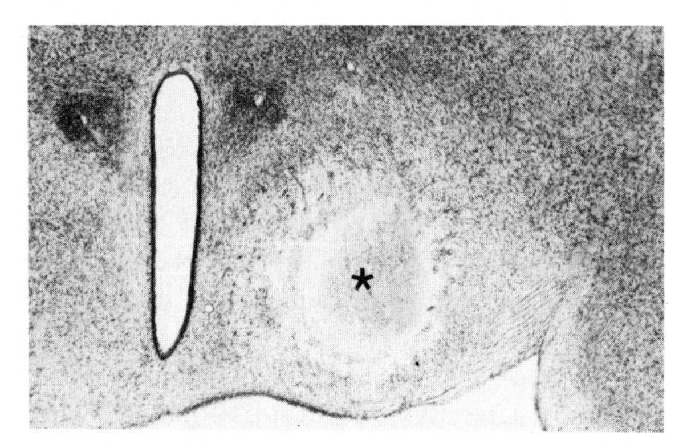

Fig. 7. Electric lesioned site (*) of the brain of the same animal as in Fig. 6. Thus, the enkephalinergic neurons in this area project to the septal area. (From Sakanaka et al., 1982.)

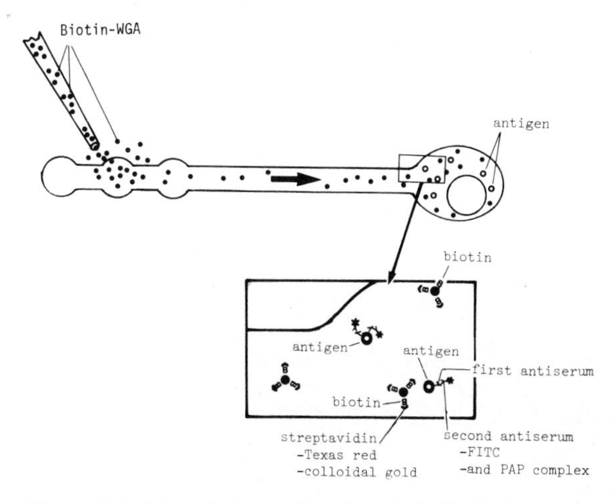

Fig. 8. Principles of the combination method. Black dots represent biotinized tracers visualized with streptavidin–Texas red conjugate (for light microscopy) or streptavidin–colloidal gold (for electronmicroscopy). Open circles represent the antigen visualized with the immunofluorescent method (for light microscopy) or PAP method (for electronmicroscopy).

of avidin–biotin linking and the use of the same fixative for immunohistochemistry allow the maximum sensitivity of both the tracer procedures and the immunocytochemistry (Heitzman and Richards, 1974). (5) Texas red and FITC do not overlap in their emission of fluorescein (Titus et al., 1982), making discrimination between retrograde tracer and immunoreaction easy. (6) Because the detection systems for the immunoreactivity (an immunoreaction) and for the tracer (the avidin–biotin reaction) are separate, there is no interference between the two. (7) By using PAP and diaminobenzidine reaction for immunocytochemistry and colloidal gold particles for detection of the tracer instead of fluorochrome, it is possible to identify double-labeled cells, dendrites, and axons under the electronmicroscope. The protocol used is given in Appendix 6.

Procedure

(a) Preparation of retrograde tracers. Biotinized HRP purchased from Vector laboratory (U.S.A.), is dissolved in Millipore-filtered distilled water. The solution is put into a collodion bag (Sartorius, GmBH) and dialyzed in the 0.05 M Tris-HCl buffer (pH 8.6) to replace the buffer that it originally contained. Then the solution is concentrated to approximately 5% under reduced pressure and used as the tracer. WGA (Vector, U.S.A. or Hounen Co., Japan) is conjugated with biotin and used as a tracer. This biotinization is done using biotin-*N*-hydroxysuccimide (NHS-biotin; Pierce Chemical Co., U.S.A.) by the method of Guesdon et al. (1979). The procedure is as follows. Five milligrams of WGA is dissolved in 0.5 ml of 0.1 M $NaHCO_3$, to which 50 μl of 0.1 M NHS-biotin in dimethyl formamide is added. After 1 h of reaction, the mixture is dialyzed against saline for 24 h at 4°C to remove the non-reacted NHS-biotin. Biotinized WGA (B-WGA) should be stored in the freezer (−20°C). Biotinized HRP (B-HRP) can be stored in a refrigerator (4°C) for at least 6 months.

(b) Injection of biotinized tracer. One microlitre of a solution of B-WGA in saline or B-HRP in buffer is injected into the specific portion of interest with a 5 μl Hamilton syringe. To avoid diffusion of the tracer into other areas, an iontophoretic injection can be made into the brain. 5% B-HRP solution in Tris-HCl buffer is placed in a glass micropipette and connected to a high-voltage positive constant current source (4 μA, pulse current of 7 s on, 7 s off). A small oval injection 200–300 μm in diameter can be made (Fig. 9A, B).

(c) Tissue preparation. The animals injected with the tracer are kept alive for 24–48 h, then perfused through the ascending aorta with 50 ml of saline, followed by 400 ml of 4% paraformaldehyde–0.1 M sodium phosphate buffer (pH 7.4) for 30 min. The brains are quickly removed, immersed in the same fixative for 2 h at 4°C, then immersed for 24 h in 0.1 M sodium phosphate buffer containing 30% sucrose (pH 7.4) at 4°C. The brains are sectioned in a cryostat set at 20–30 μm and rinsed in 0.02 M PBS (pH 7.4). When sections are mounted on a slideglass, the section thickness is reduced to 7–15 μm because antibody and streptavidin are less permeable.

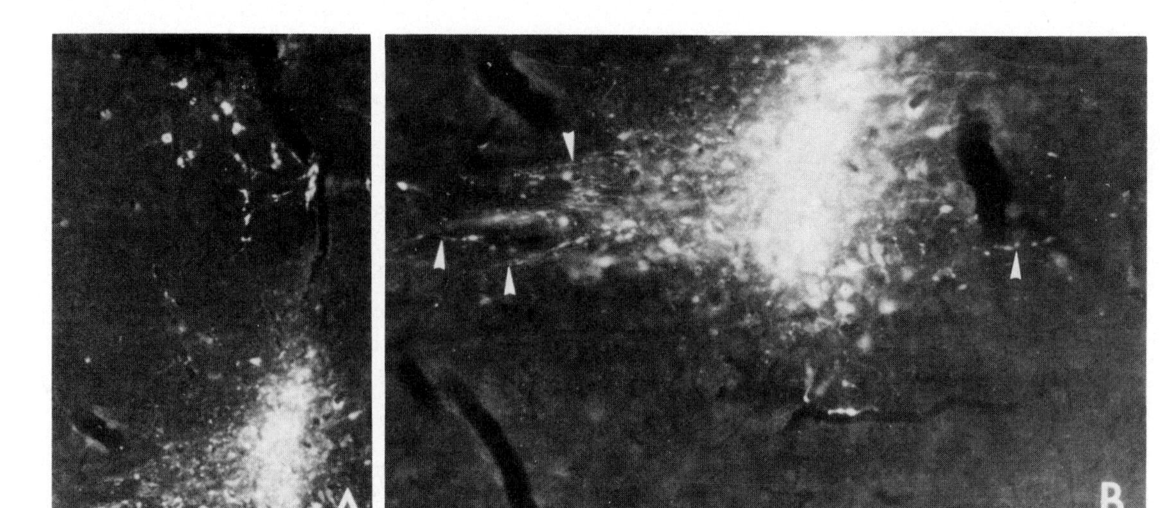

Fig. 9. Injection site of iontophoretically applied biotinized tracer (B-HRP).

(d) Pretreatment of an animal. If necessary, in order to raise the immunoreaction in the cell soma, colchicine solution is injected at 8–12 h before killing. The details have been described earlier (see p. 4).

(e) Staining procedures. The sections are first incubated in a PBS solution with specific antiserum (rabbit) and streptavidin–Texas red (1:250, v/v; Amersham, U.K.) for 24 h at room temperature (15–20°C). After being rinsed with PBS for 10–20 min, the sections are then incubated in PBS with anti-rabbit IgG conjugated with FITC (Coons, 1958). The sections are rinsed with PBS and mounted with a PBS–glycerin mixture (1:1) or Entellan (Merck, U.S.A.). When Entellan is used, the tissue mounted on the slide is air dried, a drop of the glycerin solution is put on the slide, and then the section is coverslipped.

(f) Observation. Observations are made with a Nikon EF microscope. The biotinized tracer that had accumulated in the soma is visualized as a complex of biotinized tracer–streptavidin–Texas red (Titus et al., 1982). This complex possesses red fluorescence when a G dichroic mirror and 580-nm absorption filter system are used (Fig. 10A). Immunoreaction has green fluorescence of FITC under a B dichroic mirror and a 520–545-nm interference filter (Fig. 10B). If an interference filter is not used, a weak yellow fluorescence derived from leakage of Texas red is seen. An example of the staining is shown in Figs. 10A, B and 11. In this experiment, 1 μl of B-WGA was injected into nucleus caudatus putamen. After a day the rat was perfused with 4% paraformaldehyde–0.1 M phosphate buffer (pH 7.4) and then immunocytochemistry using anti-tyrosine hydroxylase antibody and tracer location were carried out. Among the labeled soma, the complex of B-WGA and streptavidin–Texas red looked granular (Figs. 10A and 11). On the other hand, immunoreaction appeared as a dif-

Fig. 10. Double-labeled neurons in the substantia nigra after injection of B-WGA into the nucleus caudatus putamen and incubation with anti-tyrosine hydroxylase (TH) antiserum. (A) Neuronal cell bodies containing B-WGA visualised by the streptavidin–Texas red conjugate. (Observed using a G dichroic mirror and 580-nm absorption filter.) (B) TH immunoreactive neurons in the same field as A. (Observed using a B dichroic mirror and a 520–545-nm interference filter.) Arrowheads indicate double-labeled cells.

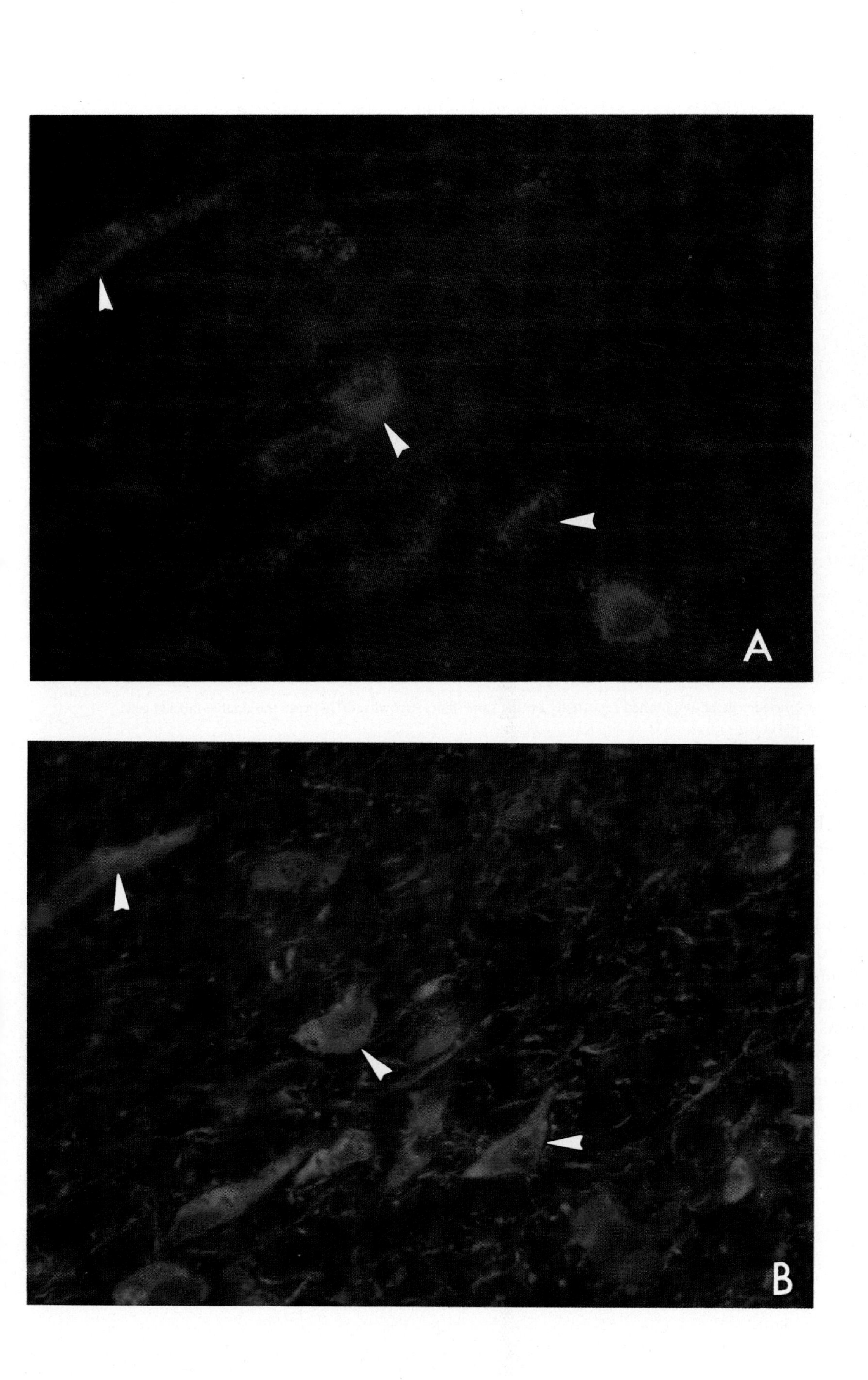
A
B

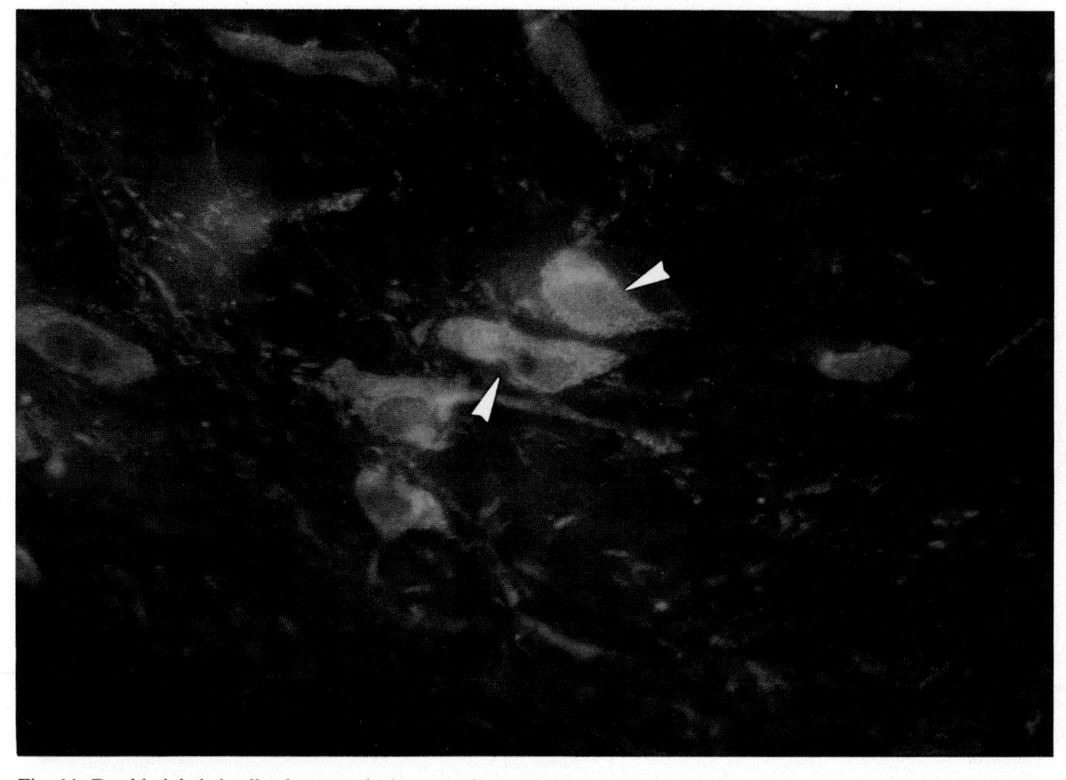

Fig. 11. Double-labeled cells photographed repeatedly on the same film. Arrowheads indicate the double-labeled cells.

fuse green fluorescein (Figs. 10B and 11). When the same area was repeatedly photographed on the same film by using different filters (B and G excitation filters), identification of double-labeled cells was considerably facilitated (Fig. 11).

Fig. 12 compares the sensitivity of the B-WGA and the B-HRP tracer technique. When tracers at the same concentration were injected into the spinal cord of different rats in the same volume, a number of tracer labeled neurons appeared in the reticular formation, vestibular nuclei and raphe nuclei of the medulla oblongata in both cases. The number of labeled neurons was greater with B-WGA injection than with B-HRP.

Electronmicroscopy

The most remarkable feature of the combination method involving a biotinized tracer is that both the reactions can be seen under an electronmicroscope. Fig. 8 shows the principle involved. Staining for electronmicroscopic observation is carried out by using a PAP–diaminobenzidine reaction for immunohistochemistry and biotin–streptavidin gold linking for detection of the biotinized tracer.

Procedure

(a) Tissue preparation. Rats injected with the tracer are perfused with saline and with 400 ml of 0.1 M phosphate buffer containing 4% paraformaldehyde, 0.05% glutaraldehyde, and 0.2% picric acid. The brains are removed, cut into blocks approximately 2 × 2 × 3 mm and immersed in the same fixative for 2 h at 4°C. The blocks are then immersed in 0.1 M phosphate buffer containing 10% sucrose (pH 7.4) for 1 h at 4°C, and then in the same buffer containing 20% sucrose for 12 h at 4°C. The blocks are frozen in liquid nitrogen, immediately thawed in 20% sucrose buffer (see also p. 9), sectioned using a Vibratome (Oxford,

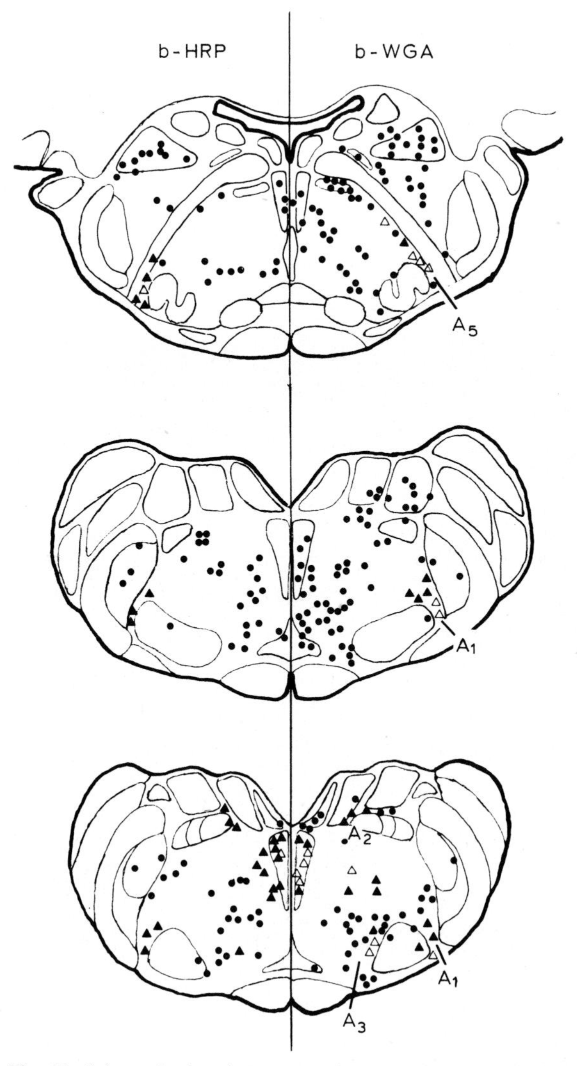

Fig. 12. Schematic drawings comparing sensitivity of B-WGA (right side) and B-HRP (left side) tracer technique. Black dots represent cells labeled retrogradely when the tracers were injected into the spinal cord. Black triangles represent the tyrosine hydroxylase immunoreactive cells and open triangles show the double-labeled cells of tracer and tyrosine hydroxylase immunoreactivity. A1–3 and A5 represent the medullary noradrenaline neuron systems.

U.S.A.), 50 μm thick, and rinsed in PBS to remove the fixative and sucrose buffer.

(b) Staining procedures (electronmicroscopy). (i) Preembedding staining. This step is identical to that described earlier (p. 9). The sections are incubated in a PBS solution with the primary antiserum (1:1000–5000, v/v) for 24 h at 4°C. Then they are washed three times, 10 min each wash, in ice-cold PBS, followed by incubation in PBS with the secondary antibody (1:100; v/v) for 12 h at 4°C. The sections are rinsed three times (10 min each) in ice-cold PBS, and incubated in PBS with the PAP complex (1:100; v/v) (Sternberger, 1970) for 3 h at room temperature. They are rinsed in ice-cold PBS, followed by reaction with diaminobenzidine. After block-staining in 70% ethanol solution with 1% uranyl acetate and dehydration, the sections are mounted with one drop of Epon on silicon-coated slides and coverslips (see also p. 10). A small area that contains immunoreactive structures is cut out and stuck onto a prepared cylinder of Epon with an adhesive. Ultrathin sections are made and collected on collodion-coated nickel grids (150 mesh, Veco, U.K.).

(ii) Postembedding staining. The postembedding reaction with streptavidin-labeled colloidal gold (SCG) is as follows. Preparation of colloidal gold–streptavidin complex is described in Appendix 4. The ultrathin sections on the grids are etched with 10% hydrogen peroxide–PBS solution for 20 min and incubated in a solution of 2% bovine serum albumin and PBS for 30 min at room temperature. Then, they are incubated in a 1:10 mixture of SCG and bovine serum albumin–PBS for 1 h at room temperature, washed with double distilled water, and stained with aqueous lead citrate for 2 min, and examined under a JEOL 100CX electronmicroscope. Fig. 8 shows the principles of this method.

Fig. 13A, B shows an example of the combination method in electronmicroscopy. After B-WGA was injected into the nucleus caudatus putamen, double-labeled cells of tracer and tyrosine hydroxylase immunoreactivity appeared in the pars compacta of the substantia nigra (Figs. 10, 11). When examined by electronmicroscopy, the immunolabeling was distributed throughout the cytoplasm, especially in the endoplasmic reticulum (Fig. 13A). In this same ultrathin section, the neurons were double-labeled by B-WGA, which was

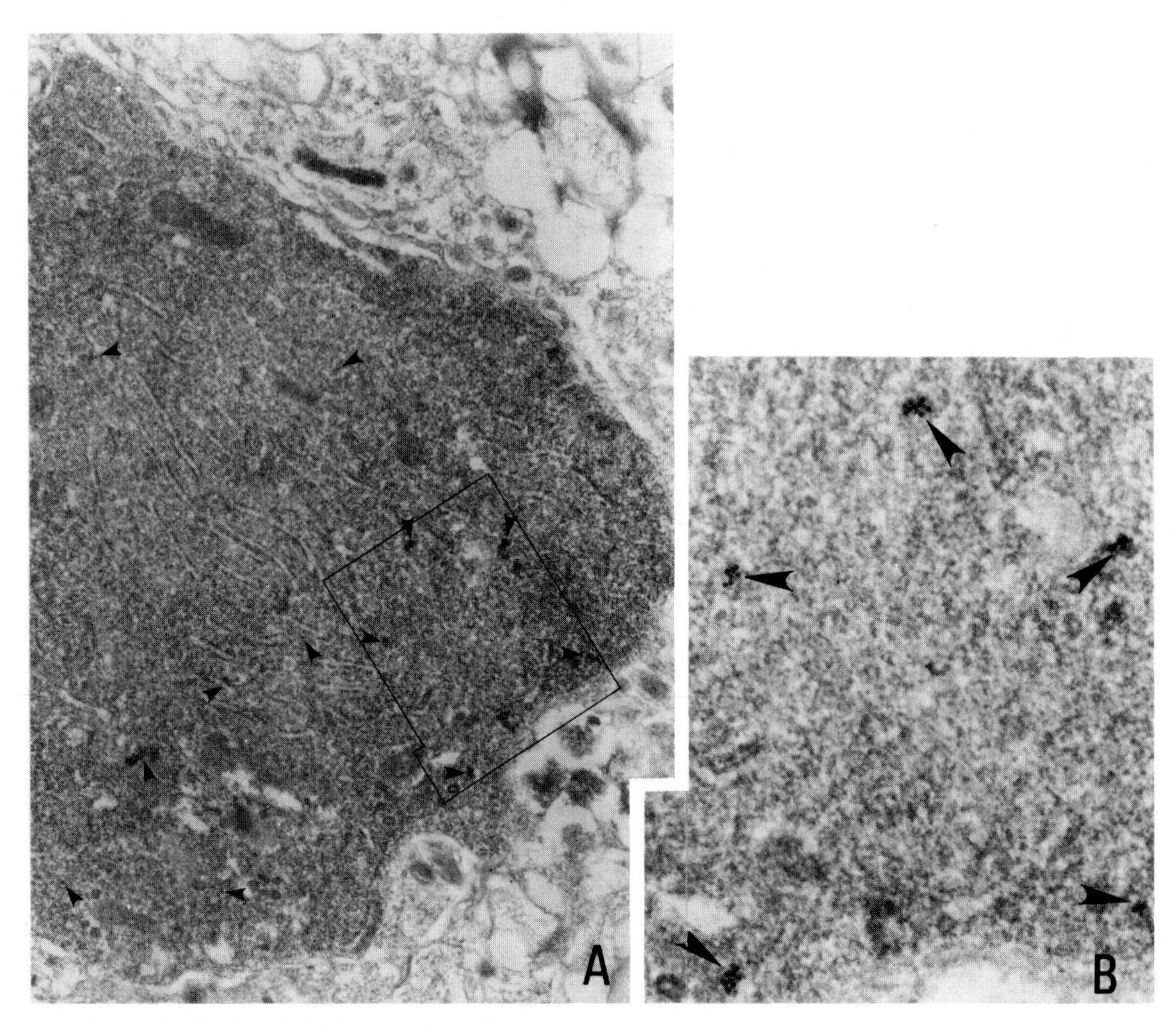

Fig. 13. (A) Electronmicrograph showing a double-labeled neuron in the substantia nigra. B-WGA was injected into the nucleus caudatus putamen, and the tracer-labeled and tyrosine hydroxylase immunoreactive nigral cells were observed under an electronmicroscope. Immunoreaction was found throughout the cytoplasm of the cell. B-WGA, determined by the streptavidin–colloidal gold complex, is shown by arrowheads. (B) Higher magnification of the area framed in A. The clusters of colloidal gold are clearly visible (arrowheads).

assayed by the clear black dots of colloidal gold (Fig. 13A, B, arrowheads). Thus, the complex of B-WGA and streptavidin–colloidal gold was seen in the cytoplasm with a number of small to big clusters. Double-labeling was also confirmed by using streptavidin–ferritin complex as the marker for tracing and PAP–diaminobenzidine reaction as the marker for immunoreactivity (Fig. 14). The process is quite similar to the streptavidin–gold technique apart from the substitution of streptavidin–ferritin–PBS solution. Though both reactions are clearly visible, the ferritin molecule is quite difficult to observe under the electronmicroscope even at high magnification. Therefore, we use the colloidal gold as the marker of the biotinized tracer.

In this section, we studied the nigrostriatal dopamine system and the immunoreaction and accumulation of retrograde biotinized tracer occurred in the same neurons. However, this combination method for electronmicroscopy could be used more

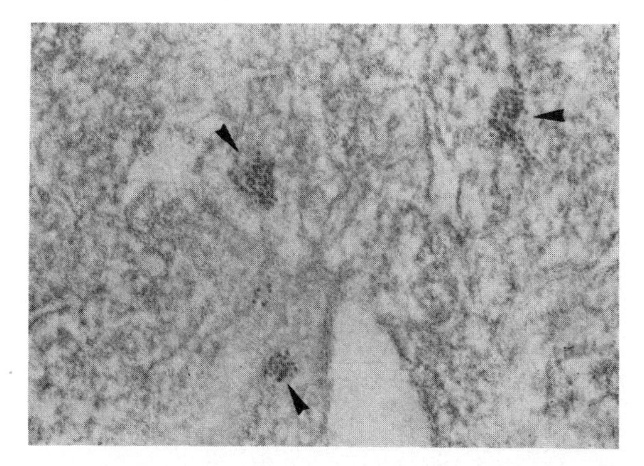

Fig. 14. Electronmicrograph showing double-labeled cells in the substantia nigra. B-WGA was visualized with ferritin–streptavidin and antigen (tyrosine hydroxylase) is visualised by PAP technique.

widely. For example, when we observe immunoreactive terminals, the most frequent profile is an axo-dendritic contact or axo-somatic contact. However, in the preparations in which only one antigen is demonstrated, no information is available concerning where the postsynaptic neuronal elements project. Accordingly, when we observe immunoreactive structures in the cerebral cortex, we can identify the immunoreactive terminals which make the synaptic contact with the cerebral cortical pyramidal cells projecting to, say, the spinal cord, by injecting a biotinized tracer into the spinal cord before killing the animal and processing it for histochemistry. In addition, since the biotinized tracer, particularly B-WGA, is transported "anterogradely", we can demonstrate the origins of the nerve terminals forming synapses on immunoreactive cells before injecting B-WGA into the area where it is already known to project to the area which is under examination.

(c) Control experiment. Control experiments for the "Combination" immunohistochemistry are the same as those used for the conventional PAP technique. Sofroniew et al. (1983) provide a critical discussion of the relevant controls. In our laboratory, the specificity of the primary antiserum is first checked by radioimmunoassay. However, the working dilution of the antiserum is different between immunoassay and immunohistochemistry, and so an additional check by an absorption test is necessary (Sofroniew et al., 1983). Antiserum is pretreated with excess synthetic polypeptide (10–1000 μM) or purified protein for two days at 4°C, and serves as the control serum. When sections are incubated with the control serum, the structures stained by that antiserum should disappear.

Specificity checks of biotinized tracers were done, first, by injecting B-WGA into the third, lateral, and fourth ventricles, and then by injecting B-WGA into the blood stream via the femoral artery. In neither case were labeled cells found in the brain.

Other combination techniques of immunohistochemistry and retrograde tracing (light microscopy)

Takeda et al. (1984) detected a hypothalamo-cortical histaminergic pathway by injecting HRP into the cerebral cortex, and making two serial frontal sections in which one section was stained with histidine decarboxylase (marker enzyme of the histaminergic system) antiserum and the other reacted with HRP–tetramethyl benzidine (Mesulam, 1978). Fig. 15A, B shows the same neurons labeled by immunoreaction and tetramethyl benzidine reaction in neighbouring sections (arrowheads).

The techniques used by Bowker et al. (1981) and Priestley et al. (1981) are similar. HRP was used as a retrograde tracer and immunohistochemistry by PAP reaction was combined with the technique. Sensitivity of the diaminobenzidine reaction of the tracer was enhanced by intensification with cobalt chloride of the brown reaction (Bowker et al., 1981) or by observing under interference contrast illumination microscope (Priestley et al., 1981), while immunoreaction was identified by a diaminobenzidine reaction showing a brown precipitate. As tracer, WGA-conjugated HRP, which is more sensitive than HRP itself, can be used (for review see Hökfelt et al., 1983). HRP and PAP combination techniques were also reported by Beitz (1982), Ritchie et al. (1982), Yazierski et al. (1982), Rye et al. (1984) and Wainer and Rye (1984).

There are some approaches based upon a dif-

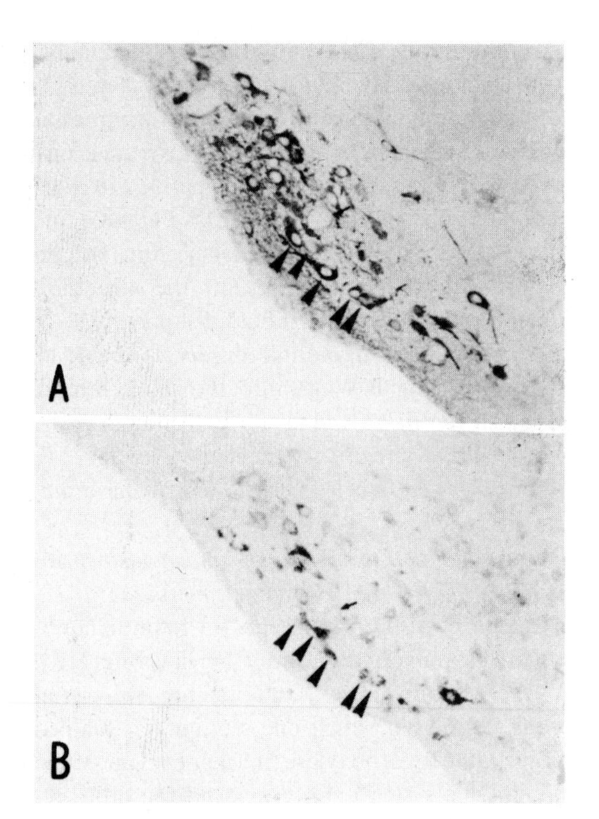

Fig. 15. Two serial frontal sections showing histidine decarboxylase immunoreactive (A) cells containing HRP-granules as tracer (B) in the caudal magnocellular nucleus of the rat. Arrows show the vessels and arrowheads show the identical neurons containing HRP and immunoreaction. (From Takeda et al., 1984.)

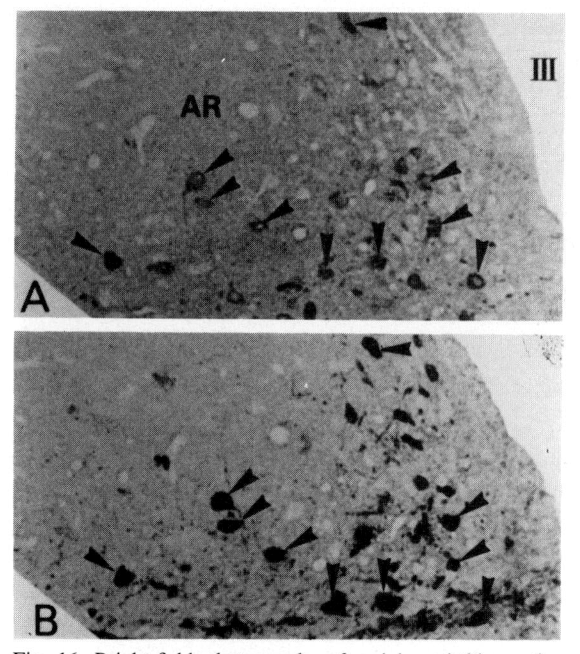

Fig. 16. Bright-field photographs of serial semi-thin sections show a single cell contains two separate antigens. Immunostained with alpha melanocyte stimulating hormone (MSH) antiserum (A) and with gamma MSH antiserum (B). Arrowheads indicate the same neurons containing alpha and gamma MSH. AR: arcuate nucleus; III: third ventricle.

ferent principle. Lechan et al. (1981) used WGA as a retrograde tracer and identified the molecule by PAP immunohistochemistry, while the antigen was detected by the immunofluorescent technique or the fluorescein avidin–biotin complex technique. The advantage of this method is that the markers for the immunoreaction and tracer are different, and WGA as a tracer has high sensitivity.

Several authors have also combined retrograde tracer, HRP and immunocytochemistry with electronmicroscopy (see Oldfield et al., 1983, 1985).

Multiple antigen immunohistochemistry

Light microscopy

Since Hökfelt et al. (1977) found that noradrenaline and somatostatin coexist in some sympathetic neurons, a number of peptide–classical neurotransmitter and peptide–peptide co-localizations have been reported. To investigate co-localization, two main techniques have been employed. In one technique, semithin sectioning of 2–5 μm is done serially. One section is stained with an antiserum, and a neighbouring section is stained with another antiserum (Fig. 16). The other is the double-labeling technique on a single tissue section. Both techniques can also be used for cell identification with staining by a cell or tissue marker antigen and a desired antigen.

In the double-labeling technique, there are many different procedures. However, in our laboratory, only one technique has been shown to have the necessary specificity and sensitivity. The principle of this technique is illustrated in Fig. 17. The technique depends upon specific differences in the two primary antisera and absence of crossreactivities between the primary and secondary antisera and

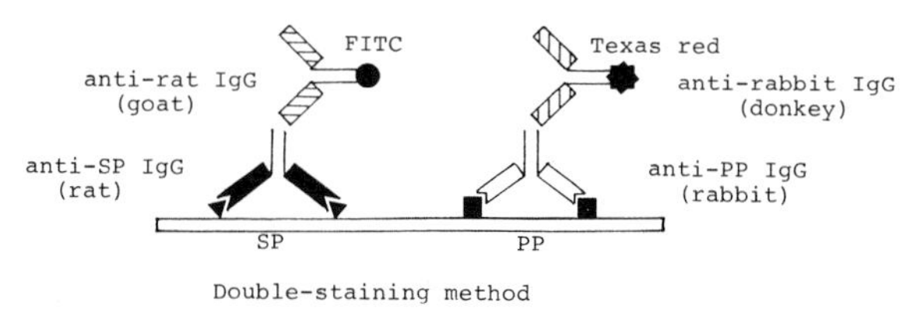

Fig. 17. Principle of the double-staining method. FITC has a green fluorescence and is visible under a fluorescent microscope equipped with B dichroic mirror and 520 nm absorption filter. Texas red has a strong red fluorescence and is visible under a fluorescent microscope with G dichroic mirror and 580 nm absorption filter.

between each of the secondary antisera. As an example, double-staining for substance P (SP) and pancreatic polypeptide (PP) immunoreactivities is described. For visualization of SP immunoreactivity, the monoclonal anti-SP antibody was raised in the rat and FITC-conjugated goat anti-rat IgG was used. For PP immunoreactivity, polyclonal antibody raised in the rabbit and Texas red-conjugated donkey anti-rabbit IgG are used. Because the secondary antiserum used in this technique does not crossreact with the IgG molecule of the other species (species-specific), there is no false staining. Crossreaction of antisera (specifically secondary antisera) must be checked strictly (Fig. 18). If there is any crossreaction against the IgG of another species which is to be used in double immunostaining, an alternative secondary antibody must be found which is species-specific (the IgG fractions crossreacting with the IgG of other species can be eliminated by differential absorption — see Appendix 8). The procedure is described below.

Procedure

The procedure for tissue preparation is identical to that described earlier (p. 4).

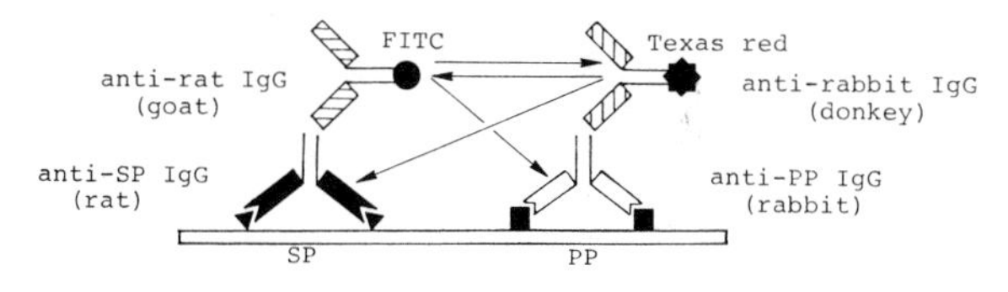

Fig. 18. Possible crossreaction among primary and secondary antibodies.

The tissue for double-staining is first incubated in a mixture containing both SP monoclonal antibody raised in rat and PP polyclonal antibody raised in the rabbit. After overnight incubation of the first antisera mixture in a humid atmosphere at room temperature, the tissues are then rinsed three times for 10 min each with PBS. Next, the tissue sections are incubated with a mixture of the second antisera, FITC-conjugated goat anti-rat IgG and Texas red-conjugated donkey anti-rabbit IgG for 24 h in a humid atmosphere at room temperature. Then the tissue is mounted in a glycerine–PBS mixture. The tissues are examined with a fluorescent microscope with FITC under a B-dichroic mirror illumination system, whereas PP in the same tissue

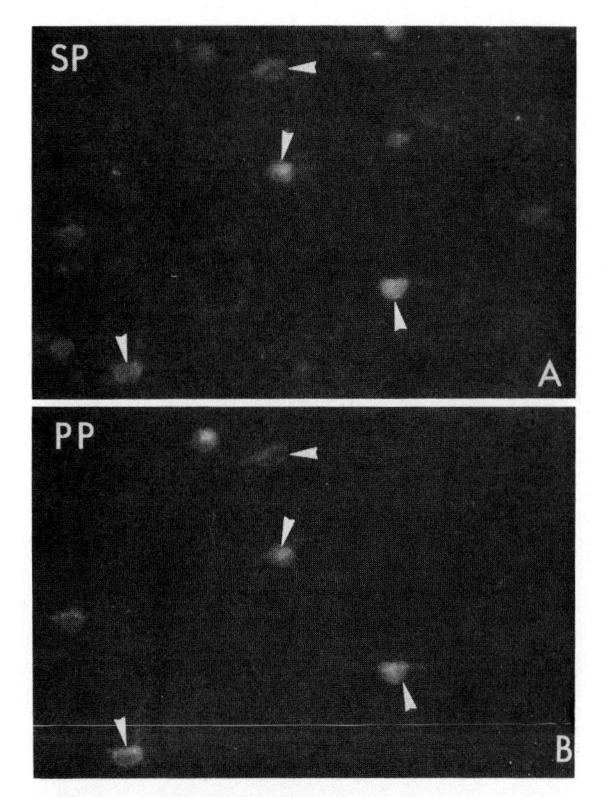

Fig. 19. Fluorescent photomicrographs showing the coexistence of substance P (SP) and pancreatic polypeptide (PP) in the chick retina revealed by the double-staining method (flat preparation). PP is observed by a Texas red immunofluorescence (red fluorescein) and SP by FITC immunofluorescence technique (green fluorescein). (From Katayama-Kumoi et al., 1985.)

exhibit red fluorescence with Texas red under a G dichroic mirror system illumination (Fig. 19A, B).

Specificity check of antisera. Specificity of primary antisera has already been described (see p. 21). Specificity of the second antisera was tested as follows: (1) the sections, incubated with SP antiserum, were reacted with Texas red-conjugated donkey IgG; (2) the sections, incubated with PP antiserum, were incubated with FITC-conjugated goat anti-rat IgG; (3) sections incubated with SP antiserum were incubated with both secondary antisera and observed by a B dichroic mirror filter; and (4) the sections, incubated with PP antiserum were incubated with both secondary antisera and observed by a G dichroic mirror filter (Fig. 18).

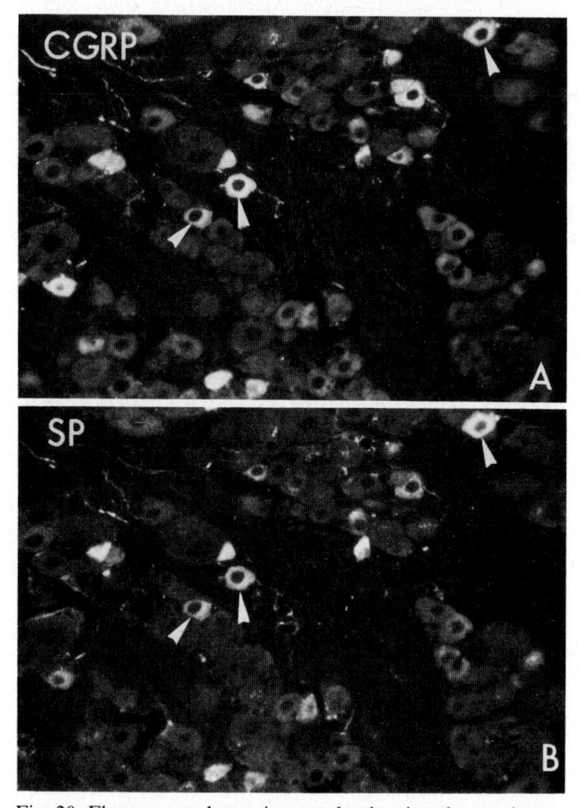

Fig. 20. Fluorescent photomicrographs showing the coexistence of calcitonin gene-related peptide (CGRP) and substance P (SP) in the trigeminal ganglion revealed by the double-staining method. CGRP is observed by a Texas red immunofluorescence and SP is by FITC immunofluorescence. (From Lee et al., 1985.)

Fig. 20 also shows an example of double-labeled neurons containing both calcitonin gene-related peptide (CGRP)- and SP-like immunoreactivities in the trigeminal ganglion.

Electronmicroscopic observation of double-staining technique

The gold-labeled antigen detection method of Larsson et al. (1983) is useful for this purpose. The principle is shown in Fig. 21. Ultrathin sections are stained by primary antisera. Antibodies are allowed to react with tissue-bound antigen with one of their two antigen-combining sites and with a marker-bound pure antigen with the remaining site. As the marker, colloidal gold particles with different sizes (18 and 30 nm) are used, and antigen-coated colloidal gold is prepared by coating of gold granules with albumin-conjugated pure antigen (where a synthetic peptide might be the antigen). Small peptides do not absorb well to gold granules and must be conjugated to a larger protein such as albumin before they can be absorbed by colloidal gold. Detailed procedures are described in the review of Larsson et al. (1983).

Recently, Hisano et al. (1984) presented a new technique for the double-labeling. They fixed tissues by the freeze-substitution technique, embedded and cut ultrathin sections. Then they incubated the sections on the grid with both a prepared protein A–small colloidal gold–antibody complex and protein A–large colloidal gold–antibody complex (to an alternative antigen). However, they noted that the Fc

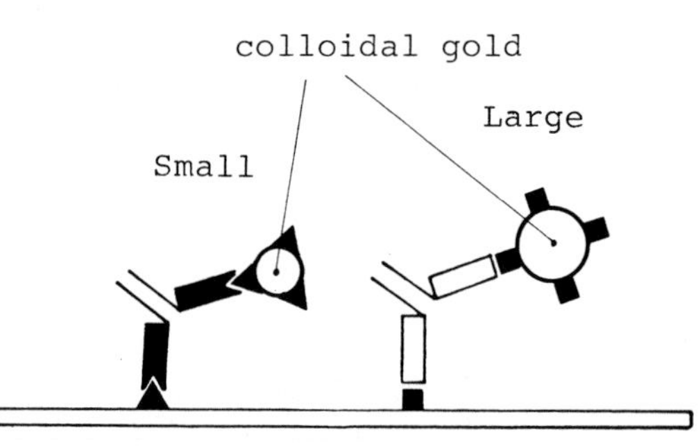

Fig. 21. Principle of the gold-labeled antigen detection method of Larsson.

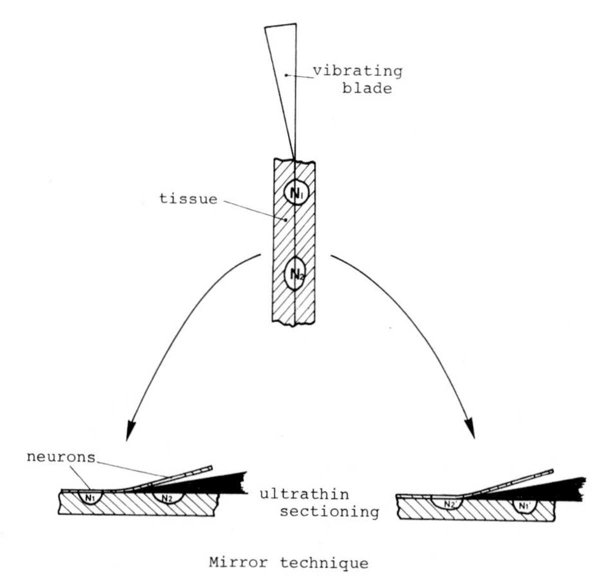

Fig. 22. Principle of mirror technique of electronmicroscopic level. N1 and N2 show the neurons shown in Fig. 23.

binding sites of protein A–gold had to be saturated by the use of excess amounts of antibodies.

Associated techniques

Mirror technique at ultramicroscopic level (Kubota et al., 1985)

The technique is depicted in a schematic drawing (Fig. 22). Fifty to 100-μm Vibratome sections are prepared. Serial ultrathin sections are cut from the surface of a pair of matched sections, one of which has been incubated with an antibody and the other with an alternative antibody to give immunostaining in the superficial (5–10 μm) layers. The immunostaining is carried out by the preembedding method using the PAP technique. The ultrathin sections are then mounted on Formvar-coated single slot (2 × 1 mm) grids. Then, sections are stained and observed by electron microscopy (Fig. 23).

Immunocytochemistry combined with false transmitter histochemistry (Fig. 24)

Before developing the immunocytochemistry, fine structures of amine terminals have been examined by using "false" transmitters such as 5-hydroxydopamine (5-OHDA), 6-hydroxydopamine, 5,6-dihydroxytryptamine or 5,7-dihydroxytryptamine. The first two have been used for the demonstration of catecholamine (CA) terminals and the other for serotonin (5-HT) terminals (for review see Richards, 1983). Recently immunocytochemistry has been combined with "false" transmitter histochemistry, and the relationship between CA terminals and other bioactive substances containing terminals or neurons examined (Matsuyama et al., 1985; Yamano et al., 1985).

Colchicine (3.5 mg/ml, 10 μl) was injected into the ventricle of rats weighing about 100 g. After 12 h, they were further injected with 5-OHDA (15 mg/kg) into the ventricle. After an additional 6–12 h, the animals were perfused with saline, followed by a fixative developed by Somogyi and Takagi (1982). This procedure can be applied to the examination of CA terminals and immunoreactive fibres or neurons. When the relationship between CA terminals and immunoreactive fibres of other bioactive substances is to be determined, colchicine injection can be omitted. The peripheral nervous system 5-OHDA (100 mg/kg) was repeatedly injected intraperitoneally for three days (once per day) before the animal was killed. Compared with conventional "false" transmitter histochemistry, the number of cored vesicles demonstrated in the CA terminals is much less. This is due to the fixatives used: for demonstration only of "false" transmitter, a conventional method (Karnovsky, 1965) can be used, whilst for immunocytochemistry, the content of glutaraldehyde was much lower than that in Karnovsky's fixative, resulting in reduced histochemical sensitivity.

Comments

Several important points must be remembered in order to obtain good immunostaining. (1) Because fixation affects the retention of antigenicity, a good fixative for that antigen must be selected. In addition the perfusion must be adequate, and if the material cannot be perfused it should be immersed for

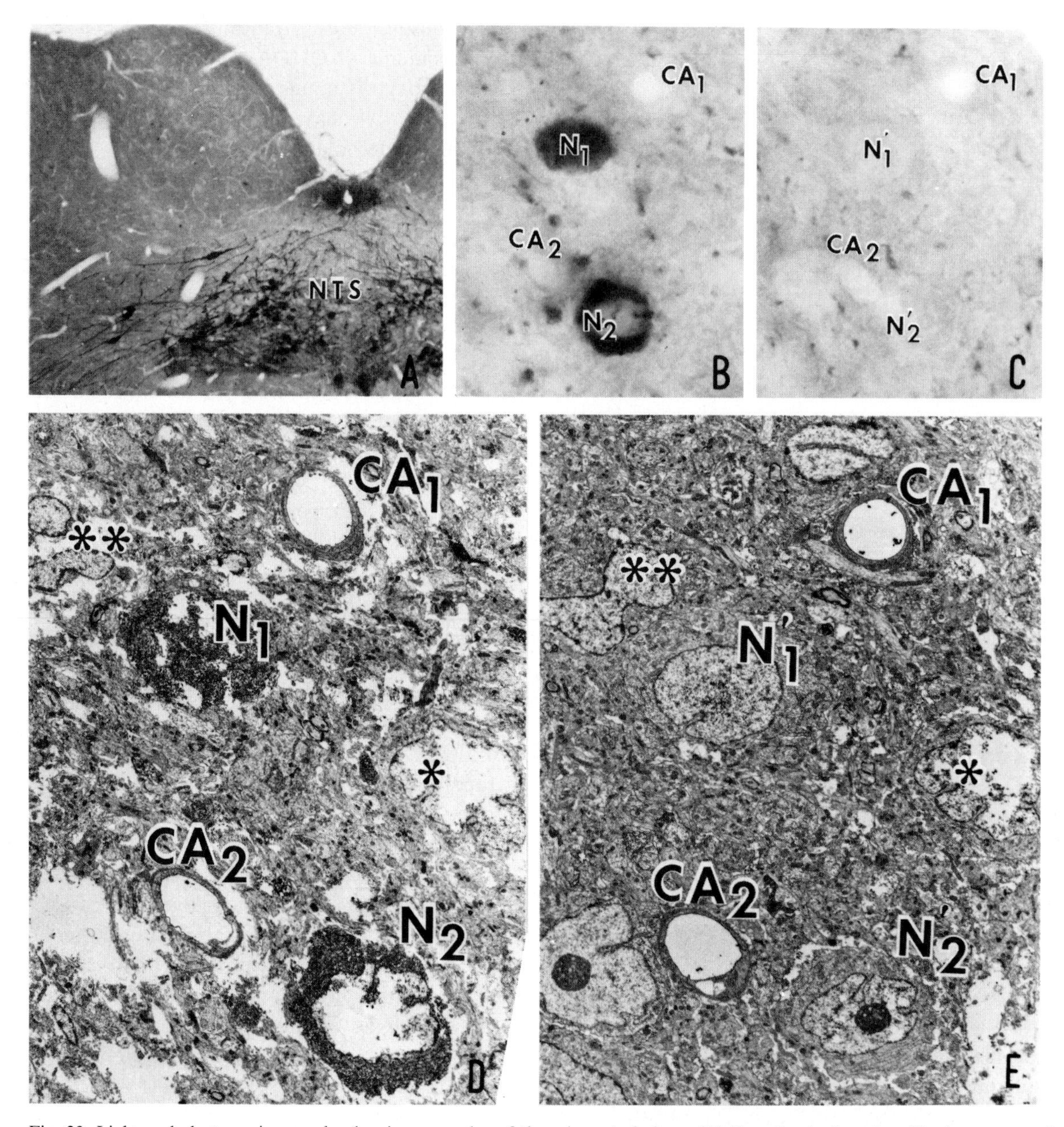

Fig. 23. Light and electronmicrographs showing examples of the mirror technique. (A) Tyrosine hydroxylase-like immunoreactive neurons in the nucleus tractus solitarii; (B) higher magnification of A; (C) adjacent section of A and B was immunostained with anti-substance P antiserum. (D, E) Low magnification electronmicrographs showing the pair-matched and ultrathin-sectioned tyrosine hydroxylase immunoreactive (D) and substance P immunoreactive (E) structures. N_1 and N'_1, and N_2 and N'_2 show the same neurons, respectively. CA_1 and CA_2, capillaries; * and **, non-immunoreactive cells. (From Kubota et al., 1985.)

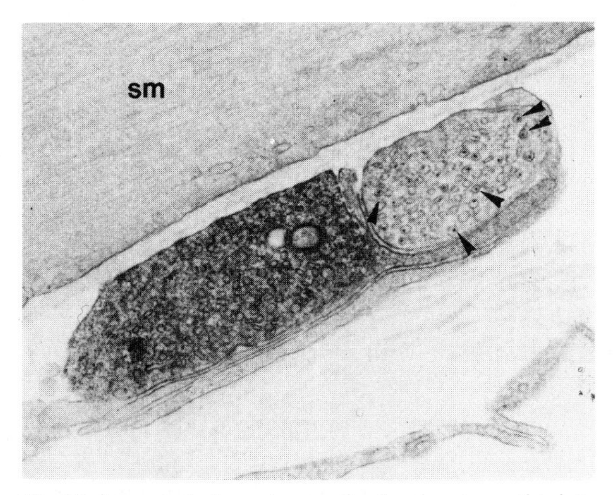

Fig. 24. Immunoelectronmicrographs showing vasoactive intestinal polypeptide (VIP) immunoreactive nerve terminals in the anterior cerebral artery of the rat with 5-hydroxydopamine treatment. VIP immunoreactive terminal and catecholamine terminal form a cluster. Arrowheads show small granulated vesicles. (From Matsuyama et al., 1985.)

a long period. (2) Characteristics of antibodies seriously influence the immunoreaction. When a commercial antibody is used, reactivity changes within the same lot number may be found even though the agent may be the same. (3) In the various steps of frozen sectioning, "gradual freezing" must be avoided. We recommend freezing of the tissue rapidly in solid CO_2 and acetone, or solid CO_2.

To obtain good results with the combination of light microscope for immunohistochemistry and retrograde tracing, the substrate for avidin–biotin reaction must be selected carefully. In addition, when avidin D, the egg white protein (Woolley and Longsmith, 1942), is used for detection of the biotinized tracer, care should be taken to minimize non-specific reactions, because avidin D readily attaches to membrane glycoproteins. The usual cautions for the specificity in an immunoreaction described above also apply.

In double labeling, particular attention must be paid to the crossreactivity of secondary antisera. This point has been discussed earlier (see p. 21).

Acknowledgements

The authors would like to thank Dr. H. Takagi for valuable advice and providing electronmicrographs and Drs. Y. Katayama-Kumoi and Y. Lee for providing micrographs of double-labeling. We also thank Drs. N. Takeda and T. Matsuyama for providing photographs of combinations of retrograde tracing and immunohistochemistry. We are most grateful to Prof. Shiotani for valuable advice.

Appendix

(1) Preparation of gelatin-coated slideglass

(a) Dissolve 0.5 g of gelatin powder with 100 ml of warm distilled water.
(b) Cool the mixture and add 0.05 g of chromium potassium sulfate ($CrK(SO_4)_2 \cdot 12H_2O$).
(c) Dip slides in this mixture for a short time.
(d) Dry the slides at 60°C for 12 h.

(2) Enhancement of DAB reaction by nickel ammonium sulfate

The usual DAB reaction is replaced by the following protocol.
(a) Sections which have reacted with antisera are rinsed with 0.05 M Tris-HCl (pH 7.6) for 10 min.
(b) Reaction with the following reaction mixture for 10–30 min:

0.05 M Tris-HCl buffer	(200 ml)
DAB	(40 mg)
Nickel ammonium sulfate	
$(NH_4)_2Ni\ (SO_4)_2 \cdot 6H_2O$	(1.2 g)
10% H_2O_2	(100 μl)

(c) Wash with distilled water.

(3) Protocol for preembedding staining (Somogyi and Takagi, 1982)

Step 1

(a) Perfusion by 50 ml of saline followed by fixative (300–400 ml):

4% paraformaldehyde aqueous solution
0.05% glutaraldehyde
0.2% picric acid
0.1 M phosphate buffer (pH 7.2–7.6)

(b) Immersion in the same fixative for 2–3 h at 4°C.
(c) Cut into blocks (ca. 2 × 4 × 5 mm) at 4°C.
(d) Wash in 0.1 M phosphate buffer at 4°C.
(e) 10% sucrose in 0.1 M phosphate buffer for 1 h at 4°C.
(f) 20% sucrose in 0.1 M phosphate buffer for 24 h at 4°C.
(g) Freeze in liquid nitrogen and thaw in 0.1 M phosphate buffer.
(h) 50–100 μm Vibratome sectioning.
(i) Rinse in 0.1 M phosphate buffer.

Step 2

(a) Wash two times in PBS for 10 min each at 4°C.
(b) 20% normal goat serum in PBS for 30 min at room temperature.
(c) Incubation with primary antiserum diluted with PBS containing 1% normal goat serum (PBS–NGS) at 4°C, overnight.

Step 3

(a) Wash three times in PBS–NGS for 30 min each at 4°C.
(b) Incubation with second antiserum in PBS–NGS at 4°C, overnight.

Step 4

(a) Wash three times in PBS–NGS for 30 min each at 4°C.
(b) Incubation with PAP complex in PBS–NGS for 3 h at room temperature.
(c) Wash in PBS for 30 min at 4°C.

Step 5

(a) Wash in 0.05 M Tris-HCl buffer (pH 7.6) for 30 min at 4°C.
(b) Incubation with DAB solution (50 mg in 100 ml of Tris-HCl buffer).
(c) Incubation with DAB solution with hydrogen peroxide added (final concentration = 0.01%) for 3–20 min at room temperature.
(d) Wash in Tris-HCl buffer.
(e) Incubation in the 1% osmium tetroxide in 0.1 M phosphate buffer at room temperature.
(f) Wash in phosphate buffer.

Step 6

(a) Block staining by 1% uranyl acetate in 70% ethanol.
(b) Dehydration with series of ethanol and propylene oxide.
(c) Immersion in the propylene oxide–Epon (1:1) for 1–2 h at room temperature.
(d) Immersion in Epon for 2 h at room temperature.
(e) Mount sections on the silicon-coated slide with one drop of Epon.
(f) Cover with silicon-coated coverslip and put in an incubator at 60°C for 2 days.
(g) Observe under the light microscope (sketch or photograph).
(h) Cut out the desired portion under an operating microscope.
(i) Mount on a prepared cylinder of Epon.
(j) Ultrathin sectioning.
(k) Mount on coated grids.

(4) Preparation of colloidal gold (Frens, 1973)

(a) 100 ml of a 0.01% tetrachloroauric acid solution is heated.
(b) 1% sodium citrate solution is added and boiled until the solution becomes wine-red colour (particle size of colloidal gold depends on the volume of the reducing agent, i.e. 4, 2, 1.5, 1.0 ml citrate produces 15, 18, 32, and 50 nm of colloidal gold, respectively).
(c) Adjust the gold solution to pH 6.9 with 0.2 M K_2CO_3.

(5) Preparation of colloidal gold–protein complex (Roth et al., 1978)

Determine the minimum amount of protein for full stabilization of the colloidal gold (Horisberger et

al., 1975; Roth and Bender, 1978).
(a) Serial dilutions of protein (0.1 ml) are mixed with 0.5 ml of colloidal gold suspension and allowed to stand for 5 min.
(b) Add 10% NaCl solution.
(c) Flocculation (colour change from red to blue) is judged visually.
(d) 20% excess of protein is added to colloidal gold suspension.
(e) After a few minutes, polyethylene glycol (MW 20 000) is added at final concentration, 0.05%.
(f) Ultracentrifuge at 100 000 × g for 1 h at 4°C.
(g) Dissolve the precipitate with 0.05% polyethylene glycol–PBS solution.
Note: When a constant size of the colloidal gold particle is necessary (as in double-labeling), centrifugation in a continuous density gradient of 10–30% glycerine–PBS with 0.05% polyethylene glycol is carried out (10 000 × g, 15 min).

(6) Combination technique of immunohistochemistry and retrograde tracing (light microscopy) (Shiosaka et al., 1984)

Injection of tracer

(a) Iontophoretic injection: a 5% B-HRP solution in Tris-HCl buffer (pH 8.6) is placed in a glass micropipette connected to a high voltage constant current (4 μA, 7 s on and 7 s off) for 10–30 min.
(b) Injection by pressure: 0.2–1 μl of biotinized tracer is injected by using 5 μl of Hamilton's syringe.
(c) Leave animal to survive for 1–2 days.
(d) Inject a colchicine solution (5 μl/100 g body weight of 8 mg/ml saline), if necessary.
(e) Leave animal to survive for a further 12 h.
(f) Perfusion with 50 ml of saline followed by 400 ml of the fixative (4% paraformaldehyde in 0.1 M phosphate buffer).
(g) Postfixation for 2 h.
(h) Immersion in 30% sucrose in 0.1 M phosphate buffer for 1 day at 4°C.
(i) Cut frozen sections of 20 μm thick and rinse in PBS.

Reaction

(j) Incubate in reaction medium at room temperature for 24 h:
primary antiserum (usual 1:1000–1:5000 dilution)
streptavidin–Texas red (Amersham, final dilution 1:250–1:500) PBS
(k) Wash three times with PBS at 4°C for 10 min each.
(l) Incubate with second antiserum (1:1000 dilution) for 1 day at room temperature.
(m) Observation with B dichroic mirror and 520–540 nm interference filter (FITC for immunoreaction) and G dichroic mirror and 580 nm absorption filter (Texas red for the tracer).

(7) Combination technique of immunohistochemistry and retrograde tracing (electronmicroscopy)

Preembedding staining

(a) Injection of B-WGA (do not use B-HRP) into CNS region.
(b) Leave animal to survive for 24–48 h.
(c) Colchicine treatment, if necessary.
(d) Tissue preparation is same as section 3, though step 5 (e) is omitted.

Postembedding staining

The operations are carried out at room temperature.
(e) Ultrathin-section tissue on the nickel grid is etched with 10% -H_2O_2–PBS solution for 20 min.
(f) Rinse in a 2% bovine serum albumin–PBS solution (BSA–PBS) for 20 min.
(g) Incubate in a droplet of streptavidin–gold solution diluted by BSA–PBS for 1 h.
(h) Wash in Millipore filtered water.
(i) Grid is fixed by floating on a drop of 4% osmium tetroxide in 0.1 M phosphate buffer for 4 s.
(j) Stain with aqueous lead citrate for 2 min.
(k) Observe under electronmicroscope.

(8) Elimination of crossreactivity

We have described the system for the elimination of the crossreaction against a rabbit IgG from anti-goat IgG antiserum (Takeda et al., 1984a).
(a) Dissolve 20 mg of a rabbit IgG in 2 ml of 0.2 M Hepes buffer (pH 7.4) — solution (a).
(b) Wash 2 ml of the Affi-gel 10 (Bio-Rad) once with isopropanol and three times with distilled water.
(c) Add solution (a) to washed Affi-gel 10 and incubate overnight at 4°C with agitation.
(d) Add 200 μl of 1 M ethanolamine to this solution.
(e) Make 1 ml of column and wash with PBS.
(f) The eluate is obtained and used for immunohistochemistry.

References

Altman, L. G., Schneider, B. G. and Papermaster, D. S. (1984) Rapid embedding of tissue in Lewcryl K4M for immunoelectron microscopy. *J. Histochem. Cytochem.*, 32: 1217–1223.

Armbruster, B. L., Carlemalm, E., Chiovetti, R., Garavito, R. M., Hobot, J. A., Kellenberger, K. and Villiger, W. (1982) Specimen preparation for electron microscopy using low temperature embedding resins. *J. Microsc.*, 126: 77–85.

Bauer, H., Gerber, H. and Horisberger, M. (1975) Morphology of colloidal gold, ferritin and anti-ferritin antibody complexes. *Experientia*, 15: 1149–1151.

Beitz, A. J. (1982) The sites of origin of brain stem neurotensin and serotonin projections to the rodent nucleus raphe magnus. *J. Neurosci.*, 2: 829–842.

Bishop, A. E., Polak, J. M., Bloom, S. R. and Pearse, A. G. E. (1978) A new universal technique for the immunocytochemical localization of peptidergic innervation. *J. Endocrinol.*, 77: 25P.

Bowker, R. M., Steinbusch, H. W. M. and Coulter, J. D. (1981) Serotonergic and peptidergic projections to the spinal cord demonstrated by a combined retrograde HRP histochemical and immunocytochemical staining method. *Brain Res.*, 211: 412–417.

Bowker, R. M., Westlund, K. N., Sullivan, M. C. and Coulter, J. D. (1982) A combined retrograde transport and immunocytochemical staining method for demonstrating the origins of serotonergic projections. *J. Histochem. Cytochem.*, 30: 805–810.

Carlemalm, E., Garavito, M. and Villiger, W. (1982) Resin development for electron microscopy and an analysis of embedding at low temperature. *J. Microsc.*, 126: 123–143.

Coons, A. H. (1958) Fluorescent antibody methods. In J.F. Danielli (Ed.), *General Cytochemical Methods*. Academic Press, New York, pp. 399–422.

Coons, A. H. and Kaplan, M. H. (1950) Localization of antigens in tissue cells. II. Improvement in a method for the detection of antigen by means of fluorescent antibody. *J. Exp. Med.*, 91: 1–13.

Costa, M., Buffa, R., Furness, J. B. and Salcia, E. L. (1980) Immunohisto-chemical localization of polypeptides in peripheral autonomic nerves using whole mount preparations. *Histochemistry*, 65: 157–165.

Dahlstrom, A. (1968) Effect of colchicine on transport of amine storage granules in sympathetic nerve of the rat. *Eur. J. Pharmac.*, 5: 111–112.

De May, J., Moeremans, M., Geuens, R., Nuydens, R. and De Brabander, M. (1981) High resolution light and electron microscopic localization of tubulin with the IGS (Immuno Gold Staining) Method. *Cell Biol. Int. Rep.*, 5: 889–899.

Eldred, W. D., Zucker, C., Karten, H. J. and Yazulla, S. (1983) Comparison of fixation and penetration enhancement technique for use in ultrastructural immunocytochemistry. *J. Histochem. Cytochem.*, 31: 285–292.

Emson, P. C. (Ed.) (1983) *Chemical Neuroanatomy*. Raven Press, New York.

Faulk, W. P. and Taylor, G. M. (1971) An immunocolloid method for the electron microscope. *Immunochemistry*, 8: 1081–1083.

Forsgreen, A. and Sjoquist, J. (1966) "Protein A" from *S. aureus*. I. Pseudoimmune reaction with human r-globulin. *J. Immunol.*, 97: 822–827.

Forsgreen, A. and Sjoquist, J. (1967) "Protein A" from *S. aureus*. III. Reaction with rabbit r-globulin. *J. Immunol.*, 99: 19–24.

Frens, G. (1973) Controlled nucleation for the regulation of the particle size in monodisperse gold solutions. *Nature Phys. Sci.*, 241: 20–21.

Gerfen, C. and Sawchenko, P. E. (1984) An anterograde neuroanatomical tracing method that shows the detailed morphology of neurons, their axons and terminals: Immunohistochemical localization of an axonally transported plant lectin, *Phaseolus vulgaris* Leucoagglutinin (PHA-L). *Brain Res.*, 290: 219–238.

Graham, R. C. and Karnovsky, M. J. (1966) The early stages of absorption of injected horseradish peroxidase in the proximal tubules of mouse kidney. Ultrastructural cytochemistry by a new technique. *J. Histochem. Cytochem.*, 14: 291–302.

Green, N. M. (1963) Avidin. The use of ^{14}C biotin for kinetic studies and for assay. *Biochem. J.*, 89: 585–591.

Green, N. M. and Toms, E. J. (1970) Purification and crystallization of avidine. *Biochem. J.*, 118: 67–70.

Guesdon, J. L., Ternynck, T. and Avrameas, S. (1979) The use of avidin–biotin interaction in immunoenzymatic techniques. *J. Histochem. Cytochem.*, 27: 1131–1139.

Hayashi, H., Takagi, H., Takeda, N., Kubota, Y., Tohyama,

M., Watanabe, T. and Wada, H. (1984) Fine structure of histaminergic neurons in the caudal magnocellular nucleus of the rat as demonstrated by immunohistochemistry using histidine decarboxylase as a marker. *J. Comp. Neurol.*, 229: 233–241.

Heitzman, H. and Richards, F. M. (1974) Use of the avidin–biotin complex for specific staining of biological membranes in electronmicroscopy. *Proc. Natl. Acad. Sci. U.S.A.*, 71: 3537–3541.

Hisano, S., Adachi, T. and Daikoku, S. (1984) Immunolabeling of adenohypophysial cells with protein A–colloidal gold–antibody complex for electron microscopy. *J. Histochem. Cytochem.*, 32: 705–711.

Hofmann, K. (1980) Iminobiotin affinity columns and their application to retrieval of streptavidin. *Proc. Natl. Acad. Sci. U.S.A.*, 77: 4666–4668.

Hökfelt, T. and Dahlstrom, A. (1971) Effects of two mitotic inhibitors (colchicine and vinblastine) on the distribution and axonal transport of noradrenaline storage particles, studied by fluorescence and electron microscopy. *Z. Zellforsch.*, 119: 460–482.

Hökfelt, T., Fuxe, K. and Goldstein, M. (1975) Application of immunohistochemistry to studies on monoamine cell systems with special references to nervous tissues. *Ann. NY Acad. Sci.*, 254: 407–432.

Hökfelt, T., Elfvin, G., Elde, M., Schultzberg, M., Goldstein, M. and Luft, R. (1977) Occurrence of somatostatin-like immunoreactivity in some peripheral sympathetic noradrenergic neurons. *Proc. Natl. Acad. Sci. U.S.A.*, 74: 3587–3591.

Hökfelt, T., Skagerberg, G., Skirboll, L. and Bjorklund, A. (1983) Combination of retrograde tracing and neurochemistry histochemistry. In A. Björklund and T. Hökfelt (Eds.), *Handbook of Chemical Neuroanatomy, Vol. 1. Methods in Chemical Neuroanatomy*. Elsevier, New York, pp. 228–284.

Hökfelt, T., Johanson, O. and Goldstein, M. (1984) Chemical anatomy of the brain. *Nature*, 225: 1326–1334.

Horisberger, M., Rosset, J. and Bauer, H. (1975) Colloidal gold granules as a marker for cell surface receptors in the scanning electron microscope. *Experientia*, 31: 1147.

Hsu, S. M. and Reine, L. (1981) Protein A, avidin and biotin in immunohistochemistry. *J. Histochem. Cytochem.*, 29: 1349–1353.

Hsu, S. M., Reine, L. and Fanger, H. (1981) The use of avidin–biotin–peroxidase complex (ABC) in immunoperoxidase techniques: A comparison between ABC and unlabeled antibody (PAP) procedures. *J. Histochem. Cytochem.*, 29: 577–580.

Ishimoto, I., Shiosaka, S., Shimizu, Y., Kuwayama, Y., Fukuda, M., Inagaki, S., Takagi, H., Sakanaka, M., Takatsuki, K., Senba, E. and Tohyama, M. (1982) Two types of substance P-containing cells and their uneven distribution in the chicken retina: an immunohistochemical analysis with flat-mounts. *Brain Res.*, 240: 171–176.

Karnovsky, M. J. (1965) A formaldehyde-glutaraldehyde fixative of high osmolality for use in electronmicroscopy. *J. Cell Biol.*, 27: 137A.

Katayama-Kumoi, Y., Kiyama, H., Emson, P. C., Kimmel, J. R. and Tohyama, M. (1985) Coexistence of pancreatic polypeptide and substance P in the chicken retina. *Brain Res.*, 361: 25–35.

King, J. C., Lechan, R. M., Kugel, G. and Anthony, E. (1983) A fixative for immunocytochemical localization of peptides in the central nervous system. *J. Histochem. Cytochem.*, 31: 62–68.

Kreutzberg, G. (1969) Neuronal dynamics and flow-IV. Blockage of intra-axonal enzyme transport of colchicine. *Proc. Natl. Acad. Sci. U.S.A.*, 62: 722–728.

Kristenson, K. (1975) Retrograde axonal transport of protein tracers. In W. M. Cowan and M. Cuenod (Eds.), *The Use of Axonal Transport for Studies of Neuronal Connectivity*. Elsevier, New York, pp. 69–82.

Kronvall, G., Seal, U. S., Finstad, J. and Williams, R. C. (1970) Phylogenic insight into evolution of mammalian Fc fragment of rG globulin using staphylococcal protein A. *J. Immunol.*, 104: 140–147.

Kubota, Y., Takagi, H., Morishita, Y., Powell, J. F. and Smith, A. D. (1985) Synaptic interaction between catecholaminergic neurons and substance P-immunoreactive axons in the caudal part of the nucleus of the solitary tract of the rat: demonstration by the electron microscopic mirror technique. *Brain Res.*, 333: (in press).

Kuypers, H. G. J. M., Catsman-Berrevoets, C. E. and Padt, R. E. (1977) Retrograde axonal transport of fluorescent substances in the rat's forebrain. *Neurosci. Lett.*, 6: 127–135.

Larsson, L.-I. (1977) Ultrastructural localization of a new neuronal peptide (VIP). *Histochemistry*, 54: 173–176.

Larsson, L.-I. (1983) Methods for immunocytochemistry of neurohormonal peptides. In A. Björklund and T. Hökfelt (Eds.), *Handbook of Chemical Neuroanatomy. Vol. 1. Methods in Chemical Neuroanatomy*. Elsevier, New York, pp. 228–284.

LaVail, J. H. and LaVail, M. M. (1972) Retrograde axonal transport in the central nervous system. *Science*, 176: 1416–1417.

Lechago, J., Sun, N. C. J. and Weinstein, W. M. (1979) Simultaneous visualization of two antigens in the same tissue section by combining immunoperoxidase with immunofluorescence techniques. *J. Histochem. Cytochem.*, 27: 1221–1225.

Lechan, D. M., Nestler, J. and Jacobson, S. J. (1981) Immunohistochemical localization of retrogradely and anterogradely transported wheat germ agglutinin (WGA) within the central nervous system of the rat: application to immunostaining of a second antigen within the same neuron. *J. Histochem. Cytochem.*, 29: 255–262.

Lee, Y., Kawai, Y., Shiosaka, S., Takami, K., Kiyama, H., Hillyard, C. J., Girgis, S., MacIntyre, I., Emson, P. C. and Tohyama, M. (1985) Coexistence of calcitonin gene-related peptide and substance P-like peptide in single cells of the trigeminal ganglion of the rat: immunohistochemical analysis. *Brain Res.*, 330: 194–196.

Ljungdahl, A., Hokfelt, T. and Nilson, G. (1978) Distribution of substance P-like immunoreactivity in the central nervous system of the rat-I. Cell bodies and nerve terminals. *Neuroscience*, 3: 861–943.

Matsuyama, T., Shiosaka, S., Wanaka, A., Yoneda, S., Kimura, K., Hayakawa, T., Emson, P. C. and Tohyama, M. (1985) Fine structure of peptidergic and catecholaminergic nerve fibers in the anterior cerebral artery and their interrelationship: An immunoelectron microscopic study. *J. Comp. Neurol.*, 235: 268–276.

McLean, I. and Nakane, P. K. (1974) Periodate-lysin-paraformaldehyde fixative. A new fixative for immunoelectron-microscopy. *J. Histochem. Cytochem.*, 22: 1077–1083.

Mesulam, M.-M. (1978) Tetramethyl benzidine for horseradish peroxidase neurohistochemistry; a non-carcinogenic blue reaction product with superior sensitivity for visualising neural afferents and efferents. *J. Histochem. Cytochem.*, 26: 106–117.

Nakane, P. K. and Pierce, G. B. (1966) Enzyme-labeled antibodies: preparation and application for the localization of antigens. *J. Histochem. Cytochem.*, 14: 929–931.

Nakane, P. K. and Pierce, G. B. (1967) Enzyme-labeled antibodies for the light and electron microscopic localization of tissue antigens. *J. Cell Biol.*, 32: 307–318.

Nauta, H. J. W., Pritz, M. B. and Lasek, R. J. (1974) Afferents to the rat caudoputamen studied with horseradish peroxidase. An evaluation of a retrograde neuroanatomical research method. *Brain Res.*, 67: 219–238.

Oldfield, B. J., Hou-Yu, A. and Silverman, A.-J. (1983) Technique for simultaneous ultrastructural demonstration of anterogradely transported horseradish peroxidase and an immunocytochemically identified neuropeptide. *J. Histochem. Cytochem.*, 31: 1145–1150.

Oldfield, B. J., Hou-Yu, A. and Silverman, A.-J. (1985) A combined electron microscopic HRP and immunocytochemical study of the limbic projections to rat hypothalamic nuclei containing vasopressin and oxytocin neurons. *J. Comp. Neurol.*, 231: 221–231.

Pearse, A. G. E. (1962) Buffered formaldehyde as a killing agent and primary fixative for electron-microscopy. *Anat. Rec.*, 142: 342.

Pearse, A. G. E. and Polak, J. M. (1975) Bifunctional reagents as vapour- and liquid-phase fixatives for immunohistochemistry. *Histochem. J.*, 7: 179–186.

Petrutz, P., Dimeo, P., Ordronneau, P., Weaver, C. and Keefer, D. A. (1975) Improved immunoglobulin enzyme bridge method for light-microscopic demonstration of hormone-containing cells of rat adenohypophysis. *Histochemistry*, 46: 9–26.

Pickel, V. M., Joh, T. H. and Reis, D. J. (1975) Ultrastructural localization of tyrosine hydroxylase in noradrenergic neurons of brain. *Proc. Natl. Acad. Sci. U.S.A.*, 72: 659–663.

Priestley, J. V., Somogyi, P. and Cuello, C. (1981) Neurotransmitter-specific projection neurons revealed by combined immunohistochemistry with retrograde transport of HRP. *Brain Res.*, 220: 231–240.

Richard, G. (1983) Ultrastructural visualization of biogeniamine. In A. Björklund and A. Hökfelt (Eds.), *Handbook of Chemical Neuroanatomy, Vol. 1. Methods in Chemical Neuroanatomy*. Elsevier, New York, pp. 122–146.

Riggs, J. L., Seiwald, R. J., Burcholter, J. H., Downs, C. M. and Metcalf, T. G. (1958) Isothiocyanate compounds as fluorescent labeling agents for immune serum. *Am. J. Clin. Pathol.*, 34: 1081–1097.

Ritchie, T. C., Westlund, K. N., Bowker, R. M., Coultier, J. D. and Leonard, R. B. (1982) The relationship of the medullary catecholamine containing neurons to the vagal motor nuclei. *Neuroscience*, 7: 1471–1482.

Romano, E. L. and Romano, M. (1977) Staphylococcal protein A bound to colloidal gold: A useful reagent to label antigen–antibody sites in electron-microscopy. *Immunochemistry*, 14: 711–715.

Romano, E. L., Stolinski, C. and Hughes-Jones, N. C. (1974) An antiglobulin reagent labeled with colloidal gold for use in electron microscopy. *Immunochemistry*, 11: 521–522.

Roth, J. and Bender, M. (1978) Colloidal gold, ferritin and peroxidase as markers for electron microscopic double labeling lectin techniques. *J. Histochem. Cytochem.*, 26: 163–169.

Roth, J., Bendayan, M. and Orci, L. (1978) Ultrastructural localization of intracellular antigens by the use of protein A–gold complex. *J. Histochem. Cytochem.*, 26: 1074–1081.

Roth, J., Bendayan, M., Carlmalm, E., Villiger, W. and Garavito, M. (1981a) Enhancement of structural preservation and immunocytochemical staining in low temperature embedding pancreatic tissue. *J. Histochem. Cytochem.*, 29: 663–761.

Roth, J., Ravazzola, M., Bendayan, M. and Orci, L. (1981b) Application of the protein A–gold technique for electron microscopic demonstration of polypeptide hormones. *Endocrinology*, 108: 247–252.

Rye, D. B., Wainer, B. H., Mesulam, M.-M., Mufson, E. J. and Saper, C. B. (1984) Cortical projections arising in the basal forebrain: A study of cholinergic and noncholinergic components employing combined retrograde tracing and immunohistochemical localization of choline acetyltransferase. *Neuroscience*, 13: 627–643.

Sakanaka, M., Senba, E., Shiosaka, S., Takatsuki, K., Inagaki, S., Takagi, H., Kawai, Y., Hara, Y. and Tohyama, M. (1982) Evidence for the existence of an enkephalin-containing pathway from the area just ventrolateral to the anterior hypothalamic nucleus to the lateral septal area of the rat. *Brain Res.*, 239: 240–244.

Sawchenko, P. E. and Swanson, L. W. (1981) A method for tracing biochemically defined fluorescence retrograde transport and immunohistochemical techniques. *Brain Res.*, 210: 31–41.

Schwab, M. E., Javoy-Agid, F. and Agid, Y. (1978) Labeled wheat germ agglutinin (WGA) as new, highly sensitive retrograde tracer in the rat brain hippocampal system. *Brain Res.*, 152: 145–150.

Shimada, S., Shiosaka, S., Takami, K., Yamano, M. and Tohyama, M. (1985) Somatostatinergic neurons in the insular cortex project to the spinal cord: combined retrograde axonal transport and immunohistochemical study. *Brain Res.*, 326: 197–200.

Shiosaka, S. and Tohyama, M. (1984) Evidence for an α-MSH-ergic hippocampal commissural connection in the rat, revealed by a double-labeling technique. *Neurosci. Lett.*, 49: 213–216.

Shiosaka, S. and Shibasaki, T. and Tohyama, M. (1984) Bilateral α-melanocyte stimulating hormonergic fiber system from zona incerta to cerebral cortex: combined retrograde axonal transport and immunohistochemical study. *Brain Res.*, 309: 350–353.

Shiosaka, S., Kawai, Y., Shibasaki, T. and Tohyama, M. (1985) The descending α-MSH (α-melanocyte-stimulating hormonergic) projection from the zona incerta and lateral hypothalamic area to the inferior colliculus and spinal cord in the rat. *Brain Res.*, 381: 371–375.

Skirboll, L. and Hökfelt, T. (1983) Transmitter specific mapping of neuronal pathways by immunohistochemistry combined with fluorescent dyes. In A.C. Cuello (Ed.), *IBRO Handbook Series: Methods in the Neurosciences. Immunohistochemistry.* Wiley, Chichester.

Skirboll, L., Hökfelt, T., Norell, G., Phillipson, O., Kuypers, H. G. J. M., Bentivoglio, M., Catsman-Berrevoets, C. E., Visser, T. J., Steinbusch, H., Verhofstad, A., Cuello, A. C., Goldstein, M. and Brownstein, M. (1984) A method for specific transmitter identification of retrogradely labeled neurons: Immunofluorescence combined with fluorescence tracing. *Brain Res. Rev.*, 8: 99–127.

Smolen, A. J., Glazer, E. J. and Ross, L. L. (1979) Horseradish peroxidase histochemistry combined with glyoxylic acid-induced fluorescence used to identify brain stem catecholaminergic neurons which project to the chick thoracic spinal cord. *Brain Res.*, 160: 353–357.

Sofroniew, M. V., Couture, R. and Cuello, A. C. (1983) Immunohistochemistry: Preparation of antibodies and staining specificity. In A. Björklund and T. Hökfelt (Eds.), *Handbook of Chemical Neuroanatomy. Vol. 1. Methods in Chemical Neuroanatomy.* Elsevier, New York, pp. 210–227.

Somogyi, P. and Takagi, H. (1982) A note on the use of picric acid–paraformaldehyde–glutaraldehyde fixative for correlated light and electron microscopic immunocytochemistry. *Neuroscience,* 7: 1979–1983.

Steindler, D. A., Isaacson, L. G. and Trosko, B. K. (1983) Combined immunocytochemistry and autoradiographic retrograde axonal tracing for identification of transmitters of projection neurons. *J. Neurosci. Methods,* 9: 217–228.

Sternberger, L. A. (1967) Electron microscopic immunocytochemistry: A review. *J. Histochem. Cytochem.*, 15: 139–159.

Sternberger, L. A. (1970) The unlabeled antibody enzyme method of immunohistochemistry. *J. Histochem. Cytochem.*, 8: 315–325.

Sternberger, L. A. (1979) The unlabeled antibody peroxidase–antiperoxidase (PAP) method. In L. A. Sternberger (Ed.), *Immunocytochemistry,* Wiley, New York, pp. 104–169.

Swanson, L. W., Wyss, J. M. and Cowan, W. M. (1978) An autoradiographic study of the organization of intrahippocampal association pathways in the rat. *J. Comp. Neurol.*, 181: 681–716.

Takeda, N., Inagaki, S., Shiosaka, S., Taguchi, Y., Oertel, W. H., Tohyama, M., Watanabe, T. and Wada, H. (1984a) Immunohistochemical evidence for the coexistence of histidine decarboxylase-like and glutamate decarboxylase-like immunoreactivities in nerve cells of the magnocellular nucleus of the posterior hypothalamus of rats. *Proc. Natl. Acad. Sci. U.S.A.*, 81: 7647–7650.

Takeda, N., Inagaki, S., Taguchi, Y., Tohyama, M., Watanabe, T. and Wada, H. (1984b) Origins of histamine-containing fibers in the cerebral cortex of rats studied by immunohistochemistry with histidine decarboxylase as a marker and transection. *Brain Res.*, 323: 55–63.

Titus, J. A., Haugland, R., Sharrow, S. O. and Segal, D. M. (1982) Texas red, a hydrophilic, red-emitting fluorophore for use with fluorescein in dual parameter flow microfluorochrometric and fluorescence microscopic studies. *J. Immunol. Methods.*, 50: 193–204.

Tramu, G., Pillez, A. and Leonardelli, J. (1978) An efficient method of antibody elution for the successive or simultaneous localization of two antigens by immunocytochemistry. *J. Histochem. Cytochem.*, 26: 322–324.

Van der Kooy, D. and Sawchenko, P. E. (1982) Characterization of serotonergic neurons using concurrent fluorescent retrograde axonal tracing and immunohistochemistry. *J. Histochem. Cytochem.*, 30: 794–798.

Van der Kooy, D. and Steinbusch, H. W. M. (1980) Simultaneous fluorescent retrograde axonal tracing and immunofluorescent characterization of neurons. *J. Neurosci. Res.*, 5: 479–484.

Van der Kooy, D. and Wise, R. A. (1980) Retrograde fluorescent tracing of substantia nigra neurons combined with catecholamine histofluorescence. *Brain Res.*, 183: 447–452.

Vandesande, F. and Dierickx, K. (1975) Identification of the vasopressin-producing and of the oxytocin-producing neurons in the hypothalamic magnocellular neurosecretory system of the rat. *Cell Tissue Res.*, 164: 153–162.

Varndell, I. M., Tapia, F. J., Probert, L., Buchan, A. M. J., Gu, J., De May, J., Bloom, S. R. and Polak, J. M. (1982) Immunogold staining procedure for the localization of regulatory peptides. *Peptide,* 3: 259–272.

Varndell, I. M., Harris, A., Tapia, F. J., Yanaihara, N., De May, J., Bloom, S. R. and Polak, J. M. (1983) Intracellular topography of immunoreactive gastrin demonstrated using electron immunocytochemistry. *Experientia,* 39: 713–717.

Wainer, B. H. and Rye, D. B. (1984) Retrograde horseradish peroxidase tracing combined with the localization of choline-acetyltransferase immunoreactivity. *J. Histochem. Cytochem.*, 32: 439–443.

Wilchek, M. and Bayerm, E. A. (1984) The avidin–biotin com-

plex in immunology. *Immunol. Today,* 5: 39–43.
Woolley, D. W. and Longsmith, L. G. (1942) Isolation of an antibiotin factor from egg white. *J. Biol. Chem.,* 142: 285–290.
Yamamoto, K., Matsuyama, T., Shiosaka, S., Inagaki, S., Senba, E., Shimizu, Y., Ishimoto, I., Hayakawa, T., Matsumoto, M. and Tohyama, M. (1983) Overall distribution of substance P-containing nerves in wall of the cerebral arteries of the guinea pig and its origins. *J. Comp. Neurol.,* 215: 421–426.
Yamano, M., Bai, F.-L., Tohyama, M. and Shiotani, Y. (1985) Ultrastructural evidence of direct synaptic contact of catecholamine terminals with oxytocin-containing neurons in the parvocellular portion of the rat hypothalamic paraventricular nucleus. *Brain Res.,* 336: 176–179.
Yamazoe, M., Shiosaka, S., Emson, P. C. and Tohyama, M. (1985) Distribution of neuropeptide Y in the lower brainstem: an immunohistochemical analysis. *Brain Res.,* 335: 109–120.
Yazierski, R. P., Bowker, R. M., Kevetter, G. A., Westlund, K. N., Coulter, J. D. and Willis, W. D. (1982) Serotonergic projections to the caudal brain stem: a double label study using horseradish peroxidase and serotonin immunocytochemistry. *Brain Res.,* 239: 258–264.
Zamboni, L. and De Martino, C. (1967) Buffered picric acid–formaldehyde; a new rapid fixative for electron microscopy. *J. Cell Biol.,* 35: 148A.

SECTION II

Basal Ganglia

P. C. Emson, M. N. Rossor and M. Tohyama (Eds.),
Progress in Brain Research, Vol. 66.

CHAPTER 2

Distribution of peptides in basal ganglia

Hiroshi Takagi

2nd Department of Anatomy, Kinki University School of Medicine, Sayama-cho, Minami-Kawachi-gun, Osaka 589, Japan

Introduction

There is some confusion over the definition of the term "basal ganglia". Almost all investigators agree that the caudate nucleus and the lentiform nucleus (the putamen and globus pallidus), i.e. the corpus striatum, are the major components of the basal ganglia. The claustrum is sometimes included. Although the amygdaloid nuclei are often included in the basal ganglia, the fibre connections and functions of the nuclei are very different from those of the corpus striatum. Recently, the areas called "ventral striatum", i.e., nucleus accumbens, olfactory tubercle and ventral part of the striatum, and those called "ventral pallidum", i.e., ventral extension of the globus pallidus including the rostral or subcommissural part of the substantia innominata have been shown to be closely related in the rat (Heimer and Wilson, 1975; Heimer, 1976). The present chapter will focus on the distribution of the peptides in the major components of the "basal ganglia", such as the striatum, globus pallidus, nucleus accumbens and olfactory tubercle. The peptide immunoreactivities found in the claustrum will be summarised in Table II. The detailed localization of the peptides in the amygdaloid nuclei was summarized by Shiosaka et al. (1984).

The striatum (or neostriatum) has been one of the areas on which many investigators have focused their interest, because this area is abundant in dopamine fibres and acetylcholinesterase positive structures (for a review see Graybiel and Ragsdale, 1983). Recently, immunocytochemistry and radioimmunoassay studies have demonstrated that the striatum is also rich in various kinds of neuropeptides, and these findings strongly suggest that neuropeptides play some important role in the regulation of the function of the striatum in addition to dopamine and acetylcholine. Furthermore, abnormal levels of peptides in the striatum may relate to psychiatric and movement disorders. Thus, identification of the regional distribution of various kinds of peptides including their fibre connections and fine structure, especially in the striatum, can help us to understand the physiological significance of peptides in the basal ganglia and provide fundamental knowledge of the involvement of peptides in neurological disorders.

Distribution of neuropeptides in the striatum

Neurons in the striatum have been divided into several categories using different techniques such as Golgi impregnation methods, intracellular markers and ultrastructural analysis (for reviews see Pasik et al., 1979; Graybiel and Ragsdale, 1983). Immunocytochemistry has revealed the presence of various kinds of peptide-containing cells in the striatum. Some of the peptide-containing structures are shown to be localized to specific cell types. Thus, to understand the significance of peptidergic cells in the striatum, it is important to elucidate in which cell type a particular immunoreactivity is localized. The present findings are summarized in Table I.

TABLE I

Correspondence of peptide-immunoreactive cells to the category of cell types in the striatum

Immunoreactive cells	Cell types
ENKs	medium-size spiny neuron
DYNs	?
SP	medium-size spiny and aspiny neuron
SRIF	medium-size aspiny neuron
NPY	medium-size aspiny neuron (possibly)
CCK	medium-size aspiny neuron
NT	?
VIP	?
FMRF-amide	?

Opioid peptides

Since Hughes et al. (1975) first isolated from the brain two pentapeptides, methionine-enkephalin (MET-ENK) and leucine-enkephalin (LEU-ENK) with potent opiate agonist activity, many other opioid peptides have been identified in the brain. Among them are (1) enkephalins (ENKs) (MET-ENK and LEU-ENK) and their related peptides (MET-ENK-arg^{6}-gly^{7}-leu^{8} and MET-ENK-arg^{6}-phe^{7}), (2) pro-opiomelanocortin compounds, such as β-endorphin (β-END) and α-melanocyte stimulating hormone (α-MSH), and (3) dynorphins (DYNs) and α-neo-endorphin (α-neo-END), all of which have been demonstrated in the striatum, using radioimmunoassay and immunocytochemistry.

Enkephalins and enkephalin-related peptides

Studies on biosynthesis in the adrenal medulla, demonstrated that pre-pro-enkephalin A may be cleaved into intermediate size peptides containing one or several copies of ENK sequences which in turn give rise to smaller end-products such as MET-ENK and LEU-ENK (Lewis et al., 1979; Jones et al., 1982; Noda et al., 1982). The same precursor as in the adrenal medulla, has been suggested as the source of ENKs in the brain (Bloch et al., 1983). The striatum contains a large amount of each ENK and also contains its possible transient precursors, i.e., BAM-12P, BAM-22P and synenkephalin (Bloch et al., 1983; Pittius et al., 1983; Liston et al., 1984; Zamir et al., 1984, 1985). Many species contain much more MET-ENK than other ENKs, including MET-ENK-arg^{6}-gly^{7}-leu^{8} and MET-ENK-arg^{6}-phe^{7} (Rossier et al., 1980; Ikeda et al., 1983; Pittius et al., 1983; Zamir et al., 1984). For example, human striatum (both caudate nucleus and putamen) contains about 4–6 times more MET-ENK than the other ENKs. The existence of large amounts of ENKs in the striatum is consistent with the evidence that opiate receptors are highly concentrated in this region (Kuhar et al., 1973; Pert et al., 1976; Atweh and Kuhar, 1977; Young and Kuhar, 1979; Herkenham and Pert, 1981; Hamel and Beaudet, 1984).

(a) Enkephalin-like immunoreactive (ENK-LIr) cells. Immunocytochemical studies indicate that MET-ENK, LEU-ENK, MET-ENK-arg^{6}-phe^{7} and BAM-22P (possible transient precursor of ENKs) are contained within the medium- (or small-) size striatal cells in many species (Pickel et al., 1980; Finley et al., 1981a; DiFiglia et al., 1982; Williams and Dockray, 1982; Bloch et al., 1983; Aronin et al., 1984; Graybiel and Chesselet, 1984; Beckstead and Kersey, 1985; Inagaki and Parent, 1985). Whether MET-ENK-arg^{6}-gly^{7}-leu^{8} is also contained in the striatal cells or not is unknown. MET-ENK and LEU-ENK have been suggested to be present in the medium-size spiny neurons in the rat and monkey, from light- and electron-microscopic observations (Pickel et al., 1980; DiFiglia et al., 1982). The perikarya always have an unindented nucleus of round or oval shape which is surrounded by a thin rim of cytoplasm (Fig. 1A,B). The unindented nucleus is a unique feature of medium-size spiny neurons (Somogyi and Smith, 1979; DiFiglia et al., 1980; Dimova et al., 1980; Wilson and Groves, 1980; Bishop et al., 1982).

MET-ENK-LIr cells are distributed throughout the rostro-caudal extent of the striatum in many species (Pickel et al., 1980; DiFiglia et al., 1982; Graybiel and Chesselet, 1984; Beckstead and Kersey, 1985; Inagaki and Parent, 1985), though the

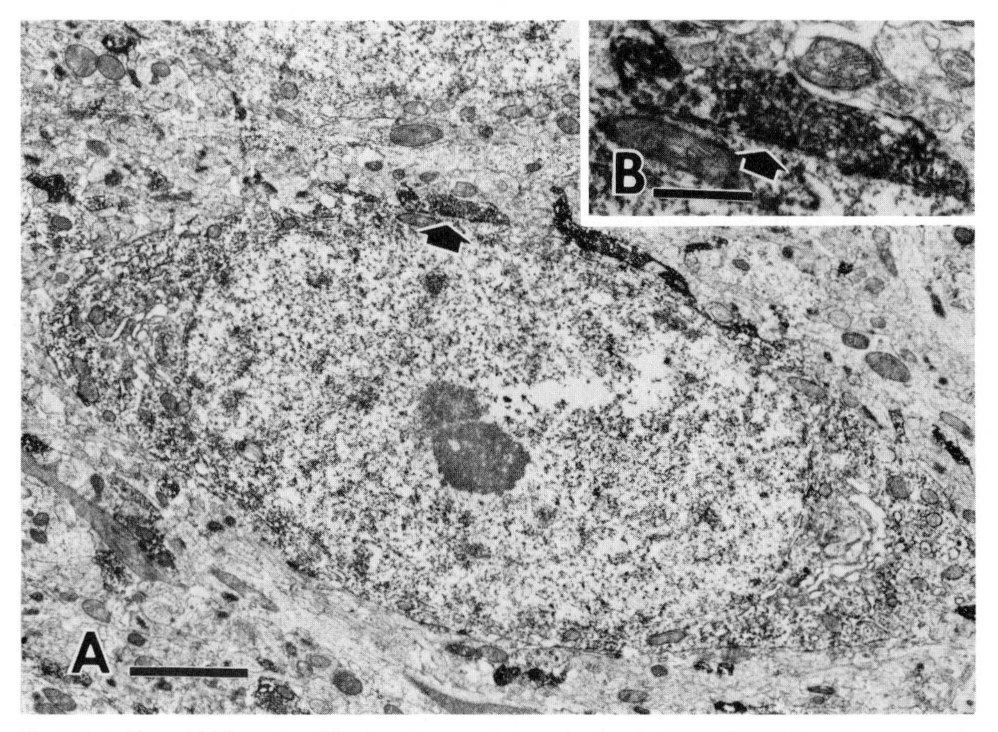

Fig. 1. Low (A) and high (B) magnification electron micrographs showing MET-ENK-LIr perikaryon and boutons in the rat striatum. The morphological features of immunoreactive neurons are similar to those of medium-size spiny neuron; smooth surfaced nucleus and thin rim of cytoplasm. Note a synaptic contact between an ENK-LIr perikaryon and ENK-LIr bouton (arrow). Scales: (A) 2.0 μm; (B) 0.5 μm.

distribution may differ with the species. In the squirrel monkey, the MET-ENK-LIr cells are prominent rostrally in the caudate nucleus and putamen as well as the ventro-medial aspect of the head of the caudate nucleus which is adjacent to the lateral ventricle (Fig. 2A,B) and along the same part of the putamen (Fig. 2A–F) (Inagaki and Parent, 1985). Furthermore, MET-ENK-LIr cells are conspicuous throughout the tail of the caudate nucleus (Fig. 2E,F) and in the ventral part of the putamen at the most caudal level (Fig. 2D–F). MET-ENK-LIr cells in other species such as rat and cat also show a somewhat heterogeneous distribution; they tend to occur more frequently in the medial and/or ventral or the caudal parts (Finley et al., 1981a; Beckstead and Kersey, 1985). Graybiel and Chesselet (1984) reported in the cat that there are MET-ENK-LIr cell-poor zones which are more prominent in the caudate nucleus than the putamen. These areas appear to match acetylcholinesterase-poor zones (striosomes) (see below). Detailed comparative topographical studies on the distribution of MET-ENK-LIr and LEU-ENK-LIr neurons in the striatum have not been made. MET-ENK-like immunoreactivity (MET-ENK-LI) and LEU-ENK-like immunoreactivity (LEU-ENK-LI) appear to coexist in some striatal cells (Aronin et al., 1984). However, a careful analysis must be done, because of the possible crossreactivity between both ENKs, caused by the similarities in amino acid sequences. Larsson et al. (1979), by selectively demonstrating either MET-ENK-LI or LEU-ENK-LI provided evidence that they are localized in two different neuron systems in the guinea pig striatum as well as in the peripheral nervous system.

Williams and Dockray (1982) suggested that

MET-ENK-arg^6-phe^7 may be a transient precursor form of MET-ENK. In the rat striatum, the MET-ENK-arg^6-phe^7 antiserum revealed numerous immunoreactive cell bodies that were not shown by the MET-ENK antiserum, whereas in the globus pallidus abundant MET-ENK-LIr nerve fibres and terminals were revealed, but the MET-ENK-arg^6-phe^7 antiserum did not reveal any immunoreactive structures. Therefore, they suggested that MET-ENK-arg^6-phe^7 may play a role different from that of MET-ENK in the basal ganglia.

Subsequent section analysis has shown that about one half of the MET-ENK-LIr or LEU-ENK-LIr cells in the rat striatum also display glutamic acid decarboxylase (GAD)-like immunoreactivity (GAD-LI), and that MET-ENK-LIr or LEU--ENK-LIr cells account for one half of the GAD-like immunoreactive (GAD-LIr) cells (Aronin et al., 1984). This study, therefore, indicates that there are three populations of striatal neurons, those containing either ENK-LI or GAD-LI alone or both ENK-LI and GAD-LI. GABAergic cells may belong to either medium-size spiny neurons (presumed projection neurons) or medium-size aspiny neurons (presumed interneurons) (Ribak et al., 1979; Bolam et al., 1983a). GABA neurons containing ENK-LI may be identical to the former type of neurons, because all the ENK-LIr neurons seem to belong to this type. On the other hand, GABA neurons lacking ENK-LI may be composed of both types or only of the latter type.

ENK-LIr cells receive synaptic input from non-immunoreactive and immunoreactive axon terminals (Pickel et al., 1980; DiFiglia et al., 1982; Somogyi et al., 1982) (Fig. 1). Recently, Kubota et al. (1986) demonstrated, by using the electron microscopic mirror technique (Kubota et al., 1985) for the detection of two different antigens, that some of the non-immunoreactive axon terminals are tyrosine hydroxylase (catecholamine synthesizing enzyme) -like immunoreactive.

(b) Enkephalin-like immunoreactive striatal afferents. MET-ENK-LIr fibres display heterogeneous distribution patterns in many species, including man (Simantov et al., 1977; Sar et al., 1978; Finley et al., 1981a; Graybiel et al., 1981; Haber and Elde, 1982; Chesselet and Graybiel, 1983; Bouras et al., 1984; Inagaki and Parent, 1985). In monkey, numerous MET-ENK-LIr fibres are present in the rostral portion of the caudate nucleus and putamen where they show a densely stained patchy pattern. This patchy staining is found throughout the head of the caudate nucleus with the highest density being found in the medial and ventral portions (Fig. 2A), and tending to decrease in the caudal direction (Fig. 2B–F). In the caudal portion of the caudate nucleus and putamen, MET-ENK-LI is sparse to moderate, although there are some abundantly stained areas at the most caudal levels (Fig. 2E,F) (Haber and Elde, 1982; Inagaki and Parent, 1985). In the rat, dense accumulation of immunoreactive fibres is found in the medial and ventral parts of the striatum at the level of preoptic anterior hypothalamus (Finley et al., 1981a). The rat striatum also shows a patchy pattern which appears to be less dense and less widely spread (Haber and Elde,

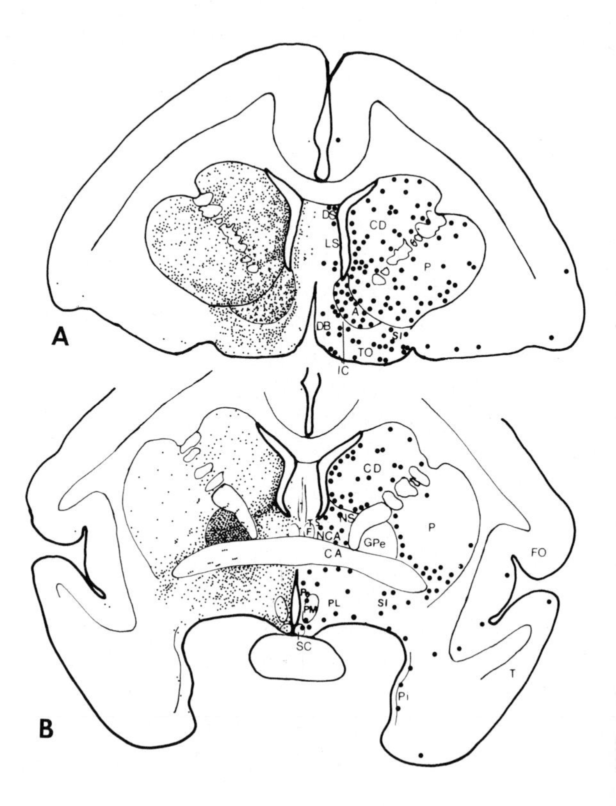

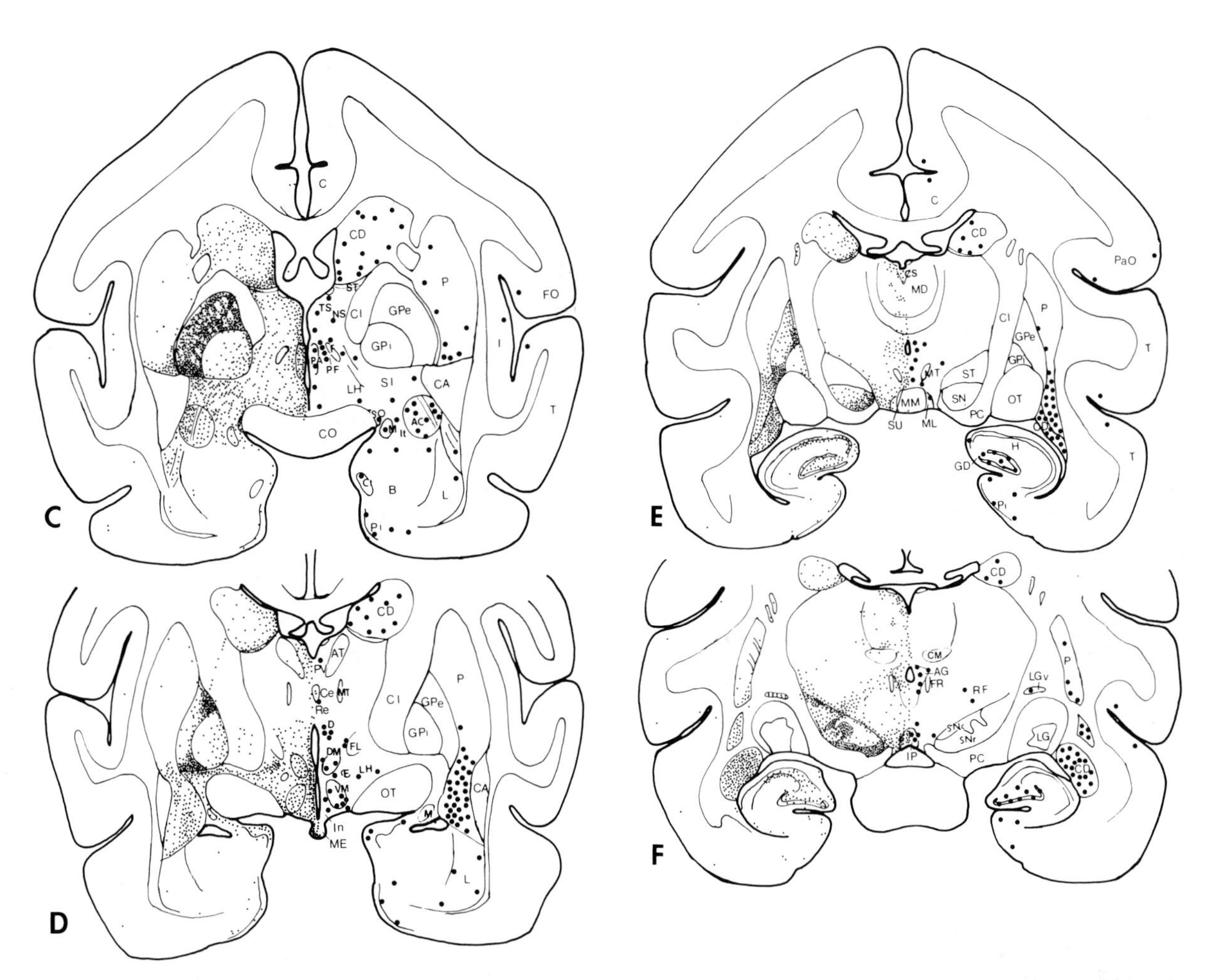

Fig. 2. (A–F) Series of drawings of frontal sections through the forebrain and upper brain stem of squirrel monkey illustrating the topographical distribution of MET-ENK-LIr neurons (●) on the right side and fibres (·) on the left side. Abbreviations: A, nucleus accumbens; CD, caudate nucleus; GPe, globus pallidus, external segment; GPi, globus pallidus, internal segment; P, putamen; SI, substantia innominata; OT, olfactory tubercle; others, see Inagaki and Parent (1985). (Illustrations kindly supplied by Dr. S. Inagaki.)

1982). Graybiel et al. (1981) reported that in the cat striatum these densely stained ENK-LIr patches correspond well to the acetylcholinesterase-poor zones (striosomes) (Graybiel and Ragsdale, 1978). To a certain extent these patches also correspond to the areas of high opioid binding sites (Young and Kuhar, 1979; Herkenham and Pert, 1981; Hamel and Beaudet, 1984). This chemoarchitectural compartmentalization of the striatum may be related to the organization of striatal afferent and efferent connections (Graybiel and Ragsdale, 1978; Graybiel et al., 1981; Gerfen, 1984). Recently, Graybiel and Chesselet (1984) have demonstrated that different immunocytochemical protocols for the MET-ENK-LIr structures result in remarkably different staining patterns with respect to compartmental distribution. MET-ENK-rich patches are stained as zones corresponding to the striosomes in the sections treated with a protocol favouring fibre immunostaining. In contrast to this, the striosomes appear as MET-ENK-poor patches surrounded by MET-ENK-LIr cells and neuropils in the sections treated with an alternative protocol favouring perikaryal staining. These data suggest that the immunostaining of ENK-LIr neuropil achieved with a single immunocytochemical protocol may not fully represent the immunoreactive elements.

The human striatum also shows a heterogeneous distribution of MET-ENK-LI with predominant staining in the rostral portions (Bouras et al., 1984). The distribution of MET-ENK-LIr and LEU-ENK-LIr fibres appears to be very similar in the guinea pig striatum, but MET-ENK-LIr fibres are about 4–5 times as numerous as LEU-ENK-LIr fibres (Larsson et al., 1979).

Electron microscopic examinations have shown that ENK-LIr striatal afferents establish synaptic contacts with heterogeneous (ENK-negative and ENK-positive) neuronal elements on perikarya and dendrites. Two types of postsynaptic neurons were identified in the monkey and rat striatum, by using LEU-ENK antiserum (DiFiglia et al., 1982; Somogyi et al., 1982). The first type of neurons may belong to medium-size spiny neurons which are either ENK-negative or ENK-positive with the former occurring more frequently than the latter. The second type of neurons may belong to Golgi-classified aspiny type 1 neurons of DiFiglia et al. (1980) in the monkey striatum, or to the large aspiny neurons of Bolam et al. (1981b) in the rat striatum, both of which are non-immunoreactive. LEU-ENK-LIr axon terminals also form synapses with the axon hillocks and initial segments of non-immunoreactive cells (DiFiglia et al., 1982). No axo-axonic synaptic contacts have been identified, although there are many direct appositions between ENK-LIr axon terminals and non-immunoreactive terminals (Pickel et al., 1980; DiFiglia et al., 1982; Somogyi et al., 1982). Bouyer et al. (1984) demonstrated two kinds of relationship between MET-ENK-LIr axon terminals and degenerating terminals after cortical lesions. One was a direct apposition without synaptic membranous specialization; the functional significance of this remains to be elucidated. The other was that the same ENK-negative dendrites and spines made synaptic contacts with both degenerating and ENK-LIr axon terminals, suggesting the convergence of cortical efferents and enkephalinergic afferents on the same striatal neurons. Pickel et al. (1980) reported that MET-ENK-LIr and LEU-ENK-LIr axon terminals make asymmetrical synaptic contacts with the other neuronal elements in the rat striatum, whereas other authors have shown that a great majority of LEU-ENK-LIr axon terminals form symmetrical type synapses in the rat and monkey (DiFiglia et al., 1982; Somogyi et al., 1982). This disagreement is puzzling, since one of the LEU-ENK antisera used by Pickel et al. (1980) and Somogyi et al. (1982) was obtained from the same source. Another interesting finding is that both types of synapse, i.e. symmetrical and asymmetrical, were formed by some of the same LEU-ENK-LIr terminals (in fig. 26 of DiFiglia et al., 1982).

The origin of the ENK-LIr afferents might be primarily if not entirely, intrinsic, since there are abundant ENK-LIr cells in the striatum and they are considered to be medium-size spiny neurons that give rise to many local axon collaterals (Preston et al., 1980; for a review see Pasik et al., 1979). The findings that these collaterals always form symmetrical synapses (Wilson and Groves, 1980; Somogyi et al., 1981; Bishop et al., 1982) may favour the abundance of a symmetrical type formed by ENK-LIr axon terminals in the striatum.

Since numerous ENK-LIr neurons also exhibit GAD-LI (Aronin et al., 1984), many of the ENK-LIr afferents may be GABAergic.

An extrinsic source of ENK-LIr afferents has not yet been identified.

(c) Enkephalin-like immunoreactive striatal efferents. The ENK-containing striato-pallidal pathway has been unequivocally demonstrated with immunocytochemistry combined with lesion or retrograde tracer labeling method in the rat (Cuello and Paxinos, 1978; Brann and Emson, 1980; Corrêa et al., 1981; DelFiacco et al., 1982). These observations are supported by the fact that medium-size spiny neurons have been observed to project to the globus pallidus (Chang et al., 1981; Bishop et al., 1982). The globus pallidus may be innervated by

diffusely distributed striatal enkephalinergic cells in an overlapping fashion (Corrêa et al., 1981). In the primate, a very dense network of ENK-LIr fibres is distributed in the external segment of the globus pallidus, homologous to the globus pallidus of the nonprimate. By contrast, only a few or moderate number of ENK-LIr fibres are found in the internal segment of the globus pallidus, homologous to the entopeduncular nucleus of nonprimates (Haber and Elde, 1982; Bouras et al., 1984; Grafe et al., 1985; Inagaki and Parent, 1985) (Fig. 2B–E). There is no direct evidence for enkephalinergic connections between the striatum and the internal segment of the globus pallidus (or entopeduncular nucleus) although there may be collaterals from ENK-LIr fibres descending to the substantia nigra.

Whether an enkephalinergic striato-nigral pathway exists is not clear. Only a few ENK-LIr fibres have been reported in the substantia nigra of rat (Sar et al., 1978; Wamsley et al., 1980; Finley et al., 1981a; Khachaturian et al., 1983). However, Inagaki and Parent (1984) demonstrated a progressive increase from the rodent to primates in both the number and the complexity of organizational features of MET-ENK-LIr fibres in the substantia nigra (see also Haber and Elde, 1982; Beckstead and Kersey, 1985) (Fig. 2F). Furthermore, levels of ENK-LI decrease significantly in the substantia nigra of patients with Huntington's disease (Emson et al., 1980a; Marshall et al., 1983) or with striato-pallidal infarction (Pioro et al., 1984). These findings argue in favour of this pathway in man. (See Emson, this volume.)

GABA is considered to be contained both in the striato-pallidal and striato-nigral pathways (Brownstein et al., 1977; Gale et al., 1977; Fonnum et al., 1978; Jessell et al., 1978; Nagy et al., 1978; Ribak et al., 1980; Staines et al., 1980). The coexistence of GAD-LI and ENK-LI in a proportion of the striatal cells (Aronin et al., 1984; also see above) suggests that ENK-LI may be partly, if not exclusively, associated with these GABAergic striatofugal pathways.

ENK-LIr fibres in the striatum are sometimes myelinated (Pickel et al., 1980; DiFiglia et al., 1982). These myelinated fibres may be some of the striatofugal axons.

(d) Summary. A large number of ENK-LIr cells are seen in the striatum, which may belong to the medium-sized spiny neurons. The distribution pattern of these cells differs amongst species. About half of the ENK-LIr cells contain GABA. ENK-LIr neurons seem to project to the globus pallidus in the cat and rat, and in the primates to the external segment of the globus pallidus homologous to the globus pallidus of nonprimates. In addition, in primates, these neurons appear to project to the substantia nigra. ENK-LIr neurons receive both immunoreactive and non-immunoreactive afferents. Some of the latter inputs have been shown to contain catecholamines.

A dense ENK-LIr fibre plexus is also seen in the striatum, though there do exist species differences. Most of them seem to originate from intrinsic ENK-LIr neurons. Two types of neurons, postsynaptic to ENK-LIr terminals, have been identified in the monkey and rat striatum: one belonging to a medium-sized spiny neuron (both ENK-negative and ENK-positive), and the other to an aspiny type 1 neuron (or large size aspiny neuron; always lacking ENK-LI).

β-Endorphin

The striatum contains a small amount of β-endorphin-like immunoreactivity (β-END-LI) (Verhoef et al., 1982). β-END may exert some effects on dopamine metabolism in the striatum (Van Loon and Kim, 1978; George and Van Loon, 1982; Kameyama et al., 1982). It is suggested that β-END-LI derives from the ascending hypothalamic pathway originating in or near the arcuate nucleus (Watson et al., 1978a; Pelletier et al., 1980). Oertel and Mugnaini (1984) and Oertel et al. (1983), by using GAD antiserum and a monoclonal antibody to the N-terminal of β-END, demonstrated that GAD-LI and opioid peptide-like immunoreactivity coexist in the medium-size, striatal neurons. The staining with their β-END antibody, however, may not represent β-END itself, but may be crossreactivity with other opioid peptide(s).

Some of the other pro-opiomelanocortin compounds, adrenocorticotropin (ACTH) and β-lipotropin have not been found in the striatum (Watson et al., 1978a; Pelletier and LeClerc, 1979), although these two peptides and β-END could be localized in the same cell bodies in the hypothalamus (Watson et al., 1978a; Pelletier et al., 1980).

α-Melanocyte stimulating hormone

In good agreement with radioimmunoassay data showing low concentrations of α-MSH-like immunoreactivity (α-MSH-LI) in the striatum (Eskay et al., 1979; O'Donohue et al., 1979), a few α-MSH-like immunoreactive (α-MSH-LIr) fibres were found in the rat striatum (Guy et al., 1981; Umegaki et al., 1983). α-MSH-LIr fibres may originate in the arcuate nucleus and/or dorsolateral hypothalamus (Eskay et al., 1979; Pelletier et al., 1980; Watson and Akil, 1980). The autoradiographic localization of ^{125}I-α-MSH suggests the presence of α-MSH receptors in the rat striatum (Pelletier et al., 1975).

Dynorphins and α-neo-endorphin

Radioimmunoassay and immunocytochemical studies indicate that the striatum contains a moderate amount of DYN-like immunoreactivity (DYN-LI) such as DYN_{1-17} (DYN-A), DYN_{1-13} (DYN-B) and DYN_{1-18}, and α-neo-END (Höllt et al., 1980; Weber et al., 1982; Graybiel and Chesselet, 1984; Zamir et al., 1984; Quirion et al., 1985). DYNs and α-neo-END have been also shown to occur in high concentrations in the substantia nigra of the rat and of primates (Vincent et al., 1982a; Weber et al., 1982; Haber and Watson, 1983; Zamir et al., 1984). The striatum has been found to contain many immunoreactive cell bodies, using antiserum raised either to DYN-A (Vincent et al., 1982a) or DYN-B (Graybiel and Chesselet, 1984). Vincent et al. (1982a) provided further evidence for a DYNergic striato-nigral pathway by showing that DYN-like immunoreactive (DYN-LIr) fibres in the substantia nigra markedly decrease after striatal lesions. DYN-LIr cell bodies are of medium size and it is therefore probable that DYN-LIr projecting cells belong to medium-size spiny neurons. DYN-B-LIr cells are selectively concentrated in the striosomes (Graybiel and Chesselet, 1984). Both DYN-A-LIr and DYN-B-LIr fibres are also found in the globus pallidus and entopeduncular nucleus to various extents depending on the animal species (Chesselet and Graybiel, 1983; Haber and Watson, 1983). It may be that DYNs are also associated with the striato-pallidal pathway, although there is as yet no evidence. Morphological features of the neuronal structures containing α-neo-END-like immunoreactivity have not been established.

Tachykinins

Substance P

Substance P (SP) is a tachykinin that shares the common C-terminal amino acid sequence Phe-X-Gly-Leu-Met-NH_2 (Erspamer, 1981). SP (X = Phe) was first discovered by Von Euler and Gaddum (1931). Radioimmunoassay studies showed that a low or moderate amount of SP-like immunoreactivity (SP-LI) is contained in the striatum (Buck et al., 1981; Aronin et al., 1983; Tenovuo et al., 1984), and a large number of SP-like immunoreactive (SP-LIr) neurons and fibres have been found in the striatum (Cuello and Kanazawa, 1978; Ljungdahl et al., 1978; Bolam et al., 1983b; Beach and McGeer, 1984; Kohno et al., 1984; Beckstead and Kersey, 1985). This peptide plays an important role in the functioning of the basal ganglia (Del Río et al., 1983; Melis and Gale, 1984; Sagar et al., 1984). Rothman et al. (1984) revealed the heterogeneity of detectable SP receptors in the rat striatum, although whether this implies the existence of physiological SP receptor subtypes is not clear (Sandberg and Iversen, 1982).

(a) Substance P-like immunoreactive cells. SP-LIr cell bodies were reported to be distributed predominantly in the rostral part of the striatum of the rat (Cuello and Kanazawa, 1978; Ljungdahl et al., 1978). However, recent studies have demonstrated that SP-LIr cells occur throughout the rostro-caudal extent of the striatum not only in the rat

(Kohno et al., 1984), but also in other species such as the cat (Beckstead and Kersey, 1985) and baboon (Beach and McGeer, 1984). In the rat, SP-LIr cells in the posterior part of the striatum could be detected only after in situ colchicine injection, while in the cat and baboon, they could be found even without colchicine pretreatment. This may be due to the difference in the concentration of the peptides contained in the cell bodies between the two species (Beach and McGeer, 1984). In the cat, many SP-LIr cells are in the medial half of the caudal three-quarters of the putamen and in clusters of irregular sizes and shapes in the head of the caudate nucleus (Beckstead and Kersey, 1985). Clusters of SP-LIr cells are also found in the baboon neostriatum (Beach and McGeer, 1984) and in the dorsal part of the rat caudate-putamen (Kohno et al., 1984). SP-LIr cell bodies appear to be heavily concentrated within the dense patches of SP-LI (Beach and McGeer, 1984). Graybiel et al. (1981) first reported SP-LIr dense patches in the caudate nucleus of the cat. SP-LIr dense patches or SP-LIr poor zones in the caudate nucleus correspond in many cases to LEU-ENK-LI-rich and acetylcholinesterase-poor zones (striosomes). The staining pattern of SP-LIr dense patches seems to represent clusters of SP-LIr cell bodies with arborizations of their local axon collaterals in the materials which are good enough to resolve individual immunoreactive cell bodies (Graybiel and Chesselet, 1984; Beckstead and Kersey, 1985).

The neurons have medium-size perikarya with round, oval, triangular or polymorphic shape (Cuello and Kanazawa, 1978; Ljungdahl et al., 1978; Bolam et al., 1983b; Beach and McGeer, 1984; Kohno et al., 1984; Beckstead and Kersey, 1985). In human brain, Beach and McGeer (1984) reported the presence of abundant medium-size SP-LIr neurons in the caudate nucleus and putamen. Some authors further reported a few large SP-LIr neurons with polymorphic perikarya (Ljungdahl et al., 1978; Beach and McGeer, 1984). The electron microscopic examination of SP-LIr cells demonstrated two different types (type I and II) of cell bodies in the rat striatum (Bolam et al., 1983b). Both types of cell bodies are round or oval, and of medium size, but type I has a smooth surfaced nucleus and has ultrastructural features typical of the medium-size spiny neurons. On the other hand, type II has a deeply indented nucleus with a thin rim of cytoplasm and seems to belong to the medium-size aspiny neurons. Medium-size cells with triangular and pleomorphic perikarya, or large-size cells have not yet been analyzed at the electron microscopic level. There is no evidence that SP-LIr cells contain other known bioactive substances, at least in the striatum. It has been suggested that SP and GABA are contained in separate striatonigral pathways (Brownstein et al., 1977; Gale et al., 1977; Jessell et al., 1978).

(b) Substance P-like immunoreactive striatal afferents. SP-LIr nerve fibres and terminals are homogeneously distributed throughout the rat striatum except for SP-LIr bands in the caudal part (Ljungdahl et al., 1978), whereas, the striatum of the cat, baboon and man shows SP-LI-rich patches which may be associated with the location of SP-LIr striatal afferents as well as intrinsic cells (Graybiel et al., 1981; Beach and McGeer, 1984; Beckstead and Kersey, 1985).

Both intrinsic and extrinsic neuronal sources are suspected to be the source of SP-LIr afferents. Bolam et al. (1983b) identified axon collaterals emerging from the cell bodies of their SP-LIr type I neurons. These collaterals were found to form symmetrical synaptic contact with a non-immunoreactive dendritic shaft. These findings indicate that at least some of the SP-LIr striatal axons are supplied by this type of neuron. However, other intrinsic sources, including the SP-LIr type II neurons of Bolam and colleagues may be possible. As to the extrinsic origin, Sugimoto et al. (1984) demonstrated that the thalamostriatal pathway is one candidate, by sequential section analysis with the retrograde horseradish peroxidase (HRP) labeling method and immunocytochemistry. SP-LIr cells located in the globus pallidus may also be a possible candidate, although there is no direct evidence for this (Cuello and Kanazawa, 1978; Marshall et al., 1983).

The synapses formed by SP-LIr boutons in the

rat striatum have been so far identified to belong only to the symmetrical type (Bolam et al., 1983b). Their postsynaptic targets are dendritic shafts, some of which possess spines or correspond to the proximal dendrites of non-immunoreactive neurons which are believed from ultrastructural evidence to be medium-size spiny neurons. A few SP-LIr boutons also form synapses with the initial segments of SP-LIr cells.

(c) Substance P-like immunoreactive striatal efferents. Lesion studies indicate that SP-LI is contained in the striato-nigral pathway (Brownstein et al., 1977; Gale et al., 1977; Hong et al., 1977; Kanazawa et al., 1977b; Mroz, 1977; Jessell et al., 1978; Ben-Ari et al., 1979; Pioro et al., 1984) and striato-entopeduncular pathway (Paxinos et al., 1978; Kanazawa et al., 1980). The presence of a striato-nigral SP pathway has been confirmed by in vivo radiolabeling procedure with ^{35}S- and ^{3}H-SP (Sperk and Singer, 1982; Torrens et al., 1982; Krause et al., 1984).

A recent experimental immunocytochemical study has revealed that there exist two subpathways in the striato-nigral SP tract: SP-LIr cells located in the rostral part of the striatum of the rat, except its dorsal region, project to the medial two-thirds of the substantia nigra, whereas the SP-LIr cells in the caudal striatum innervate preferentially the lateral third of the substantia nigra (Kohno et al., 1984). Medium-sized spiny neurons classified by Golgi studies are the major types of the two discrete striato-nigral projection neurons identified in the rat (Somogyi and Smith, 1979; Somogyi et al., 1979, 1981; Bolam et al., 1981a,b). SP-LIr type I cells defined by Bolam et al. (1983b) using immunocytochemistry may correspond to the neurons mentioned above. Whether another type (large aspiny) striato-nigral projection neurons (Bolam et al., 1981a,b), do or do not contain SP, is unknown, although some authors reported that large SP-LIr neurons are occasionally present in the striatum (Ljungdahl et al., 1978; Beach and McGeer, 1984).

In primates, including man, a dense SP-LIr staining of nerve fibres occurs in the internal segment in contrast to the weak staining in the external segment except for a few small areas (Haber and Elde, 1981; Beach and McGeer, 1984; Grafe et al., 1985). This strongly suggests that there exists a striato-pallidal (internal segment) SP pathway in the primate.

In Huntington's and Parkinson's diseases, impairments of SP striato-pallidal and striato-nigral pathways are thought to be related to the pathogenesis of these basal ganglia diseases (Kanazawa et al., 1977a; Emson et al., 1980a; Buck et al., 1981; Aronin et al., 1983; Mauborgne et al., 1983; Tenovuo et al., 1984). However, it is not clear whether or not the abnormal level of SP-LI in this system is a primary or secondary phenomenon. (See Agid et al., this volume, and Emson, this volume.)

Other tachykinins

Recent biochemical studies have shown that two tachykinins other than SP are contained in the striatum; substance K (neurokinin α) and kassinin (neurokinin β) (Kanazawa et al., 1984). In contrast to the high density of SP binding sites, only a low density of substance K or kassinin (a potent structural analogue of substance K) binding sites was found (Mantyh et al., 1984). Localization of these two novel peptides in the striatum has not yet been explored.

Summary

A number of SP-LIr neurons are distributed in the striatum in various mammals. These neurons are composed of striato-nigral and striato-entopeduncular SP tracts. Medium-sized spiny neurons (SP-LIr type I cells) seem to be the origin of these tracts. In addition, medium-sized aspiny neurons (SP-LIr type II cells) were also detected.

Numerous SP-LIr fibres were also detected in the striatum, and were considered to be both intrinsic and extrinsic to the striatum.

Somatostatin

Somatostatin (SRIF)-like immunoreactivity (SRIF-LI) has been shown to be present in the striatum by radioimmunoassay (Brownstein et al., 1975; Kobayashi et al., 1977; Barden et al., 1981; Beal et al., 1983) and by immunocytochemical studies (Hökfelt et al., 1978; Bennett-Clarke et al., 1980; Finley et al., 1981b; Graybiel et al., 1981; DiFiglia and Aronin, 1982; Shiosaka et al., 1982; Vincent et al., 1982b,c, 1983a; Vincent and Johansson, 1983). SRIF in the brain exists mainly in two different molecular weight forms: SRIF-14, composed of 14 amino acids, and SRIF-28, of 28 amino acids (Schally et al., 1976; Lauber et al., 1979; Rorstad et al., 1979; Patel et al., 1981; Aronin et al., 1983; Benoit et al., 1984). The striatum contains both SRIF-14 and SRIF-28; the former is the major form (Beal et al., 1983). The striatum contains a moderate or high density of SRIF binding sites (Srikant and Patel, 1981; Tapia-Arancibia et al., 1981; Epelbaum et al., 1982).

Although the role of SRIF in the basal ganglia has not been fully understood, radioimmunoassay studies have demonstrated that SRIF is increased in the striatum and globus pallidus in Huntington's disease (Aronin et al., 1983; Nemeroff et al., 1983b). Also it is suggested that SRIF is involved in the local regulation of striatal dopaminergic transmission (Garcia-Sevilla et al., 1978; Chesselet and Reisine, 1983; Beal and Martin, 1984b). Furthermore, animal experiments have shown that administration of SRIF can alter locomotor activity (Cohn and Cohn, 1975).

(a) Somatostatin-like immunoreactive cells. SRIF-like immunoreactive (SRIF-LIr) cell bodies are almost uniformly distributed throughout the striatum at all levels with somewhat more cells in the medial-ventral portions in the rat (Johansson et al., 1984), or in the dorsal cap of the caudate nucleus of the cat (Graybiel et al., 1981). SRIF-LIr cells usually occur individually, but sometimes form small groups (Graybiel et al., 1981; Takagi et al., 1983). Johansson et al. (1984) calculated that there are about 50–60 (up to 100) immunoreactive cell bodies on one side of a 8–14 μm-thick section in the colchicine-treated rat.

The SRIF-LIr cells have perikarya of various shapes; round, ovoid, spindle and triangular. The

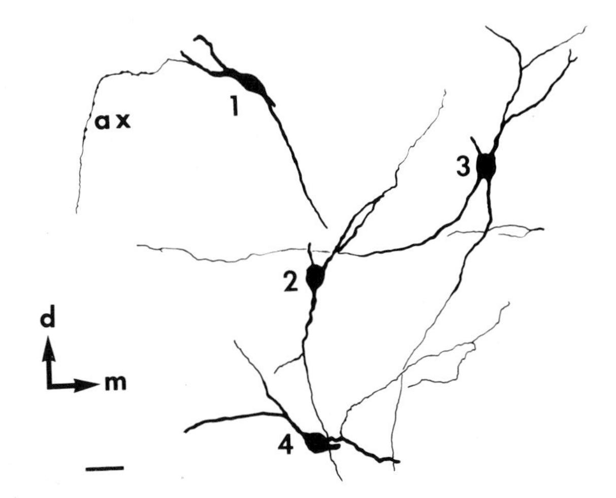

Fig. 3. Camera lucida drawing of SRIF-LIr neurons in the rat striatum. All cells are drawn from the animals without colchicine pretreatment. An axon (ax) with many varicosities can be seen arising from cell 1. d, dorsal; m, medial. Scale: 20 μm.

diameter of the perikarya varies according to their shape, but in many species, all of the SRIF-LIr cells may belong to the medium-size neurons on the basis of light microscopy (Bennett-Clarke et al., 1980; Finley et al., 1981b; Graybiel et al., 1981; DiFiglia and Aronin, 1982; Vincent et al., 1982b,c, 1983a; Vincent and Johansson, 1983; Johansson et al., 1984; Marshall and Landis, 1985). They give rise to two or more dendrite-like processes without apparent spines at the light microscopic level (DiFiglia and Aronin, 1982; Takagi et al., 1983; Vincent et al., 1983a) (Fig. 3). The electron microscopic studies using the rat confirmed the absence (Takagi et al., 1983) or paucity of dendritic spines (DiFiglia and Aronin, 1982), and further demonstrated that the cell nucleus is always indented (DiFiglia and Aronin, 1982; Takagi et al., 1983; Vincent and Johansson, 1983). Therefore, these results suggest that SRIF-LIr cells correspond to the medium-size aspiny neurons. Although detailed comparison of the immunoreactive neurons with subtypes of neurons classified according to the Golgi method is limited (Takagi et al., 1983), some of the SRIF-LIr cells seem to correspond to medium-sized (aspiny) type III nerve cells as described by Dimova et al. (1980) in the rat, based upon the beaded-like appearance

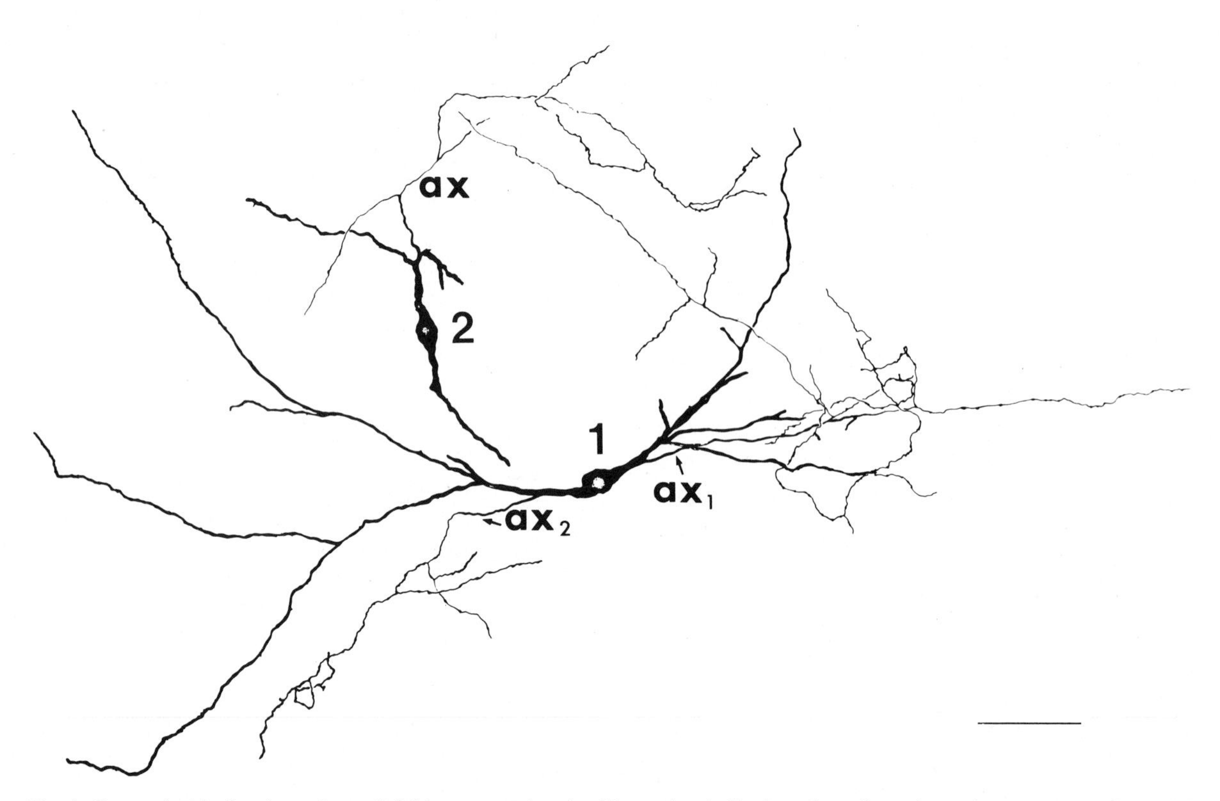

Fig. 4. Camera lucida drawings of two Golgi-impregnated and gold-toned, spindle-shaped, medium-size aspiny neurons (cell 1 and 2) in the rat striatum. The axons (ax, ax_1 and ax_2) of each neuron originate from the base of a primary dendrite as it emerges from the perikaryon. Note that one of the neurons (cell 1) has two axons (ax_1 and ax_2) projecting from the opposite poles of the cell. Scale: 50 μm.

of their dendrites and the fine structural characteristics of the cell bodies (DiFiglia and Aronin, 1982). On the other hand, some other cells (cell 1 in Fig. 3) well resemble the spindle shaped, medium-size aspiny (MA1) neurons of Takagi et al. (1984b) (Figs. 4, 5A). Furthermore, some SRIF-LIr neurons resemble an aspiny neuron intracellularly filled with HRP and electrophysiologically excited following stimulation of the cortex or substantia nigra (Bishop et al., 1982). Since SRIF-LIr neurons receive both asymmetrical and symmetrical synapses from non-immunoreactive boutons on their perikarya and dendrites (Takagi et al., 1983), some of the non-immunoreactive boutons forming the former type of synapse may belong to cortical or nigral afferents.

It is still not known whether SRIF-LIr neurons are interneurons, or send their axons to other areas outside the striatum. It is likely that SRIF-LIr neurons belong to the former type, judging from the comparative analysis with Golgi studies as described above. However, some of the SRIF-LIr fibres are present in both pallidal structures and substantia nigra (Finley et al., 1981b; Johansson et al., 1984). Further studies are needed to decide whether there also exist SRIF striato-fugal pathways.

Vincent et al. (1982b,c) demonstrated the coexistence of SRIF-LI and avian pancreatic polypeptide (APP)-like immunoreactivity (APP-LI) in rat striatal neurons. APP-like immunoreactive (APP-LIr) neurons may be identical to neuropeptide Y-like immunoreactive (NPY-LIr) ones in the rat brain (Allen et al., 1983; also see below). It is not clear

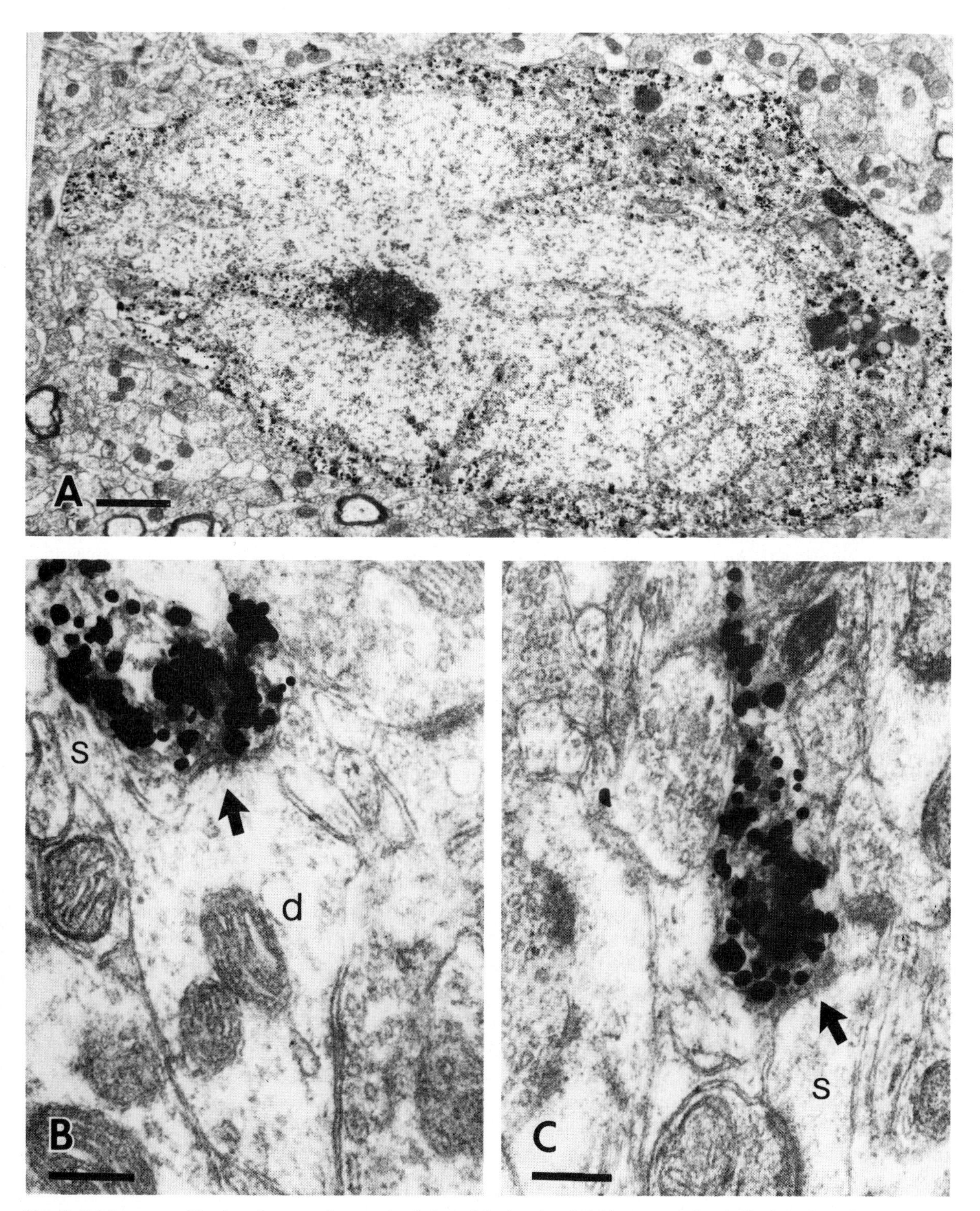

Fig. 5. (A) Low magnification electron micrograph of the cell body of a Golgi-impregnated, spindle-shaped, medium-size aspiny neuron (cell 2 in Fig. 4). Note the highly indented nucleus. (B, C) Electron micrographs showing symmetrical synapse (arrows) established by Golgi-impregnated boutons of local axon collaterals of cell 1 (A) and cell 2 (B) (Fig. 4) with unimpregnated dendritic profiles. The impregnated boutons are in synaptic contact with a dendritic shaft (d) emitting a spine (s) (A) or with a spine (B). Scales: (A) 1 μm; (B, C) 0.2 μm.

whether all of the SRIF-LIr striatal cells contain APP-LI, or presumed NPY-LI. Vincent et al. (1983a) also reported that all of the striatal neurons showing nicotinamide adenosine dinucleotide phosphate (NADPH)-diaphorase activity contain both SRIF-LI and APP-LI. The functional significance of the coexistence of these three different substances in a single neuron is not known. It is interesting to note that in the other brain areas, NADPH-diaphorase-positive cells do not always contain these two peptide-like immunoreactivities (Vincent et al., 1983a). No other reports suggest the coexistence of SRIF and the other bioactive substances except for NPY in single striatal cells. Cholinergic cells or GABAergic cells seem to belong to a separate population from SRIF-LIr cells (Vincent et al., 1983b; Araki et al., 1985).

(b) Somatostatin-like immunoreactive striatal afferents. Fine fibres with numerous varicosities are present in the striatum. They increase in density towards the medial-ventral parts of the striatum with patches of high density in the ventral aspects extending ventrally into the nucleus commissuralis interstitialis anterior (Elde and Hökfelt, 1979; Bennett-Clarke et al., 1980; Johansson et al., 1984). Differences in topographical distribution of SRIF-LIr fibres correspond to the radioimmunoassay data (Beal et al., 1983).

Electron microscope studies have demonstrated that the type of synapses formed by SRIF-LIr boutons are always symmetrical and their postsynaptic targets are mainly dendritic shafts, followed by spines and less frequently perikarya (DiFiglia and Aronin, 1982, 1984; Takagi et al., 1983) (Fig. 6). This suggests that some of their postsynaptic targets are spiny neurons.

The origin of SRIF-LIr fibres is uncertain. The presence of numerous SRIF-LIr cells in the striatum suggests that the afferents are of intrinsic origin. In fact, Golgi-classified medium-size aspiny neurons that closely resemble SRIF-LIr cells give rise to their local axon collaterals with boutons that always form symmetrical synapses (Takagi et al., 1984b) (Fig. 5B,C). However, radioimmunoassay determinations following neurotoxic lesions of the striatum suggest that about half of the SRIF-LI may be of extrinsic origin, although the other half is localizable to interneurons (Beal and Martin, 1983a). There is, however, no further evidence demonstrating the extrinsic source(s) of SRIF afferents in the striatum. The fact that lesions made in the globus pallidus, substantia nigra, frontal cortex and amygdaloid nuclei had no effect on biochemical quantities of SRIF-LI in the striatum suggests that these areas at least do not contribute to striatal SRIF-LI (Beal and Martin, 1983a,b).

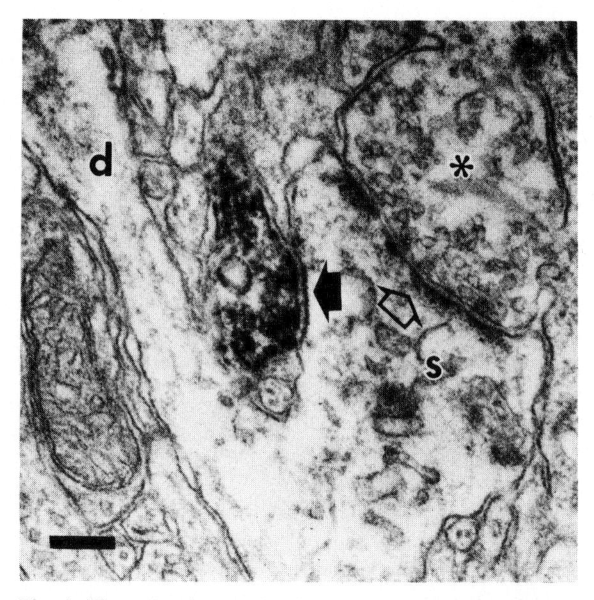

Fig. 6. Electron micrograph of a striatal SRIF-LIr bouton in symmetrical synaptic contact (solid arrow) with a non-immunoreactive dendritic spine (s). The spine is also in asymmetrical synaptic contact (open arrow) with a nonreactive bouton (asterisk). d, dendritic shaft. Scale: 0.2 μm.

(c) Summary. SRIF-LIr cells are almost uniformly distributed in the striatum. These cells belong to the medium-sized aspiny neurons. The projecting fields of these cells are obscure. Mny SRIF-LIr cells are shown to contain APP-LI.

Dense fibre plexuses of SRIF-LIr fibres are also seen in the striatum. The striatal neurons on which SRIF-LIr fibres terminate have been suggested to belong partly to the spiny neurons. The possibility that some of the SRIF-LIr fibres in the striatum originate from intrinsic SRIF-LIr neurons has been shown.

Neuropeptide Y

NPY, composed of 36 amino acid residues, was isolated from extracts of porcine brain using a chemical method to detect peptide amides (Tatemoto et al., 1982). This peptide is structurally similar to the pancreatic polypeptide (PP) from avian, porcine, bovine and human pancreas (Kimmel et al., 1975; Tatemoto et al., 1982). There is evidence to indicate that NPY may be the true peptide of the PP family in the rat brain (Allen et al., 1983). Therefore, it is likely that striatal structures stained with antiserum against PP such as APP or bovine PP, are identical to NPY-like immunoreactive (NPY-LIr) ones (Lorén et al., 1979b; Olschowka et al., 1981; Vincent et al., 1982b,c, 1983a). A large amount of NPY-LI is contained in the rat striatum (Allen et al., 1983).

(a) Neuropeptide Y-like immunoreactive cells. NPY-LIr cells have medium-sized perikaraya and are widely distributed throughout the rat striatum (Allen et al., 1983; Nakagawa et al., 1985). These cells well resemble APP-LIr neurons (Vincent et al., 1982b,c, 1983a) and SRIF-LIr neurons (DiFiglia and Aronin, 1982; Vincent et al., 1982b,c, 1983a; Takagi et al., 1983) in shape and size of the cell body, and in dendritic profiles. NPY-LIr cells also have an indented nucleus (unpublished data). Taken together with the possible identity of NPY-LI and APP-LI, at least some of the NPY-LIr cells may well correspond to medium-size aspiny neurons.

(b) Neuropeptide Y-like immunoreactive striatal afferents. NPY-LIr fibres in the rat striatum show very heterogeneous (sometimes in a patchy fashion) distributions predominantly occurring in the lateral half (Nakagawa et al., 1985). NPY-LIr afferents may be of intrinsic origin. However, other sources cannot be excluded, since NPY-LIr cell bodies are present in the regions which have the cells of origins of striatal afferents, such as the globus pallidus, amygdaloid complex, claustrum and cortex, including piriform cortex (Nakagawa et al., 1985).

Cholecystokinin

In the brain, there are at least five molecular forms of cholecystokinin (CCK)-like immunoreactivity (CCK-LI) with 39, 33, 12, 8 and 4 amino acids residues. Of these, CCK-8 is the most abundant form (Dockray et al., 1978; Larsson and Rehfeld, 1979; Emson et al., 1980b). By radioimmunoassay, the striatum has been shown to contain a high concentration of CCK-LI (Larsson and Rehfeld, 1979; Emson et al., 1980b; Beinfeld et al., 1981; Meyer et al., 1982; Gilles et al., 1983). This area also contains binding sites for this peptide (Saito et al., 1980; Zarbin et al., 1983). Interaction between CCK and dopamine in the striatum has been reported by pharmacological (Fuxe et al., 1980b, 1981; Fekete et al., 1981; Kovacs et al., 1981; Agnati et al., 1983; Mashal et al., 1983; Meyer and Krauss, 1983; Conzelmann et al., 1984; Markstein and Hökfelt, 1984; Meyer et al., 1984) and anatomical (Fallon et al., 1983; see also Hökfelt et al., 1980) studies. The latter studies demonstrated the coexistence of both substances in the cells of origin of the dopaminergic striatal afferents.

(a) Cholecystokinin-like immunoreactive (CCK-LIr) cells. Since the immunocytochemical demonstration that CCK-LI is present in nerve fibres, but not in cell bodies in the striatum (Innis et al., 1979; Lorén et al., 1979a; Hökfelt et al., 1980; Vanderhaeghen et al., 1980; Cho et al., 1983), CCK-LIr fibres found in the striatum have been considered to be supplied exclusively from extrinsic cells (Meyer et al., 1982; Fallon et al., 1983; Gilles et al., 1983). Recently we have found a few CCK-LIr cells mainly in the ventral half of the rat caudal striatum (but not in the ventral striatum) (Takagi et al., 1984a).

The CCK-LIr cell bodies are of medium size, and

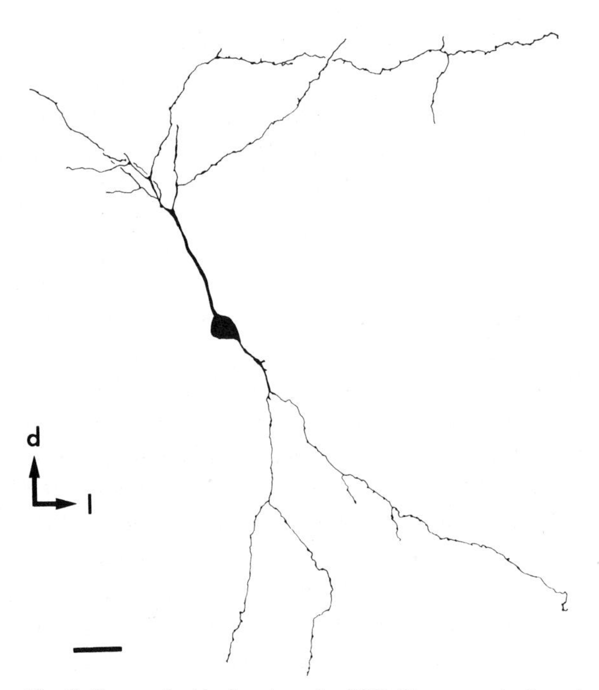

Fig. 7. Camera lucida drawing of a CCK-LIr neuron in the rat striatum. The cell gives rise to two straight immunoreactive dendrites. d, dorsal; l, lateral. Scale: 20 μm.

round or fusiform in shape. About two dendrite-like processes emerge from the cell body (Figs. 7, 8A). These CCK-LIr cells have characteristic features similar to medium-size aspiny neurons, i.e., an indented nucleus and paucity of dendritic spines (Fig. 8B). Some of the CCK-LIr cells share both light and electron microscopic features with spindle shaped, medium-size aspiny neurons (Takagi et al., 1984b) in the same species (Figs. 4, 5A). This type of Golgi-classified neuron gives rise to local axon collaterals and may be an interneuron (Takagi et al., 1984b) (Fig. 5B,C). However, radioimmunoassay studies suggest that a CCK striatofugal pathway is present, since in addition to the reduction of CCK-LI levels in the globus pallidus and substantia nigra in Huntington's disease, injections of kainic acid into the rat striatum resulted in a decrease in CCK-LI in the striatum and substantia nigra (Emson et al., 1980b). If this is the case, then questions arise, such as what category of cells the CCK projection neurons are or whether CCK-LIr medium-size aspiny neurons are really nonprojection neurons. Further studies are needed to answer these questions.

(b) Cholecystokinin-like immunoreactive striatal afferents. CCK-LIr nerve fibres show a heterogeneous distribution in the striatum. They are localized especially to the ventral, medial and dorsomedial parts of the striatum (Lorén et al., 1979a). Towards the caudal part of the striatum, CCK-LIr fibres tend to increase and occupy the medial and ventral parts, and at the most caudal part of the striatum, they are found almost in the whole striatum. Furthermore, CCK-LIr fibres often appear in patches, mainly in the medial part of the striatum (Hökfelt et al., 1980; Cho et al., 1983; Záborszky et al., 1985).

Heterogeneous sources, including the intrinsic striatal cells, have been suggested as the origins of CCK-LIr afferents to the striatum. Combined CCK immunocytochemistry and retrograde fluorescence tracer studies have shown that CCK-LIr neurons in the substantia nigra and ventral tegmental area send a significant number of their axons to the ipsilateral striatum and a few of them to the contralateral striatum (Fallon et al., 1983). Many of these CCK-LIr neurons also contain dopamine (Hökfelt et al., 1980; Fallon et al., 1983). Overlap of CCK-LIr and dopamine fibres is found in the periventricular and caudal parts of the striatum, suggesting the coexistence of CCK-LI and dopamine in the same fibres. However, CCK-LIr fibres in patches may not contain dopamine, since the latter fibres do not disappear after treatment with 6-hydroxydopamine, a neurotoxin which destroys dopamine and noradrenaline neurons (Hökfelt et al., 1980). A radioimmunoassay study demonstrated the complicated result that 6-hydroxydopamine injections into the ventral mesencephalon do not cause a significant decrease in CCK-LI in the striatum (Gilles et al., 1983). This suggests that CCK-LIr cells in the ventral mesencephalon are not the major source of CCK afferents to striatum. Some of the other sources, possibly the major, may be the piriform cortex or claustrum, and amygdaloid complex,

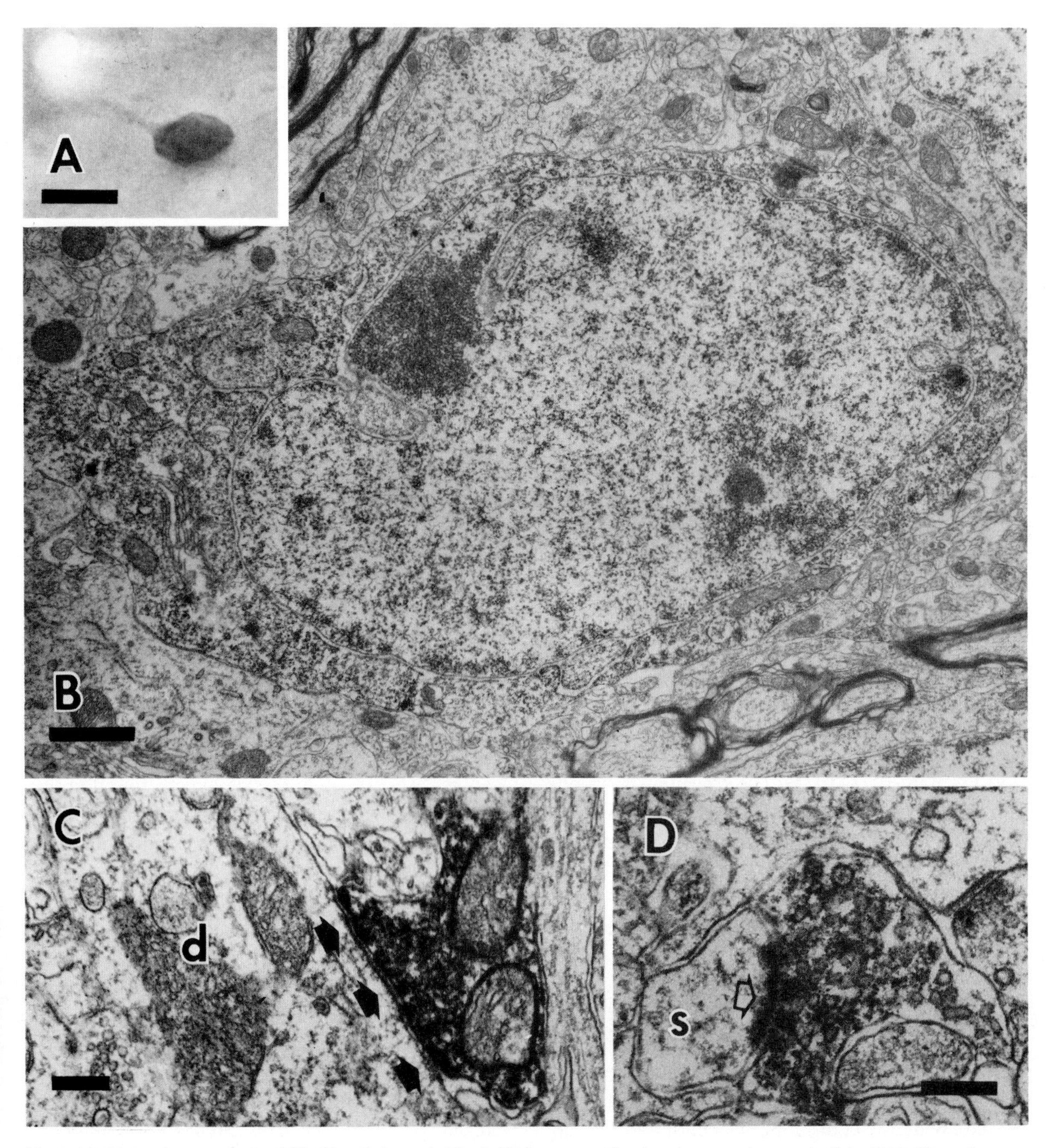

Fig. 8. (A) Light micrograph of a CCK-LIr cell shown in Fig. 7. (B) Low magnification electron micrograph of the CCK-LIr perikaryon shown in A. Note an indented nucleus. Electron micrographs showing (C) symmetrical synaptic contact (solid arrows) between a CCK-LIr bouton and nonreactive dendritic shaft (d) and (D) asymmetrical contact (open arrow) between CCK-LIr bouton and nonreactive dendritic spine (s) in the rat striatum. Scales: (A) 10 μm; (B) 1.0 μm; (C, D) 0.25 μm.

since in addition to the fact that these areas contain many CCK-LIr cell bodies, a knife cut severing the first two areas from the striatum caused a 70% decrease in CCK-LI, and a knife cut separating the last area from the striatum caused a 30% decrease (Meyer et al., 1982). However, the precise origin has not yet been fully determined (Záborszky et al., 1985). Although there are also CCK-LIr cells in the dorsal raphe nucleus, they may not project to the striatum (Steinbusch et al., 1981; Van der Kooy et al., 1981). On the other hand, CCK-LIr striatal neurons may participate in the supply of a part of the striatal afferents. The presence of both asymmetrical and symmetrical types of synapse may reflect the heterogeneity of the cells of origins. The former type occurs more frequently than the latter one (Takagi et al., 1984a) (Fig. 8C,D). Some of the latter type of synapse seem to belong to the CCK-LIr striatal cells, since the medium-sized aspiny neurons closely resembling some CCK-LIr cells have axon collaterals forming the symmetrical synapse (Takagi et al., 1984b) (Fig. 5B,C).

Neurotensin

Neurotensin (NT) is a tridecapeptide which was first isolated from bovine hypothalamus (Carraway and Leeman, 1973, 1975). The striatum shows one of the highest concentrations of NT-like immunoreactivity (NT-LI) in the cat brain and the immunoreactive material co-elutes with synthetic NT on gel chromatography (Goedert and Emson, 1983). Jennes et al. (1982) demonstrated the existence of NT-like immunoreactive (NT-LIr) neurons in the rat, as well as fibres in the rat and other species (Uhl et al., 1977; Hara et al., 1982; Goedert et al., 1983). This may be due to the different antisera used by each author. A possible interaction between NT and nigrostriatal dopamine system has been suggested (Widerlöv et al., 1982; DeQuidt and Emson, 1983; Nemeroff et al., 1983a; Okuma et al., 1983; Goedert et al., 1984b). The level of this peptide is increased in the striatum in Huntington's disease, although its pathological significance is not known (Manberg et al., 1982; Nemeroff et al., 1983b).

(a) Neurotensin-like immunoreactive cells. A few NT-LIr cells have so far been identified in the ventromedial part of the rat striatum close to the bed nucleus of the stria terminalis, but no NT-LIr cells have been found in the remaining striatum (Jennes et al., 1982). No other information is available on the characteristic features of these cells.

(b) Neurotensin-like immunoreactive striatal afferents. The cat and rat appear to have a different pattern of distribution of NT-LIr fibres in the striatum. In the cat striatum, NT-LIr fibres are widely distributed throughout the entire striatum, with densely stained patchy zones of irregular shape being present in the caudate nucleus (Goedert et al., 1983). These NT-rich patches match well with the acetylcholinesterase-poor zones (striosomes). On the other hand, the distribution of NT receptors has been reported to be inverse to that of NT-LI (Goedert et al., 1984a). In contrast to the cat, in the rat, a high density of NT-LIr fibres is seen in the medial, ventral and dorsal portions almost throughout the entire rostro-caudal extension of the striatum without showing any apparent patchy pattern (Jennes et al., 1982).

The origin of the NT-LIr striatal afferents is unknown. NT-LIr cell bodies have been found in several regions which project to the striatum, such as the amygdaloid complex, ventral tegmental area and dorsal raphe nucleus (Uhl et al., 1979; Hara et al., 1982; Jennes et al., 1982). However, no direct evidence has yet been obtained for the source of NT-LIr afferents. The possibility that part, if not all, of the NT-LIr fibres come from their intrinsic neurons cannot be excluded.

Lesion studies combined with the autoradiographic receptor binding technique indicated that more than 50% of rat striatal NT receptors are localized to intrinsic neurons, 30% to dopaminergic nerve afferents and 20% to cortico-striatal nerve afferents (Goedert et al., 1984b). These data may necessitate a detailed ultrastructural analysis of the interrelationship between NT-LIr and the other neuronal elements in the striatum.

Vasoactive intestinal polypeptide

Vasoactive intestinal polypeptide (VIP), a 28 amino acid peptide, was originally isolated from porcine small intestine (Said and Mutt, 1970; Mutt and Said, 1974). The striatum contains a small or moderate amount of VIP-like immunoreactivity (VIP-LI) (Besson et al., 1979; Emson et al., 1979; Lorén et al., 1979c). Immunocytochemistry has revealed that VIP-like immunoreactive (VIP-LIr) nerve fibres are only sparsely scattered throughout this region (Lorén et al., 1979c) with some patchy staining patterns (Fuxe et al., 1977; Roberts et al., 1982). In contrast to these immunocytochemical findings, the striatum has a very high density of specific binding sites for VIP (Taylor and Pert, 1979) and shows a significant increase in local glucose utilization after intrastriatal injection of VIP (McCulloch et al., 1983). The latter authors also found an increase in glucose utilization in the entopeduncular nucleus and lateral habenula, suggesting that VIP may activate the striato-entopeduncular-lateral habenular circuit.

The cells of origin of the striatal VIP afferents are unknown at present. Medial forebrain bundle lesions produced a significant decrease of VIP-LI in the nucleus accumbens, but no change in the striatum (Marley et al., 1981). Occasional occurrence of VIP-LIr cells was reported in the striatum of several mammalian species, although precise location and morphological features of these cells were obscure (Fukui et al., 1982), and an intrinsic source remains possible.

FMRF-amide (phe-met-arg-phe-NH_2)

FMRF-amide is found within the C-terminal of MET-ENK-arg[6]-phe[7] and accounts for the C-terminal part of the gene that codes for MET-ENK and LEU-ENK in bovine adrenal medulla (Gubler et al., 1982; Noda et al., 1982). However, this peptide seems to be distinct from the ENK-related peptides (Dockray et al., 1981a,b; Weber et al., 1981).

FMRF-amide-like immunoreactive cell bodies are mostly localized to dorsomedial, medial and ventral parts of rostral striatum in the rat (Williams and Dockray, 1983). The cells have medium-sized perikarya and sometimes give rise to bipolar-like dendrites. FMRF-amide-like immunoreactive fibres are scattered in the striatum as well as in the globus pallidus. Anatomical and physiological properties of an FMRF-amide-containing neuron system in the basal ganglia remain to be elucidated.

Thyrotropin-releasing hormone

Thyrotropin-releasing hormone (TRH) is a tripeptide, which was first isolated and structurally characterized as a hypothalamic hormone (Burgus et al., 1970). Radioimmunoassay studies have shown that the striatum in the rat and primates contains small amounts of TRH-like immunoreactivity (TRH-LI) (Spindel et al., 1980, 1981; Mori et al., 1982a,b; Nemeroff et al., 1983b). The increase in TRH-LI in the striatum in Huntington's disease may suggest the involvement of TRH in the dysfunction of the basal ganglia in this disease (Spindel et al., 1980; Nemeroff et al., 1983b). There is no detailed description of the immunocytochemical localization of striatal TRH-LI, although the nucleus accumbens was reported to have a network of high density of TRH-like immunoreactive (TRH-LIr) fibres (Hökfelt et al., 1975; Johansson and Hökfelt, 1980). The origin of the TRH innervation of the striatum is not known. Radioimmunoassay studies combined with lesions suggest that some of the TRH-LI is supplied by striatal intrinsic cells (Spindel et al., 1981), although there is no morphological evidence supporting this. The striatum shows a low concentration of TRH binding sites (Rostène et al., 1984; Manaker et al., 1985).

Angiotensin II and angiotensin converting enzyme

Angiotensin II (AII) and angiotensin converting enzyme (ACE) which converts AI to AII have been shown to be present in the brain (Changaris et al., 1978; Phillips et al., 1979; Fuxe et al., 1980a; Arregui et al., 1982; Brownfield et al., 1982). ACE has been used as a possible marker for the detection of

AII in the brain. The striatum contains very large amounts of ACE, followed by the globus pallidus and substantia nigra in human and rat (Arregui et al., 1982; Saavedra et al., 1982).

ACE-like immunoreactive (ACE-LIr) fibres are widely scattered in the striatum (Brownfield et al., 1982). Although ACE-LIr cell bodies have not yet been immunocytochemically identified in this region, biochemical studies have suggested these cells exist within the striatum because: (1) intrastriatal injections of kainic acid caused a remarkable decrease of ACE in the rat striatum and substantia nigra (Arregui et al., 1978; Singh and McGeer, 1978), and (2) a marked decrease of ACE content was found in the striatum and substantia nigra in Huntington's disease (Arregui et al., 1977, 1978), suggesting an ACE projection from the striatum to the substantia nigra. An autoradiographic study supported the presence of an ACE striato-nigral pathway by means of visualization of ACE with ^{3}H-captopril, a selective ACE inhibitor (Strittmatter et al., 1984).

There is no immunocytochemical evidence supporting the existence of AII-positive cells in the striatum. One radioimmunoassay study demonstrated that the striatum contains some AII-like immunoreactivity (Sirret et al., 1981). AII-like immunoreactive fibres are widely scattered in the striatum (Brownfield et al., 1982). The AII receptors are few (Mendelsohn et al., 1984) or non-detectable (Strittmatter et al., 1984).

Corticotropin-releasing factor

Corticotropin-releasing factor (CRF) is a 41 amino acid peptide, first isolated from ovine hypothalamus (Vale et al., 1981). Recently, this peptide has also been shown to be present in various extrahypothalamic areas (Joseph and Knigge, 1983; Swanson et al., 1983).

No CRF-like immunoreactive (CRF-LIr) cell bodies, but a few nerve fibres have been found in the rat striatum (Swanson et al., 1983). Exceptionally, relatively dense wedge-shaped plexuses of CRF-LIr fibres were found in the ventral and medial part of the striatum at the most caudal level, and small patches of densely packed fibres occasionally in the medial-most parts at the middle level.

Calcitonin gene-related peptide

Calcitonin gene-related peptide (CGRP), a peptide composed of 37 amino acids, was recently demon-

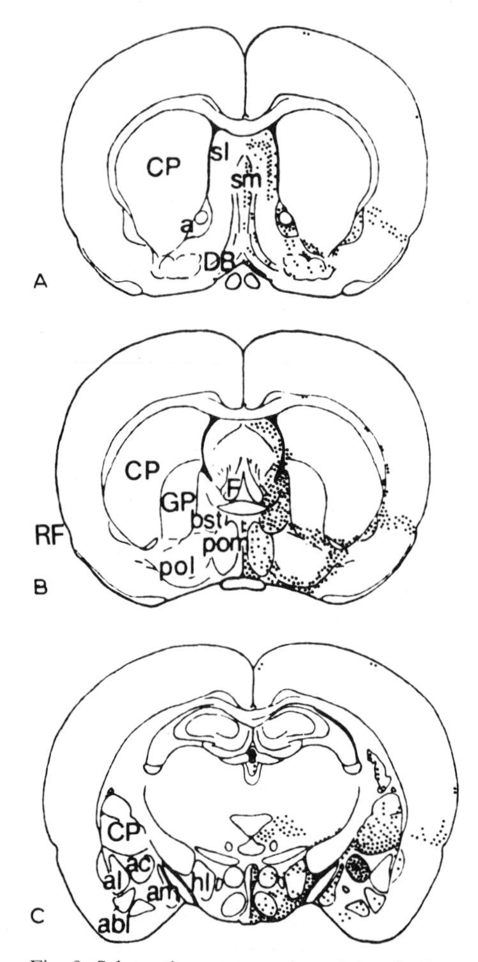

Fig. 9. Schematic representation of the distribution of CGRP-LIr nerve fibres (dots) in the rat forebrain. Arrangement from rostral to caudal in the frontal plane (A–C). a, nucleus accumbens; abl, ac, al, am, amygdaloid complex; bst, bed nucleus of stria terminalis; CP, caudate-putamen; DB, diagonal band; F, fornix; GP, globus pallidus; hl, nucleus lateralis hypothalami; pol, nucleus preopticus lateralis; pom, nucleus preopticus medialis; RF, rhinal fissure; sl, nucleus lateralis septi; sm, nucleus medialis septi. (Illustrations kindly supplied by Dr. Y. Kawai.)

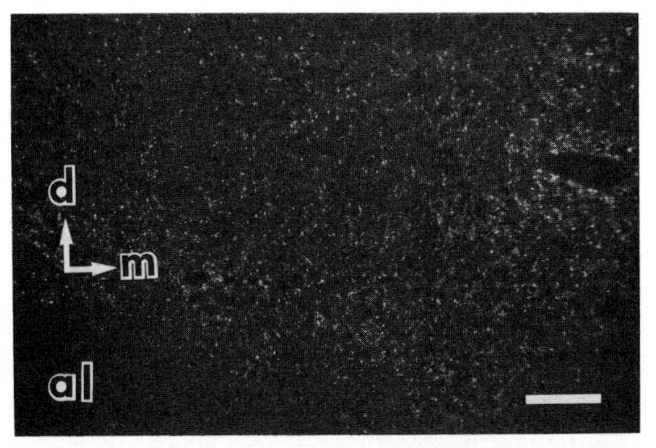

Fig. 10. Fluorescent photomicrograph showing CGRP-LIr fibre plexus in the ventro-caudal portion of the rat striatum as shown in Fig. 9C. al, nucleus amygdaloideus lateralis; d, dorsal; m, medial. Scale: 100 μm. (Micrograph kindly supplied by Dr. Y. Kawai.)

strated in nerve tissue by recombinant DNA and molecular biological techniques (Amara et al., 1982; Rosenfeld et al., 1983).

In the rat, the detailed localization of CGRP-like immunoreactive (CGRP-LIr) structures has been reported by Kawai et al. (1985) using immunocytochemistry. No CGRP-LIr cell bodies have been identified, but a high density of CGRP-LIr fibre plexus is seen in the ventrocaudal portions of the striatum (Figs. 9, 10), a part of which seems to be relevant to the ventral striatum (see also Rosenfeld et al., 1983). These fibres can be traced medially to the bed nucleus of the stria terminalis and the hypothalamus via the area along the ventral surface of the globus pallidus. In addition, CGRP-LIr fibre plexuses with moderate densities are seen in the caudal striatum (Fig. 9C).

Vasopressin

No vasopressin-like immunoreactivity (vasopressin-LI) is detectable in human striatum (Rossor et al., 1981) or few in rat striatum (Hawthorn et al., 1980), although moderate amounts are present in other basal ganglia regions, such as the globus pallidus, amygdaloid nuclei and claustrum (Hawthorn et al., 1980; Rossor et al., 1981, 1982).

Distribution of neuropeptides in the ventral striatum

The ventral striatum includes the nucleus accumbens, olfactory tubercle and the ventral part of the striatum. Since the distribution of neuropeptides in the ventral striatum is covered in the former section, here we will focus on the distribution of peptides in the nucleus accumbens and olfactory tubercle.

Opioid peptides

Enkephalin and related peptides

Radioimmunoassay studies have demonstrated MET-ENK-LI, LEU-ENK-LI or MET-ENK-arg^{6}-gly^{7}-leu^{8}-LI in the nucleus accumbens of the rat and primates with similar amounts in the striatum (Fonnum and Walaas, 1981; Zamir et al., 1984, 1985). Opioid binding sites have been reported to be labeled moderately to strongly in the rat nucleus accumbens (Hamel and Beaudet, 1984). A large number of MET-ENK-LIr or LEU-ENK-LIr cell bodies, which are medium or small in size, occur in the nucleus accumbens and olfactory tubercle (Finley et al., 1981a; Johansson and Hökfelt, 1981; Inagaki and Parent, 1985) (Fig. 2A). Both nucleus accumbens and olfactory tubercle exhibit a medium density of ENK-LIr fibres and nerve terminals which surround patchy, bundle-like or finger-like dense staining areas in the monkey (Haber and Elde, 1982; Inagaki and Parent, 1985) (Fig. 2A), although in the human and rat, the latter region displays only a few fibres (Finley et al., 1981a; Bouras et al., 1984). The highest density of ENK-LIr fibre networks is seen in the vicinity of the anterior commissure (Elde et al., 1976; Simantov et al., 1977; Watson et al., 1977a; Sar et al., 1978; Wamsley et al., 1980).

Little is known of the fibre connections of ENK neuron systems in the ventral striatum. Recently, Zahm et al. (1985) demonstrated, by using adjacent section analysis of post-embedding materials, that MET-ENK-LI and GAD-LI coexist in axon terminals of rat ventral pallidum, one of the presumed terminal fields of GABAergic afferents from the nucleus accumbens (Pycock and Horton, 1976; Waddington and Cross, 1978; Walaas and Fonnum, 1980). This suggests that ENK-LIr cells in the nucleus accumbens contain GABA, as demonstrated in the striatum, and project to the ventral pallidum.

Pro-opiomelanocortin compounds

O'Donohue et al. (1979) reported that the nucleus accumbens and olfactory tubercle in the rat contain small amounts of α-MSH-LI, respectively 2.5 times larger, and about the same amount as that in the striatum. Jacobowitz and O'Donohue (1978) demonstrated that α-MSH-LIr fibres are arranged along the outside or along the medial edge of the nucleus accumbens. In contrast to this view, Umegaki et al. (1983) have shown that many of the α-MSH-LIr fibres are located mostly within the medial half of this region in the same species.

Kawai et al. (1984) examined the localization and origin of γ-MSH-LIr fibres in the rat brain. They showed a low to medium density of γ-MSH-LIr fibre plexus in the nucleus accumbens, while the striatum is devoid of them. These fibres originate from γ-MSH-LIr neurons located in the arcuate nucleus. In addition, ACTH and β-lipotropin-like immunoreactivities are present in the nucleus accumbens (Watson et al., 1977b, 1978b) but detailed information is not available.

Dynorphins and α-neo-endorphin

The level of DYN-B-LI and α-neo-END-LI in the nucleus accumbens of monkey is, respectively, 2–3 times and 1.4–2 times higher than that in the striatum (Zamir et al., 1984). Very weak DYN-A-LIr staining has been reported in the rat, monkey and human (Haber and Watson, 1983).

Substance P

A close interaction between SP and dopamine has been suggested in the nucleus accumbens of the rat (Kalivas and Miller, 1984b). In the rat nucleus accumbens, the level of SP-LI is similar to that in the striatum, whereas the monkey accumbens has about 2.5–3 times more SP-LI than the caudate nucleus or putamen (Fonnum and Walaas, 1981; Zamir et al., 1984).

Single SP-LIr cells are present in the middle part of the rat nucleus accumbens (Ljungdahl et al., 1978; Johansson and Hökfelt, 1981). Their perikarya are of medium size (15–25 μm in diameter) and are occasionally large (above 25 μm). A high density of SP-LIr fibres is found in the rostral part of the nucleus, whereas a medium density of immunoreactive fibres occupies the caudal part with the exception of the small medial area which still shows a high density (Johansson and Hökfelt, 1981).

The olfactory tubercles also have some SP-LIr cells and a medium density of SP-LIr fibres (Ljungdahl et al., 1978).

Somatostatin

Although little is known of the functional roles of SRIF in the ventral striatum (Beal and Martin, 1984a), high concentrations of SRIF-LI have been detected. The nucleus accumbens contains 3.5, 2 and 4 times more SRIF-LI than the human caudate nucleus (Nemeroff et al., 1983b), monkey striatum (Zamir et al., 1984) and rat striatum (Fonnum and Walaas, 1981), respectively. On the other hand, Beal and Martin (1984a) have reported that the rat nucleus accumbens contains only slightly more SRIF-LI than the ventrolateral striatum. These differences in the ratio seem to reflect heterogeneous regional distributions of SRIF-LIr structures in the striatum.

Numerous medium-sized SRIF-LIr cell bodies are present in the middle part of the nucleus accumbens and they tend to decrease in number both rostrally and caudally. A high density of SRIF-LIr fibres is found evenly throughout the middle and caudal parts, and in a patchy fashion in the rostral part of the rat nucleus accumbens (Johansson and Hökfelt, 1981; Johansson et al., 1984).

The olfactory tubercle also has both many SRIF-LIr (medium- or large-sized) cells and a high density of fibre networks (Johansson et al., 1984).

Both nucleus accumbens and olfactory tubercle

contain a high or medium density of SRIF binding sites (Epelbaum et al., 1982).

Neuropeptide Y

The rat nucleus accumbens and olfactory tubercle contain a small group of NPY-LIr cells (Nakagawa et al., 1985). They have medium-sized, fusiform or ovoid perikarya (Nakagawa et al., 1985). Vincent et al. (1982c) demonstrated that the coexistence of SRIF-LI and APP-LI occurs in many cells located in both regions, as well as in the striatum. Therefore, the SRIF-LIr cells in the former two regions may also be identical to NPY-LIr cells which were presumably visualized by the crossreaction with APP antiserum.

Relatively low or medium density of NPY-LIr fibres is found in the nucleus accumbens or the olfactory tubercle, respectively, although their origin is obscure (Nakagawa et al., 1985).

Cholecystokinin

Radioimmunoassay has shown high concentrations of CCK-LI in the nucleus accumbens and olfactory tubercle (Marley et al., 1982; Studler et al., 1984). Heterogeneous localization of CCK-LI within the two regions is roughly in accordance with the immunocytochemical data (Záborszky et al., 1985).

The middle and caudal parts of the nucleus accumbens and olfactory tubercle contain numerous CCK-LIr fibres with patchy profiles almost at the medial halves, which decrease in number rostrally (Lorén et al., 1979a; Hökfelt et al., 1980; Cho et al., 1983; Záborszky et al., 1985). The ventral part of the caudate nucleus also contains a high density of CCK-LIr fibre networks. No CCK-LIr cell bodies have been identified in the ventral striatum.

Radioimmunoassay studies (Williams et al., 1981; Studler et al., 1981, 1984; Marley et al., 1982; Gilles et al., 1983; Záborszky et al., 1985) and immunocytochemical studies (Hökfelt et al., 1980; Fallon et al., 1983; Záborszky et al., 1985) combined with experimental procedures have indicated that a proportion of the CCK-LI in limbic structures, including posteromedial parts of the nucleus accumbens and olfactory tubercle, and the subcommissural part of caudate-putamen, are supplied by CCK-LIr cells in the ventral tegmental area and substantia nigra. Hökfelt et al. (1980) demonstrated that CCK-LI coexists with dopamine in a subpopulation of mesencephalic dopaminergic neurons and that these CCK/dopamine cells project to the posteromedial parts of the nucleus accumbens. Besides this ascending CCK pathway, Záborszky et al. (1985) suggested the presence of a descending CCK pathway to the ventral striatum; the lateral part of the nucleus accumbens, the olfactory tubercle and the subcommissural part of caudate-putamen receive CCK afferents from the pyriform and medial prefrontal cortices and/or the amygdaloid complex. Therefore, the ventral striatum may receive heterogeneous types of afferents containing either CCK-LI only or CCK-LI and dopamine.

Nucleus accumbens and olfactory tubercle show high concentrations of CCK binding sites which colocalize with dopamine binding sites (Zarbin et al., 1983), suggesting that this is part of the neuronal mechanism involved in the CCK/dopamine interaction.

Pharmacological interaction between CCK and dopamine has been reported in the nucleus accumbens (Fuxe et al., 1980b; Chang et al., 1983; Vaccarino and Koob, 1984; White and Wang, 1984). (See also Fuxe et al., this volume.)

Immuno-electron microscopic study has revealed that CCK-LIr axon terminals form both asymmetrical and symmetrical synapses in the nucleus accumbens with the former type occurring more frequently (Baali-Cherif et al., 1984).

Neurotensin

The nucleus accumbens has been reported to contain about three times or more NT than the striatum in man and rat (Kobayashi et al., 1977; Govoni et al., 1980; Fonnum and Walaas, 1981; Emson et al., 1982; Manberg et al., 1982; Nemeroff et al., 1983b). Jennes et al. (1982) have reported that both the nucleus accumbens and olfactory tubercle contain a high density of NT-LIr fibres with a patchy appearance that is also a characteristic of their distribution in the striatum. The lateral part of the nucleus accumbens appears to exhibit more dense immunostained fibres (Uhl et al., 1977).

Combined retrograde HRP tracing and immunocytochemistry demonstrated that the medial part of the nucleus accumbens is innervated by some NT-LIr neurons in the ventral and medial parts of the ventral tegmental area (Kalivas and Miller, 1984a). These NT-LIr afferents may also contain dopamine since there is evidence for coexistence of the two substances in some neurons located in the ventral tegmental area (Hökfelt et al., 1983). Pharmacological interactions between NT and dopamine have been reported (Kalivas et al., 1983, 1984).

Vasoactive intestinal polypeptide

Both nucleus accumbens and olfactory tubercle contain large amounts of VIP-LI, which is 2–6 times higher in the nucleus accumbens than in the striatum (Fonnum and Walaas, 1981; Marley et al., 1981) and in the olfactory tubercle is about half of that in the striatum of the mouse (Lorén et al., 1979c).

No VIP-LIr cell bodies have been found in either region. The localization of VIP-LIr fibres in the nucleus accumbens is limited to certain areas. They are present in the nucleus accumbens along the medial border of the nucleus as well as the area close to the lateral ventricle (Fuxe et al., 1977; Lorén et al., 1979c; Johansson and Hökfelt, 1981), although there is another report describing a wider distribution of VIP-LIr fibres within the nucleus (Sims et al., 1980).

The nucleus accumbens, but not the olfactory tubercle, may receive VIP afferents via the medial forebrain bundle, since lesion of this bundle resulted in about 50% decrease of VIP-LI in the unilateral side of the nucleus accumbens, but no change in the olfactory tubercle (Marley et al., 1981).

Both nucleus accumbens and olfactory tubercle have many VIP binding sites (Staun-Olsen et al., 1985).

FMRF-amide

A few FMRF-amide immunoreactive cell bodies are present in the rat nucleus accumbens, mostly in the dorsal aspects of the caudal region. A dense immunoreactive fibre plexus is also found in the medial half of the caudal region (Williams and Dockray, 1983). On the other hand, Weber et al. (1981) described strong immunostaining of cell bodies and moderately dense labeling in the olfactory tubercle.

Thyrotropin releasing hormone

The content of TRH-LI in the nucleus accumbens is three times higher than in the caudate nucleus in human brain (Nemeroff et al., 1983b). Monkey accumbens also showed a similar (2–3 times higher) ratio to the striatum, except that the former contains a little less TRH-LI than the caudate body (Mori et al., 1982a). Different concentrations of TRH binding sites in the nucleus accumbens have been reported; both very high (Rostène et al., 1984) and low (Manaker et al., 1985). This discrepancy may be caused by the difference in the procedures used. Numerous TRH-LIr fibres, but no cell bodies have been found in the nucleus accumbens. Dense networks of fibres are present, especially in the dorsal part (Hökfelt et al., 1975; Johansson and Hökfelt, 1981). The cells of origin are not known.

TRH may modify dopaminergic transmission in the nucleus accumbens (Narumi and Nagawa, 1983). Involvement of TRH in the locomotor activity at the level of the nucleus accumbens has been suggested (Miyamoto and Nagawa, 1977; Heal and Green, 1979) although some results are inconsistent (Costall et al., 1979).

Angiotensin II and angiotensin converting enzyme

Radioimmunoassay has shown that the nucleus accumbens contains about half of the ACE in the striatum of the man and rat (Arregui et al., 1982; Saavedra et al., 1982). Immunocytochemically densely stained AII-LIr fibres are found in this nucleus (Changaris et al., 1978). Mendelsohn et al. (1984) demonstrated low level of AII receptors.

Corticotropin releasing factor

A number of CRF-LIr cell bodies and nerve fibres are present in the nucleus accumbens at the medial part of the caudal region. These cells appear to be continuous with a CRF-LIr cell group in the nucleus of the stria terminalis (Joseph and Knigge, 1983; Swanson et al., 1983). Immunoreactive structures are almost undetectable in the olfactory tubercle.

Calcitonin gene-related peptide

A high density of CGRP-LIr fibre plexus is found in the caudal part of the nucleus accumbens, and the ventral part of the striatum (Kawai et al., 1985) (Fig. 9). No other data are available.

Vasopressin

The nucleus accumbens in human brain shows a low content of vasopressin (Rossor et al., 1981).

Distribution of neuropeptides in the pallidal structures

The globus pallidus, one of the main targets from the striatum, also contains many peptides. These peptides could be divided into three groups based upon their contents: (i) MET- and LEU-ENK, MET-ENK-arg^6-gly^7-leu^8, SP and NPY with a high density of immunoreactive structures; (ii) DYNs, α-neo-END, CCK and TRH with a moderate density of immunoreactive structures; and (iii) α-MSH, SRIF, NT, VIP, FMRF-amide, CRF, CGRP and vasopressin with a low or very low density of immunoreactive structures (O'Donohue et al., 1979; Emson et al., 1979, 1980b; Beinfeld et al., 1981; Buck et al., 1981; Rossor et al., 1981; Manberg et al., 1982; Mori et al., 1982a; Allen et al., 1983; Aronin et al., 1983; Beal et al., 1983; Chesselet and Graybiel, 1983; Pittius et al., 1983; Swanson et al., 1983; Williams and Dockray, 1983; Zamir et al., 1984, 1985; Kawai et al., 1985; Quirion et al., 1985).

There is good evidence for striato-pallidal ENK and SP neuron systems (see above). However, there is a prominent topographic separation between the ENK-LIr and SP-LIr fibres in the pallidal structures; in the primates, ENK-LIr fibres are abundant in the external segment, whereas SP-LIr fibres are prominent in the internal segment. Judging from the fibre connections of these areas (Graybiel and Ragsdale, 1979), enkephalinergic afferents are more involved in the intrinsic corpus striatal circuit through the subthalamic nucleus than SP afferents, whereas SP afferents involve more the long pallidal efferent systems (Beckstead and Kersey, 1985). In accordance with the results in primates, the globus pallidus of non-primates, which is homologous to the external segment of the primate, contains a number of ENK-LIr fibres but fewer SP-LIr fibres, while in the entopeduncular nucleus of the non-primates, homologous to the internal segment of the primates, SP-LIr fibres were much more numerous than ENK-LIr fibres (see above). Thus the topographic differences between ENK-LIr and SP-LIr fibres in the dorsal pallidum (or conventionally defined globus pallidus) are fundamentally similar among the mammals. However, in the ventral pallidum, no common profiles could be obtained. For example, in the rat subcommissural pallidum, many of both ENK-LIr and SP-LIr fibres are present, while in the primate, ENK-LIr fibres are more numerous than SP-LIr fibres.

ENK-LIr and SP-LIr fibres show the unique pattern in the pallidal structures of both non-primates and primates including man, i.e. ribbon-like fibres or 'woolly fibers' (Haber and Elde, 1981; Haber and Nauta, 1983; Beach and McGeer, 1984) (Fig. 11). The ENK-LIr and SP-LIr woolly fibres are frequently seen both in the dorsal and ventral pallidum with some regional differences. Although the significance of these woolly fibres in the pallidal structures is obscure, it has been suggested that unstained pallidal neuronal elements (cell bodies and dendrites) are enmeshed by a plexus of ENK-LIr and SP-LIr striato-pallidal fibres (Haber and Elde, 1981).

Besides ENK and SP, there is still little evidence to indicate that the other peptides in the globus pal-

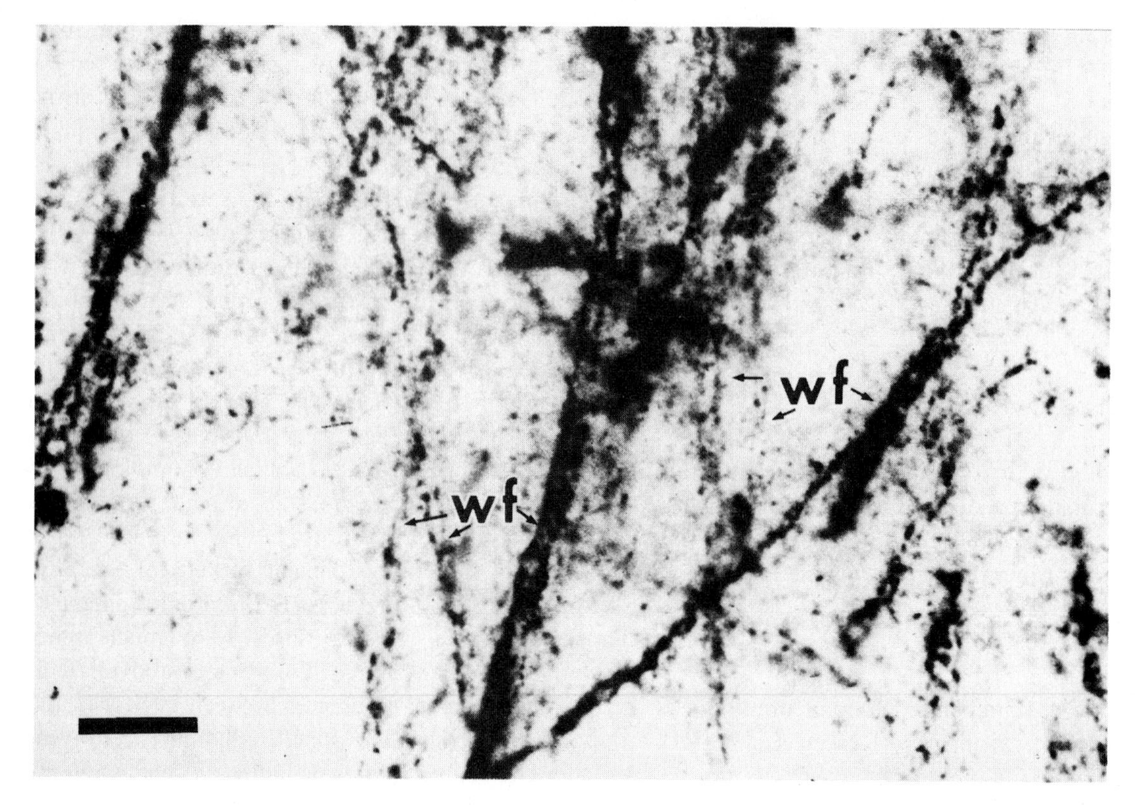

Fig. 11. Light micrograph showing LEU-ENK-LIr fibres in the external segment of human globus pallidus. Many woolly-fibres (wf) are seen. The two parallel lines appear to be made up of structures resembling beaded nerve fibres running in parallel. Scale: 50 μm. (Micrograph kindly supplied by Dr. J. Pearson.)

lidus are also present within the striato-pallidal neuron systems. However, this is likely, because almost all of the peptides described above have been identified in nerve fibres but not in cell bodies, and the globus pallidus is one of the main targets from the striatum. Also, the possibility could not be excluded that the cells of origin lie in other areas than the striatum. Further analysis will be needed on the peptidergic neuronal connections.

As mentioned above, almost all immunoreactive structures identified in the globus pallidus are nerve fibres. However, the presence of immunoreactive cells in the pallidal structures has also been reported; SP-LIr cells in the rat globus pallidus (Cuello and Kanazawa, 1978), in the entopeduncular nucleus of the rat (Shinoda et al., 1984) and external segment of human globus pallidus (Marshall et al., 1983), and a few NPY-LIr cells in the rat globus pallidus (Nakagawa et al., 1985) and numerous SRIF-LIr cells in the rat entopeduncular nucleus (Johansson et al., 1984). SP-LIr cells located in the rat entopeduncular nucleus were shown to project to the lateral part of the lateral habenular nucleus (Shinoda et al., 1984). However, little is known of the projection fields of other immunoreactive cells.

Distribution of neuropeptides in the claustrum

The claustrum contains many peptides. Few radioimmunoassay data are available. Therefore, peptide-containing structures have been mainly obtained by immunocytochemical studies. Localization of the peptides in the claustrum is summarized in Table II.

TABLE II

Peptide-immunoreactive structures in the claustrum

Peptides	Cells	Fibres	References
ENKs	−	+	Roberts et al. (1982)
β-END	?	?	
ACTH	?	?	
α,γ-MSH	−	−	Kawai et al. (1984); O'Donohue et al. (1979)
DYNs	?	?	
α-neo-END	?	?	
SP	−	+	Roberts et al. (1982)
SRIF	+	+	Finley et al. (1981b); Johansson et al. (1984)
NPY	+	+	Nakagawa et al. (1985)
CCK	+	+	Meyer et al. (1982); Záborszky et al. (1985)
NT	−	+	Jennes et al. (1982)
VIP	+	+	Sims et al. (1980)
FMRF-amide	−	−	Williams and Dockray (1983)
TRH[a]	?	?	
AII	?	?	
CRF	−	−	Swanson et al. (1983)
CGRP	−	+	Kawai et al. (1985)
Vasopressin[b]	?	?	

[a] A high concentration of TRH binding sites (Rostène et al., 1984).
[b] Low content of vasopressin (Rossor et al., 1981).

Comments

As described above, a variety of neuropeptides are present in the basal ganglia, particularly in the striatum. The first approach to elucidation of the function of these peptides in the basal ganglia is to map their topographies, and next, if immunoreactive cell bodies are identified, to classify these cells into subtypes based upon their morphological characteristics both by light and electron microscopy. These attempts are now in progress in many laboratories. Another powerful approach is to elucidate the fibre connections of peptides in the basal ganglia. Several peptidergic tracts have been established such as striato-nigral SP tract, striato-entopeduncular SP tract, entopedunculo-lateral habenular SP tract and striato-pallidal SP or ENK tract. Further information will certainly be accumulated within a few years.

Subsequent steps will be to establish the interaction among the neuronal structures which contain different peptides. Double staining techniques or "mirror" technique at the electron microscopic level can be useful for this purpose (see Shiosaka and Tohyama, this volume). Finally, the coexistence of multiple putative transmitters in single cells which seems to be common in the peripheral nervous system (see Inagaki and Kitoh, this volume) must be examined more precisely.

Acknowledgements

The author wishes to thank Dr. M. Tohyama for his helpful advice and Drs. J. Pearson, S. Inagaki, Y. Kawai, N. Sakamoto and Y. Kubota for providing illustrations of their materials. Thanks are also extended to Miss Y. Morishima for her skilled photographic work and to Mrs. M. Nimura for her efforts in typing the manuscript.

References

Agnati, L., Fuxe, K., Benfenati, F., Celani, M. F., Battistini, N., Mutt, V., Cariccioli, L., Galli, G. and Hökfelt, T. (1983) Differential modulation by CCK-8 and CCK-4 of [^{3}H]spiperone binding sites linked to dopamine and 5-hydroxytryptamine receptors in the brain of the rat. *Neurosci. Lett.,* 35: 179–183.

Allen, Y. S., Adrian, T. E., Allen, J. M., Tatemoto, K., Crow, T. J., Bloom, S. R. and Polak, J. M. (1983) Neuropeptide Y distribution in the rat brain. *Science,* 221: 877–879.

Amara, S. G., Jones, V. and Rosenfeld, M. G. (1982) Alternative RNA processing in calcitonin gene expression generates mRNAs encoding different polypeptide products. *Nature,* 298: 240–244.

Araki, M., McGeer, P. L. and McGeer, E. G. (1985) Differential effect of kainic acid on somatostatin, GABAergic and cholinergic neurons in the rat striatum. *Neurosci. Lett.,* 53: 197–202.

Aronin, N., Cooper, P. E., Lorenz, L. J., Bird, E. D., Sagar, S. M., Leeman, S. E. and Martin, J. B. (1983) Somatostatin is increased in the basal ganglia in Huntington's disease. *Ann. Neurol.,* 13: 519–526.

Aronin, N., DiFiglia, M., Graveland, G. A., Schwartz, W. J.

and Wu, J.-Y. (1984) Localization of immunoreactive enkephalins in GABA synthesizing neurons of the rat neostriatum. *Brain Res.,* 300: 376–380.

Arregui, A., Bennett, J. P., Bird, E. D., Yamamura, H. I., Iversen, L. L. and Snyder, S. H. (1977) Huntington's chorea: Selective depletion of activity of angiotensin converting enzyme in the corpus striatum. *Ann. Neurol.,* 2: 294–298.

Arregui, A., Emson, P. C. and Spokes, E. G. (1978) Angiotensin-converting enzyme in substantia nigra: Reduction of activity in Huntington's disease and after intrastriatal kainic acid in rats. *Eur. J. Pharmacol.,* 52: 121–124.

Arregui, A., Perry, E. K., Rossor, M. and Tomlinson, B. E. (1982) Angiotensin converting enzyme in Alzheimer's disease: Increased activity in caudate nucleus and cortical areas. *J. Neurochem.,* 38: 1490–1492.

Atweh, S. F. and Kuhar, M. (1977) Autoradiographic localization of opiate receptors in rat brain. III. The telencephalon. *Brain Res.,* 134: 393–405.

Baali-Cherif, H., Arluison, M. and Tramu, G. (1984) Ultrastructural study of cholecystokinin-like immunoreactive nerve terminals in the rat nucleus accumbens. *Neurosci. Lett.,* 49: 331–335.

Barden, N., Merand, Y., Rouleau, D., Moore, S., Dockray, G. J. and Dupont, A. (1981) Regional distributions of somatostatin and cholecystokinin-like immunoreactivities in rat and bovine brain. *Peptides,* 2: 299–302.

Beach, T. G. and McGeer, E. G. (1984) The distribution of substance P in the primate basal ganglia: An immunohistochemical study of baboon and human brain. *Neuroscience,* 13: 29–52.

Beal, M. F. and Martin, J. B. (1983a) Effects of lesions on somatostatin-like immunoreactivity in the rat striatum. *Brain Res.,* 266: 67–73.

Beal, M. F. and Martin, J. B. (1983b) Effects of lesions on somatostatin-like immunoreactivity in the rat striatum. *Neurology,* 33: 177.

Beal, M. F. and Martin, J. B. (1984a) Effects of neuroleptic drugs on brain somatostatin-like-immunoreactivity. *Neurosci. Lett.,* 47: 125–130.

Beal, M. F. and Martin, J. B. (1984b) The effect of somatostatin on striatal catecholamines. *Neurosci. Lett.,* 44: 271–276.

Beal, M. F., Domesick, V. B. and Martin, J. B. (1983) Regional somatostatin distribution in the rat striatum. *Brain Res.,* 278: 103–108.

Beckstead, R. M. and Kersey, K. S. (1985) Immunohistochemical demonstration of differential substance P-, met-enkephalin-, and glutamic-acid decarboxylase-containing cell body and axon distributions in the corpus striatum of the cat. *J. Comp. Neurol.,* 232: 481–498.

Beinfeld, M. C., Meyer, D. K., Eskay, R. L., Jensen, R. T. and Brownstein, M. J. (1981) The distribution of cholecystokinin immunoreactivity in the central nervous system of the rat as determined by radioimmunoassay. *Brain Res.,* 212: 51–57.

Ben-Ari, Y., Pradelles, P., Oros, C. and Dray, F. (1979) Identification of authentic substance P in striatonigral and amygdaloid nuclei using combined high performance liquid chromatography and radioimmunoassay. *Brain Res.,* 173: 360–363.

Bennett-Clarke, C., Romagnano, M. A. and Joseph, S. H. (1980) Distribution of somatostatin in the rat brain: telencephalon and diencephalon. *Brain Res.,* 188: 473–486.

Benoit, R., Böhlen, P., Esch, F. and Ling, N. (1984) Neuropeptides derived from prosomatostatin that do not contain the somatostatin-14 sequence. *Brain Res.,* 311: 23–29.

Besson, J., Rotszien, W., LaBurthe, M., Epelbaum, J., Beaudet, A., Kordon, C. and Rosselin, G. (1979) Vasoactive intestinal peptide (VIP): Brain distribution subcellular localization and effect of deafferentation of the hypothalamus in male rats. *Brain Res.,* 165: 79–85.

Bishop, G. A., Chang, H. T. and Kitai, S. T. (1982) Morphological and physiological properties of neostriatal neurons: An intracellular horseradish peroxidase study in the rat. *Neuroscience,* 7: 179–191.

Bloch, B., Baird, A., Ling, N., Benoit, R. and Guillemin, R. (1983) Immunohistochemical evidence that brain enkephalins arise from a precursor similar to adrenal preproenkephalin. *Brain Res.,* 263: 251–257.

Bolam, J. P., Powell, J. F., Totterdell, S. and Smith, A. D. (1981a) The proportion of neurons in the rat neostriatum that project to the substantia nigra demonstrated using horseradish peroxidase conjugated with wheat-germ agglutinin. *Brain Res.,* 220: 339–343.

Bolam, J. P., Somogyi, P., Totterdell, S. and Smith, A. D. (1981b) A second type of striatonigral neuron: A comparison between retrogradely labeled and Golgi-stained neurons at the light and electron microscopic levels. *Neuroscience,* 6: 2141–2157.

Bolam, J. P., Clarke, D. J., Smith, A. D. and Somogyi, P. (1983a) A type of aspiny neuron in the rat neostriatum accumulates $[^3H]\gamma$-aminobutyric acid: combination of Golgi-staining, autoradiography and electron microscopy. *J. Comp. Neurol.,* 213: 121–134.

Bolam, J. P., Somogyi, P., Takagi, H., Fodor, I. and Smith, A. D. (1983b) Localization of substance P-like immunoreactivity in neurons and nerve terminals in the neostriatum of the rat: A correlated light and electron microscopic study. *J. Neurocytol.,* 12: 325–344.

Bouras, C., Taban, C. H. and Constantinidis, J. (1984) Mapping of enkephalins in human brain. An immunohistofluorescence study on brains from patients with senile and presenile dementia. *Neuroscience,* 12: 179–190.

Bouyer, J. J., Miller, R. J. and Pickel, V. M. (1984) Ultrastructural relation between cortical efferents and terminals containing enkephalin-like immunoreactivity in rat neostriatum. *Regul. Peptides,* 8: 105–115.

Brann, M. R. and Emson, P. C. (1980) Microiontophoretic injection of fluorescent tracer combined with simultaneous immunofluorescent histochemistry for the demonstration of efferents from the caudate-putamen projecting to the globus pallidus. *Neurosci. Lett.*, 16: 61–65.

Brownfield, M. S., Reid, I. A., Ganten, D. and Ganong, W. F. (1982) Differential distribution of immunoreactive angiotensin and angiotensin-converting enzyme in rat brain. *Neuroscience*, 7: 1759–1769.

Brownstein, M. J., Arimura, A., Sato, H., Schally, A. V. and Kizer, J. S. (1975) The regional distribution of somatostatin in the rat brain. *Endocrinology*, 96: 1456–1461.

Brownstein, M. J., Mroz, E. A., Tappaz, M. L. and Leeman, S. E. (1977) On the origin of substance P and glutamic acid decarboxylase (GAD) in the substantia nigra. *Brain Res.*, 135: 315–323.

Buck, S. H., Burks, T. F., Brown, M. R. and Yamamura, H. I. (1981) Reduction in basal ganglia and substantia nigra substance P levels in Huntington's disease. *Brain Res.*, 209: 464–469.

Burgus, R., Dunn, T. F., Desiderio, D., Ward, D. N., Vale, W. and Guillemin, R. (1970) Characterization of ovine hypothalamic hypophysiotropic TSH-releasing factor. *Nature*, 226: 321–325.

Carraway, R. and Leeman, S. (1973) The isolation of a new hypotensive peptide, neurotensin, from bovine hypothalami. *J. Biol. Chem.*, 248: 6854–6861.

Carraway, R. and Leeman, S. (1975) The amino acid sequence of a hypothalamic peptide neurotensin. *J. Biol. Chem.*, 250: 1907–1911.

Chang, H. T., Wilson, C. J. and Kitai, S. T. (1981) Single neostriatal efferent axons in the globus pallidus: a light and electron microscopic study. *Science*, 213: 915–918.

Chang, R. S. L., Lotti, V. J., Martin, G. E. and Chen, T. B. (1983) Increase in ^{125}I-cholecystokinin receptor binding following chronic haloperidol treatment, intracisternal 6-hydroxydopamine or ventral tegmental lesions. *Life Sci.*, 32: 871–878.

Changaris, D. C., Keil, L. C. and Severs, W. B. (1978) Angiotensin II immunohistochemistry of the rat brain. *Neuroendocrinology*, 25: 257–274.

Chesselet, M. F. and Graybiel, A. M. (1983) Met-enkephalin-like and dynorphin-like immunoreactivities of the basal ganglia of the cat. *Life Sci.*, Suppl. 1, 33: 37–40.

Chesselet, M. F. and Reisine, D. (1983) Somatostatin regulates dopamine release in rat striatal slices and cat caudate nuclei. *J. Neurosci.*, 3: 232–236.

Cho, H. J., Shiotani, Y., Shiosaka, S., Inagaki, S., Kubota, Y., Kiyama, H., Umegaki, K., Tateishi, K., Hashimura, E., Hamaoka, T. and Tohyama, M. (1983) Ontogeny of cholecystokinin-8-containing neuron system of the rat: an immunohistochemical analysis-I. Forebrain and upper brainstem. *J. Comp. Neurol.*, 218: 25–41.

Cohn, M. L. and Cohn, M. (1975) "Barrel rotation" induced by somatostatin in the non-lesioned rat. *Brain Res.*, 96: 138–141.

Conzelmann, U., Holland, A. and Meyer, D. K. (1984) Effects of selective dopamine D_2-receptor agonists on the release of cholecystokinin-like immunoreactivity from rat neostriatum. *Eur. J. Pharmacol.*, 101: 119–125.

Corrêa, F. M. A., Innis, R. B., Hester, L. D. and Snyder, S. H. (1981) Diffuse enkephalin innervation from caudate to globus pallidus. *Neurosci. Lett.*, 25: 63–68.

Costall, B., Hui, S. G., Metcalf, G. and Naylor, R. J. (1979) A study of the changes in motor behaviour caused by TRH on intracerebral injection. *Eur. J. Pharmacol.*, 53: 143–150.

Cuello, A. C. and Kanazawa, I. (1978) The distribution of substance P immunoreactive fibers in the rat central nervous system. *J. Comp. Neurol.*, 178: 129–156.

Cuello, A. C. and Paxinos, G. (1978) Evidence for a long Leu-enkephalin striopallidal pathway in rat brain. *Nature*, 271: 178–180.

DelFiacco, M., Paxinos, G. and Cuello, A. C. (1982) Neostriatal enkephalin-immunoreactive neurones project to the globus pallidus. *Brain Res.*, 231: 1–17.

Del Río, J., Naranjo, J. R., Yang, H. Y. and Costa, E. (1983) Substance P-induced release of met-enkephalin from striatal and periaqueductal gray slices. *Brain Res.*, 279: 121–126.

De Quidt, M. E. and Emson, P. C. (1983) Neurotensin facilitates dopamine release in vitro from rat striatal slices. *Brain Res.*, 274: 376–380.

DiFiglia, M. and Aronin, N. (1982) Ultrastructural features of immunoreactive somatostatin neurons in the rat caudate nucleus. *J. Neurosci.*, 2: 1267–1274.

DiFiglia, M. and Aronin, N. (1984) Quantitative electron microscopic study of immunoreactive somatostatin axons in the rat neostriatum. *Neurosci. Lett.*, 50: 325–331.

DiFiglia, M., Pasik, T. and Pasik, P. (1980) Ultrastructure of Golgi-impregnated and gold-toned spiny and aspiny neurons in the monkey neostriatum. *J. Neurocytol.*, 9: 471–492.

DiFiglia, M., Aronin, N. and Martin, J. B. (1982) Light and electron microscopic localization of immunoreactive leu-enkephalin in the monkey basal ganglia. *J. Neurosci.*, 2: 303–320.

Dimova, R., Vuillet, J. and Seite, R. (1980) Study of the rat neostriatum using a combined Golgi-electron microscope technique and serial sections. *Neuroscience*, 5: 1581–1596.

Dockray, G. J., Gregory, R. A. and Hutchinson, J. B. (1978) Isolation, structure and biological activity of two cholecystokinin octapeptides from sheep brain. *Nature*, 274: 711–713.

Dockray, G. J., Vaillant, C. and Williams, R. G. (1981a) New vertebrate brain–gut peptide related to a molluscan neuropeptide and an opioid peptide. *Nature*, 293: 656–657.

Dockray, G. J., Vaillant, C., Williams, R. G., Gayton, R. J. and Osborne, N. N. (1981b) Vertebrate brain–gut peptides related to FMRFamide and Met-enkephalin Arg^{6}Phe^7. *Peptides*, Suppl. 2, 2: 25–30.

Elde, R. and Hökfelt, T. (1979) Localization of hypophysiotrophic peptides and other biologically active peptides within the brain. *Annu. Rev. Physiol.*, 41: 587–602.

Elde, R., Hökfelt, T., Johansson, O. and Terenius, L. (1976) Immunohistochemical studies using antibodies to leucine-enkephalin: Initial observations on the nervous system of the rat. *Neuroscience,* 1: 349–351.

Emson, P. C., Fahrenkrug, J. and Spokes, E. G. (1979) Vasoactive intestinal polypeptide (VIP): Distribution in normal human brain and in Huntington's disease. *Brain Res.,* 173: 174–178.

Emson, P. C., Arregui, A., Clement-Jones, V., Sandberg, B. E. B. and Rossor, M. (1980a) Regional distribution of methionine-enkephalin and substance P-like immunoreactivity in normal human brain and in Huntington's disease. *Brain Res.,* 199: 147–160.

Emson, P. C., Rehfeld, J. F., Langevin, H. and Rossor, M. (1980b) Reduction in cholecystokinin-like immunoreactivity in the basal ganglia in Huntington's disease. *Brain Res.,* 198: 497–500.

Emson, P. C., Goedert, M., Benton, H., St-Pierre, S. and Rioux, F. (1982) The regional distribution and chromatographic characterization of neurotensin-like immunoreactivity in the rat. In E. Costa and M. Trabucchi (Eds.), *Regulatory Peptides: From Molecular Biology to Function,* Raven, New York, pp. 477–485.

Epelbaum, J., Tapia-Arancibia, L., Kordon, C. and Enjalbert, A. (1982) Characterization, regional distribution, and subcellular distribution of ^{125}I-Tyr$_1$-somatostatin binding sites in rat brain. *J. Neurochem.,* 38: 1515–1523.

Erspamer, V. (1981) The tachykinin peptide family. *Trends Neurosci.,* 4: 267–269.

Eskay, R. L., Giraud, P., Oliver, C. and Brownstein, M. J. (1979) Distribution of α-melanocyte-stimulating hormone in the rat brain. Evidence that α-MSH-containing cells in the arcuate region send projections to extrahypothalamic areas. *Brain Res.,* 178: 55–67.

Fallon, J. H., Hicks, R. and Loughlin, S. E. (1983) The origin of cholecystokinin terminals in the basal forebrain of the rat: evidence from immunofluorescence and retrograde tracing. *Neurosci. Lett.,* 37: 29–35.

Fekete, M., Kadar, T., Penke, B., Kovacs, K. and Telegdy, G. (1981) Influence of cholecystokinin octapeptide sulfate ester on monoamine metabolism in rats. *J. Neural Transm.,* 50: 81–88.

Finley, J. C. W., Manderdrut, J. L. and Petrusz, P. (1981a) The immunocytochemical localization of enkephalin in the central nervous system of the rat. *J. Comp. Neurol.,* 198: 541–565.

Finley, J. C. W., Maderdrut, J. L., Roger, L. J. and Petrusz, P. (1981b) The immunocytochemical localization of somatostatin-containing neurons in the rat central nervous system. *Neuroscience,* 6: 2173–2192.

Fonnum, F. and Walaas, I. (1981) Localization of neurotransmitters in nucleus accumbens. In R. B. Chronister and J. F. DeFrance (Eds.), *The Neurobiology of the Nucleus Accumbens,* Haer Institute for Electrophysiological Research, Brunswick, ME, pp. 259–272.

Fonnum, F., Gottesfeld, Z. and Grofová, I. (1978) Distribution of glutamate decarboxylase, choline acetyl-transferase and aromatic amino acid decarboxylase in the basal ganglia of normal and operated rats. Evidence for striatopallidal, striatoento-penduncular and striatonigral GABAergic fibers. *Brain Res.,* 143: 125–138.

Fukui, K., Kato, N., Kimura, H., Tange, A., Okamura, H., Obata, H. L., Ibata, Y. and Yanaihara, N. (1982) The distribution of vasoactive intestinal polypeptide (VIP) in central nervous system, studied by immunohistochemistry. *Neurosci. Lett.,* Suppl., 9: S96.

Fuxe, K., Hökfelt, T., Said, S. I. and Mutt, V. (1977) Vasoactive intestinal polypeptide and the nervous system: Immunohistochemical evidence for localization in central and peripheral neurons, particularly intracortical neurons of the cerebral cortex. *Neurosci. Lett.,* 5: 241–246.

Fuxe, K., Anderson, K., Ganten, D., Hökfelt, T. and Enroth, P. (1980a) Evidence for the existence of an angiotensin II-like immunoreactive central neuron system and its interactions with the central catecholamine pathways. In F. Gross and G. Vogel (Eds.), *Enzymatic Release of Vasoactive Peptides,* Raven, New York, pp. 161–170.

Fuxe, K., Andersson, K., Locatelli, V., Agnati, L. F., Hökfelt, T., Skirboll, L. and Mutt, V. (1980b) Cholecystokinin peptides produce marked reduction of dopamine turnover in discrete areas of rat brain following intraventricular injection. *Eur. J. Pharmacol.,* 67: 329–331.

Fuxe, K., Agnati, L. F., Benfenati, F., Cimmino, M., Algeri, S., Hökfelt, T. and Mutt, V. (1981) Modulation by cholecystokinins of ^{3}H-spiroperidol binding in rat striatum: Evidence for increased affinity and reduction in the number of binding sites. *Acta Physiol. Scand.,* 113: 567–569.

Gale, K., Hong, J. S. and Guidotti, A. (1977) Presence of substance P and GABA in separate striatonigral neurons. *Brain Res.,* 136: 371–375.

Garcia-Sevilla, J. A., Magnusson, T. and Carlsson, A. (1978) Effect of intracerebroventricularly administered somatostatin on brain monoamine turnover. *Brain Res.,* 155: 159–164.

George, S. R. and Van Loon, G. R. (1982) β-Endorphin alters dopamine uptake by the dopamine neurons of the hypothalamus and striatum. *Brain Res.,* 248: 293–303.

Gerfen, C. R. (1984) The neostriatal mosaic: compartmentalization of corticostriatal input and striatonigral output systems. *Nature,* 311: 461–464.

Gilles, C., Lotstra, F. and Vanderhaeghen, J. J. (1983) CCK nerve terminals in the rat striatal and limbic areas originate partly in the brainstem and partly in the telencephalic structures. *Life Sci.,* 32: 1683–1690.

Goedert, M. and Emson, P. C. (1983) The regional distribution of neurotensin-like immunoreactivity in central and peripheral tissues of the cat. *Brain Res.,* 272: 291–297.

Goedert, M., Mantyh, P. W., Hunt, S. P. and Emson, P. C. (1983) Mosaic distribution of neurotensin-like immunoreactivity in the cat striatum. *Brain Res.,* 274: 176–179.

Goedert, M., Mantyh, P. W., Emson, P. C. and Hunt, S. P. (1984a) Inverse relationship between neurotensin receptors and neurotensin-like immunoreactivity in cat striatum. *Nature,* 307: 543–546.

Goedert, M., Pittaway, K. and Emson, P. C. (1984b) Neurotensin receptors in the rat striatum: lesion studies. *Brain Res.,* 299: 164–168.

Govoni, S., Hong, J. S., Yang, H.-Y. T. and Costa, E. (1980) Increase of neurotensin content elicited by neuroleptics in nucleus accumbens. *J. Pharmacol. Exp. Ther.,* 215: 413–417.

Grafe, M. R., Forno, L. S. and Eng, L. F. (1985) Immunocytochemical studies of substance P and met-enkephalin in the basal ganglia and substantia nigra in Huntington's, Parkinson's and Alzheimer's diseases. *J. Neuropathol. Exp. Neurol.,* 44: 47–59.

Graybiel, A. M. and Chesselet, M.-F. (1984) Compartmental distribution of striatal cell bodies expressing [Met]enkephalin-like immunoreactivity. *Proc. Natl. Acad. Sci. U.S.A.,* 81: 7980–7984.

Graybiel, A. M. and Ragsdale, C. W. (1978) Histochemically distinct compartments in the striatum of human, monkey and cat demonstrated by acetylcholinesterase staining. *Proc. Natl. Acad. Sci. U.S.A.,* 75: 5723–5726.

Graybiel, A. M. and Ragsdale, C. W. (1979) Fiber connections of the basal ganglia. In M. Cuénod, G. W. Kreutzberg and F. E. Bloom (Eds.), *Development and Chemical Specificity of Neurons,* Elsevier, Amsterdam, pp. 239–283.

Graybiel, A. M. and Ragsdale, C. W. (1983) Biochemical anatomy of the striatum. In P.C. Emson (Ed.), *Chemical Neuroanatomy,* Raven, New York, pp. 427–504.

Graybiel, A. M., Ragsdale, C. W., Yoneoka, E. S. and Elde, R. P. (1981) An immunohistochemical study of enkephalins and other neuropeptides in the striatum of the cat with evidence that the opiate peptides are arranged to form mosaic patterns in register with the striosomal compartments visible by acetylcholinesterase staining. *Neuroscience,* 6: 377–397.

Gubler, U., Seeburg, P., Hoffman, B. J., Gage, L. P. and Udenfriend, S. (1982) Molecular cloning establishes pro-enkephalin as precursor of enkephalin-containing peptides. *Nature,* 295: 206–208.

Guy, J., Vaudry, H. and Pelletier, G. (1981) Differential projections of two immunoreactive α-melanocyte stimulating hormone (α-MSH) neuronal systems in the rat brain. *Brain Res.,* 220: 199–202.

Haber, S. and Elde, R. (1981) Correlation between met-enkephalin and substance P immunoreactivity in the primate globus pallidus. *Neuroscience,* 6: 1291–1297.

Haber, S. and Elde, R. (1982) The distribution of enkephalin immunoreactive fibers and terminals in the monkey central nervous system: An immunohistochemical study. *Neuroscience,* 7: 1049–1095.

Haber, S. N. and Nauta, W. J. (1983) Ramifications of the globus pallidus in the rat as indicated by patterns of immunohistochemistry. *Neuroscience,* 9: 245–260.

Haber, S. N. and Watson, S. J. (1983) The comparison between enkephalin-like and dynorphin-like immunoreactivity in both monkey and human globus pallidus and substantia nigra. *Life Sci.,* Suppl. I, 33: 33–36.

Hamel, E. and Beaudet, A. (1984) Electron microscopic autoradiographic localization of opioid receptors in rat neostriatum. *Nature,* 312: 155–157.

Hara, Y., Shiosaka, S., Senba, E., Sakanaka, M., Inagaki, S., Takagi, H., Kawai, Y., Takatsuki, K., Matsuzaki, T. and Tohyama, M. (1982) Ontogeny of the neurotensin-containing neuron system of the rat: Immunohistochemical analysis. I. Forebrain and diencephalon. *J. Comp. Neurol.,* 208: 177–195.

Hawthorn, J., Ang, V. T. Y. and Jenkins, J. S. (1980) Localization of vasopressin in the rat brain. *Brain Res.,* 197: 75–81.

Heal, D. J. and Green, A. R. (1979) Administration of TRH to rats releases dopamine in n. accumbens but not in n. caudatus. *Neuropharmacology,* 18: 23–31.

Heimer, L. (1976) The olfactory cortex and the ventral striatum. In K. E. Livingston and O. Hornykiewicz (Eds.), *Limbic Mechanisms: The Continuing Evolution of the Limbic System Concept,* Plenum, New York, pp. 95–187.

Heimer, L. and Wilson, R. D. (1975) The subcortical projections of the allocortex: Similarities in the neural associations of the hippocampus, the piriform cortex, and the neocortex. In M. Santini (Ed.), *Golgi Centennial Symposium,* Raven, New York, pp. 177–193.

Herkenham, M. and Pert, C. B. (1981) Mosaic distribution of opiate receptors, parafascicular projections and acetylcholinesterase in rat striatum. *Nature,* 291: 415–418.

Hökfelt, T., Fuxe, K., Johansson, O., Jeffcoate, S. and White, N. (1975) Distribution of thyrotropin releasing hormone (TRH) in the central nervous system as revealed with immunocytochemistry. *Eur. J. Pharmacol.,* 34: 389–392.

Hökfelt, T., Elde, R., Johansson, O., Ljungdahl, Å., Schultzberg, M., Fuxe, K., Goldstein, M., Nilsson, G., Pernow, B., Terenius, L., Ganten, D., Jeffcoate, S. L., Rehfeld, J. and Said, S. (1978) Distribution of peptide-containing neurons. In M. A. Lipton, A. DiMascio and K. F. Killam (Eds.), *Psychopharmacology: A Generation of Progress,* Raven, New York, pp. 39–66.

Hökfelt, T., Skirboll, L., Rehfeld, J. F., Goldstein, M., Markey, K. and Dann, O. (1980) A subpopulation of mesencephalic dopamine neurons projecting to limbic areas contains a cholecystokinin-like peptide: Evidence from immunohistochemistry combined with retrograde tracing. *Neuroscience,* 5: 2093–2124.

Hökfelt, T., Everitt, E. J., Norheim, E., Rossel, S. and Goldstein, M. (1983) Neurotensin-like immunoreactivity in dopamine neurons in the arcuate nucleus and ventral mesencephalon. In *Proc. 5th Int. Catecholamine Symp.,* Goteborg, pp. 171.

Höllt, V., Haarmann, I., Bovermann, K., Jerlicz, M. and Herz, A. (1980) Dynorphin-related immunoreactive peptides in rat brain and pituitary. *Neurosci. Lett.,* 18: 149–153.

Hong, J. S., Yang, H.-Y. T., Racagni, G. and Costa, E. (1977) Projections of substance P containing neurons from neostriatum to substantia nigra. *Brain Res.,* 122: 541–544.

Hughes, J., Smith, T. W., Kosterlitz, H. W., Fothergill, L. A., Morgan, B. A. and Morris, H. R. (1975) Identification of two related pentapeptides from the brain with potent opiate agonist activity. *Nature,* 256: 577–579.

Ikeda, Y., Nakao, K., Yoshimasa, T., Sakamoto, M., Suda, M., Yanaihara, N. and Imura, H. (1983) Parallel distribution of methionine-enkephalin-arg^6-gly^7-leu^8 with methionine-enkephalin, leucine-enkephalin and methionine-enkephalin-arg^6-phe^7 in human and bovine brains. *Life Sci.,* Suppl. I, 33: 65–68.

Inagaki, S. and Parent, A. (1984) Distribution of substance P and enkephalin-like immunoreactivity in the substantia nigra of rat, cat and monkey. *Brain Res. Bull.,* 13: 319–329.

Inagaki, S. and Parent, A. (1985) Distribution of enkephalin-immunoreactive neurons in the forebrain and upper brain stem of the squirrel monkey. *Brain Res.,* 359: 267–280.

Innis, R. B., Correa, F. M., Uhl, G. R., Schneider, B. and Snyder, S. H. (1979) Cholecystokinin octapeptide-like immunoreactivity: Histochemical localization in rat brain. *Proc. Natl. Acad. Sci. U.S.A.,* 76: 521–525.

Jacobowitz, D. M. and O'Donohue, T. L. (1978) α-Melanocyte stimulating hormone: Immunohistochemical identification and mapping in neurons of rat brain. *Proc. Natl. Acad. Sci. U.S.A.,* 78: 6300–6304.

Jennes, L., Stumpf, W. E. and Kalivas, P. W. (1982) Neurotensin: topographical distribution in rat brain by immunohistochemistry. *J. Comp. Neurol.,* 210: 211–224.

Jessell, T. M., Emson, P. C., Paxinos, G. and Cuello, A. C. (1978) Topographic projections of substance P and GABA pathways in the striato- and pallido-nigral system: A biochemical and immunohistochemical study. *Brain Res.,* 152: 487–498.

Johansson, O. and Hökfelt, T. (1980) Thyrotropin releasing hormone, somatostatin, and enkephalin: distribution studies using immunohistochemical techniques. *J. Histochem. Cytochem.,* 28: 364–366.

Johansson, O. and Hökfelt, T. (1981) Nucleus accumbens: transmitter histochemistry with special reference to peptide-containing neurons. In R. B. Chronister and J. F. DeFrance (Eds.), *The Neurobiology of the Nucleus Accumbens,* Haer Institute for Electrophysiological Research, Brunswick, ME, pp. 147–172.

Johansson, O., Hökfelt, T. and Elde, R. P. (1984) Immunohistochemical distribution of somatostatin-like immunoreactivity in the central nervous system of the adult rat. *Neuroscience,* 13: 265–339.

Jones, B. N., Shively, J. E., Kilpatrick, D. L., Stern, A. S., Lewis, R. V., Kojima, K. and Udenfriend, S. (1982) Adrenal opioid proteins of 8600 and 12 600 daltons: intermediates in proenkephalin processing. *Proc. Natl. Acad. Sci. U.S.A.,* 79: 2096–2100.

Joseph, S. A. and Knigge, K. M. (1983) Corticotropin releasing factor: Immunocytochemical localization in rat brain. *Neurosci. Lett.,* 35: 135–141.

Kalivas, P. W. and Miller, J. S. (1984a) Neurotensin neurons in the ventral tegmental area project to the medial nucleus accumbens. *Brain Res.,* 300: 157–160.

Kalivas, P. W. and Miller, J. S. (1984b) Substance P modulation of dopamine in the nucleus accumbens. *Neurosci. Lett.,* 48: 55–59.

Kalivas, P. W., Burgess, P. W., Nemeroff, C. B. and Prange, A. J. (1983) Behavioral and neurochemical effects of neurotensin microinjection into the ventral tegmental area. *Neuroscience,* 8: 495–505.

Kalivas, P. W., Nemeroff, C. B. and Prange, A. J. (1984) Neurotensin microinjection into the nucleus accumbens antagonizes dopamine-induced increase in locomotion and rearing. *Neuroscience,* 11: 919–930.

Kameyama, T., Ukai, M., Noma, S. and Hiramatsu, M. (1982) Differential effects of α-, β- and γ-endorphins on dopamine metabolism in the mouse brain. *Brain Res.,* 244: 305–309.

Kanazawa, I., Bird, E., O'Connell, R. and Powell, D. (1977a) Evidence for a decrease in substance P content of substantia nigra in Huntington's chorea. *Brain Res.,* 120: 387–392.

Kanazawa, I., Emson, P. C. and Cuello, A. C. (1977b) Evidence for the existence of substance P-containing fibres in striato-nigral and pallido-nigral pathways in rat brain. *Brain Res.,* 119: 447–453.

Kanazawa, I., Mogaki, S., Muramoto, O. and Kuzuhara, S. (1980) On the origin of substance P-containing fibers in the entopeduncular nucleus and the substantia nigra of the rat. *Brain Res.,* 184: 481–485.

Kanazawa, I., Ogawa, I., Kimura, S. and Munekata, E. (1984) Regional distributin of substance P, neurokinin α and neurkinin β in rat central nervous system. *Neurosci. Res.,* 2: 111–120.

Kawai, Y., Inagaki, S., Shiosaka, S., Shibasaki, T., Ling, N., Tohyama, M. and Shiotani, Y. (1984) Distribution and projection of γ-melanocyte stimulating hormone in the rat brain: An immunohistochemical analysis. *Brain Res.,* 297: 21–32.

Kawai, Y., Takami, K., Shiosaka, S., Emson, P. C., Hillyard, C. J., Girgis, S. I., MacIntyre, I. and Tohyama, M. (1985) Topographic localization of calcitonin gene-related peptide in the rat brain: An immunohistochemical analysis. *Neuroscience,* 15: 747–763.

Khachaturian, H ., Lewis, M. E. and Watson, S. J. (1983) Enkephalin systems in diencephalon and brainstem of the rat. *J. Comp. Neurol.,* 220: 310–320.

Kimmel, J. R., Hayden, L. J. and Pollock, H. G. (1975) Isolation and characterization of a new pancreatic polypeptide hormone. *J. Biol. Chem.,* 250: 9369–9376.

Kobayashi, R. M., Brown, M. and Vale, W. (1977) Regional distribution of neurotensin and somatostatin in the non-lesioned rat. *Brain Res.,* 126: 584–588.

Kohno, J., Shiosaka, S., Shinoda, K., Inagaki, S. and Tohyama, M. (1984) Two distinct strio-nigral substance P pathways in the rat: An experimental immunohistochemical study. *Brain Res.,* 308: 309–317.

Kovacs, G. A., Szabo, G., Penke, B. and Telegdy, G. (1981) Effects of cholecystokinin on striatal dopamine metabolism and on apomorphine-induced stereotyped cage climbing in mouse. *Eur. J. Pharmacol.*, 69: 313–319.

Krause, J. E., Reiner, A. J., Advis, J. P. and McKelvy, J. F. (1984) In vivo biosynthesis of [^{35}S]- and [^{3}H] substance P in the striatum of the rat and their axonal transport to the substantia nigra. *J. Neurosci.*, 4: 775–785.

Kubota, Y., Takagi, H., Morishima, Y., Powell, J. F. and Smith, A. D. (1985) Synaptic interaction between catecholaminergic neurons and substance P-immunoreactive axons in the caudal part of the nucleus of the solitary tract of the rat: Demonstrated by the electron microscopic mirror technique. *Brain Res.*, 333: 188–192.

Kubota, Y., Inagaki, S., Kito, S., Takagi, H. and Smith, A. D. (1986) Ultrastructural evidence of dopaminergic input to enkephalinergic neurons in rat neostriatum. *Brain Res.* (in press).

Kuhar, M. J., Pert, C. B. and Snyder, S. H. (1973) Regional distribution of opiate receptor binding in monkey and human brain. *Nature*, 245: 447–450.

Larsson, L.-I. and Rehfeld, J. P. (1979) Localization and molecular heterogeneity of cholecystokinin in the central and peripheral nervous system. *Brain Res.*, 165: 201–218.

Larsson, L.-I., Childers, S. and Snyder, S. H. (1979) Met- and Leu-enkephalin immunoreactivity in separate neurons. *Nature*, 282: 407–410.

Lauber, M., Camier, M. and Cohen, P. (1979) Higher molecular weight forms of immunoreactive somatostatin in mouse hypothalamic extracts: evidence of processing in vitro. *Proc. Natl. Acad. Sci. U.S.A.*, 76: 6004–6008.

Lewis, R. V., Stern, A. S., Rossier, J., Stein, S. and Udenfriend, S. (1979) Putative enkephalin precursors in bovine adrenal medulla. *Biochem. Biophys. Res. Commun.*, 89: 822–829.

Liston, D., Böhlen, P. and Rossier, J. (1984) Purification from brain of synenkephalin, the N-terminal fragment of proenkephalin. *J. Neurochem.*, 43: 335–341.

Ljungdahl, A., Hökfelt, T. and Nilsson, G. (1978) Distribution of substance P-like immunoreactivity in the central nervous system of the rat. I. Cell bodies and nerve terminals. *Neuroscience*, 3: 861–943.

Lorén, I., Alumets, J., Håkanson, R. and Sundler, F. (1979a) Distribution of gastrin and CCK-like peptides in rat brain. An immunocytochemical study. *Histochemistry*, 59: 249–257.

Lorén, I., Alumets, J., Håkanson, R. and Sundler, F. (1979b) Immunoreactive pancreatic polypeptide (PP) occurs in the central and peripheral nervous system: Preliminary immunocytochemical observations. *Cell. Tissue Res.*, 200: 179–186.

Lorén, I., Emson, P. C., Fahrenkrug, J., Björklund, A., Alumets, J., Håkanson, R. and Sundler, F. (1979c) Distribution of vasoactive intestinal polypeptide in the rat and mouse brain. *Neuroscience*, 4: 1953–1976.

Manaker, S., Winokur, A., Rostène, W. H. and Rainbow, T. C. (1985) Autoradiographic localization of thyrotropin-releasing hormone receptors in the rat central nervous system. *J. Neurosci.*, 5: 167–174.

Manberg, P. J., Nemeroff, C. B., Iversen, L. L., Rossor, M. N., Kizer, J. S. and Prange, A. J. (1982) Human brain distribution of neurotensin in normals, schizophrenics, and Huntington's choreics. *Ann. N.Y. Acad. Sci.*, 400: 354–367.

Mantyh, P. W., Maggio, J. E. and Hunt, S. P. (1984) The autoradiographic distribution of kassinin and substance K binding sites is different from the distribution of substance P binding sites in rat brain. *Eur. J. Pharmacol.*, 102: 361–364.

Markstein, R. and Hökfelt, T. (1984) Effect of cholecystokinin-octapeptide on dopamine release from slices of cat caudate nucleus. *J. Neurosci.*, 4: 570–575.

Marley, P. D., Emson, P. C., Hunt, S. P. and Fahrenkrug, J. (1981) A long ascending projection in the brat brain containing vasoactive intestinal polypeptide. *Neurosci. Lett.*, 27: 261–266.

Marley, P. D., Emson, P. C. and Rehfeld, J. F. (1982) Effect on 6-hydroxy-dopamine lesions of the medial forebrain bundle on the distribution of cholecystokinin in rat forebrain. *Brain Res.*, 252: 382–385.

Marshall, P. E. and Landis, D. M. D. (1985) Huntington's disease is accompanied by changes in the distribution of somatostatin-containing neuronal processes. *Brain Res.*, 329: 71–82.

Marshall, P. E., Landis, D. M. and Zalneraitis, E. L. (1983) Immunocytochemical studies of substance P and leucine-enkephalin in Huntington's disease. *Brain Res.*, 289: 11–26.

Mashal, R. D., Owen, F., Deakin, J. F. and Poulter, M. (1983) The effects of cholecystokinin on dopaminergic mechanisms in rat striatum. *Brain Res.*, 277: 375–376.

Mauborgne, A., Javoy-Agid, F., Legrand, J. C., Agid, Y. and Cesselin, F. (1983) Decrease of substance P-like immunoreactivity in the substantia nigra and pallidum of Parkinsonian brains. *Brain Res.*, 268: 167–170.

McCulloch, J., Kelly, P. A., Uddman, R. and Edvinsson, L. (1983) Functional role for vasoactive intestinal polypeptide in the caudate nucleus: A 2-deoxy [^{14}C]glucose investigation. *Proc. Natl. Acad. Sci., U.S.A.*, 80: 1472–1476.

Melis, M. R. and Gale, K. (1984) Intranigral application of substance P antagonists prevents the haloperidol-induced activation of striatal tyrosine hydroxylase. *Naunyn Schmiedeberg's Arch. Pharmacol.*, 326: 83–86.

Mendelsohn, F. A. O., Quirion, R., Saavedra, J. M., Aguilera, G. and Catt, K. J. (1984) Autoradiographic localization of angiotensin II receptors in rat brain. *Proc. Natl. Acad. Sci. U.S.A.*, 81: 1575–1579.

Meyer, D. K. and Krauss, J. (1983) Dopamine modulates cholecystokinin release in neostriatum. *Nature*, 301: 338–340.

Meyer, D. K., Beinfeld, M. C., Oertel, W. H. and Brownstein, M. J. (1982) Origin of the cholecystokinin-containing fibers in the rat caudatoputamen. *Science*, 215: 187–188.

Meyer, D. K., Holland, A. and Conzelmann, U. (1984) Dopamine D_1-receptor stimulation reduces neostriatal cholecystokinin release. *Eur. J. Pharmacol.*, 104: 387–388.

Miyamoto, M. and Nagawa, Y. (1977) Mesolimbic involvement in the locomotor stimulant action of thyrotropin-releasing hormone (TRH) in rats. *Eur. J. Pharmacol.*, 44: 143–152.

Mori, M., Jayaraman, A., Prasad, C., Pegues, J. and Wilber, J. F. (1982a) Distribution of histidyl-proline diketopiperazine [cyclo(His-Pro)] and thyrotropin-releasing hormone (TRH) in the primate central nervous system. *Brain Res.*, 245: 183–186.

Mori, M., Prasad, C. and Wilber, J. F. (1982b) Regional dissociation of histidyl-proline diketopiperazine (cyclo-(His-Pro)) and thyrotropin-releasing hormone (TRH) in the rat brain. *Brain Res.*, 231: 451–453.

Mroz, E. A., Brownstein, M. J. and Leeman, S. E. (1977) Evidence for substance P in the striato-nigral tract. *Brain Res.*, 125: 305–311.

Mutt, V. and Said, S. I. (1974) Structure of the porcine vasoactive intestinal octacosapeptide: the amino acid sequence. Use of kallikrein in its determination. *Eur. J. Biochem.*, 42: 581–589.

Nagy, J. I., Carter, D. A. and Fibiger, H. C. (1978) Anterior striatal projections to the globus pallidus, entopeduncular nucleus and substantia nigra in the rat: The GABA connection. *Brain Res.*, 158: 15–29.

Nakagawa, Y., Shiosaka, S., Emson, P. C. and Tohyama, M. (1985) Distribution of neuropeptide Y in the forebrain and diencephalon: An immunohistochemical analysis. *Brain Res.*, 361: 52–60.

Narumi, S. and Nagawa, Y. (1983) Modification of dopaminergic transmission by thyrotropin-releasing hormone. *Adv. Biochem. Psychopharmacol.*, 36: 185–197.

Nemeroff, C. B., Luttinger, D., Hernandez, D. E., Mailman, R. B., Mason, G. A., Davis, S. D., Wilderlöv, E., Frye, G. D., Kilts, C. A., Beaumont, K., Breese, G. R. and Prange, A. J. (1983a) Interactions of neurotensin with brain dopamine systems: Biochemical and behavioral studies. *J. Pharmacol. Exp. Ther.*, 225: 337–345.

Nemeroff, C. B., Youngblood, W. M., Manberg, P. J., Prange, A. J. and Kizer, J. S. (1983b) Regional brain concentrations of neuropeptides in Huntington's chorea and schizophrenia. *Science*, 221: 972–975.

Noda, M., Furutani, Y., Takahashi, H., Toyosato, M., Hirose, T., Inayama, S., Nakanishi, S. and Numa, S. (1982) Cloning and sequence analysis of cDNA for bovine adrenal preproenkephalin. *Nature*, 295: 202–206.

O'Donohue, T. L., Miller, R. L. and Jacobowitz, D. M. (1979) Identification, characterization and stereotaxic mapping of intraneuronal α-melanocyte stimulating hormone-like immunoreactive peptides in discrete regions of the rat brain. *Brain Res.*, 176: 101–123.

Oertel, W. H. and Mugnaini, E. (1984) Immunocytochemical studies of GABAergic neurons in rat basal ganglia and their relations to other neuronal systems. *Neurosci. Lett.*, 47: 233–238.

Oertel, W. H., Riethmuller, G., Mugnaini, E., Schmechel, D. E., Weindl, A., Gramsch, C. and Herz, A. (1983) Opioid peptide-like immunoreactivity localized in GABAergic neurons of rat neostriatum and central amygdaloid nucleus. *Life Sci.*, Suppl. 1, 33: 73–76.

Okuma, Y., Fukuda, Y. and Osumi, Y. (1983) Neurotensin potentiates the potassium-induced release of endogenous dopamine from rat striatal slices. *Eur. J. Pharmacol.*, 93: 27–33.

Olschowka, J. A., O'Donohue, T. L. and Jacobowitz, D. M. (1981) The distribution of bovine pancreatic polypeptide-like immunoreactive neurons in rat brain. *Peptides*, 2: 309–331.

Pasik, P., Pasik, T. and DiFiglia, M. (1979) The internal organization of the neostriatum in mammals. In I. Divac and R. G. E. Oberg (Eds.), *The Neostriatum*, Pergamon, New York, pp. 5–36.

Patel, Y. C., Wheatley, Y. and Ning, C. (1981) Multiple forms of immunoreactive somatostatin: comparison of distribution in neural and non-neural tissue and portal plasma of the rat. *Endocrinology*, 109: 1943–1949.

Paxinos, G., Emson, P. C. and Cuello, A. C. (1978) Substance P projections to the entopeduncular nucleus, the medial preoptic area and the lateral septum. *Neurosci. Lett.*, 7: 133–136.

Pelletier, G. and LeClerc, R. (1979) Immunohistochemical localization of adrenocorticotropin in the rat brain. *Endocrinology*, 104: 1426–1433.

Pelletier, G., Labrie, F., Kastin, A. J. and Schally, A. V. (1975) Radioautographic localization of radioactivity in rat brain after intracarotid injection of [^{125}I]-α-melanocyte-stimulating hormone. *Pharmacol. Biochem. Behav.*, 3: 671–674.

Pelletier, G., LeClerc, R., Saavedra, J. M., Brownstein, M. J., Vaudry, H., Ferland, L. and Labrie, F. (1980) Distribution of β-lipotropin (β-LPH), adrenocorticotropin (ACTH) and α-melanocyte-stimulating hormone (α-MSH). *Brain Res.*, 192: 433–440.

Pert, C. B., Kuhar, M. J. and Snyder, S. H. (1976) Opiate receptors: Autoradiographic localization in rat brain. *Proc. Natl. Acad. Sci. U.S.A.*, 73: 3729–3733.

Phillips, M. I., Wehenmeyer, J., Felix, D., Ganten, D. and Hoffman, W. E. (1979) Evidence for an endogenous brain renin-angiotensin system. *Fed. Proc. Fed. Am. Soc. Exp. Biol.*, 38: 2260–2266.

Pickel, V. M., Sumal, K. K., Beckley, S. C., Miller, R. J. and Reis, D. J. (1980) Immunocytochemical localization of enkephalin in the neostriatum of rat brain: a light and electron microscopic study. *J. Comp. Neurol.*, 189: 721–740.

Pioro, E. P., Hughes, J. T. and Cuello, A. C. (1984) Loss of substance P and enkephalin immunoreactivity in the human substantia nigra after striato-pallidal infarction. *Brain Res.*, 292: 339–347.

Pittius, C. W., Seizinger, B. R., Mehraein, P., Pasi, A. and Herz, A. (1983) Proenkephalin-A-derived peptides are present in human brain. *Life Sci.*, Suppl. I, 33: 41–44.

Preston, R. J., Bishop, G. A. and Kitai, S. T. (1980) Medium spiny neuron projection from the rat striatum: An intracellu-

lar horseradish peroxidase study. *Brain Res.*, 183: 253–263.
Pycock, C. and Horton, R. (1976) Evidence for an accumbens-pallidal pathway in the rat and its possible GABA-minergic control. *Brain Res.*, 110: 629–634.
Quirion, R., Gaudreau, P., Martel, J.-C., St.-Pierre, S. and Zamir, N. (1985) Possible interactions between dynorphin and dopaminergic systems in rat basal ganglia and substantia nigra. *Brain Res.*, 331: 358–362.
Ribak, C. E., Vaughn, J. E. and Roberts, E. (1979) The GABA neurons and their axon terminals in rat corpus striatum as demonstrated by GAD immunocytochemistry. *J. Comp. Neurol.*, 187: 261–284.
Ribak, C. E., Vaughn, J. E. and Roberts, E. (1980) GABAergic nerve terminals decrease in the substantia nigra following hemitransections of the striatonigral and pallidonigral pathways. *Brain Res.*, 192: 413–420.
Roberts, G. W., Woodhams, P. L., Polak, J. M. and Crow, T. J. (1982) Distribution of neuropeptides in the limbic system of the rat: The amygdaloid complex. *Neuroscience*, 7: 99–131.
Rorstad, O. P., Epelbaum, J., Brazeau, P. and Martin, J. B. (1979) Chromatographic and biological properties of immunoreactive somatostatin in hypothalamic and extrahypothalamic brain regions of the rat. *Endocrinology*, 105: 1083–1092.
Rosenfeld, M. G., Mermod, J.-J., Amara, S. G., Swanson, L. W., Sawchenko, P. E., River, J., Vale, W. W. and Evans, R. M. (1983) Production of a novel neuropeptide encoded by the calcitonin gene via tissue-specific RNA processing. *Nature*, 304: 129–135.
Rossier, J., Audigier, Y., Ling, N., Cros, J. and Udenfriend, S. (1980) Met-enkephalin-Arg^6-Phe^7, present in high amounts in brain of rat, cattle and man, is an opioid agonist. *Nature*, 288: 88–90.
Rossor, M. N., Iversen, L. L., Hawthorn, J., Ang, V. T. Y. and Jenkins, J. S. (1981) Extrahypothalamic vasopressin in human brain. *Brain Res.*, 214: 349–355.
Rossor, M. N., Hunt, S. P., Iversen, L. L., Bannister, R., Hawthorn, J., Ang, V. T. Y. and Jenkins, J. S. (1982) Extrahypothalamic vasopressin is unchanged in Parkinson's disease and Huntington's disease. *Brain Res.*, 253: 341–343.
Rostène, W. H., Morgat, J. L., Dussaillant, M., Rainbow, T. C., Sarrieau, A., Vial, M. and Rosselin, G. (1984) In vitro biochemical characterization and autoradiographic distribution of ^{3}H-thyrotropin-releasing hormone bindings sites in rat brain sections. *Neuroendocrinology*, 39: 81–86.
Rothman, R. B., Herkenham, M., Pert, C. B., Liang, T. and Cascieri, M. A. (1984) Visualization of rat brain receptors for the neuropeptide, substance P. *Brain Res.*, 309: 47–54.
Saavedra, J. M., Fernandez-Pardal, J. and Chevillard, C. (1982) Angiotensin-converting enzyme in discrete areas of the rat forebrain and pituitary gland. *Brain Res.*, 245: 317–325.
Sagar, S. M., Beal, M. F., Marshall, P. E., Landis, D. M. D. and Martin, J. B. (1984) Implications of neuropeptides in neurological diseases. *Peptides*, Suppl. 1, 5: 255–262.
Said, S. and Mutt, V. (1970) Polypeptide with broad biological activity: isolation from small intestine. *Science*, 169: 1217–1218.
Saito, A., Sankaran, H., Goldfine, I. D. and Williams, J. A. (1980) Cholecystokinin receptors in brain: characterization and distribution. *Science*, 208: 1155–1156.
Sandberg, B. E. B. and Iversen, L. L. (1982) Substance P. *J. Med. Chem.*, 25: 1009–1015.
Sar, M., Stumpf, W. E., Miller, R. J., Chang, K. J. and Cuatrecasas, P. (1978) Immunohistochemical localization of enkephalin in rat brain and spinal cord. *J. Comp. Neurol.*, 182: 17–38.
Schally, A. V., Dupont, A., Arimura, A., Redding, T. W., Nishi, N., Linthicum, G. L. and Schlesinger, D. H. (1976) Isolation and structure of somatostatin from porcine hypothalamus. *Biochemistry*, 15: 509–514.
Shinoda, K., Inagaki, S., Shiosaka, S., Kohno, J. and Tohyama, M. (1984) Experimental immunohistochemical studies on the substance P neuron system in the lateral habenular nucleus of the rat: Distribution and origins. *J. Comp. Neurol.*, 222: 578–588.
Shiosaka, S., Takatsuki, K., Sakanaka, M., Inagaki, S., Takagi, H., Senba, E., Kawai, Y., Iida, H., Minagawa, H., Hara, Y., Matsuzaki, T. and Tohyama, M. (1982) Ontogeny of somatostatin-containing neuron system of the rat: Immunohistochemical analysis. II. Forebrain and diencephalon. *J. Comp. Neurol.*, 204: 211–224.
Shiosaka, S., Sakanaka, M., Inagaki, S., Senba, E., Hara, Y., Takatsuki, K., Takagi, H., Kawai, Y. and Tohyama, M. (1984) Putative neurotransmitters in the amygdaloid complex with special reference to peptidergic pathways. In P. C. Emson (Ed.), *Chemical Neuroanatomy*, Raven, New York, pp. 359–389.
Simantov, R., Kuhar, M. J., Uhl, G. R. and Snyder, S. H. (1977) Opioid peptide enkephalin: Immunohistochemical mapping in rat central nervous system. *Proc. Natl. Acad. Sci. U.S.A.*, 74: 2167–2171.
Sims, K. B., Hoffman, D. L., Said, S. I. and Zimmerman, E. A. (1980) Vasoactive intestinal polypeptide (VIP) in mouse and rat brain: an immunocytochemical study. *Brain Res.*, 186: 165–183.
Singh, E. A. and McGeer, E. G. (1978) Angiotensin converting enzyme in kainic acid-injected striata. *Ann. Neurol.*, 4: 85–86.
Sirett, N. E., Bray, J. J. and Hubbard, J. I. (1981) Localization of immunoreactive angiotensin in the hippocampus and striatum of the brain. *Brain Res.*, 217: 405–411.
Somogyi, P. and Smith, A. D. (1979) Projection of neostriatal spiny neurons to the substantia nigra. Application of a combined Golgi-staining and horseradish peroxidase transport procedure at both light and electron microscopic levels. *Brain Res.*, 178: 3–15.
Somogyi, P., Hodgson, A. J. and Smith, A. D. (1979) An approach to tracing neuron networks in the cerebral cortex and basal ganglia. Combination of Golgi staining, retrograde transport of horseradish peroxidase and anterograde degener-

ation of synaptic boutons in the same material. *Neuroscience,* 4: 1805–1852.

Somogyi, P., Bolam, J. P. and Smith, A. D. (1981) Monosynaptic cortical input and local axon collaterals of identified striatonigral neurons. A light and electron microscopic study using the Golgi-peroxidase transport-degeneration procedure. *J. Comp. Neurol.,* 195: 567–584.

Somogyi, P., Priestley, J. V., Cuello, A. C., Smith, A. D. and Takagi, H. (1982) Synaptic connections of enkephalin-immunoreactive nerve terminals in the neostriatum: A correlated light and electron microscopic study. *J. Neurocytol.,* 11: 779–807.

Sperk, G. and Singer, E. A. (1982) In vivo synthesis of substance P in the corpus striatum of the rat and its transport to the substantia nigra. *Brain Res.,* 238: 127–135.

Spindel, E. R., Wurtman, R. J. and Bird, E. D. (1980) Increased TRH content of the basal ganglia in Huntington's disease. *N. Engl. J. Med.,* 303: 1235–1236.

Spindel, E. R., Pettibone, D. J. and Wurtman, R. J. (1981) Thyrotropin-releasing hormone (TRH) content of rat striatum: Modification by drugs and lesions. *Brain Res.,* 216: 323–331.

Srikant, C. B. and Patel, Y. C. (1981) Somatostatin receptors. Identification and characterization in rat brain membranes. *Proc. Natl. Acad. Sci. U.S.A.,* 78: 3930–3934.

Staines, W. A., Nagy, J. I., Vincent, S. R. and Fibiger, H. C. (1980) Neurotransmitters contained in the efferents of the striatum. *Brain Res.,* 194: 391–402.

Staun-Olsen, P., Ottesen, B., Gammeltoft, S. and Fahrenkrug, J. (1985) The regional distribution of receptors for vasoactive intestinal polypeptide (VIP) in the rat central nervous system. *Brain Res.,* 330: 317–321.

Steinbusch, H. W. M., Nieuwenhuys, R., Verhofstad, A. A. J. and van der Kooy, D. (1981) The nucleus raphe dorsalis of the rat and its projection upon the caudatoputamen: A combined cytoarchitechtonic, immunohistochemical and retrograde transport study. *J. Physiol. (Paris),* 77: 157–174.

Strittmatter, S. M., Lo, M. M. S., Javitch, J. A. and Snyder, S. H. (1984) Autoradiographic visualization of angiotensin-converting enzyme in rat brain with [^{3}H]captopril: Localization to a striatonigral pathway. *Proc. Natl. Acad. Sci. U.S.A.,* 81: 1599–1603.

Studler, J. M., Simon, H., Cesselin, F., Legrand, J. C., Glowinski, J. and Tassin, J. P. (1981) Biochemical investigation on the localization of the cholecystokinin octapeptide in dopaminergic neurons originating from the ventral tegmental area of the rat. *Neuropeptides,* 2: 131–139.

Studler, J. M., Reinbaud, M., Tramu, G., Blanc, G., Glowinski, J. and Tassin, J. P. (1984) Pharmacological study on the mixed CCK8/DA meso-nucleus accumbens pathway: Evidence for the existence of storage sites containing the two transmitters. *Brain Res.,* 298: 91–97.

Sugimoto, T., Takeda, M., Kaneko, T. and Mizuno, N. (1984) Substance P-positive thalamocaudate neurons in the centromedian-parafascicular complex in the cat. *Brain Res.,* 323: 181–189.

Swanson, L. W., Sawchenko, P. E., Rivier, J. and Vale, W. W. (1983) Organization of ovine corticotropin-releasing factor immunoreactive cells and fibers in the rat brain: An immunohistochemical study. *Neuroendocrinology,* 36: 165–186.

Takagi, H., Somogyi, P., Somogyi, J. and Smith, A. D. (1983) Fine structural studies on a type of somatostatin-immunoreactive neuron and its synaptic connections in the rat neostriatum: a correlated light and electron microscopic study. *J. Comp. Neurol.,* 214: 1–16.

Takagi, H., Mizuta, H., Matsuda, T., Inagaki, S., Tateishi, K. and Hamaoka, T. (1984a) The occurrence of cholecystokinin-like immunoreactive neurons in the rat neostriatum: light and electron microscopic analysis. *Brain Res.,* 309: 346–349.

Takagi, H., Somogyi, P. and Smith, A. D. (1984b) Aspiny neurons and their local axons in the neostriatum of the rat: a correlated light and electron microscopic study of Golgi-impregnated materials. *J. Neurocytol.,* 13: 239–265.

Tapia-Arancibia L., Epelbaum, J., Enjalbert, A. and Kordon, C. (1981) Somatostatin binding sites in various structures of the rat brain. *Eur. J. Pharmacol.,* 71: 523–526.

Tatemoto, K., Carlquist, M. and Mutt, V. (1982) Neuropeptide Y — a novel brain peptide with structural similarities to peptide YY and pancreatic polypeptide. *Nature,* 296: 659–660.

Taylor, D. P. and Pert, C. B. (1979) Vasoactive intestinal polypeptide: Specific binding to rat brain membranes. *Proc. Natl. Acad. Sci. U.S.A.,* 76: 660–664.

Tenovuo, O., Rinne, U. K. and Viljanen, M. K. (1984) Substance P immunoreactivity in the post-mortem Parkinsonian brain. *Brain Res.,* 303: 113–116.

Torrens, Y., Michelot, R., Beaujouan, J. C., Glowinski, J. and Bockaert, J. (1982) In vivo biosynthesis of ^{35}S-substance P from ^{35}S-methionine in the rat striatum and its transport to the substantia nigra. *J. Neurochem.,* 38: 1728–1734.

Uhl, G., Kuhar, M. and Snyder, S. (1977) Neurotensin: immunohistochemical localization in rat central nervous system. *Proc. Natl. Acad. Sci. U.S.A.,* 74: 4059–4063.

Uhl, G. R., Goodman, R. R. and Snyder, S. H. (1979) Neurotensin-containing cell bodies, fibers and nerve terminals in the brain stem of the rat: Immunohistochemical mapping. *Brain Res.,* 167: 77–91.

Umegaki, K., Shiosaka, S., Kawai, Y., Shinoda, K., Yagura, A., Shibasaki, T., Ling, N. and Tohyama, M. (1983) The distribution of α-melanocyte stimulating hormone (α-MSH) in the central nervous system of the rat: An immunohistochemical study-I. Forebrain and upper brain stem. *Cell. Mol. Biol.,* 29: 377–386.

Vaccarino, F. J. and Koob, G. F. (1984) Microinjections of nanogram amounts of sulfated cholecystokinin octapeptide into the rat nucleus accumbens attenuates brain stimulation reward. *Neurosci. Lett.,* 52: 61–66.

Vale, W., Spiess, J., Rivier, C. and Rivier, J. (1981) Characterization of a 41-residue ovine hypothalamic peptide that stimulates secretion of corticotropin and β-endorphin. *Science,* 213: 1394–1397.

Vanderhaeghem, J. J., Lotstra, F., De Mey, J. and Gilles, C.

(1980) Immunohistochemical localization of cholecystokinin and gastrin-like peptides in the brain and hypophysis of the rat. *Proc. Natl. Acad. Sci. U.S.A.*, 77: 1190–1194.

Van der Kooy, D., Hunt, S. P., Steinbusch, H. W. M. and Verhofstad, A. A. M. (1981) Separate populations of cholecystokinin and 5-hydroxytryptamine-containing neuronal cells in the rat dorsal raphe, and their contribution to the ascending raphe projections. *Neurosci. Lett.*, 26: 25–30.

Van Loon, G. R. and Kim, C. (1978) β-Endorphin-induced increase in striatal dopamine turnover. *Life Sci.*, 23: 961–970.

Verhoef, J., Wiegant, V. M. and De Wied, D. (1982) Regional distribution of α- and γ-type endorphins in rat brain. *Brain Res.*, 231: 454–460.

Vincent, S. R. and Johansson, O. (1983) Striatal neurons containing both somatostatin- and avian pancreatic polypeptide (APP)-like immunoreactivities and NADPH-diaphorase activity: A light and electron microscopic study. *J. Comp. Neurol.*, 217: 264–270.

Vincent, S. R., Hökfelt, T., Christensson, I. and Terenius, L. (1982a) Immunohistochemical evidence for a dynorphin immunoreactive striatonigral pathway. *Eur. J. Pharmacol.*, 85: 251–252.

Vincent, S. R., Johansson, O., Skirboll, L. and Hökfelt, T. (1982b) Coexistence of somatostatin- and avian pancreatic polypeptide-like immunoreactivities in striatal neurons which are selectively stained for NADPH-diaphorase activity. In E. Costa and M. Trabucchi (Eds.), *Regulatory Peptides: From Molecular Biology to Function*, Raven Press, New York, pp. 453–462.

Vincent, S. R., Skirboll, L., Hökfelt, T., Johansson, O., Lundberg, J. M., Elde, R. P., Terenius, L. and Kimmel, J. (1982c) Coexistence of somatostatin- and avian pancreatic polypeptide (APP)-like immunoreactivity in some forebrain neurons. *Neuroscience*, 7: 439–446.

Vincent, S. R., Johansson, O., Hökfelt, T., Skirboll, L., Elde, R. P., Terenius, L., Kimmel, J. and Goldstein, M. (1983a) NADPH-diaphorase: a selective histochemical marker for striatal neurons containing both somatostatin- and avian pancreatic polypeptide (APP)-like immunoreactivities. *J. Comp. Neurol.*, 217: 252–263.

Vincent, S. R., Staines, W. A. and Fibiger, H. C. (1983b) Histochemical demonstration of separate populations of somatostatin and cholinergic neurons in the rat striatum. *Neurosci. Lett.*, 35: 111–114.

Von Euler, U. S. and Gaddum, J. H. (1931) An unidentified depressor substance in certain tissue extracts. *J. Physiol. (London)*, 72: 74–87.

Waddington, J. L. and Cross, A. J. (1978) Neurochemical changes following kainic acid lesions of the nucleus accumbens: Implications for a GABAergic accumbal-ventral tegmental pathway. *Life Sci.*, 22: 1011–1014.

Walaas, I. and Fonnum, F. (1980) Biochemical evidence for γ-aminobutyrate containing fibers from the nucleus accumbens to the substantia nigra and ventral tegmental area in the rat. *Neuroscience*, 5: 63–72.

Wamsley, J. K., Young, W. S. III and Kuhar, M. (1980): Immunohistochemical localization of enkephalin in rat forebrain. *Brain Res.*, 190: 153–174.

Watson, S. J. and Akil, H. (1980) α-MSH in rat brain: occurrence within and outside of β-endorphin neurons. *Brain Res.*, 182: 217–223.

Watson, S. J., Akil, H., Sullivan, S. and Barchas, J. D. (1977a) Immunocytochemical localization of methionine enkephalin: Preliminary observations. *Life Sci.*, 21: 733–738.

Watson, S. J., Barchas, J. D. and Li, C. H. (1977b) β-Lipotropin: Localization of cells and axons in rat brain by immunocytochemistry. *Proc. Natl. Acad. Sci. U.S.A.*, 74: 5155–5158.

Watson, S. J., Akil, H., Richard, C. W. III and Barchas, J. D. (1978a) Evidence for two separate opiate peptide neuronal systems. *Nature*, 275: 226–228.

Watson, S. J., Richard, C. W. III and Barchas, J. D. (1978b) Adrenocorticotropin in rat brain: Immunocytochemical localization in cells and axons. *Science*, 200: 1180–1181.

Weber, E., Evans, C. J., Samuelsson, S. J. and Barchas, J. D. (1981) Novel peptide neuronal system in rat brain and pituitary. *Science*, 214: 1248–1251.

Weber, E., Roth, K. A. and Barchas, J. D. (1982) Immunohistochemical distribution of α-neo-endorphin/dynorphin neuronal systems in rat brain: Evidence for colocalization. *Proc. Natl. Acad. Sci. U.S.A.*, 79: 3062–3066.

White, F. J. and Wang, R. Y. (1984) Interactions of cholecystokinin octapeptide and dopamine on nucleus accumbens neurons. *Brain Res.*, 300: 161–166.

Widerlöv, E., Kilts, C. D., Mailman, R. G., Nemeroff, C. B., Macown, T. J., Prange, A. J. and Breese, G. R. (1982) Increase in dopamine metabolites in rat brain by neurotensin. *J. Pharmacol. Exp. Ther.*, 222: 1–6.

Williams, R. G. and Dockray, G. J. (1982) Differential distribution in rat basal ganglia of met-enkephalin- and met-enkephalin Arg^6-Phe^7-like peptides revealed by immunohistochemistry. *Brain Res.*, 240: 167–170.

Williams, R. G. and Dockray, G. J. (1983) Immunohistochemical studies of FMRF-amide-like immunoreactivity in rat brain. *Brain Res.*, 276: 213–229.

Williams, R. G., Gayton, R., Zhu, W.-Y. and Dockray, G. J. (1981) Changes in brain cholecystokinin octapeptide following lesions of the medial forebrain bundle. *Brain Res.*, 213: 227–230.

Wilson, C. J. and Groves, P. M. (1980) Fine structure and synaptic connections of the common spiny neuron of the rat neostriatum: A study employing intracellular injection of horseradish peroxidase. *J. Comp. Neurol.*, 194: 599–615.

Young, W. S. III and Kuhar, M. J. (1979) A new method for receptor autoradiography: [^{3}H]opioid receptors in rat brain. *Brain Res.*, 179: 255–270.

Záborszky, L., Alheid, G. F., Beinfeld, M. C., Eiden, L. E., Heimer, L. and Palkovits, M. (1985) Cholecystokinin innervation of the ventral striatum: A morphological and radioimmunological study. *Neuroscience*, 14: 427–453.

Zahm, D. S., Záborszky, L., Alones, V. E. and Heimer, L. (1985)

Evidence for the coexistence of glutamate decarboxylase and met-enkephalin immunoreactivities in axon terminals of rat ventral pallidum. *Brain Res.,* 325: 317–321.

Zamir, N., Skofitsch, G., Bannon, M. J., Helke, C. J., Kopin, I. and Jacobowitz, D. M. (1984) Primate model of Parkinson's disease: alterations in multiple opioid systems in the basal ganglia. *Brain Res.,* 322: 356–360.

Zamir, N., Palkovits, M. and Brownstein, M. (1985) Distribution of immunoreactive met-enkephalin-arg^{6}-gly^{7}-leu^{8} and leu-enkephalin in discrete regions of the rat brain. *Brain Res.,* 326: 1–8.

Zarbin, M. A., Innis, R. B., Wamsley, J. K., Snyder, S. H. and Kuhar, M. J. (1983) Autoradiographic localization of cholecystokinin receptors in rodent brain. *J. Neurosci.,* 3: 877–906.

P. C. Emson, M. N. Rossor and M. Tohyama (Eds.),
Progress in Brain Research, Vol. 66.

CHAPTER 3

Physiology of peptides in basal ganglia

Shinichi Hosokawa and Motohiro Kato

Department of Neurophysiology, Neurological Institute, Faculty of Medicine, Kyushu University, Fukuoka 812, Japan

Introduction

Recent advances in the techniques of radioimmunoassay and immunohistochemistry have identified more than 20 peptides in neurons of the brain, spinal cord and peripheral nervous system (Emson, 1979; Hökfelt et al., 1980a). The discovery of these neuropeptides has attracted much interest in the possible functional roles of these peptides in the nervous system, and has already opened new and fruitful approaches to the study of memory, pain and satiety. Although radioimmunoassay and immunohistochemical studies have increasingly revealed peptide-containing nerve fibres in the central nervous system (CNS), one would still be hard placed to state exactly how any single peptidergic system operates. This partly reflects the formidable difficulty of doing the appropriate experiment in the labyrinthine maze of the CNS. In spite of this, experimental evidence has accumulated which suggests that at least some of these peptides, for example substance P (SP), may act as neurotransmitters. The alternative possibility is that the presence, or the release, of peptides from nerve terminals is not directly related to synaptic transmission, but rather to trophic, or longer-term events; this has only rarely been considered (Harkins et al., 1980).

In the basal ganglia, more than 10 neuropeptides (or "peptide immunoreactivities") have already been found (see Takagi, this volume). These include the enkephalins, SP, cholecystokinin, neurotensin, thyrotropin releasing hormone, somatostatin, vasoactive intestinal polypeptide, dynorphin and so on. Some of these peptides, for example SP in the striatonigral fibres and enkephalin (ENK) in the striatopallidal fibres, fulfill most of what are usually considered to be criteria for a neurotransmitter. These criteria are that the "transmitter" (1) should show synaptosomal localization in a specific neuronal system; (2) be released from presynaptic endings following nerve stimulation; (3) show presence of a specific receptor; (4) iontophoretic application should produce changes in neuronal activity; and (5) the transmitter candidate should mimic the action of the natural transmitter. Studies of changes in the content of neuropeptides in diseases of basal ganglia such as Parkinson's disease or Huntington's disease also suggest the pathophysiological role of neuropeptides in the basal ganglia.

In this paper, we will review the evidence for neuropeptides as neurotransmitters and discuss the possible physiological roles of these peptides in the basal ganglia.

Opioid peptides

The basal ganglia shows one of the highest levels of enkephalin-like immunoreactivity (ENK-LI) in the CNS (Table I), comparable to those in the spinal cord and brain stem (Hong et al., 1977b; Hughes et al., 1977; Emson et al., 1980a). On the other hand, most mammals seem to have little or no immunologically detectable endorphins in the basal ganglia (Rossier et al., 1977). In the basal ganglia, the concentration of ENK-LI is highest in the lateral segment of the globus pallidus. However, the medial segment of globus pallidus, striatum, nucleus accumbens and substantia nigra also contain signifi-

TABLE I

Levels of neuropeptides in the brain

	MET-ENK (pmol/g) human[a]	SP (pmol/g) human[b]	CCK (pmol/g) guinea pig[c]	VIP (pmol/g) human[d]	SRIF (ng/mg) rat[e]	TRH (ng/g) rat[f]	ACE (pmol/min per mg) human[g]	NT (pmol/g) human[h]
Cerebral cortex	42	14	214	8	–	–	18	1
Caudate	116	138	167	5	0.6	21	307	3
Globus pallidus			–	2	–	–		10
lateral segment	1163	197					155	
medial segment	675	877					149	
Substantia nigra	661	1535	–	2	0.3	–	105	–
Amygdala	26	25	280	21	3.7	23	–	5
Hypothalamus	141	112	74	23	73.1	489	–	43
Cerebellum	–	–	1	0	0.4	2	13	1

MET-ENK, methionine enkephalin; SP, substance P; CCK, cholecystokinin; VIP, vasoactive intestinal polypeptide; SRIF, somatostatin; TRH, thyrotropin releasing hormone; ACE, angiotensin converting enzyme; NT, neurotensin.
[a,b] Emson et al. (1980); [c] Larsson and Rehfeld (1979); [d] Emson et al. (1979); [e] Kobayashi et al. (1977); [f] Brownstein et al. (1975); [g] Arregui et al. (1979); [h] Cooper et al. (1981).

cant amounts of ENK-LI. Immunohistochemical studies have shown that there is an extremely dense mesh of enkaphalin-like immunoreactive (ENK-LIr) nerve terminals in the globus pallidus (Elde et al., 1976; Simantov et al., 1978). ENK-LIr neuronal cell bodies, as well as nerve terminals, have also been demonstrated in the striatum (DelFiacco et al., 1982). Immunohistochemical studies also show that the distribution of ENK-LIr fibres coincides with the regional distribution of opiate receptors as determined by biochemical techniques and autoradiography (Elde et al., 1976; Simantov et al., 1978).

The distribution of two subtypes of ENK, leucine enkephalin (LEU-ENK) and methionine enkephalin (MET-ENK), has been demonstrated. Although the ratios of the MET-ENK- and LEU-ENK-LI are similar in the whole brain, the ratio varies considerably in different brain regions (Hughes et al., 1977). Whether MET-ENK-LI and LEU-ENK-LI always coexist in the same neurotransmitter system or whether there may exist two separate neurotransmitter systems with diverse functions remains to be resolved. It is, however, relevant to note that MET-ENK-LI and LEU-ENK-LI are always present in the same precursor so that for separate MET--ENK and LEU-ENK cells to exist would imply selective degradation of either peptide.

Chronic kainate lesions in the striatum produced topographic depletion of ENK-LIr terminals in the globus pallidus. However, the intrapallidal injection of kainate could not modify the ENK-LI immunofluorescence of the globus pallidus. All these observations indicate that most of the enkephalinergic fibres come from the striatum and terminate in the globus pallidus (Hong et al., 1977b; Childers et al., 1978). Intrastriatal injection of kainate also reduced the striatal content of ENK-LI by at least 50%, suggesting that the majority of striatal ENK-LI is present in small interneurons.

The presence of small but significant amounts of ENK-LI in the substantia nigra raises the possibility that this may be stored in nigral neurons or in the terminals of axons projecting to the substantia nigra. The loss of ENK-LI in the substantia nigra in Huntington's disease suggests that this depletion is associated with the descending ENK-containing striatonigral projections (Emson et al., 1980a). Kainate lesions and immunohistochemical studies confirm the presence of the expected ENK-containing striatonigral pathway (see Agid et al., this volume).

Since the discovery of ENK as an endogenous ligand at the opiate receptor, an important question has been raised as to whether ENK fulfills the criteria of a normal neurotransmitter. One such criterion is that ENK should be localized in nerve endings. Subcellular localization studies have shown that opioid activity in rat brain is concentrated in the synaptosomal fraction (Simantov et al., 1976). Electronmicroscope studies have shown that ENK-LI is localized in both large and small vesicles, axons and dendrites in the striatum.

If ENK does have a neurotransmitter role we should expect that it would be released from presynaptic endings following nerve stimulation. Several studies have demonstrated in vitro release of ENK-LI from the striatum (Henderson et al., 1978) and the globus pallidus (Iversen et al., 1978a) slices. The release of ENK-LI was evoked by potassium and was calcium dependent.

The effects of ENK on CNS neurons have been studied electrophysiologically. In general, the actions of microiontophoretically applied ENK are inhibitory. The inhibitory actions of ENK on the firing rate of single neurons in the CNS have been widely documented by extracellular recordings (Zieglgansberger and Fry, 1976). Intracellular recordings, using a slice preparation from the guinea pig pons, also showed that the application of opiates and opioids evoked a stereospecific, naloxone reversible hyperpolarization, accompanied by a decrease in membrane resistance. The hyperpolarizations were dose dependent and accompanied by a marked reduction or abolition of spontaneous firing. This effect was probably direct, since normorphine was still effective in low Ca^{2+}, high Mg^{2+} solutions. Naloxone-reversible depression of neuronal firing by ENK has also been observed in the striatum (Fredrickson and Norris, 1976; Hill et al., 1976; Zieglgansberger and Fry, 1976). On the other hand, naloxone-sensitive, morphine-evoked excitation has also been described in two exceptional cases, Renshaw cells in the spinal cord (Davies and Dray, 1976b) and pyramidal cells in the hippocampus (Nicoll et al., 1977) where it has been suggested that it acts presynaptically to reduce the action of an inhibitory transmitter.

These various electrophysiological experiments have generated considerable debate amongst physiologists concerning the mechanism of action of ENK. Since ENK produced hyperpolarization in a direct manner, its major action may be on postsynaptic receptors. However, Jan et al. (1980) have shown, using the guinea pig inferior mesenteric autonomic ganglion. that the depressive effect of ENK on the cholinergic fast excitatory postsynaptic potential (EPSP) and the SP-mediated slow EPSP could be attributed to a presynaptic effect, since (1) the postsynaptic response to the agonists was unchanged, (2) the number of transmitter quanta making up the fast EPSP was reduced, and (3) the direct effect of acetylcholine application was undiminished. These findings suggest that one of the main roles of ENK is to modulate other neurotransmitter actions, and this action would be distinct from the action of "classical" neurotransmitters which normally directly open voltage sensitive ion channels.

A number of observations suggest that opioid peptides affect the activity of the dopaminergic nigro-striatal system. Intraventricular injection of morphine or opioid peptides induced the behavioral symptoms of decreased dopaminergic transmission, characterized by catalepsy (Izumi et al., 1977) and morphine or opioid peptides inhibited K^+-induced $[^3H]$dopamine ($[^3H]$DA) release from striatal slices (Celsens and Kuschinsky, 1974; Loh et al., 1976). However, systemic injection of opiates produced an increase in firing rate of nigro-striatal neurons (Iwatsubo and Clouet, 1977), and intraventricular and intrastriatal administration induced a naloxone-reversible increase in dopamine (DA) synthesis in caudate nucleus (Biggio et al., 1978). The mechanisms of these rather paradoxical and contradictory observations are still not completely resolved. It is possible that the primary action of opiates is the activation of opioid receptors at the nigro-striatal DA nerve terminals, which results in a decreased release of DA. This primary effect would trigger the second phase, i.e. a compensatory increase in DA synthesis and increase in firing of nigro-striatal DA neurons, which may be partly due to the removal of the normal inhibitory effect of DA on presynaptic receptors. Pollard et al. (1978) demonstrated that the degeneration of nigro-striatal DA neurons, produced by intranigral injection of 6-hydroxydo-pamine (6OH-DA), induced a 30% decrease in receptors binding LEU-ENK (δ receptors) in the striatum, which suggests that about one-third of striatal opiate receptors are localized on the nerve terminals of the nigro-striatal dopaminergic neurons. This presynaptic localization of opiate receptors is consistent with the view that these receptors could provide a substrate for presynaptic inhibition in the striatum.

Recently, a new family of opioid peptides, the dynorphins (DYN), have been shown to occur in high concentrations in the rat substantia nigra (Vincent et al., 1982; Weber et al., 1982) and dynorphin-like immunoreactive (DYN-LIr) cell bodies have been found in the neostriatum (Vincent et al., 1982). DYN-LIr fibres in the substantia nigra disappeared following ibotenic acid lesions in the striatum, which strongly suggests that the striatal DYN-LIr cells are the source of the DYN-LIr fibres in the substantia nigra. This descending, DYN-LI-containing striato-nigral system provides a further, chemically defined projection in addition to the earlier described γ-aminobutyric acid (GABA) and SP pathways. Intranigral injection of DYN induced dose-dependent contralateral rotation. This rotatory behaviour was markedly enhanced by the DA-releasing agent D-amphetamine. This suggests that DYN activates DA neurons directly or indirectly at the site of injection.

Substance P

SP is one of the most extensively studied peptides in the brain, and was detected by von Euler and Gaddum (1931) on the basis of its hypotensive and smooth muscle-contracting properties. Its role in sensory neurotransmission of pain has been extensively studied. High levels of the peptide are also found in the basal ganglia (Table I). In the basal ganglia, the substantia nigra and globus pallidus internal segment contain the highest content of the peptide (Cuello and Kanazawa, 1978; Ljungdahl et al., 1978a). Within the nigra, SP-LI is more concentrated in the pars reticulata than in the pars compacta (Brownstein et al., 1976; Kanazawa et al., 1977a). Evidence has accumulated that indicates that SP is a neurotransmitter candidate of the striato-nigral fibres.

Most of SP-LI is located in nerve-ending particles (synaptosomes) in the substantia nigra, which suggests that SP is likely to be closely involved in neurotransmission (Duffy et al., 1975). Lesion studies indicated that most of the SP-LIr neurons in the rat substantia nigra have their cell bodies in the anterior striatum (Brownstein et al., 1977; Gale et al., 1977; Hong et al., 1977a; Kanazawa et al., 1977b; Mroz et al., 1977). SP immunofluorescent cell bodies have indeed been observed in the anterior striatum in the rat (Kanazawa et al., 1977b; Ljungdahl et al., 1978b). However, another recent study suggested two distinct striatonigral SP pathways in the rat, one from the posterior portion of striatum to the substantia nigra pars lateralis, and another from the lateroventral part of the anterior portion of striatum to the substantia nigra pars compacta and pars reticulata (Kohno et al., 1984). The existence of a pallidonigral SP pathway is still a matter of debate (Brownstein et al., 1977; Ljungdahl et al., 1978a). Kanazawa et al. (1977b) showed a large reduction in the SP-LI concentration in the substantia nigra following pallidal lesions, and found that some pallidal neurons contain SP-LI. They suggested that SP-LIr fibres terminating in the substantia nigra may also originate from the globus pallidus as well as the striatum. However, since the striatonigral pathway passes through the pallidum, it is almost impossible to destroy selectively the pallidonigral fibres by pallidal lesions. The high concentration of SP-LI in the medial segment of globus pallidus may represent its presence in terminals of collaterals of striatonigral axons. However, the existence of such collaterals has not been confirmed in detail.

Ca^{2+}-dependent release of SP-LI from tissue

stores in the isolated rat substantia nigra has been demonstrated in response to depolarizing stimuli (Schenker et al., 1976; Jessell, 1978). Davies and Dray (1976a) reported the effects of iontophoretic application of SP on the neuronal firing rate in the substantia nigra. SP induced a small increase in the firing rate in most of the neurons. The neurons in the pars compacta as well as in the pars reticulata responded to SP.

Because a transmitter candidate must "mimic" the response of natural transmitter, it must first be determined which synaptic response SP must mimic. So far, little evidence exists with respect to the role of endogenous SP in the substantia nigra, since it is not possible to lesion selectively SP-containing nigral afferents. The lack of specific antagonists of SP has also hampered progress in this area. It has been demonstrated that the slow EPSP which lasts several minutes after nerve stimulation has similar characteristics to the SP-induced slow EPSP found in both autonomic ganglia (Jan et al., 1979) and nociceptive primary afferent neurons, in terms of its time course and the increased conductance of neuronal membrane. Such a slow EPSP produced in nigral neurons by striatal stimulation has not been demonstrated. Although most of the responses of nigral neurons following striatal stimulation are inhibitory, probably mediated by GABA, some neurons in the substantia nigra have also been shown to respond to striatal stimulation by slow excitation and EPSP-IPSP sequences in intracellular studies (Feger and Ohye, 1975; Dray et al., 1976). In order to determine whether these excitatory effects are really mediated by SP, it would be necessary to know whether these effects are blocked by SP antagonists or not.

Interactions of SP with other neurotransmitters in the substantia nigra have been extensively reported. Unilateral intranigral application of SP induces a behavioural activation characterized by contralateral turning and by increased rearing and grooming (Olpe and Koella, 1977; James and Starr, 1979; Kelley and Iversen, 1979). This behavioural activation was not observed in animals with 6OH-DA destruction of striatal DA terminals, and pretreatment by DA antagonist abolished the behavioural responses to SP. Intranigral application of SP also resulted in increased DA turnover in the striatum, i.e. an increase in homovanillic acid (HVA) and decrease in DA content in the striatum (Waldmeier et al., 1978; James and Starr, 1979). However, it did not affect the levels of serotonin (5-HT), 5-hydroxyindolacetic acid (5HIAA) or amino acids such as glutamate and GABA. The change in the activity of the dopaminergic nigrostriatal system by intranigral SP application has also been estimated from [^{3}H]DA release in the caudate nucleus by means of push-pull cannulae (Nieoullon et al., 1977). Unilateral infusion of SP into the substantia nigra via push-pull cannulae led to an increase in [^{3}H]DA release in the ipsilateral caudate, which was accompanied by a decrease in [^{3}H]DA release in the ipsilateral nigra (Cheramy et al., 1977; Michelot et al., 1979). However, infusion of anti-SP gamma globulins decreased the [^{3}H]DA release in the ipsilateral caudate. These observations all strongly suggest that SP increases the firing rate of dopaminergic neurons. Whether SP acts directly upon these neurons or whether its action is mediated through the local circuitry remains to be determined.

The role of SP in the regulation of nigral GABAergic activity has also been studied by Melis and Gale (1984). The unilateral intranigral application of SP analogues with antagonist properties induced a significant decrease in GABA turnover in the target region of the nigrotectal projections, namely the deep layers of superior colliculus on the homolateral side. This action was not prevented by coinjection of the GABA antagonist biculline-methiodide. Melis and Gale suggested that endogenous nigral SP exerts a tonic excitatory action on efferent GABAergic projections from pars reticulata of substantia nigra.

A functional interaction of SP with ENK was also suggested, because immunohistochemical and radioimmunocytochemical mapping have revealed that these two peptides exhibit a remarkable overlap in many cerebral structures such as some cortical areas, caudate nucleus and nucleus accumbens.

However, whether this overlap represents a functional interaction of these peptides or simply a coincidence in their distribution requires further physiological and ultrastructural studies (McLean et al., 1985). It may also be relevant to note that recent studies suggest that ENK-LI and SP-LI coexist with GABA in the striato-pallidal and striato-nigral projections.

Cholecystokinin

Cholecystokinin (CCK), which was first isolated from the small intestine on the basis of its ability to cause gallbladder contraction and pancreatic enzyme secretion, is also widely distributed in the CNS (Table I). There are a number of molecular forms of CCK, but the predominant form in the brain is the carboxy-terminal octapeptide (CCK-8), most of which probably exists in the sulphated form (CCK-8S). Many areas, e.g. cerebral cortex, hippocampus, amygdala, hypothalamus and several brainstem nuclei, contain both CCK-LIr cell bodies and terminal fields, whereas other regions including the striatum, thalamus, septum and most of the spinal cord, contain principally fibres and terminal fields. The striatum has one of the highest concentrations of CCK-LI in the rat brain (Beinfeld et al., 1981). Hökfelt et al. (1980a) have shown that the peptide can be visualized most strongly by immunocytochemistry in the ventral striatum, i.e. nucleus accumbens, olfactory tubercle and ventromedial parts of the caudate-putamen.

Lesion studies, using radioimmunoassay (Studler et al., 1981; Williams et al., 1981) and immunohistochemical technique (Hökfelt et al., 1980b; Záborszky et al., 1985), have indicated that a proportion of striatal CCK-containing afferents originate from the ventral tegmental area and substantia nigra CCK-LIr cells. The ascending CCK pathway seems to terminate primarily in the postero-medial parts of the olfactory tubercle and nucleus accumbens, and in the most ventral and medial aspects of the caudate-putamen. A second major CCK-containing pathway, with a possible origin in the piriform and medial prefrontal cortices and the amygdala, projects to the subcommissural caudate-putamen, the olfactory tubercle and the lateral part of the nucleus accumbens. It has been demonstrated that CCK is synthesized de novo (Golterman et al., 1980), localized in synaptosomal fractions after tissue disruption, released from synaptic vesicles in a calcium-dependent manner (Emson et al., 1980b; Meyer and Kraus, 1983), and binds to specific high affinity CCK receptors (Innis and Snyder, 1980; Saito et al., 1980). Local microiontophoretic application of CCK increased the firing rates of neurons in the cerebral cortex, thalamus (Ishibashi et al., 1979a), and also in the nucleus accumbens (White and Wang, 1984). The excitatory effects of CCK were selectively antagonized by the CCK antagonist proglumide. These above findings strongly suggest that CCK acts as a neurotransmitter or neuromodulator.

Recently, it has been shown that a subpopulation of mesolimbic DA neurons contain CCK-LI in nerve cell bodies and in nerve terminals (Hökfelt et al., 1980a). 6OH-DA lesions of the DA-containing A10 neurons result in consistent reductions in both CCK-8-LI and DA content only in the medial posterior nucleus accumbens (Studler et al., 1981; Marley et al., 1982), suggesting that these CCK-8/DA neurons project primarily to this region. The functional significance of the coexistence of CCK-8-LI and DA is unclear at this time since there are few reports within the literature which have examined the interaction of CCK-8 and DA within the brain areas where this coexistence occurs. However, a recent study (Wang and White, 1983) utilizing electrophysiological techniques demonstrated that CCK-8 and DA interact to influence the activity of neurons within the medial nucleus accumbens. Iontophoretic application of CCK-8 onto medial nucleus accumbens neurons resulted in increased neuronal firing rates, whereas the application of DA resulted in the inhibition of this CCK-8 induced increase. These results are consistent with the report by Fuxe et al. (1980) which states that CCK-8 reduced DA turnover in the anterior nucleus accumbens. Voigt and Wang (1984) also observed that CCK suppressed K^+-induced release of DA from the nucleus accumbens in a concentration-dependent manner. These data are all consistent with the suggestion that CCK may act as a functional antagonist to DA within this structure.

Vasoactive intestinal polypeptide

Vasoactive intestinal polypeptide (VIP), which was originally isolated from the small intestine as a substance that causes vasodilatation, has been found to be distributed in the mammalian CNS (Table I), especially in forebrain areas. As with CCK, highest levels of VIP-LI occur in the cerebral cortex, although cortical VIP-LI levels are only one-tenth those of CCK-LI. VIP is amongst the most recent to be added to the list of possible CNS neurotransmitter candidates. Some of the VIP-LI is localized in synaptosomal fractions from which it could be released by depolarizing stimuli (Emson et al., 1978). Iontophoretic application of VIP to the cerebral cortex and hippocampal slice produces potent excitatory effects (Phillips et al., 1978; Dodd et al., 1979).

Small but significant amounts of VIP-LI are present in the basal ganglia (Emson et al., 1979). Nerve cell terminals containing VIP-LI are scattered sparsely throughout the striatum (Loren et al., 1979). In contrast to its low endogenous level of VIP-LI, the striatum contains one of the highest densities of specific binding sites for VIP of all regions examined (Taylor et al., 1979).

The functional role of VIP in the basal ganglia is completely unknown. However, it has been demonstrated that intrastriatal administration of VIP significantly increased metabolic activity in the injected striatum, where the activated area was localized in small punctate areas scattered throughout the nucleus at considerable distances from the injected side. This indicates that VIP can modify functional processes in the striatum (McCulloch et al., 1983).

Somatostatin

Somatostatin (SRIF), a tetradecapeptide, which was originally isolated from bovine hypothalamic extracts, was shown to inhibit growth hormone release from rat pituitary cells (Vale et al., 1972). Although the hypothalamus contains the highest concentration of somatostatin-like immunoreactivity (SRIF-LI), the majority of brain SRIF-LI (90%) is situated outside the hypothalamus (Kobayashi et al., 1977). In the basal ganglia, quantitations of SRIF-LI level by radioimmunoassay have revealed appreciable amounts of SRIF-LI in nucleus accumbens, striatum and substantia nigra (Table I). Immunohistochemical studies have shown SRIF-LIr neurons in the striatum, and SRIF-LIr nerve terminals in the striatum and nucleus accumbens. The striatum shows a mosaic-like appearance of two interdigitating, neurochemically distinct compartments, the "patch" and the "matrix". SRIF-LIr fibres form a dense plexus in the "matrix", which also shows a high acetylcholinesterase activity (Gerfen, 1984). The origin of the SRIF-LI found in these areas remains to be determined although the presence of SRIF-LI in intrinsic cells suggests an interneuronal origin. (See also chapters 2 and 4.)

Neurochemical studies have shown that the majority of CNS SRIF-LI is found in synaptosomal fractions (Martin et al., 1978) and in synaptic vesicles (Berelowitz et al., 1978). In addition, studies have shown that SRIF-LI is released from slices of rat hypothalamus and amygdala in a calcium-dependent fashion (Iversen et al., 1978b). Furthermore, SRIF-LI can also be released from hypothalamic synaptosomes by high potassium.

Interaction of SRIF with the nigrostriatal dopaminergic system has also been demonstrated. Intraventricularly injected SRIF induced an increase in DA synthesis and turnover in several brain areas including the striatum (Garcia-Sevilla et al., 1978). SRIF also increased DA release both in vitro from superfused striatal slices and in vivo from caudate nucleus and substantia nigra (Chesselet and Reisine, 1983). This suggests that SRIF in the striatum may play a role in the local control of DA release from the terminals of dopaminergic nigro-striatal neurons.

Recently, some investigators, using radioimmunoassay, have found that the concentration of SRIF-LI is significantly increased throughout the atrophic basal ganglia in Huntington's disease (Aronin et al., 1983; Nemeroff, 1983; see also Emson, this volume). A recent immunohistochemical study demonstrates that this increase in SRIF-LI (and NPY-LI) content in Huntington's disease is associated with an increase in the number of immunostained fibres and cells, and probably reflects the selective survival of SRIF-LIr striatal neurons (Marshall and Landis, 1985; see Emson, this volume). Although the pathophysiological significance of the change in SRIF-LI in Huntington's disease remains to be determined, it might be possible that a developing imbalance in the levels of neuropeptides in the basal ganglia may contribute to some of the clinical manifestations of Huntington's disease.

Thyrotropin releasing hormone

Thyrotropin releasing hormone (TRH), originally found as a hormone in the hypothalamus to facilitate the release of thyrotropin and prolactin, has been known to have variable actions on the CNS, even outside the hypothalamus, and has been postulated as a neurotransmitter or a neuromodulator. It has been found that TRH-like immunoreactivity (TRH-LI) is distributed widely in the CNS (Jackson and Reichlin, 1974; Oliver et al., 1974) (Table I).

A large amount of information has been accumulated about the action of TRH on the CNS. It modifies a wide variety of behavioural activities, including an increase in spontaneous motor activity (Miyamoto and Nagawa, 1977), a suppression of feeding and drinking activity (Vijayan and McCann, 1977), and antagonism of hypothermia induced by pentobarbital and reserpine (Miyamoto et al., 1981). TRH has been suspected to exert a predominantly inhibitory effect on the neuronal activities in various parts of the CNS (Renaud and Martin, 1975; Winokur and Beckman, 1978), whereas it was found to depolarize and excite frog spinal motoneurons (Nicoll, 1977). In the hypothalamus TRH has been reported to have both excitatory and inhibitory effects (Ishibashi et al., 1979b).

In the neurons of the CNS, TRH-LI is found primarily in the synaptosomal fractions, and its Ca^{2+}-dependent release from synaptosomes has been reported (Emson, 1979).

Studies of the effects of TRH upon the basal ganglia are rather limited in number. It has been demonstrated that TRH injected intraperitoneally (i.p.) or directly into nucleus accumbens bilaterally causes a clear locomotor hyperactivity, characterized by rearing, mild sniffing and grooming (Miyamoto and Nagawa, 1977). DA, injected either i.p. or into the accumbens, produces similar effects. This hyperactivity can be blocked by haloperidol or pimozide, DA receptor blockers. On the other hand, the direct injection of TRH and DA to the caudate did not produce hyperactivity. These observations suggest that TRH acts on the basal ganglia through the DA system, to produce locomotor hyperactivity. The effects of TRH on locomotion, however, are mediated through the mesolimbic dopaminergic system rather than the nigro-striatal dopaminergic system.

It has also been reported that TRH increases the release of [^{3}H]DA from slices prepared from nucleus accumbens, but not from caudate nucleus (Heal and Green, 1979; Kerwin and Pycock, 1979; Miyamoto et al., 1979). TRH, however, had no effect on basal or DA-stimulated adenylate cyclase, and did not displace [^{3}H]spiperone binding in membranes from the nucleus accumbens. These results indicate that the action of TRH on the locomotor hyperactivity is likely to be mediated through the presynaptic release of DA in the nucleus accumbens, but not through direct postsynaptic interactions with DA receptors in the nucleus (Green et al., 1976; Kerwin and Pycock, 1979).

Fukuda et al. (1979) observed that TRH facilitated circling behaviour induced by apomorphine or L-DOPA in mice with unilateral striatal lesions due to injection of 6OH-DA or aspiration of the caudate nucleus. The stereotypic movements resulting have been presumed to be due predominantly to dysfunctions of the striatum rather than nucleus accumbens. The facilitatory effect of TRH on these movements, however, was much weaker than those on the locomotor activity, because the dose required to elicit these effects was 5–10 times larger. They also suggested that low doses of TRH facilitated postsynaptic DA transmission if the DA-receptors were hypersensitized, whereas high dose caused a presynaptic stimulation, releasing [^{14}C]DA from the striatal slices in rats.

These postulated actions of TRH on the basal ganglia through dopaminergic mechanism still remain controversial. Some investigators question the DA-releasing action of TRH in the accumbens-mesolimbic system (Costall et al., 1979; Ervin et al., 1981a). Malouin and Bedard (1982) reported that intrastriatal injection of TRH in cats caused similar head turning responses to those induced by intrastriatal DA. However, haloperidol blocked the DA-induced head turning, but not the TRH-induced head turning. They suggested that the TRH-induced head turning was unlikely to be mediated by the postulated "mesolimbic DA releasing" effect, although the precise mechanism was unknown. They also suggested an involvement of a cholinergic mechanism in the effects of TRH on the head turning response.

Even though the mechanisms of action of TRH in the basal ganglia still need further investigation, TRH clearly alters behaviour, at least in part, by influencing the function of the basal ganglia. Although more than one neurotransmitter mechanism is likely to be involved, and a whole variety of possible interactions exist, the DA-system is certainly implicated in the behavioural effects produced by TRH.

One study of Huntington's disease reported that the concentration of TRH-LI in the caudate and putamen was approximately three times higher than in controls (Spindel et al., 1980). However, the significance of this change remains obscure (Emson, this volume).

Angiotensin

Angiotensin I, a decapeptide, is formed in the circulating blood, and is almost devoid of biological activity in the periphery and the brain. Angiotensin II (AII), an octapeptide, is produced by the action of angiotensin converting enzyme (ACE) on angiotensin I. AII as a peripheral hormone plays a major role in maintenance of extracellular fluid volume and blood pressure. It also produces a variety of central effects on intraventricular administration, many of which appear linked to its peripheral effects. These include increased drinking, elevation of blood pressure, increased vasopressin release, and increased sympathetic outflow. Many of the central actions of AII are likely to be mediated via the circumventricular organs, small areas of the brain that lack a blood brain barrier, which include the area postrema, subfornical organ, and organum vasculosum of the lamina terminalis. The proposal that AII is formed locally within the brain and acts as a neurotransmitter is based on the following findings. First, the brain contains a complete renin–angiotensin system, including angiotensinogen, renin, angiotensin I, converting enzyme, and AII-like immunoreactivity (AII-LI) (Phillips et al., 1979; Ganten et al., 1982). However, the major problem is that it has proven extremely difficult to demonstrate the presence of AII in the brain. Although immunohistochemistry has revealed a sparse but widely distributed AII-LI in the CNS (Quinlan and Phillips, 1981), some workers reported that endogenous AII-LI in the brain appears to be very low or absent (Reid, 1979; Meyer et al., 1982). In contrast to the very low levels of endogenous angiotensin, the brain possesses abundant ACE with the highest activity in the striatum and in the cerebellum (Yang and Neff, 1972) (Table I). Arregui et al. (1979) reported the existence of a striatonigral neuronal system containing ACE activity, since the levels of ACE activity in the substantia nigra are significantly reduced in Huntington's disease and after intrastriatal kainate lesion in rats. However, the significance of such a system remains obscure, since the possible cellular localization of striatal ACE activity has not been determined and, moreover, ACE is a nonspecific peptidase which also inactivates bradykinin, SP, neurotensin, ENK and other peptides.

AII produces a wide variety of biological responses when injected into the CNS. Iontophoretic administration of AII increased firing rates of neurons in some cerebral structures such as subfornical organ (Felix and Akert, 1974), supraoptic neurosecretory cells (Nicoll and Barker, 1971) and hippocampus. In the basal ganglia, Simonnet et al. (1981) have shown that microiontophoretic application of AII modified the spontaneous activity of some neurons in the neostriatum and that the action of AII was generally inhibitory. AII also stimulated the spontaneous release of [^{3}H]DA continuously synthetized from [^{3}H]tyrosine in striatal slices. They suggested that there are angiotensin-containing nerve endings in the neostriatum, and that the cell bodies of such nerves are located outside the neostriatum.

Neurotensin

Neurotensin (NT) is a tridecapeptide which exists in specific regions of the CNS, pituitary, and the gastrointestinal tract (Uhl and Snyder, 1976; Kobayashi et al., 1977). In the CNS, neurotensin-like immunoreactivity (NT-LI) is mainly distributed in the substantia gelatinosa of the spinal cord and trigeminal nucleus, central amygdaloid nucleus, median eminence, and preoptic and basal hypothalamic areas (Table I). In the basal ganglia NT-LIr cells and fibres are found in the substantia nigra and ventral tegmental area in the midbrain (Uhl et al., 1979; Jennes et al., 1982) and modest densities of NT-LIr fibres, terminals, and NT receptors are noted in the caudate, putamen, and globus pallidus.

NT-LI is concentrated in synaptosomal subcellular fractions (Uhl and Snyder, 1976), and can be released from hypothalamic tissue in a Ca^{2+}-dependent manner (Emson, 1979). [^{3}H]NT binding has also been demonstrated recently (Kitabgi et al., 1977). These findings suggest a neurotransmitter role for NT in the brain, although other criteria must also be satisfied before full transmitter status can be imputed to this peptide.

Recent immunohistochemical studies have shown that NT and DA possess a similar distribution; the ventral tegmental area containing a moderate density of NT perikarya, and the nucleus accumbens having a moderate density of NT fibres (Uhl et al., 1977). The ventral tegmental area and the nucleus accumbens also contain NT receptors (Young and Kuhar, 1981), and recently it was found that autoradiographically identified NT receptors in the ventral tegmental area are located on DA neurons (Quirion et al., 1982).

Several lines of evidence also suggest an interaction between NT and DA in the mammalian CNS. It has been observed that iontophoretic application of NT increased the firing frequency of neurons in the ventral tegmental area and substantia nigra (Andrade and Aghajanian, 1981) and that NT microinjection into the ventral tegmental area produced a mesolimbic DA-dependent increase in spontaneous motor activity (Kalivas et al., 1983) and increases in the concentration of DA metabolites in the nucleus accumbens (Nemeroff et al., 1982), a major termination site of the circuit. However, bilateral injection of NT into the substantia nigra, the site of origin of the nigroneostriatal system, produced no overt behavioural effects, which suggests that NT may modulate the activity of the mesolimbic DA system but not the nigroneostriatal DA system. In apparent contrast to the facilitatory effect on the mesolimbic DA system after NT injection into the ventral tegmental area, a number of studies have concluded that intraventricular application of NT can exert a "neuroleptic-like" action in the mammalian CNS (Nemeroff et al., 1982). More specifically, NT injection into the nucleus accumbens blocks the behavioural response following pharmacologically induced release of mesolimbic DA. Thus, Ervin et al. (1981b) found that intra-accumbens injection with NT blocked amphetamine--induced locomotion and Kalivas et al. (1982, 1984) demonstrated that intra-accumbens NT blocked the behavioural hyperactivity produced by NT injection into the ventral tegmental area and that produced by DA injection into the nucleus accumbens. The mechanism of the above rather contradictory findings remains to be resolved.

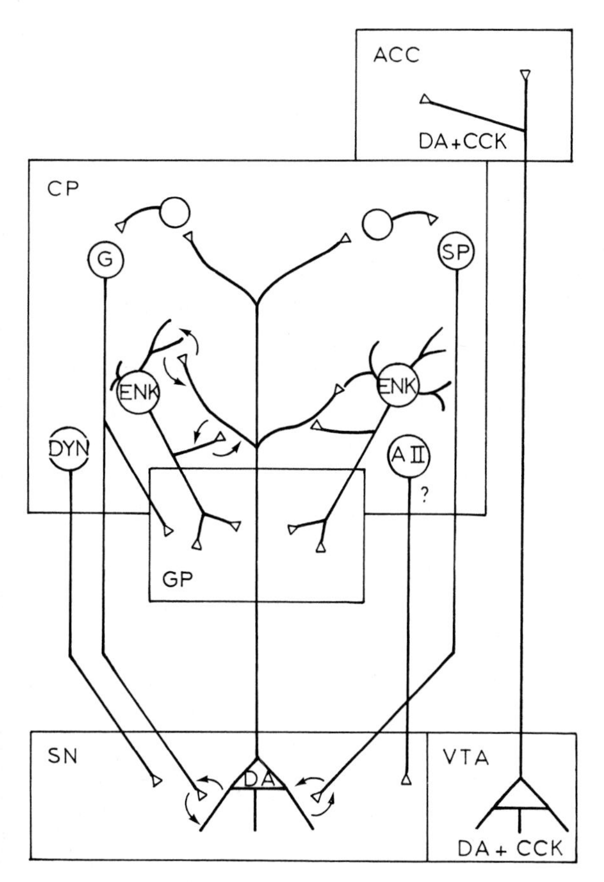

Fig. 1. Schematic representation of the postulated synaptic connections of peptide-containing neurons in the basal ganglia. Bent arrows indicate possible axo-dendritic, axo-axonic and dendro-axonic interactions. SP, substance P; ENK, enkephalin; AII, angiotensin II; DYN, dynorphin; CCK, cholecystokinin; DA, dopamine; G, γ-aminobutyric acid; CP, caudate putamen; ACC, nucleus accumbens; GP, globus pallidus; SN, substantia nigra; VTA, ventral tegmental area (modified from Cuello et al., 1981).

Some considerations about actions of peptides in the basal ganglia

It is clear from this review that a considerable number of peptides exist in the basal ganglia and that at least some of these peptides can be considered as potential neurotransmitters or neuromodulators (Fig. 1). At the same time, the existence of so many peptides raises several questions. First, why are so many peptides necessary? One excitatory and one inhibitory transmitter would apparently be sufficient to operate the nervous system. Secondly, are the actions of peptides qualitatively different from those of the "classical" transmitters such as acetylcholine and monoamines? It is probably too early to answer these questions. Even though the specific actions of the peptides in the basal ganglia remain obscure, certain characteristics of the neuropeptides may be of crucial relevance to their actions in the basal ganglia.

Time course of actions of peptide

Some of the peptides such as SP and ENK have been reported to act slowly (i.e. their postsynaptic effects peak slowly) and their actions persist for relatively long periods of time often exceeding 5 min (Jan and Jan, 1983). This slow and prolonged action of the neuropeptides may have crucial relevance to functions of the basal ganglia. This possibility is worth emphasizing because so many of the clinical symptoms of basal ganglia diseases are characterized by episodic or relatively long lasting changes in the quality of movement or affect-state. Examples of such episodes are the diurnal fluctuation of symptoms in parkinsonism and the "on-off" phenomena seen during treatment. Physiological studies also suggested that the basal ganglia are more relevant to slow movements (DeLong and Georgopoulos, 1979).

Coexistence and corelease of peptides and "classical" transmitters

The question as to whether a neuron can produce and release more than one transmitter has attracted considerable interest. Since Hökfelt et al. (1978) discovered that SP coexists with 5-HT in a part of the medulla oblongata, several examples have been observed in a relatively short time (Hökfelt et al., 1980b), which may indicate that coexistence is ubiquitous in the CNS. Although it is still not certain whether all of the instances of immunohistochemically identified "coexistence" have physiological significance, some studies suggest strongly that a single neuron can release two different transmitters

(Jan and Jan, 1983). If two transmitters are released from the same nerve terminals they may act in a variety of ways but available examples suggest that they usually act synergistically. The most studied examples are the neurons of the cat sympathetic ganglia (Lundberg et al., 1979) innervating exocrine glands which contain both VIP-LI and acetylcholinesterase. It is well known that acetylcholine produces sweat secretion, and this effect is atropine-sensitive. On the other hand, the concomitant vasodilatation, which also occurs on stimulation of the postganglionic nerves is atropine-resistant. Fahrenkrug et al. (1978) suggested that VIP may be the mediator of atropine-resistant vasodilatation. It is now postulated that these two effects, the secretion and vasodilatation in exocrine glands, are induced by acetylcholine and VIP, respectively, released from the same nerve terminals to act synergistically (Lundberg et al., 1979). Clearly, in other systems different mechanisms may operate. There is, for example, evidence that SP may block cholinergic receptors (Belcher and Ryall, 1977).

It has also been demonstrated that a subpopulation of DA neurons contain CCK-LI; these neurons are found in the most medial parts of the pars compacta of the substantia nigra (A9 group) and the ventral tegmental area (A10 group) and project to nucleus accumbens and other limbic areas in the forebrain. The occurrence of a peptide in a population of DA neurons is interesting in view of our extensive knowledge of the functional role of these DA systems. In particular, the mesolimbic DA systems have been associated with higher mental functions and, according to the so-called "DA hypothesis", disturbances in this system may represent one component in the pathogenesis of schizophrenia. If a CCK-like peptide is released together with DA, the peptide could also be involved in the etiology and symptomatology of schizophrenia. It has recently been demonstrated that CCK fragments can inhibit DA release in the nucleus accumbens–tuberculum olfactorium region. Hökfelt et al. (1980b) have speculated that an imbalance between peptide and amine could be an aetiologic factor in schizophrenia, whereby a loss or decrease in peptide would lead to an "overactive" DA system.

Local interactions of transmitters in striatal mosaic

The striatum of the basal ganglia has a mosaic of two interdigitating, neurochemically distinct compartments (Gerfen, 1984; Graybiel, 1984). One type, the "patch" compartments, which are 300–600 μm-wide pockets, is identified by patches of dense opiate receptor binding, and is enriched in GAD-, SP- and ENK-like immunoreactivity. The other compartment, the "matrix", has a high acetylcholinesterase activity. These chemically specified compartments raise the possibility that they represent mechanisms whereby groups of neurons and their process can be modulated in a coordinated way. Such group modulation could occur through a high density of conventional synaptic contacts established by synchronized inputs. A second possibility is that the "patch" may form sites for local extracellular diffusion of neuroactive substances. There is already evidence in the peripheral nervous system for non-synaptic effects being exerted by luteinizing hormone-releasing hormone at distances approaching the diameter of the "patch" (Jan and Jan, 1983). Since the DA island fibres are spatially confined to the "patch", interactions between various peptides and dopaminergic fibres could be mediated by such local extracellular actions (Graybiel, 1984).

Conclusions

Recent advance in radioimmunoassay and immunohistochemical technique has revealed the distribution of a variety of peptides in the basal ganglia. Furthermore, physiological and pharmacological studies have shown that for at least some of these peptides there is good evidence for a neurotransmitter or a neuromodulator role. However, our understanding of the physiological role of these peptide "transmitters" is still primitive. Further detailed physiological studies will be necessary to understand the functional characters of the peptides. In particular development of drugs which specifically modify the actions of peptides will be critical to any understanding of the specific physiological roles of peptides in the basal ganglia. Investigations concentrating on changes of peptide content in the basal ganglia in disease may also provide knowledge useful in clinical therapeutic approaches as well as providing insights into the physiological role of peptides in the brain.

References

Andrade, R. and Aghajanian, G. K. (1981) Neurotensin selectively activates dopaminergic neurons in the substantia nigra. *Soc. Neurosci. Abstr.*, 7: 573.

Aronin, N., Cooper, P. E., Lorenz, L. J., Bird, E. D., Sagar, S. M., Leeman, S. E. and Martin, J. B. (1983) Somatostatin is increased in the basal ganglia in Huntington's disease. *Am. Neurol.*, 13: 519–526.

Arregui, A., Iversen, L. L., Spokes, E. G. S. and Emson, P. C. (1979) Alterations in postmortem brain angiotensin-converting enzyme activity and some neuropeptides in Huntington's disease. In T. N. Chase (Ed.), *Advances in Neurology, Vol. 23*, Raven Press, New York, pp. 517–525.

Beinfeld, M. C., Meyer, D. K., Eskay, R. L., Jensen, R. T. and Brownstein, M. J. (1981) The distribution of cholecystokinin immunoreactivity in the central nervous system of the rat as determined by radioimmunoassay. *Brain Res.*, 212: 51–57.

Belcher, G. and Ryall, R. W. (1977) Substance P and Renshaw cells: a new concept of inhibitory synaptic interactions. *J. Physiol. (London)*, 272: 105–119.

Berelowitz, M., Hudson, A., Pimstone, B., Kronheim, S. and Bennett, G. W. (1978) Subcellular localization of growth hormone release inhibiting hormone in rat hypothalamus, cerebral cortex, striatum and thalamus. *J. Neurochem.*, 31: 751–753.

Biggio, G., Casu, M., Corda, M. G., DiBello, C. and Gessa, G. L. (1978) Stimulation of dopamine synthesis in caudate nucleus by intrastriatal enkephalins and antagonism by naloxone. *Science*, 200: 552–554.

Brownstein, M. J., Utiger, R. D., Palkovits, M. and Kizer, J. S. (1975) Effect of hypothalamic deafferentation on thyrotropin-releasing hormone levels in rat brain. *Proc. Natl. Acad. Sci. U.S.A.*, 72: 4177–4179.

Brownstein, M. J., Mroz, E. A., Kizer, J. S., Palkovits, M. and Leeman, S. E. (1976) Regional distribution of substance P in the brain of the rat. *Brain Res.*, 116: 299–305.

Brownstein, M. J., Mroz, E. A., Tappaz, M. L. and Leeman, S. E. (1977) On the origin of substance P and glutamic acid decarboxylase (GAD) in the substantia nigra. *Brain Res.*, 135: 315–323.

Celsens, B. and Kuschinsky, K. (1974) Effects of morphine on kinetics of ^{14}C-dopamine in rat striatal slices. *Naunyn Schmiedebergs Arch. Exp. Pathol. Pharmakol.*, 284: 159–165.

Cheramy, A., Nieoullon, A., Michelot, R. and Glowinski, J. (1977) Effects of intranigral application of dopamine and substance P on the in vivo release of newly synthesized ^{3}H-dopamine in the ipsilateral caudate nucleus of the cat. *Neurosci. Lett.*, 4: 105–109.

Chesselet, M.-F. and Reisine, D. (1983) Somatostatin regulates dopamine release in rat striatal slices and cat caudate nuclei. *J. Neurosci.*, 3: 232–236.

Childers, R. R., Schwarcz, R., Coyle, J. T. and Snyder, S. H. (1978) Radioimmunoassay of enkephalins: levels of methionine- and leucine-enkephalin in morphine-dependent and kainic acid lesioned rat brains. In E. Costa and M. Trabucchi (Eds.), *Endorphins: Advances in Biochemical Psychopharmacology, Vol. 18*, Raven Press, New York, pp. 161–174.

Costall, B. S., Hui, C. G., Metcalf, G. and Naylor, R. J. (1979) A study of the changes in motor behaviour caused by TRH on intracerebral injection. *Eur. J. Pharmacol.*, 53: 143–150.

Cuello, A. C. and Kanazawa, I. (1978) The distribution of substance P immunoreactive fibers in the rat central nervous system. *J. Comp. Neurol.*, 178: 129–156.

Cuello, A. C., DelFiacco, M., Paxinos, G., Somogyi, P. and Priestley, J. V. (1981) Neuropeptides in striato-nigral pathways. *J. Neural Transm.*, 51: 83–96.

Davies, J. and Dray, A. (1976a) Substance P in the substantia nigra. *Brain Res.*, 107: 623–627.

Davies, J. and Dray, A. (1976b) Effects of enkephalin and morphine on Renshaw cells in feline spinal cord. *Nature*, 262: 603–604.

DelFiacco, M., Paxinos, G. and Cuello, A. (1982) Neostriatal enkephalin-immunoreactive neurones project to the globus pallidus. *Brain Res.*, 231: 1–17.

DeLong, R. and Georgopoulos, A. P. (1979) Motor functions of the basal ganglia as revealed by studies of single cell activity in the behaving primate. In L. J. Poirier, T. L. Sourkes and P. J. Bedard (Eds.), *Advances in Neurology, Vol. 24*, Raven Press, New York, pp. 131–140.

Dodd, J., Kelly, J. S. and Said, S. J. (1979) Excitation of CA1 neurones of the rat hippocampus by the octacosapeptide, vasoactive intestinal polypeptide (VIP). *Br. J. Pharmacol.*, 66: 126.

Dray, A., Gonye, T. J. and Oakley, N. R. (1976) Caudate stimulation and substantia nigra activity in the rat. *J. Physiol. (London)*, 259: 825–849.

Duffy, M. J., Mulhall, D. and Powell, D. (1975) Subcellular distribution of substance P in bovine hypothalamus and substantia nigra. *J. Neurochem.*, 25: 305–307.

Elde, R., Hökfelt, T., Johannson, O. and Terenius, L. (1976) Immunohistochemical studies using antibodies to Leu-enkephalin: initial observations on the nervous system of the rat. *Neuroscience*, 1: 349–355.

Emson, P. C. (1979) Peptides as neurotransmitter candidates in the mammalian CNS. *Prog. Neurobiol.*, 13: 61–116.

Emson, P. C., Fahrenkrug, J., Schaffalitzky de Muckadell, O. B., Jessell, T. M. and Iversen, L. L. (1978) Vasoactive intestinal polypeptide (VIP): Vesicular localization and potassium evoked release from rat hypothalamus. *Brain Res.*, 140: 174–178.

Emson, P. C., Fahrenkrug, J. and Spokes, E. G. S. (1979) Vasoactive intestinal polypeptide (VIP): distribution in normal human brain and in Huntington's disease. *Brain Res.*, 173: 174–178.

Emson, P. C., Arregui, A., Clement-Jones, V., Sandberg, B. E. B. and Rossor, M. (1980a) Regional distribution of methionine-enkephalin and substance P-like immunoreactivity in normal human brain and in Huntington's disease. *Brain Res.*, 199: 147–160.

Emson, P. C., Lee, C. M. and Rehfeld, J. E. (1980b) Cholecystokinin octapeptide: vesicular localization and calcium dependent release from rat brain in vitro. *Life Sci.*, 26: 2157–2163.

Ervin, G. N., Schmitz, S. A., Nemeroff, C. B. and Prange, A. J. (1981a) Thyrotropin-releasing hormone and amphetamine produce different patterns of behavioral excitation in rats. *Eur. J. Pharmacol.*, 72: 35.

Ervin, G. N., Birkemo, L. S., Nemeroff, C. B. and Prange, A. J., Jr. (1981b) Neurotensin blocks certain amphetamine-induced behaviors. *Nature*, 291: 73–76.

Fahrenkrug, J., Haglund, U., Jodal, M., Lundgren, O., Olbe, L. and Schaffalitzky de Muckadell, O. B. (1978) Nervous release of vasoactive intestinal polypeptide in the gastrointestinal tract of cats: possible physiological implications. *J. Physiol. (London)*, 284: 291–305.

Feger, J. and Ohye, C. (1975) The unitary activity of the substantia nigra following stimulation of the striatum in the awake monkey. *Brain Res.*, 89: 155–159.

Felix, D. and Akert, K. (1974) Effect of angiotensin-II on neurones of cat subfornical organ. *Brain Res.*, 76: 350–353.

Fredrickson, R. C. A. and Norris, F. H. (1976) Enkephalin-induced depression of single neurons in brain areas with opiate receptors: antagonism by naloxone. *Science*, 194: 440–442.

Fukuda, N., Miyamoto, M., Narumi, S., Nagai, Y. and Shima, T. (1979) Thyrotropin-releasing hormone (TRH): Enhancement of dependent circling behavior and its own circling-inducing effect in unilateral striatal lesioned animals. *Folia Pharmacol. Jpn.*, 75: 251–270.

Fuxe, K., Andersson, K., Locatelli, V., Agnati, L. F., Hökfelt, T., Skirboll, L. and Mutt, V. (1980) Cholecystokinin peptides produce marked reduction of dopamine turnover in discrete areas in the rat brain following intraventricular injection. *Eur. J. Pharmacol.*, 67: 325–331.

Gale, K., Hong, J. S. and Guidotti, A. (1977) Presence of substance P and GABA in separate striatonigral neurons. *Brain Res.*, 136: 371–375.

Ganten, D., Printz, M., Phillips, M. I. and Scholkens, B. A. (Eds.) (1982) *The Renin–Angiotensin System in the Brain*, Springer-Verlag, Berlin.

Garcia-Sevilla, J. A., Magnusson, T. and Carlsson, A. (1978) Effect of intra-cerebroventricularly administered somatostatin on brain monoamines turnover. *Brain Res.*, 155: 159–164.

Gerfen, R. C. (1984) The neostriatal mosaic: compartmentalization of corticostriatal input and striatonigral output systems. *Nature*, 311: 461–464.

Golterman, N. R., Rehfeld, J. F. and Roigaard-Petersen, H. (1980) In vivo biosynthesis of cholecystokinin in rat cerebral cortex. *J. Biol. Chem.*, 255: 6181–6185.

Graybiel, A. M. (1984) Neurochemically specified subsystems in the basal ganglia. In *Ciba Foundation Symposium 107: Functions of the Basal Ganglia*, Pitman, London, pp. 114–149.

Green, A. R., Heal, D. J., Grahame-Smith, D. G. and Kelly, P. H. (1976) The contrasting actions of TRH and cycloheximide in altering the effects of centrally acting drugs. *Neuropharmacology*, 17: 265–270.

Harkins, J. A., Stewart, J. M. and Krivoy, W. A. (1980) Multiple modes of communication by substance P. *Int. J. Neurol.*, 2–3–4: 233–238.

Heal, D. J. and Green, A. R. (1979) Administration of thyrotrophin releasing hormone (TRH) to rats releases dopamine in n. accumbens but not in n. caudates. *Neuropharmacology*, 18: 23–31.

Henderson, G., Hughes, J. and Kosterlitz, H. W. (1978) In vitro release of leu- and met-enkephalin from the corpus striatum. *Nature*, 271: 677–679.

Hill, R. G., Pepper, C. M. and Mitchell, J. F. (1976) Depression of nociceptive and other neurones in the brain by iontophoretically applied met-enkephalin. *Nature*, 262: 604–606.

Hökfelt, T., Ljungdahl, A., Steinbusch, H., Verhofstad, A., Nilsson, G., Brodin, E., Pernow, B. and Goldstein, M. (1978) Immunohistochemical evidence of substance-P-like immunoreactivity in some 5 hydroxtryptamine-containing neurons in the rat central nervous system. *Neuroscience*, 3: 517–538.

Hökfelt, T., Skirboll, L., Rehfeld, J. F., Goldstein, M., Markey, K. and Dann, O. (1980a) A subpopulation of mesencephalic dopamine neurons projecting to limbic areas contains a cholecystokinin-like peptide evidence from immunohistochemistry combined with retrograde tracing. *Neuroscience*, 5: 2093–2124.

Hökfelt, T., Johansson, O., Ljungdahl, A., Lundberg, J. M. and Schultzberg, M. (1980b) Peptidergic neurones. *Nature*, 284: 515–521.

Hong, J. S., Yang, H.-Y. T., Racagni, G. and Costa, E. (1977a) Projections of substance P containing neurons from neostriatum to substantia nigra. *Brain Res.*, 122: 541–544.

Hong, J.-S., Yang, H.-Y. T. and Costa, E. (1977b) On the location of methionine enkephalin neurons in rat striatum. *Neuropharmacology*, 16: 451–453.

Hughes, J., Kosterlitz, H. W. and Smith, T. W. (1977) The distribution of methionine-enkephalin and leucine-enkephalin in the brain and peripheral tissues. *Br. J. Pharmacol.*, 61: 639–648.

Innis, R. B. and Snyder, S. H. (1980) Distinct cholecystokinin receptors in brain and pancreas. *Proc. Natl. Acad. Sci. U.S.A.*, 77: 6917–6921.

Ishibashi, S., Oomura, Y., Okajima, T. and Shibata, S. (1979a) Cholecystokinin, motilin and secretin effects on the central nervous system. *Physiol. Behav.*, 23: 401–403.

Ishibashi, S., Oomura, Y. and Okajima, T. (1979b) Facilitatory and inhibitory effects of TRH on lateral hypothalamic and ventro medial neurons. *Physiol. Behav.*, 22: 785–787.

Iversen, S. D., Bloom, F. E., Vargo, T. and Guillemin, R. (1978a) Release of enkephalin from rat globus pallidus in vitro. *Nature*, 271: 679–681.

Iversen, L. L., Iversen, S. D., Bloom, F. E., Douglas, C., Brown, M. and Vale, W. (1978b) Calcium-dependent release of somatostatin and neurotensin in vitro. *Nature*, 273: 161–163.

Iwatsubo, K. and Clouet, D. H. (1977) Effects of morphine and haloperidol on the electrical activity of rat nigrostriatal neurons. *J. Pharmacol. Exp. Ther.*, 202: 429–436.

Izumi, K., Motomatsu, T., Chretien, M., Butterworth, R. F., Lis, M., Seidan, N. and Barbeau, A. (1977) β-Endorphin induced akinesia in rats: effect of apomorphine and α-methyl-*p*-tyrosine and related modifications of dopamine turnover in the basal ganglia. *Life Sci.*, 20: 1149–1156.

Jackson, I. and Reichlin, S. (1974) Thyrotropin-releasing hormone (TRH): distribution in hypothalamic and extrahypothalamic brain tissues of mammalian and submammalian chordates. *Endocrinology*, 95: 854–862.

James, T. A. and Starr, M. S. (1979) Effects of substance P injected into the substantia nigra. *Br. J. Pharmacol.*, 65: 423–429.

Jan, Y. N. and Jan, L. Y. (1983) Some features of peptidergic transmission. *Prog. Brain Res.*, 58: 49–59.

Jan, Y. N., Jan, L. Y. and Kuffler, S. W. (1979) A peptide as a possible transmitter in sympathetic ganglia of the frog. *Proc. Natl. Acad. Sci. U.S.A.*, 76: 423–429.

Jan, Y. N., Jan, L. Y. and Kuffler, S. W. (1980) Further evidence for peptidergic transmission in sympathetic ganglia. *Proc. Natl. Acad. Sci. U.S.A.*, 77: 5008–5012.

Jennes, L., Stumpf, W. E. and Kalivas, P. W. (1982) Neurotensin: topographical distribution in rat brain by immunohistochemistry. *J. Comp. Neurol.*, 210: 211–224.

Jessell, T. M. (1978) Substance P release from the rat substantia nigra. *Brain Res.*, 151: 469–478.

Kalivas, P. W., Nemeroff, C. B. and Prange, A. J., Jr. (1982) Neuroanatomical site-specific modulation of spontaneous motor activity by neurotensin. *Eur. J. Pharmacol.*, 78: 471–474.

Kalivas, P. W., Burgess, P. W., Nemeroff, C. B. and Prange, A. J., Jr. (1983) Behavioral and neurochemical effects of neurotensin microinjection into the ventral tegmental area. *Neuroscience*, 8: 495–505.

Kalivas, P. W., Nemeroff, C. B. and Prange, A. J. (1984) Neurotensin microinjection into the nucleus accumbens antagonizes dopamine-induced increase in locomotion and rearing. *Neuroscience*, 11: 919–930.

Kanazawa, I., Bird, E., O'Connell, R. and Powell, D. (1977a) Evidence for the decrease of substance P content in the substantia nigra of Huntington's Chorea. *Brain Res.*, 120: 387–392.

Kanazawa, I., Emson, P. C. and Cuello, A. C. (1977b) Evidence for the existence of substance P-containing fibres in striato-nigral and pallidonigral pathways in rat brain. *Brain Res.*, 119: 447–453.

Kelley, A. E. and Iversen, S. D. (1979) Substance P infusion into substantia nigra of the rat: behavioural analysis and involvement of striatal dopamine. *Eur. J. Pharmacol.*, 60: 171–179.

Kerwin, R. W. and Pycock, C. J. (1979) Thyrotropin releasing hormone stimulates release of [^{3}H]dopamine from slices of rat nucleus accumbens in vitro. *Br. J. Pharmacol.*, 67: 323–325.

Kitabgi, P., Carraway, R., Rietschoten, J. V., Granier, C., Morgat, J. L., Menez, A., Leeman, S. and Freychet, P. (1977) Neurotensin: specific binding to synaptic membranes from rat brain. *Proc. Natl. Acad. Sci. U.S.A.*, 74: 1846–1850.

Kobayashi, R. M., Brown, M. and Vale, W. (1977) Regional distribution of neurotensin and somatostatin in rat brain. *Brain Res.*, 126: 584–588.

Kohno, J., Shiosaka, S., Shinoda, K., Inagaki, S. and Tohyama, M. (1984) Two distinct strio-nigral substance P pathways in the rat: an experimental immunohistochemical study. *Brain Res.*, 308: 309–317.

Kuhar, M. J., Pert, C. B. and Snyder, S. H. (1973) Regional distribution of opiate receptor binding in monkey and human brain. *Nature*, 245: 447–451.

Larsson, L. I. and Rehfeld, J. F. (1979) Localization and molecular heterogeneity of cholecystokinin in central and peripheral nervous system. *Brain Res.*, 165: 201–218.

Ljungdahl, A., Hökfelt, T. and Nilsson, G. (1978a) Distribution of substance P-like immunoreactivity in the central nervous system of the rat. I. Cell bodies and nerve terminals. *Neuroscience*, 3: 861–943.

Ljungdahl, A., Hökfelt, T., Nilsson, G. and Goldstein, M. (1978b) Distribution of substance P-like immunoreactivity in the central nervous system of the rat. II. Light microscopic localization in relation to catecholamine-containing neurons. *Neuroscience*, 3: 945–976.

Loh, H. H., Tseng, L. F., Wei, E. and Li, C. H. (1976) β-Endorphin is a potent analgesic agent. *Proc. Natl. Acad. Sci. U.S.A.*, 83: 2895–2898.

Loren, I., Alumets, J., Hakanson, R. and Sundler, F. (1979) Immunoreactive pancreatic polypeptide (PP) occurs in the central and peripheral nervous system: preliminary immunocytochemical observations. *Cell Tissue Res.*, 200: 179–196.

Lundberg, J. M., Hökfelt, T., Schultzberg, M., Uvnas-Wallensten, K., Kohler, C. and Said, S. I. (1979) Occurrence of vasoactive intestinal polypeptide (VIP)-like immunoreactivity in certain cholinergic neurons of the cat: evidence from combined immunohistochemistry and acetylcholinesterase staining. *Neuroscience*, 4: 1539–1559.

Malouin, F. and Bedard, P. J. (1982) Head turning induced by unilateral intracaudate thyrotropin-releasing hormone (TRH) injection in the cat. *Eur. J. Pharmacol.*, 81: 559–567.

Marley, P. D., Emson, P. C. and Rehfeld, J. F. (1982) Effect of 6-hydroxydopamine lesions of the medial forebrain bundle on the distribution of cholecystokinin in rat forebrain. *Brain Res.*, 252: 382–385.

Marshall, P. E. and Landis, D. M. D. (1985) Huntington's disease is accompanied by changes in the distribution of somatostatin-containing neuronal processes. *Brain Res.*, 329: 71–82.

Martin, J. B., Brazeau, P., Tannenbaum, G. S., Willoughby, J. O., Epelbaum, J., Cass Terry, L. and Durand, D. (1978) Neuroendocrine organization of growth hormone regulation. In S. Reichlin, R. J. Baldessarini and J. B. Martin (Eds.), *The Hypothalamus*, Raven Press, New York, pp. 329–358.

McCulloch, J., Kelly, P. A. T., Uddman, R. and Edvinsson, L. (1983) Functional role for vasoactive intestinal polypeptide in the caudate nucleus: A 2-deoxy[^{14}C]glucose investigation. *Proc. Natl. Acad. Sci. U.S.A.*, 80: 1472–1476.

McLean, S., Skirboll, L. R. and Pert, C. B. (1985) Comparison of substance P and enkephalin distribution in rat brain: an overview using radioimmunocytochemistry. *Neuroscience*, 14: 837–852.

Melis, M. R. and Gale, K. (1984) Evidence that nigral substance P controls the activity of the nigrotectal GABAergic pathway. *Brain Res.*, 295: 389–393.

Meyer, D. K. and Kraus, J. (1983) Cholecystokinin modulates dopamine release in neostriatum. *Nature*, 301: 338–340.

Meyer, D. K., Phillips, M. I. and Eiden, L. (1982) Studies on the presence of angiotensin II in rat brain. *J. Neurochem.*, 38: 816–820.

Michelot, R., Leviel, V., Giorguieff-Chesselet, M. F., Cheramy, A. and Glowinski, J. (1979) Effects of the unilateral nigral modulation of substance P transmission on the activity of the two nigro-striatal dopaminergic pathways. *Life Sci.*, 24: 715–724.

Miyamoto, M. and Nagawa, Y. (1977) Mesolimbic involvement in the locomotor stimulant action of thyrotropin-releasing hormone (TRH) in rats. *Eur. J. Pharmacol.*, 44: 143–152.

Miyamoto, M., Narumi, S., Nagai, Y., Shima, T. and Nagawa, Y. (1979) Thyrotropin-releasing hormone: Hyperactivity and mesolimbic dopamine system in rats. *Jpn. J. Pharmacol.*, 29: 335–347.

Miyamoto, M., Fukuda, N., Narumi, S., Nagai, Y., Saji, Y. and Nagawa, Y. (1981) -Butyrolactone- -carbonyl-histidyl-prolinamide citrate (DN-1417): A novel TRH analog with potent effects on the central nervous system. *Life Sci.*, 28: 861–869.

Mroz, E. A., Brownstein, M. J. and Leeman, S. E. (1977) Evidence for substance P in the striato-nigral tract. *Brain Res.*, 125: 305–311.

Nemeroff, C. B., Hernandez, D. E., Luttinger, D., Kalivas, P. W. and Prange, A. J., Jr. (1982) Interactions of neurotensin with brain dopamine systems. *Ann. N.Y. Acad. Sci.*, 400: 330–334.

Nemeroff, C. B., Youngblood, W. W., Manberg, P. J., Prange, A. J., Jr. and Kizer, J. S. (1983) Regional brain concentrations of neuropeptides in Huntington's chorea and schizophrenia. *Science*, 221: 972–975.

Nicoll, R. A. (1977) Excitatory actions of TRH on spinal motorneurons. *Nature*, 265: 242–243.

Nicoll, R. A. and Barker, J. L. (1971) Excitation of supraoptic neuro-secretory cells by angiotensin II. *Nature New Biol.*, 233: 172–174.

Nicoll, R., Siggins, G. and Bloom, F. (1977) Neuronal actions of endorphins and enkephalin among brain regions: a comparative microiontophoretic study. *Proc. Natl. Acad. Sci. U.S.A.*, 74: 2584–2588.

Nieoullon, A., Cheramy, A. and Glowinski, J. (1977) An adaptation of the push-pull cannula method to study the in vivo release of [^{3}H]dopamine synthesized from [^{3}H]tyrosine in the cat caudate nucleus: Effects of various physical and pharmacological treatments. *J. Neurochem.*, 28: 819–828.

Oliver, C., Eskay, R. L., Ben-Jonathan, N. and Porter, J. C. (1974) Distribution and concentration of thyrotropin-releasing hormone in rat brain. *Endocrinology*, 95: 540–546.

Olpe, H. R. and Koella, W. P. (1977) Rotary behavior in rats by intranigral application of substance P and an eledoisin fragment. *Brain Res.*, 126: 576–579.

Phillips, J. W., Kirkpatrick, J. R. and Said, S. I. (1978) Vasoactive intestinal polypeptide excitation of central neurons. *Can. J. Physiol. Pharmacol.*, 56: 337–340.

Phillips, M. I., Weyhenmeyer, J., Felix, D., Ganten, D. and Hoffman, W. E. (1979) Evidence for an endogenous brain renin-angiotensin system. *Fed. Proc.*, 38: 2260–2266.

Pollard, H., Llorens, C., Schwartz, J. C., Gros, C. and Dray, F. (1978) Localization of opiate receptors and enkephalins in the rat striatum in relationship with the nigrostriatal dopaminergic system: lesion studies. *Brain Res.*, 151: 392–398.

Quinlan, J. T. and Phillips, M. I. (1981) Immunoreactivity for an angiotensin II-like peptide in the human brain. *Brain Res.*, 205: 212–218.

Quirion, R., Everist, H. D. and Pert, A. (1982) Nigrostriatal dopamine terminals bear neurotensin receptors but mesolimbic do not. *Soc. Neurosci. Abs.*, 8: 582.

Reid, I. A. (1979) The brain renin-angiotensin system: a critical analysis. *Fed. Proc.*, 38: 2255–2259.

Renaud, L. P. and Martin, J. B. (1975) Thyrotropin releasing hormone (TRH) – Depressant action on central neural activity. *Brain Res.*, 86: 150–154.

Rossier, J., Vargo, T. M., Minick, S., Ling, N., Bloom, F. E. and Guillemin, R. (1977) Regional distribution of beta-endorphin and enkephalin contents in rat brain and pituitary. *Proc. Natl. Acad. Sci. U.S.A.*, 74: 5162–5165.

Saito, A., Sankaran, H., Goldfine, I. D. and Williams, J. A. (1980) Cholecystokin receptors in brain: characterization and distribution. *Science*, 208: 1155–1156.

Schenker, C., Mroz, E. A. and Leeman, S. E. (1976) Release of substance P from isolated nerve endings. *Nature*, 264: 790–792.

Simantov, R. and Snyder, S. H. (1976) Brain–pituitary opiate mechanisms: pituitary opiate receptor binding, radioimmunoassays for methionine enkephalin and leucine enkephalin and ^{3}H-enkephalin interactions with the opiate receptor. In H. W. Kosterlitz (Ed.), *Endogenous Opioid Peptides*, Amsterdam, North Holland, pp. 41–48.

Simantov, R., Snowman, A. M. and Snyder, S. H. (1978) A morphine-like factor "enkephalin" in rat brain: subcellular localization. *Brain Res.*, 107: 650–657.

Simonnet, G., Giorguieff-Chesselet, M. G., Carayon, A., Bioulac, B., Cesselin, F., Glowinski, J. and Vincent, J. D. (1981) Angiotensine II et systeme nigro-neostriatal. *J. Physiol. (Paris)*, 77: 71–79.

Spindel, E. R., Wurtman, R. J. and Bird, E. D. (1980) Increased TRH content of the basal ganglia in Huntington's disease. *N. Engl. J. Med.*, 303: 1235–1236.

Studler, J. M., Simon, H., Cesselin, F., Legrand, J. C., Glowinski, J. and Tassin, J. P. (1981) Biochemical investigation on the localization of the cholecystokinin octapeptide in dopaminergic neurons originating from the ventral tegmental area of the rat. *Neuropeptides*, 2: 131–139.

Taylor, D. P. and Pert, C. B. (1979) Vasoactive intestinal polypeptide: Specific binding to rat brain membranes. *Proc. Natl. Acad. Sci. U.S.A.*, 76: 660–664.

Uhl, G. R. and Snyder, S. H. (1976) Regional and subcellular distribution of brain neurotensin. *Life Sci.*, 19: 1827–1832.

Uhl, G. R., Kuhar, M. J. and Snyder, S. H. (1977) Neurotensin: Immunohistochemical localization in rat central nervous system. *Proc. Natl. Acad. Sci. U.S.A.*, 74: 4059–4063.

Uhl, G. R., Goodman, R. R. and Snyder, S. H. (1979) Neurotensin-containing cell bodies, fibres and nerve terminals in the brain stem of the rat: immunohistochemical mapping. *Brain Res.*, 167: 77–91.

Vale, W., Brazeau, P., Grant, G., Nusey, A., Burgus, R., Rivier, J., Ling, N. and Guillemin, R. (1972) Premieres observations sur le mode d'action de la somatostatine un facteur hypothalamique qui inhibe la secretion de l'hormone de croissance. *C. R. Acad. Sci. Paris*, 275: 2913.

Vijayan, E. and McCann, S. M. (1977) Suppression of feeding and drinking activity in rats following intraventricular injection of thyrotropin releasing hormone (TRH). *Endocrinology*, 100: 1727–1730.

Vincent, S., Hökfelt, T., Christensson, I. and Terenius, L. (1982) Immunohistochemical evidence for a dynorphin immunoreactive striato-nigral pathway. *Eur. J. Pharmacol.*, 85: 251–252.

Voigt, M. M. and Wang, R. Y. (1984) In vivo release of dopamine in the nucleus accumbens of the rat: modulation by cholecystokinin. *Brain Res.*, 296: 189–193.

Von Euler, U. S. and Gaddum, J. H. (1931) An unidentified depressor substance in certain tissue extracts. *J. Physiol. (London)*, 72: 74–87.

Waldmeier, P. C., Kam, R. and Stocklin, K. (1978) Increased dopamine metabolism in rat striatum after infusions of substance P into the substantia nigra. *Brain Res.*, 159: 223.

Wang, R. Y. and White, F. J. (1983) Cholecystokinin octapeptide excitation and dopamine inhibition of nucleus accumbens neurons in the rat. Abstract, *5th Int. Catecholamine Symp.*, pp. 314–315.

Weber, E., Roth, K. A. and Barchas, J. D. (1982) Immunohistochemical distribution of α-neo-endorphin/dynorphin neuronal systems in rat brain: Evidence for colocalization. *Proc. Natl. Acad. Sci. U.S.A.*, 79: 3062–3066.

White, F. J. and Wang, R. Y. (1984) Interactions of cholecystokinin octapeptide and dopamine on nucleus accumbens neurons. *Brain Res.*, 300: 161–166.

Williams, R. G., Gayton, R. J., Zhu, W.-Y. and Dockray, G. J. (1981) Changes in brain cholecystokinin octapeptide following lesions of the medial forebrain bundle. *Brain Res.*, 213: 227–230.

Winokur, A. and Beckman, A. L. (1978) Effects of thyrotropin releasing hormone, norepinephrine and acetylcholine on the activity of neurons in the hypothalamus, septum and cerebral cortex of the rat. *Brain Res.*, 150: 205–209.

Yang, H.-Y. T. and Neff, N. H. (1972) Distribution and properties of angiotensin converting enzyme of rat brain. *J. Neurochem.*, 19: 2443–2450.

Young, W. S. III and Kuhar, M. J. (1981) Neurotensin receptor localization by light microscopic autoratiography in rat brain. *Brain Res.*, 206: 273–285.

Zaborszky, L., Alheid, G. F., Beinfeld, M. C., Eiden, L. E., Heimer, L. and Palkovits, M. (1985) Cholecystokinin innervation of the ventral striatum: A morphological and radioimmunological study. *Neuroscience*, 14: 427–453.

Zieglgansberger, W. and Fry, J. P. (1976) Actions of enkephalin on cortical and striatal neurones of naive and morphine tolerant/dependent rats. In H. W. Kosterlitz (Ed.), *Opiates and Endogenous Opioid Peptides*, North Holland, Amsterdam, pp. 231–238.

P. C. Emson, M. N. Rossor and M. Tohyama (Eds.),
Progress in Brain Research, Vol. 66.

CHAPTER 4

Neuropeptides and the pathology of Huntington's disease

P. C. Emson

MRC Group, Institute of Animal Physiology, Babraham, Cambridge CB2 4AT, U.K.

Introduction

In the early 1970s the discovery of the enkephalins (methionine and leucine enkephalin) by Hughes and Kosterlitz (Hughes, 1975; Hughes et al., 1977) and the sequencing of substance P (Chang et al., 1971) and the first releasing factor (Boler et al., 1969; Burgus et al., 1969) led to a flood of research aimed at localising, characterising and identifying central nervous system (CNS) neuropeptides. There is now an ever-expanding list of neuropeptides, that is peptides with a localisation in specific CNS neurones, which may be neurotransmitters or neurohormones. These discoveries have also revolutionised the study of CNS organisation as it is now possible to develop antibodies to specific neuropeptide sequences and using immunological techniques to localise or measure the peptide in specific brain areas or pathways.

Technical advances have now made it possible to pick out low abundance brain-specific messenger RNA and using the complementary DNA technique to predict a peptide sequence to which antibodies can be raised and a specific CNS localisation demonstrated, without ever having to attempt a purification of the putative peptide transmitter (Sutcliffe et al., 1983). Such studies will undoubtedly lead to the recognition of further neurone-specific antigens, but the most impressive feature of neuropeptide histochemistry has been the complexity and specificity of CNS organisation revealed. These techniques applied to the human CNS have demonstrated that the human brain, like other mammalian brains, is equally full of neuropeptides and with few exceptions much of the organisation of neuropeptide systems in lower mammals is retained in the human. In consequence, it is possible to use neuropeptides as markers of specific pathways, or neuronal cell types and to use measurements of peptide content as an index of the integrity of peptide-containing cells or pathways. Examples of this type of approach are found throughout this book but examples include substance P as a marker for human striatonigral pathway (this chapter) or the use of vasoactive intestinal polypeptide and cholecystokinin content as an index of selectivity of cortical neuronal loss in Alzheimer's disease.

However, before considering the degeneration of neuropeptide-containing pathways in the basal ganglia in Huntington's disease, it is appropriate to review briefly the evidence that peptides are stable post-mortem (despite initial predictions) and to illustrate the type of biochemical technique used, such as high pressure liquid chromatography (HPLC) or gel chromatography, to establish the integrity of the peptide after death.

Post-mortem stability and characterisation of neuropeptides

It is usually difficult to obtain human brain tissue less than 4–6 h after death. Biopsy samples are rarely available, are confined to surgically accessible tissue and are often taken from tissue adjacent to tumours or epileptic foci. As a consequence of these constraints it has been important to establish if peptides remain intact post-mortem and for how long this stability persists. Thus in an initial experiment, Kanazawa and Jessell (1976) showed that substance P in the rat brain was stable for up to 24 h after death. This initial observation suggested that the post-mortem stability of substance P was high and indeed assays of substance P in this laboratory (Gale et al., 1978; Emson et al., 1980) indicated that the content of substance P in normal human brains was not significantly affected by the mode of death (agonal status), drug treatment (neuroleptics etc.), age at death (although there is a tendency for many peptides to decline with age) or to autopsy delay. Subsequent studies have established that methionine–enkephalin, substance P, cholecystokinin (CCK), vasoactive intestinal polypeptide (VIP) and somatostatin, for example, are all relatively stable after death (Emson et al., 1981). The reason for this stability is unknown but may lie in the packaging of neuropeptides in vesicles to avoid cytoplasmic peptidase activity.

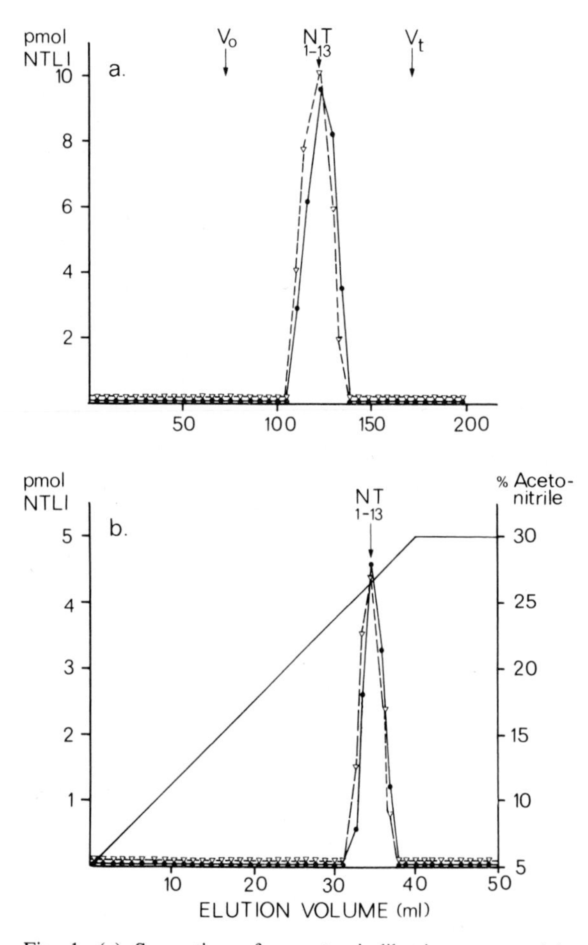

Fig. 1. (a) Separation of neurotensin-like immunoreactivity (NTLI) by Sephadex G25 column chromatography. V_o, void volume of column; V_t, total column volume; N_{1-13} indicates the expected elution of synthetic bovine neurotensin. Sequence directed radioimmunoassay detects carboxy-terminal (●——●) or amino-terminal immunoreactivity (▽- - -▽). Note that all the immunoreactivity separates at the position expected for intact neurotensin. (b) Separation of neurotensin-like immunoreactivity by reverse phase HPLC. Note the elution of neurotensin-like immunoreactivity at the position of the synthetic standard.

Given the apparent stability of many peptides post-mortem it is also important to establish the identity of the immunoreactivity as the authentic peptide rather than a "degraded" immunoreactive fragment. The usual technique of choice for such studies is reverse phase HPLC which allows the separation of even complex mixtures of peptides with relative ease (Fig. 1 shows the separation of neurotensin). A combination of HPLC and sequence directed radioimmunoassay should give reasonable evidence that the peptide is intact (i.e. it separates at the position expected for the intact peptide) and is immunologically intact (i.e. it contains the appropriate amino acids to react with appropriate sequence-specific antiserum (Fig. 1 shows such a separation and assay of brain neurotensin). All of these various data are consistent with the assumption that the peptide detected contains the complete amino acid sequence. These HPLC studies are not always, however, the method of choice as gel chromatography is sometimes more appropriate. A case in point here is with CCK-octapeptide (CCK-8), which occurs in a number of molecular forms (sulphated, desulphated, oxidized, unoxidized etc.) which makes HPLC traces extremely complicated, however, gel chromatography resolves all these forms into one peak of CCK-8 based on molecular weight. It should be noted that CCK-8 with the possibility of four oxidized forms and sulphated and desulphated forms is rather unusual, and most peptides can be readily resolved by HPLC. CCK-8 is

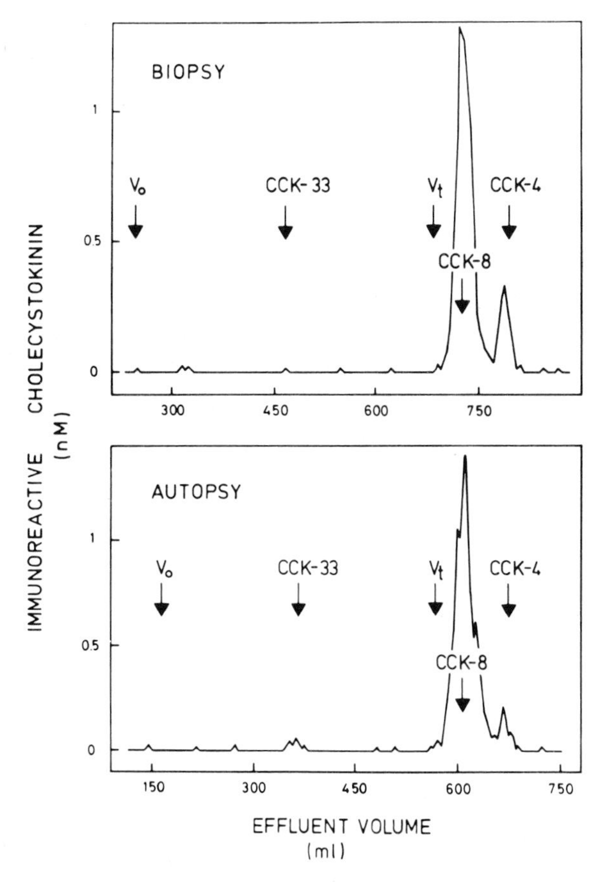

Fig. 2. Comparison of the chromatographic profiles of cholecystokinin immunoreactivity from a human brain cerebral cortical biopsy sample and a sample obtained after usual post-mortem handling (autopsy). Note the similar amounts of CCK-8 in autopsy and biopsy samples.

a major cortical peptide and with this peptide we are fortunate enough to have chromatographic evidence that the material from biopsy samples of human brain (Fig. 2) is identical to the material extracted from post-mortem brain (Fig. 2).

Much less is known of the stability of receptors and enzymes after death. In some cases, for example, the enzymes glutamic acid decarboxylase and choline acetyltransferase, there are both agonal and post-mortem changes which influence enzyme content measured after death. Some receptors, including peptide receptors, have been measured in human brain but few systematic studies have been done of receptor stability (exceptions are dopamine, muscarinic and GABA receptors; Lloyd and Davidson, 1979; Reisine et al., 1979). As for peptidase enzymes (enzymes which may be concerned with degradation of neuropeptides) no systematic studies of enzyme stability have been carried out. The information available for the only two enzymes which have been studied, angiotensin converting enzyme and proline endopeptidase, suggests they are not affected by agonal or post-mortem changes. The enzymes' stability in model systems (such as the mouse "cooling" model, Fig. 3) has not been tested (Arregui et al., 1978; Pittaway et al., 1984).

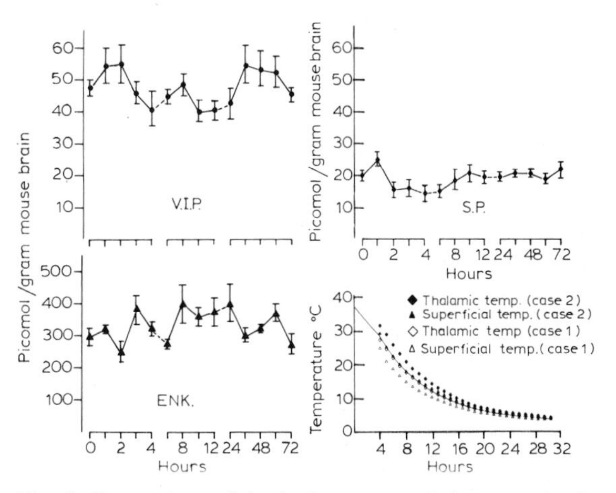

Fig. 3. Comparison of the brain content of three neuroactive peptides (vasoactive intestinal polypeptide (VIP), substance P (SP), and methionine enkephalin (MET-ENK)) in mouse cadavers killed at time 0 and then kept in an incubator programmed to mimic the normal cooling curve of a human cadaver. Note the relative stability of the brain content even up to 72 h post-mortem.

The mouse "cooling" model has been used extensively in our laboratory and it consists of programming an incubator to cool mouse cadavers over a period of 72 h to a programme that matches the normal cooling curve of the human brain. The stability of many neuropeptides in this model has been assessed and so far the stability of all the peptides studied in this model has been impressive although the content of thyrotropin-releasing hormone (TRH) actually increased.

Neuropathology of Huntington's disease and the organisation of the neostriatum

The regions particularly affected in Huntington's disease are the caudate nucleus and putamen (Bruyn, 1968, 1973). These regions correspond to the neo or dorsal striatum (for detailed discussion see Graybiel, 1984 and Takagi, this volume). The ventral striatum, which includes the nucleus accumbens and olfactory tubercle, has not been so extensively studied in Huntington's disease but it is clear that pathological changes, although less marked are found in these areas (Bruyn, 1968, 1973). It would not be appropriate to consider here the detailed organisation of the neostriatum which is discussed in some detail by Takagi (this volume), however, it is appropriate to note briefly that the major output neurones of the striatum are believed to be the spiny neurones and it is these that degenerate in Huntington's disease (Bruyn, 1973). In contrast, the aspiny neurones (characterized by local axons and deeply indented nuclei) are apparently relatively preserved in Huntington's disease (Marshall and Landis, 1982). The probable localisation of neuropeptides in these different cell types will be discussed briefly when considering the changes observed in different peptides.

Substance P and the tachykinins

Initial observations by Powell et al. (1973) on human brain material revealed that the substantia nigra had a high content of substance P. Subsequent studies (Kanazawa et al., 1977; Gale et al., 1978; Emson et al., 1980a) confirmed this observation and showed that the pars reticulata had the highest substance P content of the brain areas assayed. These observations, together with the parallel immunohistochemical studies showing the localisation of substance P-like immunoreactivity in the human substantia nigra (Emson et al., 1981) are consistent with the localisation of substance P in striatal efferent neurones. This suggestion has been confirmed by studies, primarily in the rat, that confirm substance P is present in descending striatonigral projections (Kanazawa et al., 1977; Brownstein et al., 1977; Jessell et al., 1978) and by anatomical studies that show substance P-like immunoreactivity can be localised in populations of medium-sized spiny neurones (Bolam et al., 1983).

Given the probable anatomical localisation of substance P in the human striatum (although substance P-positive neurones have not been visualized in human material) it is not surprising to find a dramatic loss of substance P-like immunoreactivity in Huntington's disease (Kanazawa et al., 1977; Gale et al., 1978; Emson et al., 1980) (Table I, Fig. 4).

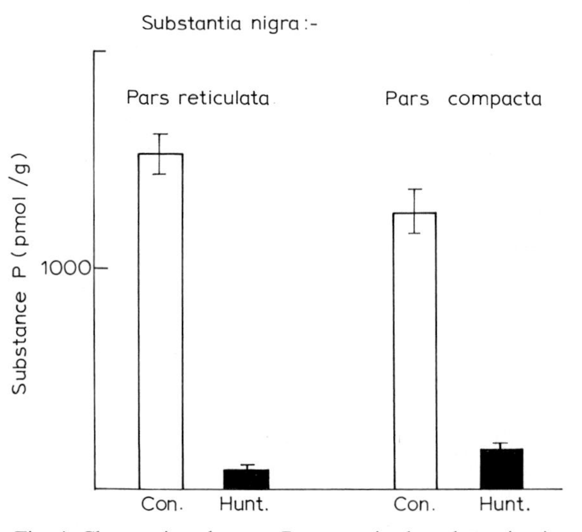

Fig. 4. Changes in substance P content in the substantia nigra pars compacta and pars reticulata of patients with Huntington's disease (black bars) and normal controls (open bars). Values are means ± S.E.M. for the number of determinations listed in Table IV.

The loss of substance P is not restricted to the substantia nigra, but includes both segments of the globus pallidus. The absence of any dramatic loss of substance P content from the remaining caudate-putamen might be explained by tissue shrinkage, masking the loss of substance P-containing neurones. Paralleling these biochemical changes in substance P and opiate-like peptides (see later) there is a dramatic loss of GABA and glutamic acid decarboxylase, indicating a degeneration of striatal-nigral GABA-ergic fibres. Recent histochemical evi-

TABLE I

Substance P in Huntington's disease

Region	Control	Huntington's disease
Caudate nucleus	138 ± 14 (13)	158 ± 28 (10)
Putamen	112 ± 29 (6)	77 ± 20 (10)
Globus pallidus (lateral)	197 ± 99 (6)	18 ± 9* (10)
Globus pallidus (medial)	877 ± 253 (6)	178 ± 102* (10)
Substantia nigra (pars compacta)	1264 ± 239 (27)	158 ± 41* (24)
Substantia nigra (pars reticulata)	1535 ± 177 (30)	86 ± 25* (25)

Each value is expressed as pmol/g wet weight and is the mean ± S.E.M. of the number of determinations in parentheses.
* $P < 0.05$.

dence (W. Oertel, personal communication) suggests that all medium-sized spiny neurones in the neostriatum are GABA-ergic and if this is the case the various peptides in the striatum will co-exist with GABA in different populations of spiny neurones.

Apart from substance P, recent work has identified two other mammalian tachykinins, which have been termed neurokinin α (substance K, or neuromedin L) and neurokinin β (neuromedin K) (Kanagawa et al., 1983; Kimura et al., 1983). Both these peptides contain the C-terminal tachykinin sequence but contain a valine at position 8 instead of the phenylalanine at position 8 found in substance P. The substitution of valine for phenylalanine modifies the properties of these molecules and separate substance K receptors (originally designated as "E" type receptors; Lee et al., 1982) may be found in the mammalian brain (Mantyh et al., 1984). Determinations of peptide immunoreactivity which may correspond to these two peptides, using an antiserum raised against kassin (which has a valine at position 8) indicates that kassin immunoreactivity is found in the substantia nigra and so neurokinin α and β may be additional striato-nigral transmitters. The presence of neurokinin α in a tachykinin gene (Nawa et al., 1983) suggests that both neurokinin α and β may co-exist with substance P in tachykinin-containing neurones of the striatum. It is interesting to note that the rat substantia nigra seems to contain primarily substance K receptors (Mantyh et al., 1984) so it may be that neurokinin α and β may be more important than substance P in influencing nigral cell activity.

Opiate-like peptides

The mammalian opioid peptides are derived from three precursors, pro-opiomelanocortin, pro-enkephalin A and pro-enkephalin B (prodynorphin; Höllt, 1983). Development of radioimmunoassays and immunohistochemical procedures showed initially that enkephalin-like immunoreactivity was most concentrated in the spinal cord (dorsal horn), trigeminal nucleus and the globus pallidus of the rat (Hong et al., 1977a,b; Hökfelt et al., 1977; Hughes et al., 1977). The presence of high concentrations of enkephalin-like peptides in the basal ganglia of the rat led several groups to study their distribution in the human basal ganglia (Gramsch et al., 1979; Emson et al., 1980a). Our studies showed that Met-enkephalin-like immunoreactivity was particularly concentrated in the lateral segment of globus pallidus with lower concentrations in the medial segment and substantia nigra (Table II). The probable presence of Met-enkephalin in the human striato-nigral pathway was confirmed when determinations carried out on Huntington's disease material showed substantial losses of Met-enkephalin immunoreactivity (Table II) from both segments of globus pallidus and the substantia nigra (Table II) (Fig. 5).

TABLE II

Met-enkephalin in Huntington's disease

Region	Control	Huntington's disease
Caudate nucleus	116 ± 40 (10)	156 ± 54 (10)
Putamen	200 ± 71 (12)	184 ± 68 (12)
Globus pallidus (lateral)	1163 ± 216 (20)	527 ± 102* (39)
Globus pallidus (medial)	675 ± 168 (20)	317 ± 48* (40)
Substantia nigra (pars compacta)	577 ± 103 (15)	229 ± 71* (15)
Substantia nigra (pars reticulata)	661 ± 145 (15)	230 ± 54* (16)

Each value is expressed as pmol/g wet weight and is the mean ± S.E.M. of the number of determinations in parentheses.
* $P < 0.05$.

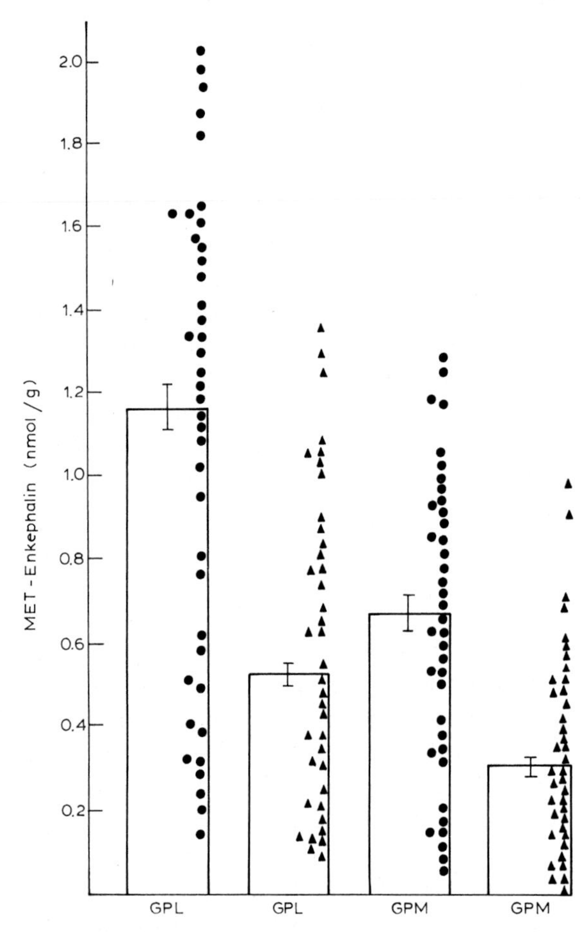

Fig. 5. Distribution of Met-enkephalin immunoreactivity in the lateral segment of the globus pallidus (GPL) and medial segment of the globus pallidus (GPM). ●, normal controls; ▲, Huntington's disease cases. Bars are means ± S.E.M.

The observation of significant quantities of Met-enkephalin-like immunoreactivity in the human substantia nigra is in contrast to observations in the rat where only small amounts of enkephalin-like material are found in the nigra of the albino rat (Hökfelt et al., 1977). Histochemical studies on the human brain show convincingly that the lateral globus pallidus contains the majority of enkephalin-like immunoreactivity and that in Huntington's disease there is a dramatic loss of enkephalin staining from the globus pallidus and substantia nigra. Met-enkephalin staining also nicely reveals that the human caudate and putamen contains the patches or striosomes described by Graybiel (Graybiel, 1984).

The distribution of other peptides from the Pro-enkephalin A precursor Met-enkephalin-Arg^6-Phe^7, and Met-enkephalin-Arg^6-Gly^7-Leu^8 has been determined in the human brain (Pittius et al., 1984) (Table III). The ratio of the content of these peptides to Met-enkephalin was in reasonable agreement with the expected ratio from the precursor (i.e. Met-enkephalin/Met-enkephalin-Arg^6-Phe^7/Met-enkephalin-Arg^6-Gly^7-Leu^8, 4:1:1), except that content of Met-enkephalin-Arg^6-Gly^7-Leu^8 was slightly lower than expected. Determination of post-mortem stability of these peptides by Pittius et al. (1984) suggested that these peptides are also stable post-mortem.

Turning to other opioid peptide precursors, pro-opiomelanocortin contains β-endorphin, how-

TABLE III

Distribution of pro-enkephalin A derived peptides in human brain

	Immunoreactivity (pmol/g)		
	MERF[a]	MERGL[b]	BAM = 12P[c]
Frontal lobe cortex	2.54 ± 0.40	1.24 ± 0.21	0.29 ± 0.06
Occipital lobe cortex	5.95 ± 0.86	1.26 ± 0.31	0.37 ± 0.06
Temporal lobe cortex	1.85 ± 0.31	1.00 ± 0.24	0.33 ± 0.06
Cerebral white matter	0.54 ± 0.25	0.40 ± 0.13	< 0.15
Amygdala, medial	14.7 ± 3.61	5.12 ± 1.27	2.27 ± 0.28
Amygdala, lateral	10.4 ± 1.74	2.91 ± 0.67	2.87 ± 0.64
Hippocampus	3.79 ± 0.48	1.23 ± 0.13	0.65 ± 0.11
Septum pellucidum	1.21 ± 3.12	4.45 ± 1.13	1.02 ± 0.33
Anterior hypothalamus	66.1 ± 6.23	25.3 ± 2.64	12.1 ± 2.03
Posterior hypothalamus	47.6 ± 7.14	14.4 ± 2.51	5.26 ± 1.27
Mammillary bodies	15.6 ± 3.28	2.85 ± 0.99	0.98 ± 0.23
Thalamus, medial nuclei	6.98 ± 0.90	1.90 ± 0.30	0.55 ± 0.19
Thalamus, lat. nuclei	2.13 ± 0.58	0.72 ± 0.19	0.72 ± 0.19
Pulvinar	4.03 ± 0.92	1.07 ± 0.15	0.54 ± 0.14
Caudate nucleus	255 ± 36.3	84.1 ± 8.85	13.6 ± 1.57
Globus pallidus, medial	458 ± 60.1	137 ± 23.5	16.4 ± 2.45
Globus pallidus, lateral	521 ± 33.8	162 ± 16.5	17.6 ± 2.89
Olfactory bulb	28.4 ± 4.60	7.49 ± 1.62	1.79 ± 0.43
Olfactory region	7.58 ± 2.55	2.56 ± 0.91	2.14 ± 0.72
Periaqueductal gray	62.1 ± 8.30	18.5 ± 2.39	7.59 ± 2.05
Substantia nigra	215 ± 16.4	53.4 ± 3.81	8.49 ± 1.75
Colliculi superiores	38.0 ± 6.64	7.45 ± 1.32	1.14 ± 0.34
Colliculi inferiores	40.5 ± 8.19	12.0 ± 3.53	1.19 ± 0.48
Red nucleus	28.6 ± 8.66	7.56 ± 2.99	< 1.1
Pons dors/loc. cerul.	41.4 ± 5.31	15.9 ± 1.16	0.85 ± 0.21
Medulla obl. dors.	54.2 ± 6.18	17.7 ± 2.14	3.59 ± 1.14
Medulla obl. ventr.	2.23 ± 0.59	0.64 ± 0.07	< 0.12
Area postrema	85.0 ± 4.98	22.5 ± 4.00	< 3.0
Olivary nucl. inf.	13.0 ± 3.97	2.09 ± 0.42	< 1.0
Geniculate body lat.	1.93 ± 0.56	2.15 ± 0.74	< 1.5
Pineal gland	33.4 ± 11.0	2.63 ± 0.79	< 2.0
Cerebellum, vermis	12.3 ± 1.87	9.42 ± 4.04	0.66 ± 0.06
Cereb., lobe quadrang.	7.16 ± 1.15	3.11 ± 0.56	0.27 ± 0.04
Dentate nucleus	2.23 ± 0.91	0.84 ± 0.14	< 0.4
Spinal cord:			
Cervical ventral	33.9 ± 7.45	10.8 ± 23.13	< 0.3
Cervical dorsal	32.9 ± 5.22	8.72 ± 1.45	< 0.3
Thoracic ventral	32.8 ± 1.65	11.8 ± 1.75	< 0.4
Thoracic dorsal	24.5 ± 9.50	6.69 ± 1.96	< 0.4
Lumbar ventral	27.9 ± 5.91	7.62 ± 1.87	< 0.35
Lumbar dorsal	46.6 ± 8.40	9.60 ± 1.32	0.81 ± 0.34

[a] MERF = Met-enkephalin-Arg^6-Phe^7.
[b] MERGL = Met-enkephalin-Arg^6-Gly^7-Leu^8.
[c] BAM = 12P = BAM-12P-BAM-22P.

ever, there is no evidence that neurones containing this peptide are localized in the striatum, whilst the prodynorphin precursor contains dynorphin A, dynorphin B (rimorphin) and α- and β-neoendorphin. Development of dynorphin-directed radioimmunoassays has established that dynorphin_{1-8} and dynorphin B are found in approximately equal quantities in the human substantia nigra (Table IV) (Gramsch et al., 1982; Pittius, 1984). Examination of the data in Table IV indicates interesting differences in content through the basal ganglia and contrasts sharply with the distribution of Met-enkephalin-like immunoreactivity (Table II). The distribution of dynorphin-like peptides resembles more closely that found for substance P than Met-enkephalin, suggesting that if the dynorphin peptides are found together with Met-enkephalin, this must be only a small percentage of all Met-enkephalin-containing striatal neurones. Indeed, histochemical studies (Vincent et al., 1982; Graybiel, 1984) show clearly that dynorphin-positive immunoreactivity is distributed differently in the striatum and its outputs. The concentration of dynorphin peptides in the medial globus pallidus and substantia nigra (which parallels the distribution of substance P) suggests that there will be dynorphin--containing striato-nigral and striato-pallidal projections and indeed, these have been reported in the rat (Vincent et al., 1982b; Palkovits et al., 1984).

TABLE IV

Distribution of dynorphin in human brain

	Immunoreactivity (pmol/g)	
	Dyn_{1-8}	Dyn B
Amygdala, medial part	9.6 ± 1.1	17.6 ± 1.4
Ant. hypothalamus	5.3 ± 0.7	9.6 ± 1.4
Post. hypothalamus	3.8 ± 0.9	8.4 ± 2.4
Caudate nucleus	20.6 ± 1.9	5.0 ± 1.4
Pallidum, medial part	61.8 ± 8.1	19.0 ± 7.7
Pallidum, lateral part	5.1 ± 1.3	6.0 ± 0.8
Putamen	7.5 ± 0.4	2.1 ± 0.3
Substantia nigra	100.0 ± 10.4	116.9 ± 30.9
Cerebellum	0.8 ± 0.6	< 0.06
Spinal cord, ventral	0.8 ± 0.1	0.9 ± 0.2
Spinal cord, dorsal	1.6 ± 0.3	2.0 ± 0.5

Acidic acetone extracts were used to measure Dyn_{1-8} and acetic acid extracts for Dyn B (pmol/g), means ± S.E.M. of 5–9 tissues.

Somatostatin and neuropeptide Y

These two peptides are considered together because recent experimental work suggests that in some striatal neurones these two peptides co-exist. Somatostatin-like immunoreactivity has been reported in medium-sized aspiny neurones of the striatum (DiFiglia and Aronin, 1982; Takagi et al., 1983, 1984b; Bolam, 1984). Parallel histochemical studies using antisera directed against the neuropeptide Y-related peptide (avian pancreatic polypeptide) show that neurones containing this peptide are also medium-sized aspiny neurones and that in some, possibly all, of these neurones the two peptides somatostatin and neuropeptide Y co-exist (see Graybiel and Ragsdale, 1983, for discussion). Another interesting feature of these neurones is that they are apparently preserved in Huntington's disease. This was originally demonstrated for somatostatin by Aronin et al. (1983) and we have subsequently confirmed that neuropeptide Y immunoreactivity and neuropeptide Y-positive neurones are preserved in parallel with somatostatin (Table V and Fig. 6) (Dawbarn et al., 1985), the density of neuropeptide Y cells increasing some three-fold in the caudate nucleus of patients with Huntington's disease. Interpretation of such changes may not be straightforward as increases in peptide content could be a reflection of reduced turnover, reduced release etc., but the most likely explanation for the increased peptide content and cell density is that these aspiny neurones are less damaged in Huntington's disease.

Neither somatostatin nor neuropeptide Y are decreased in the substantia nigra in Huntington's disease, so it is likely that the small amount of somatostatin and neuropeptide Y in this area is not derived from the striatum.

TABLE V

Concentration of neuropeptide Y and somatostatin in different brain areas in control and Huntington's disease

Brain region	Neuropeptide Y concentration (pmol/g)		Somatostatin concentration (pmol/g)	
	Control	Huntington's disease	Control	Huntington's disease
Brodmann area 10	14.6 ± 1.3 (9)	16.5 ± 1.8 (9)	44.6 ± 6.6 (9)	51.2 ± 5.0 (9)
Caudate	24.4 ± 3.1 (9)	47.6 ± 6.0 (9)*	71.2 ± 12.3 (9)	125.3 ± 16.1 (9)*
Globus pallidus medial	0.8 ± 0.2 (6)	1.5 ± 0.5 (7)	27.2 ± 8.4 (4)	21.0 ± 5.9 (5)
Hippocampus	10.9 ± 1.4 (5)	16.1 ± 4.0 (5)	68.6 ± 10.9 (5)	67.3 ± 13.7 (5)
Putamen	11.0 ± 1.7 (12)	28.1 ± 2.9 (12)*	95.6 ± 16.6 (12)	119.9 ± 28.2 (12)*
Hypothalamus	5.2 ± 1.2 (8)	4.8 ± 1.7 (8)	686.0 ± 88.0 (8)	793.0 ± 118.0 (8)
Substantia nigra pars compacta	–	–	26.4 ± 4.7 (4)	31.9 ± 12.2 (3)

Values are means ± S.E.M. (values in brackets indicate number of samples).
* $P < 0.05$ by *t*-test.

Cholecystokinin and vasoactive intestinal polypeptide

Both these gut peptides were sequenced in Victor Mutt's laboratory (Jorpes and Mutt, 1966; Said and Mutt, 1969). In the mammalian brain both peptides are markers for the population of cerebral cortical interneurones (Emson and Hunt, 1981) and are mostly concentrated in the fore-brain. In the rat brain there is little VIP in the basal ganglia, however, CCK has been shown to be a marker for a particular population of dopamine neurones' (Hökfelt et al., 1980) projection to the nucleus accumbens. The co-existence of CCK with dopamine, together with the recent report that there are CCK-containing neurones in the rat striatum (see Takagi, this volume) has meant that determinations of CCK in Parkinson's disease (Agid et al., this volume) and Huntington's disease (Emson et al., 1980) are relevant.

In Huntington's disease the content of VIP in the striatum is essentially unchanged, whilst there is a reduction of CCK content in the globus pallidus and substantia nigra (Emson et al., 1980b) (Table VI). Interpretation of this change is complicated; it could perhaps relate to a loss of a cortico-nigral projection. Meyer et al. (1982) suggested that much of striatal CCK originates from the pyriform cortex or claustrum. Alternatively it may reflect a loss of CCK-containing striatal output neurones. The recent report of Takagi et al. (1984a) of CCK-containing striatal neurones in the rat suggests this as a possibility, however, the neurone type described by Takagi et al. (1984a) is believed to be an interneurone (medium-sized aspiny neurones; Bolam, 1984; Takagi et al., 1984a) at least in the rat.

Other peptides

The peptides so far discussed are those known from immunohistochemical studies to be associated with defined striatal neurones or striatal inputs. Other peptides have been looked at in Huntington's disease and these include neurotensin, TRH and homocarnosine (Table VII). Both neurotensin and TRH are increased in Huntington's disease. In the case of neurotensin this may reflect the relative survival of neurotensin/dopamine-containing fibres (neurotensin and dopamine have been reported to co-exist in the rat; Hökfelt et al., 1984), or alternatively survival of neurotensin-containing striatal neurones which have been reported (Jennes et al., 1982). As for TRH, it is not known if there are striatal TRH-containing neurones, although there are TRH-positive cells in the septum and nucleus accumbens (Johanssen et al., 1983). It is of course

TABLE VI

Regional distribution of cholecystokinin and vasoactive intestinal polypeptide in normal human brain and in Huntington's disease

Region	CCK		VIP	
	Control	Huntington's disease	Control	Huntington's disease
Caudate nucleus	316 ± 94 (10)	284 ± 24 (10)	4.6 ± 1.3 (12)	4.2 ± 2.0 (12)
Putamen	268 ± 24 (10)	308 ± 52 (10)	2.3 ± 0.5 (20)	2.5 ± 2.0 (20)
Globus pallidus	84 ± 16 (10)	40 ± 8* (10)	2.4 ± 1.4 (12)	3.1 ± 1.9 (10)
Substantia nigra	260 ± 40 (10)	100 ± 12* (10)	1.9 ± 1.3 (6)	2.8 ± 1.6 (10)
Frontal cortex	548 ± 56 (10)	552 ± 64 (10)	17.3 ± 2.3 (21)	14.2 ± 1.5 (11)

Each value is expressed as pmol/g wet weight and is the mean ± S.E.M. of the number of determinations in parentheses.
* $P < 0.05$.

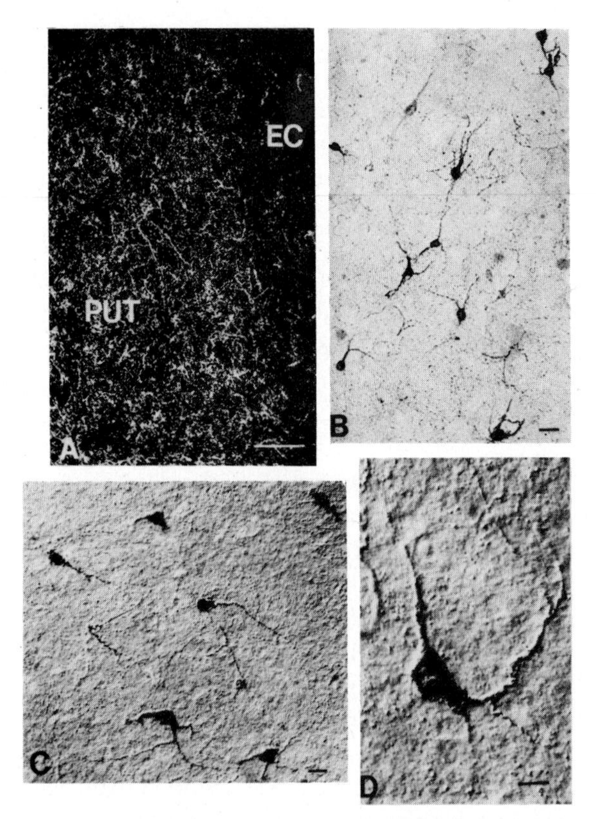

Fig. 6. Neuropeptide Y-like immunoreactivity in the corpus striatum of Huntington's disease brain, stained by the peroxidase–antiperoxidase method. (A) Dark field photomicrograph of the putamen (PUT) and external capsule (EC). Note the dense fibre staining in the putamen. (B) Phase contrast photomicrograph of neurones in the caudate nucleus. (C, D) Differential interference photomicrographs of cells in the putamen. Bars indicate (A) 50 μm; (B–D) 10 μm.

likely that as further neuropeptide markers are discovered some of these will be localised in the basal ganglia.

Peptide receptors

Paralleling the loss of medium-sized spiny neurons from the human caudate and putamen in Huntington's disease, it would be logical to expect parallel loss of transmitter receptors expressed on these neurones. In agreement with this expectation measurements of acidic amino acid, dopamine and GABA receptor binding sites all show a reduction in number (Lloyd and Davidson, 1979; Olsen et al., 1979; Reisine et al., 1979). In contrast to these assays for "classical" transmitter receptors only one systematic study of a peptide receptor in Huntington's disease has been carried out (Hays et al., 1981). Hays and colleagues report a loss of a large number of striatal CCK binding sites consistent with the loss of CCK "receptors" on degenerating striatal neurones (Table VIII). Parallel studies by Hays et al. (1981) showed that kainic acid injections into the rat striatum produced a substantial loss of CCK binding sites mimicking the effects found in Huntington's disease. These two results strongly suggest that most of the striatal CCK receptors are on the intrinsic neurones rather than axons of projection neurones. Hays et al. (1981) did not carry out a systematic study of peptide receptor stability post-

TABLE VII

Content of neurotensin, thyrotropin-releasing hormone and homocarnosine in control and Huntington's disease brains

	Neurotensin (pmol/g)		TRH (pg/mg)		Homocarnosine (μmol/g)	
	Control	Huntington's disease	Control	Huntington's disease	Control	Huntington's disease
Caudate nucleus	5.37 ± 0.99 (8)	7.56 ± 1.0 (7)	1.0 ± 0.1 (8)	2.7 ± 0.4* (21)	–	–
Putamen	3.12 ± 0.92 (12)	3.07 ± 0.54 (9)	0.9 ± 0.1 (21)	2.6 ± 0.3* (23)	–	–
Globus pallidus (lateral segment)	12.12 ± 1.82 (8)	24.18 ± 2.7* (10)	–	–	–	–
Globus pallidus (medial segment)	17.18 ± 3.4 (8)	29.18 ± 3.50* (10)	–	–	–	–
Substantia nigra (compacta)	54.5 ± 6.7 (12)	67.18 ± 8.9 (10)	–	–	–	–
Substantia nigra (reticulata)	36.3 ± 4.0	40.0 ± 7.4 (10)	–	–	0.90 ± 0.09 (8)	0.57 ± 0.09* (8)

From Emson et al. (unpublished); Spindel, Wurtman and Bird (1980); and Perry, Hansen and Kloster (1973). Values in parentheses indicate number of samples.
* $P < 0.05$.

TABLE VIII

CCK-Receptor binding in brains of patients with Huntington's disease and matched controls

Brain region	Control (fmol/mg protein)	Huntington's disease (fmol/mg protein)
Basal ganglia	5.2 ± 0.4	1.0 ± 0.3*
Cerebral cortex	3.2 ± 0.2	1.8 ± 0.3*

* $P < 0.05$, from Hays et al. (1981).

mortem and it seems sensible that this should be done before attempts are made to measure opiate, substance P and other appropriate receptors in Huntington's disease.

Peptide-degrading enzymes

Following the realisation that many neuropeptides may be putative transmitters, attention has turned to possible mechanisms for inactivation of peptide transmitters. Following the example of other amino acid or monoamine transmitters two possibilities were obvious: one was re-accumulation (uptake) into the neurone releasing the peptide and the second was inactivation by specific peptidases localised in the synaptic cleft. A number of studies have indicated that most peptides are not actively reaccumulated, so that attention was transferred to degrading enzymes. Initial studies suggested that at least reasonably specific peptidases might exist (Malfroy et al., 1978), but subsequent work revealed this to be something of an oversimplification and terms such as "enkephalinase" or "neurotensinase" need to be used with caution. Current work suggests that there are a number of neuropeptide-degrading endopeptidases localised on neuronal membranes, however, these are likely to be capable of degrading a range of substrates depending on the specificity of the active site of the peptidase. Despite these reservations about substrates for peptidases, there is evidence that some peptidases are regionally distributed in the CNS and may contribute to the degradation of appropriate neuropeptides.

One such peptidase is angiotensin-converting enzyme (EC 3.4.15.1) so named because of its involvement in the periphery with the formation of angiotensin II from angiotensin I. Histochemical and biochemical studies in the rat show clearly that this enzyme is localised in the striato-nigral pathway

TABLE IX

Activity of angiotensin-converting enzyme and proline endopeptidase in brain regions of control subjects and of patients dying with Huntington's disease

Region	Angiotensin converting enzyme activity			Proline endopeptidase activity		
	Control[a]	Huntington's[a]	Reduction (%)	Control[b]	Huntington's[b]	Reduction (%)
Caudate nucleus	307 ± 43 (10)	125 ± 27 (9)*	60	4.3 ± 0.3	0.9 ± 0.3*	78
Putamen	205 ± 28 (10)	98 ± 10 (11)*	52	–	–	–
Globus pallidus (lateral)	155 ± 6 (11)	44 ± 3 (13)*	72	2.9 ± 0.3	1.1 ± 0.2*	63
Globus pallidus (medial)	149 ± 11 (11)	55 ± 5 (12)*	63	3.1 ± 0.3	1.2 ± 0.3*	60
Substantia nigra (reticulata)	105 ± 6 (16)	23 ± 4 (10)*	78	1.7 ± 0.3	1.0 ± 0.2*	42
Substantia nigra (compacta)	55 ± 6 (16)	29 ± 5 (10)	48	1.8 ± 0.4	0.8 ± 0.2*	56

[a] Each value is expressed as pmol/mg per min and is the mean ± S.E.M. of the number of determinations in parentheses.
[b] Expressed as nmol/mg protein per min.
* $P < 0.05$, Student's t-test (two tailed).

and in this location it probably functions as an endopeptidase cleaving substance P and other neuropeptides present in this pathway (Arregui et al., 1978; Iversen, 1984). Consistent with this suggestion there is a substantial loss of angiotensin-converting enzyme in Huntington's disease (Arregui et al., 1978, 1979) (Table IX).

Whilst angiotensin-converting enzyme is a predominantly membrane-bound enzyme, other soluble peptidases may be to some extent at least neuronal or glial markers. One enzyme that is primarily a neuronal marker enzyme is proline endopeptidase (E.C. 3.4.21.26). Determinations of the content of this enzyme in Huntington's disease also show that this enzyme is depleted in Huntington's disease, presumably paralleling the loss of neostriatal neurones (Table IX).

Discussion

The results presented in this review show clearly that neuropeptides can be used as markers for the damaged striatal neurones found in Huntington's disease. In some cases the biochemical and immunohistochemical results reveal complexities of striatal organisation that were not obvious with classical anatomical techniques. Such complexities include the presence of striosomes (Graybiel, 1984) or the chemical distinction between striato-nigral (substance P-containing) and striato-pallidal (enkephalin-containing) efferents. None of these observations bring us any nearer to understanding the biochemical deficit in Huntington's disease, however, it may be relevant that some neuronal types (aspiny neurones) seem to be more resistant than others (the spiny neurones) to the genetic defect of Huntington's disease. It is tempting to speculate that the reason for the differential sensitivity of neurones to the defect of Huntington's disease may lie in receptors the various neurones carry. This might occur, for example, if the spiny but not all aspiny neurones carried receptors for the acidic amino acids (glutamate/aspartate), so that differential sensitivity might relate to the sensitivity to excess acidic amino acid. This so-called excitotoxin hypothesis (Coyle et al., 1977) is attractive, however, we will have to await the elucidation of the exact genetic defect before this question can be resolved. It is to be hoped that this discovery will not be long delayed (Gusella et al., 1983).

References

Aronin, N., Cooper, P. E., Lorenz, L. J., Bird, E. D., Sagar, S. M., Leeman, S. E. and Martin, J. B. (1983) Somatostatin is increased in the basal ganglia in Huntington's disease. *Ann. Neurol.*, 13: 519–526.

Arregui, A., Emson, P. C. and Spokes, E. G. S. (1978) Angiotensin-converting enzyme in substantia nigra. Reduction of activity in Huntington's disease and after intrastriatal kainic acid in the rat. *Eur. J. Pharmacol.*, 52: 121–124.

Arregui, A., Iversen, L. L., Spokes, E. G. S. and Emson, P. C. (1979) Alterations in postmortem brain angiotensin-converting enzyme activity and some neuropeptides in Huntington's disease. In T. N. Chase, N. S. Wexler and A. Barbea (Eds.), *Advances in Neurology, Vol. 23,* Raven Press, New York, pp. 517–525.

Bolam, J. P. (1984) Synapses of identified neurons in the neostriatum. In D. Evered and M. O'Connor (Eds.), *Ciba Found. Symp., Functions of the Basal Ganglia,* pp. 30–41.

Bolam, J. P., Somogyi, P., Takagi, H., Fodor, I. and Smith, A. D. (1983) Localisation of substance P-like immunoreactivity in neurons and nerve terminals in the neostriatum of the rat: a correlated light and electron microscopic study. *J. Neurocytol.*, 12: 325–344.

Boler, J., Enzmann, F., Folkers, K., Bowers, C. Y. and Schally, A. V. (1969) The identity of chemical or hormonal properties of the thyrotropin releasing hormone and pyroglutamylhistidy-prolineamide. *Biochem. Biophys. Res. Commun.*, 37: 705–710.

Brownstein, M. J., Mroz, E. A., Toppaz, M. L. and Leeman, S. E. (1977) On the origin of substance P and glutamic acid decarboxylase (GAD) in the substantia nigra. *Brain Res.*, 135: 315–323.

Bruyn, G. W. (1968) Huntington's chorea; historical, clinical and laboratory synopsis. In P. J. Vinken and G. W. Bruyn (Eds.), *Handbook of Clinical Neurology, Vol. 6,* North Holland, Amsterdam, pp. 379–396.

Bruyn, G. (1973) Neuropathological changes in Huntington's chorea. In A. Barbeau, T. N. Chase and G. W. Paulson (Eds.), *Huntington's Chorea 1872–1972,* Raven Press, New York, pp. 399–403.

Burgus, R., Dunn, T., Desiderio, D. and Guillemin, R. (1969) Structure moleculaire de facteur hypothalamique hypophysiotrope TRF d'origine ovine: Mise en evidence par spectrometrie de masse de la sequence PCA-His-Pro-NH_2. *C.R. Acad. Sci. (Paris),* 269: 1870–1873.

Chang, M. M., Leeman, S. E. and Niall, H. D. (1971) Aminoacid sequence of substance P. *Nature New Biol.*, 232: 86–87.

Coyle, J. T., Schwartz, R., Bennett, J. P. and Campochiaro, P. (1977) Clinical, neuropathologic and pharmacologic aspects of Huntington's disease: correlates with a new animal model. *Prog. Neuro-Psychopharmacol.*, 1: 13–30.

Dawbarn, D., DeQuidt, M. E. and Emson, P. C. (1985) Survival of basal ganglia neuropeptide Y/somatostatin neurones in Huntington's disease. *Brain Res.* (in press).

DiFiglia, M. and Aronin, N. (1982) Ultrastructural features of immunoreactive somatostatin neurons in the rat caudate-putamen. *J. Neurosci.*, 2: 1267–1274.

Emson, P. C. and Hunt, S. P. (1981) Anatomical chemistry of the cerebral cortex. In F. G. Warden, G. Adelman and S. G. Dennis (Eds.), *The Organisation of the Cerebral Cortex,* F. O. Schmitt, M.I.T. Press, Cambridge, MA, pp. 325–346.

Emson, P. C., Arregui, A., Clement-Jones, V., Sandberg, B. E. and Rossor, M. N. (1980a) Regional distribution of methionine enkephalin and substance P-like immunoreactivity in normal human brain and in Huntington's disease. *Brain Res.*, 199: 147–160.

Emson, P. C., Rehfeld, J. F., Rossor, M. N. and Langevin, H. (1980b) Reductions in cholecystokinin-like immunoreactivity in the basal ganglia in Huntington's disease. *Brain Res.*, 198: 497–500.

Emson, P. C., Rossor, M. N., Hunt, S. P., Marley, P. D., Clement-Jones, V., Rehfeld, J. F. and Fahrenkrug, J. (1981) Distribution and post-mortem stability of substance P, metenkephalin, vasoactive intestinal polypeptide and cholecystokinin in normal human brain and in Huntington's disease. In F. C. Rose (Ed.), *Metabolic Disorders of the Nervous System,* Pitman, London, pp. 312–321.

Gale, J. S., Bird, E. D., Spokes, E. G., Iversen, L. L. and Jessell, T. M. (1978) Human brain substance P: distribution in controls and Huntington's disease. *J. Neurochem.*, 30: 633–634.

Gramsch, C., Höllt, V., Mehraein, P., Pasi, A. and Herz, A. (1979) Regional distribution of methionine enkephalin and β-endorphin-like immunoreactivity in human brain and pituitary. *Brain Res.*, 171: 261–270.

Gramsch, C., Höllt, V., Pasi, A., Mehraein, P. and Herz, A. (1982) Immunoreactive dynorphin in human brain and pituitary. *Brain Res.*, 233: 65–74.

Graybiel, A. M. (1984) Neurochemically specified subsystems in the basal ganglia. In D. Evered and M. O'Connor (Eds.), *Ciba Found. Symp., Functions of the Basal Ganglia,* pp. 114–143.

Graybiel, A. M. and Ragsdale, C. W., Jr. (1984) Biochemical anatomy of the striatum. In P. C. Emson (Ed.), *Chemical Neuroanatomy,* Raven Press, New York, pp. 427–504.

Gusella, J. F., Wexler, N. S., Conneally, P. M., Naylor, S. L., Anderson, M. A., Tanzi, R. E., Watkins, P. C., Ottina, K., Wallace, M. R., Sakaguchi, A. Y., Young, A. B., Shoulson, I., Bonilla, E. and Martin, J. B. (1983) A polymorphic DNA marker genetically linked to Huntington's disease. *Nature,* 306: 234–238.

Hays, S. E., Meyer, D. K. and Paul, S. M. (1981) Localisation of cholecystokinin receptors to neuronal elements in the rat caudate nucleus. *Brain Res.*, 219: 208–213.

Hökfelt, T., Elde, R., Johansson, O., Terenius, L. and Stein, L. (1977) The distribution of enkephalin-immunoreactive cell bodies in the rat central nervous system. *Neurosci. Lett.*, 5: 25–31.

Hökfelt, T., Skirboll, L., Rehfeld, J. F., Goldstein, M., Markey,

K. and Dann, O. (1980) A subpopulation of mesencephalic dopamine neurons projecting to limbic areas contain a cholecystokinin-like peptide. Evidence from immunohistochemistry combined with retrograde tracing. *Neuroscience,* 5: 2093–2124.

Hökfelt, T., Everitt, B. J., Theodorsson-Norheim, E. and Goldstein, M. (1984) Occurrence of neurotensin-like immunoreactivity in subpopulations of hypothalamic, mesencephalic and medullary catecholamine neurons. *J. Comp. Neurol.* (in press).

Höllt, V. (1983) Multiple endogenous opioid peptides. *Trends Neurosci.,* 6: 24–26.

Hong, J. S., Yang, H. Y. and Costa, E. (1977a) On the location of methionine–enkephalin neurons in the rat striatum. *Neuropharmacology,* 16: 451–453.

Hong, J. S., Yong, H. Y., Fratta, W. and Costa, E. (1977b) Determination of methionine–enkephalin in discrete regions of rat brain. *Brain Res.,* 134: 383–386.

Hughes, J. (1975) Isolation of an endogenous compound from the brain with pharmacological properties similar to morphine. *Brain Res.,* 88: 295–306.

Hughes, J., Kosterlitz, H. W. and Smith, T. W. (1977) The distribution of methionine and leucine enkephalin in the brain and peripheral tissue. *Br. J. Pharmacol.,* 61: 639–648.

Iversen, L. L. (1984) Amino acids and peptides: fast and slow chemical signals in the nervous system. *Proc. R. Soc. Lond. B.,* 221: 245–260.

Jennes, L., Stumpf, W. E. and Kalivas, P. W. (1982) Neurotensin topographical distribution in the rat brain by immunohistochemistry. *J. Comp. Neurol.,* 210: 211–234.

Jessell, T. M., Emson, P. C., Paxinos, A. G. and Cuello, A. C. (1978) Topographic projections of substance P and GABA pathways in the striato-pallidal-nigral system: a biochemical and immunohistochemical study. *Brain Res.,* 152: 487–498.

Johansson, O., Hökfelt, T., Jeffcoate, S. L., White, N. and Spindel, E. (1983) Light and electron microscopic immunohistochemical studies on TRH in the central nervous system of the rat. In E. C. Griffiths and G. W. Bennett (Eds.), *Thyrotropin-Releasing Hormone,* Raven Press, New York.

Jorpes, E. and Mutt, V. (1966) Cholecystokinin and paraeozymin: one single hormone. *Acta Physiol. Scand.,* 66: 196–202.

Kanagawa, K., Minamino, M., Fukuda, A. and Matsuo, H. (1983) Neuromedin K: a novel mammalian tachykinin identified in porcine spinal cord. *Biochem. Biophys. Res. Commun.,* 114: 533–540.

Kanazawa, I. and Jessell, T. M. (1976) Postmortem changes and regional distribution of substance P in the rat and mouse nervous system. *Brain Res.,* 117: 362–367.

Kanazawa, I., Bird, E., O'Connell, R. and Powell, D. (1977) Evidence for the decrease of substance P content in the substantia nigra of Huntington's chorea. *Brain Res.,* 120: 387–392.

Kimura, S., Okada, M., Sugita, Y., Kanazawa, I. and Murekata, E. (1983) Novel neuropeptides, neurokinin α and β isolated from porcine spinal cord. *Proc. Jpn. Acad. Ser. B.,* 59: 101–104.

Lee, C. M., Iversen, L. L., Hanley, M. R. and Sandberg, B. E. B. (1982) The possible existence of multiple receptors for substance P. *Naunyn-Schmiedebergs Arch. Pharmakol.,* 318: 281–287.

Lloyd, K. G. and Davidson, L. (1979) Alterations in ^{3}H-GABA binding in Huntington's disease: A phospholipid component. In T. Chase, N. Wexler and A. Barbeau (Eds.), *Advances in Neurology, Vol. 23,* Raven Press, New York, pp. 705–716.

Maggio, J. E. and Hunter, J. C. (1984) Regional distribution of kassinin-like immunoreactivity in rat central and peripheral tissues and the effect of capsaicin. *Brain Res.,* 307: 370–373.

Malfroy, B., Swertz, J. P., Gayon, A., Rogues, B. P. and Schwartz, J. C. (1978) High affinity of enkephalin-degrading peptidase in brain is increased after morphine. *Nature,* 276: 523–526.

Mantyh, P. W., Maggio, J. E. and Hunt, S. P. (1984) The autoradiographic distribution of kassinin and substance K binding sites is different from the distribution of substance P binding sites in the rat brain. *Eur. J. Pharmacol.,* 102: 361–364.

Marshall, P. and Landis, D. M. D. (1982) Somatostatin in the neostriatum and substantia nigra in normal human brain and in Huntington's disease. *Proc. Am. Soc. Neurosci.,* 12: 507.

Meyer, D. K., Beinfeld, M. C., Oertel, W. H. and Brownstein, M. J. (1982) Origin of the cholecystokinin-containing fibers in the rat caudate-putamen. *Science,* 215: 187–188.

Nawa, H., Hirose, T., Takashima, H., Inayama, S. and Nakanishi, S. (1983) Nucleotide sequence of cloned cDNA for types of bovine substance P precursor. *Nature,* 306: 32–36.

Olsen, R. W., Van Ness, P. C. and Tourtellotte, W. W. (1979) Gamma-aminobutyric acid receptor binding curves for human brain regions: comparison of Huntington's disease and normal. In T. N. Chase, N. S. Wexler and A. Barbeau (Eds.), *Advances in Neurology, Vol. 23,* Raven Press, New York, pp. 697–704.

Palkovits, M., Brownstein, M. J. and Zamir, N. (1984) On the origin of dynorphin A and α neo-endorphin in the substantia nigra. *Neuropeptides,* 4: 193–199.

Perry, T. L., Hansen, S. and Kloster, M. (1973) Huntington's chorea, deficiency of gamma-aminobutyric acid in the brain. *N. Engl. J. Med.,* 288: 337–356.

Pittaway, K. M., Reynolds, G. P. and Emson, P. C. (1984) Decreased proline endopeptidase activity in the basal ganglia in Huntington's disease. *J. Neurochem.,* 43: 878–880.

Pittius, C. W., Seizing, B. R., Pasi, A., Mehraein, P. and Herz, A. (1984) Distribution and characterisation of opioid peptides derived from pro-enkephalin A in human and rat central nervous system. *Brain Res.,* 304: 127–136.

Powell, D., Leeman, S., Tregear, G. W., Niall, H. D. and Potts, J. T., Jr. (1973) Radioimmunoassay for substance P. *Nature New Biol.,* 241: 252–254.

Reisine, T. D., Beaumont, K., Bird, E. D., Spokes, E. G. S. and Yamamura, H. I. (1979) Huntington's disease: Alterations in neurotransmitter receptor binding in the human brain. In T. N. Chase, N. Wexler and A. Barbeau (Eds.), *Advances in Neurology, Vol. 23,* Raven Press, New York, pp. 177–726.

Said, S. and Mutt, v. (1969) Long-acting vasodilator peptide from lung tissue. *Nature,* 224: 699–700.

Spindel, E. R., Wurtman, R. J. and Bird, E. D. (1980) Increased TRH content of the basal ganglia in Huntington's disease. *N. Engl. J. Med.,* 303: 1235–1236.

Sutcliffe, J. T., Milner, R. J., Schinnick, T. M. and Bloom, F. E. (1983) Identifying the protein products of brain-specific genes with antibodies to chemically synthesized peptides. *Cell,* 33: 671–682.

Takagi, H., Somogyi, P., Somogyi, J. and Smith, A. D. (1983) Fine structural studies on a type of somatostatin-immunoreactive neuron and its synaptic connections in the rat neostriatum: a correlated light and electron microscopic study. *J. Comp. Neurol.,* 214: 1–16.

Takagi, H., Mizuta, H., Matsuda, T., Inagaki, S., Tateishi, K. and Hamaoka, T. (1984a) The occurrence of cholecystokinin-like immunoreactive neurons in the rat neostriatum: light and electron microscopic analysis. *Brain Res.,* 309: 346–349.

Takagi, H., Somogyi, P. and Smith, A. D. (1984b) Aspiny neurons and their local axons in the neostriatum of the rat: a correlated light and electron microscopic study of Golgi-impregnated material. *J. Neurocytol.,* 13: 239–265.

Vincent, S. R., Hökfelt, T., Christensson, I. and Terenius, L. (1982a) Dynorphin-immunoreactive neurons in the central nervous system of the rat. *Neurosci. Lett.,* 33: 185–190.

Vincent, S., Hökfelt, T., Christensson, I. and Terenius, L. (1982b) Immunohistochemical evidence for a dynorphin immunoreactive striato-nigral pathway. *Eur. J. Pharmacol.,* 85: 251–252.

P. C. Emson, M. N. Rossor and M. Tohyama (Eds.),
Progress in Brain Research, Vol. 66.

CHAPTER 5

Neuropeptides and Parkinson's disease

Y. Agid[1], H. Taquet[2], F. Cesselin[2], J. Epelbaum[3] and F. Javoy-Agid[1]

[1]*Laboratoire de Médecine Expérimentale, Unité 289 de l'INSERM,* [2]*Laboratoire de Biochimie Médicale, CHU Pitié-Salpêtrière, 91, Bd de l'Hôpital, 75634 Paris Cedex 13, and* [3]*Unité 159 de l'INSERM, 2 ter, rue d'Alésia, 75014 Paris, France*

Introduction

Parkinson's disease is a unique degenerative neurological disease since the symptoms, mainly akinesia, are improved by drug treatment, i.e. levodopa or dopamine agonists. The clinical improvement of the patients is probably related to the restoration of dopamine (DA) transmission in the brain. Indeed, from a neurochemical point, the disease is essentially characterised by a generalized central DA deficiency: not only the nigro-striatal (Hornykiewicz, 1966; Bernheimer et al., 1973), but the mesocorticolimbic (Price et al., 1978; Javoy-Agid et al., 1982b, 1983; Scatton et al., 1983) and the hypothalamic (Ehringer and Hornykiewicz, 1960; Rinne and Sonninen, 1973; Javoy-Agid et al., 1984b) DA systems are also damaged, though to different degrees (Javoy-Agid et al., 1983b). Thus, Parkinson's disease represents a rather specific model of a central DA denervation in man, of interest for the study of the relationships between neuropeptides and DA-containing neurons. Such interactions have been evidenced from investigations on the corresponding animal model, the 6-hydroxydopamine-induced degeneration of the rat DA ascending pathways. In addition, in consideration of the rich and well-classified symptomatology characteristic of the disease, the possible implication of alterations in neuropeptide transmission in the appearance of parkinsonian symptoms may be sought. At present, little is known concerning the role of neuropeptides; most central peptidergic deficits produced in animal studies or evidenced in human pathology (Spokes, 1981) are not clearly related to specific functional disturbances. In order to give some insight on these points, the main data obtained for a number of neuropeptides (cholecystokinin-8 (CCK-8), substance-P (SP), Met-enkephalin (MET-ENK), Leu-enkephalin (LEU-ENK), somatostatin (SRIF) and thyrotropin-releasing hormone (TRH)) within the parkinsonian brain post mortem will be reviewed.

Regional distribution of neuropeptides in the brain post mortem of patients with Parkinson's disease

Several neuropeptides have been studied in Parkinson's disease to date. The characteristics of a number of systems (CCK-8, SP, MET-ENK, LEU-ENK, SRIF), but not all (TRH, vasopressin), are modified in discrete brain areas of patients.

Alterations in peptidergic systems

Cholecystokinin-8

In the human brain, cholecystokinin-like immunoreactivity (CCK-LI) is concentrated mainly in the amygdala and the cerebral cortex but also in subcortical areas, particularly DA-rich areas such as the striatum (Emson et al., 1980b; Studler et al., 1982; Vanderhaeghen et al., 1982). Interestingly, according to studies in animals, striatal CCK-LI is contained in nerve terminals distributed in patches close to the median line, but the peptide is also present within some DA-containing terminals innervating the medial and posterior part of the nucleus accumbens (Hökfelt et al., 1980; Studler et al., 1981; Marley et al., 1982). The CCK/DA containing nerve terminals arise from the ventral tegmental area of the mesencephalon and the external and medial substantia nigra pars compacta where CCK-LI is partly localized within DA cell bodies (Hökfelt et al., 1980; Vanderhaeghen et al., 1980). The mesencephalic DA cell bodies are known to degenerate in Parkinson's disease (Javoy-Agid et al., 1982b). A decrease in CCK-LI levels could thus be expected in the ventral tegmental area and substantia nigra of parkinsonian brains. The peptide concentrations are reduced in the substantia nigra pars compacta and reticulata but not in other structures (ventral tegmental area, caudate nucleus, putamen, nucleus accumbens) when compared to control values (Fig. 1) (Studler et al., 1982). Therefore most CCK-like immunoreactive (CCK-LIr) neurons are preserved in the disease. The reduction in

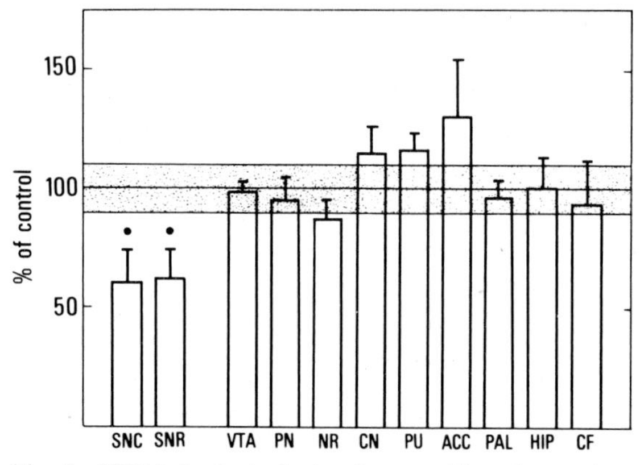

Fig. 1. CCK-8 levels in brains from parkinsonian patients. CCK-8 levels were estimated by radioimmunoassay of 8–13 control and 6–11 parkinsonian brains (Studler et al., 1982). Data (mean ± S.E.M.) are expressed as percentage of respective control values (horizontal bars). Several structures were examined: nucleus accumbens (ACC), cerebral frontal cortex (Brodmann area 9) (CF), caudate nucleus (CN), hippocampus (HIP), red nucleus (NR), external pallidum (PAL), nucleus paranigralis (PN), putamen (PU), substantia nigra pars compacta (SNC) and pars reticulata (SNR), ventral tegmental area (VTA) of the mesencephalon.

* $P < 0.01$, significantly different when compared to respective control.

the nigral peptide levels may result from the degeneration of the CCK-LIr cell bodies, and/or of CCK inputs of unknown origin. The absence of detectable changes in ventral tegmental area CCK--LI levels does not exclude the possibility that some CCK/DA neurons are damaged, but they could represent a subpopulation of neurons too small to be detected. The latter hypothesis is all the more satisfying since the DA content decreases by 60% in this area, in contrast to the more than 80% reduction in the substantia nigra (Javoy-Agid et al., 1983b), indicating that not all DA neurons degenerate, but that some are spared, which could be those containing CCK-LI. Finally, the data are compatible with, but do not definitely demonstrate, the coexistence of CCK-LI in DA neurons visualized in animals (Hökfelt et al., 1980). The clinical significance of the nigral CCK-LI abnormality remains unknown. Experimental pharmacology in animals demonstrates an excitatory effect of the peptide on nigral dopamine neurons (Skirboll et al., 1981). It might be proposed that administration of CCK-8 agonists could potentiate the effects of DA agonists and further help parkinsonian patients. However, intravenous administration of ceruletide, a synthetic CCK-8 analogue, in man does not seem to modify motoneuronal excitability (Delwaïde et al., 1983).

Substance P

Substance P-like immunoreactivity (SP-LI) is heterogeneously distributed in the human brain with highest levels in both parts of the substantia nigra and in the internal pallidum (Kanazawa et al., 1977a; Gale et al., 1978; Emson et al., 1980a). The high peptide concentration detected in the nigra very likely corresponds to a massive SP innervation originating in the striatum, as demonstrated in the rat (Kanazawa et al., 1977b). The presence of the peptide in the vicinity of the DA neurons suggests that the SP afferents modulate the activity of the DA neurons in the human substantia nigra, as shown by electrophysiological, behavioural and pharmacological experiments in animals (for review see Glowinski et al., 1980), and thereby interfere

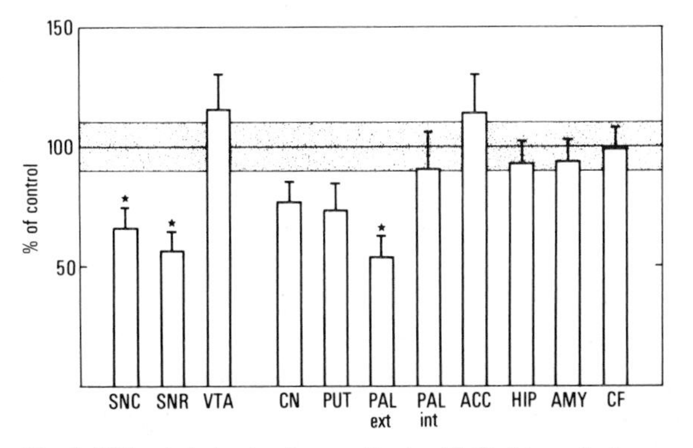

Fig. 2. SP levels in brains from patients with Parkinson's disease. SP immunoreactivities were determined by radioimmunoassay of brains from 9–15 controls and 10–15 parkinsonian patients (Mauborgne et al., 1983). Data (mean ± S.E.M.) are expressed as percentage of respective control values (horizontal bars). Abbreviations are detailed in the legend to Fig. 1. Amygdala (AMY); external (ext) and internal (int) pallidum.

with behaviour. In Parkinson's disease, a 30–40% decrease in SP-LI levels is observed in the substantia nigra (pars compacta and reticulata), the putamen and the pallidum (Fig. 2) (Mauborgne et al., 1983). In other brain structures, SP-LI values are not different from controls, providing evidence that most SP systems are spared in the disease (A. Mauborgne et al., unpublished data). The decline in pallidal and nigral SP-LI indicates that striato-pallidal and/or striato-nigral SP-LIr neurons, as described in animals (Kanazawa, 1977b; Staines et al., 1980), are affected as proposed on the basis of similar biochemical observations in Huntington's disease (Emson et al., 1980a). Classical neuropathology does not favour the existence of a severe striatal neuronal loss in Parkinson's disease, although lesions of the striatum and the pallidum have been observed in some cases (Blackwood and Corsellis, 1977). The physiological significance of such a biochemical deficiency remains unclear: neuronal loss or metabolic changes? The intense immunoreactivity for SP visualised in pallidum and substantia nigra of parkinsonian brains (Grafe et al., 1985) would support the latter hypothesis. Since SP has an excitatory effect on DA neurons (for review see Glowinski et al., 1980), the functional consequences may be opposite, depending on the significance of the biochemical abnormality. (1) The degeneration or the reduced metabolic activity of the nigral SP nerve terminals would accentuate the already deficient DA nigro-striatal transmission, and aggravate symptoms (i.e. akinesia); (2) on the other hand, a metabolic change consisting of an increased SP turnover, not fully compensated by synthesis, would activate DA transmission in the surviving DA neurons and represent a mechanism to counterbalance the DA deficiency.

Met-enkephalin

Behavioural, biochemical and pharmacological studies in animals (for review see Javoy-Agid et al., 1982b) indicate that some effects of MET-ENK systems are mediated through DA systems and vice versa. Such relationships might be predicted in humans as well, considering the parallel distribution of the peptide and the amine within the brain (Gramsch et al., 1979; Emson et al., 1980a; Taquet et al., 1982a,b, 1983). According to animal studies, striatal MET-ENK concentrations are increased by blockade of DA receptors with chronic neuroleptic treatment, unilateral hemisection of the brain (Costa et al., 1978) or chronic nigro-striatal DA denervation (Thal et al., 1983). A chronic lesion of the DA nigro-striatal pathway is also present in Parkinson's disease. However, in parkinsonian brains the peptide concentrations are reduced massively in the mesencephalon (substantia nigra and ventral tegmental area), less in the putamen and the pallidum but are not different from control values in other areas (Fig. 3) (Taquet et al., 1982a,b, 1983). The normal MET-ENK-like immunoreactivity (MET-ENK-LI) content found in most brain structures suggests that the corresponding neurons are anatomically intact. Since a putamino-pallidal enkephalin system has been demonstrated in animals (Cuello and Paxinos, 1978), the reduced MET-ENK-LI concentrations in putamen and pallidum of parkinsonian brains could reflect alteration of transmission in this neuronal pathway, though intrastriatal neurons might be implicated (Pickel et al., 1980). In the ventral mesencephalon the reduction in MET-ENK-LI content is in the same order

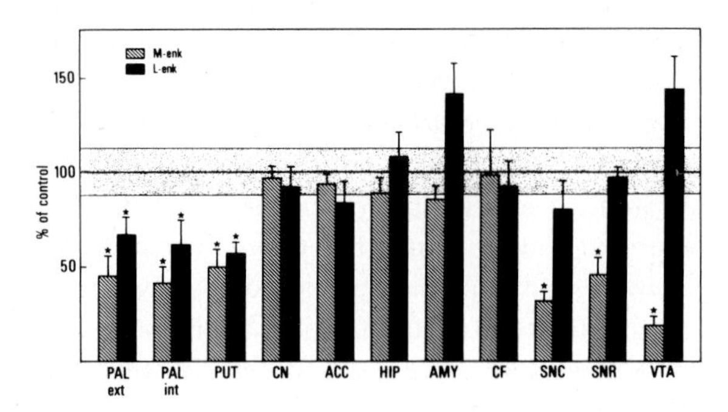

Fig. 3. MET-ENK and LEU-ENK content in brains of patients with Parkinson's disease. MET- and LEU-ENK were estimated in 15 control and 15 parkinsonian brains (Taquet et al., 1983). Data (mean ± S.E.M.) are expressed as percentage of respective control values (horizontal bar). Abbreviations are detailed in the legend to Fig. 1.

of magnitude as that of DA (Taquet et al., 1982b). In addition, in the normal human brain, the rostro--caudal patterns of distribution of DA and MET-ENK-LI in both parts of the substantia nigra are superimposable (Taquet et al., 1982b). Together, these two observations suggest that DA and MET--ENK could coexist within the same neurons in the human mesencephalon. In fact, according to immuno-histochemical studies in man, which corroborate data obtained in the rat (Johnson et al., 1980), the mesencephalic melanin-containing neurons, also stained with a Tyr-hydroxylase antibody (Pearson et al., 1979; Gaspar et al., 1983) are surrounded by MET-ENK-LIr nerve terminals (Gaspar et al., 1983). The close contact between DA neurons and MET-ENK-LIr nerve terminals does not necessarily imply a functional connection between the two systems. Yet, an interaction is strongly suggested in the substantia nigra. The number of opiate receptor sites (estimated by the B_{max}, using [^{3}H](D-Ala^2-Met^5)-enkephalinamide as ligand) is reduced by 30–40% in the substantia nigra of patients with Parkinson's disease with no modification of the K_d (Javoy-Agid et al., 1982a; Llorens-Cortes et al., 1984), in agreement with results obtained in the rat (Llorens-Cortes et al., 1979). Enkephalinase activity was reduced to the same extent, indicating that the specific degradative enzyme of MET-ENK may be located postsynaptically (Javoy-Agid et al., 1982a; Llorens-Cortes et al., 1984). The biochemical and immunohistochemical data suggest the existence of close relationships between DA and MET-ENK systems in the human brain and support the hypothesis of a connection between MET-ENK fibres and DA cell bodies in the ventral mesencephalon. Although the dramatic decrease in peptide levels in the substantia nigra and in the ventral tegmental area may reflect neuronal damage (a reduced MET-ENK innervation has been visualized in the substantia nigra of one case of Parkinson's disease (Hunt et al., 1983), it may rather indicate a modification of MET-ENK metabolism. Indeed, immunohistochemical studies evidence a normal nigral MET-ENK staining in most cases of Parkinson's disease (Grafe et al., 1985). The biochemical data obtained in the striatum suggest a loss in striatal MET-ENK inputs to nigra in Parkinson's disease similar to that proposed in Huntington's chorea (Emson et al., 1980a). The implication of the mesencephalic and putamino-pallidal MET-ENK neuronal loss in the occurrence of Parkinsonian symptoms is not known yet.

Leu-enkephalin

Interest in LEU-ENK systems in the brain of patients with Parkinson's disease is related to two main observations: first, the regional distribution of LEU-ENK-like immunoreactivity (LEU-ENK-LI) in human brain resembles that of MET-ENK-LI (Kubek and Wilber, 1980; Taquet et al., 1983) which is itself very similar to that of DA and thus anatomical relationships may exist between the DA and LEU-ENK systems; second, the ratio of MET- to LEU-ENK levels throughout the brain varies (from 2 to 5 in most brain areas, but reaches 1 and 12 in the hippocampus and the nucleus accumbens, respectively (Taquet et al., 1983)), suggesting that the two peptides are located in different neuronal systems and might thus have distinct functions.

In Parkinson's disease, levels of LEU-ENK-LI are reduced by 30–40% in the putamino-pallidal complex (Fig. 3) (Taquet et al., 1983). The data indicate that a LEU-ENK system might exist between the putamen and the pallidum in human brain as demonstrated in the rat (Cuello and Paxinos, 1978); the LEU-ENK, like the MET-ENK, putamino-pallidal pathway is affected in Parkinson's disease, suggesting that MET- and LEU-ENK may coexist within the same neurons in these structures. In the substantia nigra and ventral tegmental area, the normal level of LEU-ENK-LI suggests that LEU- in contrast to MET-ENK neurons are spared. The destruction of a subpopulation of LEU-ENK neurons is not excluded, but this neuronal damage could be masked by the presence of a bulk of LEU-ENK neurons of a distinct origin (Taquet et al., 1985). This hypothesis is supported by the recent discovery of the neo-endorphin/dynorphin precursor, or pro-enkephalin-B which contains three copies of LEU-ENK (Kakidani et al., 1982). Dynorphin and neo-endorphin are highly concentrated in the substantia nigra (Maysinger et al., 1982; Taquet et al., 1985). On the other hand, MET- and LEU-ENK-LI (both contained in the precursor pro-enkephalin A) (Comb et al., 1982) predominate in the striatum. It may be put forward that pro-enkephalin-A neurons degenerate while pro-enkephalin-B neurons are spared. Consequently, a decrease in LEU-ENK due to a pro-enkephalin-A neuronal loss could be masked by the relatively higher amounts of LEU-ENK from pro-enkephalin-B origin.

Finally, MET- and LEU-ENK-containing neurons in the putamino-pallidal complex are partly altered (functionally or anatomically), while only MET-ENK neurons seem to be damaged in the ventral mesencephalon. There again, the clinical implication of the local LEU-ENK-LI deficiency is not understood, and no behavioural data are yet available from animal studies.

Somatostatin

Somatostatin-like immunoreactivity (SRIF-LI) is mainly localized in the hypothalamus but is also present in substantial amounts in extra-hypothalamic areas such as cortical structures. A systematic study on the regional distribution of SRIF-LI levels in the brain of parkinsonian patients was undertaken since: (1) a severe reduction in the neocortical SRIF-LI levels has been described in patients with Alzheimer's disease (Davies, 1979; Rossor et al., 1982a), and it is now quite clear that at least one third of the parkinsonian population presents signs of intellectual deterioration (Martilla and Rinne, 1976; Lieberman et al., 1979); (2) a marked decrease in SRIF-LI levels has been observed in the cerebrospinal fluid of parkinsonian patients (Dupont et al., 1982), although no difference was documented between demented and non-demented patients.

No difference in SRIF-LI concentrations is observed in several brain areas between control and parkinsonian subjects as long as the whole population of parkinsonian patients is considered (Epelbaum et al., 1983). However, a significant reduction in SRIF-LI levels is detectable in the cerebral cortex exclusively in demented parkinsonian patients (Fig. 4): SRIF-LI levels are decreased in the frontal, cingulate and entorhinal cortex; in the hippocampus, the biochemical deficit is only observed in severely demented patients; the peptide content in the occipital cortex, caudate nucleus and hypothalamus is not significantly different from controls (Epelbaum

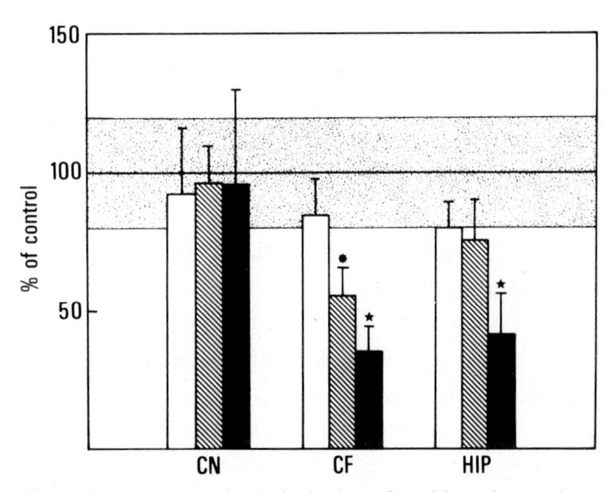

Fig. 4. Somatostatin levels in brains of parkinsonian patients. Somatostatin concentrations (mean ± S.E.M.) are expressed as percentage of respective control values (horizontal bar) (Epelbaum et al., 1983). Data were obtained on control (13–15), and 7 non-demented (□), 9 demented (▨) and 4 severely demented (■) parkinsonian brains. Abbreviations are detailed in the legend to Fig. 1.

et al., 1983). The reduction in SRIF-LI levels, restricted to the cerebral cortex seems to represent a common biochemical characteristic of dementia in patients with Alzheimer's or Parkinson's disease. Since SRIF-LI has been shown to be contained in intrinsic cortical neurons (Sorensen, 1982), intracortical neurons are most likely damaged in demented parkinsonian patients. This abnormality may play a role in the genesis of intellectual impairment.

Neuropeptides apparently unaffected in the brain of parkinsonian patients

A number of peptide systems seem undamaged in the brain of parkinsonian patients. This is the case for TRH (Javoy-Agid et al., 1983a). The concentrations of this peptide are not different from controls in many brain areas where the peptide is detectable (caudate nucleus, putamen, nucleus accumbens, pallidum, substantia nigra, ventral tegmental area, hypothalamus, amygdala, periaqueductal grey, red nucleus). Similarly, no change is reported for vasopressin-like immunoreactivity in substantia nigra, periaqueductal grey matter, locus coeruleus, and pallidum of the parkinsonian brain as compared to controls (Rossor et al., 1982b).

The existence of apparently intact peptide systems in the brain of patients is important since it demonstrates some specificity in the peptide deficiencies found in Parkinson's disease.

Functional significance of brain peptide deficiencies in Parkinson's disease

In Parkinson's disease, several peptidergic systems are affected in discrete brain areas, primarily the basal ganglia (CCK-8, SP, MET- and LEU-ENK) and the cerebral cortex (SRIF), whereas some are probably anatomically intact (TRH and perhaps vasopressin).

(1) A crucial question remains, and that concerns the physiological significance of these peptide abnormalities: do they reflect a degeneration of the corresponding neurons or a metabolic change, possibly related to the DA denervation?

These peptide deficiencies may reflect neuronal degeneration. In animals, for example, electrolytic lesions of the putamen are followed by a loss in LEU-ENK-LIr nerve terminals in the pallidum (Cuello and Paxinos, 1978). It must be kept in mind that neuronal damage in Parkinson's disease is extensive, and not restricted to DA-containing neurons, indicating that other biochemical abnormalities may be predicted in the disease. Besides the catecholamine-containing areas well described for their neuronal loss (Escourolle et al., 1970), one should also consider brain areas in which Lewy bodies (possible markers of degenerative processes) have been visualized. Some lesions may be responsible for the degeneration of peptide-containing systems. Combined immunohistochemical and biochemical studies should help to answer the question of a neuronal degeneration (although the former technique probably does not at present allow a suitable quantitative estimation). The recent demonstration of a massive diminution in the nigral MET-ENK-LI innervation of a parkinsonian patient, using anti-MET-ENK-antibody staining, supports this point of view (Hunt et al., 1983).

An alteration in peptide turnover responsible for the change in peptide levels must be considered. In animal brain, peptide concentrations are modified by pharmacological or surgical manipulations, for instance of the DA nigro-striatal system (Costa et al., 1978; Thal et al., 1983). Metabolic changes could account for moderate reductions in peptide contents, as observed for SP and MET-ENK-LI in the pallidum (Fig. 2) since immunohistochemical staining is not different from control brains in Parkinson's disease (Grafe et al., 1985). The changes may be related to the DA deficiency. Along with this idea, the reduction in putamen of MET- and LEU-ENK-LI which contrasts with the normal caudate nucleus content might be consequent to the more severe DA denervation in the former than in the latter structure (Bernheimer et al., 1973; Javoy-Agid et al., 1983b; Scatton et al., 1983). The data fit well with the clear hypersensitivity of the DA D2 receptor sites which develops in the putamen, and not in the caudate nucleus of these patients (Bokobza et al., 1984). However, hypersensitive caudate nucleus D2 receptors have also been observed by others (Rinne et al., 1981). In terms of pathophysiology, the data indicate a difference between the putamen and the caudate nucleus function in Parkinson's disease. If some of the peptide deficiencies observed in parkinsonian brains may be taken for metabolic changes, several questions still arise. (a) Is this metabolic change a consequence of the degeneration of other neurotransmitter systems (DA or others)? (b) Does it signify an increase or a decrease in the peptide turnover? By analogy with other rapid turnover-rate neurotransmitters, such as acetylcholine, a reduction in peptide concentrations would preferentially indicate an increased turnover (with enhanced release not fully compensated by synthesis). On the other hand, the reduced levels could be an indication of a lowered metabolic activity. (c) What is the physiological significance of a peptide deficiency? If one refers to the hypothetical modulating function of peptides (coexisting or not with a classical neurotransmitter) the deficit may disrupt the smooth functional synchrony between the mediator and the peptide, thus being a factor of decompensation at the synaptic level.

(2) Another important problem concerns the implication of these peptide deficits in the occurrence of parkinsonian symptoms, which, in addition, raises the question of the functional role of peptidergic systems. Alterations in peptidergic pathways may contribute to the appearance of clinical disorders in patients through different mechanisms. (a) Changes in peptide transmission may, per se, be responsible for symptoms, i.e. the cortical SRIF deficiency and dementia (Epelbaum et al., 1983). At present, this relation seems quite selective, since intellectual deterioration in patients is not correlated to modifications in other neuropeptides levels (SP, CCK-8, MET- and LEU-ENK, vasointestinal peptide) (Jegou et al., 1985). SRIF-LI concentrations are also reduced in the cerebral cortex of patients with Alzheimer's disease (Davies, 1979; Rossor et al., 1982a) indicating that damage to the cortical SRIF-LIr neurons probably represents a clue to the appearance of dementia. From a biochemical point of view, the analogy between dementia in Parkinson's and Alzheimer's disease must be extended to the cortical cholinergic systems as well. Choline acetyltransferase activity (an index of the cholinergic innervation) is reduced in cortical areas of parkinsonian patients when compared to control subjects, the reduction being maximal in demented patients (Ruberg et al., 1982; Dubois et al., 1983). Such results emphasize the nosological similarity between Alzheimer's and Parkinson's type of dementia as far as neurotransmitters are concerned and prompt further investigations on anatomo-clinical and biochemical correlations. (b) Peptidergic systems may mediate physiological changes via other neurotransmitter systems. In Parkinson's disease, for example, SP-, MET-ENK- and CCK-LIr neurons are affected within the substantia nigra, in addition to the massive DA neuronal loss. The peptidergic abnormality most likely interferes with DA transmission and may contribute to the symptomatology. According to studies in animals, these three peptides have an excitatory effect on the nigral DA neurons (Davies and Dray, 1976; Waldmeier et al., 1978; Pert et al., 1979; Skirboll et al., 1981). From a pathophysiological point, depending on their significance (discussed above), these peptide deficiencies may contribute to the aggravation of symptoms (reduced activity) or be a regulatory process to overcome the deficiency in DA transmission (increased utilization). The selectivity in the biochemical changes [many other neurotransmitter systems are intact in the mesencephalon of patients (i.e. choline acetyltransferase, glutamic acid decarboxylase, TRH, histidine decarboxylase, LEU-ENK containing neurons (Javoy-Agid et al., 1982a, 1984a))] further suggests an involvement of these three peptides in the parkinsonian symptomatology. (c) Peptidergic systems may be indirectly implicated in parkinsonian symptoms as a result of a functional imbalance between anatomically intact (i.e. normal peptide levels) and damaged (i.e. DA deficiency) systems. The hypothalamus of parkinsonian patients may represent, from this point of view, an interesting example: in contrast to the

50–60% reduction in DA levels, many other classical neurotransmitters and peptides are found in normal concentrations in the hypothalamus (taken as a whole) of parkinsonian patients (Javoy-Agid et al., 1984b). The resulting functional imbalance might play a role in the genesis of various endocrine abnormalities and autonomic symptoms.

In conclusion, Parkinson's disease is characterized biochemically by the degeneration of the DA ascending neurons consequent to the lesion of the substantia nigra and ventral tegmental area and of noradrenergic, serotoninergic and cholinergic ascending systems which results from lesions in the locus coeruleus, raphe nuclei and substantia innominata, respectively (for review see Javoy-Agid et al., 1984). This biochemical map of the parkinsonian brain must be extended to the neuropeptide deficiencies described above. The significance of most of these peptide deficiencies remains and will remain a mystery as long as the function of peptide systems is not fully understood, unless the observed abnormalities per se contribute to an understanding of the role of peptides in the brain.

Acknowledgements

We are indebted to Drs. Beck, Bouchon, Forette, Laurent, Moulias, Piette (Charles Foix Hospital, Ivry s/Seine); Escourolle, Gray, Hauw (Salpêtrière Hospital, Paris); Morel-Maroger (Corbeil-Essonnes Hospital); Rascol, Montastruc (Purpan Hospital, Toulouse); Pollak (CHU La Tronche, Grenoble) for providing human material. The authors wish to thank Drs. Hamon and Grouselle (College de France, Paris) and Legrand, Mauborgne and Studler (CHU Pitié-Salpêtrière, Paris) for their collaboration, and Mrs. J. Sage for typing the manuscript. The study was supported by INSERM (PRC Santé Mentale et Cerveau) and CNRS (Jeune Equipe).

References

Bernheimer, H., Birkmayer, W., Hornykiewicz, O., Jellinger, K. and Seitelberger, F. (1973) Brain dopamine on the striatum of Parkinson and Huntington. Clinical, morphological and neurochemical correlations. *J. Neurol. Sci.*, 20: 415–455.

Blackwood, W. and Corsellis, J. A. N. (Eds.) (1977) *Greenfield's Neuropathology*, Edward Arnold, London, pp. 608–650.

Bokobza, B., Ruberg, M., Scatton, B., Javoy-Agid, F. and Agid, Y. (1984) ^{3}H-Spiperone binding, dopamine and HVA concentrations in Parkinson's disease and supranuclear palsy. *Eur. J. Pharmacol.* (in press).

Comb, M. T., Seeburg, P. H., Adelman, J., Eiden, L. and Herbert, E. (1982) Primary structure of the human Met- and Leu-enkephalin precursor and its mRNA. *Nature*, 295: 663–666.

Costa, E., Fratta, W., Hong, J. S., Moroni, F. and Yang, H. Y. T. (1978) Interactions between enkephalinergic and other neuronal systems. In E. Costa and M. Trabucchi (Eds.), *Advances in Biochemical Psychopharmacology, Vol. 18*, Raven Press, New York, pp. 217–226.

Cuello, A. C. and Paxinos, G. (1978) Evidence for a long leu-enkephalin striopallidal pathway in rat brain. *Nature*, 271: 178–180.

Davies, P. (1979) Neurotransmitter related enzymes in senile dementia of the Alzheimer type. *Brain Res.*, 171: 319–327.

Davies, J. and Dray, A. (1976) Substance P in the substantia nigra. *Brain Res.*, 107: 623–627.

Delwaide, P. J., Maertens, A., Dubois, V. and Schoenen, J. (1983) Neuropeptides modify motoneuronal excitability in man. In *Peptides and Neurological Disease*, August 7–10, Cambridge (abstract).

Dubois, B., Ruberg, M., Javoy-Agid, F., Ploska, A. and Agid, Y. (1983) Degeneration of a subcortico-cortical cholinergic system in Parkinson's disease. *Brain Res.*, 288: 213–218.

Dupont, E., Christensen, S. E., Hansen, A. P., Olivarius, B. de F. and Orskov, H. (1982) Low cerebrospinal fluid somatostatin in Parkinson disease: an irreversible abnormality. *Neurology*, 32: 312–314.

Ehringer, H. and Hornykiewicz, O. (1960) Verteilung von Noradrenalin und Dopamin (3-hydroxytyramin) in Gehirn des Menschen und ihr Verhalten bei Erkrankungen des Extrapyramidalen Systems. *Wien. Klin. Wochenschr.*, 38: 1236–1239.

Emson, P. C., Arregui, A., Clement-Jones, V., Sanberg, B. E. B. and Rossor, M. (1980a) Regional distribution of methionine-enkephalin and substance P-like immunoreactivity in normal human brain and in Huntington's disease. *Brain Res.*, 199: 147–160.

Emson, P. C., Rehfeld, J. F., Langevin, H. and Rossor, M. (1980b) Reduction in cholecystokinin-like immunoreactivity in the basal ganglia in Huntington's disease. *Brain Res.*, 198: 497–500.

Epelbaum, J., Ruberg, M., Moyse, E., Javoy-Agid, F., Dubois, B. and Agid, Y. (1983) Somatostatin and dementia in Parkinson's disease. *Brain Res.*, 278: 376–379.

Escourolle, R., De Recondo, J. and Gray, F. (1970) Etude anatomopathologique de syndromes parkinsoniens. In J. de

Ajuriaguerra (Ed.), *Monoamines, Noyaux Gris Centraux et Syndrome de Parkinson. Symposium Bel Air IV*, Masson, Paris, pp. 173–229.

Gale, J. S., Bird, E. D., Spokes, E. G., Iversen, L. L. and Jessel, T. (1978) Human brain substance P: distribution in controls and Huntington's chorea. *J. Neurochem.*, 30: 633–634.

Gaspar, P., Berger, B., Gay, M., Hamon, M., Cesselin, F., Vigny, A., Javoy-Agid, F. and Agid, Y. (1983) Tyrosine hydroxylase and methionine-enkephalin in the human mesencephalon: immunohistochemical localization and relationships. *J. Neurol. Sci.*, 58: 247–267.

Glowinski, J., Michelot, R. and Cheramy, A. (1980) Role of striato-nigral substance P in the regulation of the activity of the nigro-striatal dopaminergic neurons. In E. Costa and M. Trabucchi (Eds.), *Neural Peptides and Neuronal Communication*, Raven Press, New York, pp. 51–63.

Grafe, M. R., Forno, L. S. and Eng, L. F. (1985) Immunocytochemical studies of substance-P and Met-enkephalin in the basal ganglia and substantia nigra in Huntington's, Parkinson's and Alzheimer's diseases. *J. Neuropathol. Exp. Neurol.*, 44: 47–59.

Gramsch, C., Hollt, V., Mehraein, P., Pasi, A. and Herz, A. (1979) Regional distribution of methionine-enkephalin and beta-endorphin like immunoreactivity in human brain and pituitary. *Brain Res.*, 171: 261–270.

Hökfelt, T., Skirboll, L., Rehfeld, J. F., Goldstein, M., Markey, K. and Dann, O. (1980) A subpopulation of mesencephalic dopamine neurons projecting to limbic areas contains a cholecystokinin-like peptide: evidence from immunohistochemistry combined with retrograde tracing. *Neuroscience*, 5: 2093–2124.

Hornykiewicz, O. (1966) Dopamine (^{3}H-hydroxytyramine) and brain function. *Pharmacol. Rev.*, 18: 925–964.

Hunt, S. P., Peck, R. W. and Rossor, M. N. (1983) Patterns of neuropeptide immunoreactivity in the human brain: changes associated with Huntington's chorea and Parkinson's disease. In *Peptides and Neurological Diseases*, August 7–10, Cambridge (abstract).

Javoy-Agid, F., Ruberg, M., Taquet, H., Studler, J. M., Lloyd, K. G., Garbarg, M., Llorens Cortes, C., Grouselle, D. and Agid, Y. (1982a) Biochemical neuroanatomy of the human substantia nigra (pars compacta) in normal and parkinsonian subjects. In A. J. Friedhoff and T. N. Chase (Eds.), *Gilles de la Tourette Syndrome*, Raven Press, New York, pp. 151–163.

Javoy-Agid, F., Taquet, H., Berger, B., Gaspar, P., Morel-Maroger, A., Montastruc, J. L., Scatton, B., Ruberg, M. and Agid, Y. (1982b) Relations between dopamine and methionine-enkephalin systems in control and parkinsonian brains. In *Proceedings of the 12th World Congress of Neurology, September 20–25, 1981, Kyoto*, Excerpta Medica, Amsterdam, Neurology 568, pp. 187–202.

Javoy-Agid, F., Grouselle, D., Tixier-Vidal, A. and Agid, Y. (1983a) Thyrotropin releasing hormone in brain of patients with Parkinson's disease. *Neuropeptides*, 3: 405–410.

Javoy-Agid, F., Ruberg, M., Taquet, H., Bokobza, B., Agid, Y., Gaspar, P., Berger, B., N'guyen-Legros, J., Alvarez, C., Gray, F., Hauw, J. J., Scatton, B. and Rouquier, L. (1983b) Biochemical neuropathology of Parkinson disease. In *Advances in Neurology. Proceedings of the VII International Symposium on Parkinson Disease, June 1982, Frankfurt*, Raven Press, New York, pp. 189–198.

Javoy-Agid, F., Taquet, H., Cesselin, F., Epelbaum, J., Grouselle, D., Mauborgne, A., Studler, J. M. and Agid, Y. (1984a) Neuropeptides in Parkinson's disease. 5th International Catecholamine Symposium. Goteborg, June 1983. In E. Usdin, A. Carlsson, A. Dahlstrom and J. Engel (Eds.), *Catecholamine, Vol. 3*, Alan Liss, Inc., SMS Publications, New York (in press).

Javoy-Agid, F., Ruberg, M., Pique, L., Bertagna, X., Taquet, H., Studler, J. M., Cesselin, F., Epelbaum, J. and Agid, Y. (1984b) Biochemistry of the hypothalamus in Parkinson disease. *Neurology*, 34: 672–675.

Jegou, S., Javoy-Agid, F., Delbende, C., Ruberg, M., Vaudry, H. and Agid, Y. (1985) Cortical vasoactive intestinal peptide in relation to dementia in Parkinson's disease. *J. Neurol. Neurosurg. Psychiatry* (in press).

Johnson, R. P., Sar, M. and Stumpf, W. E. (1980) A topographic localization of enkephalin on the dopamine neurons of the rat substantia nigra and ventral tegmental area demonstrated by combined histofluorescence immunocytochemistry. *Brain Res.*, 194: 566–571.

Kakidani, H., Furutani, Y., Takahashi, H., Noda, M., Morimoto, Y., Hirose, T., Asai, M., Inayama, S., Nakanischi, S. and Numa, S. (1982) Cloning and sequence analysis of cDNA for porcine β-neo-endorphin/dynorphin precursor. *Nature*, 298: 245–249.

Kanazawa, I., Bird, E., O'Connel, R. and Powell, D. (1977a) Evidence for a decrease in substance P content of substantia nigra in Huntington's chorea. *Brain Res.*, 120: 387–392.

Kanazawa, I., Emson, P. C. and Cuello, A. C. (1977b) Evidence for the existence of substance P-containing fibres in striato-nigral and pallido-nigral pathways in rat brain. *Brain Res.*, 119: 447–453.

Kubek, M. J. and Wilber, J. F. (1980) Regional distribution of leucine-enkephalin in hypothalamic and extrahypothalamic loci of the human nervous system. *Neurosci. Lett.*, 18: 155–161.

Lieberman, A., Dziatolowski, M., Kupersmith, M., Serby, M., Goodgold, A., Korein, J. and Goldstein, M. (1979) Dementia in Parkinson's disease. *Ann. Neurol.*, 6: 355–359.

Llorens-Cortes, C., Pollard, H. and Schwartz, J. C. (1979) Localization of opiate receptors in substantia nigra. *Neurosci. Lett.*, 17: 165–170.

Llorens-Cortes, C., Javoy-Agid, F., Agid, Y., Taquet, H. and Schwartz, J. C. (1984) Enkephalinergic markers in substantia nigra and caudate nucleus from parkinsonian subjects. *J. Neurochem.*, 47: 874–877.

Marley, P. D., Emson, P. C. and Rehfeld, J. F. (1982) Effect of

6-hydroxydopamine lesions of the medial forebrain bundle on the distribution of cholecystokinin in rat forebrain. *Brain Res.*, 252: 382–385.

Martilla, R. J. and Rinne, U. K. (1976) Dementia in Parkinson's disease. *Acta Neurol. Scand.*, 54: 431–441.

Mauborgne, A., Javoy-Agid, F., Legrand, J. C., Agid, Y. and Cesselin, F. (1983) Decrease of substance P-like immunoreactivity in the substantia nigra and pallidum of parkinsonian brains. *Brain Res.*, 268: 167–170.

Maysinger, D., Hollt, V., Seizinger, B. R., Menraein, P., Pasi, A. and Herz, A. (1982) Parallel distribution of immunoreactive α-neo-endorphin and dynorphin in rat and human tissue. *Neuropeptides*, 2: 211–225.

Pearson, J. M., Goldstein, M. and Brandas, L. (1979) Tyrosine hydroxylase immunohistochemistry in the human brain. *Brain Res.*, 165: 333–337.

Pert, A., Dewald, A., Liao, H. and Sivit, C. (1979) Effects of opiates and opioid peptides on motor behaviors. Sites and mechanisms of action. In E. Usdin, W. E. Bunney and N. S. Kline (Eds.), *Endorphins in Mental Health Research*, Oxford University Press, Oxford, pp. 45–61.

Pickel, V. M., Sumal, K. K., Beckley, S. C., Miller, R. J. and Reis, D. J. (1980) Immunocytochemical localization of enkephalin in the neostriatum of rat brain: a light and electron microscopic study. *J. Comp. Neurol.*, 189: 721–740.

Price, K. S., Farley, I. J. and Hornykiewicz, O. (1978) Neurochemistry of Parkinson's disease: relation between striatal and limbic dopamine. In P. J. Roberts, G. N. Woodruff and L. L. Iversen (Eds.), *Dopamine. Advances in Biochemical Psychopharmacology, Vol. 19*, Raven Press, New York, pp. 293–309.

Rinne, U. K. and Sonninen, V. (1973) Brain catecholamines and their metabolites in parkinsonian patients. *Arch. Neurol.*, 28: 107–110.

Rinne, U. K., Koskinen, V. and Lonnberg, P. (1980) Neurotransmitter receptors in the parkinsonian brain. In U. K. Rinne, M. Klinger and M. Stamm (Eds.), *Parkinson's Disease. Current Progress, Problems and Management*, Elsevier/North Holland, Amsterdam, pp. 93–107.

Rossor, M. N., Emson, P. C., Iversen, L. L., Mountjoy, C. Q., Roth, M., Fahrenkrug, J. and Rehfeld, J. F. (1982a) Neuropeptides and neurotransmitters in cerebral cortex in Alzheimer's disease. In S. Corkin et al. (Eds.), *Alzheimer's Disease: a Report of Progress*, Raven Press, New York, pp. 15–23.

Rossor, M. N., Hunt, S. P., Iversen, L. L., Bannister, R., Hawthorn, J., Ang, V. T. Y. and Jenkins, J. S. (1982b) Extrahypothalamic vasopressin is unchanged in Parkinson's disease and Huntington's disease. *Brain Res.*, 253: 341–343.

Ruberg, M., Ploska, A., Javoy-Agid, F. and Agid, Y. (1982) Muscarinic binding and choline acetyltransferase in parkinsonian subjects with reference to dementia. *Brain Res.*, 232: 129–139.

Scatton, B., Javoy-Agid, F., Rouquier, L., Dubois, B. and Agid, Y. (1983) Reduction of cortical dopamine, noradrenaline, serotonin and their metabolites in Parkinson's disease. *Brain Res.*, 275: 321–328.

Skirboll, L. R., Grace, A. A., Hommer, D. W., Rehfeld, J. F., Goldstein, M., Hökfelt, T. and Bunney, B. S. (1981) Peptide-monoamine coexistence: studies of the actions of cholecystokinin-like peptide on the electrical activity of midbrain dopamine neurons. *Neuroscience*, 6: 2111–2124.

Sorensen, K. V. (1982) Somatostatin: localization and distribution in the cortex and sub-cortical white matter of human brain. *Neuroscience*, 7: 1227–1232.

Spokes, E. G. S. (1981) The neurochemistry of Huntington's chorea. *TINS*, 35: 115–118.

Staines, W. A., Nagy, J. J., Vincent, S. R. and Fibiger, H. C. (1980) Neurotransmitters contained in the efferents of the striatum. *Brain Res.*, 194: 391–402.

Studler, J. M., Simon, H., Cesselin, F., Legrand, J. C., Glowinski, J. and Tassin, J. P. (1981) Biochemical investigation on the localization of the cholecystokinin octapeptide in dopaminergic neurons originating from the ventral tegmental area of the rat. *Neuropeptides*, 2: 131–139.

Studler, J. M., Javoy-Agid, F., Cesselin, F., Legrand, J. C. and Agid, Y. (1982) CCK-8 immunoreactivity distribution in human brain: selective decrease in the substantia nigra from parkinsonian patients. *Brain Res.*, 243: 176–179.

Taquet, H., Javoy-Agid, F., Cesselin, F. and Agid, Y. (1982a) Methionine-enkephalin deficiency in brains of patients with Parkinson's disease. *Lancet*, 1: 1367–1368.

Taquet, H., Javoy-Agid, F., Cesselin, F., Hamon, M., Legrand, J. C. and Agid, Y. (1982b) Microtopography of methionine-enkephalin dopamine and noradrenaline in the ventral mesencephalon of human control and parkinsonian brains. *Brain Res.*, 235: 303–314.

Taquet, H., Javoy-Agid, F., Hamon, M., Legrand, J. C., Agid, Y. and Cesselin, F. (1983) Parkinson's disease affects differently Met5 and Leu5-enkephalin in the human brain. *Brain Res.*, 280: 379–382.

Taquet, H., Javoy-Agid, F., Giraud, P., Legrand, J. C., Agid, Y. and Cesselin, F. (1985) Dynorphin levels in parkinsonian patients: Leu5-enkephalin production from either proenkephalin A or prodynorphin in human brain. *Brain Res.*, 341: 390–392.

Thal, L. J., Sharpless, N. S., Hirschhorn, I. D., Horowitz, S. G. and Makman, M. H. (1983) Striatal Met-enkephalin concentration increases following nigrostriatal denervation. *Biochem. Pharmacol.*, 32: 3297–3301.

Vanderhaeghen, J. J., Lostra, F., De Mey, J. and Gilles, C. (1980) Immunohistochemical localization of cholecystokinin- and gastrin-like peptides in the brain and hypophysis of the rat. *Proc. Natl. Acad. Sci. U.S.A.*, 77: 1190–1194.

Vanderhaeghen, J. J., Lostra, F., Vierendeels, G., Deschepper, C., Verhas, M., Verbanck, P. and Gilles, C. (1982) Cholecystokinin in the central nervous system: relationship with cerebral cortex, dopaminergic and limbic systems, spinal cord and hypothalamo-hypophyseal pathways. In S. Katsuki, T. Tsubaki and Y. Toyokura (Eds.), *Neurology, Proceedings 12th World Congress of Neurology, Kyoto, 1981*, Excerpta Medica, Amsterdam, pp. 298–311.

Waldmeier, P. C., Kam, R. and Stocklink, K. (1978) Increased dopamine metabolism in rat striatum after infusions of substance P into substantia nigra. *Brain Res.*, 159: 223–227.

SECTION III

Cerebral Cortex

P. C. Emson, M. N. Rossor and M. Tohyama (Eds.),
Progress in Brain Research, Vol. 66.

CHAPTER 6

Morphology and distribution of peptide-containing neurones in the cerebral cortex

John G. Parnavelas

Department of Anatomy and Embryology, University College London, London WC1E 6BT, U.K.

Introduction

The discovery in the last 10 years of a number of small biologically active peptides in the cerebral cortex has come somewhat as a surprise. At present, their physiological role in this area of the brain is unknown. The recent development of microassay methods and the generation of antibodies to a number of these peptides have provided possibilities for studying their precise anatomical distribution. The aim of this chapter is to bring together our current knowledge of the cytology of the cortex with information derived from peptide immunocytochemistry. It would be a monumental task to attempt an adequate description of the morphology and peptide histochemistry of the entire cerebral cortex. I have, therefore, concentrated on the visual cortex of the rat, an area that has been the subject of extensive anatomical, physiological and, recently, neurochemical investigations. I shall briefly mention observations on the localisation of peptides in the same cortical region of other mammals. It is hoped that a comparative approach will provide some clues to the function of these neuroactive substances.

Morphological description of cortical neurones

The primary visual cortex (area 17) of the rat, like the neocortex of other mammals, appears stratified in Nissl-stained coronal sections. Cells in each of six layers tend to have distinct packing density, size, shape and pattern of connections (Lorente de Nó, 1949; Peters, 1981; Parnavelas and McDonald, 1983). This lamination has long been thought to be important in cortical function. Although lamination is clearly noticeable in Nissl-stained material, it tends to be obscured in Golgi preparations by the staining of processes. The Golgi stains, although capricious in nature, have been utilised extensively in studies of cortical cytomorphology since they were first discovered nearly one hundred years ago (Golgi, 1886; Cajal, 1911; Lorente de Nó, 1949). They selectively impregnate a small proportion (approximately 1–5%) of cells in minute detail. Based on observations of Golgi preparations, it is customary to subdivide neurones of the neocortex into two categories: *pyramidal* and *nonpyramidal*. A strict correlation has not yet been found between cell morphology and function in the visual cortex (Kelly and Van Essen, 1974; Gilbert and Wiesel, 1979; Lin et al., 1979; Parnavelas et al., 1983; Parnavelas, 1984).

Pyramidal cells comprise the majority of neurones in the visual cortex and are found in all cor-

All micrographs (except Figs. 1 and 14) and camera lucida drawings are from coronal sections of rat visual cortex. In all light microscope illustrations the pial surface is at the top of the page.
Note added in proof. This chapter reviews literature up to June, 1984.

Fig. 1. Pyramidal neuron in layer II of the rabbit visual cortex. This gold-toned Golgi impregnated cell was photographed with Nomarski optics. × 300. (Courtesy of Dr. L. Müller.)

tical layers except layer I. They are defined by the characteristic and consistent appearance of their perikarya, dendrites and axons. Their features are too well known to require further description (O'Leary, 1941; Lorente de Nó, 1949; Szentágothai, 1973; Parnavelas et al., 1977a) (Fig. 1). Traditionally the pyramidal cell has been regarded as the projection or output neuron of the cortex, although recent evidence in cat visual cortex shows that the axons of some pyramidal cells do not leave the cortex (Gilbert and Wiesel, 1983). Nonpyramidal cells comprise a heteromorphic group of cells distributed in all cortical layers. They are regarded as intrinsic neurones because their axons remain within the visual cortex and form Gray's type II synapses with the perikarya and dendrites of pyramidal and nonpyramidal neurones (Parnavelas et al., 1977b; Feldman and Peters, 1978; Peters and Fairén, 1978). Feldman and Peters (1978) have used dendritic geometry to classify nonpyramidal neurones in the visual cortex of the rat as *multipolar, bitufted* and *bipolar*. The dendrites of these cells may have a modest number of spines or none at all, and on this basis they are sometimes distinguished as spinous or spine-free nonpyramidal neurones.

The preponderance of nonpyramidal cells in the rat visual cortex are multipolar. Their somata vary considerably in shape (spherical, ovoid, polygonal) and their dendrites, which may be spine-free (Fig. 2) or spinous (Fig. 3), display variable patterns. Most are stellate with three or more primary dendrites radiating in all orientations from the cell body, but an appreciable number of multipolar cells exhibit a degree of dendritic polarisation (Parnavelas et al., 1977a; Feldman and Peters, 1978).

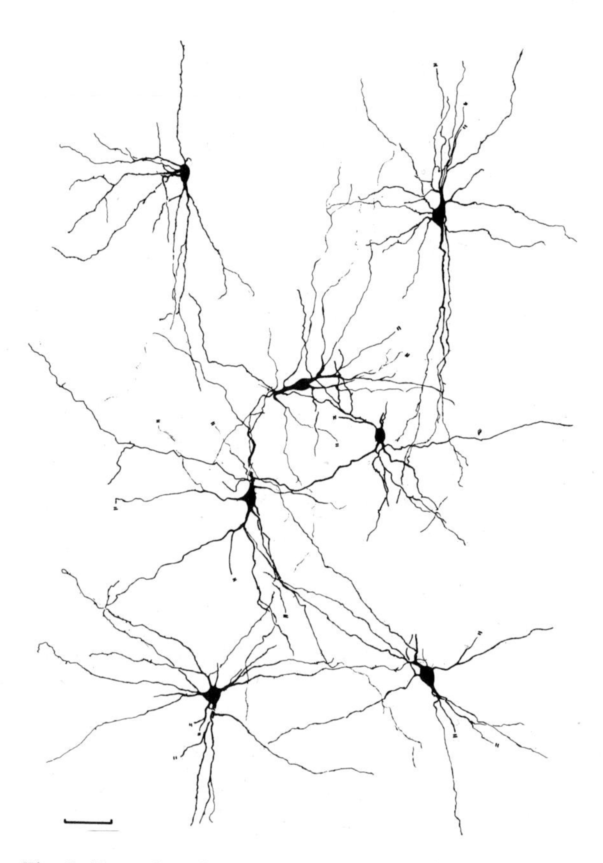

Fig. 2. Examples of spine-free multipolar nonpyramidal neurones from various cortical layers. Bar, 50 μm. (From Feldman and Peters (1978) with permission.)

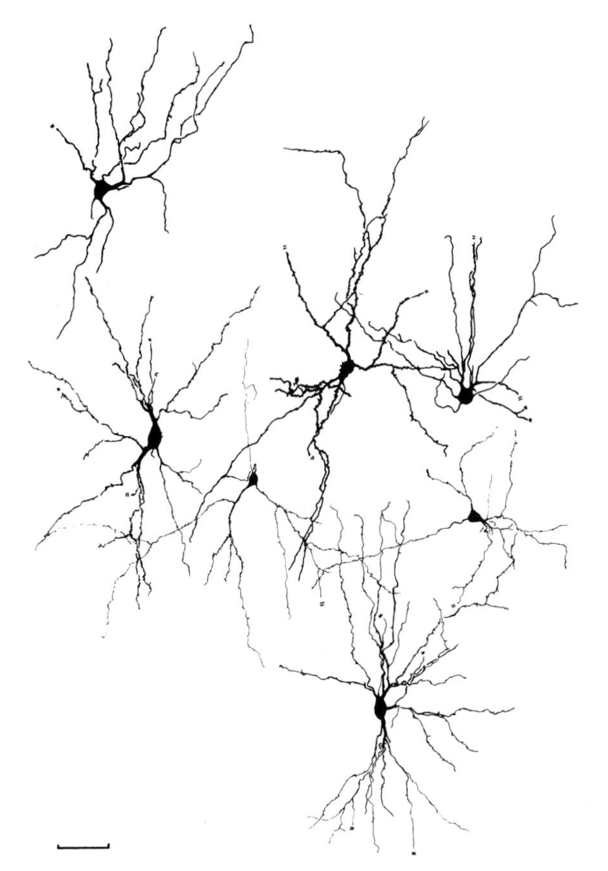

Fig. 3. Examples of sparsely spinous multipolar nonpyramidal neurones from various cortical layers. Bar, 50 μm. (From Feldman and Peters (1978) with permission.)

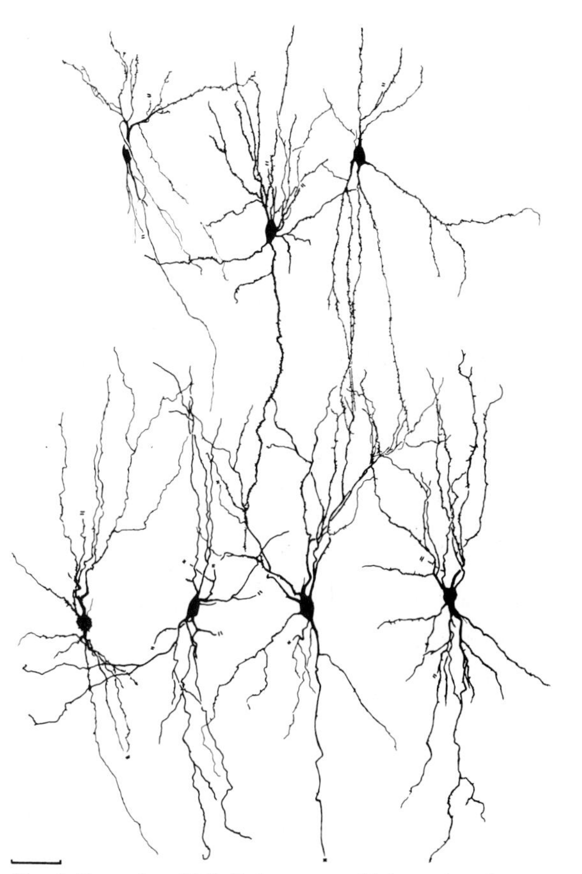

Fig. 4. Examples of bitufted nonpyramidal neurones from various cortical layers. Bar, 50 μm. (From Feldman and Peters (1978) with permission.)

Common among this group are neurones, concentrated in layer VI, with horizontally oriented dendrites. The axonal distribution of multipolar nonpyramidal cells is highly diverse with the majority forming an elaborate local axonal plexus.

Bitufted cells are also present throughout the thickness of the visual cortex. The dendrites, which bear a few spines or none at all, arise from each pole of a vertically elongated soma and branch extensively to form superficial and deep dendritic tufts (Fig. 4). One subgroup of bitufted cells are known as chandelier cells. Their name is derived from the appearance of their terminal axonal arbors which look like candles in a chandelier. Chandelier cells have been shown to form axo-axonal synapses with the axon initial segments of pyramidal neurones (Peters, 1981). Finally, bipolar cells represent the smallest subpopulation of nonpyramidal neurones in the visual cortex and are concentrated in layers II, III and IV. They are distinguished by elongated perikarya and vertically oriented, narrow dendritic fields (Fig. 5).

Rockel et al. (1980) have recently made a comparison of the number of cells in a strip through the entire thickness of the cortex in many structural and functional areas and in a variety of species ranging from mouse to man. They reported that, with the exception of area 17 of primates, the absolute number of neurones through the thickness of the cortex is the same in all areas and in all species, and the

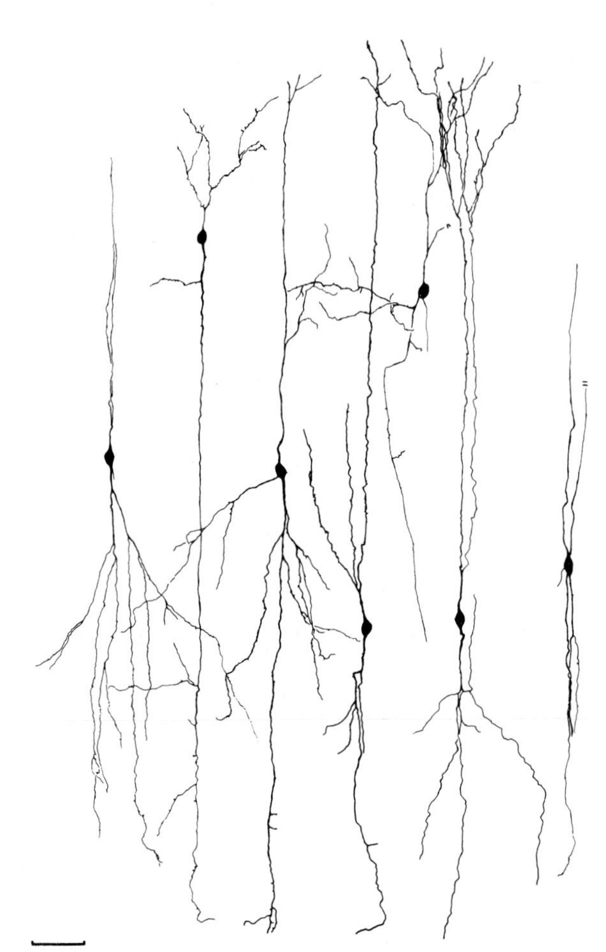

Fig. 5. Examples of bipolar nonpyramidal neurones from various cortical layers. Bar, 50 μm. (From Feldman and Peters (1978) with permission.)

proportions of the two main cell types, pyramidal and nonpyramidal, are similar. However, in area 17 of the primates the number of neurones is approximately 2.5 times greater than in other cortical areas. Although the proportions of the pyramidal and nonpyramidal neurones are roughly the same in a number of diverse cortical areas and in a variety of species, these authors suggest that the details of the form of the cells need not be identical. Comparison of Golgi-stained neurones in the visual cortex of the rat and in the corresponding cortex of more "visual" mammals such as cat and primates reveals no clear differences in the distribution and morphology of pyramidal cells (O'Leary, 1941; Szentágothai, 1973; Parnavelas et al., 1977a). However, significant differences become apparent when comparing the types of nonpyramidal neurones identified in various mammals. For example, large multipolar neurones with dendrites richly endowed with spines and axons that descend to the white matter (category I stellate cells of Szentágothai, 1973) have not been identified in the rat visual cortex but have been observed in cat (Szentágothai, 1973), monkey (Valverde, 1971) and human (Cajal, 1911). Also absent from the rat visual cortex are "basket" cells, a subclass of small ("midget") nonpyramidal cells and probably "double-bouquet" cells, all of which appear frequently in Golgi preparations of cat, monkey and human visual cortex (Cajal, 1911; Valverde, 1971; Szentágothai, 1973; Meyer, 1983). It appears, therefore, that the rat visual cortex contains nonpyramidal neurones whose dendritic and axonal fields are less elaborate than those in the visual cortex of mammals with more highly developed visual systems.

Peptides

Although a large number of peptides have been detected in the rat visual cortex by radioimmunoassay or immunohistochemistry (see Parnavelas and McDonald, 1983), only four have so far been localised unequivocally to intracortical neurones. They are: somatostatin, vasoactive intestinal polypeptide, cholecystokinin and neuropeptide Y.

Somatostatin

The tetradecapeptide somatotropin release-inhibiting factor or somatostatin (SRIF), derived from large molecular weight precursors (Lauber et al., 1979), was first isolated from sheep hypothalamus and shown to inhibit the release of growth hormone from rat pituitary cells (Krulich et al., 1968; Vale et al., 1972; Brazeau et al., 1973). Subsequently it has also been localised by radioimmunoassay and immunohistochemical analyses to a variety of per-

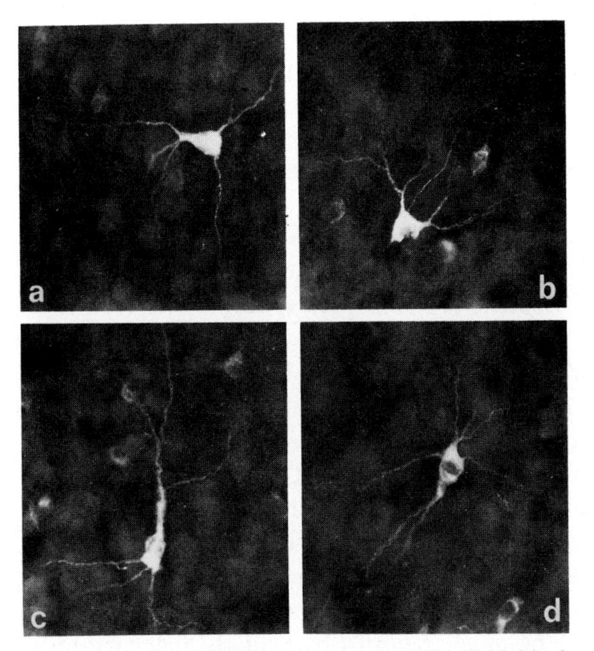

Fig. 6. Examples of SRIF-labelled neurones. (a, b) Multipolar cells in layers V (a) and II (b); (c, d) bitufted cells in layer V. Immunofluorescence technique. × 280. (From McDonald et al. (1982a) with permission.)

ipheral tissues and extrahypothalamic brain including the neocortex (Hökfelt et al., 1974; Brownstein et al., 1975; Epelbaum et al., 1977; Patel and Reichlin, 1978; Finley et al., 1981; Bennett-Clarke et al., 1980; Geola et al., 1981; Shiosaka et al., 1982). Immunohistochemical studies have provided divergent views regarding the cellular localisation and distribution of SRIF-like immunoreactivity (SRIF-LI) in the cerebral cortex of the rat (Bennett-Clarke et al., 1980; Krisch, 1980; Finley et al., 1981; Takatsuki et al., 1981; McDonald et al., 1982a). It is likely that the contradictory findings are due to various antisera utilised which may recognise different forms of SRIF-LI (Rorstad et al., 1979; Patel et al., 1981).

Our observations in the visual cortex of the rat (McDonald et al., 1982a) show that SRIF-like-immunoreactive (SRIF-LIr) neurones comprise approximately 2–3% of the neuronal population. They are almost exclusively nonpyramidal cells whose perikarya are present in layers II through VI. The majority are scattered in layers II and III, and layer IV consistently contains the smallest number of labelled neurones. The large majority of SRIF-LIr cells exhibit morphologies typical of multipolar and bitufted forms of nonpyramidal neurones (Fig. 6), but a very small number of bipolar cells are also present. Various multipolar forms may be recognised: (1) stellate cells of varying sizes (Fig. 6a); (2) cells, chiefly present in layers II and III, which lack descending dendrites (Fig. 6b); and (3) cells, predominantly present in layers V and VI, with horizontally elongated perikarya and dendritic envelopes.

Examination of SRIF-LIr cells with the electron microscope shows that they possess features which are typical of nonpyramidal neurones as described in previous ultrastructural studies of the cerebral cortex of rats (Parnavelas et al., 1977b; Peters and Fairén, 1978) and other mammals (Colonnier, 1968; Sloper, 1973). Their perikarya are round or oval, span a considerable spectrum of sizes, and display abundant organelle-rich cytoplasm. The nuclei, which sometimes show invaginations, are more electron dense than those of pyramidal cells and display a fairly even distribution of chromatin. Particularly characteristic of all SRIF-stained cells, irrespective of soma size, laminar position or dendritic form, is the large concentration of granular endoplasmic reticulum consisting of orderly arrays of often long cisternae (Fig. 7). These cells receive both Gray's type I and type II synapses (the former predominating) on their cell bodies and dendrites, and their axons establish type II synaptic contacts with dendritic shafts of unknown origin (Fig. 13a).

We were interested to examine the distribution and morphology of SRIF-LIr neurones in the cortices of mammals which utilise vision to different extents. It was intriguing to find, using the same antiserum we utilised in our study of the rat visual cortex, that in the hedgehog SRIF-labelled cells are present in all layers except layer I but show a rather pronounced concentration in layers V and VI and in the white matter (Fig. 14a). In the sheep, SRIF-labelled cells are present in all layers with a slight increase in the lower layers, and in the monkey,

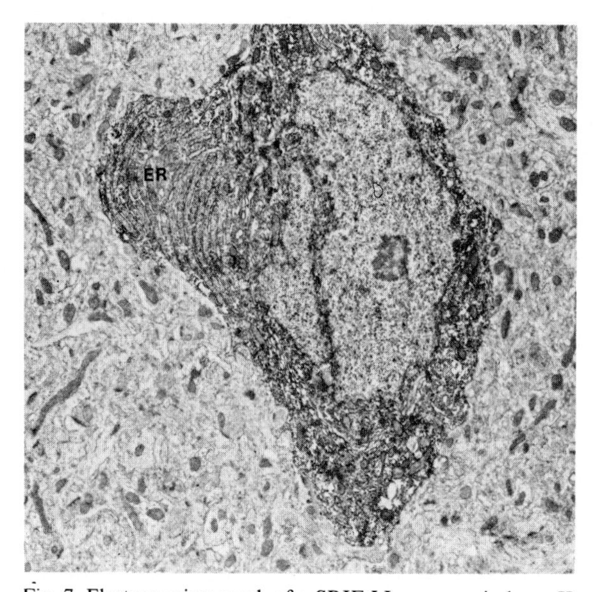

Fig. 7. Electron micrograph of a SRIF-LIr neurone in layer II. The organelle-rich cytoplasm contains an accumulation of organised cisternae of granular endoplasmic reticulum (ER). ×4800.

most labelled cells are distributed in layers II and III and deep in the cortex including the white matter. These observations show that SRIF-LIr neurones in the visual cortices of various animals display significant differences in distribution. These differences may be due to the antiserum recognising different molecular forms of the peptide in various animals or due to differences in intracortical neuronal organisation between animals. However, the overall morphologies of the labelled cells are comparable, i.e. primarily multipolar and bitufted nonpyramidal neurones with many cells in the lower cortical layers displaying horizontal orientation.

Although the role of SRIF in the cerebral cortex is unknown, evidence suggests that it may affect synaptic transmission. It has been shown that SRIF-LI is localised in synaptosomes and its release from isolated cortical neurones in tissue culture, as with neurotransmitters, is calcium dependent (Iversen et al., 1978; Bennett et al., 1979). In addition, cortical receptors have been demonstrated for SRIF and the large precursor form SRIF-28 (Reubi et al., 1981; Strikant and Patel, 1981). The physiological role of SRIF in the visual cortex is uncertain at present. Recent studies have reported both excitatory and inhibitory postsynaptic action in several cortical areas of the rat (Renaud et al., 1975; Olpe et al., 1980; Phillis and Kirkpatrick, 1980; Kelly, 1982).

Vasoactive intestinal polypeptide

Vasoactive intestinal polypeptide (VIP) is a 28-amino acid peptide originally isolated from porcine duodenum by Said and Mutt (1970, 1972). VIP-like immunoreactivity (VIP-LI) is widely distributed in gut, pancreas, genito-urinary tract (Said, 1980) and in the central nervous system (Emson, 1979). Radioimmunoassay and immunohistochemical studies have localised VIP-LI to several sites throughout the brain and reported particularly high concentrations in the neocortex (Fuxe et al., 1977; Fahrenkrug and Schaffalitzky de Muckadell, 1978; Besson et al., 1979; Lorén et al., 1979b; Samson et al., 1979; Hökfelt et al., 1982).

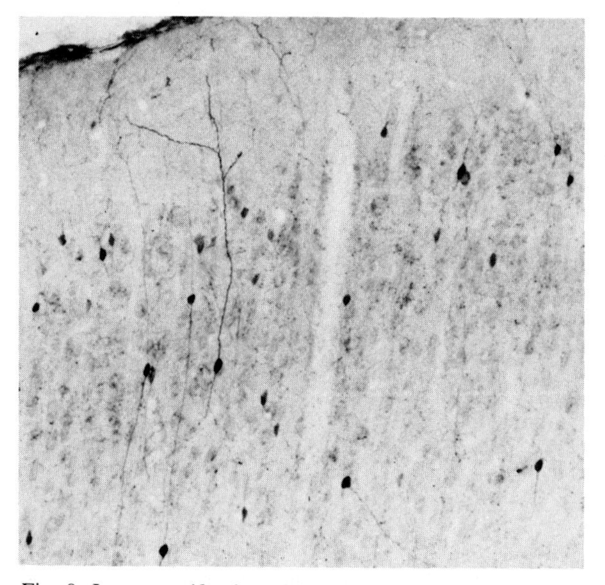

Fig. 8. Low magnification photomicrograph of VIP-LIr neurones in layers I–IV. Note the overall vertical orientation of several dendrites. Peroxidase–antiperoxidase technique. ×110.

Our immunocytochemical studies in the rat visual cortex (McDonald et al., 1982b) have shown that VIP-labelled neurones comprise approximately 3% of the neuronal population. Labelled perikarya are found in layers II through VI with the highest number in layers II and III. They are all nonpyramidal neurones displaying chiefly bipolar morphology (see Feldman and Peters, 1978) (Fig. 8). The ascending dendrites of bipolar cells commonly give rise to a bouquet of varicose branches which course obliquely through the upper portion of layer II and through layer I to form a subpial tuft (Fig. 8). These branches are often seen in close association with pial and cerebral blood vessels. The descending dendrites give rise to a number of oblique and horizontal branches. Other VIP-like immunoreactive (VIP-LIr) neurones are small, multipolar cells located primarily in layers II and VI. A similar picture has been described for the localisation of VIP-LI in other cortical areas of the rat and mouse (Fuxe et al., 1977; Emson et al., 1979; Lorén et al., 1979b; Sims et al., 1980; Emson and Hunt, 1981; Morrison et al., 1984).

Using the electron microscope, we have recently observed that VIP-LIr neurones possess large nuclei and a fairly thin rim of perinuclear cytoplasm. Contained in the cytoplasm are clusters of ribosomes, abundant mitochondria and a rather sparse granular endoplasmic reticulum consisting of short individual cisternae (Fig. 9). The nuclei are usually deeply invaginated and contain small clumps of heterochromatin. The perikarya and dendrites receive primarily type I synapses as well as a small number of type II synaptic contacts. The axons of VIP-LIr neurones, which at times elaborate in the vicinity of the same or other VIP-stained cells, form type II synapses with dendrites and somata of pyramidal and nonpyramidal neurones (Fig. 13b).

Examination of the visual cortex of the squirrel reveals a similar picture for the distribution and morphology of VIP-labelled cells as that of the corresponding cortex of rat and mouse. In the hedgehog visual cortex, VIP-stained somata are situated in layers II through VI with the majority found in layers II and III (Fig. 14b). In the sheep almost all labelled cells are found in the upper three layers, and in the cat visual cortex they are found in all layers with the majority situated in layers II and III (see also Obata-Tsuto et al., 1983). Similar to the rat, most VIP-stained neurones in these species are bipolar in form with dendrites extending perpendicular to the cortical surface. It is interesting that although there are similarities in the distribution and morphology of VIP-LIr cells in the visual cortices of a variety of species, there are also some differences. These include the rather variable proportion of stained cells in the lower cortical layers in

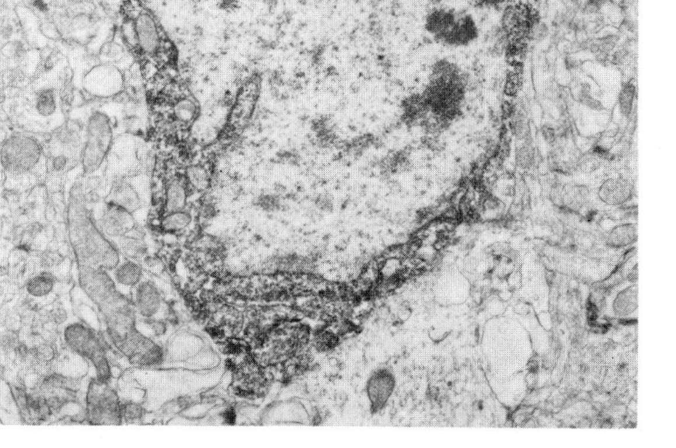

Fig. 9. Electron micrograph of a typical VIP-LIr neurone in layer II. ×4600.

different species and the presence in layer I of the cat visual cortex of a prominent group of cells with long processes directed parallel to the pial surface.

Eckenstein and Baughman (1984) have recently localised choline acetyltransferase (ChAT), the acetylcholine-synthesising enzyme, in nonpyramidal neurones of the rat cerebral cortex whose distribution and morphology were similar to VIP-containing cells. Further analysis of sections alternately processed for ChAT and VIP-LI showed that a substantial proportion of neurones stained for ChAT also contained VIP-LI. These results strongly suggest that ChAT- and VIP-like-immunoreactivities co-exist in cells of the rat cerebral cortex. It has not yet been determined whether all VIP-labelled cells also contain ChAT. These authors also reported an intricate association of fibres containing VIP- or ChAT-like-immunoreactivities with cortical blood vessels. Although their data do not distinguish whether the ChAT-stained fibres originate in the basal forebrain — a rich source of cholinergic projections to the cortex — or from intrinsic ChAT-VIP cells, the observed association suggests that both substances may influence cortical blood flow. The innervation of cortical blood vessels by cholinergic and VIP fibres has been described in detail (Edvinsson et al., 1972, 1980; Larsson et al., 1976), and recent studies have localised specific VIP and muscarinic binding sites in cortical microvessels (Grammas et al., 1983; Huang and Rorstad, 1983). It is now established that VIP is a strong vasodilatory agent (Edvinsson, 1982; Lee et al., 1984). Its presence in cortical interneurones strongly suggests that it may participate in the neural regulation of cerebral blood flow and metabolism (Magistretti et al., 1981; McCulloch and Kelly, 1983).

A number of studies provide evidence that VIP satisfies many of the criteria for a neurotransmitter candidate in the central nervous system. The evidence has been reviewed recently by Fahrenkrug (1980) and Marley and Emson (1982). Briefly, VIP-LI has been found in synaptosomes prepared from cortical tissue and this immunoreactivity can be released by calcium-dependent potassium depolarisation (Giachetti et al., 1977; Besson et al., 1982). Furthermore, VIP receptors have been described and these receptors are linked to an adenylate cyclase (Quick et al., 1978; Borghi et al., 1979; Taylor and Pert, 1979; Staun-Olsen et al., 1982). Finally, recent electrophysiological studies have reported an excitatory action of iontophoretically applied VIP on cortical neurones (Phillis et al., 1978; Phillis and Kirkpatrick, 1980; Kelly, 1982; Lamour et al., 1983).

Cholecystokinin

Cholecystokinin (CCK) is a gastrointestinal hormone discovered in 1928 and later purified as a 33-amino acid polypeptide from porcine intestine (Mutt and Jorpes, 1968). The carboxy terminal of CCK is structurally similar to that of gastrin and, consequently, antibodies raised against gastrin frequently cross react with CCK. The presence of gastrin-like immunoreactivity was demonstrated in the central nervous system in recent years (Vanderhaeghen et al., 1975). This substance was later shown to be predominantly the sulphated cholecystokinin octapeptide (CCK-8) (Dockray, 1976; Rehfeld, 1978; Straus and Yalow, 1978; Larsson and Rehfeld, 1979). Radioimmunoassay and immunohistochemical studies have examined the regional distribution of CCK-LI in the central nervous system of various mammals and established that it is present in particularly high concentrations in the cerebral cortex (Rehfeld, 1978; Emson, 1979; Larsson and Rehfeld, 1979; Barden et al., 1981; Beinfeld et al., 1981; Emson and Hunt, 1981; Emson et al., 1982).

Our studies in the rat visual cortex (McDonald et al., 1982c) have shown that CCK-like immunoreactive (CCK-LIr) cells comprise approximately 1% of the neuronal population in this region. They are nonpyramidal cells whose perikarya are distributed in all cortical layers with the majority identified in layers II and III. Most CCK-LIr cells are of the bitufted variety although a substantial number of multipolar cells and a small number of bipolar cells are also present (Fig. 10). Labelled bitufted cells are present in all layers but predominate

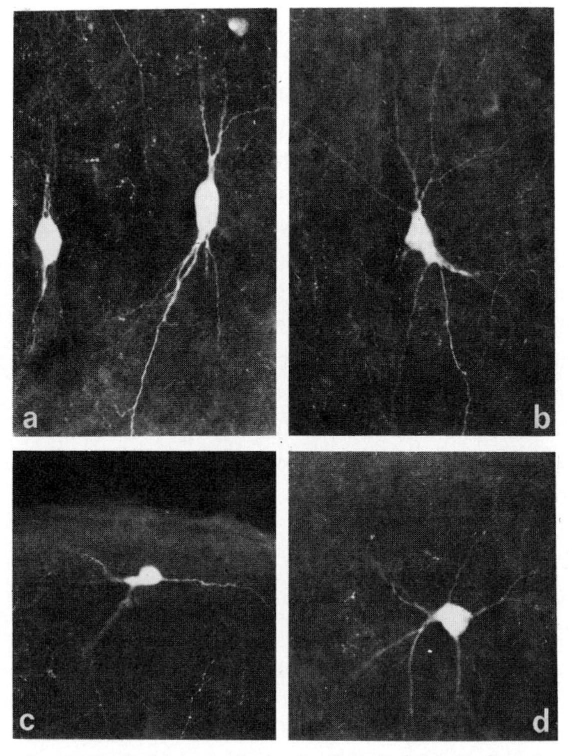

Fig. 10. Bitufted (a), multipolar (b, d) and a horizontally oriented neurone in layer I (c) stained with an antibody against CCK. Immunofluorescence technique. ×280. (From McDonald et al. (1982c) with permission.)

in layers II and III (Fig. 10a). Their somata are typically elongate and large in size and give rise to one or two primary dendrites from each pole. These dendrites usually divide into branches close to the soma to form superficial and deep dendritic tufts of almost equal length. In our studies, cells which displayed such superficial and deep tufts were considered bitufted irrespective of the ratio of the major and minor axes of the dendritic envelope. This differs from the scheme of Feldman and Peters (1978) who arbitrarily classified cells as bipolar if this ratio was 3:1 or greater. Differences in the classification of cells as bipolar or bitufted may account for the differences in the forms of CCK-labelled cells described by various investigators (Emson and Hunt, 1981, 1984; Hendry et al., 1983; Peters et al., 1983), although this may be attributed to a variety of antisera utilised or to staining which lacks sufficent detail to characterise cell forms accurately. CCK-labelled neurones are observed routinely in layer I which is sparsely populated by nerve cells. The multipolar (Fig. 10c) and bitufted varieties of labelled cells are present in this layer as well as other forms unique to layer I (see Bradford et al., 1977). It was interesting to note that the distributions and morphologies of CCK-labelled neurones in the visual cortices of hedgehog, squirrel, sheep and cat are comparable to those described for the rat (Fig. 14c).

Recent examination with the electron microscope of a number of CCK-LIr cells taken from throughout the depth of the visual cortex revealed that these cells possess fairly large nuclei which are frequently

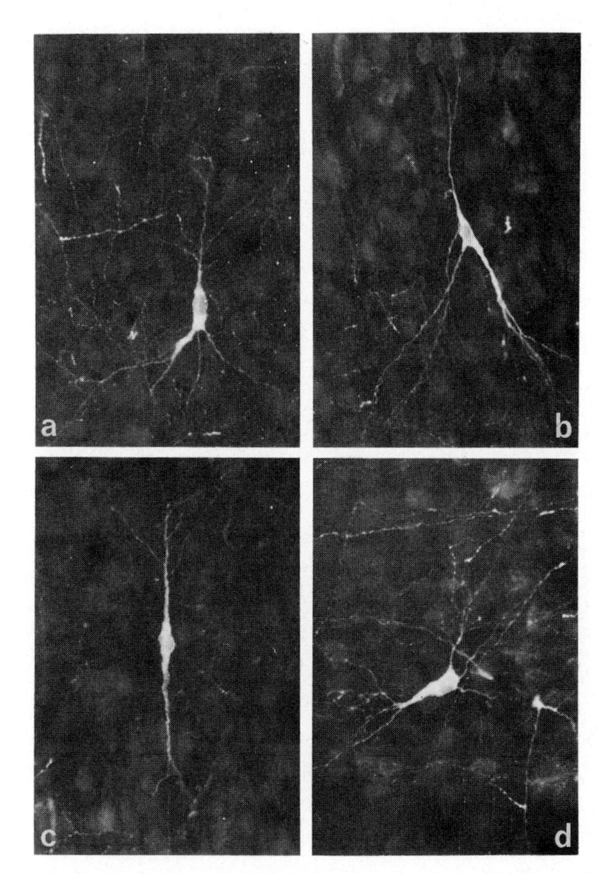

Fig. 11. Examples of NPY-labelled neurones from an 8-day old animal displaying various dendritic forms. (a) Bitufted, layer V; (b, d) multipolar, layer VI; (c) bipolar, layer IV. Immunofluorescence technique. ×280.

deeply invaginated. The cytoplasm typically contains a complement of organelles that usually includes one to three long cisternae of granular endoplasmic reticulum organised concentrically around the nucleus (Fig. 12a). These cells receive both type I and type II axosomatic synapses but no more than three or four in any section through the soma. Particularly notable is the close association and frequent formation of synapses between CCK-stained axons and the perikarya and apical dendrites of cortical pyramidal cells.

Increasing evidence in recent years suggests a neurotransmitter role for CCK in the cerebral cortex, but its precise effects remain elusive. Briefly, CCK-8 appears to be synthesised in neurones and transported to axonal endings from which it can be released by mechanisms similar to those determining the release of other transmitters (Pinget et al., 1979; Dodd et al., 1980; Emson et al., 1980; Emson and Hunt, 1984). Furthermore, investigators have reported a high concentration of CCK receptors in the cortex (Innis and Snyder, 1980; Saito et al., 1980; Hays et al., 1981; Gaudreau et al., 1983). Finally, iontophoretic application of CCK has produced excitatory action on some neurones in a variety of cortical areas in anaesthetised rats (Phillis and Kirkpatrick, 1980; Kelly, 1982; Chiodo and Bunney, 1983; Lamour et al., 1983). Similar effects have been reported in the hippocampus in vitro (Dodd and Kelly, 1981).

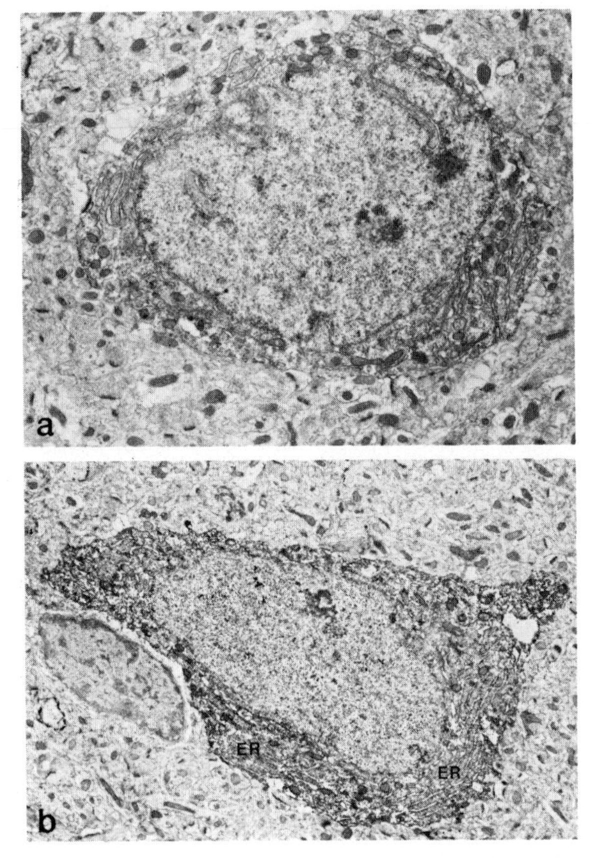

Fig. 12. CCK- (a) and NPY-labelled (b) neurones in layer III. Note the prominent array of granular endoplasmic reticulum (ER) contained in the cytoplasm of the NPY-containing cell. (a) × 5300; (b) × 3900.

Neuropeptide Y

Avian pancreatic polypeptide (APP) is a 36-amino acid peptide isolated from chicken pancreas and characterised by Kimmel et al. (1975). APP and one of its mammalian homologues, bovine pancreatic polypeptide (BPP), have been localised by immunocytochemistry in various sites throughout the brain and in the peripheral nervous system (Larsson et al., 1974; Lorén et al., 1979a; Lundberg et al., 1980; Hunt et al., 1981; Hökfelt et al., 1981; Olschowka et al., 1981; Jacobowitz and Olschowka, 1982; Card et al., 1983). However, neither APP-like nor BPP-like immunoreactivities have been detected in significant quantities in mammalian brain by assay. Recently, another 36-amino acid peptide, neuropeptide Y (NPY), was isolated from porcine brain (Tatemoto, 1982; Tatemoto et al., 1982). It shares major sequence homologies with APP and with other pancreatic polypeptides (PP), and is most likely the endogenous PP in the mammalian brain (Allen et al., 1984; Emson and Hunt, 1984). A number of immunohistochemical studies have clearly shown that the pattern of immunoreactivity observed in the brain with an antiserum to APP is identical to that obtained with NPY antisera (Allen et al., 1984; Moore et al., 1984; also our observations in the rat visual cortex). Consequently, the PP-like immunoreactivity described in the literature will be referred to here as NPY-like immunoreactivity (NPY-LI).

Several studies have described NPY-LI in the cerebral cortex (Lorén et al., 1979a; McDonald et al., 1982d; Vincent et al., 1982a,b; Dawbarn et al., 1984). Our observations in the rat visual cortex (McDonald et al., 1982d) have shown that NPY-labelled cells comprise approximately 1–2% of the neuronal population and are distributed in layers II to VI with a slight increase in concentration in the deeper layers. The overwhelming majority of labelled cells exhibit morphologies characteristic of multipolar, bitufted and bipolar varieties of non-pyramidal neurones (Fig. 11). However, a few immunoreactive pyramidal neurones are also present. It is interesting to note that the morphology of some NPY-LIr cells located throughout the cortex, and particularly a group with horizontally oriented dendrites in the lower layers, closely resembles SRIF-labelled cells in the visual cortex (McDonald et al., 1982a). This is expected in view of the fact that NPY and SRIF have been shown to coexist in some neurones in the rat and human cerebral cortex (Vincent et al., 1982a,b). Furthermore, our electron microscopic observations have revealed that the nuclear, cytoplasmic and synaptic features exhibited by NPY-labelled cells are similar to those displayed by SRIF-containing neurones (Fig. 12b). As expected, the axons of these cells make predominantly type II synapses within the visual cortex (Fig. 13d). It is not known at present what percentage of NPY-containing cells also stain with antiserum directed against SRIF.

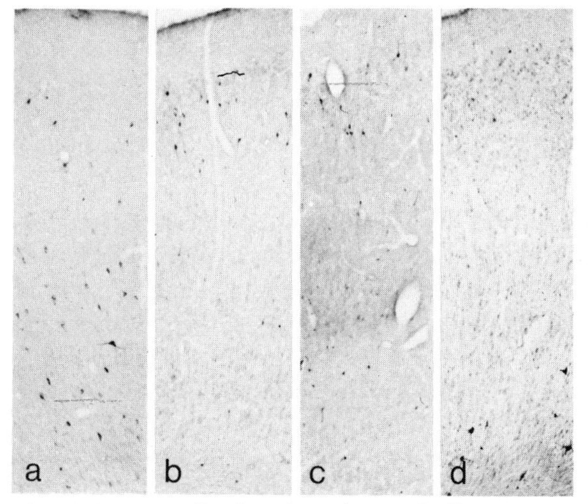

Fig. 14. Low magnification photomicrographs illustrating the distribution of SRIF- (a), VIP- (b), CCK- (c) and NPY- (d) LIr cells in coronal sections through the visual cortex of hedgehog. × 36. (Courtesy of Dr. G. Papadopoulos.)

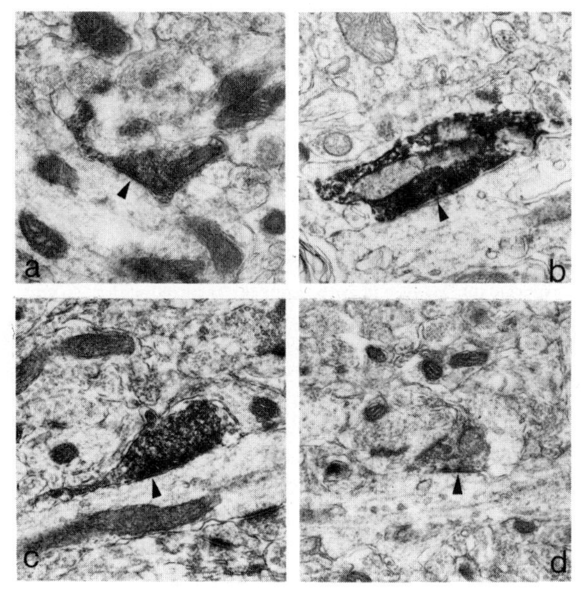

Fig. 13. SRIF- (a), VIP- (b), CCK- (c) and NPY- (d) LIr axon-terminals forming Gray type II axo-dendritic synaptic contacts. × 15 700.

In the visual cortex of the hedgehog, NPY-LIr neurones are present in layers II through VI but tend to be concentrated in layer VI and in the white matter (Fig. 14d). This differs from the distribution observed in the sheep where cells are encountered in all layers but predominantly in layers II and III. As for the distribution in cat and monkey visual cortex, labelled cells are predominantly present in layer VI and white matter with an occasional cell seen in the superficial layers including layer I. It appears, therefore, that the distribution of NPY-LIr cells varies significantly among species. In the peripheral nervous system NPY has been shown to inhibit pancreatic secretion and have vasoconstrictor actions (Tatemoto et al., 1982). Its function(s) in the central nervous system remains to be elucidated.

Acknowledgements

The NPY-LIr cells illustrated in Fig. 11 were obtained from tissue prepared in collaboration with Dr. J. K. McDonald using an anti-NPY serum kindly provided by Dr. P. C. Emson. The electron micrographs were taken from tissue prepared in collaboration with Wendy Kelly and Eva Franke. Drs. G. Papadopoulos and N. Brecha kindly provided information and tissue from a number of mammalian species. I wish to thank Drs. L. Müller and G. Papadopoulos for providing illustrations, Shihan Jayasuriya for typing the manuscript and the MRC for financial support (Project grant G8312953N).

References

Allen, Y. S., Adrian, T. E., Allen, J. M., Tatemoto, K., Crow, T. J., Bloom, S. R. and Polak, J. M. (1984) Neuropeptide Y distribution in the rat brain. *Science*, 221: 877–879.

Barden, N., Mérand, Y., Rouleau, D., Moore, S., Dockray, G. J. and Dupont, A. (1981) Regional distributions of somatostatin and cholecystokinin-like immunoreactivities in rat and bovine brain. *Peptides*, 2: 299–302.

Beinfeld, M. C., Meyer, D. K., Eskay, R. L., Jensen, R. T. and Brownstein, M. J. (1981) The distribution of cholecystokinin immunoreactivity in the central nervous system of the rat as determined by radioimmunoassay. *Brain Res.*, 212: 51–57.

Bennett, G. W., Edwardson, J. A., Marcano de Cotte, D., Berelowitz, M., Pimstone, B. L. and Kronheim, S. (1979) Release of somatostatin from rat brain synaptosomes. *J. Neurochem.*, 32: 1127–1130.

Bennett-Clarke, C., Romagnano, M. A. and Joseph, S. A. (1980) Distribution of somatostatin in the rat brain: telencephalon and diencephalon. *Brain Res.*, 188: 473–486.

Besson, J., Rotsztejn, W., Laburthe, M., Epelbaum, J., Beaudet, A., Kordon, C. and Rosselin, G. (1979) Vasoactive intestinal peptide (VIP): brain distribution, subcellular localization and effect of deafferentation of the hypothalamus in male rats. *Brain Res.*, 165: 79–85.

Besson, J., Rotsztejn, W., Poussin, B., Lhiaubet, A. M. and Rosselin, G. (1982) Release of vasoactive intestinal peptide from rat brain slices by various depolarizing agents. *Neurosci. Lett.*, 28: 281–285.

Borghi, C., Nicosia, S., Giachetti, A. and Said, S. I. (1979) Vasoactive intestinal polypeptide (VIP) stimulates adenylate cyclase in selected areas of rat brain. *Life Sci.*, 24: 65–70.

Bradford, R., Parnavelas, J. G. and Lieberman, A. R. (1977) Neurons in layer I of the developing occipital cortex of the rat. *J. Comp. Neurol.*, 176: 121–132.

Brazeau, P., Vale, W., Burgus, R., Ling, N., Butcher, M., Rivier, J. and Guillemin, R. (1973) Hypothalamic polypeptide that inhibits the secretion of immunoreactive pituitary growth hormone. *Science*, 179: 77–79.

Brownstein, M., Arimura, A., Sato, H., Schally, A. V. and Kizer, J. S. (1975) The regional distribution of somatostatin in the rat brain. *Endocrinology*, 96: 1456–1461.

Cajal, S. R. (1911) *Histologie du Systéme Nerveux de l'Homme et des Vertébrés, Vol. II* (translated by S. Azoulay), Maloine, Paris, pp. 599–618.

Card, J. P., Brecha, N. and Moore, R. Y. (1983) Immunohistochemical localization of avian pancreatic polypeptide-like immunoreactivity in the rat hypothalamus. *J. Comp. Neurol.*, 217: 123–136.

Chiodo, L. A. and Bunney, B. S. (1983) Proglumide: selective antagonism of excitatory effects of cholecystokinin in central nervous system. *Science*, 219: 1449–1451.

Colonnier, M. (1968) Synaptic patterns on different cell types in the different laminae of the cat visual cortex. An electron microscope study. *Brain Res.*, 9: 268–287.

Dawbarn, D., Hunt, S. P. and Emson, P. C. (1984) Neuropeptide Y: regional distribution, chromatographic characterization and immunohistochemical demonstration in post-mortem human brain. *Brain Res.*, 296: 168–173.

Dockray, G. J. (1976) Immunochemical evidence of cholecystokinin-like peptides in brain. *Nature*, 264: 568–570.

Dodd, J. and Kelly, J. S. (1981) The actions of cholecystokinin and related peptides on pyramidal neurones of the mammalian hippocampus. *Brain Res.*, 205: 337–350.

Dodd, P. R., Edwardson, J. A. and Dockray, G. J. (1980) The depolarization-induced release of cholecystokinin C-terminal octapeptide (CCK-8) from rat synaptosomes and brain slices. *Regul. Peptides*, 1: 17–29.

Eckenstein, F. and Baughman, R. W. (1984) Two types of cholinergic innervation in cortex, one co-localized with vasoactive intestinal polypeptide. *Nature*, 309: 153–155.

Edvinsson, L. (1982) Vasoactive intestinal polypeptide and the cerebral circulation. In S. I. Said (Ed.), *Vasoactive Intestinal Polypeptide*, Raven Press, New York, pp. 149–168.

Edvinsson, L., Nielsen, K. C., Owman, Ch. and Sporrong, B. (1972) Cholinergic mechanisms in pial vessels. Histochemistry, electron microscopy and pharmacology. *Z. Zellforsch.*, 134: 311–325.

Edvinsson, L., Fahrenkurg, J., Hanko, J., Owman, C., Sundler, F. and Uddman, R. (1980) VIP (vasoactive intestinal polypeptide)-containing nerves of intracranial arteries of mammals. *Cell Tiss. Res.*, 208: 135–142.

Emson, P. C. (1979) Peptides as neurotransmitter candidates in the mammalian CNS. *Prog. Neurobiol.*, 13: 61–116.

Emson, P. C. and Hunt, S. P. (1981) Anatomical chemistry of the cerebral cortex. In F. O. Schmitt, F. C. Worden, G. Adelman and S. G. Dennis (Eds.), *The Organization of the Cerebral Cortex,* MIT Press, Cambridge, MA, pp. 325–345.

Emson, P. C. and Hunt, S. P. (1984) Peptide containing neurones of the cerebral cortex. In E. G. Jones and A. Peters (Eds.), *The Cerebral Cortex, Vol. 2,* Plenum Press, New York, pp. 145–169.

Emson, P. C., Gilbert, R. F. T., Lorén, I., Fahrenkrug, J., Sundler, F. and Schaffalitzky de Muckadell, O. B. (1979) Development of vasoactive intestinal polypeptide (VIP) containing neurones in the rat brain. *Brain Res.*, 177: 437–444.

Emson, P. C., Lee, C. M. and Rehfeld, J. F. (1980) Cholecystokinin octapeptide: vesicular localization and calcium dependent release from rat brain in vitro. *Life Sci.*, 26: 2157–2163.

Emson, P. C., Rehfeld, J. F. and Rossor, M. N. (1982) Distribution of cholecystokinin-like peptides in the human brain. *J. Neurochem.*, 38: 1177–1179.

Epelbaum, J., Brazeau, P., Tsang, D., Brawer, J. and Martin, J. B. (1977) Subcellular distribution of radioimmunoassayable somatostatin in rat brain. *Brain Res.*, 126: 309–323.

Fahrenkrug, J. (1980) Vasoactive intestinal polypeptide. *Trends Neurosci.*, 3: 1–2.

Fahrenkrug, J. and Schaffalitzky de Muckadell, O. B. (1978) Distribution of vasoactive intestinal polypeptide (VIP) in the porcine central nervous system. *J. Neurochem.*, 31: 1445–1451.

Feldman, M. L. and Peters, A. (1978) The forms of non-pyramidal neurons in the visual cortex of the rat. *J. Comp. Neurol.*, 179: 761–794.

Finley, J. C. W., Maderdrut, J. L., Roger, L. J. and Petrusz, P. (1981) The immunocytochemical localization of somatostatin-containing neurons in the rat central nervous system. *Neuroscience,* 6: 2173–2192.

Fuxe, K., Hökfelt, T., Said, S. I. and Mutt, V. (1977) Vasoactive intestinal polypeptide and the nervous system: immunohistochemical evidence for localization in central and peripheral neurons, particularly intracortical neurons of the cerebral cortex. *Neurosci. Lett.*, 5: 241–246.

Gaudreau, P., Quirion, R., St.-Pierre, S. and Pert, C. B. (1983) Tritium-sensitive film autoradiography of [^{3}H]cholecystokinin-5/pentagastrin receptors in rat brain. *Eur. J. Pharmacol.*, 87: 173–174.

Geola, F. L., Yamada, T., Warwick, R. J., Tourtelotte, W. W. and Hershman, J. M. (1981) Regional distribution of somatostatin-like immunoreactivity in the human brain. *Brain Res.*, 229: 35–42.

Giachetti, A., Said, S. I., Reynolds, R. C. and Koniges, F. C. (1977) Vasoactive intestinal polypeptide in brain: localization in and release from isolated nerve terminals. *Proc. Natl. Acad. Sci. U.S.A.*, 74: 3424–3428.

Gilbert, C. D. and Wiesel, T. (1979) Morphology and intracortical projections of functionally characterised neurones in the cat visual cortex. *Nature,* 280: 120–125.

Gilbert, C. D. and Wiesel, T. N. (1983) Clustered intrinsic connections in cat visual cortex. *J. Neurosci.*, 3: 1116–1133.

Golgi, C. (1886) Recherches sur l'histologie des centres nerveux. *Arch. Ital. Biol.*, 3: 285–317.

Grammas, P., Diglio, C. A., Marks, B. H., Giacomelli, F. and Wiener, J. (1983) Identification of muscarinic receptors on rat cerebral cortical microvessels. *J. Neurochem.*, 40: 645–651.

Hays, S. E., Houston, S. H., Beinfeld, M. C. and Paul, S. M. (1981) Postnatal ontogeny of cholecystokinin receptors in rat brain. *Brain Res.*, 213: 237–241.

Hendry, S. H. C., Jones, E. G. and Beinfeld, M. C. (1983) Cholecystokinin-immunoreactive neurons in rat and monkey cerebral cortex make symmetric synapses and have intimate associations with blood vessels. *Proc. Natl. Acad. Sci. U.S.A.*, 80: 2400–2404.

Hökfelt, T., Efendic, S., Johansson, O., Luft, R. and Arimura, A. (1974) Immunohistochemical localization of somatostatin (growth hormone release-inhibiting factor) in the guinea pig brain. *Brain Res.*, 80: 165–169.

Hökfelt, T., Lundberg, J. M., Terenius, L., Jancsó, G. and Kimmel, J. R. (1981) Avian pancreatic polypeptide (APP) immunoreactive neurons in the spinal cord and spinal trigeminal nucleus. *Peptides,* 2: 81–87.

Hökfelt, T., Schultzberg, M., Lundberg, J. M., Fuxe, K., Mutt, V., Fahrenkrug, J. and Said, S. I. (1982) Distribution of vasoactive intestinal polypeptide in the central and peripheral nervous system as revealed by immunocytochemistry. In S. I. Said (Ed.), *Vasoactive Intestinal Polypeptide,* Raven Press, New York, pp. 65–90.

Huang, M. and Rorstad, O. P. (1983) Effects of vasoactive intestinal polypeptide, monoamines, prostaglandins, and 2-chloroadenosine on adenylate cyclase in rat cerebral microvessels. *J. Neurochem.*, 40: 719–726.

Hunt, S. P., Emson, P. C., Gilbert, R., Goldstein, M. and Kimmel, J. R. (1981) Presence of avian pancreatic polypeptide-like immunoreactivity in catecholamine and methionine-enkephalin-containing neurones within the central nervous system. *Neurosci. Lett.*, 21: 125–130.

Innis, R. B. and Snyder, S. H. (1980) Distinct cholecystokinin receptors in brain and pancreas. *Proc. Natl. Acad. Sci. U.S.A.*, 77: 6917–6921.

Iversen, L. L., Iversen, S. D., Bloom, F., Douglas, C., Brown, M. and Vale, W. (1978) Calcium-dependent release of somatostatin and neurotensin from rat brain in vitro. *Nature,* 273: 161–163.

Jacobowitz, D. M. and Olschowka, J. A. (1982) Bovine pancreatic polypeptide-like immunoreactivity in brain and peripheral nervous system: coexistence with catecholaminergic nerves. *Peptides,* 3: 569–590.

Kelly, J. S. (1982) Electrophysiology of peptides in the central nervous system. *Br. Med. Bull.*, 38: 283–290.

Kelly, J. P. and Van Essen, D. C. (1974) Cell structure and function in the visual cortex of the cat. *J. Physiol.*, 238: 515–547.

Kimmel, J. R., Hayden, L. J. and Pollock, H. G. (1975) Isolation and characterization of a new pancreatic polypeptide hormone. *J. Biol. Chem.*, 250: 9369–9376.

Krisch, B. (1980) Differing immunoreactivities of somatostatin in the cortex and the hypothalamus of the rat. *Cell Tiss. Res.*, 212: 457–464.

Krulich, L., Dhariwal, A. P. S. and McCann, S. M. (1968) Stimulatory and inhibitory effects of purified hypothalamic extracts on growth hormone release from rat pituitary in vitro. *Endocrinology*, 83: 783–790.

Lamour, Y., Dutar, P. and Jobert, A. (1983) Effects of neuropeptides on rat cortical neurons: laminar distribution and interaction with the effect of acetylcholine. *Neuroscience*, 10: 107–117.

Larsson, L.-I. and Rehfeld, J. F. (1979) Localization and molecular heterogeneity of cholecystokinin in the central and peripheral nervous system. *Brain Res.*, 165: 201–218.

Larsson, L.-I., Sundler, F., Håkanson, R., Pollock, H. G. and Kimmel, J. R. (1974) Localization of APP, a postulated new hormone, to a pancreatic endocrine cell type. *Histochemistry*, 42: 377–382.

Larsson, L.-I., Edvinsson, L., Fahrenkrug, J., Håkanson, R., Owman, Ch., Schaffalitzky de Muckadell, O. B. and Sundler, F. (1976) Immunocytochemical localization of a vasodilatory polypeptide (VIP) in cerebrovascular nerves. *Brain Res.*, 113: 400–404.

Lauber, M., Camier, M. and Cohen, P. (1979) Higher molecular weight forms of immunoreactive somatostatin in mouse hypothalamic extracts: evidence of processing in vitro. *Proc. Natl. Acad. Sci. U.S.A.*, 76: 6004–6008.

Lee, T. J.-F., Saito, A. and Berezin, I. (1984) Vasoactive intestinal polypeptide-like substance: the potential transmitter for cerebral vasodilation. *Science*, 224: 898–901.

Lin, C.-S., Friedlander, M. J. and Sherman, S. M. (1979) Morphology of physiologically identified neurons in the visual cortex of the cat. *Brain Res.*, 172: 344–348.

Lorén, I., Alumets, J., Håkanson, R. and Sundler, F. (1979a) Immunoreactive pancreatic polypeptide (PP) occurs in the central and peripheral nervous system: preliminary immunocytochemical observations. *Cell Tiss. Res.*, 200: 179–186.

Lorén, I., Emson, P. C., Fahrenkrug, J., Björklund, A., Alumets, J., Håkanson, R. and Sundler, F. (1979b) Distribution of vasoactive intestinal polypeptide in the rat and mouse brain. *Neuroscience*, 4: 1953–1976.

Lorente de Nó, R. (1949) Cerebral cortex: architecture, intracortical connections, motor projections. In J. F. Fulton (Ed.), *Physiology of the Nervous System*, 3rd edn., Oxford University Press, Oxford, pp. 288–330.

Lundberg, J. M., Hökfelt, T., Anggaard, A., Kimmel, J., Goldstein, M. and Markey, K. (1980) Coexistence of an avian pancreatic polypeptide (APP) immunoreactive substance and catecholamines in some peripheral and central neurons. *Acta Physiol. Scand.*, 110: 107–109.

Magistretti, P. J., Morrison, J. H., Schoemaker, W. J., Sapin, V. and Bloom, F. E. (1981) Vasoactive intestinal polypeptide induces glycogenolysis in mouse cortical slices: a possible regulatory mechanism for the local control of energy metabolism. *Proc. Natl. Acad. Sci. U.S.A.*, 78: 6535–6539.

Marley, P. and Emson, P. (1982) VIP as a neurotransmitter in the central nervous system. In S. I. Said (Ed.), *Vasoactive Intestinal Polypeptide*, Raven Press, New York, pp. 341–360.

McCulloch, J. and Kelly, P. A. T. (1983) A functional role for vasoactive intestinal polypeptide in anterior cingulate cortex. *Nature*, 304: 438–440.

McDonald, J. K., Parnavelas, J. G., Karamanlidis, A. N., Brecha, N. and Koenig, J. I. (1982a) The morphology and distribution of peptide-containing neurons in the adult and developing visual cortex of the rat. I. Somatostatin. *J. Neurocytol.*, 11: 809–824.

McDonald, J. K., Parnavelas, J. G., Karamanlidis, A. N. and Brecha, N. (1982b) The morphology and distribution of peptide-containing neurons in the adult and developing visual cortex of the rat. II. Vasoactive intestinal polypeptide. *J. Neurocytol.*, 11: 825–837.

McDonald, J. K., Parnavelas, J. G., Karamanlidis, A. N., Rosenquist, G. and Brecha, N. (1982c) The morphology and distribution of peptide-containing neurons in the adult and developing visual cortex of the rat. III. Cholecystokinin. *J. Neurocytol.*, 11: 881–895.

McDonald, J. K., Parnavelas, J. G., Karamanlidis, A. N. and Brecha, N. (1982d) The morphology and distribution of peptide-containing neurons in the adult and developing visual cortex of the rat. IV. Avian pancreatic polypeptide. *J. Neurocytol.*, 11: 985–995.

Meyer, G. (1983) Axonal patterns and topography of short-axon neurons in visual areas 17, 18 and 19 of the cat. *J. Comp. Neurol.*, 220: 405–438.

Moore, R. Y., Gustafson, E. L. and Card, J. P. (1984) Identical immunoreactivity of afferents to the rat suprachiasmatic nucleus with antisera against avian pancreatic polypeptide, molluscan cardioexcitatory peptide and neuropeptide Y. *Cell Tiss. Res.*, 236: 41–46.

Morrison, J. H., Magistretti, P. J., Benoit, R. and Bloom, F. E. (1984) The distribution and morphological characteristics of the intracortical VIP-positive cell: an immunohistochemical analysis. *Brain Res.*, 292: 269–282.

Mutt, V. and Jorpes, J. E. (1968) Structure of porcine cholecystokinin-pancreozymin. 1. Cleavage with thrombin and with trypsin. *Eur. J. Biochem.*, 6: 156–162.

Obata-Tsuto, H. L., Okamura, H., Tsuto, T., Terubayashi, H., Fukui, K., Yanaihara, N. and Ibata, Y. (1983) Distribution of the VIP-like immunoreactive neurons in the cat central nervous system. *Brain Res. Bull.*, 10: 653–660.

O'Leary, J. L. (1941) Structure of the area striata of the cat. *J. Comp. Neurol.*, 75: 131–164.

Olpe, H.-R., Balcar, V. J., Bittiger, H., Rink, H. and Sieber, P. (1980) Central actions of somatostatin. *Eur. J. Pharmacol.*, 63: 127–133.

Olschowka, J. A., O'Donahue, T. L. and Jacobowitz, D. M. (1981) The distribution of bovine pancreatic polypeptide-like immunoreactive neurons in rat brain. *Peptides,* 2: 309–331.

Parnavelas, J. G. (1984) Physiological properties of identified neurons. In E. G. Jones and A. Peters (Eds.), *The Cerebral Cortex, Vol. 2,* Plenum Press, New York, pp. 205–239.

Parnavelas, J. G. and McDonald, J. K. (1983) The cerebral cortex. In P. C. Emson (Ed.), *Chemical Neuroanatomy,* Raven Press, New York, pp. 505–549.

Parnavelas, J. G., Burne, R. A. and Lin, C.-S. (1983) Distribution and morphology of functionally identified neurons in the visual cortex of the rat. *Brain Res.,* 261: 21–29.

Parnavelas, J. G., Lieberman, A. R. and Webster, K. E. (1977a) Organization of neurons in the visual cortex, area 17, of the rat. *J. Anat.,* 124: 305–322.

Parnavelas, J. G., Sullivan, K., Lieberman, A. R. and Webster, K. E. (1977b) Neurons and their synaptic organization in the visual cortex of the rat. Electron microscopy of Golgi preparations. *Cell Tiss. Res.,* 183: 499–517.

Patel, Y. C. and Reichlin, S. (1978) Somatostatin in hypothalamus, extrahypothalamic brain, and peripheral tissues of the rat. *Endocrinology,* 102: 523–530.

Patel, Y. C., Wheatley, T. and Ning, C. (1981) Multiple forms of immunoreactive somatostatin: comparison of distribution in neural and nonneural tissues and portal plasma in the rat. *Endocrinology,* 109: 1943–1949.

Peters, A. (1981) Neuronal organization in rat visual cortex. In R. J. Harrison (Ed.), *Progress in Anatomy, Vol. 1,* Cambridge University Press, Cambridge, pp. 95–121.

Peters, A. and Fairén, A. (1978) Smooth and sparsely-spined stellate cells in the visual cortex of the rat: a study using combined Golgi-electron microscope technique. *J. Comp. Neurol.,* 181: 129–172.

Peters, A., Miller, M. and Kimerer, L. M. (1983) Cholecystokinin-like immunoreactive neurons in rat cerebral cortex. *Neuroscience,* 8: 431–448.

Phillis, J. W. and Kirkpatrick, J. R. (1980) The actions of motilin, luteinizing hormone releasing hormone, cholecystokinin, somatostatin, vasoactive intestinal peptide, and other peptides on rat cerebral cortical neurons. *Can. J. Physiol. Pharmacol.,* 58: 612–623.

Phillis, J. W., Kirkpatrick, J. R. and Said, S. I. (1978) Vasoactive intestinal polypeptide excitation of central neurons. *Can. J. Physiol. Pharmacol.,* 56: 337–340.

Pinget, M., Straus, E. and Yalow, R. S. (1978) Release of cholecystokinin peptides from a synaptosome-enriched fraction of rat cerebral cortex. *Life Sci.,* 25: 339–342.

Quick, M., Iversen, L. L. and Bloom, S. R. (1978) Effect of vasoactive intestinal peptide (VIP) and other peptides on cAMP accumulation in rat brain. *Biochem. Pharmacol.,* 27: 2209–2213.

Rehfeld, J. F. (1978) Immunochemical studies on cholecystokinin. II. Distribution and molecular heterogeneity in the central nervous system and small intestine of man and dog. *J. Biol. Chem.,* 253: 4022–4030.

Renaud, L. P., Martin, J. B. and Brazeau, P. (1975) Depressant action of TRH, LH-RH and somatostatin on activity of central neurones. *Nature,* 255: 233–235.

Reubi, J.-C., Rivier, J., Perrin, M., Brown, M. and Vale, W. (1981) Somatostatin receptors in brain and pancreas: different pharmacological properties. *Soc. Neurosci. Abstr.,* 7: 431.

Rockel, A. J., Hiorns, R. W. and Powell, T. P. S. (1980) The basic uniformity in structure of the neocortex. *Brain,* 103: 221–244.

Rorstad, O. P., Epelbaum, J., Brazeau, P. and Martin, J. B. (1979) Chromatographic and biological properties of immunoreactive somatostatin in hypothalamic and extrahypothalamic brain regions of the rat. *Endocrinology,* 105: 1083–1092.

Said, S. I. (1980) Peptides common to the nervous system and the gastrointestinal tract. In L. Martini and W. F. Ganong (Eds.), *Frontiers in Neuroendocrinology, Vol. 6,* Raven Press, New York, pp. 293–331.

Said, S. I. and Mutt, V. (1970) Polypeptide with broad biological activity: isolation from small intestine. *Science,* 169: 1217–1218.

Said, S. I. and Mutt, V. (1972) Isolation from porcine-intestinal wall of a vasoactive octacosapeptide related to secretin and glucagon. *Eur. J. Biochem.,* 28: 199–204.

Saito, A., Sankaran, H., Goldfine, I. D. and Williams, J. A. (1980) Cholecystokinin receptors in the brain: characterization and distribution. *Science,* 208: 1155–1156.

Samson, W. F., Said, S. I. and McCann, S. M. (1979) Radioimmunologic localization of vasoactive intestinal polypeptide in hypothalamic and extrahypothalamic sites in the rat brain. *Neurosci. Lett.,* 12: 265–269.

Shiosaka, S., Takatsuki, K., Sakanaka, M., Inagaki, S., Takagi, H., Senba, E., Kawai, Y., Iida, H., Minagawa, H., Hara, Y., Matsuzaki, T. and Tohyama, M. (1982) Ontogeny of somatostatin-containing neuron system of the rat: immunohistochemical analysis. II. Forebrain and diencephalon. *J. Comp. Neurol.,* 204: 211–224.

Sims, K. B., Hoffman, D. L., Said, S. I. and Zimmerman, E. A. (1980) Vasoactive intestinal polypeptide (VIP) in mouse and rat brain: an immunocytochemical study. *Brain Res.,* 106: 165–183.

Sloper, J. J. (1973) An electron microscopic study of the neurons of the primate motor and somatic sensory cortices. *J. Neurocytol.,* 2: 351–359.

Srikant, C. B. and Patel, Y. C. (1981) Somatostatin receptors: identification and characterization in rat brain membranes. *Proc. Natl. Acad. Sci. U.S.A.,* 78: 3930–3934.

Staun-Olsen, P., Ottesen, B., Bartels, P. D., Neilsen, M. H., Gammeltoft, S. and Fahrenkrug, J. (1982) Receptors for vasoactive intestinal polypeptide on isolated synaptosomes from rat cerebral cortex. Heterogeneity of binding and desensitization of receptors. *J. Neurochem.,* 39: 1242–1251.

Straus, E. and Yalow, R. S. (1978) Species specificity of cholecystokinin in gut and brain of several mammalian species.

Proc. Natl. Acad. Sci. U.S.A., 75: 486–489.

Szentágothai, J. (1973) Synaptology of the visual cortex. In R. Jung (Ed.), *Handbook of Sensory Physiology, Vol. VII/3, Central Processing of Visual Information, Part B, Visual Centers of Brain,* Springer-Verlag, Berlin, pp. 269–324.

Takatsuki, K., Shiosaka, S., Sakanaka, M., Inagaki, S., Senba, E., Takagi, H. and Tohyama, M. (1981) Somatostatin in the auditory system of the rat. *Brain Res.,* 213: 211–216.

Tatemoto, K. (1982) Neuropeptide Y: complete amino acid sequence of the brain peptide. *Proc. Natl. Acad. Sci. U.S.A.,* 79: 5485–5489.

Tatemoto, K., Carlquist, M. and Mutt, V. (1982) Neuropeptide Y — a novel brain peptide with structural similarities to peptide YY and pancreatic polypeptide. *Nature,* 296: 659–660.

Taylor, D. P. and Pert, C. B. (1979) Vasoactive intestinal polypeptide: specific binding to rat brain membranes. *Proc. Natl. Acad. Sci. U.S.A.,* 76: 660–664.

Vale, W., Brazeau, P., Grant, G., Nussey, A., Burgus, R., Rivier, J., Ling, N. and Guillemin, R. (1972) Premières observations sur le mode d'action de la somatostatine, un facteur hypothalamique qui inhibi la sécrétion de l'hormone de croissance. *C.R. Acad. Sci. (Paris),* 275: 2913–2916.

Valverde, F. (1971) Short axon neuronal subsystems in the visual cortex of the monkey. *Int. J. Neurosci.,* 1: 181–197.

Vanderhaeghen, J. J., Signeau, J. C. and Gepts, W. (1975) New peptide in the vertebrate CNS reacting with antigastrin antibodies. *Nature,* 257: 604–605.

Vincent, S. R., Johansson, O., Hökfelt, T., Meyerson, B., Sachs, C., Elde, R. P., Terenius, L. and Kimmel, J. (1982a) Neuropeptide coexistence in human cortical neurones. *Nature,* 298: 65–67.

Vincent, S. R., Skirboll, L., Hökfelt, T., Johansson, O., Lundberg, J. M., Elde, R. P., Terenius, L. and Kimmel, J. (1982b) Coexistence of somatostatin- and avian pancreatic polypeptide (APP)-like immunoreactivity in some forebrain neurons. *Neuroscience,* 7: 439–446.

P. C. Emson, M. N. Rossor and M. Tohyama (Eds.),
Progress in Brain Research, Vol. 66.

CHAPTER 7

Cerebral changes in Alzheimer's disease

Fred Plum

Department of Neurology, New York Hospital–Cornell Medical Center, New York, NY 10021, U.S.A.

Epidemiology

Alzheimer's disease, called here "SDAT" (senile dementia, Alzheimer type), if it remains unprevented and untreated, is the coming plague of the 21st century. Its emergence as a major health problem is directly related to improved medical care of organs outside the central nervous system and the resulting increased longevity of populations in Western society creating a population increasingly at risk for senile dementia. Fig. 1 emphasizes this by illustrating what has happened to the United States population since the turn of the 20th century. Gradually but steadily not only has our population increased, but the percentage who live beyond the age of 65 years has proportionately increased by far more. Presently about 11% of Americans are over age 65, and ca. 2.3% are over the age of 80. By 2025, however, we predict that with a population of 300 million, one-fifth will be more than 65 years of age, and almost 5% will be over the age of 80. Given an estimated incidence of socially incapacitating dementia of about 6.5–7% of the population over the age of 65 and ca. 20% of the population over age 80, these figures have devastating social, economic, and ethical implications (Plum, 1979).

Presently, the National Institute of Aging estimates that there are about one and a half million persons in nursing homes in the U.S.A. Not all have SDAT. Nevertheless, the Institute estimates that approximately one-half of those receiving custodial care have lost the cognitive capacity to function independently. One must assume that most of this group has SDAT since all surveys show the disorder to represent more than half of all dementia (Tomlinson et al., 1970). Furthermore, best estimates are that an equal number of patients with advanced SDAT receive care at home by their families and custodians. Placing even a low dollar cost on the care of these unfortunates, this calculates out at an

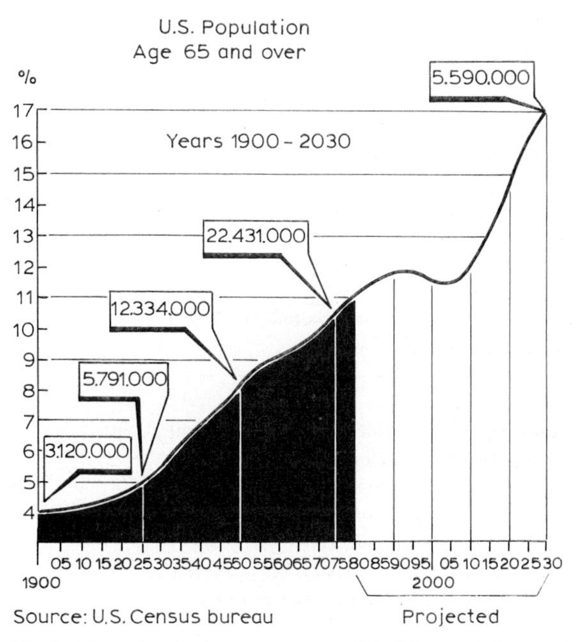

Fig. 1. Chart showing the percentage of the US population aged 65 years and older from 1900 to 1975, with predictions for 1980 to 2030. Reprinted from Plum (1979), with permission of the publisher.

TABLE I

Predicted total and aged population U.S.A. (millions)

	Total	Over 65 years	Over 80 years
1980	227	25 (11%)	5.2 (2.3%)
2000	268	34 (13%)	7.5 (3.8%)
2025	301	60 (20%)	14.5 (4.8%)

annual cost of at least 20 billion dollars. If we take our previous incidence figures, they predict that by the year 2025 we can expect some 4 million persons requiring custodial treatment (Table I) with a predicted cost in 1980 dollars exceeding 60 billion dollars. To place these figures in perspective, the rather generous United States health research budget for 1980, counting all sources, was a little less than 4 billion dollars and had not increased in dollar terms for the previous decade. We are literally in the position of providing a pound of treatment for the ounce that is being spent on preventive research.

Clinical aspects

Clinically, the signs and symptoms of SDAT provide a fascinating repertoire of evidence for relatively selective impairment of the association cortex of the brain (Table II). The onset can be at any age after 9 years but is something of a curiosity before about age 50 to 55, after which it rises in steadily increasing numbers. The onset is insidious, but sometimes families of the patients provide one with hints that behavior had been abnormal in nonincapacitating ways for many years before clinical changes became sufficient to identify them as a disease. In many patients, episodic early warnings can be seen at least several years before more florid symptoms appear, sometimes taking the form of a delirium or psychotic reaction following cataract extractions or general anesthesia. Once clinical symptoms begin they are almost invariably progressive, although pauses in the advance sometimes may last from months to years. Nonetheless, once the illness begins to interfere with work or social compensation, not much more than a year or so goes by without the development of serious social distress.

TABLE II

SDAT — a disease of association cortex: insidious onset, gradual progression, early warnings

Early:
Amnesia-dementia with selective, focal predilections:
- Spatial disorientation
- Constructional–dressing apraxias
- Failure of out-of-context learned acts ("apraxia")
- No sensorimotor defects
- Social amenities and sphincters preserved
- Awareness often present, insight often not

Late:
Global loss of language and cognition;
incontinence of indifference

The first unequivocal symptoms of SDAT are those of impaired memory often selecting distinct and focal psychological functions. In the earlier stages patients forget names, ideas and numbers. Some lose their sense of spatial place. Faces slip from recollection in a manner not easy to distinguish from absentmindedness. Nevertheless, the fact remains in the mind and the initial defect is mainly one of recalling, since a tiny cue usually brings forth the correct answer. Indeed, the only difference I can discern between "benign forgetfulness" and the signs of more malignant Alzheimer's disease is one of degree rather than kind. Some patients show remarkably focal early disturbances. Nominal memory loss occasionally may be so prominent as to suggest an isolated aphasia, whereas others at an early stage develop spatial disorientation or constructional-dressing apraxias. As part of the early symbolic memory failure, acalculia is frequent as is agnosia for body parts. Patients have characteristic and curious difficulties with commands to carry out instructions involving crossing the midline of the body. At the same time early learned skills, such as competitive games, bicycle riding, skiing, or skating, may be well preserved. One of our patients, a former tennis champion, still won the over-60's regional title in his community but did so without any capacity to remember the score or even recall who had won the last point or served. Specific signs of involvement of either corticospinal tract dysfunction or somatosensory or special sensory functions are, in our experience, always lacking unless the patient has had complicating strokes or other diseases. Similarly, seizures have no place in the illness. Patients with Alzheimer's disease preserve their social amenities and sphincters to a remarkable degree in the early stages of the illness. They are hyperalert rather than drowsy, and while commonly aware of their disturbances they often appear to suffer only a shallow emotional response to the awareness. Later on one observes a global loss of language and cognition coupled with the urinary incontinence of indifference. Even at this stage, however, early learned automatic movements appear to be retained for almost as long as one can detect any cognitive capacity whatsoever.

The main problems in differential diagnosis come when neither CT scans nor EEGs show a specific abnormality. The clinical spectrum is not pathognomonic and atypical cases often occur. As alternative diagnoses, multiinfarct dementia, hypertensive vascular disease, psychogenic depression, therapeutic drug toxicity, some of the diffuse angiopathies, or rare examples of Pick's Disease, metabolic deficiency, meningitis, or diffuse brain tumor most often give problems. Most of the time, however, our experience closely parallels that recently reported by Sulkava et al. (1983), who, on clinical grounds, accurately predicted Alzheimer's disease in 22 out of 27 patients who came to autopsy, with several of the undiagnosed five having clinically or pathologically inconsistent findings. The few instances in which we have been either unable to reach a diagnosis or possibly reached an incorrect one in recent years consisted of one instance of autopsy-demonstrated idiopathic diencephalic angiopathy, one of limbic encephalitis associated with cancer, one chronic arteritis confined to the nervous system, and two patients with typical clinical Alzheimer's disease in whom the brain contained no specific pathological changes; of this, more later.

Human metabolic studies

The results of positron emission tomographic studies of glucose and oxidative metabolism in the brain support the concept that SDAT is a disorder of the association cortex. Studies to date find a generalized reduction of cerebral metabolism of 20–30% in early cases which spares only the region of the pre- and postcentral gyrus and possibly the occipital pole. Early cases may show side-to-side differences in hemispheric metabolism. However, averaging among patients shows general metabolic reductions with focal exceptions and accentuations. Thus in the 20 early cases studied at NIH (Chase et al., 1984), the parieto-occipital region underwent the greatest metabolic reduction, whereas the frontal, supplementary motor, and association areas showed a less profound loss. Kuhl et al. (1983) found a similar distribution of maximal hypometabolism. The finding is well diagrammed in Fig. 2 from the NIH group which illustrates the parieto-occipital distribution of the maximal metabolic reduction in their 20 minimal-to-moderately affected patients. The figure illustrates relative sparing of the pre- and postcentral gyrus, the inferior frontal area, and the possibly somewhat greater impairment in

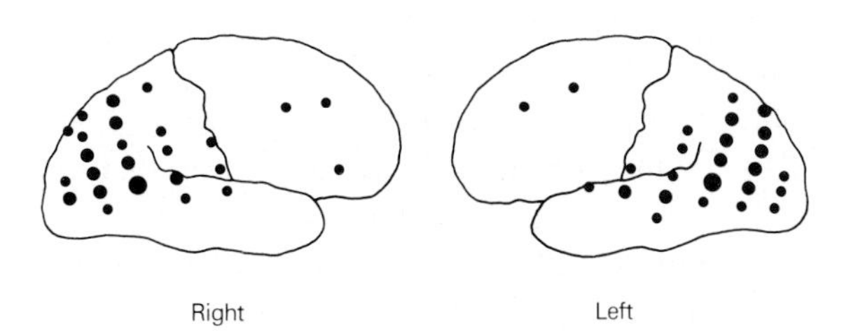

Fig. 2. Reductions in local cortical glucose metabolism in Alzheimer's disease. Filled circles indicate the magnitude of the mean metabolic decrease in each region for 17 patients with Alzheimer's disease as compared with 5 normal controls. Small circles, ≧30% decrease; medium circles, ≧40% decrease; large circles, ≧50% decrease. Reprinted from Chase et al. (1984) with permission of the author and publisher.

the language area on the dominant hemisphere. Actual positron images of the reduced glucose uptake lying in the parieto-occipital distribution can be observed in the Kuhl et al. (1983) study. Frackowiak et al. (1981) similarly found more severe temporo-parietal reduction of metabolism in the early stages of presumed SDAT but noted a profound reduction in frontal metabolism when cognitive loss became severe.

Having outlined the prominent clinical and metabolic evidence for Alzheimer's disease as a disorder of the association cortex, let me briefly review some recent findings on the pathology, pharmacology, and chemistry of the disorder which appear to raise questions or suggest leads that the clinical investigator might pursue in attempting to determine its fundamental cause and cure.

Neuropathology

Structurally, Alzheimer's disease produces abnormalities at various levels of the brain, both cortical and subcortical, neither of which is fully understood. Plaques which at least some researchers (Struble et al., 1982) now consider to be the neuropathological tomb stones of amyloid-cored, dying, pinched-off cholinergic nerve endings plus additional assembled detritus from the decaying cerebral cortex, are observed most prominently in the outer layers of all four lobes of the cortex, the hippocampus, the amygdala, the hypothalamus, and the granular layer of the cerebellum. Neurofibrillary tangles, the silver-staining twisted cords that one finds in larger neurons, are more prominent in younger than in aged patients with SDAT (Terry and Katzman, 1983). When present, they are particularly prominent in the cerebral cortex, the hippocampus, the basal nucleus of Meynert, the amygdala, the hypothalamus, and the pontine tegmentum. Neuropathologists are inconsistent in their reports of grossly visible cerebral atrophy in Alzheimer's disease (Tomlinson et al., 1970). Perhaps because the patients live longer, atrophy has almost always been reported in the brains of the presenile patients who died under age 65 years, usually after a long illness. Over that age the temporal lobe is most consistently described as relatively atrophic, usually having a volume about 18% less than controls (Miller et al., 1980; Brun and Englund, 1981; Hubbard and Anderson, 1981). Even there, however, between 10 and 20% of brains fail to show gross differences from control. Neuronal loss has been described in the cerebral cortex, especially in the large pyramidal cells of the fronto-parietal temporal association areas (Terry et al., 1981), but the laminar distribution of the cell loss is not precisely known. One would suspect that the damaged neurons come mainly from layers 3, 5, and 6, since that is where the large pyramidal cells largely reside in the association cortex, as shown in von Economo's classic diagram (1927). Cell loss is also prominent in the amygdala, the hippocampus, the basal nucleus, and inconsistently in the locus ceruleus, especially in younger cases (Herzog and Kemper, 1980; Tomlinson et al., 1981; Bondareff et al., 1982; Coyle et al., 1983).

The neuropathology of Alzheimer's disease presents several unsolved problems. Neither plaques nor tangles are specific to this disorder, both occurring at least to some degree in aged persons with brains functioning normally for their age (Tomlinson et al., 1970). Tangles and cell loss are also found in boxers following multiple head injuries, particularly as they age past 55 years (Corsellis et al., 1973). Plaques, but not tangles, are encountered in scrapie (Brice and Fraser, 1975) and have been reported in Creutzfeldt–Jakob disease (Masters et al., 1981). No one yet has been able to report whether the tangled cells in Alzheimer's and other diseases are predestined for an early death, although this seems likely since the tangles occur primarily in the large pyramidal cells of the cortex which selectively die out in SDAT. They are most prominent in the large cholinergic cells of areas such as the basal nucleus and amygdala where confident estimates of greater than 50% cell loss have been made post mortem. What has not been established is what the biochemical differences between cortical neurons in association cortex and elsewhere might be or, for that matter, in what quantitative way the cortico-cortical connectivity of these cells differs from less vulnerable areas. A great puzzle is why 8–16% of brains of patients diagnosed clinically as Alzheimer's disease are histologically normal even when, as in the recent study by the Cambridge group, substantial reductions in choline acetyltransferase can be found (Tomlinson et al., 1970; Hubbard and Anderson, 1981; Rossor et al., 1982).

Pharmacology and biochemistry

The pathopharmacology of Alzheimer's disease equally raises some questions. The case for SDAT being a disorder of cholinergic innervation has been made strongly (Coyle et al., 1983). Choline acetyltransferase is reduced in all tested areas of the cortex plus the hippocampus, the caudate, the nucleus accumbens, the amygdala, the septal area, the region of the anterior perforated substance, the anteromedial nucleus of the thalamus, probably the hypothalamus, and the cerebellar cortex. Postsynaptic M_1 receptors remain intact, while recent studies by Mash et al. (1985) indicate that cortical M_2 receptors decline in number. GABA appears to be reduced moderately but only in the temporal cortex (Rossor et al., 1982).

Davies et al. (1980) first observed that somatostatin was significantly reduced in cortex and several laboratories have subsequently confirmed the findings (Rossor et al., 1980; Morrison et al., 1985; Roberts et al., 1985). Furthermore, Morrison et al. found somatostatin-immunoreactive material in the cortical plaques of Alzheimer's specimens while Roberts et al. observed that degenerating neurons containing neurofibrillary tangles also stained immunoreactively for somatostatin. These latter findings appear to implicate somatostatin-containing cells directly in the degenerative process of Alzheimer's disease. Whether or not somatostatin co-localizes with cortical cholinergic neurons, as postulated by Antuono and Coyle (personal communication, 1985) remains to be definitely answered.

Substance P was reported as moderately reduced in the cortex of all lobes of the hemispheres by Crystal and Davies (1982). Several laboratories have reported that other neuropeptides, including cholecystokinin-octapeptide, vasoactive intestinal peptide, and neuropeptide Y, although present in cortical neurons, are generally unaffected in Alzheimer brains as are levels of dopamine and norepinephrine (see Roberts et al., 1985 and Bissette et al., this volume for details and references).

The data on cholinergic changes in forebrain have represented the single strongest stimulus for the scientific study of Alzheimer's disease (Bowen et al., 1976; Coyle et al., 1983). Nevertheless, important and obvious questions immediately arise. Studies in lower animals by Armstrong et al. (1983) using immunocytochemical techniques show no intrinsic cholinergic neurons in the cerebral cortex. If true in man as well, this means that all cholinergic projections on cortex come from deeper sources. What is the normal cortical cytoarchitecture of muscarinic receptor cells? Does cell loss in the cortex coincide with the muscarinic receptors? If choline acetyltransferase declines as much in areas 4 and 6 as it does in other areas of the cortex (Rossor et al., 1982), why does one see so little clinical evidence of corticospinal tract and sensory dysfunction? Finally, are cholinergic degeneration and plaque formation associated with or are they independent of the fundamental pathological abnormality which causes cortical neuron and subcortical degeneration in Alzheimer's disease?

Advances in biochemical understanding of the mechanisms of Alzheimer's disease have come less quickly than the accelerated findings by so many investigators in the United Kingdom and United States on the cholinergic and somatostatin changes. Table III indicates some of the major findings, none of which presently are resolved in a manner which allow one to speculate on how they contribute to the cause or pathogenesis of Alzheimer's disease. Even the nature of the paired helical filaments that make up the neurofibrillary tangles (Wisniewski et al., 1976) has escaped analysis. This difficulty is probably explained by Selkoe and colleagues' (1982) recent discovery that the protein in the filaments is essentially insoluble by standard methods. The finding implies that past protein assays have

TABLE III

SDAT — biochemical leads and unknowns

Paired helical filaments comprise tangles
Filament protein insoluble and unsolved
Nature and importance of amyloid uncertain
CHO altered:
$CMRO_2$ ↓; PFK ↓ brain (Bowen, 1979), platelets (Ksiezak-Reding, 1983), fibroblasts (Sorbi, 1983)
Protein metabolism altered:
Tangles; nuclear volume ↓, cyto RNA ↓ (Mann, 1981)
Chromatin-histones abnormal (Crapper McLachlin, 1984)
Aluminum uncertain

examined everything else in the cell without it being realized that the pathological structure was firmly resisting even the first steps of analysis. At least at first finding, the paired helical filaments of the tangles in SDAT are not immunologically identical to any normal brain protein (Ihara et al., 1983). Other leads that appear less immediately important are listed in Table III.

What causes Alzheimer's disease? No one appears to have a strong lead. Most clinicians find in their patients that the disorder carries a heavy familial factor, perhaps being transmitted as an autosomal dominant trait with a high rate of penetrance, provided the subject lives long enough. Many patients relate a strong history of siblings and forebears becoming demented at similar ages (Heyman et al., 1983). The illness shows a far higher than chance association with Down's syndrome as well as with immunocytic malignancies (Heston et al., 1981). SDAT has been observed in both members of identical twinships although sometimes at widely spaced ages (Cook et al., 1981). The available evidence thus suggests a heavy genetic predisposition but leaves open the additional factors which may either predispose or precipitate the illness in those capable of having it.

Treatment

Efforts to treat Alzheimer's disease have been many, but none so far discovered fundamentally change the social incapacity of the patient or the apparent natural progression of the disease. Almost all who care for these patients agree that the non-specific aspects of care and particularly a compassionate and informed doctor's advice to the family are invaluable. Mace and Rabins (1981) have written a useful book for families of patients with this disease. For those few patients in whom serious depression appears to be a problem, tricyclic anti-depressants may alleviate that symptom. Liquid tasteless haloperidol, given in tiny titrations of as little as 0.1–0.2 mg per dose in food or drink, often helps to control the restless agitation of home-bound victims. A number of efforts have been made to replace the known deficiency of forebrain acetylcholine, but up until the present none of these have achieved practical success. The same conclusion, unfortunately, must be drawn about the attempted use of various peptides to relieve symptoms. Trials are underway using gangliosides and certain other relatively unlikely agents, such as interferon and aluminum chelating agents. It is too soon to know the results, but the theoretical basis for such approaches does not give much hope. In nonsurgical medicine it is often the patient for whom there is no specific treatment who most needs the doctor's attention. Sadly, the increasing numbers of patients needing care for Alzheimer's disease epitomize this principle.

Is Alzheimer's disease simply another form of brain aging? Acetylcholine turnover declines in experimental animals during normal senescence (Gibson et al., 1981) and aged human beings without any decline in mental capacity show decreased choline acetyltransferase, plaques, occasional tangles, and granulovaculo-degeneration, only the quantity differing from SDAT (Tomlinson et al., 1970; Rossor et al., 1982). But perhaps aging itself is a disease: Down's syndrome and progeria suggest that it is.

Any condition which debilitates 25% of the population if they succeed in reaching the age of 85 years and selectively eliminates, by processes that are as yet not understood at the molecular level, the essence of humanness deserves every bit of scientific effort we can muster. It seems unimportant whether we call this a process or a disease. Either way, whatever the mechanisms are, they must operate through specific and potentially understandable pathways to produce this most devastating natural biological threat of our time.

References

Armstrong, D. M., Saper, C. B., Levery, A. I., Wainer, B. H. and Terry, R. D. (1983) Distribution of cholinergic neurons in rat brain: demonstrated by the immunocytochemical localization of choline acetyltransferase. *J. Comp. Neurol.*, 216: 53–68.

Bondareff, W., Mountjoy, C. Q. and Roth, M. (1982) Loss of neurons of origin of the adrenergic projection to cerebral cortex (nucleus locus ceruleus) in senile dementia. *Neurology*, 32: 164–168.

Bowen, D. M., Smith, C. B., White, P. and Davison, A. N. (1976) Neurotransmitter-related enzymes and indices of hypoxia in senile dementia and other abiotrophies. *Brain*, 99: 459–496.

Bowen, D. M., White, P., Spillane, J. A., Goodhart, M. J., Curzon, G., Iwangoff, P., Meier-Ruge, W. and Davison, A. N. (1979) Accelerated aging or selective neuronal loss as an important cause of dementia. *Lancet*, 1: 11–13.

Brice, M. E. and Fraser, H. (1975) Amyloid plaques in the brains of mice infected with scrapie. *Neuropathol. Appl. Neurobiol.*, 1: 189–202.

Brun, A. and Englund, E. (1981) Regional pattern of degeneration in Alzheimer's disease, neuronal loss, and histopathological grading. *Histopathology*, 5: 549–564.

Chase, T. N., Foster, N. L., Fedio, P. et al. (1984) Regional cortical dysfunction in Alzheimer's disease as determined by positron emission tomograph. *Ann. Neurol.* (in press).

Cook, R. H., Schneck, S. A. and Clark, D. B. (1981) Twins with Alzheimer's disease. *Arch. Neurol.*, 38: 300–301.

Corsellis, J. A. N., Bruton, C. J. and Freeman-Browne, D. (1973) The aftermath of boxing. *Psychol. Med.*, 3: 270–303.

Coyle, J. T., Price, D. L. and DeLong, M. R. (1983) Alzheimer's disease: a disorder of cortical cholinergic innervation. *Science*, 219: 1184–1190.

Crystal, H. A. and Davies, P. (1982) Cortical substance P-like immunoreactivity in cases of Alzheimer's disease and senile dementia of Alzheimer type. *J. Neurochem.*, 1781–1784.

Davies, P., Katzman, R. and Terry, R. D. (1980) Reduced somatostatin-like immunoreactivity in cerebral cortex from cases of Alzheimer's disease and Alzheimer senile dementia. *Nature*, 288: 279–280.

Economo, C. von (1927) *Zellaufbau der Grosshirnrinde des Menschen*, Julius Springer, Berlin.

Frackowiak, R. S. J., Pozzilli, C., Legg, N. J., DuBoulay, G. H., Marshall, J., Lenzi, G. L. and Jones, T. (1981) Regional cerebral oxygen supply and utilization in dementia. A clinical and physiological study with oxygen-15 and positron tomography. *Brain*, 104: 753–778.

Gibson, G. E., Peterson, C. and Jenden, D. J. (1981) Brain acetylcholine synthesis declines with senescence. *Science*, 213: 674–676.

Herzog, A. G. and Kemper, T. L. (1980) Amygdaloid changes in aging and dementia. *Arch. Neurol.*, 37: 625–629.

Heston, L. L., Mastri, A. R., Anderson, E. and White, J. (1981) Dementia of the Alzheimer type. Clinical genetics, natural history, and associated conditions. *Arch. Gen. Psychiatry*, 38: 1085–1090.

Heyman, A., Wilkinson, W. E., Hurwitz, B. J., Schmechel, D., Sigmon, A. H., Weinberg, T., Helms, M. J. and Swift, M. (1983) Alzheimer's disease: genetic aspects and associated clinical disorders. *Ann. Neurol.*, 14: 507–515.

Hubbard, B. M. and Anderson, J. M. (1981) A quantitative study of cerebral atrophy in old age and senile dementia. *J. Neurol. Sci.*, 50: 135–145.

Ihara, Y., Abraham, C. and Selkoe, D. (1983) Antibodies to paired helical filaments in Alzheimer's disease do not recognize normal brain protein. *Nature*, 304: 727–729.

Ksiezak-Reding, H., Murphy, C. and Blass, J. P. (1983) Enzyme activities in platelets from patients with Alzheimer's disease. *Aging*, 6: 11 (abstract).

Kuhl, D. E., Metter, E. J., Riege, W. H., Hawkins, R. A., Mazziotta, J. C., Phelps, M. E. and Kling, A. S. (1983) Local cerebral glucose utilization in patients with depression, multiple infarct dementia, and Alzheimer's disease. *J. Cereb. Blood Flow Metab.*, 3: S494.

Mace, N. L. and Rabins, P. V. (1981) *The 36-Hour Day. A Family Guide to Caring for Patients with Alzheimer's Disease, Related Dementing Illnesses, and Memory Loss in Later Life*, Johns Hopkins, Baltimore.

Mann, D. M. A., Neary, D., Yates, P. O., Lincoln, J., Snowden, J. S. and Stanworth, P. (1981) Neurofibrillary pathology and protein synthetic capability in nerve cells in Alzheimer's disease. *Neuropathol. Appl. Neurobiol.*, 7: 37–47.

Mash, D. C., Flynn, D. D. and Potter, L. T. (1985) Loss of M_2 muscarine receptors in the cerebral cortex in Alzheimer's disease and experimental cholinergic denervation. *Science*, 228: 1115–1117.

Masters, C. L., Gajdusek, D. C. and Gibbs, C. J., Jr. (1981) Creutzfeldt–Jakob disease virus isolation from the Gerstmann–Straussler syndrome with an analysis of the various forms of amyloid plaque deposition in the virus-induced spongiform encephalopathies. *Brain*, 104: 559–588.

McLachin, D. R. Crapper, Lewis, P. N., Lukiw, W. J., Sima, A., Bergeron, C. and DeBoni, U. (1984) Chromatin structure in dementia. *Ann. Neurol.,* 15: 329–334.

Miller, A. K. H., Alston, R. L. and Corsellis, J. A. N. (1980) Variation with age in the volumes of gray and white matter in the cerebral hemispheres of man: measurements with an image analyzer. *Neuropathol. Appl. Neurobiol.,* 6: 119–132.

Morrison, J. H., Rogers, J., Scherr, S., Benoit, R. and Bloom, F. E. (1985) Somatostatin immunoreactivity in neuritic plaques of Alzheimer's patients. *Nature,* 314: 90–92.

Plum, F. (1979) Dementia: an approaching epidemic. *Nature,* 279: 372–373.

Roberts, G. W., Crow, T. J. and Polak, J. M. (1985) Location of neuronal tangles in somatostatin neurons in Alzheimer's disease. *Nature,* 314: 92–94.

Rossor, M. N., Emson, P. C., Mountjoy, C. Q., Roth, M. and Iversen, L. L. (1980) Reduced amounts of immunoreactive somatostatin in the temporal cortex in senile dementia of Alzheimer type. *Neurosci. Lett.,* 20: 373–377.

Rossor, M. N., Garrett, N. J., Johnson, A. L., Mountjoy, C. O., Roth, M. and Iversen, L. (1982) A post-mortem study of the cholinergic and GABA systems in senile dementia. *Brain,* 105: 313–330.

Selkoe, D. J., Ihara, Y. and Salazar, F. J. (1983) Alzheimer's disease: insolubility of partially purified paired helical filaments in sodium dodecyl sulfate and urea. *Science,* 215: 1243–1245.

Sorbi, S. and Blass, J. P. (1983) Fibroblast phosphofructokinase in Alzheimer's disease and Down's syndrome. *Banbury Rep.,* 15: 297–307.

Struble, R. G., Cork, L. C., Whitehouse, P. J. and Price, D. L. (1982) Cholinergic innervation in neuritic plaques. *Science,* 216: 413–415.

Sulkava, R., Haltia, M., Paeteu, A., Wikstrom, J. and Palo, J. (1983) Accuracy of clinical diagnosis in primary degenerative dementia: correlation with neuropathological findings. *J. Neurol. Neurosurg. Psychiatry,* 46: 9–13.

Terry, R. D. and Katzman, R. (1983) Senile dementia of the Alzheimer type. *Ann. Neurol.,* 14: 497–506.

Terry, R. D., Peck, A., DeTeresa, R., Schecter, P. and Horoupian, D. S. (1981) Some morphometric aspects of the brain in senile dementia of the Alzheimer type. *Ann. Neurol.,* 10: 184–192.

Tomlinson, B. E., Blessed, G. and Roth, M. (1970) Observations on the brains of demented old people. *J. Neurol. Sci.,* 11: 205–242.

Tomlinson, B. E., Irving, D. and Blessed, G. (1981) Cell loss in the locus coeruleus in senile dementia of Alzheimer type. *J. Neurol. Sci.,* 49: 419–428.

Wisniewski, H. M., Narang, H. K. and Terry, R. D. (1976) Neurofibrillary tangles of paired helical filaments. *J. Neurol. Sci.,* 27: 173–181.

P. C. Emson, M. N. Rossor and M. Tohyama (Eds.),
Progress in Brain Research, Vol. 66.

CHAPTER 8

Neuropeptides and dementia

M. N. Rossor[1], P. C. Emson[2], D. Dawbarn[2], C. Q. Mountjoy[3] and M. Roth[3]

[1]*National Hospital, Queen Square, London WC1N 3BG,* [2]*MRC Group, Institute of Animal Physiology, Babraham, Cambridge, and* [3]*Dept. Psychiatry, Addenbrookes Hospital, Cambridge, U.K.*

Introduction

Recent neurochemical studies of dementia have focused on the cholinergic system following the demonstration in 1976 of reduced activity of the biosynthetic enzyme, choline acetyltransferase (ChAT) in cerebral cortex of patients dying with a diagnosis of Alzheimer's disease (Davies and Maloney, 1976; Bowen et al., 1976; Perry et al., 1977a). This observation has since been confirmed in subsequent post-mortem (Davies, 1979; Perry and Perry, 1980; Rossor et al., 1982b; Wilcock et al., 1982; Bird et al., 1983; Wood et al., 1983) and biopsy studies (Sims et al., 1980; Francis et al., 1985) (for reviews see Rossor, 1982; Coyle et al., 1983; Hardy et al., 1985). The loss of cortical enzyme activity reflects damage to the ascending projection from basal forebrain (Whitehouse et al., 1982; Rossor et al., 1982d; Coyle et al., 1983; Nagai et al., 1983; Pearson et al., 1983) and can be correlated with the severity of both the clinical and histopathological features (Perry et al., 1978; Wilcock et al., 1982; Mountjoy et al., 1984). The other ascending projections to cerebral cortex which utilise the classical neurotransmitters noradrenaline and 5-hydroxytryptamine have also been implicated in the pathophysiology of Alzheimer's disease (Forno, 1978; Adolfsson et al., 1979; Cross et al., 1981, 1983; Tomlinson et al., 1981; Bondareff et al., 1982; Bowen et al., 1983; Arai et al., 1984).

With few exceptions ignorance of the potential role of peptides in cognition precludes analysis based on their functional role in dementia. However, analysis of peptides as neuronal markers may provide information on selective vulnerability of neurons in degenerative disease. This is of particular relevance to the cerebral cortex in Alzheimer's disease since the characteristic histological features are most obvious in this area (see Plum, this volume).

In contrast to the changes in cholinergic and noradrenergic systems which reflect damage to ascending projections, the neurofibrillary tangles and cell loss reflect perikaryal damage within the cortex itself. The neurotransmitter identity of these cells remains unknown although reduced somatostatin-like immunoreactivity (SRIF-LI) is found within cerebral cortex of patients with Alzheimer's disease (Davies et al., 1980; Rossor et al., 1980) with sparing of the other cortical neuropeptides vasoactive intestinal polypeptide (VIP) and cholecystokinin (CCK). Moreover, recent immunohistochemical studies have demonstrated SRIF-LI within senile plaques (Morrison et al., 1985) and within neurons containing tangles (Roberts et al., 1985). Most of the peptide studies in dementia relate to Alzheimer's disease and this will form the focus of this chapter. However, the peptide changes observed in Parkinson's and Huntington's disease (see Emson, this volume, and Agid et al., this volume) may also contribute to the cognitive impairment that may occur, and limited data are available on peptide changes in Pick's disease and Down's syndrome.

Alzheimer's disease

Changes in cortical peptides

A number of peptides have now been examined in the cerebral cortex of patients with Alzheimer's disease (Table I). The most consistent finding has been that of reduced concentrations of SRIF-LI (Davies et al., 1980; Rossor et al., 1980; Davies and Terry, 1981; Ferrier et al., 1983; Wood et al., 1983; Rossor et al., 1984).

TABLE I

Summary of neuropeptide changes in Alzheimer's disease

Peptide	Cerebral cortex	Subcortical areas	Publication
Somatostatin	ca. 80% ↓ all areas		Davies et al. (1980, 1981)
	ca. 30–50% ↓	No change putamen, amygdala	Rossor et al. (1980)
	ca. 50–70% ↓ confined to temporal cortex in older cases		Rossor et al. (1984)
	↓ in temporal cortex severe cases		Perry et al. (1981)
	ca. 40% ↓	↓ septum ca. 100% nbm No change amygdala	Ferrier et al. (1983)
	ca. 50% ↓ frontal cortex		Wood et al. (1983)
	ca. 80% ↓ all areas	ca. 80% ↓ amygdala	Nemeroff et al. (1983)
Neuropeptide Y	No change	90% ↑ nbm	Allen et al. (1984) Dawbarn et al. (1985)
Cholecystokinin	No change overall but ↓ aqueous extracted in temporal cortex in severe cases and short autopsy delay		Perry et al. (1981)
	No change	No change putamen	Rossor et al. (1981)
	No change	No change nbm, septum, amygdala	Ferrier et al. (1983)
Vasoactive intestinal polypeptide	No change		Rossor et al. (1980)
	No change		Perry et al. (1981)
	No change	No change septum, nbm, hypothalamus	Ferrier et al. (1983)
Met-enkephalin	No change	No change amygdala	Rossor et al. (1982)
Substance P	ca. 50–60% ↓ all areas		Crystal and Davies (1982)
	Increased in hippocampus in severe cases		Perry et al. (1981)
	No change	No change	Yates et al. (1983)
	No change	No change septum, nbm, ↑ in putamen	Ferrier et al. (1983)
Neurotensin	No change	↓ 56% septum	Ferrier et al. (1983)
		No change amygdala	Biggins et al. (1983)
	No change		Emson (unpublished)
	No change	ca. 30% ↓ in amygdala	Nemeroff et al. (1983)
TRH	No change frontal, cingulate	No change	Yates et al. (1983)
		No change in amygdala	Biggins et al. (1983)
	Undetectable	No change	Nemeroff et al. (1983)
LHRH		No change amygdala, hypothalamus	Yates et al. (1983)

Somatostatin and neuropeptide Y

SRIF-LI is widely distributed throughout the human brain with high concentrations within cerebral cortex (Cooper et al., 1981; Emson et al., 1981; Sorenson, 1982). Immunohistochemistry in the rat reveals perikaryal staining predominantly of non-pyramidal neurons throughout cortex with the majority in layers II, III and VI. However, the total number of cells containing somatostatin may be small and in the rat visual cortex comprises about 3% of the neuronal population (McDonald et al., 1982; see Parnavelas, this volume). Morrison et al. (1983) using an antiserum raised against somatostatin-28 demonstrated cell bodies within rat cerebral cortex whereas an antiserum against somatostatin-1–14 produced little staining despite binding in a radioimmunoassay. An antiserum against somatostatin-1–12 revealed staining of neuronal processes. It is not yet established which molecular species are involved in the somatostatin deficit in Alzheimer's disease, and whether it reflects a predominant loss of terminals or concomitant loss of perikarya. The association of reduced cortical ChAT activity in those brains with loss of SRIF-LI raises the possibility of coexistence of these two markers within the same cells and SRIF-LI has been demonstrated in the area of the substantia innominata, the origin of the cholinergic projection, but staining is observed in terminals rather than cell bodies (Candy et al., 1982). Lesions of the substantia innominata which resulted in a 70% reduction in cortical ChAT activity produced no change in SRIF-LI (McKinney et al., 1982). This together with the observation that SRIF-LI in the substantia innominata is increased rather than decreased in Alzheimer's disease (Ferrier et al., 1984) makes it unlikely that the peptide is found in the terminals of the ascending cholinergic projection. However, the short time course of the reported studies does not preclude a post-synaptic influence of the cholinergic neurons on somatostatin cells within the cerebral cortex. Recently Delfs et al. (1984) have reported the co-existence of acetylcholinesterase and SRIF-LI within cerebral cortical neurons in cell culture. However, acetylcholinesterase is not specific for cholinergic neurons and coexistence has yet to be shown using ChAT immunostaining in cortical sections.

The loss of SRIF-LI is greatest from the temporal lobe in which reductions from control have varied from 75% (Davies and Terry, 1981) to 41% (Ferrier et al., 1983). In our own earlier study (Rossor et al., 1980), we found a loss of SRIF-LI which was confined to the temporal cortex in contrast to the more widespread changes observed by Davies et al. (1980). The probable cause for this discrepancy lies in the difference in the age of the populations studied since this has an important influence on the neurochemical profile (Rossor et al., 1984). Analysis of variance of ChAT activity in a group of 25 controls and 25 Alzheimer cases demonstrated significant disease/age at death interactions. This was based upon a division of the cases at the median age for the group of 79 years at death, and yielded a distinct pattern of ChAT activity which was less severe and spared the frontal lobe in the older cases (Rossor et al., 1981, 1982b). A similar analysis of a larger group of 54 control and 49 Alzheimer brains, which did not differ in the severity of dementia as determined by an overall rating scale (Roth and Hopkins, 1954), confirmed the pattern of ChAT deficit with sparing of the frontal cortex in older (> 79 years) patients (Rossor et al., 1984). In this same group the somatostatin deficit also spared the frontal cortex and was only apparent in the temporal cortex in the older group in contrast to more widespread changes in younger cases (Table II). Concentrations of SRIF-LI did not change with age in the control group for either frontal or temporal cortex but a paradoxical non-significant increase was apparent in the Alzheimer group (Fig. 1). Reduced concentrations of SRIF-LI are also found in cortical biopsies but K^+ evoked release is normal (Francis and Bowen, 1985).

The early studies of Perry et al. (1978) demonstrated significant associations of the mean cortical ChAT activity with cortical senile plaque count and with the severity of dementia at the time of death.

TABLE II

Somatostatin-like immunoreactivity in frontal and temporal cortex of Alzheimer and control brain

	Control	Alzheimer	% Reduction	*P* value
Whole group				
Frontal cortex	1.92 ± 0.03 (79.6)	1.77 ± 0.04 (61.2)	29	0.003
Temporal cortex	2.18 ± 0.03 (157.8)	1.81 ± 0.04 (62.8)	57	<0.001
Young (<79 years)				
Frontal cortex	1.94 ± 0.05 (79.7)	1.71 ± 0.05 (56.1)	41	0.003
Temporal cortex	2.22 ± 0.04 (161.2)	1.77 ± 0.06 (58.9)	64	<0.001
Old (>79 years)				
Frontal cortex	1.91 ± 0.04 (77.7)	1.83 ± 0.06 (71.2)	–	0.3
Temporal cortex	2.15 ± 0.05 (157.5)	1.86 ± 0.07 (80.4)	49	0.001

Values are means ± S.E.M. of logarithmically transformed data with median values of untransformed data in parentheses. The control group comprised 17 young (<79 years at death) and 25 old (>79 years at death) cases and the Alzheimer group comprised 23 young and 19 old cases. Data derived from Rossor et al. (1984).

Wilcock et al. (1982) found an association with tangles only and in our own series (Mountjoy et al., 1984) associations were found with plaque count, tangle estimate and cortical cell count in the whole group, but when the Alzheimer group were analysed alone, these associations were only apparent in temporal cortex. Similar analyses of SRIF-LI concentration also reveal associations with histological features (Fig. 2), and with severity of dementia (Fig. 3), although this is only apparent for the whole group. Concentration of SRIF-LI is also significantly correlated with ChAT activity in the Alzheimer group (Davies and Terry, 1981).

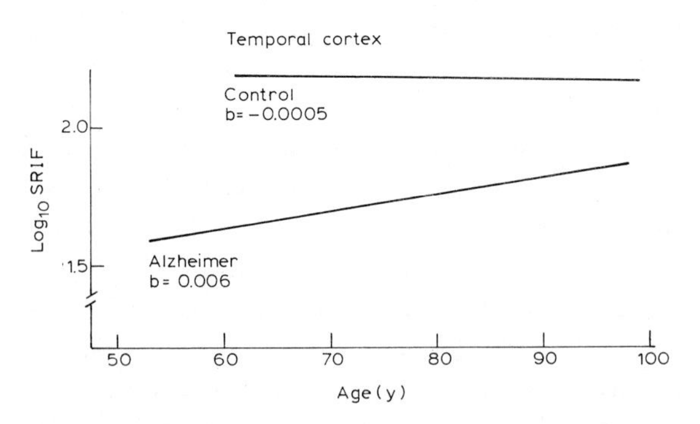

Fig. 1. Regression analysis of SRIF-LI on age for temporal cortex (Brodmann area 21) in Alzheimer's disease and controls. Concentrations of somatostatin have been logarithmically transformed in order to stabilise the variance within groups. There is a paradoxical trend towards increased concentration with age in the Alzheimer group but this is not statistically significant. Data derived from Rossor et al. (1984b).

Neuropeptide Y, a 36-amino acid peptide which shares sequence homology with avian pancreatic polypeptide, is widely distributed in human brain and is found in high concentration in basal ganglia and to a lesser extent in the cerebral cortex (Adrian et al., 1983; see Takagi, this volume; Emson, this volume; and Parnavelas, this volume). Antisera raised against avian pancreatic polypeptide have also demonstrated immunoreactivity in mammalian brain but this appears to have been due to cross

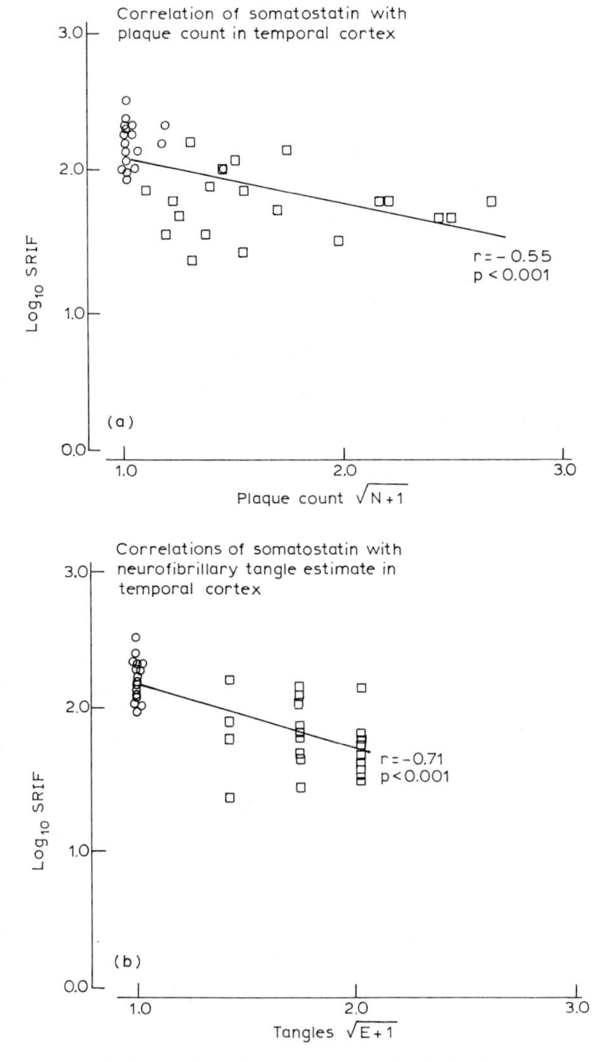

Fig. 2. (a) Correlation of concentration of SRIF-LI with numbers of senile plaques in temporal cortex (Brodmann area 21). $r = -0.55$, $P < 0.001$ for the total group. (b) Correlation of concentration of SRIF-LI with estimates of neurofibrillary tangles in temporal cortex (Brodmann area 21). $r = -0.71$, $P < 0.001$ for the total group. Somatostatin concentrations have undergone logarithmic transformation and histological counts square root transformation in order to stabilise the variance within groups. Correlations are only statistically significant if the total group is analysed. Data derived from Rossor et al. (1984a). ○, controls; □, Alzheimer's disease.

Fig. 3. Correlation of SRIF-LI in temporal cortex (Brodmann area 21) with dementia score (Roth and Hopkins, 1953) for the total group of controls and Alzheimer cases. The somatostatin concentration has been transformed logarithmically and the dementia score undergone a square root transformation to stabilise the variance within groups. $r = -0.51$, $P < 0.001$.

reaction with neuropeptide Y (see Takagi, this volume; Parnavelas, this volume). Using these antisera peptide immunoreactivity has now been demonstrated in a proportion of somatostatin-containing neurons in human cerebral cortex (Vincent et al., 1982) and within the ascending noradrenergic projection from locus coeruleus in the rat (Hunt et al., 1981). Neuropeptide Y-like immunoreactivity (NPY-LI) has also been demonstrated within a number of catecholamine neurons in the medulla oblongata of the human (Hökfelt et al., 1983). If SRIF-LIr neurons are lost from the cerebral cortex in Alzheimer's disease, one might expect a corresponding loss of NPY-LI. In addition, the loss of neurons from locus coeruleus in Alzheimer's disease (Forno, 1978; Tomlinson et al., 1981; Bondareff et al., 1982) might contribute to any loss of cortical NPY-LI. We have, therefore, determined levels of NPY-LI and SRIF-LI in the frontal and temporal cortex of 15 Alzheimer brains and 15 controls. Both peptide immunoreactivities were determined in the same tissue samples. The 50% loss of SRIF-LI from temporal cortex confirms our earlier finding (Rossor et al., 1980) but there was no corresponding decrease in NPY-LI (Table III).

Recently Allen et al. (1984) have also reported normal concentrations of NPY-LI in Alzheimer cerebral cortex together with an increase of both NPY-LI and SRIF-LI in substantia innominata.

TABLE III

Somatostatin and neuropeptide Y in Alzheimer's disease

Brain area	Control	Alzheimer
Neuropeptide Y		
Frontal cortex	1.27 ± 0.02	1.24 ± 0.02
	(18.3)	(18.4)
Temporal cortex	1.08 ± 0.04	1.19 ± 0.03
	(10.3)	(14.7)
Somatostatin		
Frontal cortex	1.84 ± 0.03	1.80 ± 0.04
	(64.4)	(65.6)
Temporal cortex	2.12 ± 0.06	1.84 ± 0.07**
	(150.0)	(68.1)

Values are means ± S.E.M. of logarithmically transformed data for peptide immunoreactivities (pmol/g tissue) with medians of untransformed data in parentheses.
$n = 20$. ** $P < 0.01$. Neuropeptide and somatostatin immunoreactivities were determined on the same samples. Data derived from Dawbarn et al. (1985).

The reason for the discrepancy between SRIF-LI and NPY-LI is uncertain. If only a small proportion of cortical neuropeptide Y is found in somatostatin neurons then the loss may be masked if other neurons are preserved. Alternatively, changes in somatostatin might reflect altered turnover rather than neuronal loss.

Cholecystokinin and vasoactive intestinal polypeptide

CCK- and VIP-like immunoreactivities (CCK-LI and VIP-LI) are also found within intrinsic cerebral cortical neurons but in contrast to somatostatin, concentrations in Alzheimer's disease do not differ substantially from controls (Rossor et al., 1980b, 1981c; Perry et al., 1981a,b; Ferrier et al., 1983).

In our own series of CCK-LI determinations (Rossor et al., 1981c) there was a non-significant ($P > 0.05$) trend towards a reduction in concentration in the temporal cortex. This finding has been confirmed in a larger series, and no age influence can be discerned when the cases are divided into younger (< 79 years) and older (> 79 years) groups. Similar results have been reported by others although Perry et al. (1981a,b) found reduced concentrations in aqueous-extracted, as opposed to acid-extracted, immunoreactivity in severe cases and those with a short autopsy delay. The normal concentrations of VIP-LI are of particular interest in light of the recent report of coexistence of VIP-LI with ChAT immunoreactivity within intrinsic cortical neurons (Eckenstein and Baughman, 1984). It is not known to what extent cortical cholinergic and VIP-LIr neurons overlap but the preservation of VIP-LI in Alzheimer's disease may imply that the loss of ChAT activity reflects damage to the ascending projection only with sparing of the intrinsic cholinergic neurons.

Other peptides in cerebral cortex

A number of other peptide immunoreactivities have been determined in cerebral cortex in Alzheimer's disease but the amounts are low and difficult to assay, and the relationship to intrinsic or ascending projections is uncertain.

Crystal and Davies (1982) reported a loss of substance P-like immunoreactivity (SP-LI) from cerebral cortex although this has not been confirmed in other studies (Ferrier et al., 1983; Yates et al., 1983a) and indeed Perry et al. (1981a) found an increase in substance P within the medial hippocampus of cases with a high density of senile plaques. The precise derivation of SP-LI within cerebral cortex is not established although an ascending projection to frontal cortex via the medial forebrain bundle (Paxinos et al., 1978) and a projection from septum to hippocampus (Vincent and McGeer, 1981) are suggested from lesioning studies. Recent evidence suggests coexistence of SP-LI within cholinergic neurons arising from caudal mesencephalon and pontine tegmentum (Vincent et al., 1983) but it is not known whether this caudal group of cells is also lost in Alzheimer's disease.

Low concentrations of Met-enkephalin-like immunoreactivity can be measured in human cerebral

TABLE IV

Met-enkephalin immunoreactivity in Alzheimer's disease (pmol/g tissue)

Brain area	Control	Alzheimer
Frontal cortex (Brodmann area 10)	1.07 ± 0.05 (10.3)	1.10 ± 0.04 (12.6)
Temporal cortex (Brodmann area 21)	1.07 ± 0.03 (11.6)	1.12 ± 0.04 (12.1)
Parietal cortex (Brodmann area 7)	1.03 ± 0.03 (10.5)	1.09 ± 0.04 (11.0)
Posterior hippocampus	1.30 ± 0.07 (19.0)	1.31 ± 0.06 (17.0)
Amygdala	1.29 ± 0.08 (17.1)	1.29 ± 0.08 (19.2)
Putamen	2.89 ± 0.10 (679)	2.78 ± 0.07 (727)
Lateral globus pallidus	3.35 ± 0.05 (2036)	3.06 ± 0.17 (1500)
Substantia nigra reticulata	2.36 ± 0.12 (362)	2.38 ± 0.12 (390)
Periaqueductal grey matter	2.02 ± 0.05 (115)	1.89 ± 0.03 (82)

Values are means ± S.E.M. for logarithmically transformed data with medians of untransformed data in parentheses. None of the differences is significant.
(n = 17–21).

cortex. Immunohistochemical studies in animals demonstrate fibre staining only (Sar et al., 1978) although there is some evidence for intrinsic hippocampal enkephalin-positive neurons (Hong and Schmid, 1981). Met-enkephalin-like immunoreactivities in both neocortex and hippocampus did not differ significantly from a control group (Rossor et al., 1982a) (Table IV).

Thyrotropin releasing hormone (TRH)-like immunoreactivity was below the limit of sensitivity in most areas of cerebral cortex in the study of Yates et al. (1983) although no difference from controls was found in cingulate and frontal cortices. Similarly, although neurotensin levels are low in cerebral cortex these are not different from control values (Ferrier et al., 1983; Nemeroff et al., 1983; Emson et al., unpublished).

Relationship of cortical peptides to histopathology

The association of changes in SRIF-LI and NPY-LI with histopathological features (Perry et al., 1983; Rossor et al., 1984; Dawbarn et al., 1985) raises the question of whether these neurons are specifically and directly involved with plaque and tangle formation or whether this only reflects the overall severity of the disease. Acetylcholinesterase staining of plaques is observed both in Alzheimer's disease (Perry et al., 1980) and in aged monkeys (Struble et al., 1982). More specific evidence for involvement of cholinergic terminals in plaques is the immunostaining for ChAT (Kitt et al., 1984a). Similar immunohistochemical studies demonstrate SRIF-LIr fibres in senile plaques (Perry et al., 1983; Armstrong et al., 1985; Morrison et al., 1985).

The SRIF-LI plaques were found predominantly in layers III and V where the majority of cortico-cortical association fibres originate (Morrison et al., 1985). The fibre staining with NPY antisera (Dawbarn et al., 1985) may involve the same neurons in view of the reported coexistence (Vincent et al., 1982). Other chemical markers are also found in association with plaques and include tyrosine hydroxylase (Kitt et al., 1984b), CCK-LI and VIP-LI (Roberts et al., 1985). This variety of immunostaining suggests that senile plaques are not neurochemically specific but rather that neurites adjacent to a developing plaque are involved indiscriminately. Consistent with this is the Golgi study of hippocampal plaques in which dystrophic neurites could be traced back to a variety of local neurons (Probst et al., 1983).

An association has also been observed between ChAT activity and neurofibrillary tangles (Wilcock et al., 1982; Mountjoy et al., 1984), and neuronal counts if the entire group is analysed (Mountjoy et al., 1984). These associations are likely, at least in part, to reflect overall severity since the number of cholinergic neurons within cerebral cortex is relatively small and insufficent to explain all of the cell loss or involvement in neurofibrillary tangle formation. Associations are also noted between SRIF-LI and neurofibrillary tangles but only if the control group is included in the analysis. Immunohistochemistry provides direct evidence of neurofibrillary tangles within SRIF-LIr neurons (Roberts et al., 1985).

An important question is the relationship between loss of peptide immunoreactivity and loss of cortical cells and whether the cortical neurons which are lost in Alzheimer's disease are neurochemically discrete. If the loss of SRIF-LI is due to loss of cells one would predict a commensurate loss of NPY-LI if this coexists in the majority of SRIF-LIr neurons and yet NPY-LI is unchanged (Allen et al., 1984; Dawbarn et al., 1985). A similar argument might apply to γ-aminobutyric acid (GABA) since a subclass of GABA neurons contain SRIF-LI (Schmechel et al., 1984) and there is recent evidence to suggest that all SRIF-LIr, VIP-LIr and CCK-LIr cortical neurons are GABAergic (Hendry et al., 1984). Early reports of reduced activity of the biosynthetic marker enzyme, glutamic acid decarboxylase (GAD), may have reflected the non-specific influence of the agonal state. In tissue from cortical biopsies both the GAD activity and potassium-evoked release of GABA are normal (Spillane et al., 1977; Smith et al., 1983). GABA is stable post-mortem and appears to be less influenced by the agonal state (Spokes et al., 1980); it is reduced in Alzheimer's disease but the change is only modest, confined to younger cases and thus does not follow the deficit of SRIF-LI (Rossor et al., 1982). Armstrong et al. (1985) formed the impression that there was not any loss of SRIF-LIr neurons but the numbers of immunoreactive perikarya were low in both control and Alzheimer's disease which precluded reliable comparisons.

Changes in subcortical peptides

Although the histopathological features of senile plaques and neurofibrillary tangles are found predominantly within the cerebral cortex, tangles are also seen in a number of subcortical nuclei such as the nucleus basalis, the raphe nucleus and the amygdala (Ishii, 1966; Kemper and Herzog, 1980; Curcio and Kemper, 1984). A variety of peptide immunoreactivities have been demonstrated within the substantia innominata, which includes the nucleus basalis. In contrast to the reduction in ChAT activity in this area (Rossor et al., 1982d) both SRIF-LI (Ferrier et al., 1983) and NPY-LI (Allen et al., 1984) are significantly increased. This may reflect tissue shrinkage with relative preservation of somatostatin containing elements although only fibre staining is seen immunohistochemically (Candy et al., 1981). Neurotensin-, substance P-, VIP and cholecystokinin-like immunoreactivities have also been measured in the substantia innominata and do not differ significantly from control values. ChAT activity is also reduced in the septal area (Rossor et al., 1982b), the origin of the cholinergic projection to hippocampus. Ferrier et al. (1983) found reduced concentrations of both SRIF-LI and neurotensin-like immunoreactivity and no statistically significant change in CCK-, VIP- or SP-LI. No difference was found in concentration of SP-, TRH- or luteinizing hormone-releasing hormone (LHRH)-like immunoreactivities in a variety of subcortical structures which included the septal area (Yates et al., 1983a).

Vasopressin has been reported to affect memory functions in both animal studies (Kovacs et al., 1979) and in humans (Weingartner et al., 1981). Although vasopressin-like immunoreactivity (VP-LI) is not found consistently in cerebral cortex it can be measured in a number of subcortical structures including the locus coeruleus (Rossor et al., 1981a) where interaction with the noradrenergic projection may underlie the behavioural actions (Kovacs et al., 1979; Olpe and Baltzer, 1981). VP-LIr cell bodies have been reported in the locus coeruleus of the rat (Caffe and van Leeuwen, 1983) but nerve terminals only have been visualised in the human and there is no loss of immunoreactivity in cases of Parkinson's disease and multisystem atrophy in which there is loss of locus coeruleus neurons (Rossor et al., 1982c). Similarly, the abnormalities in the locus coeruleus in Alzheimer's disease are not accompanied by significant changes in VP-LI and no change is observed in dissections of the entire hypothalamus (Rossor et al., 1980c). The significance of a reduction in lateral globus pallidus is uncertain (Rossor et al., 1980c). In addition to the paraventricular nucleus, VP-LIr cell bodies can be demonstrated in the suprachiasmatic nucleus of human brain. In a detailed immunohistochemical study of this nucleus an increase in cell size after 60 years was found in both controls and Alzheimer's disease cases but there was an age-related cell loss which was significantly greater in those patients dying with Alzheimer's disease (Fliers et al., 1985; Swaab et al., 1985).

Of the other subcortical structures examined in Alzheimer's disease, the amygdala shows dramatic histopathological changes, particularly in the medial nuclei (Herzog and Kemper, 1980), and the loss of ChAT activity is as great as that seen in the cerebral cortex (Rossor et al., 1982b). Interestingly, changes in SRIF-LI have not been found in the amygdala (Rossor et al., 1980a; Ferrier et al., 1983), with the exception of the study of Nemeroff et al. (1983). CCK-LI and VIP-LI are also preserved in the amygdala as in cerebral cortex (Rossor et al., 1980b, 1981c; Ferrier et al., 1983). Non-significant reductions in TRH- and neurotensin-like immunoreactivities (Biggins et al., 1983) and a small but statistically significant reduction in neurotensin-like immunoreactivity have been reported (Nemeroff et al., 1983). In an immunohistochemical study of four cases of Alzheimer's disease no Met-enkephalin-like immunoreactive staining was found in the amygdala although no control group was available (Bouras et al., 1984). Radioimmunoassay has revealed no difference when compared with a control group (Table IV). The normal concentrations of peptide immunoreactivities in the amygdala are surprising in view of the marked histopathological changes and may be due to the effect of tissue shrinkage.

TABLE V

[^{3}H]Ketanserin binding to homogenates of frontal cortex in control and Alzheimer brain

	Control	Alzheimer
Whole group	6.17 ± 0.56	3.58 ± 0.25**
Age at death <79 year	7.09 ± 0.93	3.48 ± 0.32**
Age at death >79 year	5.20 ± 0.47	3.54 ± 0.45*

Data are means ± S.E.M. for [^{3}H]ketanserin bound (pmol/g) at a ligand concentration of 1.58 nM. Binding at 0.32 nM also revealed a significant reduction in the Alzheimer group and ratios of bound ligand at the two concentrations did not differ significantly (0.48 ± 0.20 and 0.55 ± 0.18 for control and Alzheimer groups).
n = 30 for control group and 34 for Alzheimer. ** $P < 0.001$. Data derived from Reynolds et al. (1984).

Receptor studies in Alzheimer's disease

The analysis of neurotransmitter receptors in Alzheimer's disease has concentrated on binding sites for the classical neurotransmitters. In the majority of studies there has been no alteration in either the B_{max} or the K_d of the muscarinic acetylcholine receptor using tritiated quinuclidinyl benzoate ([^{3}H]QNB) as the ligand (Perry et al., 1977a; Davies and Verth, 1978; Caulfield et al., 1982). Recently a loss of M_2 with preservation of M_1 muscarinic receptor subtype has been reported (Mash et al., 1985). The M_2 receptor is located on cholinergic terminals and would seem to reflect, therefore, the loss of the ascending cholinergic projections. Adrenergic and GABA receptor binding sites do not change (Bowen et al., 1979; Cross et al., 1984) although alterations of serotonin receptors are reported (Bowen et al., 1979; Cross et al., 1984). Reduced B_{max} has been found using [^{3}H]LSD as ligand and by using spiperone displacement to define the subtype it appears to be predominantly serotonin type-2 (5-HT_2) receptors that are altered. Using [^{3}H]ketanserin to define 5-HT_2 sites, Reynolds et al. (1984) reported a loss of 42% from frontal cortex (Table V). The location of the 5-HT receptors is not known precisely but some 5-HT_2 receptors defined by [^{3}H]ketanserin may be located on the ascending cholinergic projection since they are reported to be reduced following basalis lesions (Quirion et al., 1985).

Studies of peptide receptors have been more difficult methodologically although Bowen et al. (1979) found no change in naloxone binding. Recently Beal et al. (1985) have reported a loss of somatostatin receptors from cerebral cortex.

Cerebrospinal fluid studies

Although autopsy tissue studies allow detailed analyses of peptide immunoreactivities within discrete brain areas, the relationship of post-mortem levels to in vivo changes remains uncertain. The measurement of cerebrospinal fluid (CSF) peptide immunoreactivity permits direct studies in patients but the central nervous system (CNS) source of the peptide is unknown. The general problems encountered in peptide radioimmunoassays of CSF are discussed in Bissette et al. (this volume) and Capper (this volume) but additional problems in Alzheimer's disease include changes in CSF volume due to cerebral atrophy and the lack of pathological confirmation of the diagnosis in the majority of published series.

The clinical advantage of a specific CSF marker of Alzheimer's disease would be substantial and this has led to the search for CSF correlates of the changes in the CNS observed at autopsy. Recently Soininen et al. (1984) reported reduced acetylcholinesterase activity in CSF but this had not been observed in earlier studies (Davies, 1979; Wood et al., 1982). Similarly the data on monoamine metabolites in CSF are conflicting (Gottfries et al., 1973; Mann et al., 1981; Palmer et al., 1984). More consistent has been the observation of reduced concentrations of SRIF-LI in Alzheimer's disease (Oram et al., 1981; Wood et al., 1982; Beal and Martin, 1984; Lancranjan et al., 1984; Serby et al., 1984; Soininen et al., 1984). This has also been reported in a series of histologically confirmed cases (Francis et al., 1984). However, although reduced SRIF-LI in CSF appears to be a consistent observation in Alzheimer's disease it is not specific since low concentrations have also been reported in multiple sclerosis (Sorensen et al., 1980), Parkinson's disease (Dupont et al., 1982) and in affective illness (Rubinow et al., 1983). The latter report is of particular importance since pseudodementia due to depressive illness is an important differential diagnosis to be considered in the demented patient.

A number of other peptide immunoreactivities have been reported to be reduced in the CSF of Alzheimer patients. In addition to somatostatin, TRH- and LHRH-like immunoreactivities were also reduced in the study of Oram et al. (1981), but as the authors discuss this may have related to an increase in CSF volume. Recently, reduced concentrations of adrenocorticotropic hormone (ACTH) have been found with normal levels of B-lipotropin- and B-endorphin-like immunoreactivities (Facchinetti et al., 1984).

A number of recent studies have reported low concentrations of VP-LI in CSF in Alzheimer's disease (Sundquist et al., 1983; Sorensen et al., 1983, 1985). The latter authors reported normal concentration in normal pressure hydrocephalus in which there was also ventricular enlargement, suggesting that the reduction in Alzheimer's disease is not a simple dilutional effect. Whether the change in CSF concentration reflects altered central vasopressin levels is unclear although the loss of vasopressin-containing cells from the suprachiasmatic nucleus is of interest in this regard.

Peptides in other dementias

Down's syndrome

The majority of patients with Down's syndrome who survive into their fifth decade are found at autopsy to have widespread Alzheimer-type neurofibrillary tangles. Assessment of cognitive decline may be difficult but it is believed that Down's syndrome patients also dement at this early age. Yates et al. (1983) have studied a number of cases of Down's syndrome. Cortical activity of ChAT is reduced, and this together with the reported loss of nucleus basalis neurons (Whitehouse et al., 1983) supports the view that damage to the ascending cholinergic projection is shared with Alzheimer's disease. Another common feature is a loss of noradrenaline. Peptide analyses have been limited; but no reduction in TRH-, LHRH- and SP-like immunoreactivities have been found (Yates et al., 1983a).

Parkinson's disease and Huntington's disease

A variety of cognitive deficits have been described in Parkinson's disease, but whether a significant proportion of patients develop dementia according to DSM III criteria is much less certain. A few parkinsonian patients may also share Alzheimer-type histological changes in cerebral cortex (for review see Quinn et al., 1986). Nucleus basalis counts are reduced (Nakano and Hwano, 1984) and cerebral cortical ChAT activity is low in Parkinson's disease (Dubois et al., 1983). These changes were found to be more prominent in those patients who were considered to be cognitively impaired (Ruberg et al., 1982; Whitehouse et al., 1983). Additional cortical deficits of noradrenaline and serotonin are also observed.

The peptide changes in the basal ganglia are reviewed in Agid et al. (this volume). Analysis of cortical peptides in Parkinson's disease has been less extensive. Epelbaum et al. (1983) found reduced SRIF-LI concentrations in the frontal cortex of patients who were considered retrospectively to be demented. Only a few cortical areas were examined in these studies but it is of note that reductions in both ChAT activity and SRIF-LI were found predominantly in the frontal cortex. This contrasts with the predominant temporal lobe changes of Alzheimer's disease but a similar dissociation of normal NPY-LI has been observed (Allen et al., 1985). It is not clear whether the distribution of neurochemical changes is reflected in the clinical pattern of cognitive deficit.

Dementia is also a feature of Huntington's disease and this has commonly been attributed to frontal cortical atrophy. However, this may be variable and whether the changes in the basal ganglia and their cortical connections (reviewed by Emson, this volume) may contribute to the cognitive deficits is unclear. Interestingly, despite the apparent cortical atrophy, changes in frontal cortex neurotransmitter levels are not obvious. Thus, the GABA deficit which has been the most consistent change in basal ganglia, is not found in the cerebral cortex (Bird and Iversen, 1974; Spokes et al., 1980). Similarly, the cortical peptides CCK (which is reduced in the pallidum and substantia nigra) and VIP are not altered in frontal cortex (Emson et al., 1979, 1980) although CCK receptors are reduced (Hays et al., 1981). SRIF-LI and NPY-LI which are increased in the basal ganglia, are also unaltered in cerebral cortex (Aronin et al., 1983; Dawbarn et al., 1985; see also Emson, this volume). The somatostatin loss in both Alzheimer's and Parkinson's disease appears to follow the ChAT deficit and it is of interest that ChAT activity is normal in cerebral cortex in Huntington's disease, although Spokes (1980) reported a small reduction in the septal area and hippocampus.

Other dementias

ChAT activity is normal in Pick's disease, in which there are few senile plaques (Yates et al., 1980; Wood et al., 1983) and SRIF-LI is also unchanged in cerebral cortex (Wood et al., 1983). ChAT activity was reduced in a small series of Gerstmann–Straussler disease but somatostatin levels were not reported (Wood et al., 1983). No change in ChAT activity was found in two cases of Creutzfeld–Jakob disease (Davies et al., 1982) but no peptide data have been reported.

Conclusion

As neuronal markers neuropeptides can provide useful information on the specificity of histopathological change. However, the results need to be interpreted with caution, and although it is attractive to attribute loss of peptide to neuronal degeneration this need not necessarily be so. Changes in peptide concentration may result from neuronal loss or attrition of the dendrites and axons which may be unobserved on routine histology, thus distorting relationships between quantitative histology and biochemistry. Moreover, there is no means of reliably assessing in vivo turnover in post-mortem brain and this is particularly true of peptides. Thus changes at post-mortem may reflect alterations in turnover without any loss of structural integrity of the neuron. A combination of quantitative radioimmunoassay, immunohistochemistry and specific mRNA determinations may help to resolve this problem.

The majority of peptide studies in dementia have involved Alzheimer's disease and important questions have been the relationships between change in peptide and the histopathological features of plaques, tangles and neuronal loss. The significant negative association between ChAT activity and plaque count together with acetylcholinesterase and ChAT staining indicates the involvement of cholinergic terminals in plaque formation. However, similar observations have been made with other peptides and in particular somatostatin, and the neurotransmitter specificity of plaque formation, if it exists at all, is undetermined.

Significant associations are also seen between ChAT activity and neurofibrillary tangles and neuronal counts. It is not clear whether these are two separate populations of cells or whether the development of a neurofibrillary tangle predisposes a neuron to premature death. The observed correlations need not necessarily imply a causal relationship and the relatively small number of intrinsic cortical cholinergic neurons makes it unlikely that the observed associations reflect a direct identity between the two. Similarly, it is tempting to assume that the loss of SRIF-LI is due to loss of neurons or the formation of neurofibrillary tangles. Although neurons containing tangles undoubtedly stain for somatostatin, there are probably insufficient somatostatin neurons to account for the extensive histopathological change.

In conclusion, there is evidence of neurotransmitter-specific changes involving both classical systems and peptides in the dementias and, in particular, Alzheimer's disease. Two important questions in relation to dementia remain unresolved. First, does the neurotransmitter status of a cell dictate its selective vulnerability? Second, can specific biochemical changes be related to clinically identified patterns of cognitive deficit? The analysis of peptides should help to define the biochemical changes seen in dementia and contribute to the resolution of these questions.

References

Adolfsson, R., Gottfries, C. G., Roos, B. E. and Winblad, B. (1979) Changes in brain catecholamines in patients with dementia of Alzheimer type. *Br. J. Psychiatry,* 135: 216–223.

Adrian, T. E., Allen, J. M., Bloom, S. R., Ghatei, M. A., Rossor, M. N., Roberts, G. W., Crow, T. J., Tatemotoh and Polak, J. M. (1983) Neuropeptide Y in human brain — high concentrations in basal ganglia. *Nature,* 306: 584–586.

Allen, J. M., Ferrier, I. N., Roberts, G. W., Cross, A. J., Adrian, T. E., Crow, T. J. and Bloom, S. R. (1984) Elevation of neuropeptide Y (NPY) in substantia innominata in Alzheimer's type dementia. *J. Neurol. Sci.,* 64: 325–331.

Allen, J. M., Cross, A. J., Crow, T. J., Javoy-Agid, F., Agid, Y. and Bloom, S. R. (1985) Dissociation of neuropeptide Y and somatostatin in Parkinson's disease. *Brain Res.,* 337: 197–200.

Arai, H., Kosaka, K. and Iizuka, T. (1984) Changes in biogenic amines and their metabolites in post mortem brains from patients with Alzheimer's type dementia. *J. Neurochem.,* 43: 388–393.

Armstrong, D. M., Le Roy, S., Shields, D. and Terry, R. D. (1985) Somatostatin-like immunoreactivity within neuritic plaques. *Brain Res.,* 338: 71–79.

Aronin, N., Cooper, P. E., Lorenz, L. J., Bird, E. D., Sagar, S. M., Leeman, S. E. and Martin, J. B. (1983) Somatostatin is increased in the basal ganglia in Huntington's disease. *Ann. Neurol.,* 13: 519–526.

Beal, M. F. and Martin, J. B. (1984) Somatostatin: normal and abnormal observations in the central nervous system. In R. J. Wurtman, S. H. Corkin and J. H. Growdon (Eds.), *Alzheimer's Disease: Advances in Basic Research and Therapies; Proc. IIIrd Meet. Int. Study Group on Treatment of Memory Disorders Associated with Aging, Zurich,* pp. 229–257.

Beal, M. F., Mazurek, M. F., Tran, V. H., Chattha, G., Bird, E. D. and Martin, J. B. (1985) Reduced numbers of somatostatin receptors in the cerebral cortex in Alzheimer's disease. *Science,* 229: 289–291.

Biggins, J. A., Perry, E. K., McDermott, J. R., Smith, A. I., Perry, R. H. and Edwardson, J. A. (1983) Post mortem levels of thyrotropin-releasing hormone and neurotensin in the amygdala in Alzheimer's disease, schizophrenia and depression. *J. Neurol. Sci.,* 58: 117–122.

Bird, E. D. and Iversen, L. L. (1974) Huntington's chorea, post mortem measurement of glutamic acid decarboxylase, choline acetyltransferase and dopamine in basal ganglia. *Brain,* 97: 457–472.

Bird, T. D., Stranahan, S., Sumi, S. M. and Raskind, M. (1983) Alzheimer's disease: choline acetyl transferase activity in brain tissue from clinical and pathological subgroups. *Ann. Neurol.*, 14: 284–293.

Bondareff, W., Mountjoy, C. Q. and Roth, M. (1982) Loss of neurons of origin of the adrenergic projection to cerebral cortex (nucleus locus ceruleus) in senile dementia. *Neurology*, 32: 164–168.

Bouras, C., Taban, C. H. and Constantinidis, J. (1984) Mapping of enkephalins in human brain. An immunohistofluorescence study on brains from patients with senile and presenile dementia. *Neuroscience*, 12: 179–190.

Bowen, D. M., Smith, C. B., White, P. and Davison, A. N. (1976) Neurotransmitter-related enzymes and indices of hypoxia in senile dementia and other abiotrophies. *Brain*, 99: 459–496.

Bowen, D. M., White, P., Spillane, J. A., Goodhardt, M. J., Curzon, G., Iwangoff, P., Meier-Ruge, W. and Davison, A. N. (1979) Accelerated ageing or selective neuronal loss as an important cause of dementia? *Lancet*, i: 11–14.

Bowen, D. M., Allen, S. J., Benton, J. S., Goodhardt, M. J., Haan, A., Palmer, A. M., Sims, N. R., Smith, C. C. T., Spillane, J. A., Esiri, M. M., Neary, D., Snowdon, J. S., Wilcock, G. K. and Davison, A. N. (1983) Biochemical assessment of serotonergic and cholinergic dysfunction and cerebral atrophy in Alzheimer's disease. *J. Neurochem.*, 41: 266–272.

Caffe, A. R. and van Leeuwen, F. W. (1983) Vasopressin immunoreactive cells in the dorsomedial hypothalamic region, medial amygdaloid nucleus and locus coeruleus of the rat. *Cell Tissue Res.*, 233: 23–33.

Candy, J. M., Perry, R. H., Perry, E. K. and Thompson, J. E. (1981) Distribution of putative cholinergic cell bodies and various neuropeptides in the substantia innominata region of the human brain. *J. Anat.*, 133: 123.

Caulfield, M. P., Straughan, D. W., Cross, A. J., Crow, T. J. and Birdsall, N. J. M. (1982) Cortical muscarinic receptor subtypes in Alzheimer's disease. *Lancet*, ii: 1277.

Cooper, P. E., Fernstrom, M. H., Rorstad, O. P., Leeman, S. E. and Martin, J. B. (1981) The regional distribution of somatostatin, substance P and neurotensin in human brain. *Brain Res.*, 218: 219–232.

Coyle, J. T., Price, D. L. and De Long, M. R. (1983) Alzheimer's disease: a disorder of cortical cholinergic innervation. *Science*, 219: 1184–1190.

Cross, A. J., Crow, T. J., Perry, E. K., Perry, R. H., Blessed, G. and Tomlinson, B. E. (1981) Reduced dopamine-B-hydroxylase activity in Alzheimer's disease. *Br. Med. J.*, 1: 93–94.

Cross, A. J., Crow, T. J., Johnson, J. A., Joseph, M. H., Perry, E. K., Perry, R. H., Blessed, G. and Tomlinson, B. E. (1983) Monoamine metabolism in senile dementia of Alzheimer type. *J. Neurol. Sci.*, 60: 383–392.

Cross, A. J., Crow, T. J., Johnson, J. A., Perry, E. K., Perry, R. H., Blessed, G. and Tomlinson, B. E. (1984) Studies on neurotransmitter receptor systems in neocortex and hippocampus in senile dementia of the Alzheimer type. *J. Neurol. Sci.*, 64: 109–117.

Crystal, H. A. and Davies, P. (1982) Cortical substance P-like immunoreactivity in cases of Alzheimer's disease and senile dementia of the Alzheimer type. *J. Neurochem.*, 38: 1781–1784.

Curcio, C. A. and Kemper, T. (1984) Nucleus raphe dorsalis in dementia of the Alzheimer type: neurofibrillary changes and neuronal packing density. *J. Neuropathol. Exp. Neurol.*, 43: 359–368.

Davies, P. (1979) Neurotransmitter related enzymes in senile dementia of the Alzheimer type. *Brain Res.*, 171: 319–327.

Davies, P. and Maloney, A. J. (1976) Selective loss of central cholinergic neurons in Alzheimer's disease. *Lancet*, ii: 1403.

Davies, P. and Terry, R. D. (1981) Cortical somatostatin-like immunoreactivity in cases of Alzheimer's disease and senile dementia of the Alzheimer type. *Neurobiol. Aging*, 2: 9–14.

Davies, P. and Verth, A. H. (1978) Regional distribution of muscarinic acetylcholine receptors in normal and Alzheimer's type dementia brains. *Brain Res.*, 138: 385–392.

Davies, P., Katzman, R. and Terry, R. D. (1980) Reduced somatostatin-like immunoreactivity in cerebral cortex from cases of Alzheimer's disease and Alzheimer senile dementia. *Nature*, 288: 279–280.

Davies, P., Katz, D. A. and Crystal, H. A. (1982) Choline acetyltransferase, somatostatin, and substance P in selected cases of Alzheimer's disease. In S. Corkin, K. L. Davis, J. H. Growdon, E. Usdin and R. J. Wurtman (Eds.), *Alzheimer's Disease: A Report of Progress (Aging, Vol. 19)*, Raven Press, New York, pp. 9–14.

Dawbarn, D. and Emson, P. C. (1985) Neuropeptide Y like immunoreactivity in neuritic plaques of Alzheimer's disease. *Biochem. Biophys. Res. Commun.*, 126: 289–293.

Dawbarn, D., Rossor, M. N., Mountjoy, C. Q., Roth, M. and Emson, P. C. (1985) Decreased somatostatin immunoreactivity but not neuropeptide Y immunoreactivity in cortex in senile dementia of Alzheimer type. Submitted to *Neurosci. Lett.*

Delfs, J. R., Zhu, C. H. and Dichter, M. A. (1984) Coexistence of acetylcholinesterase and somatostatin-immunoreactivity in neurons cultured from rat cerebrum. *Science*, 223: 61–63.

Dubois, B., Ruberg, M., Javoy-Agid, F., Ploska, A. and Agid, Y. (1983) A subcortico-cortical cholinergic system is affected in Parkinson's disease. *Brain Res.*, 288: 213–218.

Dupont, E., Christensen, S. E., Hansen, A. P., Olivarias, B. F. and Orskou, H. (1982) Low cerebrospinal fluid somatostatin in Parkinson's disease: an irreversible abnormality. *Neurology*, 32: 312–314.

Eckenstein, F. and Baughman, R. W. (1984) Two types of cholinergic innervation in cortex, one co-localised with vasoactive intestinal polypeptide. *Nature*, 309: 153–155.

Emson, P. C., Fahrenkrug, J. and Spokes, E. G. S. (1979) Vasoactive intestinal polypeptide (VIP): distribution in normal human brain and in Huntington's disease. *Brain Res.*, 173: 174–178.

Emson, P. C., Rehfield, J. F., Langevin, H. and Rossor, M. N. (1980) Reduction in cholecystokinin-like immunoreactivity in the basal ganglia in Huntington's disease. *Brain Res.*, 198: 497–500.

Emson, P. C., Rossor, M. N. and Lee, C. M. (1981) The regional distribution and chromatographic behaviour of somatostatin in human brain. *Neurosci. Lett.*, 22: 319–324.

Epelbaum, J., Ruberg, M., Moyse, E., Javoy-Agid, F., Dubois, B. and Agid, Y. (1983) Somatostatin and dementia in Parkinson's disease. *Brain Res.*, 278: 376–379.

Facchinetti, F., Nappi, G., Petraglia, F., Martignoni, E., Sinforiani, E. and Genazzani, A. R. (1984) Central ACTH deficit in degenerative and vascular dementia. *Life Sci.*, 35: 1691–1697.

Ferrier, I. N., Cross, A. J., Johnson, J. A., Roberts, G. W., Crow, T. J., Corsellis, J. A. N., Lee, Y. C., O'Shaughnessy, D., Adrian, T. E., McGregor, G. P., Baracese-Hamilton, A. J. and Bloom, S. R. (1983) Neuropeptides in Alzheimer type dementia. *J. Neurol. Sci.*, 62: 159–170.

Fliers, E., Swaab, D. F., Pool, C. R. and Verwer, R. W. H. (1985) The vasopressin and oxytocin neurons in the human supraoptic and periventricular nucleus: changes with aging and in senile dementia. *Brain Res.*, 342: 45–53.

Forno, L. S. (1978) The locus coeruleus in Alzheimer's disease. *J. Neuropathol. Exp. Neurol.*, 37: 614 (Abstr. 100).

Francis, P. T. and Bowen, D. M. (1985) Relevance of reduced concentrations of somatostatin in Alzheimer's disease. *Biochem. Soc. Trans.*, 13: 170–171.

Francis, P. T., Bowen, D. M., Neary, D., Palo, J., Wikstrom, J. and Olney, J. (1984) Somatostatin-like immunoreactivity in lumbar cerebrospinal fluid from neurohistologically examined demented patients. *Neurobiol. Aging*, 5: 183–186.

Francis, P. T., Palmer, A. M., Sims, N. R., Bowen, D. M., Davison, A. N., Esiri, M. M., Neary, D., Snowden, J. S. and Wilcock, G. K. (1985) Neurochemical studies of early-onset Alzheimer's disease. *N. Engl. J. Med.*, 313: 7–11.

Gottfries, C. G., Gottfries, I. and Roos, B. E. (1969) Homovanillic acid and 5-hydroxyindoleacetic acid in the cerebrospinal fluid of patients with senile dementia, presenile dementia and parkinsonism. *J. Neurochem.*, 16: 1341–1345.

Hardy, J., Adolfsson, R., Alafuzoff, I., Bucht, G., Marcusson, J., Nyberg, P., Perdahl, E., Wester, P. and Winblad, B. (1985) Transmitter deficits in Alzheimer's disease. *Neurochem. Int.*, 7: 545–563.

Hays, S. E., Goodwin, F. K. and Paul, S. M. (1981) Cholecystokinin receptors are decreased in basal ganglia and cerebral cortex of Huntington's disease. *Brain Res.*, 225: 452–456.

Hendry, S. H. C., Jones, E. G., De Felipe, J., Schmechel, D., Brandon, C. and Emson, P. C. (1984) Neuropeptide-containing neurons of the cerebral cortex are also GABAergic. *Proc. Natl. Acad. Sci. U.S.A.*, 81: 6526–6530.

Herzog, A. G. and Kemper, T. L. (1980) Amygdaloid changes in aging and dementia. *Arch. Neurol.*, 37: 625–629.

Hökfelt, T., Lundberg, J. M., Lagercrantz, M., Tatemotok, Mutt, V., Lindberg, J., Terenius, L., Everitt, B. J., Fuxe, K., Agnati, L. and Goldstein, M. (1983) Occurrence of neuropeptide Y (NPY)-like immunoreactivity in catecholamine neurons in the human medulla oblongata. *Neurosci. Lett.*, 36: 217–222.

Hong, J. S. and Schmid, R. (1981) Intrahippocampal distribution of Met-enkephalin. *Brain Res.*, 205: 415–418.

Hunt, S. P., Emson, P. C., Gilbert, R., Goldstein, M. and Kimmell, J. R. (1981) Presence of avian pancreatic polypeptide-like immunoreactivity in catecholamine and methionine-enkephalin containing neurones within the central nervous system. *Neurosci. Lett.*, 21: 125–130.

Ishii, T. (1966) Distribution of Alzheimer's neurofibrillary changes in the brain stem and hypothalamus of senile dementia. *Acta Neuropathol.*, 6: 181–187.

Kitt, C. A., Price, D. L., Struble, R. G., Cork, L. C., Wainer, B. H., Becher, M. W. and Mobley, W. C. (1984a) Evidence for cholinergic neurites in senile plaques. *Science*, 226: 1443–1445.

Kitt, C. A., Mobley, W. C., Struble, R. G., Cork, L. C., Walker, L. C., Becker, M. W., Joh, T. and Price, D. L. (1984b) Contribution of catecholaminergic systems to neurites in plaques of aged primates. *Ann. Neurol.*, 16: 118.

Kovacs, G. L., Bohus, B. and Versteeg, D. H. G. (1979) The effects of vasopressin on memory processes: the role of noradrenergic neurotransmission. *Neuroscience*, 4: 1529–1537.

Lancranjan, I., Adolfsson, R., Bucht, G., Winblad, B., Groot, K. and Arimura, A. (1984) Decrease of somatostatin-like immunoreactivity in the CSF of patients with Alzheimer's disease. In R. J. Wurtman, S. H. Corkin and J. H. Growdon (Eds.), *Alzheimer's Disease: Advances in Basic Research and Therapies, Proc. IIIrd Meet. Int. Study Group on Treatment of Memory Disorders Associated with Aging, Zurich*, p. 455.

Mann, J. J., Stanley, M., Neophytides, A., De Leon, M. J., Ferris, S. H. and Gershon, S. (1981) Central amine metabolism in Alzheimer's disease: in vivo relationship to cognitive deficit. *Neurobiol. Aging*, 2: 57–60.

Mash, D. C., Flynn, D. D. and Potter, L. T. (1985) Loss of M_2 receptors in the cerebral cortex in Alzheimer's disease and experimental cholinergic denervation. *Science*, 228: 1115–1117.

McDonald, J. K., Parnavelas, J. G., Karamanlidis, A. N., Brecha, N. and Koenig, J. I. (1982) The morphology and distribution of peptide-containing neurons in the adult and developing visual cortex of the rat. I. Somatostatin. *J. Neurocytol.*, 11: 809–824.

McKinney, M., Davies, P. and Coyle, J. T. (1982) Somatostatin is not co-localised in cholinergic neurons innervating the rat cerebral cortex-hippocampal formation. *Brain Res.*, 243: 169–172.

Morrison, J. H., Benoit, R., Magistretti, P. J. and Bloom, F. E. (1983) Immunohistochemical distribution of pro-somatostatin-related peptides in cerebral cortex. *Brain Res.*, 262: 344–351.

Morrison, J. H., Rogers, J., Scherr, S., Benoit, R. and Bloom, F. E. (1985) Somatostatin immunoreactivity in neuritic plaques of Alzheimer's patients. *Nature,* 314: 90–92.

Mountjoy, C. Q., Rossor, M. N., Iversen, L. L. and Roth, M. (1984) Correlation of cortical cholinergic and GABA deficits with quantitative neuropathological findings in senile dementia. *Brain,* 107: 507–518.

Nagai, R., McGeer, P. L., Peng, J. H., McGeer, E. G. and Dolman, C. E. (1983) Choline acetyltransferase immunohistochemistry in brains of Alzheimer's disease patients and controls. *Neurosci. Lett.,* 36: 195–199.

Nakano, I. and Harano, A. (1984) Parkinson's disease: neuron loss in the nucleus basalis without concomitant Alzheimer's disease. *Ann. Neurol.,* 15: 415–418.

Nemeroff, C. B., Bissette, G., Busby, W. H., Youngblood, W. W., Rossor, M. N., Roth, M. and Kizer, J. S. (1983) Regional brain concentrations of neurotensin, thyrotropin releasing hormone and somatostatin in Alzheimer's disease. *Neurosci. Abstr.,* 9: 1052.

Olpe, H. R. and Baltzer, V. (1981) Vasopressin activates noradrenergic neurons in the rat locus coeruleus: a microiontophoretic investigation. *Eur. J. Pharmacol.,* 73: 377–378.

Oram, J. T., Edwardson, J. A. and Millard, P. M. (1981) Investigation of cerebrospinal fluid neuropeptides in idiopathic senile dementia. *Gerontology,* 27: 216–223.

Palmer, A. M., Sims, N. S., Bowen, D. M., Neary, D., Palo, J., Wikstrom, J. and Davison, A. N. (1984) Monoamine metabolite concentrations in lumbar cerebrospinal fluid of patients with histology verified Alzheimer's dementia. *J. Neurol. Neurosurg. Psychiatry,* 47: 481–484.

Paxinos, G., Emson, P. C. and Cuello, A. C. (1978) The substance P projections to the frontal cortex and the substantia nigra. *Neurosci. Lett.,* 7: 127–131.

Pearson, R. C. A., Sofroniew, M. V., Cuello, A. C., Powell, T. P. S., Eckenstein, F., Esiri, M. M. and Wilcock, G. K. (1983) Persistence of cholinergic neurons in the basal nucleus in a brain with senile dementia of the Alzheimer's type demonstrated by immunohistochemical staining for choline acetyltransferase. *Brain Res.,* 289: 375–379.

Perry, E. K. and Perry, R. H. (1980) The cholinergic system in Alzheimer's disease. In P. J. Roberts (Ed.), *Biochemistry of Dementia,* John Wiley & Sons, Chichester, pp. 135–183.

Perry, E. K., Perry, R. H., Blessed, G. and Tomlinson, B. E. (1977a) Necropsy evidence of central cholinergic deficits in senile dementia. *Lancet,* i: 189.

Perry, E. K., Gibson, P. H., Blessed, G., Perry, R. H. and Tomlinson, B. E. (1977b) Neurotransmitter enzyme abnormalities in senile dementia. *J. Neurol. Sci.,* 34: 247–265.

Perry, E. K., Tomlinson, B. E., Blessed, G., Bergmann, K., Gibson, P. H. and Perry, R. H. (1978) Correlation of cholinergic abnormalities with senile plaques and mental test scores in senile dementia. *Br. Med. J.,* 2: 1457–1459.

Perry, R. H., Blessed, G., Perry, E. K. and Tomlinson, B. E. (1980) Histochemical observations on cholinesterase activities in the brains of elderly normal and demented (Alzheimer type) patients. *Age Aging,* 9: 9–16.

Perry, E. K., Blessed, G., Tomlinson, B. E., Perry, R. H., Crow, T. J., Cross, A. J., Dockray, G. J., Dimaline, R. and Arregui, A. (1981a) Neurochemical activities in human temporal lobe related to aging and Alzheimer-type changes. *Neurobiol. Aging,* 2: 251–256.

Perry, R. H., Dockray, G. J., Dimaline, R., Perry, E. K., Blessed, G. and Tomlinson, B. E. (1981b) Neuropeptides in Alzheimer's disease, depression and schizophrenia: A post mortem analysis of vasoactive intestinal polypeptide and cholecystokinin in cerebral cortex. *J. Neurol. Sci.,* 51: 465–472.

Perry, R. H., Candy, J. M. and Perry, E. K. (1983) Some observations and speculations concerning the cholinergic system and neuropeptides in Alzheimer's disease. In R. Katzman (Ed.), *Biological Aspects of Alzheimer's Disease. Banbury Report 15,* Cold Spring Harbor, New York, pp. 351–361.

Probst, A., Basler, V., Bron, B. and Ulrich, J. (1983) Neuritic plaques in senile dementia of Alzheimer type: A Golgi analysis in the hippocampal region. *Brain Res.,* 268: 249–254.

Quinn, N. P., Rossor, M. N. and Marsden, C. D. (1986) Dementia and Parkinson's disease — pathological and neurochemical considerations. *Br. Med. Bull.,* 42.1: 86–90.

Quirion, R., Richard, J. and Dam, T. V. (1985) Evidence for the existence of serotonin type-2 receptors on cholinergic terminals in rat cortex. *Brain Res.,* 333: 345–349.

Reynolds, G. P., Arnold, L., Rossor, M. N., Iversen, L. L., Mountjoy, C. Q. and Roth, M. (1984) Reduced binding of $[^3H]$ketanserin to cortical 5-HT2 receptors in senile dementia of the Alzheimer type. *Neurosci. Lett.,* 44: 47–51.

Roberts, G. W., Crow, T. J. and Polak, J. M. (1985) Location of neuronal tangles in somatostatin neurones in Alzheimer's disease. *Nature,* 314: 92–94.

Rossor, M. N. (1982) Neurotransmitters in CNS disease: dementia. *Lancet,* ii: 1200–1204.

Rossor, M. N., Emson, P. C., Mountjoy, C. Q., Roth, M. and Iversen, L. L. (1980a) Reduced amounts of immunoreactive somatostatin in the temporal cortex in senile dementia of Alzheimer type. *Neurosci. Lett.,* 20: 373–377.

Rossor, M. N., Fahrenkrug, J., Emson, P., Mountjoy, C., Iversen, L. and Roth, M. (1980b) Reduced cortical choline acetyltransferase activity in senile dementia of Alzheimer type is not accompanied by changes in vasoactive intestinal polypeptide. *Brain Res.,* 201: 249–253.

Rossor, M. N., Iversen, L. L., Mountjoy, C. Q., Roth, M., Hawthorn, J., Ang, V. T. Y. and Jenkins, J. S. (1980c) Arginine vasopressin and choline acetyltransferase in brains of patients with Alzheimer type senile dementia. *Lancet,* ii: 1367.

Rossor, M. N., Iversen, L. L., Hawthorn, J., Ang, V. T. Y. and Jenkins, J. S. (1981a) Extrahypothalamic vasopressin in human brain. *Brain Res.,* 214: 349–355.

Rossor, M. N., Iversen, L. L., Johnson, A. L., Mountjoy, C. Q. and Roth, M. (1981b) The cholinergic defect of the frontal cortex in Alzheimer's disease is age dependent. *Lancet,* ii: 1422.

Rossor, M. N., Rehfeld, J. F., Emson, P. C., Mountjoy, C. Q., Roth, M. and Iversen, L. L. (1981c) Normal cortical concentrations of cholecystokinin-like immunoreactivity with reduced choline acetyltransferase activity in senile dementia of the Alzheimer type. *Life Sci.*, 29: 405–410.

Rossor, M. N., Emson, P. C., Mountjoy, C. Q., Roth, M. and Iversen, L. L. (1982a) Neurotransmitters of the cerebral cortex in senile dementia of Alzheimer type. The Aging Brain. *Exp. Brain Res. Suppl.*, 5: 153–157.

Rossor, M. N., Garrett, N. J., Johnson, A. L., Mountjoy, C. Q., Roth, M. and Iversen, L. L. (1982b) A post mortem study of the cholinergic and GABA systems in senile dementia. *Brain*, 105: 313–330.

Rossor, M. N., Hunt, S. P., Iversen, L. L., Bannister, R., Hawthorn, J., Ang, V. T. Y. and Jenkins, J. S. (1982c) Extrahypothalamic vasopressin is unchanged in Parkinson's disease and Huntington's disease. *Brain Res.*, 252: 341–343.

Rossor, M. N., Svendsen, C., Hunt, S. P., Mountjoy, C. Q., Roth, M. and Iversen, L. L. (1982d) The substantia innominata in Alzheimer's disease: an histochemical and biochemical study of cholinergic marker enzymes. *Neurosci. Lett.*, 28: 217–222.

Rossor, M. N., Emson, P. C., Iversen, L. L., Mountjoy, C. Q. and Roth, M. (1984a) Patterns of neuropeptide deficits in Alzheimer's disease. In R. J. Wurtman, S. H. Corkin and J. H. Growdon (Eds.), *Advances in Basic Research and Therapies*, Center for Brain Sciences and Metabolism Charitable Trust, Zurich, pp. 29–38.

Rossor, M. N., Iversen, L. L., Reynolds, G. P., Mountjoy, C. Q. and Roth, M. (1984b) Neurochemical characteristics of early and late onset types of Alzheimer's disease. *Br. Med. J.*, 288: 961–964.

Roth, M. and Hopkins, B. (1953) Psychological test performance in patients over sixty. I. Senile psychosis and the affective disorders of old age. *J. Ment. Sci.*, 99: 439–450.

Ruberg, M., Ploska, A., Javoy-Agid, F. and Agid, Y. (1982) Muscarinic binding and choline acetyltransferase activity in parkinsonian subjects with reference to dementia. *Brain Res.*, 232: 129–139.

Rubinow, D. R., Gold, P. W., Post, R. M., Ballenger, J. C., Cowdry, R., Bollinger, J. and Reichlin, S. (1983) CSF somatostatin in affective illness. *Arch. Gen. Psychiatry*, 40: 409–412.

Sar, M., Stumpf, W. E., Miller, R. J., Chang, K. J. and Cuatracasas, P. (1978) Immunohistochemical localization of enkephalin in rat brain and spinal cord. *J. Comp. Neurol.*, 182: 17–38.

Schmechel, D. E., Vickrey, B. G., Fitzpatrick, D. and Elde, R. P. (1984) GABAergic neurons of mammalian cerebral cortex: widespread subclass in deep layers defined by somatostatin content. *Neurosci. Lett.*, 47: 227–232.

Serby, M., Richardson, S. B., Twente, S., Siekerski, J., Corwin, J. and Rotrosen, J. (1984) CSF somatostatin in Alzheimer's disease. *Neurobiol. Aging*, 5: 187–190.

Sims, N. R., Bowen, D. M., Smith, C. C. T., Flack, R. H. A., Davison, A. N., Snowden, J. S. and Neary, D. (1980) Glucose metabolism and acetylcholine synthesis in relation to neuronal activity in Alzheimer's disease. *Lancet*, i: 333–335.

Smith, C. C. T., Bowen, D. M., Sims, N. R., Neary, D. and Davison, A. N. (1983) Amino acid release from biopsy samples of temporal neocortex from patients with Alzheimer's disease. *Brain Res.*, 264: 138–141.

Soininen, H., Jolkkonen, J. T., Reinikainen, K. J., Halonen, T. O. and Riekkinen, P. J. (1984) Reduced cholinesterase activity and somatostatin-like immunoreactivity in the cerebrospinal fluid of patients with dementia of the Alzheimer type. *J. Neurol. Sci.*, 63: 167–172.

Sorensen, K. V. (1982) Somatostatin: localization and distribution in the cortex and the subcortical white matter of human brain. *Neuroscience*, 7: 1227–1232.

Sorensen, K. V., Christensen, S. E., Dupont, E., Hansen, A. P., Pedersen, E. and Orskou, A. (1980) Low somatostatin content in cerebrospinal fluid in multiple sclerosis. *Acta Neurol. Scand.*, 61: 186–191.

Sorensen, P. S., Hammer, M., Vorstrup, S. and Gjerris, F. (1983) CSF and plasma vasopressin concentrations in dementia. *J. Neurol. Neurosurg. Psychiatry*, 46: 911–916.

Sorensen, P. S., Gjerris, A. and Hammer, M. (1985) Cerebrospinal fluid vasopressin in neurological and psychiatric disorders. *J. Neurol. Neurosurg. Psychiatry*, 48: 50–57.

Spillane, J. A., White, P., Goodhardt, M. J., Flack, R. H. A., Bowen, D. M. and Davison, A. N. (1977) Selective vulnerability of neurons in organic dementia. *Nature*, 266: 558–559.

Spokes, E. G. S. (1980) Neurochemical alterations in Huntington's chorea. A study of post mortem brain tissue. *Brain*, 103: 179–210.

Spokes, E. G. S., Garrett, N. J., Rossor, M. N. and Iversen, L. L. (1980) Distribution of GABA in post mortem brain tissue from control, psychotic and Huntington's chorea subjects. *J. Neurol. Sci.*, 46: 303–313.

Struble, R. G., Cork, L. C., Whitehouse, P. J. and Price, D. L. (1982) Cholinergic innervation in neuritic plaques. *Science*, 216: 413–415.

Sundquist, J., Forsling, M. L., Olsson, J. E. and Ackerlund, M. (1983) Cerebrospinal fluid arginine vasopressin in degenerative and other neurological illnesses. *J. Neurol. Neurosurg. Psychiatry*, 45: 14–17.

Swaab, D. F., Fliers, E. and Partiman, T. S. (1985) The suprachiasmatic nucleus of the human brain in relation to sex, age and senile dementia. *Brain Res.* (in press).

Tomlinson, B. E., Irving, D. and Blessed, G. (1981) Cell loss in the locus coeruleus in senile dementia of Alzheimer type. *J. Neurol. Sci.*, 49: 419–428.

Vincent, S. R. and McGeer, E. G. (1981) A substance P projection to the hippocampus. *Brain Res.*, 215: 349–351.

Vincent, S. R., Johansson, O., Hökfelt, T., Meyerson, B., Sachs, C., Elde, R. P., Terenius, L. and Kinmel, J. (1982) Neuropeptide coexistence in human cortical neurones. *Nature*, 298: 65–67.

Vincent, S. R., Satoh, K., Armstrong, D. M. and Fibiger, H. C. (1983) Substance P in the ascending cholinergic reticular system. *Nature,* 306: 688–691.

Weingartner, H., Gold, P., Ballenger, J. C., Smallberg, S. A., Summers, R., Rubinow, D. R., Post, R. M. and Goodwin, F. K. (1981) Effects of vasopressin on human memory functions. *Science,* 211: 601–603.

Whitehouse, P. J., Price, D. L., Struble, R. G., Clark, A. W., Coyle, J. T. and De Long, M. R. (1982) Alzheimer's disease and senile dementia: loss of neurons in the basal forebrain. *Science,* 215: 1237–1239.

Whitehouse, P. J., Hedreen, J. C., White, C. L. and Price, D. L. (1983) Basal forebrain neurons in the dementia of Parkinson's disease. *Ann. Neurol.,* 13: 243–248.

Wilcock, G. K., Esiri, M. M., Bowen, D. M. and Smith, C. C. T. (1982) Alzheimer's disease: correlation of cortical choline acetyltransferase activity with the severity of dementia and histological abnormalities. *J. Neurol. Sci.,* 57: 407–417.

Wood, P. L., Etienne, P., Lal, S., Gauthier, S., Cajal, S. and Nair, N. P. V. (1982) Reduced lumbar CSF somatostatin levels in Alzheimer's disease. *Life Sci.,* 31: 2073–2079.

Wood, P. L., Etienne, P., Lal, S., Nair, N. P. V., Finlayson, M. H., Gauthier, S., Palo, J., Haltia, M., Paetau, A. and Bird, E. D. (1983) A post mortem comparison of the cortical cholinergic system in Alzheimer's disease and Pick's disease. *J. Neurol. Sci.,* 62: 211–217.

Yates, C. M., Simpson, J., Maloney, A. F. J. and Gordon, A. (1980) Neurochemical observations in a case of Pick's disease. *J. Neurol. Sci.,* 48: 257–263.

Yates, C. M., Harmar, A. J., Rosie, R., Sheward, J., Sanchez de Levy, G., Simpson, J., Maloney, A. F. J., Gordon, A. and Fink, G. (1983a) Thyrotropin-releasing hormone, luteinizing hormone-releasing hormone and substance P immunoreactivity in post-mortem brain from cases of Alzheimer-type dementia and Down's syndrome. *Brain Res.,* 258: 45–52.

Yates, C. M., Simpson, J., Gordon, A., Maloney, A. F. J., Allison, Y., Ritchie, I. M. and Urquhart, A. (1983b) Catecholamines and cholinergic enzymes in pre-senile and senile Alzheimer-type dementia and Down's syndrome. *Brain Res.,* 280: 119–126.

P. C. Emson, M. N. Rossor and M. Tohyama (Eds.),
Progress in Brain Research, Vol. 66.

CHAPTER 9

Neuropeptides and schizophrenia

Garth Bissette[1], Charles B. Nemeroff[1,2] and Angus V. P. MacKay[3]

Departments of [1]Psychiatry and [2]Pharmacology and the Center of Aging and Human Development, Duke University Medical Center, Durham, NC 27710, U.S.A. and [3]Department of Psychological Medicine, University of Glasgow, Glasgow, U.K.

Introduction

Schizophrenia is a severe psychiatric disorder that affects approximately 1% of populations across most of the cultures studied to date (Babigian, 1980). While the precise boundaries of the diagnosis of schizophrenic illness remain controversial, most clinicians associate the term with a syndrome of disordered perception, speech and behaviour which is not affectively based and which is not the expression of some demonstrable organic lesion. A major improvement in research methodology in recent years has been the increasing tendency to adopt rigorous and explicit diagnostic instruments for selecting samples of patients who are to be the subject of research. Three such sets of criteria, to which reference will be made below, are those of Wing et al. (1974), the Research Diagnostic Criteria (RDC) of Spitzer et al. (1975) and the earlier criteria of Feighner et al. (1972). The etiology of this group of diseases remains unknown despite intensive research (Crow, 1982a,b; Stevens, 1983; Weinberger et al., 1983). The lack of adequate animal models of schizophrenia (Kolpakov et al., 1983) has severely limited our understanding of the biochemical pathology of this disorder. Investigators have, therefore, been forced to seek biochemical abnormalities in accessible biological fluids, e.g. plasma or cerebrospinal fluid (CSF), and post-mortem brain tissue of schizophrenic patients. Although several theories concerning the involvement of a particular transmitter or metabolite in schizophrenia have been advanced (Snyder, 1982), only the dopamine hyperactivity hypothesis has continued to gain acceptance, despite its limitations (Lipton and Nemeroff, 1978; Haracz, 1982). The genetic diversity of human populations and the problem of misdiagnosis (or poor diagnostic reliability) are other confounding variables that hinder progress in schizophrenia research. Recently, investigators have begun to focus on the possible alteration of neuropeptidergic systems in schizophrenia. The neuropeptides vastly outnumber the conventional amine and amino acid transmitters, and several forms of potent interaction with conventional neurotransmitters have been demonstrated (MacKay, 1985). If neurochemical disturbance underlies schizophrenic phenomena, then it is not unreasonable to assume that one or more of the neuropeptides will eventually turn out to be implicated — although at the present time there is no marker to point the way as in the case of the neuroleptics and dopamine. While this relatively new research area promised to contribute to a better understanding of the neurochemical pathology and to the development of novel treatments of schizophrenia, there are several constraints that limit the information that can now be obtained. First, it is important to recognise that, with few exceptions, neuropeptides are present in brain in extremely low concentrations; in the picomolar and femtomolar range, while the "classical neurotransmitters", such as glutamate and the monoamines, are, in contrast, present in nanomolar and millimolar concentrations. The low concentrations of neuropeptides necessitates the use of sensitive measurement techniques such as radioimmunoassay (RIA). Unfortunately RIA has the potential disadvantage of lack of specificity in that peptide precursors (prohormones) and metabolites as well as the peptide under study may crossreact with the antiserum. Immunoreactivity depends on a variety of factors such as tertiary structure, charge distribution, and amino acid sequence homology. Two or more antisera directed at different sequences of the neuropeptide of interest combined with separation and purification of the immunoreactive material by high performance liquid chromatography (HPLC) are required to identify absolutely the substance being measured by RIA; unfortunately this requirement is rarely met. Both ultraviolet detection and electrochemical detections coupled with HPLC are being increasingly utilized to identify the nature of neuropeptide immunoreactivity in biological fluids. Interfering substances are often first suspected when serial dilutions of biological fluids or tissues do not, when measured by RIA, exhibit corresponding dilution of peptide concentrations. Another problem in neuropeptide research is the lack of specific peptide antagonists or neurotoxins. Recently, proglumide has been shown to be a cholecystokinin (CCK) antagonist (Collins et al., 1983) and cysteamine depletes the central nervous system (CNS) of somatostatin (SRIF) (Palkovits et al., 1982); the specificity of these compounds has not yet been fully characterized. Opioid receptor antagonists (e.g. naloxone and naltrexone) are, of course, currently available. Thus, assessment of the physiological or pathological role of a particular peptide has often been limited to the measurement of peptide concentrations; measurement of neuropeptide turnover is certainly not routinely possible at this time. With these limitations in mind, the available data on neuropeptide concentrations in CSF and post-mortem brain tissue of schizophrenic patients can be examined more closely.

In spite of the problems described above, a multitude of published studies are available on measurement of neuropeptides in schizophrenia. These studies are very difficult to compare. They often use different diagnostic criteria to define schizophrenia, different laboratory methodologies (extraction techniques, RIA vs. radioreceptor assays, etc.) or use other psychiatric or neurologic disease groups as controls. The lack of healthy volunteers as controls can be crucial, as illustrated in a recent report (Fessler et al., 1984) in which the contrast medium used in myeloencephalography was found to elevate CSF β-endorphin concentrations, thereby eliminating the use of neurologic controls undergoing this procedure in the studies of endogenous opioids (and perhaps other peptides). Almost all of the studies of post-mortem human brain tissue of schizophrenics contain patients who have received chronic neuroleptic drug treatment. This may be a significant confounding variable because several reports indicate that the concentration of several neuropeptides, including Met-enkephalin (MET-ENK) (Hong et al., 1978), substance P (SP) (Hong et al., 1978; Hanson et al., 1981), neurotensin (NT) (Govoni et al., 1980; G. Bissette, C. D. Kilts and C. B. Nemeroff, unpublished observations), and CCK octapeptide (CCK-8) (Frey, 1983), are changed in specific rat brain regions after neuroleptic drug treatment. The density of nigrostriatal NT receptors in both rat and human (Uhl and Kuhar, 1984) and CCK receptors in rat (Chang et al., 1983), have also been shown to be elevated after neuroleptic drug treatment. In many of the CSF studies in which chronic, neuroleptic-treated schizophrenics were studied, a 2–4 week drug-free interval prior to lumbar puncture was included, but the adequacy of such a short respite in reversing drug-induced changes in neuropeptide concentrations is not known. In post-mortem tissue studies this latter strategy is impossible. Finally, many of the existing studies either do not report the method of statistical analysis employed or use inappropriate statistical tests. In spite of these problems, progress has been made in determining whether certain neuropeptidergic systems are altered in schizophrenia.

CSF studies

Opioid peptides

The peptides that have undoubtedly received the most attention in neurobiology are the endorphins, and this is similarly true in schizophrenia research. Named for the endogenous opiate receptor ligand that was known to exist since 1972, the endorphins are now known to comprise a family of peptides. A plethora of reviews on opioid peptides and their potential role in psychiatric disorders is available (Terenius, 1978; Vereby et al., 1978; Davis et al., 1979; Buchsbaum, 1980; van Praag et al., 1980; MacKay, 1981; van Ree and de Wied, 1981; Berger, 1983; Koob et al., 1984).

A group of Swedish investigators, using radioreceptor assay, were the first to report changes in endogenous opioid concentrations in the CSF of schizophrenics (Terenius et al., 1976). Their radioreceptor assay used [^{3}H]dihydromorphine as the ligand and measured total endogenous opioid activity without discrimination as to which particular opioid was present. In the original report, the CSF opioid activity of four chronic schizophrenics (ill for at least 10 years, but drug-free for 4 weeks) was compared before, and 2 and 4 weeks after clozapine treatment. The diagnostic criteria used were not contained in this early report. The CSF was filtered, chromatographed on Sephadex G-10 and two fractions with opiate receptor activity were isolated (fraction I and II). MET-ENK co-eluted with fraction II, but only the concentration of fraction I was changed after neuroleptic treatment. Two of the four schizophrenic patients had elevated CSF fraction I opioid concentrations after 4 weeks of neuroleptic treatment, but no statistical test was performed on this small sample. In a later study, Lindstrom et al. (1978) reported that when CSF opioid activity, as defined by this same radioreceptor assay, was measured in nine chronic schizophrenics after a drug-free period of 1–2 months, six of the schizophrenics had higher fraction I opioid levels than the mean fraction I concentration of 19 normal volunteers. Moreover, the schizophrenics with higher CSF fraction I levels, when retested after treatment with either clozapine, flupenthixol or chlorpromazine (12 days to 2 months), exhibited values close to the normal mean control value. Again, no statistical test was employed to evaluate the data. Another study using these same methods (Rimon et al., 1980) measured CSF fraction I opioid activity in 18 drug-free acute schizophrenics (11 had never received neuroleptics, seven had discontinued neuroleptic use 4–8 weeks before this study) and 23 chronic schizophrenics who had been without neuroleptic therapy for at least 2 weeks. All schizophrenics fulfilled Feighner's criteria for definite schizophrenia and only the chronic patients whose symptoms worsened during the 2-week drug holiday were studied (9 of 12). Six of nine chronic schizophrenic patients, four of six relapsed patients and six of nine acute patients had elevated CSF fraction I opioid concentrations when compared to mean values for normal controls from the Lindstrom et al. (1978) study. Another CSF sample was obtained after 4 and 8 weeks of fluphenazine treatment. Before initiation of fluphenazine treatment, the mean CSF fraction I activity in the chronic schizophrenics was significantly lower than the acute schizophrenics ($P < 0.05$, Student's t-test) while the acute schizophrenic group had significantly lower concentrations of fraction I activity after 30, but not 60 days of fluphenazine treatment ($P < 0.05$, Mann–Whitney U test, two-tailed). In another recent report from this same group, Lindstrom et al. (1982) found that either fraction I or II CSF opioid activity was elevated above the mean normal control values of the earlier study (Lindstrom et al., 1978) in 72% of 53 neuroleptic-free (for 1 week) schizophrenic patients (using Bleuler's criteria, Berner et al., 1983). These elevations did not attain statistical significance when compared to normal controls (Wilcoxon Ranked Sum test), but within the schizophrenic group, the CSF fraction I opioid activity was significantly higher ($P < 0.01$, Mann–Whitney U test) in the hebephrenic ($n = 23$) group than in the undifferentiated category ($n = 21$). No significant correlations were noted between CSF opioid activity of either fraction I or II and duration of disease, length of neuroleptic treatment or psychotic symptoms. There are, of course, many obvious problems in comparing data between groups assayed at different times.

Dupont et al. (1978), using a radioreceptor assay employing [^{3}H]naloxone as the radioactive ligand, found decreased CSF concentrations of opioid activity in 19 chronic schizophrenic patients when compared to nine controls. No diagnostic criteria were reported and all of the schizophrenics were receiving neuroleptics. The CSF opioid(s) that displaced [^{3}H]naloxone binding co-eluted with MET- and LEU-ENK on Sephadex G-25, but the decreases seen in the schizophrenics were reported as not significant, though the method of statistical analysis was not specified. After the CSF was incubated for 5 h at 37°C, the schizophrenic group had no measurable immunoreactivity in CSF while the control CSF concentrations remained at 80% of pre-incubation levels, implying higher activity of endogenous opioid degradative enzymes in the schizophrenic group. The absolute amounts of opioid activity reported for the normal controls in this assay were approximately 10 times higher than those seen by the aforementioned Swedish investigators. Akil et al. (1978) measured CSF MET-ENK by RIA and opioid activity by radioreceptor assay in 10 healthy volunteers and expressed the results as pg MET-ENK equivalents per ml in order to compare the two methods directly. Both assays used [^{3}H]MET-ENK as the radiolabelled ligand and the CSF was extracted with acid–methanol using two chromatography columns. Both assays gave levels of MET-ENK/opioid activity that agreed with those reported by the Swedish group for fraction II. In contrast, Jeffcoate et al. (1978) reported on the measurement of β-endorphin by RIA in CSF of 20 "normal" patients undergoing diagnostic lumbar punctures or pneumoencephalograms; the CSF concentrations of β-endorphin were in the same range as the opioid activity reported by Dupont et al. (1978). Because almost all antisera to β-endorphin recognize β-lipotropin (β-LPH, the immediate precursor to β-endorphin), Jeffcoate et al. (1978) separated the two immunoreactive forms on Sephadex G-25 and used an antiserum that was directed toward β-LPH, but not β-endorphin, to assess the contribution of β-LPH to the total immunoreactivity. β-Endorphin accounted for more than 90% of the immunoreactivity in all samples tested. These two reports (Akil et al., 1978 and Jeffcoate et al., 1978) established a standard for what was considered normal concentrations of MET-ENK and β-endorphin in CSF.

In a preliminary report, Domschke et al. (1979) presented data on the CSF concentration of β-endorphin in five acute and seven chronic (10 years) schizophrenics compared to seven normal controls and 10 patients with herniated vertebral discs. No diagnostic criteria for schizophrenia were presented and all psychiatric patients were receiving neuroleptic drug therapy. The CSF was extracted in silicic acid and acetone, and β-endorphin estimated by RIA. Inappropriately using Student's t-test to analyse the data, the acute schizophrenics had increased CSF β-endorphin concentrations. The chronic schizophrenics were reported to have lower CSF β-endorphin concentrations than the controls.

Burbach et al. (1979) compared CSF concentrations of β-endorphin and MET-ENK in nine neurologically diseased "controls" and nine schizophrenic patients. No diagnostic criteria were reported for the schizophrenic patients and all were receiving neuroleptics. Both peptides were measured by RIA. No statistically significant group-related differences in CSF β-endorphin or MET-ENK concentration were detected using a paired Student's t-test. After 5 h of incubation at 37°C, there was still no significant group-related difference in CSF concentrations of either peptide. Concentrations of β-endorphin in CSF were in agreement with other reported values, but the MET-ENK concentrations were higher than those previously reported.

Naber et al. (1981a,b), using a radioreceptor assay for opioid activity and a RIA for β-endorphin, studied CSF in a variety of psychiatric patients, including schizophrenics (n = 27), schizoaffectives (n = 17), depressives (n = 35), manics (n = 13) and normal controls. Schizophrenics were diagnosed by Research Diagnostic Criteria and were medication-free for 2 weeks prior to the study. The radioreceptor assay used a MET-ENK analogue, D[^{3}H]Ala-L-Leu-enkephalin amide (D-[^{3}H]ALA) and RIA used ^{125}I-labelled β-endorphin as the radioactive ligand. Opiate receptor activity was significantly reduced in the schizophrenic males only ($P < 0.005$, Student's t-test, two-tailed), while β-endorphin concentrations were much lower than those reported by others.

Emrich et al. (1979a,b) and Hollt et al. (1982) measured the concentration of β-endorphin-like immunoreactivity by RIA in CSF and plasma of eight "normal" controls (obtained to diagnose meningitis/encephalitis and found to be negative) compared to 15 schizophrenics and a variety of other neurologic disorders. Schizophrenics were diagnosed according to the International Classification of Disease nomenclature (Berner et al., 1983) and were medication-free for 4 weeks. Plasma, but not CSF, β-endorphin was extracted with a silicic acid/acetone mixture. No significant differences between controls and schizophrenics were detected in CSF or plasma, though the statistical test employed was not described.

Van Kammen et al. (1981) measured both β-endorphin concentration and opioid activity in CSF from 30 schizophrenic patients; in addition, vasopressin and angiotensin I and II concentrations were also assayed. The results were compared to those obtained in 52 normal controls. Schizophrenic patients fulfilled Research Diagnostic Criteria and were drug-free for an average of 33 days before treatment. Concentrations of β-endorphin, vasopressin and angiotensin I and II were measured by RIA using ^{125}I-labelled peptides and opioid activity was assessed by radioreceptor assay with D-[^{3}H]ALA as the radioligand. Using ANOVA after log transformation of the data, only vasopressin was found to be reduced in the CSF of male schizophrenics, by approximately 40% ($P < 0.01$).

Recently, Wen et al. (1983) measured CSF MET-ENK levels in chronic schizophrenics ($n = 18$; 10 years or more) and neurological controls ($n = 18$; 8 stroke, 10 headache). Schizophrenics were rated before CSF withdrawal according to the Brief Psychiatric Rating Scale (BPRS); no mention was made of medication status. The RIA employed recognised MET-O-ENK, thus ^{125}I-labelled MET-O-ENK was used as the radioactive trace and all CSF samples were passed through Sep-Pak filters and oxidised with hydrogen peroxide to form MET-O-ENK. Recovery was estimated at 70% and schizophrenics were reported to have significantly less MET-ENK present in CSF than controls ($P < 0.02$, Student's t-test). No significant correlation was seen between BPRS score and MET-ENK concentration.

Other neuropeptides

There have been several recent reports in which non-opioid peptides have been measured in the CSF of schizophrenics. Widerlov et al. (1982) measured CSF NT concentrations by RIA in 21 chronic schizophrenics (diagnosed by Research Diagnostic Criteria), and twelve healthy volunteers (see Fig. 1). This study was undertaken because of a number of findings in neurochemistry, neuroanatomy and neuropharmacology which have demonstrated NT–dopamine interactions in the central nervous system (for review see Nemeroff et al., 1983b; Nemeroff and Cain, 1985). Schizophrenics were drug-free for at least 2 weeks and psychiatric symptoms were assessed with the Comprehensive Psychopathological Rating Scale (CPRS). Group mean CSF concentrations of NT-like immunoreactivity were not significantly different, but the schizophrenics were shown to consist of two subgroups (Fowlkes test, $P < 0.005$); one of these subgroups had very low CSF concentrations of NT-like immunoreactivity. After neuroleptic treatment, this latter subgroup exhibited normalized CSF NT concentrations. Only one of 16 items in the CPRS was significantly correlated with the concentration of NT in CSF from the schizophrenic group: slowness of movement ($P < 0.01$). In this regard, it is of interest that all of the catatonic schizophrenics were in the low NT subgroup. Recently, in collaboration with Widerlov and Lindstrom, we have confirmed this decrease of NT-like immunoreactivity in the CSF

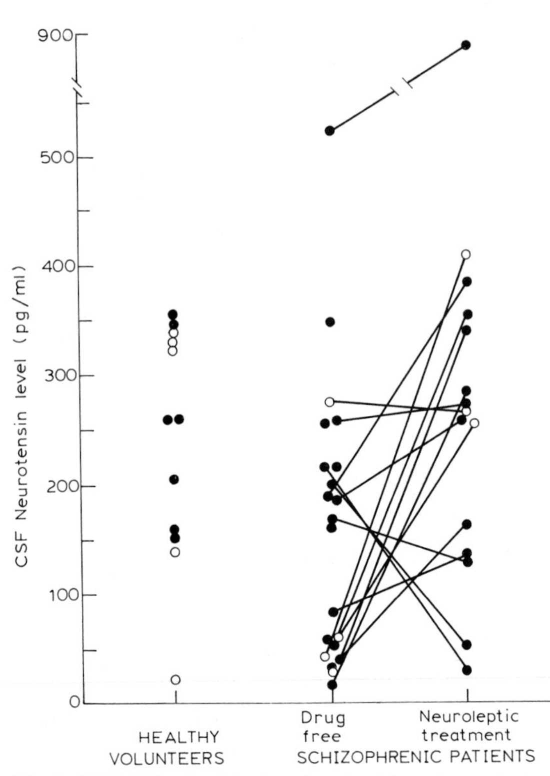

Fig. 1. CSF NT levels of 12 drug-free, healthy volunteers and 21 schizophrenic patients while drug-free and during neuroleptic treatment. The open circles represent women and the filled circles represent men. The patients were numbered according to NT content (i.e., patient 1 had the lowest level and patient 21 the highest). Only 15 of the 21 patients were tested during neuroleptic treatment. (Reproduced with permission from Widerlov et al., 1982.)

of schizophrenics, but changes after neuroleptic treatment were not found. No such CSF NT reductions were observed in patients with major depression, anorexia-bulemia or premenstrual syndrome (Manberg et al., 1983).

Gerner and Yamada (1982) and Gerner (1984) measured CSF concentrations of immunoreactive CCK, SRIF and bombesin in normal healthy volunteers (n = 29), patients with anorexia nervosa (n = 23), mania (n = 10), primary depression (n = 28) and chronic schizophrenics (n = 13, ill for more than 6 months). Diagnoses were performed using Research Diagnostic Criteria, and schizophrenics were neuroleptic-free for 14 days prior to CSF withdrawal. A small, but statistically non-significant, decrease in bombesin-like immunoreactivity was seen between the schizophrenic and control groups; no group-related differences in CSF concentrations of SRIF or CCK-like immunoreactivity were seen. In a more recent study (Gerner et al., 1984), bombesin, SRIF and CCK were measured by RIA in CSF obtained from 31 normal controls, 19 schizophrenics (from California hospitals) and 53 schizophrenics (from the National Institute of Mental Health, NIMH). All schizophrenics fulfilled Research Diagnostic Criteria and were neuroleptic-free for at least 14 days. Only the NIMH schizophrenics showed significant differences, i.e. elevations in CCK and SRIF concentrations (P < 0.001, ANOVA) and decreases in bombesin (P < 0.01, ANOVA), when compared to controls. Neither CCK, SRIF, nor bombesin showed any consistent change in CSF concentration after blockade of acid transport by probenecid or treatment of patients with the neuroleptic drug, pimozide. No significant correlation was seen between CSF concentrations of CCK, SRIF, or bombesin and those of homovanillic acid (HVA), 5-hydroxyindoleacetic acid (5-HIAA), 3-methoxy-4-hydroxyphenylglycol (MHPG), β-endorphin, tyrosine, tryptophan, gamma-aminobutyric acid or cortisol.

Verbanck et al. (1983) also measured CSF CCK concentrations in control subjects and patients with Parkinson's disease (n = 13), depression (n = 30) and schizophrenics (n = 15). Schizophrenics were diagnosed using Feighner's criteria, nine were drug-free for 6 weeks and six received haloperidol at the time CSF was obtained. The antiserum used to measure CCK also recognised gastrin and caerulein and ^{125}I-labelled gastrin was used as the radioligand in the CCK assay. The concentration of CCK was reported to be significantly decreased in the drug-free schizophrenics (P < 0.05, Snedecor's F or Student's t-test) when compared to the values of the normal controls.

Lindstrom et al. (1983) measured the CSF concentrations of immunoreactive delta sleep inducing peptide (DSIP) in healthy volunteers (n = 20), schizophrenics (n = 22) and depressed patients (n = 10). Schizophrenics fulfilled Research Diagnostic Criteria and were drug-free for 2 weeks before the first CSF sample was obtained. The DSIP RIA antisera were directed toward the N-terminus of the DSIP molecule and ^{125}I-labelled DSIP was used as the radioactive ligand. Schizophrenic patients had significantly lower ($P < 0.01$, Student's t-test, two-tailed) CSF DSIP concentrations than controls when drug-free and after 4 weeks of neuroleptic treatment.

Rimon et al. (1984) recently measured the concentration of SP in CSF from 15 controls (X-ray or urology patients), 12 depressed patients and 12 schizophrenics. Schizophrenics fulfilled Feighner's criteria and were drug-free for 2 weeks. Samples of CSF were filtered on Sep-Pak cartridges and SP was measured by RIA. Using Student's t-test, the schizophrenics were reported to have significantly increased ($P < 0.01$) levels of SP in CSF. Gel electrophoresis of the SP immunoreactivity from CSF revealed that less than 10% of the observed immunoreactivity was due to the presence of the intact SP_{1-11} molecule; fragments co-eluting with SP_{5-11} and SP_{3-11} represented the bulk of the immunoreactivity.

Recently, we (Bissette et al., 1984) have measured CSF concentrations of SRIF in 10 healthy volunteers, 29 demented patients, 23 patients with major depression and 10 schizophrenics (DSM-III criteria; Berner et al., 1983). The antiserum recognised $SRIF_{1-14}$ (cyclic or linear), $SRIF_{1-25}$, and $SRIF_{1-28}$. All three diagnostic groups had significantly reduced ($P < 0.05$, Student–Newman–Keuls test after ANOVA) CSF concentrations of SRIF when compared to the controls. Thus, decreases in CSF SRIF appear not to be specific to a particular disease state but may reflect cognitive impairment.

Post-mortem brain studies

In addition to measurement of neuropeptides in CSF, post-mortem brain tissue from schizophrenics has been studied in the hope that the neurotransmitter(s) pathogenetically involved in schizophrenia might be uncovered. As yet, only a few studies have been published regarding changes in neuropeptide concentrations in brain regions of schizophrenics. Indeed, few studies describing the concentrations of these peptides in the normal human CNS have yet been conducted.

Recently, Crow and his colleagues have measured the concentrations of several centrally active neuropeptides in brain regions from 12 controls and 14 schizophrenics (Ferrier et al., 1983; Roberts et al., 1983). All schizophrenics fulfilled Feighner's criteria and were further subclassified into type I (n = 7) and type II (n = 7) based on the presence or absence of "positive" and "negative" symptoms (Crow, 1982a). Five neuropeptides (NT, SP, CCK, SRIF and vasoactive intestinal peptide (VIP)) were assayed by RIA in four cerebrocortical regions (temporal, frontal, parietal and cingulate) and the following sub-cortical regions: hippocampus, amygdala, globus pallidus, putamen, dorso-medial thalamus and lateral thalamus. Significant alterations ($P < 0.05$, multivariate ANOVA) were seen for CCK (reduced in temporal cortex) and SP (increased in hippocampus) in the total group of schizophrenics compared to controls. Type-II schizophrenics had significantly decreased concentration of CCK in the amygdala and significantly decreased SRIF and CCK concentrations in the hippocampus; type-I schizophrenics had elevated levels of VIP in the amygdala. No significant correlation (multivariate ANOVA) was seen for any regional neuropeptide concentration with age, post-mortem delay to autopsy or presence of neuroleptic medication.

Kleinman et al. (1983) measured the concentrations of four neuropeptides (MET-ENK, SP, NT and CCK by RIA in post-mortem brain regions from normal controls (n = 18), alcoholics (n = 7), opiate users (n = 12), suicide (n = 19), and psychotic patients (n = 40). The psychotic group was sub-divided by Research Diagnostic Criteria into chronic paranoid schizophrenics (n = 11), chronic undifferentiated schizophrenics (n = 6) and "other" psychotic disorders (unspecified functional psychoses and affective psychoses). Using Newman–Keul's t-test after ANOVA, no significant differences between normal and psychotic patients were found in MET-ENK concentrations in nucleus accumbens, hypothalamus, globus pallidus or putamen. Similarly, no significant group-related differences in NT concentrations in nucleus accumbens, globus pallidus or hypothalamus were observed. Moreover, no significant differences were seen between control and psychotic patients in CCK levels in amygdala, nucleus accumbens, caudate nucleus, frontal cortex, substantia nigra, hippocampus or temporal cortex. MET-ENK concentrations were reported to be significantly ($P < 0.05$) decreased in the caudate nucleus of chronic paranoid schizophrenics compared to other diagnostic groups or controls; SP levels were significantly increased ($P < 0.05$) in the caudate nucleus of patients with psychoses when compared to diagnoses other than schizophrenia.

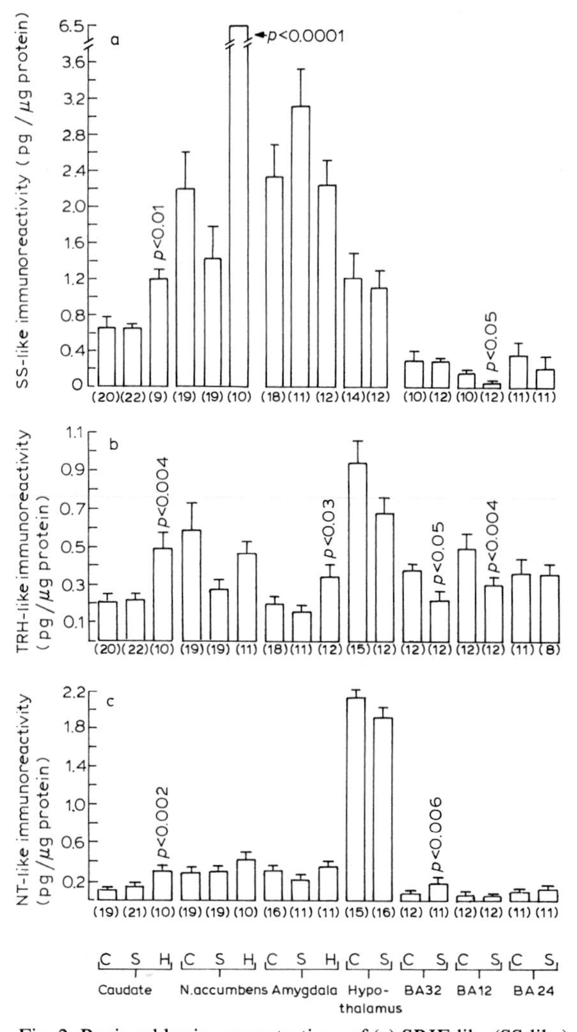

Fig. 2. Regional brain concentrations of (a) SRIF-like (SS-like), (b) TRH-like, and (c) NT-like immunoreactivity in normal controls (C), schizophrenics (S), and Huntington's chorea patients (H). Values are means ± standard errors; numbers in parentheses refer to the number of samples per group. Huntington's chorea patients had significantly higher concentrations of SRIF, TRH, NT in the caudate nucleus, SRIF in nucleus accumbens, and TRH in the amygdala. The schizophrenics had significantly lower concentrations of SRIF and TRH in one frontal cortical region (BA12) as well. In the latter region the concentration of NT was elevated. P values were derived from analysis of variance. (Reproduced with permission from Nemeroff et al., 1982.)

Our laboratory (Nemeroff et al., 1983a) has studied regional post-mortem brain concentrations of NT, SRIF and thyrotropin-releasing hormone (TRH) in controls (free of neurologic or psychiatric disease, $n = 50$), patients with Huntington's chorea ($n = 24$) and schizophrenic patients ($n = 46$). Schizophrenics were diagnosed by Research Diagnostic Criteria and were on various neuroleptic regimens before and, in some cases, up to the time of death. All peptides were measured by RIA and statistically significant differences in a particular regional peptide concentration between groups were sought by ANOVA (Fig. 2). No significant differences in NT, SRIF or TRH were seen between controls and schizophrenics in the caudate nucleus, nucleus accumbens, amygdala or hypothalamus. A significant decrease in SRIF ($P < 0.05$) and TRH ($P < 0.004$) concentrations in Brodmann's area 12 (frontal cortex) and a significant decrease ($P < 0.05$) in TRH concentration in Brodmann's area 32 (frontal cortex) were seen for schizophrenics when compared to the controls; in contrast, NT content was significantly elevated ($P < 0.006$) in Brodmann's area 32 (frontal cortex) in the schizophrenics.

Biggins et al. (1983) have recently measured the amygdala concentrations of NT and TRH by RIA in normals ($n = 7$), patients with senile dementia of the Alzheimer's type ($n = 7$), depressive illness ($n = 7$), or schizophrenia ($n = 7$). No criteria were given for the schizophrenic diagnosis and no mention was made of their neuroleptic medication status. No significant difference was seen in amygdaloid concentrations of TRH or NT between the four diagnostic groups, though the statistical test used was not specified. No correlation was seen between patient age and peptide concentration, but across diagnostic groups, males had almost twice as much NT in the amygdala as females ($P < 0.01$, test not specified). A significant positive correlation was seen for the concentration of TRH and NT ($P < 0.01$, regression analysis) in this brain region. Total NT and TRH immunoreactivity in the amygdala was characterized on HPLC; NT immunoreactivity co-eluted with the synthetic standard as a single peak while TRH immunoreactivity consisted of a major peak (50%) and several minor peaks.

A recent study conducted at the MRC Brain Bank in Cambridge, U.K., investigated the concentrations of seven neuropeptides (MET-ENK, β-endorphin, VIP, CCK, SRIF, adrenocorticotrophin and SP) in four areas of post-mortem brain obtained from 27 patients suffering from schizophrenic illness (Emson et al., 1986). Diagnosis was verified from case notes according to criteria (MacKay et al., 1982) which depended heavily upon those required for the nuclear syndrome of Wing et al. (1974). Brain areas investigated were caudate nucleus, amygdala, hippocampus and frontal cortex. No particularly clear picture emerged, although the amygdala was the site of abnormality in three peptides; SP was reduced ($P < 0.005$), as was MET-ENK ($P < 0.005$), whereas VIP was raised ($P < 0.05$, all by Mann–Whitney U-test). In the case of both MET-ENK and VIP the abnormality was confined to those cases of schizophrenia in which the illness was first documented before the age of 25 years. Neurochemical abnormality which is most marked in subjects with early onset schizophrenia had also been noted in the case of dopamine concentrations (MacKay et al., 1982).

Based on the expectation that any important and sustained alteration in the synaptic availability of the neuropeptide will result in changes in the density of membrane receptors for that peptide, investigations of opioid peptide binding sites have been undertaken in post-mortem schizophrenic brain tissues. Reisine et al. (1980) reported a 50% reduction in the density of tritiated naloxone binding sites in the caudate nucleus of schizophrenic brain tissue. Naloxone is an opioid antagonist with preferential affinity for the "μ" sub-species of opioid receptor, and this result therefore suggested a reduction in μ sites in schizophrenia. However, a more recent study failed to replicate the finding. Owen et al. (1985) examined the specific binding of tritiated naloxone, and also of tritiated etorphine, in caudate tissue from 14 cases of schizophrenia diagnosed from case-note data according to the criteria of Wing et al. (1974) and of Feighner et al. (1972). No difference from control tissue was found for naloxone binding, or for etorphine binding, which could be resolved into (δ + k) receptor components. These results suggested that no gross disturbance in opioid peptide transmission occurs in the caudate nucleus in schizophrenic illness.

Pharmacological experimentation

The administration of opioid peptide antagonists and agonists to patients suffering from schizophrenia was provoked, at least in part, by observations on the behavioural effects of such agents when injected intracerebrally into rodents. Hypokinesia induced by β-endorphin was interpreted by Bloom et al. (1976) as catatonia, but by Jacquet and Marks (1976) as neuroleptic-like bradykinesia. These imaginative interpretations of animal behaviour led to dichotomous predictions — on the one hand that opioid peptides were psychotogenic, on the other that they were neuroleptic-like. The ensuing 5 years saw a large number of reports describing the clinical effects of naloxone, β-endorphin and synthetic enkephalin analogues in patients suffering from schizophrenic illness. Neither strategy, inhibiting or stimulating central opioid peptide systems, revealed any generally useful therapeutic gain (MacKay, 1985), although it must be said that the pharmacological tools which have been available for clinical use so far leave much to be desired.

(Des-tyrosynyl[1])-γ-endorphin (DTγE) is a non-opioid derivative of γ-endorphin which has been reported by de Wied et al. (1978) to possess neuroleptic-like properties. Early studies of the administration of DTγE to schizophrenic patients were encouraging (Verhoeven et al., 1979) but negative reports have also accumulated (Casey et al., 1981; Manchanda and Hirsch, 1981; Tamminga et al., 1981), suggesting at best that DTγE may be effective in an elusive subgroup of schizophrenic responders. One theme in the studies claiming positive results has been the improvement of emotional responsiveness in chronic schizophrenia, a potentially useful extension to conventional neuroleptics.

Haemodialysis

Wagemaker and Cade (1977) revived the occasional but discarded practice of haemodialysis treatment for schizophrenic illness on the basis of an open study in which 10 out of 15 non-uremic chronic schizophrenic patients showed marked improvement. The effect of dialysis was postulated to involve the removal from plasma of an atypical psychotogenic opioid peptide, β-Leu[5]-endorphin. Subsequent controlled experiments not only failed to show any antipsychotic effect of dialysis, but rather the opposite. Twenty-four patients randomly assigned to sham or real dialysis showed that sham dialysis was associated with improvement whereas real dialysis tended to provoke deterioration (Wagemaker et al., 1984). All patients had their neuroleptic medication stopped 2 weeks prior to the study, and the deterioration in the dialysis group may have been due in part to the more rapid elimination of neuroleptic.

Discussion

At the present time, differences in the concentrations of several neuropeptides in CSF and post-mortem brain tissue have been found between schizophrenic, other diagnostic groups and controls. There is, however, considerable disagreement over the "normal" concentrations of peptides detected and the direction of the observed peptide changes in schizophrenia. Not all studies describe the criteria used to diagnose schizophrenia and few have controlled for the effects of chronic neuroleptic treatment. An embarrassing number of reports are either devoid of statistical analysis or use an inappropriate statistical test (such as Student's *t*-test in a study with more than two experimental groups). Last but not least, the virtually random investigation of many variables by large numbers of independent groups is likely to throw up statistically significant differences; as much a result of a type-I statistical error as of any real abnormality.

The problems inherent with "state of the art" neuropeptide research (RIA specificity, lack of turnover data) and the lack of any demonstrable correlation between CSF neuropeptide concentrations and functional activity of neuropeptide circuits do not, at present, allow any definitive statements to be made concerning neuropeptide involvement in schizophrenia. The absence of comprehensive studies of the "normal" neuropeptide concentration in biological fluids and tissue from healthy human controls represents a serious lacuna in the extant knowledge of what constitutes a pathological change in neuropeptide concentrations. Some neuropeptide changes, such as SRIF decreases in CSF, appear to be indicative of general cognitive dysfunction rather than being pathognomonic for any distinct diagnostic entity. While there does seem to be some degree of specificity, as far as regional neuropeptide concentration changes in post-mortem brain tissues of schizophrenics are concerned, few, if any, studies have measured the same peptide in the same region and, in these few cases, little agreement has been forthcoming. While these problems may appear disheartening at first, there are several areas where significant progress appears imminent. The close involvement of several peptides (NT, CCK, MET-ENK and SP) with dopamine systems and the changes in the concentration of these peptides and, in some cases, their receptors, with neuroleptic drug administration lends credence to the notion that neuropeptidergic systems are closely involved in the therapeutic actions of dopamine receptor antagonists in schizophrenic illness. Attempts to address some of the inconsistencies highlighted by the earlier reports should include careful experimental design that incorporates previous findings and methodologies where possible and the inclusion of such obvious information as diagnostic criteria, medication status and appropriate statistical analysis. The increasing availability of stable structural analogues of neuropeptides for radiolabeling to measure the number and affinity of neuropeptide receptors in post-mortem brain tissue should provide novel and important data.

Clinical pragmatism, rather than any compelling theoretical model, has encouraged the use of crude pharmacological probes in patients suffering from schizophrenic illness. As it became apparent in the late 1970's that opioid and other neuropeptide systems were present in areas of brain traditionally linked to pathophysiological theories for schizophrenic illness, and the modes of action of neuroleptic drugs, clinicians understandably wanted to modify these systems in an attempt to improve the chemotherapy of schizophrenia. However, only a very small group of relatively non-specific opioid peptide antagonists and agonists have been available to the clinician who wanted urgently to inhibit or to excite these systems. The results to date have been unfruitful, but these "first wave" therapeutic experiments were crude and often improperly designed. As neurobiology increases the data base of neuropeptide physiology and pharmacology, so will it become increasingly possible for more sophisticated therapeutic hypotheses to be tested in the psychotic patient. This will certainly require the availability of a range of clinically acceptable pharmacological agents capable of distinguishing between the ever-increasing range of peptide receptors. Hopefully these existing incongruities will be resolved by a combination of well-designed and well-executed studies supported by such novel neurobiological methodologies as complementary DNA probes and the use of more specific pharmacologic agents.

Acknowledgments

We are grateful to Elizabeth Brown for typing the manuscript. The authors' research is supported by NIMH MH-39415.

References

Akil, H., Watson, S. J., Sullivan, S. and Barchas, J. D. (1978) Enkephalin-like material in normal human CSF: measurement and levels. *Life Sci.*, 23: 121–126.

Berger, P. A. (1983) Endorphins in emotions, behavior and mental illness. In L. Temoshok, C. Van Dike and L. S. Vegans (Eds.), *Mind and Medicine: Emotions in Health and Illness*, Grune and Stratton, Inc., New York, pp. 153–166.

Berner, P., Gabriel, G., Katschnig, H., Kieffer, W., Lenz, G. and Simhandl, C. (1983) *Diagnostic Criteria for Schizophrenia and Affective Psychoses*, World Psychiatric Association, American Psychiatric Press, Inc., Washington, D.C.

Biggins, J., Perry, E. K., McDermott, J. R., Smith, I. A., Perry, R. H. and Edwardson, J. A. (1983) Post-mortem levels of thyrotropin-releasing hormone and neurotensin in the amygdala in Alzheimer's disease, schizophrenia and depression. *J. Neurol. Sci.*, 58: 117–122.

Bissette, G., Walleus, A., Widerlöv, E., Karlsson, I., Eklund, K., Loosen, P. T. and Nemeroff, C. B. (1984) Reductions of cerebrospinal fluid concentrations of somatostatin-like immunoreactivity in dementia, major depression and schizophrenics. *Soc. Neurosci. Abstr.*, 10: 1093.

Bloom, F., Segal, D., Ling, N. and Guillemin, R. (1976) Endorphins: profound behavioural effects in rats suggest new aetiological factors in mental illness. *Science*, 194: 630–632.

Buchsbaum, M. S., Davis, G. C. and van Kammen, D. P. (1980) Diagnostic classification and the endorphin hypothesis of schizophrenia. Individual differences and psychopharmacological strategies. In C. Baxter and T. Melnechuk (Eds.), *Perspectives in Schizophrenia Research*, Raven Press, New York, pp. 177–191.

Burbach, J. P. H., Loeber, J. G., Verhoef, J., de Kloet, E. R., van Ree, J. M. and de Wied, D. (1979) Schizophrenia and degradation of endorphins in cerebrospinal fluid. *Lancet*, ii: 480–481.

Casey, D. E., Korsgaard, S., Gerlach, J., Jorgensen, A. and Simmelsgaard, H. (1981) Effect of des-tyrosine β-endorphin in tardive dyskinesia. *Arch. Gen. Psychiatry*, 38: 158–160.

Chang, R. S. L., Lotti, V. J., Martin, G. E. and Chen, T. B. (1983) *Life Sci.*, 32: 871–878.

Collins, S., Walker, D., Forsyth, P. and Belbeck, L. (1983) The effects of proglumide on cholecystokinin-, bombesin-, and glucagon-induced satiety in the rat. *Life Sci.*, 32: 2223–2229.

Crow, T. J. (1982a) Schizophrenia. In T. J. Crow (Ed.), *Dis-*

orders of Neurohumoural Transmission, Academic Press, New York, pp. 287–340.

Crow, T. J. (1982b) The biology of schizophrenia. *Experientia,* 38: 1275–1282.

Davis, G. C., Buchsbaum, M. S. and Bunney, W. E. (1979) Research in endorphins and schizophrenia. *Schizophr. Bull.,* 5: 244–250.

De Wied, D., Bohus, B., van Ree, J. M., Kovacs, G. L. and Greven, H. M. (1978) Neuroleptic-like activity of (Des-tyr)-γ-endorphin in rats. *Lancet,* i: 1046.

Domschke, W., Dickschas, A. and Mitznegg, P. (1979) CSF β-endorphin in schizophrenia. *Lancet,* i: 1024.

Dupont, A., Villeneuve, A., Bouchard, J. P., Bouchard, R., Merand, Y., Rouleau, D. and Labrie, F. (1978) Rapid inactivation of enkephalin-like material by CSF in chronic schizophrenia. *Lancet,* ii: 1107.

Emrich, H. M., Hollt, V., Kissling, W., Fischler, M., Heinemann, H., van Zerssen, D. and Herz, A. (1979a) A measurement of β-endorphin-like immunoreactivity in CSF and plasma of neuropsychiatric patients. In Y. H. Erlich, J. Volavka, L. G. Davis and E. G. Brunngraber (Eds.), *Modulators, Mediators and Specifiers in Brain Function.* Plenum Press, New York, pp. 307–317.

Emrich, H. M., Hollt, V., Kissling, W., Fischler, M., Laspe, H., Heinemann, H., von Zerssen, D. and Herz, A. (1979b) β-Endorphin-like immunoreactivity in cerebrospinal fluid and plasma of patients with schizophrenia and neuropsychiatric disorders. *Pharmakopsychiatrie,* 12: 269–276.

Emson, P. C., Arregui, A., Rees, L. H., Besser, G. M., Lowry, P., Ratter, S., Corder, R., MacKay, A. V. P., Rossor, M., Fahrenkrug, J., Rehfeld, J. F. and Iversen, L. L. (1986) Neuropeptides in schizophrenic brain. In preparation.

Feighner, J. P., Robins, E., Guze, S. B., Woodruffe, R. A., Winokur, G. and Munoz, R. (1972) Diagnostic criteria for use in psychiatric research. *Arch. Gen. Psychiatry,* 26: 57–67.

Ferrier, I. N., Roberts, G. W., Crow, T. J., Johnstone, E. C., Owens, D. G. C., Lee, Y. C., O'Shaughnessy, D., Adrian, T. E., Polak, J. M. and Bloom, S. R. (1983) Reduced cholecystokinin-like and somatostatin-like immunoreactivity in limbic lobe is associated with negative symptoms in schizophrenia. *Life Sci.,* 33: 475–482.

Fessler, R. G., Brown, F. D., Rachlin, J. R., Mullan, S. and Fang, U. S. (1984) Elevated β-endorphin in cerebrospinal fluid after electrical brain stimulation: Artifact of contrast infusion? *Science,* 224: 1017–1019.

Frey, P. (1983) Cholecystokinin octapeptide levels in rat brain are changed after sub-chronic neuroleptic treatment. *Eur. J. Pharmacol.,* 95: 87–92.

Gerner, R. H. (1984) Cerebrospinal fluid cholecystokinin and bombesin in psychiatric disorders and normals. In R. M. Post and J. C. Ballenger (Eds.), *Neurobiology of Mood Disorders,* from Wood and Brook series, *Frontiers in Clinical Neurosciences, Vol. 1,* Williams and Wilking, Baltimore, pp. 388–392.

Gerner, R. H. and Yamada, T. (1982) Altered neuropeptide concentrations in cerebrospinal fluid of psychiatric patients. *Brain Res.,* 238: 298–302.

Gerner, R. H., van Kammen, D. P. and Ninan, P. T. (1984) Cerebrospinal fluid cholecystokinin, bombesin and somatostatin in schizophrenia and normals. *Prog. Neuropsychopharmacol. Biol. Psychiatry,* 9: 73–82.

Govoni, S., Hong, J., Yang, H.-Y. T. and Costa, E. (1980) Increase in neurotensin content elicited by neuroleptics in nucleus accumbens. *J. Pharmacol. Exp. Ther.,* 215: 413–417.

Hanson, G. R., Alpho, L., Wolf, W., Levine, R. and Lovenberg, W. (1981) Haloperidol-induced reduction of nigral substance-P-like immunoreactivity: a probe for the interactions between dopamine and substance P neuronal systems. *J. Pharmacol. Exp. Ther.,* 218: 568–574.

Haracz, J. L. (1982) The dopamine hypothesis: an overview of studies with schizophrenic patients. *Schizophr. Bull.,* 8: 438–469.

Hollt, V., Emrich, H. M., Bergmann, M., Nedopil, N., Dieterle, D., Gurland, H. J., Nussett, L., von Zerssen, D. and Herz, A. (1982) β-Endorphin-like immunoreactivity in CSF and plasma of neuropsychiatric patients. In N. S. Shah and A. G. Donald (Eds.), *Endorphins and Opiate Antagonists in Psychiatry,* Plenum Press, New York, pp. 231–243.

Hong, J. S., Yang, H.-Y. T., Fratta, W. and Costa, E. (1978a) Rat striatal methionine-enkephalin content after chronic treatment with cataleptogenic and noncataleptogenic antischizophrenic drugs. *J. Pharm. Exp. Ther.,* 205: 141–147.

Hong, J. S., Yang, H.-Y. T. and Costa, E. (1978b) Substance P content of substantia nigra after chronic treatment with antischizophrenic drugs. *Neuropharmacology,* 83–85.

Jacquet, Y. and Marks, N. (1976) The C-fragment of β-lipotropin: endogenous neuroleptic or antipsychotogen? *Science,* 194: 632–634.

Jeffcoate, W. J., McLoughlin, L., Hope, J., Rees, L. H., Ratter, S. J., Lowry, P. J. and Besser, G. M. (1978) β-Endorphin in human cerebrospinal fluid. *Lancet,* ii: 119–121.

Kleinman, J. E., Iadarola, M., Govoni, S., Hong, J., Gillin, J. C. and Wyatt, R. J. (1983) Postmortem measurements of neuropeptides in human brain. *Psychopharmacol. Bull.,* 19: 375–377.

Kolpakov, V. G., Barykina, N. N., Chepkasov, I. L., Alekhina, T. A. and Parvez, H. (1983) On animal models of schizophrenia. In S. Parvez, T. Negatsu, I. Negatsu and H. Parvez (Eds.), *Methods in Biogenic Amine Research,* Elsevier/N. Holland, Amsterdam, pp. 997–1020.

Koob, G., LeMoal, M. and Bloom, F. E. (1984) The role of endorphins in neurobiology, behavior and psychiatric disorders. In C. B. Nemeroff and A. J. Dunn (Eds.), *Peptides, Hormones and Behavior,* Spectrum Publications, New York, pp. 349–384.

Lindstrom, L. H., Widerlöv, E., Gunne, L. M., Wahlstrom, A. and Terenius, L. (1978) Endorphins in human cerebrospinal fluid: Clinical correlations to some psychotic states. *Acta Psychiat. Scand.,* 57: 153–169.

Lindstrom, L. H., Besev, G., Gunne, L. M., Sjostrom, R., Terenius, L., Wahlstrom, A. and Wistedt, B. (1982) Cerebrospinal content of endorphins in schizophrenia. In N. S. Shah and A. G. Donald (Eds.), *Endorphins and Opiate Antagonists in Psychiatry,* Plenum Press, New York, pp. 245–256.

Lindstrom, L. H., Ekman, R., Walleus, H. and Widerlöv, E. (1984) Delta-sleep inducing peptide in cerebrospinal fluid from schizophrenics, depressives and healthy volunteers. *Prog. Neuro-Psychopharmacol. Biol. Psychiatry,* 9: 83–90.

Lipton, M. A. and Nemeroff, C. B. (1978) An overview of the biogenic amine hypothesis of schizophrenia. In W. E. Fann, I. Karacan, A. D. Pokorny and R. L. Williams (Eds.), *Pharmacology and Treatment of Schizophrenia,* Spectrum Publications, New York, pp. 431–453.

MacKay, A. V. P. (1981) Endorphins and the psychiatrist. *Trends Neuro. Sci.,* 4: R9–R11.

MacKay, A. V. P. (1985) Neuropeptides and psychiatry. In K. Granville-Grossman (Ed.), *Recent Advances in Clinical Psychiatry, Vol. 5,* Churchill Livingstone, London, pp. 179–200.

MacKay, A. V. P., Iversen, L. L., Rossor, M., Spokes, E., Bird, E. D., Arregui, A., Creese, I. and Snyder, S. H. (1982) Increased brain dopamine and dopamine receptors in schizophrenia. *Arch. Gen. Psychiatry,* 39: 991–997.

Manberg, P. J., Nemeroff, C. B., Bissette, G., Prange, A. J., Jr. and Gerner, R. H. (1983) Cerebrospinal fluid levels of neurotensin-like immunoreactivity in normal controls and in patients with affective disorder, anorexia nervosa and premenstrual syndrome. *Soc. Neurosci. Abstr.,* 9: 1054.

Manchanda, R. and Hirsch, S. R. (1981) (Des-tyr)-γ-endorphin in the treatment of schizophrenia. *Psychol. Med.,* 11: 401–404.

Naber, D. (1983) Peptides. In H. Hippius and G. Winokur (Eds.), *Psychopharmacology: Part 2, Clinical Psychopharmacology,* Excerpta Medica, Amsterdam, pp. 162–177.

Naber, D., Pickar, D., Post, R. M., van Kammen, D. P., Waters, R. N., Ballenger, J. C., Goodwin, F. K. and Bunney, W. E., Jr. (1981a) Endogenous opioid activity and β-endorphin immunoreactivity in CSF of psychiatric patients and normal controls. *Am. J. Psychiatry,* 138: 1457–1462.

Naber, D., Pickar, D., Post, R. M., van Kammen, D. P., Ballenger, J., Rubinow, D., Waters, R. N. and Bunney, W. E., Jr. (1981b) CSF opioid activity in psychiatric patients. In C. Perris, G. Struwe and B. Jansson (Eds.), *Biological Psychiatry,* Elsevier, Amsterdam, pp. 372–375.

Nemeroff, C. B. and Cain, S. T. (1985) Neurotensin–dopamine interactions in the central nervous system. *Trends Pharmacol. Sci.,* 6: 201–205.

Nemeroff, C. B., Youngblood, W. W., Manberg, P. J., Prange, A. J., Jr. and Kizer, J. S. (1983a) Regional brain concentrations of neuropeptides in Huntington's chorea and schizophrenia. *Science,* 221: 972–975.

Nemeroff, C. B., Luttinger, D., Hernandez, D. E., Mailman, R. B., Mason, G. A., Davis, S. D., Widerlöv, E., Frye, G. D., Kilts, C. A., Beaumont, K., Breese, G. R. and Prange, A. J., Jr. (1983b) Interactions of neurotensin with brain dopamine systems: Biochemical and behavioral studies. *J. Pharmacol. Exp. Ther.,* 225: 337–345.

Owen, F., Bourne, R. C., Poulter, M., Crow, T. J., Paterson, S. J. and Kosterlitz, H. W. (1985) Tritiated etorphine and naloxone binding to opioid receptors in caudate nucleus in schizophrenia. *Br. J. Psychiatry,* 146: 507–509.

Palkovits, M., Brownstein, M. J., Eiden, L. E., Beinfeld, M. C., Russell, J., Arimura, A. and Szabo, S. (1982) Selective depletion of somatostatin in rat brain by cysteamine. *Brain Res.,* 240: 178–180.

Pickar, D., Naber, D., Post, R. M., van Kammen, D. P., Kaye, W., Rubinow, D. R., Ballenger, J. C. and Bunney, W. E., Jr. (1982) Endorphins in the cerebrospinal fluid of psychiatric patients. *Ann. NY Acad. Sci.,* 398: 399–412.

Reisine, T. D., Rossor, M., Spokes, E. G., Iversen, L. L. and Yamamura, H. (1980) Opiate and neuroleptic receptor alteration in human schizophrenic brain tissue. In G. Pepeu, M. J. Kuhar and S. J. Enna (Eds.), *Receptors for Neurotransmitters and Peptide Hormones,* Raven Press, New York, pp. 443–450.

Rimon, R., Terenius, L. and Kampman, R. (1980) Cerebrospinal fluid endorphins in schizophrenia. *Acta Psychiatr. Scand.,* 61: 395–403.

Rimon, R., LeGreves, P., Nyberg, F., Heikkila, L., Salmela, L. and Terenius, L. (1984) Elevation of substance P-like peptides in the CSF of psychiatric patients. *Biol. Psychiatry,* 19: 509–516.

Roberts, G. W., Ferrier, I. N., Lee, Y. C., Crow, T. J., Johnstone, E. C., Owens, D. G. C., Hamilton, A. J. B., McGregor, G., O'Shaughnessy, D., Polak, J. M. and Bloom, S. R. (1983) Peptides, the limbic lobe and schizophrenia. *Brain Res.,* 288: 199–211.

Snyder, S. H. (1982) Schizophrenia. *Lancet,* ii: 970–974.

Spitzer, R., Endicott, J. and Robins, E. (1975) *Research Diagnostic Criteria for Selected Group of Functional Disorders,* Biometrics Research Division, New York State Psychiatric Institute, New York.

Stevens, J. R. (1982) The neuropathology of schizophrenia. *Psychol. Med.,* 12: 695–700.

Tamminga, C. A., Tighe, P. J., Chase, T. N., De Fraites, E. G. and Schaffer, M. H. (1981) Des-tyrosine-γ-endorphin administration in chronic schizophrenics. *Arch. Gen. Psychiatry,* 38: 167–168.

Terenius, L. (1978) The implications of endorphins in pathological states. In J. M. van Ree and L. Terenius (Eds.), *Characteristics and Function of Opioids,* Elsevier/North Holland Medical Press, Amsterdam, pp. 143–158.

Terenius, L., Wahlstrom, A., Lindström, L. H. and Widerlöv, E. (1976) Increased CSF levels of endorphins in chronic psychosis. *Neurosci. Lett.,* 3: 157–162.

Uhl, G. M. and Kuhar, M. J. (1984) Chronic neuroleptic treatment enhances neurotensin receptor binding in human and rat substantia nigra. *Nature,* 309: 350–352.

Van Kammen, D. P., Waters, R. N., Gold, P., Sternberg, D.,

Robertson, G., Ganten, D., Pickar, D., Naber, D., Ballenger, J. C., Kaye, W. H., Post, R. M. and Bunney, W. E., Jr. (1981) Spinal fluid vasopressin, angiotensin I and II, β-endorphin and opioid activity in schizophrenia: a preliminary evaluation. In C. Perris, G. Struwe and B. Jansson (Eds.), *Biological Psychiatry*, Elsevier, Amsterdam, pp. 339–344.

Van Praag, H. M. and Verhoeven, W. M. A. (1980) Endorphins and schizophrenia. In D. de Wied and P. A. van Keep (Eds.), *Hormones and the Brain*, University Park Press, Baltimore, pp. 141–153.

Van Ree, J. M. and de Wied, D. (1981) Endorphins in schizophrenia. *Neuropharmacology*, 20: 1271–1277.

Verbanck, P. M. P., Lotstra, F., Gilles, C., Linkowski, P., Mendlewicz, J. and Vanderhaeghen, J. J. (1983) Reduced cholecystokinin immunoreactivity in the cerebrospinal fluid of patients with psychiatric disorders. *Life Sci.*, 34: 67–72.

Vereby, K., Volavka, J. and Clouet, D. (1978) Endorphins in psychiatry. *Arch. Gen. Psychiatry*, 35: 877–888.

Verhoeven, W. M. A., van Praag, H. M., van Ree, J. M. and de Wied, D. (1979) Improvement of schizophrenic patients treated with (des-tyr)-γ-endorphin (DTγE). *Arch. Gen. Psychiatry*, 36: 294–298.

Wagemaker, H. and Cade, R. (1977) The use of haemodialysis in chronic schizophrenia. *Am. J. Psychiatry*, 134: 684–685.

Wagemaker, H., Rogers, J. L. and Cade, R. (1984) Schizophrenia, haemodialysis and the placebo effect. *Arch. Gen. Psychiatry*, 41: 805–810.

Weinberger, D. R., Wagner, R. L. and Wyatt, R. J. (1983) Neuropathological studies of schizophrenia: a selective review. *Schizophr. Bull.*, 9: 193–212.

Wen, H. L., Lo, C. W. and Ho, W. K. K. (1983) Met-enkephalin level in the cerebrospinal fluid of schizophrenic patients. *Clin. Chim. Acta*, 128: 367–371.

Widerlöv, E., Lindström, L. H., Besev, G., Manberg, P. J., Nemeroff, C. B., Breese, G. R., Kizer, J. S. and Prange, A. J., Jr. (1982) Subnormal CSF levels of neurotensin in a subgroup of schizophrenic patients: Normalization after neuroleptic treatment. *Am. J. Psychiatry*, 139: 1122–1126.

Wing, J. K., Cooper, J. E. and Sartorius, N. (1974) *Measurement and Classification of Psychiatric Symptoms*, Cambridge University Press, Cambridge.

SECTION IV

Spinal Cord and Peripheral Nervous System

P. C. Emson, M. N. Rossor and M. Tohyama (Eds.),
Progress in Brain Research, Vol. 66.

CHAPTER 10

Neuropeptides in spinal cord

Masaya Tohyama and Yahē Shiotani

Department of Neuroanatomy, Institute of Higher Nervous Activity, Osaka University Medical School, 4-3-57 Nakanoshima, Kitaku, Osaka 530, Japan

Recent immunohistochemical studies have shown that the spinal cord of various animals is rich in neuropeptides. Since many of them are concentrated in the dorsal horn of the spinal cord, much interest has been drawn to the involvement of neuropeptides in sensory transmission mechanisms. However, the ventral and lateral horns also contain numerous neuropeptide-containing fibres. Furthermore, some neuropeptides are located in the neurons of the lateral and ventral horns; for example, enkephalin (ENK) in the lateral horn cells (Hökfelt et al., 1977a,b; Lundberg et al., 1980; Glazer and Basbaum, 1980, 1981; Dalsgaard et al., 1982) and calcitonin gene-related peptide (CGRP) in the ventral horn cells (Rosenfeld et al., 1983; Takami et al., 1985a). These findings suggest that neuropeptides in the spinal cord may be involved in a variety of physiological functions. In addition to mapping, current studies of the ontogeny of neuropeptides in the spinal cord show variations according to the peptide and spinal region (Senba et al., 1982; Fuji et al., 1985). A similar tendency was seen in the peptidergic fibre connections of the spinal cord. Although the projections can be divided into supraspinal (both descending and ascending), intraspinal and peripheral including both afferents and efferents, the projection patterns were different according to both the peptide and segment studied.

In this chapter, we focus on the distribution, ontogeny, fibre connections and fine structures of the individual peptides, and summarize their distribution.

Distribution of neuropeptides in the spinal cord

Somatostatin

Distribution

Distribution of somatostatin (SRIF)-like immunoreactivity (SRIF-LI) in the spinal cord has been examined in various animals from the primate to the frog (Hökfelt et al., 1975a, 1976; Elde et al., 1978; Forssmann, 1978; Burnweit and Forsmann, 1979; Dalsgaard et al., 1981; DiTirro et al., 1981; Finley et al., 1981b; Hunt et al., 1981; Inagaki et al., 1981a,b; Lorez and Kemali, 1981; Shiosaka et al., 1981; Senba et al., 1982). In most mammals, the SRIF-like immunoreactive (SRIF-LIr) fibre plexus was dense in the dorsal horn (Figs. 1, 3, 5c); the highest density being seen in lamina II or the substantia gelatinosa forming a dense band of SRIF-LIr fibres. A moderate density of SRIF-LIr fibres is found around the central canal. Some of the SRIF-LIr fibre plexus leaves the central canal and travels to the lateral horn where a low density of fibres was observed. The ventral horn contained the lowest density of SRIF-LIr fibres (Figs. 1, 3). Scattered SRIF-LIr fibres were also detected in other areas of the spinal cord (Figs. 1, 3). In the adult rats, SRIF-LIr neurons were found only in lamina II of the dorsal horn (Hunt et al., 1981; Senba et al., 1982). However, during the early ontogenetical stage, SRIF-LIr neurons occurred in other spinal regions in addition to lamina II (Senba et al., 1982). Burnweit and Forssmann (1979) reported that

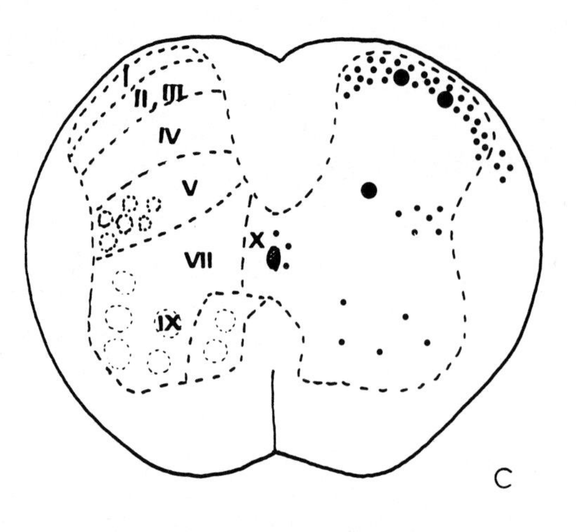
I
II, III
IV
V
VII
X
IX
C

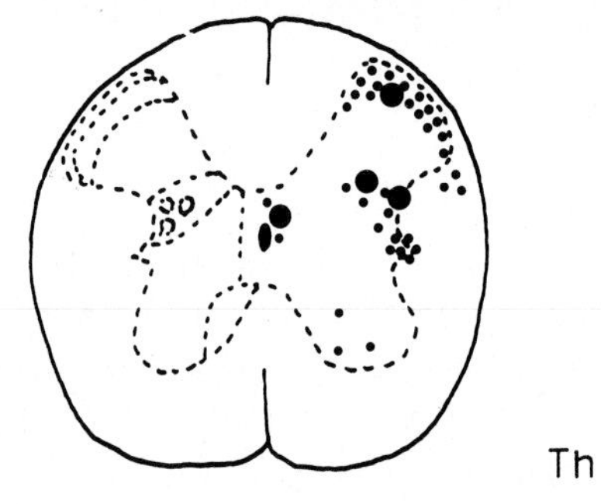
Th

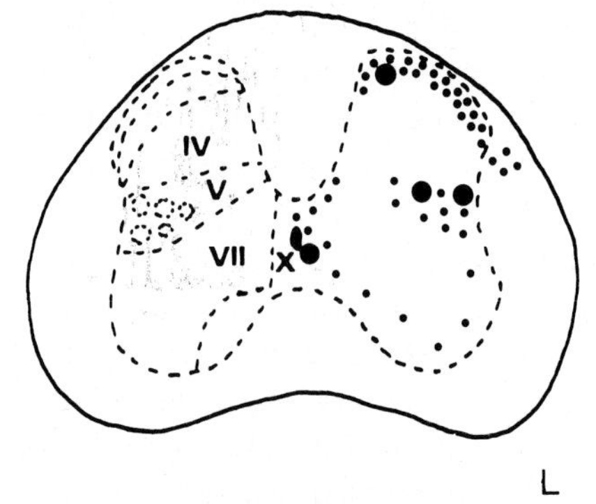
IV
V
VII
X
L

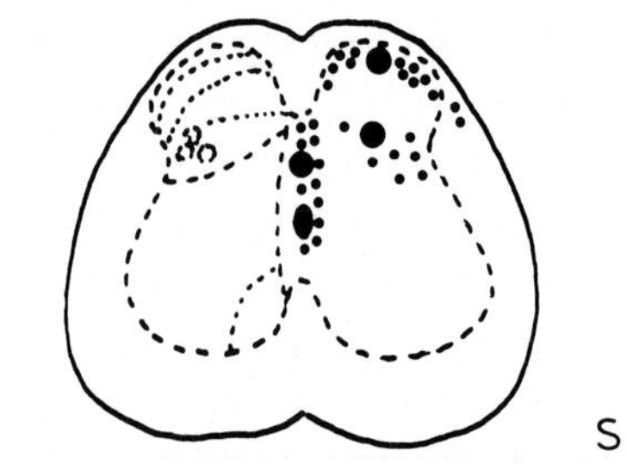
S

SRIF

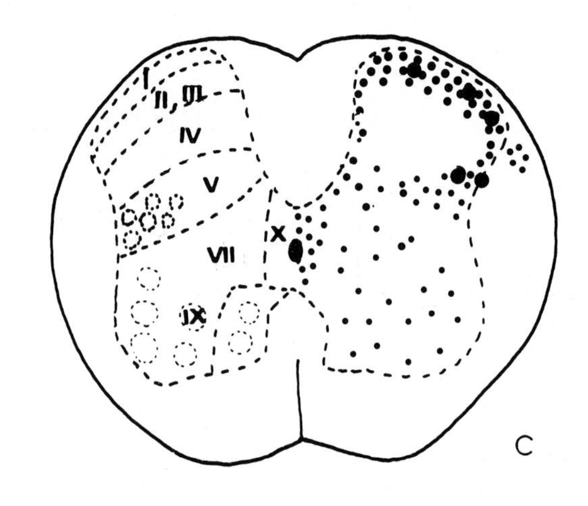
I
II, III
IV
V
VII
X
IX
C

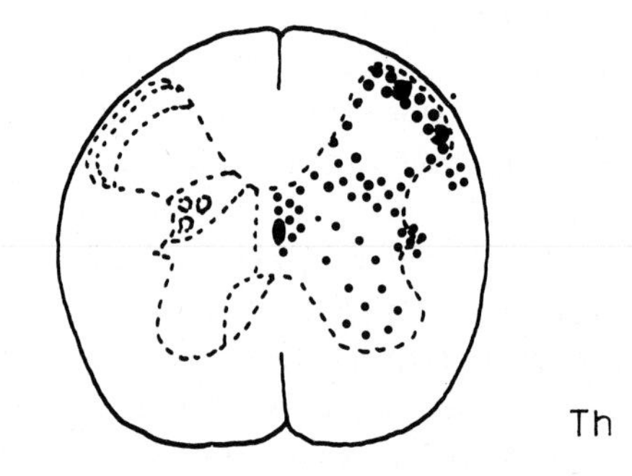
Th

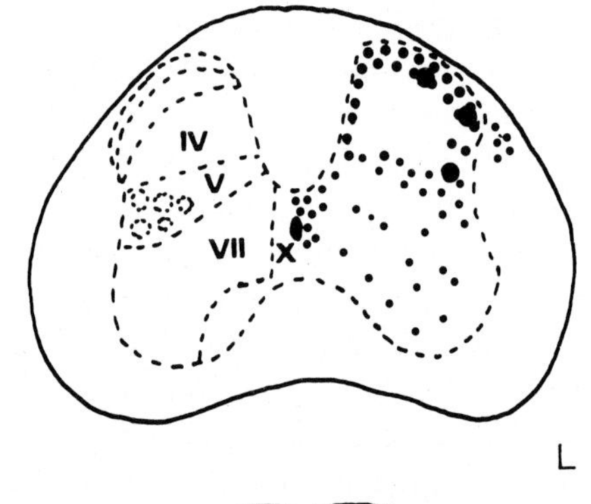
IV
V
VII
X
L

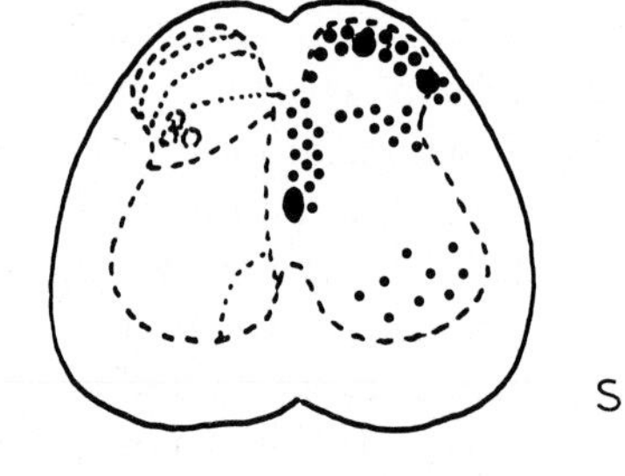
S

SP

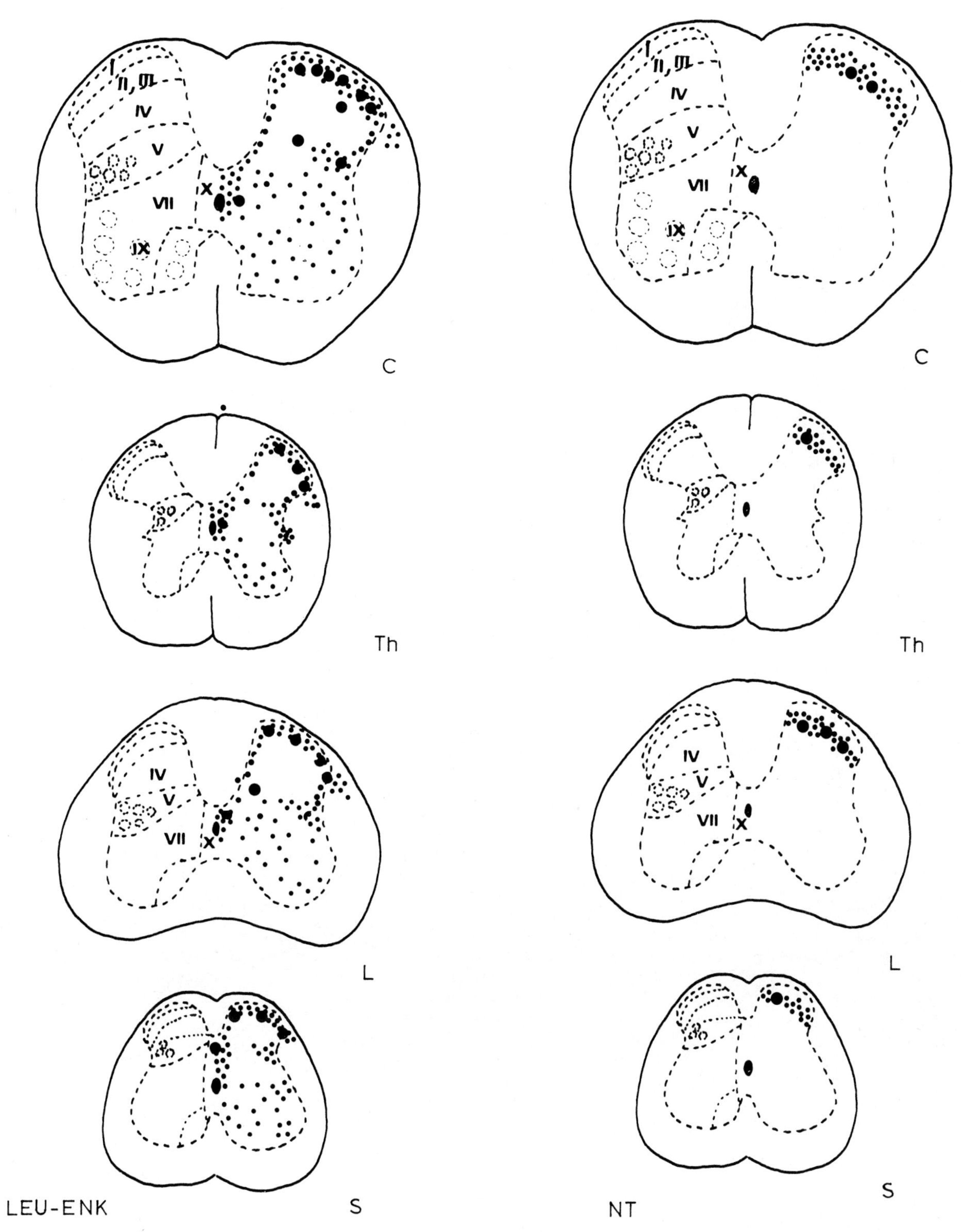

Figs. 1, 2. Schematic drawing showing the distribution of somatostatin (SRIF), substance P (SP), Leu-enkephalin (LEU-ENK) and neurotensin (NT)-like immunoreactive structures in the spinal cord through cervical (C), thoracic (Th), lumbar (L) and sacral (S) levels. Large dots indicated immunoreactive cell groups and small dots immunoreactive fibres. Abbreviations: I–X, laminae I–X. From Senba et al. (1982).

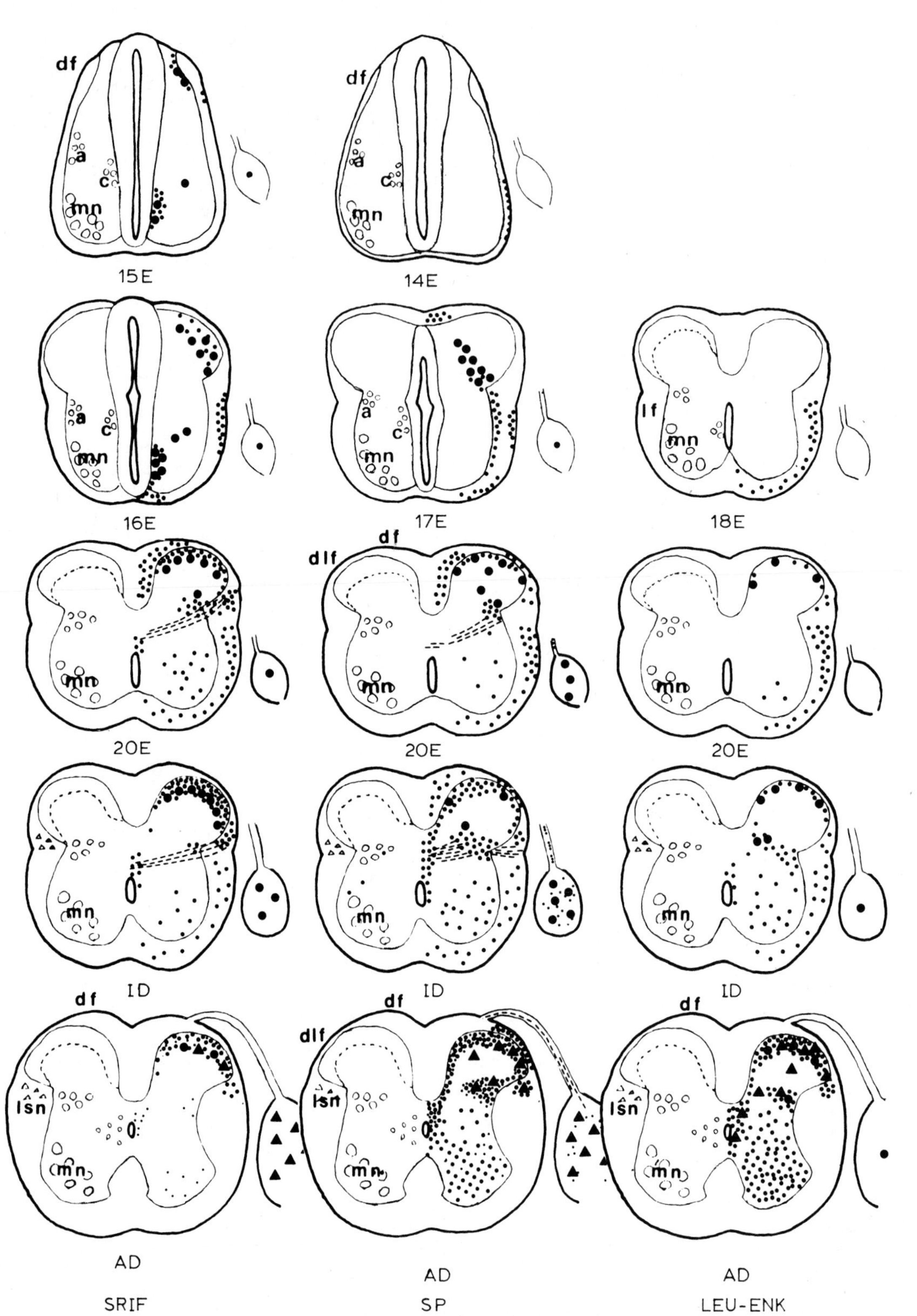

Figs. 3, 4. Schematic drawing of somatostatin (SRIF), substance P (SP), Leu-enkephalin (LEU-ENK), neurotensin (NT), cholecystokinin-8 (CCK-8) and vasoactive intestinal polypeptide (VIP)-like immunoreactive neurons (large dots) and fibres (small dots) in the rat spinal cord on gestational day 14 (14 E), 15 (15 E), 16 (16 E), 17 (17 E), 18 (18 E), 19 (19 E), 20 (20 E) new-born (0 D), postnatal

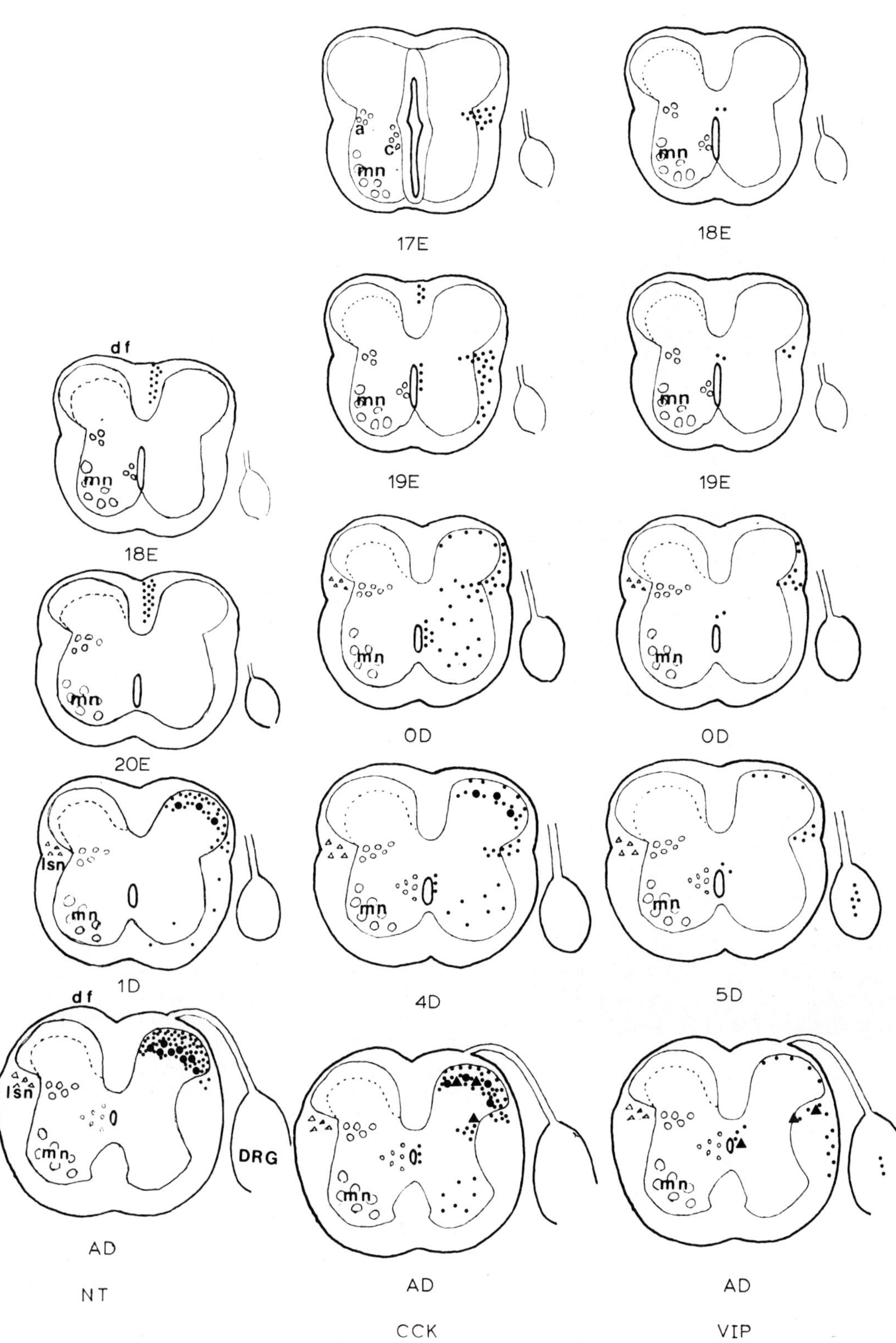

day 1 (1 D), 4 (4 D), 5 (5 D) and adult (AD). Full triangles indicate positive neurons which could be detected only when the animals were pretreated with colchicine. Abbreviations: a, association neuron; c, commissural neuron; df, dorsal funiculus; dlf, dorsolateral funiculus of Lissauer; DRG, dorsal root ganglion; lsn, a small nucleus of dorsal lateral funiculus; mn, motor neuron. From Senba et al. (1982) and Fuji et al. (1983).

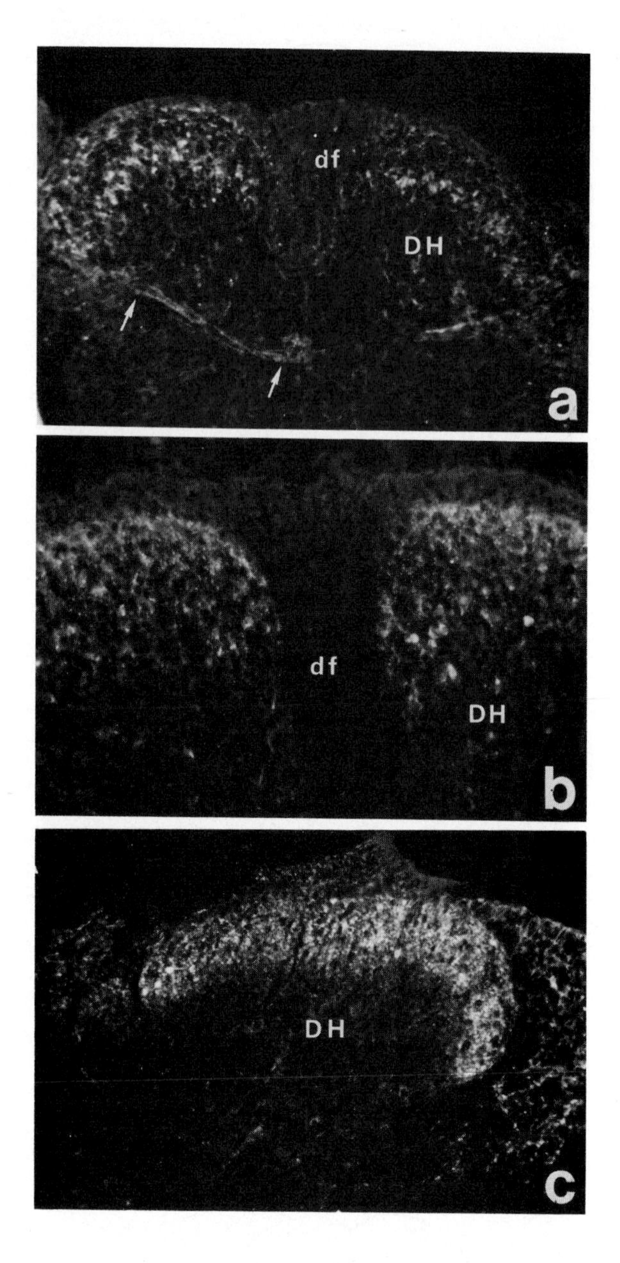

Fig. 5. Ontogeny of immunohistochemically stained SRIF-positive structures in the dorsal horn (DH) of the spinal cord. (a) Gestational day 20; (b) newborn; (c) adult. On gestational day 20 (a), small positive neurons in lamina II and scattered fibres in lamina I are observed. Note the communicating fibres between laminae V, VI and X (indicated by arrows). These positive structures increase in number in neonatal rats (b) and they are also found in the adult rat (c). a, × 125; b, × 190; c, × 160. From Senba et al. (1982).

SRIF-LIr structures increase considerably towards the junction of the spinal cord and medulla oblongata.

Fibre connections

Recently the SRIF-LIr innervation of the spinal cord has been explored in more detail (Fig. 9). Hökfelt et al. suggested in 1976 that the SRIF-LIr fibre plexus in the dorsal horn originates from SRIF-LIr neurons located in the spinal ganglia, forming a SRIF-containing primary sensory afferent system. In 1982, Kawai et al. reported the presence of an amygdalofugal descending SRIF-LIr tract, and Inagaki et al. revealed the precise terminal field of this system in 1983. These lesion studies have revealed that the SRIF-LIr neurons in the central amygdaloid nucleus send their axons ipsilaterally to various lower brain stem areas and to the upper cervical

Fig. 6. (a) Schematic drawing of the electrolytic lesions made in the amygdaloid complex of the rat. Frontal plane. Blacked area shows the lesion. Note that lesion was restricted to the central amygdaloid nucleus (ac) and the area (amc) between ac and medial amygdaloid nucleus (am). A–J: Schematic representation showing the changes of SRIF-LIr fibres in the lower brain stem 7 days after unilateral destruction of the ac and amc outlined in a stereotaxic atlas of the rat brain. Frontal plane. Large dots indicate SRIF-LIr cells and small dots indicate SRIF-LIr fibres. The lesion is shown in Fig. 6a. Note a marked decrease of SRIF--LIr fibres in the n. reticularis pontis oralis (po), n. reticularis pontis caudalis (pc) (particularly just dorsal to the superior olivary complex (os)), n. reticularis gigantocellularis (gc) (particularly just dorsal to the facial nucleus (n VII)), n. reticularis parvocellularis (pv), n. reticularis medullae oblongatae pars dorsalis (rd) et pars ventralis (pv) and hypoglossal nucleus (n XII): a less marked decrease in the central gray matter of the midbrain (MSG), midbrain reticular formation (MRF) and n VII; a small decrease in the n. reticularis paramedianus (pm) and n. reticularis lateralis (rl) on the operated side; but no reduction in other areas such as inferior olivary complex (io), n. tractus solitarii (nts) and trigeminal spinal nucleus (nV). This type of lesion also caused a reduction of SRIF-LIr fibres in the ventral horn of the spinal cord, C_{1-2}. Abbreviations: ab, basal amygdaloid nucleus; abl, basal amygdaloid nucleus, lateral part; abm, basal amygdaloid nucleus, medial part; aco, cortical amygdaloid nucleus; cp, n. caudatus putamen; ct, n. corporis trapezoidei; DP, decussatio pyramidis; ip, interpeduncular nucleus; lc, locus coeruleus; LL, lemniscus lateralis; LM, lemniscus medialis; n V, motor nucleus of the trigeminal nerve; r, n. ruber. From Inagaki et al. (1983).

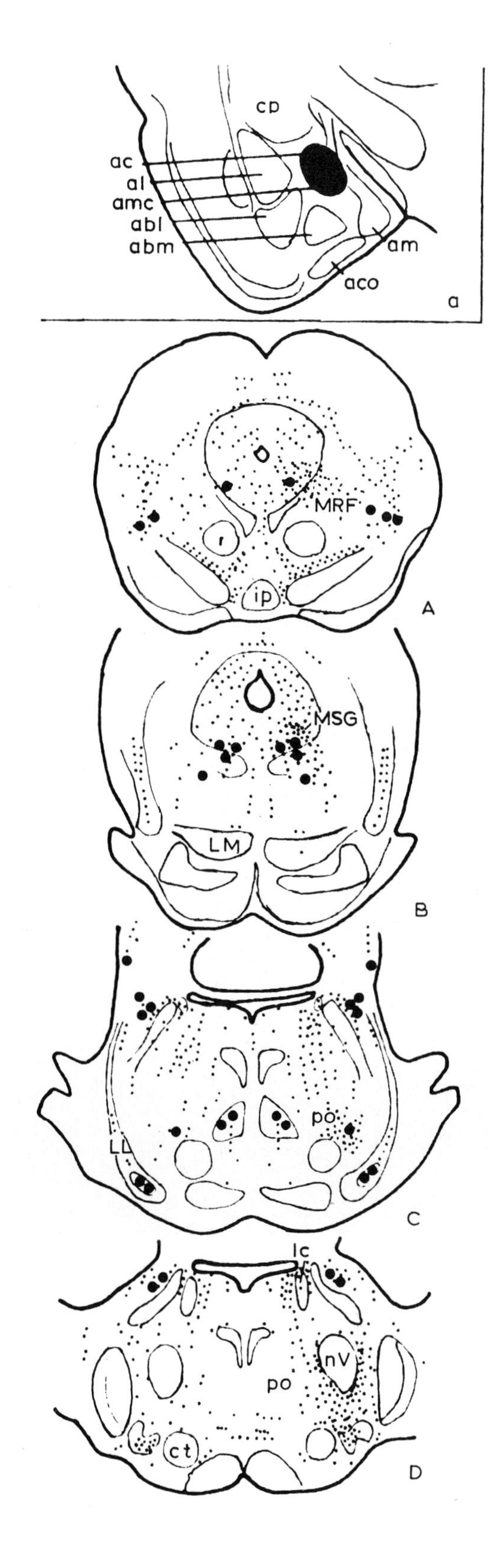
cp
ac
al
amc
abl
abm
am
aco
a
MRF
r
ip
A
MSG
LM
B
po
LL
C
lc
nV
po
ct
D

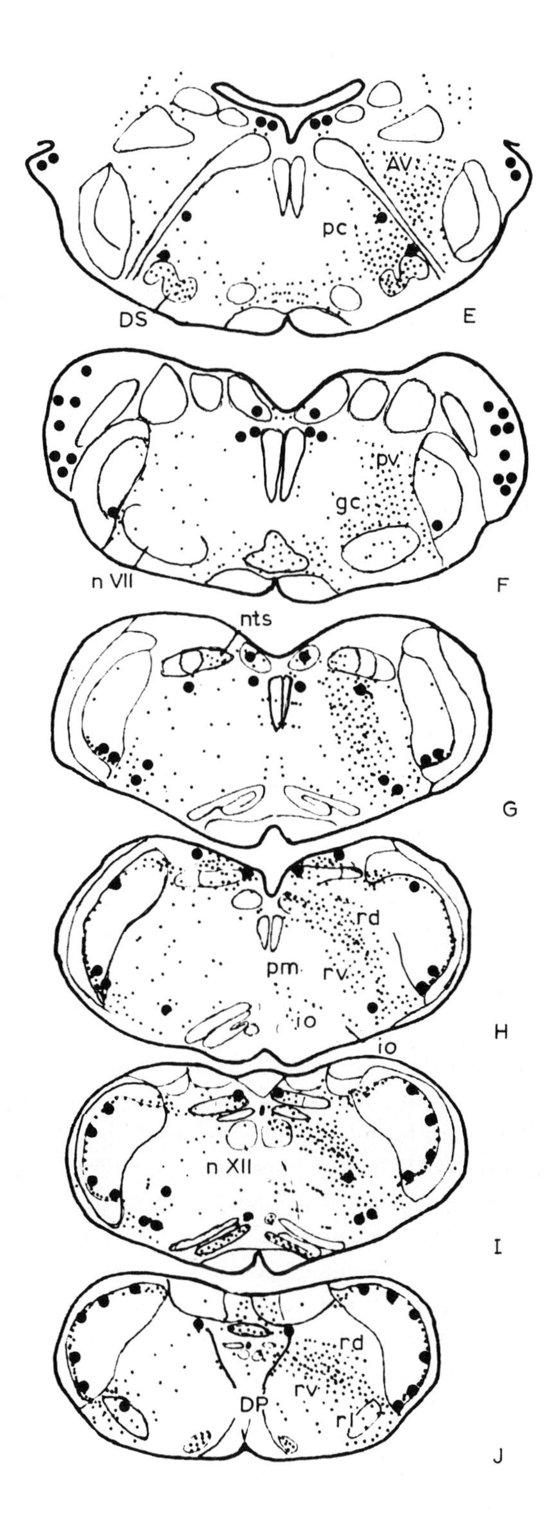
AV
pc
DS
E
pv
gc
n VII
F
nts
G
rd
pm
rv
io
io
H
n XII
I
rd
rv
DP
rl
J

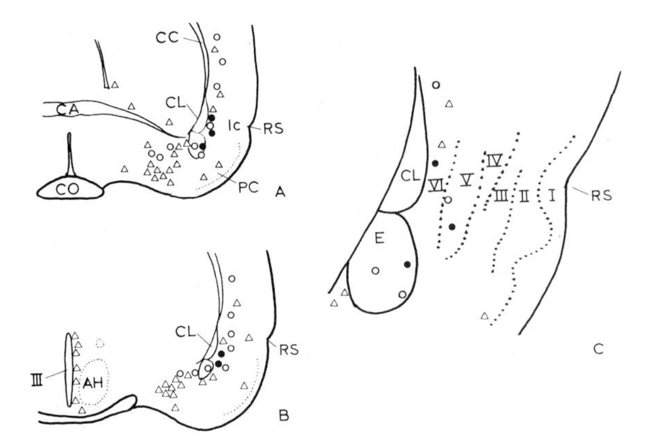

Fig. 7. Distribution of neurons labeled retrogradely by B-HRP after injection into lumbar cord (○), SRIF-positive cells (△), and double-labeled cells (●) (A, B) and (C) a precise illustration of cortical region of level (A). Abbreviations: AH, anterior hypothalamic nucleus; CA, anterior commissura; CC, corpus callosum; CL, claustrum; CO, optic chiasma; E, endopyriform nucleus; IC, insular cortex; PC, pyriform cortex; RS, rhinal sulcus; I–VI, layers of insular cortex. From Shimada et al. (1985).

cord (particularly to the ventral horn), as destruction of the central amygdaloid nucleus resulted in various degrees of ipsilateral reduction of SRIF-LIr fibres: a marked decrease in n. reticularis pontis oralis et caudalis, n. reticularis parvocellularis, n. reticularis gigantocellularis, n. reticularis medulae oblongatae pars dorsalis et ventralis and hypoglossal nucleus; a less marked decrease in the midbrain reticular formation, central gray matter of the mesencephalon, facial nucleus and lamina VII of the upper cervical cord from C_1 to C_2; and a small decrease in n. reticularis lateralis (Fig. 6). However, there was no reduction of SRIF-LIr fibres in other regions of the cord from C_1 and C_2. In addition to this descending SRIF system, another descending SRIF-LIr tract was recently demonstrated by Shimada et al. (1985) by using a double-labeling method combining immunocytochemistry and a biotinized retrograde tracer newly developed by Shiosaka et al. (Shiosaka et al., 1984, 1985a,b; Shiosaka and Tohyama, 1985) (see Shiosaka and Tohyama, this volume for the details of this method). Injection of biotin–horseradish peroxidase (B-HRP) into the spinal cord at cervical or lumbar levels resulted in the labeling of a number of neurons in the insular cortex (Figs. 7, 8). Simultaneous immunostaining revealed the existence of double-labeled neurons (both by B-HRP and SRIF antisera) in the insular cortex (Figs. 7, 8). The results provide the first direct evidence for the presence of a descending SRIF pathway from the insular cortex to the spinal cord, though the terminal fields of this system in the spinal cord have still to be elucidated.

In the past, Krisch (1981) has suggested the presence of a descending SRIF pathway from the hypothalamic periventricular nucleus to the cord. However, this possibility seems to be unlikely, because destruction of the hypothalamic periventricular zone failed to cause changes in SRIF-LIr

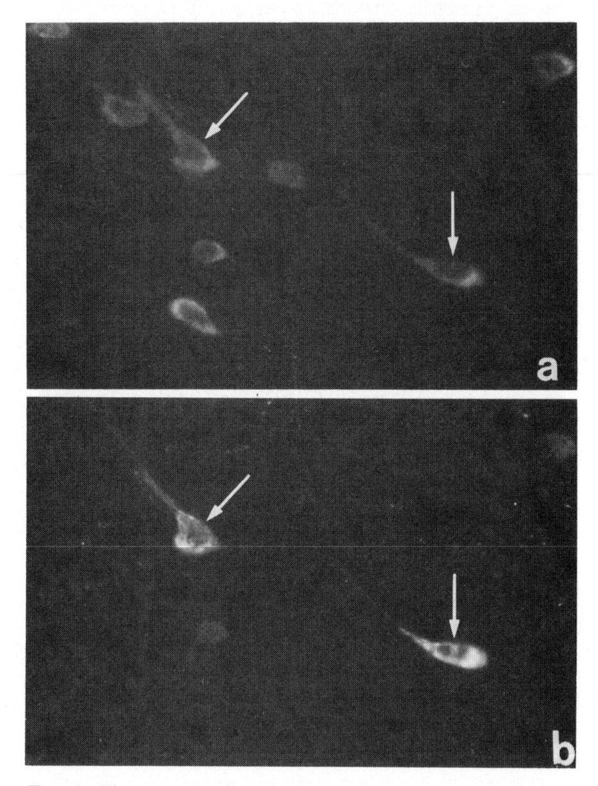

Fig. 8. Fluorescent micrographs showing double-labeled neurons in layer V of insular cortex after injection of B-HRP into the lumbar spinal cord. (a) B-HRP-labeled neurons visualized by avidin–fluorescein isothiocyanate in the insular cortex observed by a B-dichroic mirror filter and IF 520–545 nm interference filter. (b) SRIF-positive cells in the same field as a. Observation was carried out by a G-dichroic mirror and 580 nm absorption filter. Double-labeled cells were indicated by arrows. × 180. From Shimada et al. (1985).

fibres in various brain stem areas and spinal cord. Therefore, to validate the existence of this pathway, further careful experiments will be needed.

Thus, to date the SRIF innervation of the spinal cord can be classified into: (1) a descending amygdalofugal SRIF system which innervates the ventral horn of the C_1 and C_2 cord; (2) a descending cortico-spinal SRIF tract; (3) an intrinsic SRIF system and (4) a primary sensory afferent SRIF system (Fig. 9).

Ontogeny

In addition to SRIF-LIr neurons in lamina III of adult rats, several SRIF-LIr neuron groups could be seen in the rat spinal cord during development (Senba et al., 1982) (Figs. 3, 5). They could be divided into three groups according to their location and ontogenetic state. The first group of cells, which appeared on gestational day 15, was located in the dorsal horn and could be subdivided into two groups; one located in the dorsal part and the other in the ventral part (Fig. 3). SRIF-LIr cells in the ventral dorsal horn reached the maximum number on gestational day 16, after which it became difficult to detect this group of cells (Fig. 3) and only a few SRIF-LIr cells or none were found in this region on gestational day 21. On the other hand, the number of SRIF-LIr cells detected in the dorsal part of the dorsal horn increased progressively as the fetus grew (Fig. 3), and numerous SRIF-LIr cells were identified in this area at birth (Fig. 5b). On the contrary, after birth they decreased gradually in number with the rat's growth. However, colchicine treatment in the adult rat visualised numerous SRIF-LIr cells in this area, though they were less numerous than in the new born rat (Fig. 3).

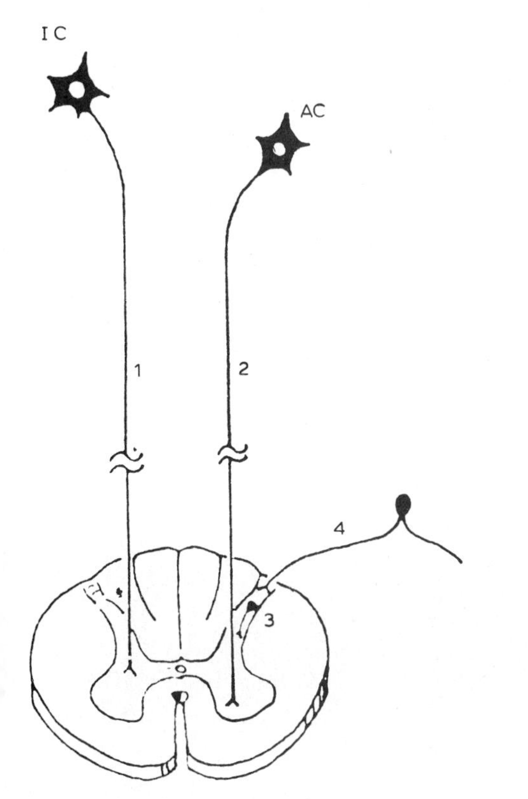

Fig. 9. Schematic drawing of SRIF spinal innervation: descending spinal projections from insular cortex (IC) (1) and central amygdaloid nucleus (AC) (2), SRIF innervation intrinsic to the spinal cord (3) and primary sensory afferent (4). SRIF-LIr fibres from AC terminate in the ventral horn of the upper cervical cord, while those of primary sensory afferent end in the dorsal horn.

The second group of cells was located in the neuropil among the association, commissural and motor neuron cell groups (Fig. 3). This group of cells appeared on gestational day 15, but became very difficult to identify by gestational day 17.

The last group of cells was located in the medial part of the ventral horn, just lateral to the ependyma and appeared on gestational day 15 (Fig. 3). SRIF-LIr cells detected in this area increased in number until gestational day 17; after which they progressively decreased in number until no SRIF-LIr cells could be seen in the adult rats. Colchicine pretreatment failed to demonstrate SRIF-LIr cells in the latter two regions.

SRIF-LIr fibres appeared in the dorsal lateral funiculus on gestational day 15 (Fig. 3). In addition, SRIF-LIr fibres were also found in the dorsal horn and ventral horn (Fig. 3) at this stage. Furthermore, numerous SRIF-LIr fibres were also found in the ventral funiculus. On gestational day 17, in addition to the SRIF-LIr fibres mentioned above, SRIF-LIr fibres occurred in the ventral lateral funiculus (Fig. 3). On gestational day 19 and

20, SRIF-LIr fibres reached their maximum extent (Figs. 1, 3a). For example, SRIF-LIr fibres found in the laminae V, VI and X formed a densely packed fibre band, most of which could be traced laterally to the lateral funiculus and medially to the contralateral side, passing the posterior commissure (Fig. 3a). It should be noted that, in the thoracic cord, in addition to this fibre band, another band which ran in lamina VII between the n. intermediolateralis and lamina X was identified. After birth, the development of the SRIF-LIr fibres could be divided into three categories. For example, SRIF-LIr fibres found in the dorsal horn of the fetus mainly occupied the marginal layer. However, after birth they continued to increase in number extending into lamina II, and in the postnatal 7th day rat, a dense plexus of SRIF-LIr fibres was found in laminae I and II. From then on, they decreased slightly in number but numerous SRIF-LIr fibres could be detected even in the adult rat (Figs. 3, 5c). SRIF-LIr fibres in the gray matter could be included in the second category. SRIF-LIr fibres in laminae V, VI and X decreased slightly in number as the rat grew, and only a few fibres were found in the adult rat (Fig. 3). The last category includes the SRIF-LIr fibres detected in the white matter. These fibres decreased remarkably in number as the rat grew and none, or only a few SRIF-LIr fibres could be detected in the adult rats (Fig. 3).

The ontogenetic development of SRIF-LIr structures in the spinal cord can be summarized as follows: SRIF-LI appears in the early fetal period before the establishment of the spinal synaptic transmission system, suggesting that SRIF may have an important trophic role in the development of the spinal cord. Furthermore, SRIF-LIr structures are found abundantly during the fetal period in the spinal cord: however, SRIF-LIr fibres in the ventral horn, laminae V, VI and X tend to decrease remarkably in number, while those found in the dorsal horn maintain their immunoreactivity even in the adult rats. These facts suggest that SRIF in the latter area might function as a neurotransmitter or neuromodulator, whereas in the former areas, SRIF might have some other, perhaps trophic, role in the development of the spinal cord.

Fine structures

Detailed histology of SRIF-LIr structures in the dorsal horn of the rat was examined by Kubota et al. (1983). SRIF-LIr fibres were of small calibre and unmyelinated. Central axons of the synaptic glomeruli often showed SRIF-LI. SRIF-LIr terminals formed mostly axo-dendritic synapses (Fig. 10) and occasionally axo-somatic synapses. Neuronal elements with which SRIF-LIr terminals make synapses mentioned above were devoid of immunoreactivity. However, in rare cases, two SRIF-LIr terminals which showed direct apposition were

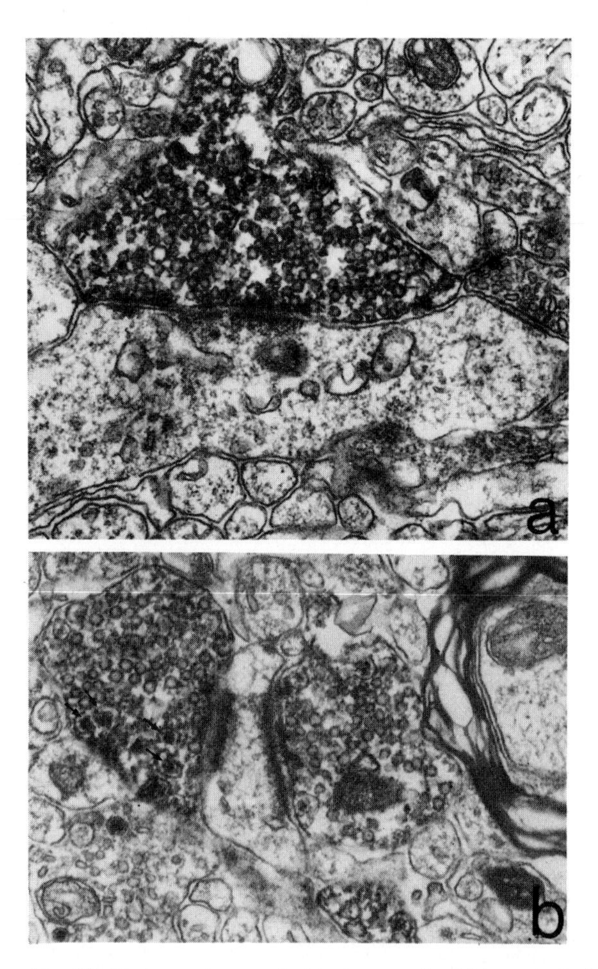

Fig. 10. Electron micrograph showing axo-dendritic contact of SRIF-LIr fibres in the dorsal horn (a). These terminals often formed synaptic contact on dendritic spines (b). Note immunoreactive large granulated vesicles in the SRIF-LIr terminal (b, arrows). ×24 300. Courtesy of Dr. Y. Kubota.

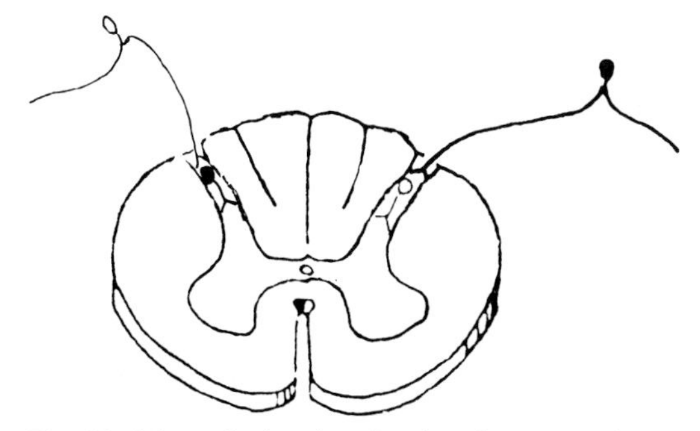

Fig. 11. Schematic drawing showing the neuronal contact of SRIF-LIr structures in the dorsal horn. SRIF-LIr primary sensory afferent formed synaptic contact with non-immunoreactive intrinsic neurons, while immunoreactive cells in the dorsal horn received non-SRIF primary sensory afferent fibres. ●, SRIF-LIr neurons; ○, non-immunoreactive cells.

seen. It should be noted that multiple SRIF-LIr terminals often make synaptic contact with a single dendrite or dendritic spine (Fig. 10b) which is devoid of immunoreactivity. On the other hand, SRIF-LIr neurons always receive non-immunoreactivity input both via axo-dendritic and axo-somatic synapses. Rhizotomy experiments revealed that some of these inputs are primary sensory afferent, because degenerating terminals formed axo-dendritic or axo-somatic synapses with SRIF-LIr neurons.

These findings indicate that (1) SRIF-LIr primary sensory afferents formed synaptic contact mostly with non-SRIF-LIr neuronal elements, though a few receive intrinsic SRIF-LIr inputs, and (2) SRIF-LIr neurons receive non-SRIF-LIr primary sensory afferents (Fig. 11).

Substance P

Distribution

Distribution of substance P (SP)-like immunoreactive (SP-LIr) structures in the spinal cord has also been examined in animals by many authors (Nilsson et al., 1974; Hökfelt et al., 1975b,c; Cuello et al., 1976; Chan-Palay and Palay, 1977; Pickel et al., 1977; Cuello and Kanazawa, 1978; Ljungdahl et al., 1978; Barber et al., 1979; Lorez and Kemali, 1981; Senba et al., 1982; Lavalley and Ho, 1983; Sasek et al., 1984).

In the adult rat, our work has shown that SP-LIr neurons were occasionally visible in laminae II, III and V. Colchicine treatment resulted in an increase in the number of immunoreactive cells in these areas; the highest density of cells was seen in lamina II, but in other areas SP-LIr cells were sparse.

SP-LIr fibres were particularly densely stained in laminae I and II where they formed a thick fibre plexus band (Figs. 1, 3, 14c). A moderate density of SP-LIr fibres was seen around the central canal. A SP-LIr fibre plexus extended in three directions from this area (Fig. 1). One was to the gray substance dorsally along the dorsal funiculus intermingling with the dense fibre band seen in laminae I and II. Other fibres went towards the midline; one group to laminae IV–VI, and the other group to lamina VII. SP-LIr fibres seen around the central canal (lamina X) were more conspicuous in the dorsal than the ventral part. In the ventral parts of lamina IV and lamina V a SP-LIr fibre plexus with a moderate density was also seen. This fibre plexus extended dorsolaterally (Fig. 1), where it invaded the lateral funiculus and dorsally intermingled with dense fibre bands located in laminae I and II at the lateral part of the dorsal horn (Fig. 1). Furthermore, it extended medially to join with the medial extension of the SP-LIr fibre plexus located in lamina X (Fig. 1). In addition to these dense fibre plexus, the ventral horn also contained a low to moderate density of SP-LIr fibres, suggesting that SP is also involved in the motor as well as the sensory function.

Fibre connections

It was originally suggested that SP may play an important role in the primary sensory afferent system following the remarkable decrease of SP-LI in the dorsal horn following dorsal rhizotomy (Hökfelt et al., 1975b; Takahashi and Otsuka, 1975) and the localization of SP-LI in small to medium sized cells in the spinal ganglion. In addition, recent studies have shown that SP-LI and thyrotropin-releasing hormone (TRH)-like immunoreactivity

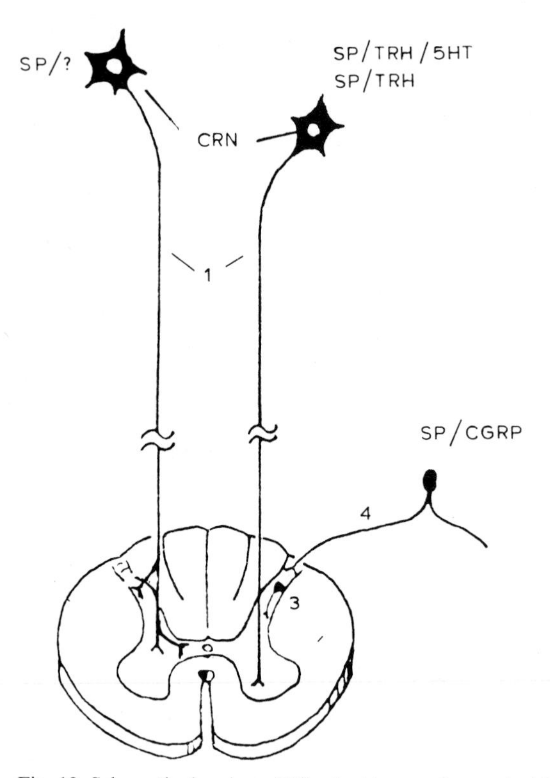

Fig. 12. Schematic drawing of SP spinal innervation: spinal SP-LIr fibres originate supraspinally from caudal raphe nuclei (CRN) (1). SP-LIr neurons containing TRH-LI project mainly to the ventral horn, while those lacking TRH-LI probably project diffusely. Most SP-LIr primary sensory afferent neurons contain CGRP-LI and end in the dorsal horn (4). In addition to these systems, an intrinsic SP system is present (3).

(TRH-LI) coexist in single serotonergic cells of the caudal raphe nuclei (Johansson et al., 1981; Gilbert et al., 1982). These cells could be divided into three categories: neurons containing both serotonin (5-HT) and SP-LI, neurons containing 5-HT, SP-LI and TRH-LI and neurons containing 5-HT alone. SP-LIr neurons in the caudal raphe nuclei which contain TRH-LI project to the ventral horn and SP-LIr neurons lacking TRH-LI probably project to the entire gray matter (Fig. 12).

Recent experimental immunohistochemical studies have demonstrated the presence of a SP neuron system intrinsic to the spinal cord as well as in supraspinal projections (Bowker et al., 1981; Hökfelt et al., 1981). In accordance with these findings, Senba et al. (1981) have demonstrated the presence of SP-LIr neurons in various parts of the dorsal horn in the adult rat after colchicine treatment and the presence of these cells both above and beneath transections of the spinal cord. The latter findings suggest that these SP-LIr cells give rise to ascending and descending axons leading to other spinal or brain stem areas. Fig. 12 gives a summary of SP spinal innervation.

Ontogeny

The development of SP-LIr structures in the spinal cord was examined in detail by Senba et al. (1982). SP-LIr cells of the spinal cord could be divided into three groups according to their localization and development. The first group of cells was located in the ventral part of the dorsal horn (Fig. 3). This group of cells first appeared on gestational day 16 and reached their maximum (as demonstrated histochemically) on gestational day 17 (Figs. 3, 13a). However, after gestational day 18, SP-LIr cells detected in this area decreased in number and only a few cells were seen at birth.

The second group of cells, located in the dorsal part of the dorsal horn, appeared on gestational day 19 and reached a maximum on gestational day 20 (Fig. 13c). The majority were located in lamina I, some in lamina II, and a few cells in lamina III. It should be noted that although the distribution and ontogeny of SP-LIr cells in the cervical and thoracic cord were generally similar, the number of SP-LIr cells in the dorsal part of the dorsal horn increased rostrally, the greatest number being identified in the upper cervical cord. From gestational day 21, the number of SP-LIr cells detected in this area decreased remarkably, and at birth only a few SP-LIr cells were detected.

The last group of cells was located in the small nucleus of the dorsal lateral funiculus. This group of cells could not be detected in the fetus or in rats which were not pretreated with colchicine. After birth, it becomes more difficult clearly to identify SP-LIr cells in the spinal cord of the rat without colchicine treatment. However, numerous SP-LIr cells can be demonstrated in laminae II, III and IV,

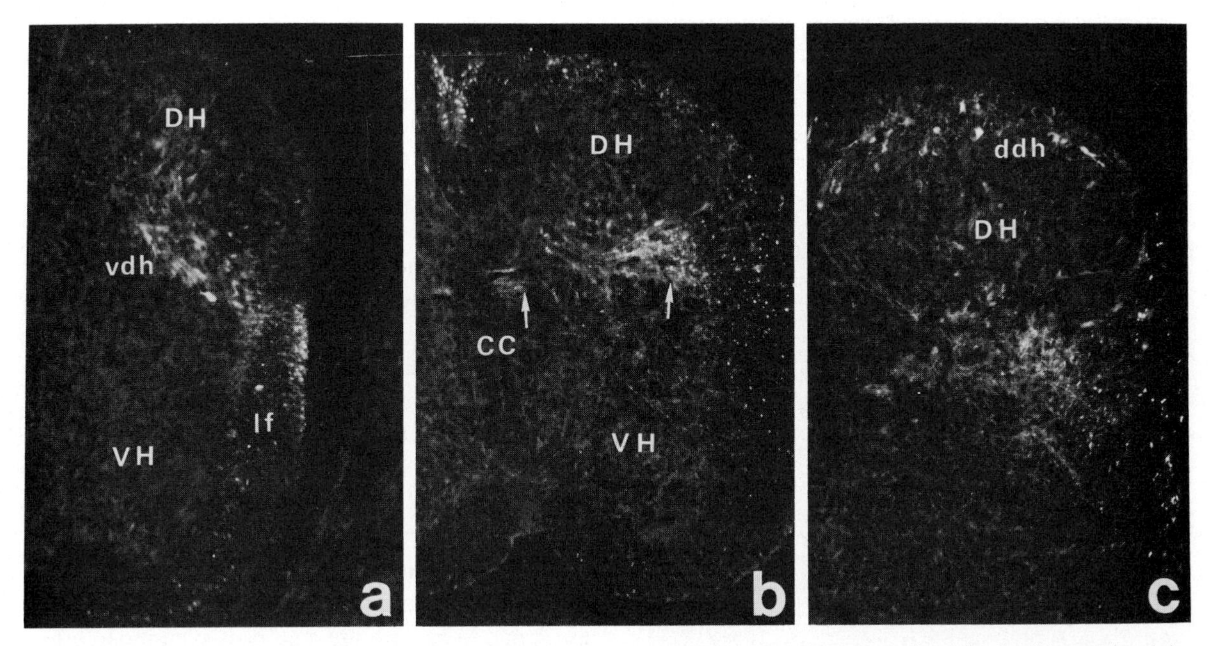

Fig. 13. Ontogeny of immunohistochemically stained SP-LIr structures in the spinal cord. (a) Gestational day 17; (b) gestational day 19; (c) gestational day 20. SP-LIr fibres first appear in the white matter on gestational day 14, whereas SP-LIr neurons first occur in the ventral part (vdh) of the dorsal horn (DH) on gestational day 16. SP-LIr structures detected in these areas increased dramatically in number and intensity on gestational day 17 (a). On gestational day 19, SP-LIr fibres are identified in the dorsal funiculus and dorsolateral fasciculus of Lissauer. Furthermore, SP-LIr fibres in the ventral horn (VH) are first observed at this stage. Note the SP-LIr fibre plexus in laminae V, VI, and X (indicated by arrows) (b). SP-LIr neurons located in the dorsal part of DH (ddh) first appear on gestational day 20 (c). (a) ×95; (b) ×95; (c) ×120. From Senba et al. (1982). CC, central canal.

with a few in the lateral spinal nucleus (Fig. 14c).

SP-LIr fibres first appeared in the ventral lateral funiculus of the fetus on gestational day 14 (Fig. 3). At this stage, SP-LIr cells could not be seen in the spinal cord. From then on, SP-LIr cells detected in the spinal cord gradually increased in number (Fig. 3). On gestational day 16, SP-LIr fibres were found in other parts of the white matter such as the ventral funiculus, dorsal lateral funiculus, and also in the gray matter around the SP-LIr cells. On gestational day 17, a small number of SP-LIr fibres were found in the dorsal lateral funiculus, which increased in number on gestational day 19, especially in fasciculus gracilis (Figs. 3, 13b). On gestational day 19–20, SP-LIr fibres in the gray matter increased remarkably in number. They formed a dense fibre plexus in the lateral part of laminae V and VI, some of which could be traced laterally to the lateral funiculus and medially to the contralateral side passing the posterior commissure (Figs. 3, 13b, c). It should be noted that in the thoracic cord another finer band in lamina VII, running between the nucleus intermediolateralis and lamina X, was observed (Fig. 3). In addition, SP-LIr fibres were also detected in the marginal layer of the dorsal horn and in the dorsolateral fasciculus of Lissauer at this stage of gestation (Fig. 3). After birth, SP-LIr fibres in the gray matter of the spinal cord increased remarkably in number. For example, in the dorsal horn, a dense fibre plexus of SP-LIr fibres was also observed in the substantia gelatinosa in addition to the marginal layer, and in the dorsolateral fasciculus of Lissauer, numerous SP-LIr fibres were seen (Figs. 3, 14a, b). Furthermore, numerous SP-LIr fibres were found in lamina X, and a moderate number were detected around the motor nuclei in the ventral horn.

In summary, SP-LI in the spinal cord of the rat

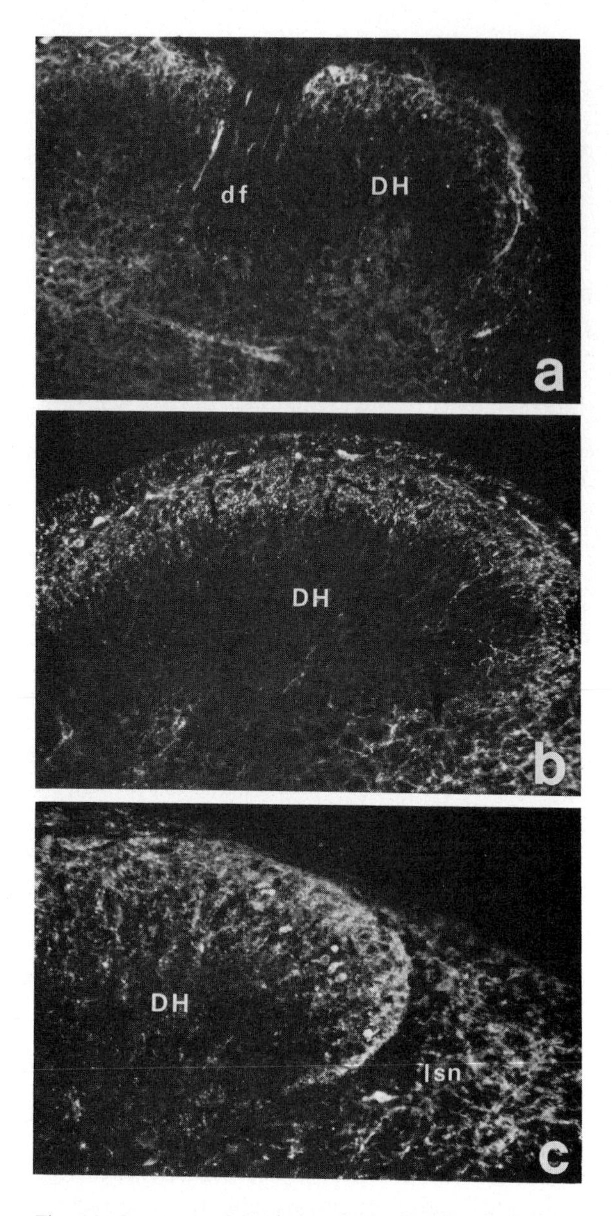

Fig. 14. Ontogeny of immunohistochemically stained SP-LIr structures in the dorsal horn (DH) of the spinal cord. (a) Postnatal day 1; (b) postnatal day 15; (c) adult. SP-LIr fibres in the dorsal part of the dorsal horn first appear on gestational day 19. After birth SP-LIr fibres continued to increase in number (compare photos). On the other hand, numerous SP-LIr neurons which have been found in the fetal period decrease in number as they grow. However, these cells are detected in the adult animals when they are pretreated with colchicine. df, dorsal funiculus; lsn, lateral spinal nucleus. (a) × 110; (b) × 110; (c) × 125. From Senba et al. (1982).

appeared in the early fetal period as seen with SRIF. SP-LIr fibres continued to increase in number after birth, and they could be seen throughout the entire spinal cord even in the adult rat.

Fine structure

Barber et al. (1979) and Di Tirro et al. (1981) examined the fine structure of the SP-LIr terminals in the dorsal horn of the rat spinal cord. They showed that SP-LIr structures are often associated with large granular vesicles and frequently formed axo-dendritic contacts, occasionally axo-somatic and in some very few cases axo-axonic contacts. These terminals formed both symmetric and asymmetric synapses. They further showed that individual SP-LIr terminals form both asymmetric and symmetric contacts with different postsynaptic elements. As to axo-axonic contact, SP-LIr terminals are found to be both pre- and post-synaptic to other axons. The authors proposed that in cases in which SP-LIr terminals are postsynaptic in structure, non--labeled terminals may belong to the ENK-containing terminals, because presynaptic inhibition by ENK of SP release at primary sensory afferent terminals has often been postulated (see ENK for details; Jessell and Iversen, 1977).

Subsequently, De Lanerolle and LaMotte (1983) examined the fine structures of SP-LIr terminals of the dorsal horn of the monkey, and they obtained similar results to those of Barber et al. (1978).

Calcitonin gene-related peptide

CGRP was recently demonstrated in nervous tissue by recombinant DNA and molecular biological techniques (Amara et al., 1982; Rosenfeld et al., 1984). Rosenfeld et al. (1984) reported that alternative processing of the mRNA transcribed from the calcitonin gene appeared to result in the production of mRNA in neural tissue distinct from that in thyroid parafollicular C cells: thyroid mRNA encoding only the calcitonin precursor but not CGRP. However, we recently found that most calcitonin cells contained CGRP (Lee et al., 1985b).

This indicates that calcitonin and CGRP are co-produced in the same C cells, and that thyroid mRNA encodes not only the calcitonin precursor but also the CGRP precursor. The distribution of CGRP-like immunoreactivity (CGRP-LI) in the brain and peripheral nervous system was recently reported in detail (Kawai et al., 1985; Kiyama et al., 1985; Lee et al., 1985a,b; Takami et al., 1985a,b).

Distribution

CGRP-LIr cells were exclusively localized to the ventral horn of the spinal cord (Gibson et al., 1984b; Rosenfeld et al., 1984; Takami et al., 1985a) (Figs. 15, 16). It is well known that ventral horn motor cells contain acetylcholine (ACH) as a neurotransmitter. Examination of serial sections through the spinal cord revealed that most of the CGRP-LIr cells in the ventral horn were ACH-containing (Takami et al., 1985a) (Fig. 17). Choline-acetyltransferase (ChAT) was used as a marker for cholinergic neurons in their study. However, it should be noted that ACH neurons located in the lateral horn lack CGRP-LIr structures (Fig. 15). This observation also applied to the brain stem (Fig. 18). ACH neurons innervating striated muscles contain CGRP-LIr structures in single cells, while other ACH neuron systems such as the basal forebrain cholinergic neurons and the autonomic cholinergic neurons lack CGRP-LI (Fig. 18).

A dense plexus of CGRP-LIr fibres was seen in laminae I and II (Figs. 15, 19). These fibres appeared much more numerous than SP-LIr fibres (Fig. 20). This may be because in the spinal ganglia which is the origin of CGRP-LIr and SP-LIr fibres in the spinal cord, there exist two types of CGRP-LIr neurons; one where CGRP-LI coexists with SP-LI in single cells (small to medium size) and the

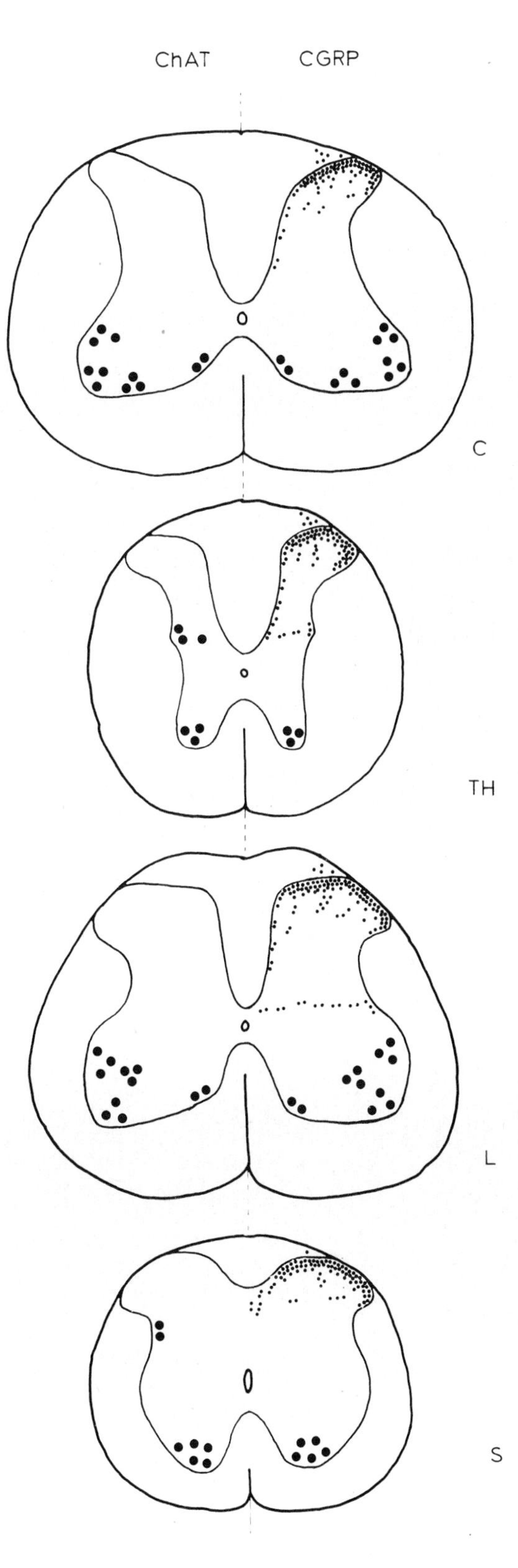

Fig. 15. Schematic drawing of CGRP-LIr structures in the spinal cord from cervical (C) to sacral (S) cord. Locations of CGRP-LIr neurons (large dots) and fibres (small dots) are represented on the right side. In addition, location of ACH neurons demonstrated by ChAT as a marker on the left by large dots. Note a similar localization of ACH and CGRP-LIr neurons in the ventral horn. L, lumbar cord; TH, thoracic cord.

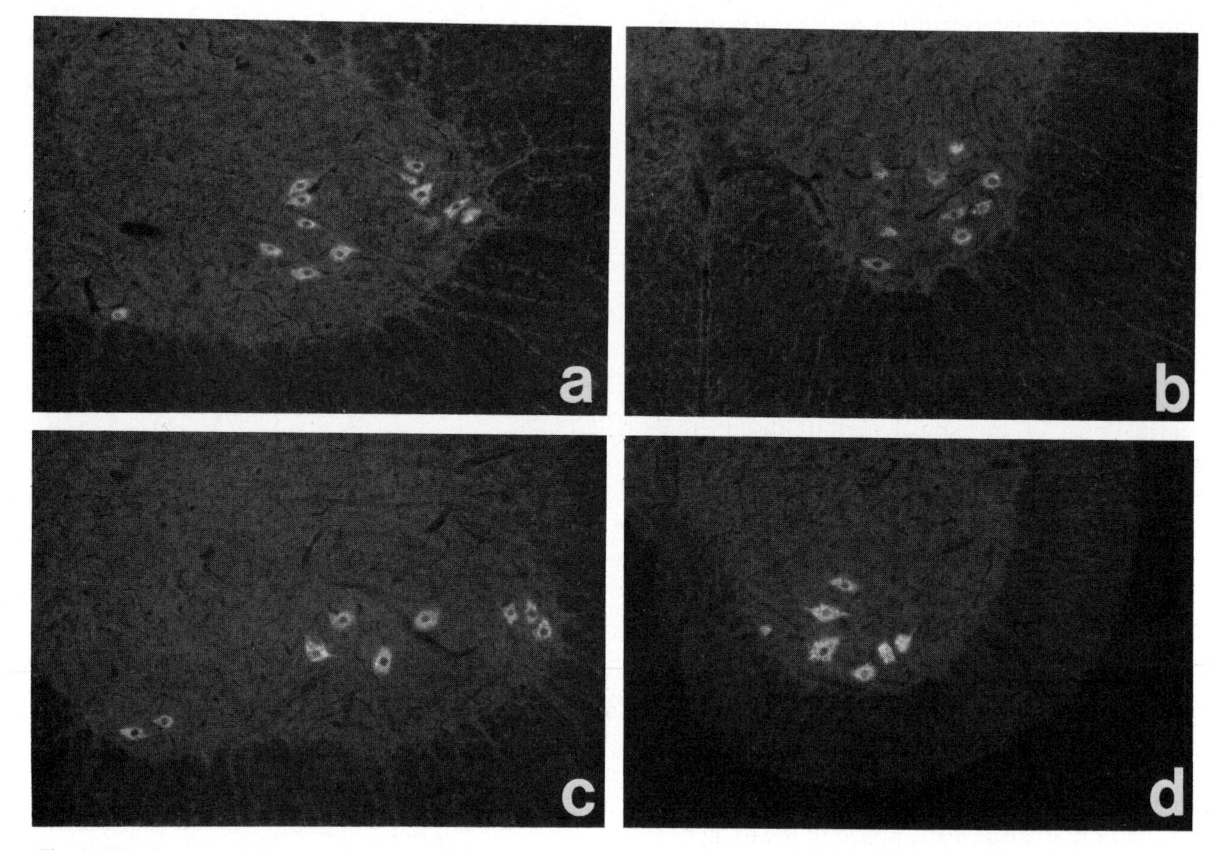

Fig. 16. Fluorescent micrographs showing CGRP-LIr neurons in the ventral horn of the spinal cord. Note that motor neuron type cells are labeled by CGRP antiserum. (a) Cervical cord; (b) thoracic cord; (c) lumbar cord; (d) sacral cord. (a–d) × 60.

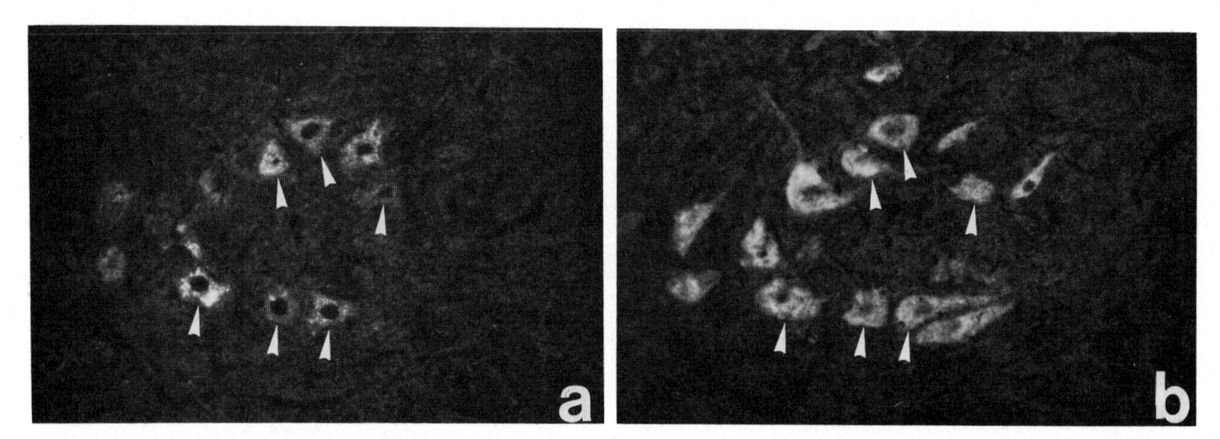

Fig. 17. Fluorescent micrographs of consecutive sections of the ventral horn at the level of the cervical cord after incubation with CGRP antiserum (a) and ChAT monoclonal antiserum (b). Most CGRP-LIr cells are simultaneously ChAT positive (arrowheads). × 185.

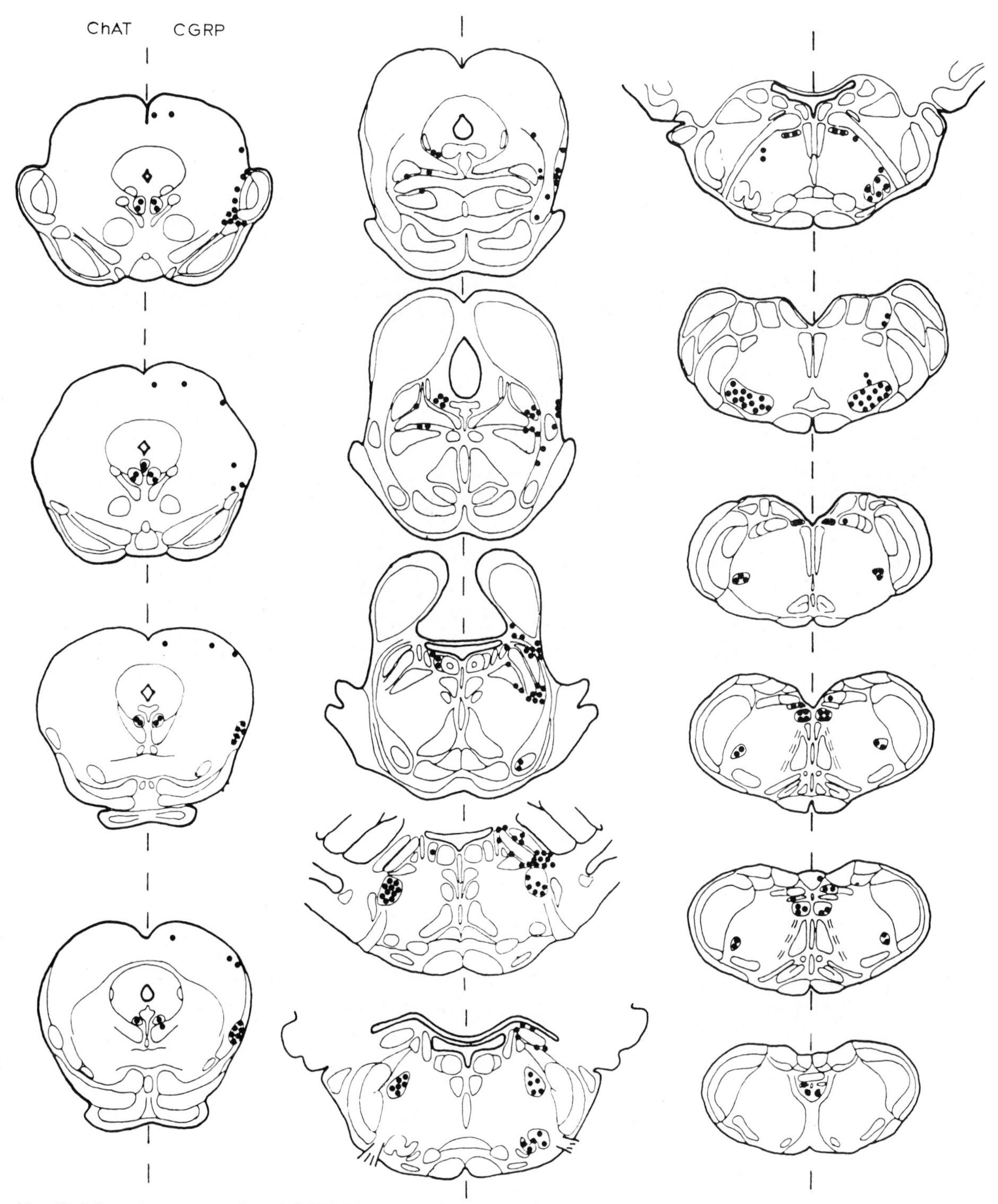

Fig. 18. Schematic representation of CGRP-LIr neurons in the lower brain stem in the right side by filled circle. Location of ACH neurons in this area is also shown on the left side by filled circle. Note that the ACH neuron group which innervates the striated muscle, contains CGRP-LIr neurons, while other ACH neuron groups lack CGRP-LIr cells. Frontal plane.

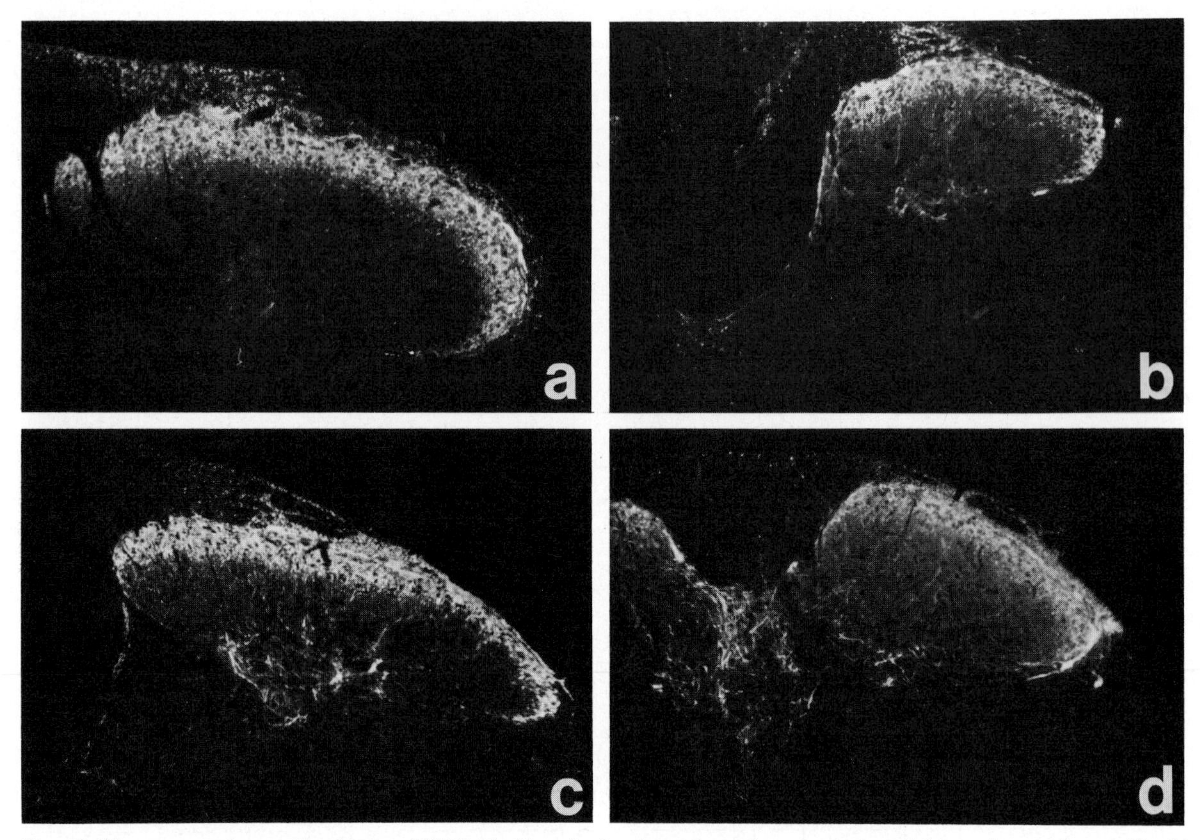

Fig. 19. Fluorescent micrographs showing CGRP-LIr fibres in the dorsal horn from cervical cord (a) to the sacral cord (d). (b) Thoracic cord; (c) lumbar cord. Note a CRGP-LIr fibre plexus with very high density in the dorsal horn.

other which lacks SP-LI; almost all the SP-LIr cells contain co-existing CGRP-LI but a substantial number of CGRP-LIr cells lack SP-LI (Lee et al., 1985b; see Shiosaka and Tohyama, this volume for details). In addition to the particularly dense CGRP-LIr fibre plexus in the lateral parts of laminae V and VI, numerous CGRP-LIr fibres run medially to lamina X. In lamina III, a low density CGRP-LIr fibre plexus was also detected. Furthermore, CGRP-LIr fibres were often seen in the gray substances along the dorsal funiculus. None or only a few CGRP-LIr fibres were seen in the ventral horn of the spinal cord.

Fibre connections

It is clear that at least two CGRP neuron systems exist in the spinal cord (Fig. 21). One is the motor neuron system where CGRP-LI coexists with ACH in single cells and terminates in the striated muscles (Takami et al., 1985a,b). The second is the primary sensory afferent CGRP neuron system. This can be further subdivided into three systems: the first is the small to medium sized CGRP-LIr neuron with coexisting SP-LI, the second is the small to medium sized CGRP-LIr neurons lacking SP-LI and the third is the large CGRP-LIr neurons lacking SP-LI. These heterogeneous subpopulations must be considered when discussing the function of primary sensory afferent CGRP and/or SP systems. In addition, the possibility that another spinal CGRP system may exist should be mentioned, as CGRP-LIr fibres are also found in laminae IV–VI and X. The fact that no CGRP-LIr neurons, except for ventral horn CGRP-LIr cells, were seen in the spinal cord suggests the presence of a descending supraspinal

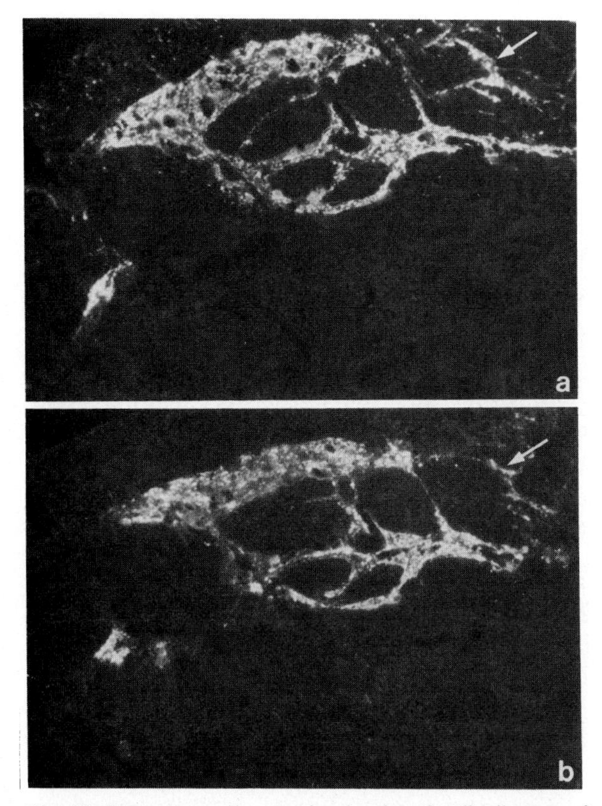

Fig. 20. Fluorescent micrographs showing the distribution of CGRP-LIr (a) and SP-LIr (b) fibres in the n. tractus spinalis nervi trigemini. Although distribution patterns of these peptides are very similar, the area where CGRP-LIr fibres are more predominant than SP-LIr fibres is identified (arrow). Frontal sections. × 140. From Lee et al. (1985a).

CGRP system, although the origins of such a system are so far obscure.

Enkephalin

Distribution

A number of reports concerning the distribution of ENK-like immunoreactivity (ENK-LI) in the spinal cord exist (Elde et al., 1976; Hökfelt et al., 1977a,b; Simantov et al., 1977; Aronin et al., 1978; Johansson et al., 1978; Sar et al., 1978; DiTirro et al., 1981; Finley et al., 1981a; Gibson et al., 1981; Glazer and Basbaum, 1981; Lorez and Kemali, 1981; Nakai et al., 1981; Haber and Elde, 1982; Senba et al., 1982; Sasek et al., 1984).

In the rat, ENK-LIr cells are abundant in lamina II, and sparse in lamina I (Figs. 1, 3). In addition to these neurons, Senba et al. (1982) have reported the presence of scattered ENK-LIr cells in laminae III, IV, V, and VII and several authors have reported cells in the sacral parasympathetic nucleus (Hökfelt et al., 1977a,b; Glazer and Basbaum, 1980, 1981; Lundberg et al., 1980; Dalsgaard, 1982).

ENK-LIr fibres are also abundant in laminae I and II, the highest fibre concentration being in the lateral region of lamina II. In lamina III, scattered ENK-LIr fibres were seen. In lamina X, a moderate density of ENK-LIr fibres was seen. ENK-LIr fibres often extended in various directions from this area (Fig. 2). One fibre group extends laterally to lamina V where a moderately dense ENK-LIr fibre plexus was also seen, another fibre group extends dorsally to pass laminae VI–III along the dorsal funiculus and joins with another ENK-LIr fibre plexus with a high density in laminae I and II, and the other ENK-LIr fibres radiate ventrolaterally or ventrally into laminae VI and VII. In the ventral

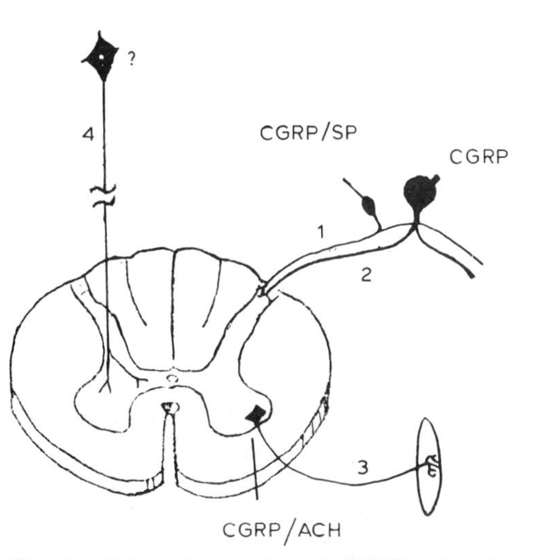

Fig. 21. Schematic drawing of CGRP spinal innervation: CGRP-LIr small to medium sized cells in the dorsal root ganglia contain SP-LIr and terminate in the dorsal horn (1). Large sized CRGP-LIr cells in this ganglion lack SP-LI but also terminate in the dorsal horn (2). CGRP-LIr cells in the ventral horn contain ACH and terminate in the striated muscles forming neuromuscular junctions (3). There may exist a supraspinal CGRP projection to the spinal cord (4).

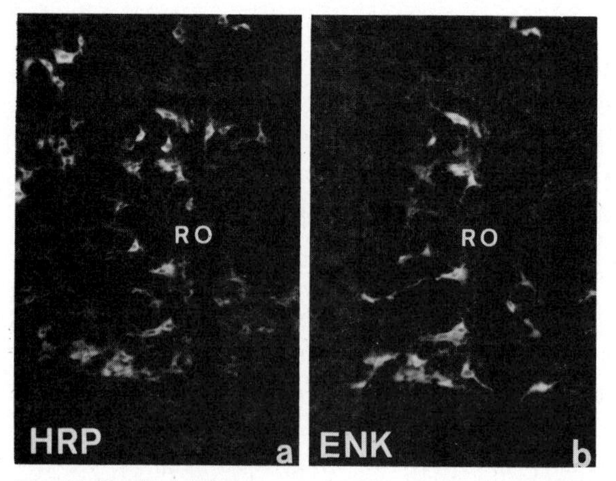

Fig. 22. Double-labeled neurons in the n. raphe obscurus (RO) after injection of B-HRP into the spinal cord and incubation with antiserum against LEU-ENK. (a) B-HRP labeled neurons visualized by avidin–fluorescein isothiocyanate; (b) ENK-LIr neurons found in the same field demonstrated by Texas-red. Note that numerous ENK-LIr neurons simultaneously contain B-HRP in the same soma. Frontal sections. × 100. Courtesy of Dr. E. Senba.

horn, a moderate density of ENK-LIr fibres can be visualised (Figs. 2, 3).

Fibre connections

A projection from ENK-LIr cells in the caudal raphe nuclei and adjacent reticular formation to the spinal cord has been reported (Hökfelt et al., 1979; Basbaum et al., 1980; Finley et al., 1981a; Hunt and Lovick, 1982) (Figs. 22, 23). ENK-LIr neurons in this area could be divided into two subpopulations; one containing 5-HT and the other not. This opiate system may contribute to the regulation of SP release from primary sensory afferents, as the K^+-evoked release of SP from trigeminal slices can be influenced by the opioid agonist Met-enkephalin-amide, an effect which is reversed by naloxone (Jessell and Iversen, 1977). In addition there are several reports which postulate a descending opiate control of nociceptive sensory information processing via SP. Macdonald and Nelson (1978) reported that opiates reduced excitatory postsynaptic potential quantal content at the dorsal root ganglion–neuron synapse without altering quantal size, suggesting that opiates may directly affect transmitter release at a presynaptic level. LaMotte and De Lanerolle (1983) have also shown the presence of a few ENK-LIr terminals forming axo-axonic contacts (ENK terminals are presynaptic in structure) in the dorsal horn of the spinal cord (see below for details). Ninkovic et al. (1981) have also shown a marked loss of opiate receptor binding sites throughout layer II following rhizotomy, consistent with the presence of the relevant receptor on sensory fibres. Thus, although it is probable that the descending ENK neuron system from the brain stem terminates in the dorsal horn and may influence primary sensory afferent inputs, there have been some doubts raised. Hunt et al. (1980), for example, examined the fine structure of ENK-LIr terminals in the dorsal horn, but they failed to ob-

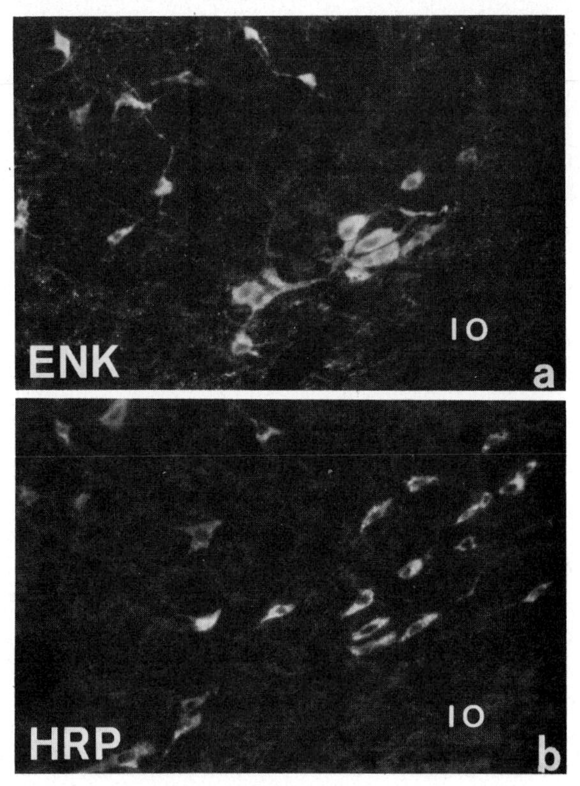

Fig. 23. Double-labeled neurons in the n. reticularis gigantocellularis just laterodorsal to inferior olivary complex (IO) after injection of B-HRP into the spinal cord. (a) B-HRP labeled neurons; (b) ENK-LIr neurons in the same field. Numerous double-labeled cells were detected. Frontal sections. × 210. Courtesy of Dr. E. Senba.

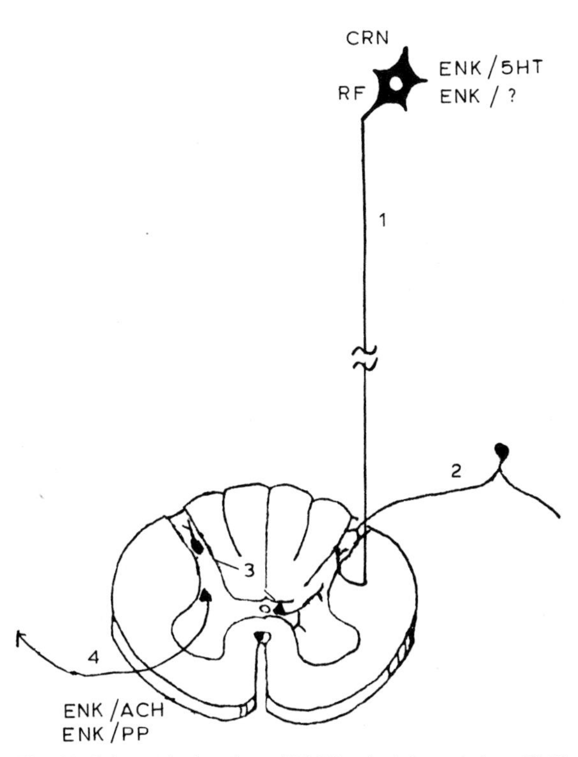

Fig. 24. Schematic drawing of ENK spinal innervation: ENK-LIr cells located in the caudal raphe nuclei (CRN) and the adjacent reticular formation (RF) project to the spinal cord. These neurons seemed to terminate in the dorsal horn (1). Some of them also contain serotonin (5-HT). In addition, there exists a primary sensory afferent system terminating in the dorsal horn (2), intrinsic ENK system (3) and preganglionic fibre system (4), some of which contain ACH or pancreatic polypeptide (PP) in single cells.

serve any ENK-LIr terminals synapsing with primary sensory afferents. The same authors (Hunt et al., 1981b) reported that hemisection of the dorsal horn failed to cause any reduction of ENK-LIr fibres in the dorsal horn, though some ENK-LI accumulated in fibres rostral to the section. These findings suggest that the majority of ENK-LIr fibres in laminae I and II do not interact with primary sensory afferent fibres. If a projection from supraspinal ENK-LIr neurons to the dorsal horn does exist, its contribution is thus likely to be very small. In support of this idea, numerous ENK-LIr neurons intrinsic to spinal cord, particularly laminae I and II, were detected. Therefore, further analysis is needed to decide whether descending ENK-LIr neurons really relate to the suppression of the inputs from primary sensory afferents or whether ENK-LIr terminals in the dorsal horn ever suppress SP release from the primary sensory afferents directly via axo-axonic contact.

In addition to the descending ENK neuron system from the brain stem, there exists an intrinsic ENK neuron system in the spinal cord as mentioned above (Hunt et al., 1981b). Furthermore, there is the possibility that part of the primary sensory afferent system contains ENK, because ENK-LIr neurons of small to medium size were detected in the dorsal root ganglion (Senba et al., 1982).

Finally, it has been shown that ENK-LIr cells in the parasympathetic nucleus project outside the spinal cord as preganglionic fibres (Hökfelt et al., 1977a,b; Glazer and Basbaum, 1980, 1981; Lundberg et al., 1980; Dalsgaard, 1982).

The ENK innervation of the spinal cord is schematically shown in Fig. 24.

Ontogeny

ENK-LIr cells could be divided into three groups based upon their localization and ontogeny (Senba et al., 1982). The first group of cells, which appeared on gestational day 20, was located in lamina I of the dorsal horn. After that, ENK-LIr cells detected in this area continued to increase in number as the rat grew and reached their maximum number on postnatal days 1–3 (Figs. 3, 25a). Thereafter, they began to decrease in number, and in the adult rats only a few ENK-LIr cells were detectable. However, colchicine treatment resulted in the visualisation of many more ENK-LIr cells in this area.

The second group of cells, which appeared at birth, was situated in lamina II. After birth, they gradually increased in number and reached a maximum on postnatal days 3–5 (Figs. 3, 25b). From that time on, detectable ENK-LIr cells began to decrease slightly in number. However, in the adult rat, detectable ENK-LIr cells were less numerous than those found in early postnatal days, but a still significant number of positive cells were identified (Fig. 3). Also after colchicine pretreatment

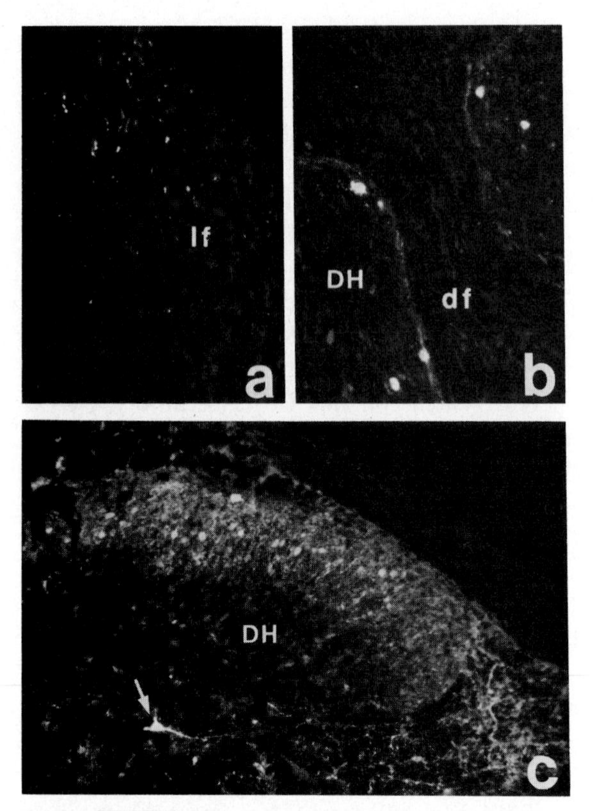

Fig. 25. Ontogeny of immunohistochemically stained ENK-LIr structures in the spinal cord. (a) Gestational day 18; (b) postnatal day 1; (c) adult. ENK-LIr fibres first appear in the lateral funiculus (lf) on gestational day 18 (a). In the perinatal stage, ENK-LIr neurons can be seen in laminae I and II (b). In the adult rat, numerous ENK-LIr structures are still found (c). Note the positive neurons in laminae I, II, and V (indicated by arrow). DH, dorsal horn; df, dorsal funiculus. (a) × 150; (b) × 150; (c) × 125. From Senba et al. (1982).

numerous ENK-LIr cells could be revealed in this area.

The third group of cells was scattered in laminae III, IV, V and VII. This group of cells first appeared in lamina V on gestational day 21. After that they appeared in laminae III, IV and VII, progressively, and reached their maximum number on postnatal days 3–5. However, after the 15th postnatal day ENK-LIr cells detected in these areas gradually decreased in number and only a few were detectable in lamina V of the adult rat. However, colchicine treatment demonstrated the existence of numerous ENK-LIr cells in laminae III, IV, V and VII, even in the adult rat (Figs. 2, 3, 25c).

ENK-LIr fibres first appeared in the white matter, both in the ventral and lateral funiculi of the fetus of gestational day 18 (Fig. 3), before the appearance of ENK-LIr fibres in the spinal cord and spinal ganglia, suggesting that the origins of ENK-LIr fibres in the dorsal and ventral horns are different and the axons reached there via dorsal and ventral funiculi, respectively. On gestational day 20, in addition to ENK-LIr fibres in the white matter, ENK-LIr fibres could be detected in the gray matter, including the ventral horn, the lateral part of lamina V and the dorsal horn, particularly lamina I (Fig. 1). After birth, the ENK-LIr fibres could be divided into two groups. One progressively decreased in number as the rat grew, and the other continued to increase in number. ENK-LIr fibres found in the gray matter belong to the latter category. For example, although in the dorsal horn of very young rats, a dense plexus of ENK-LIr fibres occupied lamina I; in the older rats, they occupied both laminae I and II (Fig. 3). Another example is in the ventral horn of adult rats where the ENK-LIr fibres were denser than in neonatal rats (Fig. 3). It should be noted that in contrast to SRIF and SP, a densely packed fibre band in laminae V and VI was not identified in the case of ENK, though in the thoracic cord a single fibre band was found in lamina VII between the n. intermediolateralis and lamina X. In summary, ENK-LIr structures in the spinal cord appear later than SRIF or SP in the perinatal stage, and they continue to increase in number after birth.

Fine structure

The fine structure of ENK-LIr terminals in the dorsal horn has been examined by several authors. Hunt et al. (1980) showed that most ENK-LIr terminals in the dorsal horn of the rat make axo-dendritic contact and they also showed closed apposition to non-labeled axon terminals as if forming an axo-axonic synapse. However, no ENK-LIr terminals forming axo-axonic synapses with the terminals of primary sensory afferents were found. An axo-somatic contact was hardly ever seen. LaMotte and De Lanerolle (1983) also examined the fine structure of the ENK-LIr terminals in the dorsal horn of the monkey. ENK-LIr terminals make synapses most frequently with various sizes of dendrites and with dendritic spines. These contacts were usually asymmetrical. ENK-LIr terminals often showed close apposition to unlabeled terminals and ENK-LIr terminals were mostly presynaptic to these terminals. These authors supposed non-labeled terminals with which the ENK-LIr terminals synapsed might belong to the primary sensory afferents. In the dorsal horn of monkey, ENK-LIr terminals form neural clusters with dendrites, dendritic spines and non-labeled terminals.

Neurotensin

Distribution

The distribution of neurotensin (NT)-like immunoreactive (NT-LIr) structures was examined by Uhl et al. (1977), Gibson et al. (1981), Hunt et al. (1981b), Senba et al. (1982) and Sasek et al. (1984).

NT-LIr cells were exclusively found in lamina II and in the dorsal part of lamina III. No NT-LIr neurons were seen in other parts (Fig. 2).

NT-LIr fibres were also concentrated in the rat dorsal horn (Figs. 2, 4). They formed a NT-LIr fibre plexus with high density in laminae I and II; in lamina III, a small number of NT-LIr fibres were seen and a few fibres in other laminae were also observed.

Fibre connections

Since no NT-LIr cells were detected in the dorsal spinal ganglion, NT-LIr structures identified in the spinal cord may originate from NT-LIr cells located in the spinal cord or supraspinal area. In addition, transection of the spinal cord failed to reduce the NT-LIr fibres in the dorsal horn of the spinal cord both above and below the section (E. Senba et al., unpublished data), suggesting that NT-LIr fibres in the dorsal horn are supplied mostly by NT-LIr neurons intrinsic to dorsal horn (Fig. 26).

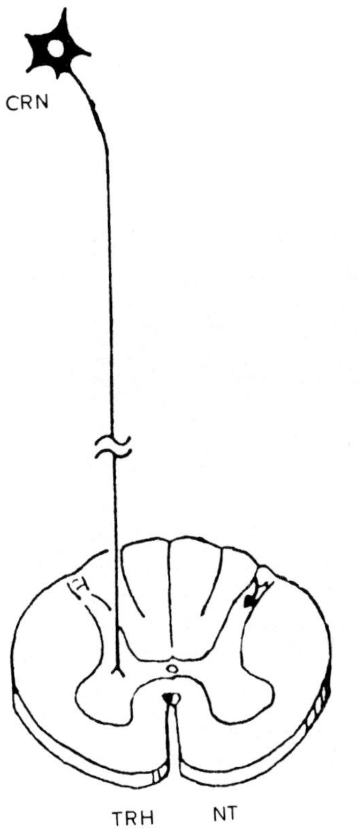

Fig. 26. Schematic drawing of NT (right side) and TRH (left side) spinal innervation: NT-LIr fibres found in the spinal cord seem to be supplied mainly by the intrinsic NT-LIr neurons, while TRH-LIr fibres originate mostly from caudal raphe nuclei (CRN) and terminate in the ventral horn (see also Fig. 14).

Ontogeny

Ontogeny of NT-LIr structures in the spinal cord has been described by Senba et al. (1982).

In contrast to SRIF, SP and ENK, no NT-LIr cells were detected in the fetus; they first appeared in the new-born rat (Fig. 4). They progressively increased in number and reached a maximum during postnatal days 7–10. After that NT-LIr cells detected in the spinal cord decreased slightly in number, but a considerable number of NT-LIr cells were still identified in the adult (Figs. 4, 27c). Colchicine pretreatment slightly increased the number of NT-LIr cells. NT-LIr cells were exclusively found in laminae I and II, and dorsal part of lamina III throughout development and no positive cells could be detected in other regions.

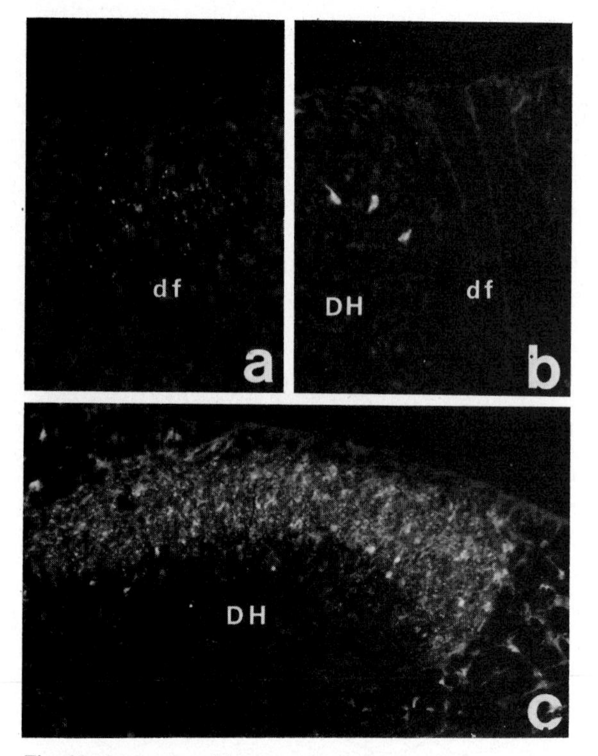

Fig. 27. Ontogeny of immunohistochemically stained NT-LIr structures in the spinal cord. (a) Gestational day 18; (b) new born; (c) adult. NT-LIr fibres first appear in the dorsal funiculus (df) on gestational day 18 (a), while NT-LIr neurons first appear at birth (b). NT-LIr structures increase in number as the rat grows. (c) This panel shows numerous NT-LIr structures in the dorsal horn (DH) of the adult rat. (a) × 150; (b) × 150; (c) × 150. From Senba et al. (1982).

NT-LIr fibres first appeared in the dorsal funiculus in the fetus of gestational day 18 (Figs. 4, 27a), and no remarkable increase of NT-LIr fibres in number or intensity was found during fetal life. NT-LIr fibres appeared at birth in lamina I. After that, they progressively increased in number to form the dense fibre plexus characteristic of laminae I and II (Figs. 4, 27c). In addition, a small number of NT-LIr fibres could be detected in the lateral spinal nucleus in the dorsal lateral funiculus with very few in laminae V and X. In summary, NT-LIr structures appeared in the spinal cord at the perinatal stage and developed markedly after birth.

Vasoactive intestinal polypeptide

Distribution

Distribution of vasoactive intestinal polypeptide (VIP)-like immunoreactive (VIP-LIr) structures in the spinal cord and spinal ganglia has been examined in various mammals including man (Emson et al., 1979; Go and Yaksh, 1980; Gibson, 1981; Anand et al., 1983; Basbaum and Glazer, 1983; Honda et al., 1983; Kawatani et al., 1983; Fuji et al., 1985). Characteristic profiles of VIP-LI occur at different levels in the spinal cord, being particularly concentrated in the sacral cord (Anand et al., 1983; Basbaum and Glazer, 1983).

Reported distribution patterns of VIP-LIr structures in the spinal cord differ slightly between authors (see Fuji et al., 1985 for a review), particularly with respect to the distribution of VIP-LIr neurons. In summary, the VIP-LIr neurons located in the spinal cord can be divided into two groups. In the first group, neurons were situated in the lateral spinal nucleus mostly near the tip of the dorsal horn (Figs. 4, 28a, c) and some medially near the base of the dorsal horn (Fig. 28b). The neurons were multipolar and had long dendritic processes, most of which passed obliquely in a medial direction towards the base of the dorsal horn. The second group was located in lamina X and was composed of oval shaped cells with a few relatively short dendrites located near the central canal. Occasionally, a few VIP-LIr neurons were observed in the superficial dorsal horn. A group of VIP-LIr neurons was also seen in the dorsal root ganglia.

VIP-LIr fibres in the spinal cord were confined to lamina I with a few being seen in the outer part of lamina II (Figs. 4, 28d). They often left this plexus and ran along the lateral curvature to terminate in lamina V. They further passed through lamina V to reach lamina X. VIP-LIr fibres were occasionally detected in the lateral funiculus but in other parts only a few or no fibres were observed.

Fibre connections

There is good evidence that VIP-LIr neurons lo-

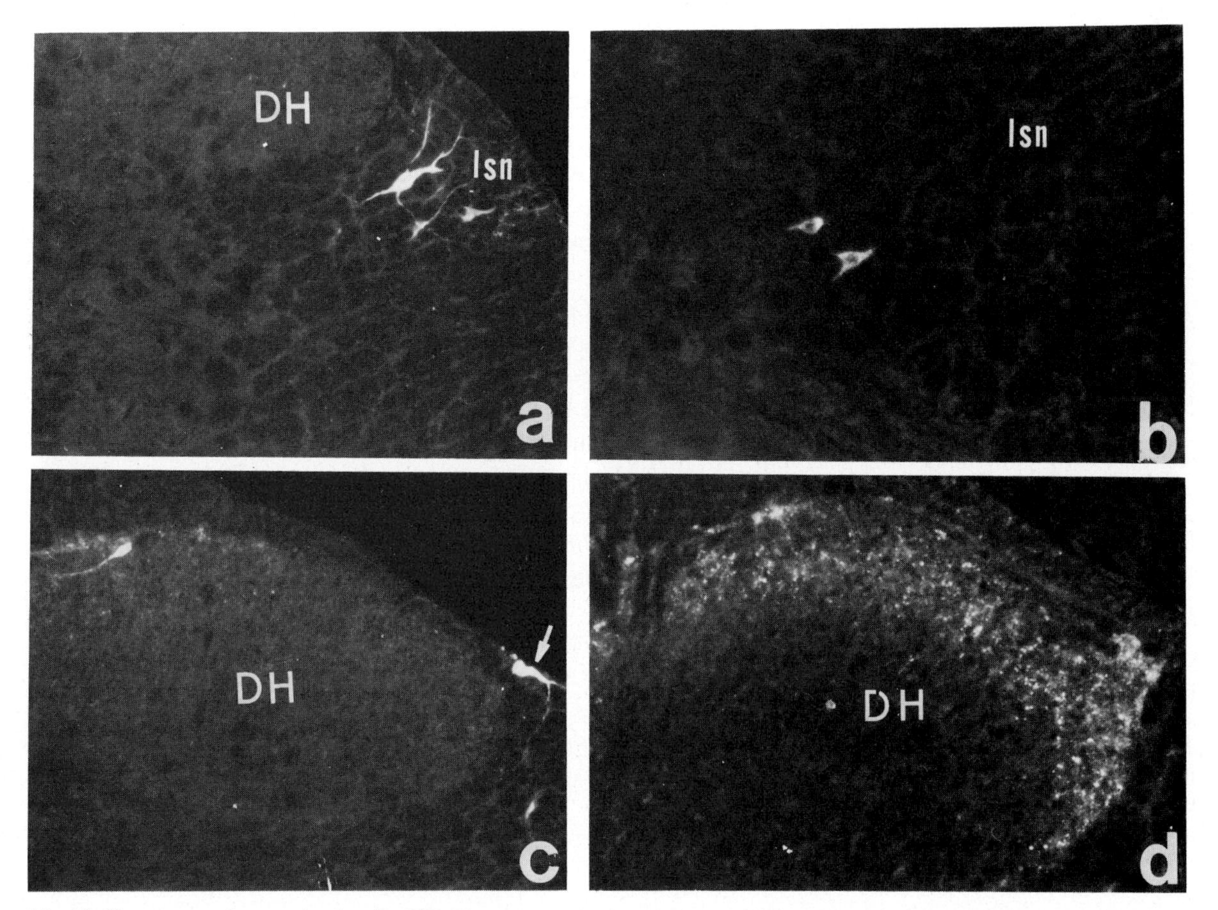

Fig. 28. Fluorescent micrographs showing VIP-LIr structures in the spinal cord of the adult rat. VIP-LIr neurons are situated in the dorsolateral part (a) and ventromedial part of the lateral spinal nucleus (lsn) (b). VIP-LIr neurons are detected in lamina I and dorsal tip of the lsn (c) (indicated by arrow). These VIP-LIr neurons are detected only in the colchicine pretreated animals. VIP-LIr fibres are distributed in laminae I and II of the dorsal horn (DH). (a) × 90; (b) × 160; (c) × 90; (d) × 160. From Fuji et al. (1985).

cated in the dorsal root ganglia are the origin of VIP-LIr fibres in the dorsal horn (Honda et al., 1983; Kawatani et al., 1983). Therefore, VIP is also one of the components of the primary sensory afferent system along with SRIF, SP, CGRP and ENK.

A recent experimental study by Fuji et al. (1985) has demonstrated that some of the VIP-LIr neurons in the spinal cord have relatively long axons which project to the supraspinal area via the lateral funiculus. This conclusion is based on the following evidence: transection of the spinal cord at the level of C_{1-2} resulted in accumulation of immunoreactive material only in those axons caudal to the lesion (Fig. 29). Most of the VIP-LI accumulating fibres were located in the lateral funiculus and a few in the ventral funiculus. After transections at the level of the lower cervical or thoracic cord, a large number of immunoreactive axons were observed in the area caudal to the lesion; these were mainly concentrated in the lateral funiculus and a few were scattered in the ventral funiculus (Fig. 29). In the lateral funiculus rostral to the lesion, only a few accumulating fibres were observed.

Fuji et al. (1983, 1985) showed that some lateral spinal nucleus neurons were labeled by VIP antiserum in the rat. The role of the lateral spinal nucleus in the rat is not fully understood, but the func-

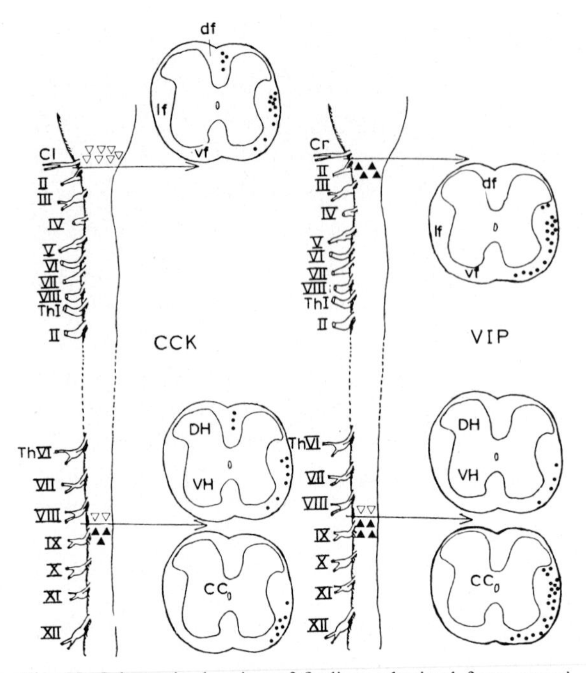

Fig. 29. Schematic drawing of findings obtained from experimental manipulations. Transections of the cord are indicated by arrows. The accumulation of the immunoreactive materials in the rostral part of the lesion (▽) and those in the caudal part to the lesion (▲) are shown. The distribution of the accumulated immunoreactive materials (●) are also shown in the scheme of coronal sections of the spinal cord. For list of abbreviations see legend to Figs. 3 and 4.

tion of the lateral cervical nucleus in the cat or other species, which is considered to be homologous to the lateral spinal nucleus of the rat (Gwyn and Waldron, 1968), has been investigated by many authors (Rexed and Strom, 1952; Morin and Catalano, 1955), although some morphological and functional differences have been described between the lateral cervical nucleus of the rat and that of the cat (Giesler et al., 1979). The lateral cervical nucleus is believed to receive cutaneous impulses and its axons ascend the dorsolateral funiculus to the medial lemniscus and so reach the thalamus (Morin and Catalano, 1955; Boivie, 1970; Blomquist et al., 1978; Craig and Burton, 1979). It is not certain whether the VIP-LIr neurons in the lateral spinal nucleus might project to the thalamus or not, because neurons projecting from the lateral spinal nucleus to the thalamus are reported to be confined to the cervical level (Giesler et al., 1979), whereas VIP-LIr neurons extend throughout the whole length of the cord. However, it is likely that they play a role in the sensory transmission system.

The VIP innervation in the spinal cord is schematically drawn in Fig. 30.

Ontogeny

Ontogeny of VIP-LIr structures in the rat spinal cord was examined by Fuji et al. (1985). No VIP-LIr neurons were detected during the fetus stage, and postnatally, no neurons were detected without pretreatment of colchicine. VIP-LIr neurons of both groups (see above) in the spinal cord appeared on postnatal day 5 and they were present at all levels of the spinal cord.

There were relatively few VIP-LIr fibres in the developing spinal cord. A few fibres in lamina X

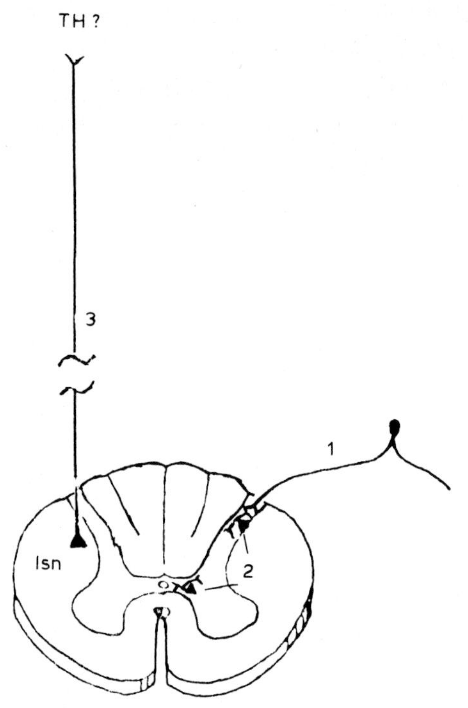

Fig. 30. Schematic drawing of VIP spinal innervation. There exist at least three kinds of VIP-LIr fibre connections: (1) primary sensory afferent VIP system, (2) intrinsic VIP system and (3) projection from spinal cord to the supraspinal area. Abbreviations: TH, thalamus; lsn, lateral spinal nucleus.

were detected on gestational day 18 and those in the lateral spinal nucleus appeared on gestational day 19, while VIP-LIr fibres in the superficial dorsal horn were first detected on postnatal day 1. Fibres increased in the growing rats and reached their maximum in adults.

Fine structures

The fine structure of the VIP-LIr terminals of the dorsal horn of the cat was examined by Honda et al. (1983). In the axon terminals, as well as other peptides, VIP-LIr structures were closely associated with large granular vesicles. Terminals often made synaptic contact with non-labeled neuronal structures. They were exclusively axo-dendritic synapses.

Cholecystokinin-8

Distribution

The location of cholecystokinin (CCK)-like immunoreactive (CCK-LIr) structures has also been examined in the spinal cord (Larsson and Rehfeld, 1979; Loren et al., 1979; Gibson et al., 1981; Vanderhaeghen et al., 1982; Schroder, 1983; Conrath-Verrier et al., 1984; Sasek et al., 1984; Fuji et al., 1985) (Fig. 4).

The majority of CCK-LIr neurons were located in laminae I and II, and the outer layer of lamina III. Their dendritic processes were difficult to identify and most of them were directed dorsoventrally on coronal sections. Medium to large sized cells

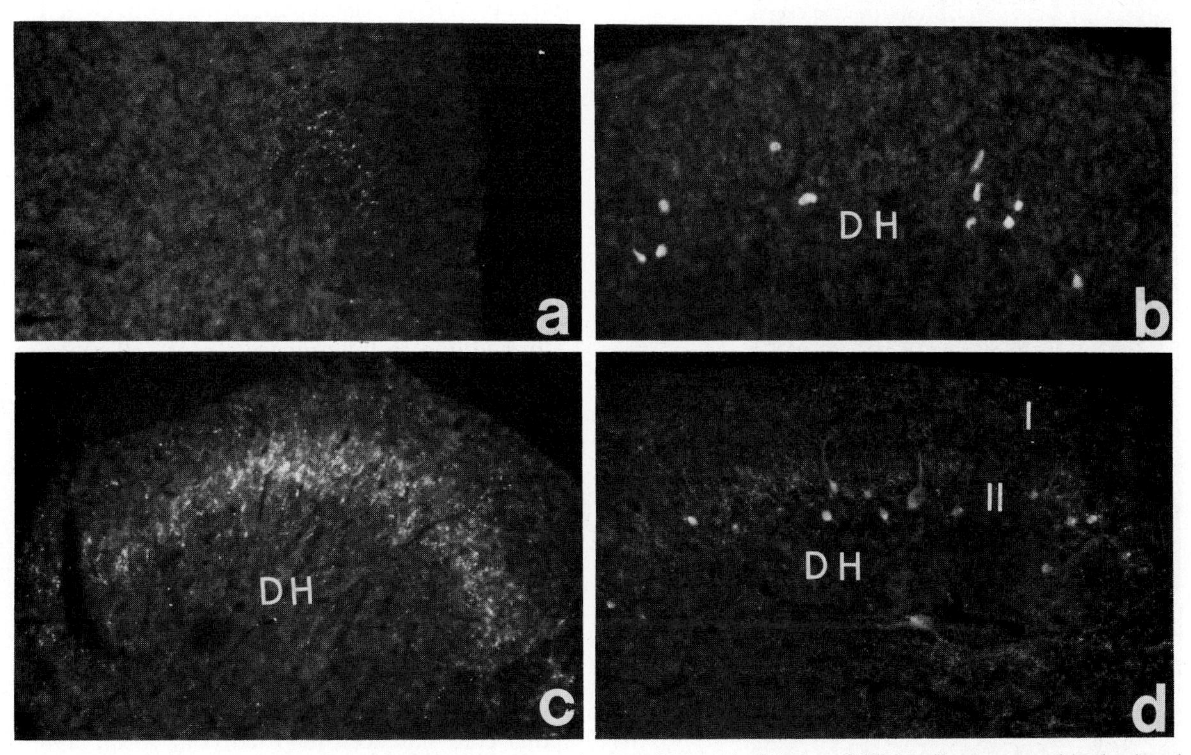

Fig. 31. Fluorescent micrographs showing CCK-LIr structures in the spinal cord of the rat: CCK-LIr fibres appeared in the dorsal part of the lateral funiculus on gestational day 17 (a). CCK-LIr neurons first appear in laminae II–III on postnatal day 4 (b). CCK-LIr fibres concentrate in lamina II of the dorsal horn (DH) of the adult rat (c) and colchicine pretreatment reveals CCK-LIr neurons distributed in laminae II, III and IV of the dorsal horn of the adult rat (d). (a) × 140; (b) × 160; (c) × 140; (d) × 140. From Fuji et al. (1985).

were occasionally detected in the same area. These cells were oval, and their dendritic processes also projected dorsoventrally (Fig. 31d). The second group of cells were large, multipolar and distributed in laminae IV and V. These cells had several long processes projecting chiefly in a mediolateral direction (Fig. 31d). In the lumbar cord additional large multipolar cells were detected in laminae X and VII, surrounding the central canal.

CCK-LIr fibres were also concentrated in the dorsal horn of the spinal cord. CCK-LIr fibres were confined to lamina II, while only a few fibres were detected in laminae I and III (Figs. 4, 31). A moderate density of fibres was also found in the dorsolateral part of the lateral funiculus, defined as the lateral spinal nucleus and the lateral part of lamina V. A few CCK-LIr fibres were scattered in the ventral horn (Fig. 4).

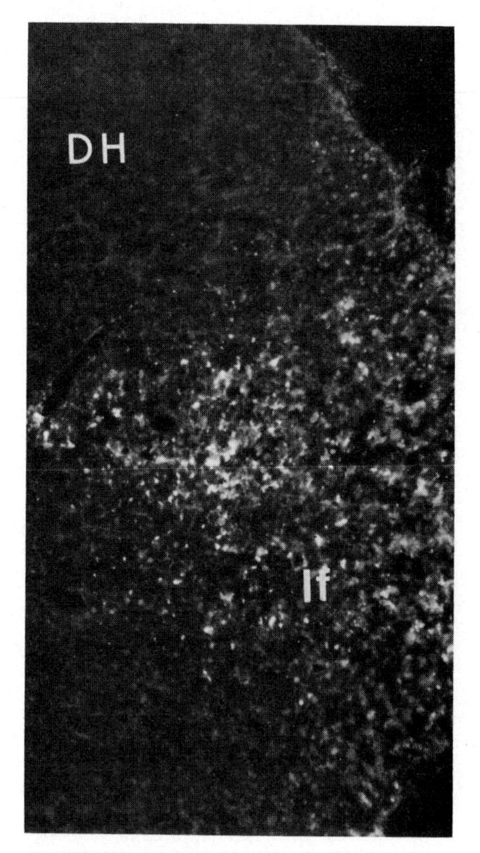

Fig. 32. Fluorescent micrograph showing the accumulation of CCK-LIr detected in the lateral funiculus (lf) after the transection of the spinal cord at the level of C_{1-2} (rostral to the lesion). DH, dorsal horn. × 110. From Fuji et al. (1985).

Fibre connections

Previously Dalsgaard et al. (1982) showed in the dorsal root ganglion that the majority of CCK-LIr neurons also contain SP. A dense concentration of CCK-LIr terminals in the dorsal horn was reported as mentioned above. A diminution in density of CCK-LIr fibres in the dorsal horn after rhizotomy or neonatal capsaicin treatment was reported (Jansco et al., 1981; Maderdrut et al., 1982; Priesley et al., 1982; Yaksh et al., 1982). Thus, it is probable that CCK may play a role as a component of primary sensory afferents. However, since many investigators have failed to demonstrate the presence of CCK-LIr neurons in the dorsal root ganglia, the possibility above must be re-examined.

Until now, two descending CCK systems to the spinal cord have been reported; one from CCK-LIr neurons located in the central gray matter of the midbrain (Skirboll et al., 1983) and the other from neurons in the caudal raphe nuclei and adjacent reticular formation (Mantyh and Hunt, 1984). These projections reached the spinal cord via lateral or dorsal funiculus, because CCK-LI accumulated in the axons of these funiculi rostral to a cervical transection of the spinal cord (Fuji et al., 1985) (Figs. 29, 32).

In addition to the primary sensory afferent and supraspinal systems, there exists an intrinsic CCK neuron system, because numerous CCK-LIr neurons have been detected in the spinal cord. Since these cells are concentrated in the dorsal horn, they contribute to CCK innervation of the dorsal horn where a dense CCK-LIr fibre plexus is present.

The organisation of CCK innervation of spinal cord is illustrated in Fig. 33.

Ontogeny

Ontogeny of CCK-LIr neuron system in the spinal cord has been reported (Fuji et al., 1985). CCK-LIr neurons in the dorsal horn were first identified in the animals on postnatal day 2–4 with colchicine treatment, while others were detected only from postnatal day 7.

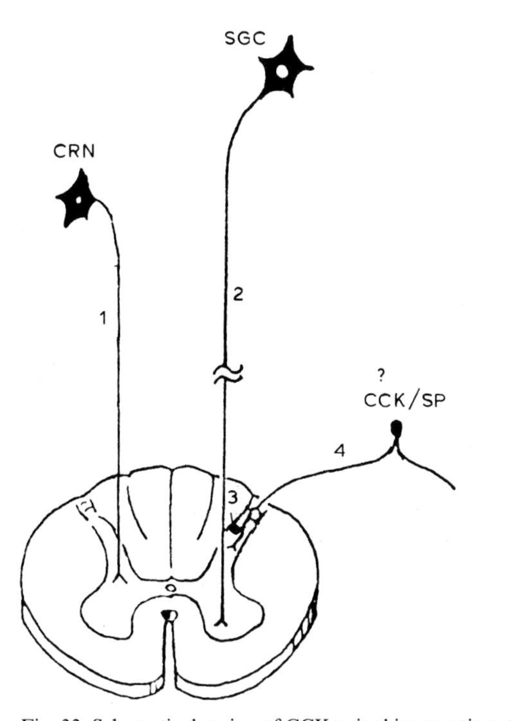

Fig. 33. Schematic drawing of CCK spinal innervation: two sites of brain stem areas are identified as the origins of spinal CCK-LIr fibres, caudal to raphe nuclei (CRN) (1), and midbrain central gray matter (SGC) (2). The intrinsic CCK neuron system is also present (3). The presence of a primary sensory afferent CCK neuron system coexisting with SP is also suggested (4).

CCK-LIr fibres could be divided into two groups based upon their changes during ontogeny. The first group includes fibres in the white matter of the dorsolateral funiculus first detected on gestational day 17 (Figs. 4, 31a) and those which appeared on gestational day 19 in the dorsal and lateral funiculus. Fibres in lamina X and the lateral part of lamina V which appeared on gestational day 19 and those in the dorsal horn and ventral horn first detected at birth belong to the second group (Fig. 4). The former group gradually disappeared after birth, while in the latter the immunoreactivity in the ventral horn remained or increased in dorsal horn as the rat grew.

Fine structure

The fine structure of CCK-LIr terminals in the dorsal horn has been examined by Conrath-Verrier et al. (1984). CCK-LIr axons were of small calibre and mostly unmyelinated. CCK-LIr terminals made asymmetric synaptic contact predominantly with non-immunoreactive dendrites (Fig. 34a) or dendritic spines. Few axo-somatic contacts but a relatively high number of axo-axonic contacts were reported (CCK-LIr terminals were both pre- and post-synaptic in structure) (see Fig. 34b).

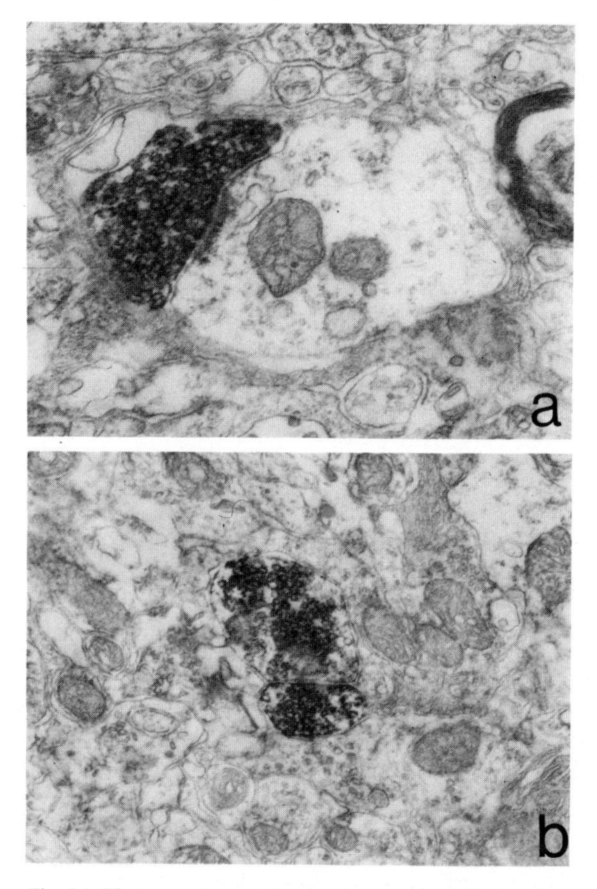

Fig. 34. Electron micrographs showing axo-dendritic asymmetric synapse of CCK-LIr terminals in the dorsal horn (a). CCK--LIr terminals in the dorsal horn most frequently make contact with non-immunoreactive dendrites. In some cases CCK-LIr terminals often showed a direct apposition with non-immunoreactive or with immunoreactive (b) axon terminals. However, no clear synaptic contact between these two axon terminals was noticed. (a) × 39 000; (b) × 24 000. Courtesy of Dr. Y. Kubota.

Neuropeptide Y

Distribution

Distribution of neuropeptide Y (NPY)-like immunoreactive (NPY-LIr) structures in the spinal cord has been examined in several mammals (Gibson et al., 1984a). In their report, Gibson et al. failed to describe the presence of NPY-LIr neurons in the dorsal horn of the spinal cord at any level. However, this is unlikely, because numerous NPY--LIr cells were detected in the trigeminal spinal nucleus with predominantly a caudal direction (Yamazoe et al., 1985) which is the homologous area to the dorsal horn of the spinal cord, and Hunt et al. (1981a) reported the presence of avian pancreatic polypeptide (APP)-like immunoreactivity (APP-LI), which cross reacts with NPY-LI in neurons in lamina I and along the border between laminae II and III of the rat spinal cord. In addition, the latter authors reported the presence of APP-LIr neurons in certain parasympathetic motor neurons which contain MET-ENK-LI in single cells (Hunt et al., 1981a).

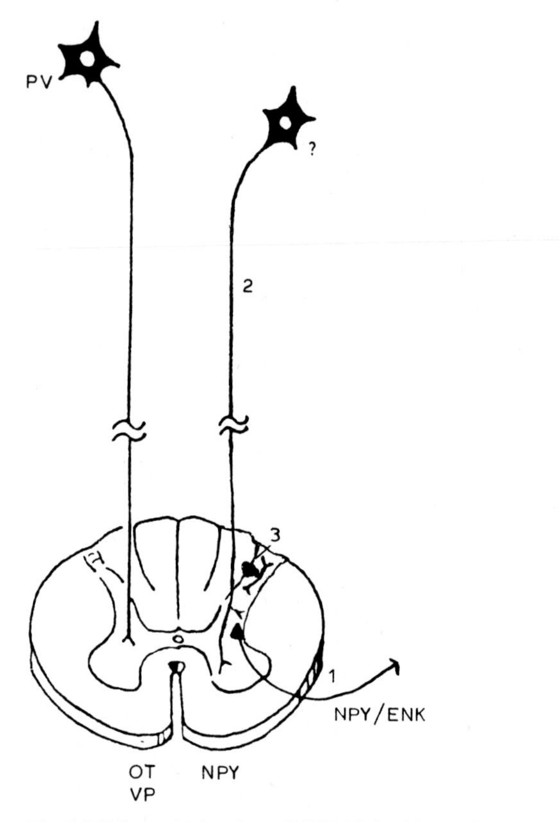

Fig. 35. Schematic drawing of NPY (right side), and OT and VP (left hand) spinal innervation. On the right side, presence of preganglionic NPY system (1), intrinsic NPY system (2) and supraspinal NPY projection (3) is illustrated. The presence of the last system is not yet well established. Coexistence of NPY-LI and ENK-LI in single preganglionic neurons was reported. On the other hand, spinal OT and VP are supplied exclusively by hypothalamic paraventricular nucleus (PV), particularly its parvocellular portion.

Numerous NPY-LIr fibres were also seen in the spinal cord, being concentrated in laminae I and II. A less numerous, but still significant, number of NPY-LIr fibres were detected in lamina X, the lateroventral part of lamina VII and the lateral part of laminae V and VI. Gibson et al. (1984a) have reported that the content of NPY-LIr structures at the sacral level is much higher than that in the other spinal regions.

Fibre connections

Little is known about the fibre connections of the spinal NPY system. The finding that NPY-LI coexists with ENK-LI in parasympathetic neurons suggests that NPY-LIr neurons project outside the spinal cord as one of the components of the preganglionic fibres of the parasympathetic nerves. The localization of APP-LIr neurons and NPY-LIr fibres in the dorsal horn suggested the presence of an intrinsic NPY system and the abundance of NPY-LIr fibres in other laminae suggested that these fibres are supplied by other NPY-LIr neurons. Little is known as to the supraspinal origins of NPY-LIr fibres in the spinal cord. However, evidence that NPY-LI and noradrenaline (NA) coexist in single cells of the lower brain stem and NA neurons in the lower brain stem project to the spinal cord (Lindvall and Bjorklund, 1983; Everitt et al., 1984), suggests that a part of the NPY-LIr fibres in the spinal cord is supplied by the descending NA neuron system although further analysis is needed.

The spinal NPY innervation is schematically drawn in Fig. 35.

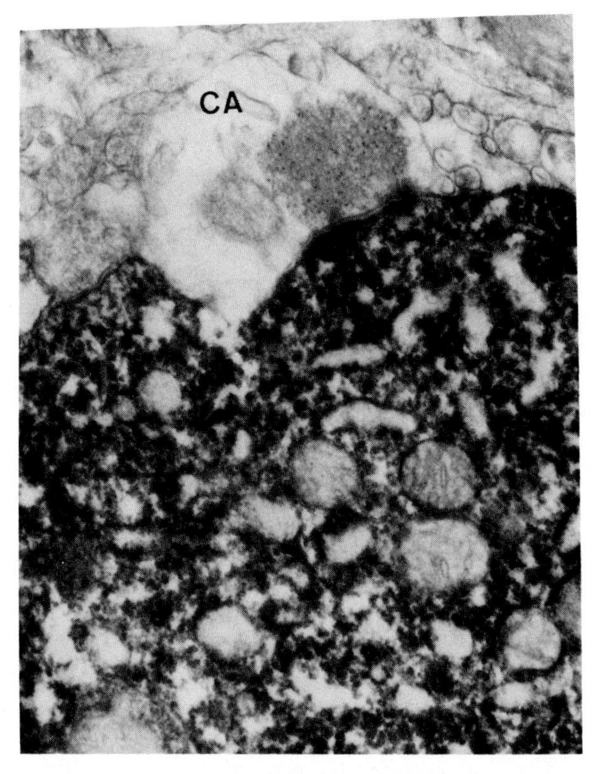

Fig. 36. Electron micrograph showing synaptic contact between catecholamine terminal (CA) and OT-LIr neurons. Axo-somatic synapse. ×42 300. From Yamano et al. (1985).

Vasopressin and oxytocin

The distribution of vasopressin (VP)-like immunoreactive (VP-LIr) and oxytocin (OT)-like immunoreactive (OT-LIr) structures in the spinal cord has been reported by many authors (see Swanson and McKellar, 1979 for summary). No immunoreactive cells have ever been detected in the spinal cord. A low density immunoreactive fibre plexus was found in laminae I and II, and also in lamina X. The dorsolateral funiculus also contained a small number of immunoreactive fibres. Millan et al. (1984) reported that the content of these peptides was greater in the lumbo-sacral than cervico-thoracic spinal cord.

Fibre connections

The fibre connections of these systems has also been examined by many authors using lesion experiments and retrograde tracers (Buijs, 1978; Swanson and McKellar, 1979; Armstrong et al., 1980; Swanson and Sawchenko, 1980; Sofroniew and Weindel, 1981; Lang et al., 1983; Millan et al., 1984). Fibres originated from immunoreactive cells located in the parvocellular portion of the hypothalamic paraventricular nucleus. Recently the direct interaction between catecholamine (CA) terminals and OT-LIr cells in the parvocellular portion of the hypothalamic paraventricular nucleus was demonstrated using immunohistochemistry combined with false transmitter (5-hydroxydopamine) histochemistry by Yamano et al. (1985). In the parvocellular portion, 5-hydroxydopamine-labeled CA terminals make synaptic contact with proximal dendrites or somata of OT-LIr cells (Fig. 36). Since these CA terminals are demonstrated to originate mostly from A_1 and A_2 NA neurons and partly from locus coeruleus NA neurons (see Lindvall and Bjorklund, 1983 for summary), it could be concluded that the ascending CA system monosynaptically regulates the extrahypothalamic OT system via axo-dendritic or axo-somatic contact (Fig. 37). Accordingly, OT-LIr terminals in the spinal cord may be influenced strongly by the activity of the ascending CA neuron system.

VP and OT fibre projections to the spinal cord are shown in Fig. 35.

Adrenocorticotropin, β-endorphin, α- and γ-melanocyte stimulating hormone

Distribution

Gibson et al. (1981) examined the distribution

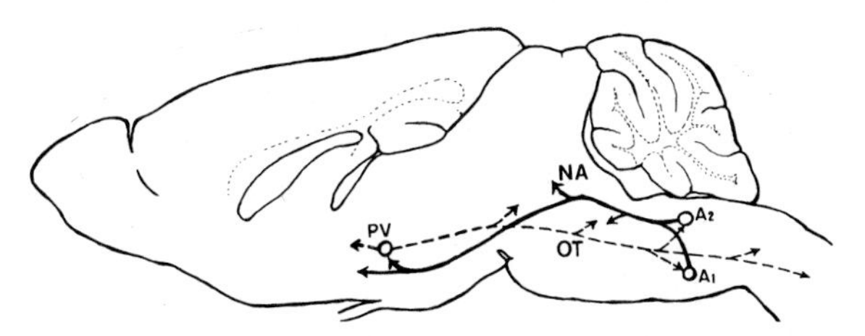

Fig. 37. Schematic drawing of interaction between ascending NA system from A_1 and A_2 NA neurons and descending (or extrahypothalamic) OT neuron system from parvocellular part of the hypothalamic paraventricular nucleus (PV). Sagittal plane.

of adrenocorticotropin (ACTH)-like immunoreactive (ACTH-LIr) fibres in the spinal cord: in laminae I, II and III, ACTH-LIr fibres are infrequently seen. In lamina X, scattered short transverse fibres can be seen. Scattered ACTH-LIr fibres were also seen in lamina VII and the dorsolateral funiculus. The intermediolateral gray contained a significant number of immunoreactive fibres. No labeled cells were detected in the spinal cord.

Since the distribution of ACTH-LIr, β-endorphin (β-END)-like immunoreactive (β-END-LIr), and γ-melanocyte stimulating hormone (γ-MSH)-like immunoreactive (γ-MSH-LIr) structures (which are all derived from the same precursor peptide) in the brain is quite similar (Watson et al., 1978; Zimmerman et al., 1978; Watson and Akil, 1980; Micevych and Elde, 1982; Osamura et al., 1982; Kawai et al., 1984), the distribution of these peptide immunoreactivities in the spinal cord is likely to be very similar to that of ACTH-LI.

No reports exist concerning the distribution of α-MSH-LI in the spinal cord. The distribution patterns of α-MSH-LIr neurons and their fibre connections in the brain are, however, those of ACTH and β-END with the exception of the arcuate nucleus and n. commissuralis (Watson et al., 1978; Watson and Akil, 1980; Umegaki et al., 1983; Kawai et al., 1984; Köhler et al., 1984; Shiosaka et al., 1984; Shiosaka and Tohyama, 1984; Yamazoe et al., 1984; Shiosaka et al., 1985a).

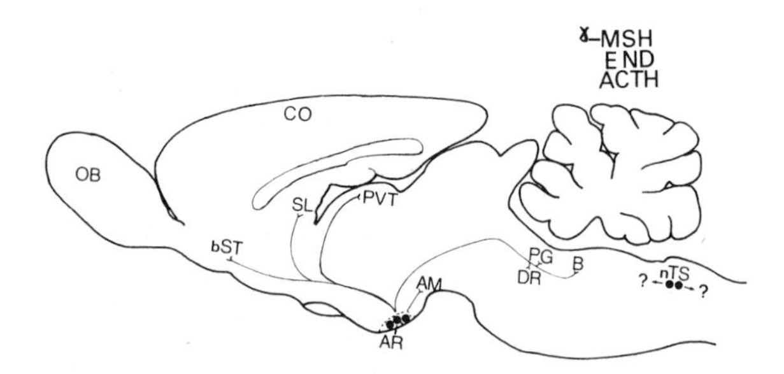

Fig. 38. Schematic drawing showing the projections of γ-MSH, β-END and ACTH neuron system in the rat. There exist two systems; one is arcuatofugal and the other is n. tractus solitarii (nTS). The former innervates forebrain, diencephalon, and upper part of the lower brain stem, such as bed nucleus of stria terminalis (bST), lateral septal area (SL), thalamic periventricular nucleus (PVT), amygdaloid complex (AM), periaqueductal gray (PG), dorsal raphe nucleus (DR) and Barrington nucleus (B), etc. Although terminal fields of the latter system are obscure, the lower part of the brain stem and spinal cords seem to be the terminal fields. Sagittal plane. AR, arcuate nucleus; CO, cerebral cortex; OB, olfactory bulb.

Fibre connections

γ-MSH, ACTH and β-END. In the central nervous system, including the spinal cord, only two sites were found to contain these peptide immunoreactivities: arcuate nucleus and commissural nucleus. Previously Kawai et al. (1984) using lesions demonstrated that γ-MSH-LIr fibres in the forebrain, diencephalon, midbrain and upper pons originate from γ-MSH-LIr neurons in the arcuate nucleus and those in the medulla oblongata from the n. commissuralis. Thus, it is likely that γ-MSH-LIr fibres in the spinal cord originate from γ-MSH-LIr neurons in the n. commissuralis as well as those seen in other lower brain stem areas. This may also apply to the β-END and ACTH positive systems.

The innervation pattern of central γ-MSH, ACTH and β-END containing neuron systems are schematically drawn in Fig. 38.

α-MSH. Köhler et al. (1984) and Shiosaka et al. (1985) examined the origins of α-MSH-LIr fibres in the spinal cord. Shiosaka et al. (1985a) explored the precise origins of these fibres in the spinal cord using a double-labeling method combining immunocytochemistry with biotinized retrograde tracer (see Shiosaka and Tohyama, this volume for details of the method). The biotinized retrograde tracer, B-HRP, injected into the thoracic cord labeled a number of neurons in the zona incerta and lateral hypothalamus (Fig. 39). In the zona incerta, labeled cells were widely distributed throughout the mediolateral area. In the lateral hypothalamus, labeled cells were concentrated in the most lateroventral portion of the ansa lenticularis and the perifornical area (Fig. 39). Some 10–20% of the neurons labeled with B-HRP were also α-MSH-like immunoreactive, and these double-labeled cells were found in the zona incerta, the lateroventral portion of the ansa lenticularis, and rarely, in the perifornical area

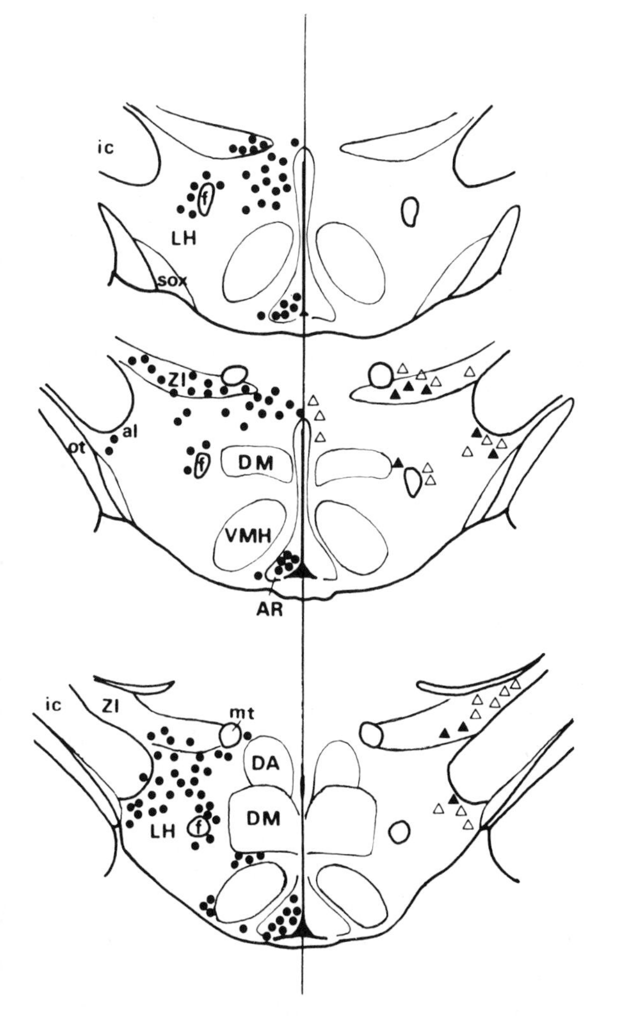

Fig. 39. Distribution of α-MSH-LIr neurons in the zona incerta (ZI), perifornical area, and lateral hypothalamus (LH) (left side, ●), and distribution of neurons labeled retrogradely by B-HRP after injection into the thoracic cord (right half, △). Double-labeled cells shown by black triangles. See the legend to Fig. 40 for list of abbreviations. Frontal plane. From Shiosaka et al. (1985a).

(Fig. 39). The morphology of these cells was very similar to the double-labeled cells identified in these areas after injection of B-HRP into the cerebral cortex, hippocampus and inferior colliculus. In the arcuate nucleus where numerous α-MSH-LIr neurons as well as γ-MSH-LIr, ACTH-LIr and β-END-LIr neurons were seen, no double-labeled cells were detected. Thus, it is clear that a large part of the α-MSH-LIr fibres in the spinal cord originate from α-MSH-LIr neurons in the zona incerta and lateral hypothalamus. However, a contribution from other α-MSH-LIr neurons such as those located in the n. commissuralis cannot be excluded, especially since this nucleus is known to project to the spinal cord and to contain α-MSH-LIr neurons. In addition, the distribution of γ-MSH-LIr, ACTH-LIr and β-END-LIr neurons in the n. commissuralis parallels that of α-MSH-LIr neurons which project to the spinal cord.

Recently, Shiosaka and his colleagues have examined the details of the α-MSH neuronal circuit in the rat brain using double-labeling methods (Shiosaka et al., 1984, 1985b; Shiosaka and Tohyama, 1984) and have shown the presence of at least two distinct α-MSH neuron systems. One was the primary melanotropin system, the neurons of which are located in the arcuate nucleus and project to the bed nucleus of stria terminalis, the periventricular nuclei of the thalamus and hypothalamus, the central gray matter of the pons and dorsal raphe nucleus. These neurons each contain γ-MSH-LI, ACTH-LI and β-END-LI as would be expected from their common precursor. α-MSH-LIr neurons located in the n. commissuralis may belong to this category, though their projection fields remain obscure. The secondary melanotropin system is another system composed of subpopulations: α-MSH--LIr neurons located in the zona incerta and lateral hypothalamus which send their axons to the cerebral cortex, hippocampus, inferior colliculus and spinal cord. α-MSH-LIr neurons located in the hippocampus send their axons to the contralateral hippocampus. These neuron systems lack γ-MSH-LI, ACTH-LI and β-END-LI and it is not established that the immunoreactivity in these cells is authentic α-MSH.

Thus, when the function of α-MSH in the spinal cord is discussed, it is necessary to consider the fact that the majority of α-MSH-LIr fibres originate from zona incerta and lateral hypothalamus, and lack γ-MSH-LI, ACTH-LI and β-END-LI, and some α-MSH-LIr fibres may be supplied by α-MSH-LIr neurons in the n. commissuralis which

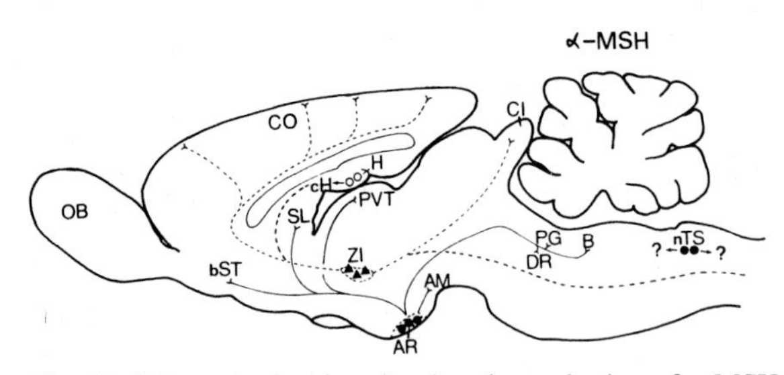

Fig. 40. Schematic drawing showing the projection of α-MSH neuron system in the rat brain. Sagittal plane. The solid lines represent the arcuatofugal α-MSH-LIr fibre system that is also γ-MSH, β-END and ACTH-like immunoreactive. The broken lines represent the bilateral α-MSH-LIr fibre system from zona incerta (ZI), perifornical area, and lateral hypothalamus (LH) to the cerebral cortex (CO), hippocampus (H) and inferior colliculus (CI). They also send their axons to the spinal cord. Open circles represent α-MSH-LIr neurons in the hippocampus that send their axons to the contralateral hippocampus. Abbreviations: al, ansa lenticularis; AM, amygdaloid complex; AR, arcuate nucleus; B, Barrington nucleus; bST, bed nucleus of stria terminalis; cH, contralateral hippocampus; DA, dorsal hypothalamic area; DM, dorsomedial hypothalamic nucleus; DR, dorsal raphe nucleus; f, fornics; ic, internal capsule; mt, mammillothalamic tract; nTS, n. tractus solitarii; OB, olfactory bulb; ot, optic tract; PG, periaqueductal gray; PVT, periventricular thalamic nucleus; sox, supraoptic decussation; VMH, ventromedial hypothalamic nucleus; ZI, zona incerta. From Shiosaka et al. (1985a).

probably contain them. Thus, α-MSH-LIr fibres are involved in two distinct systems; one which contains and one which lacks γ-MSH, ACTH and β-END.

The innervation pattern of the central α-MSH neuron system is schematically drawn in Fig. 40.

Dynorphin

Sasek et al. (1984) have examined the location of dynorphin (DYN)-like immunoreactive (DYN-LIr) structures in the spinal cord. They reported that a number of immunoreactive cells were to be found in the dorsal gray commissura (lamina X). A few DYN-LIr cells were also seen in the sacral sympathetic nucleus and dorsal horn. Numerous DYN-LIr fibres were found in the sacral parasympathetic nucleus and also lamina X. Apart from this information, detailed further data on the distribution of DYN-LIr fibres is lacking.

Fibre connections

Although numerous DYN-LIr neurons are seen in the hypothalamic paraventricular nucleus, the possibility that this nucleus is the origin of the spinal DYN-LIr fibres can be excluded, because these neurons are located in its magnocellular portion which innervates the hypophysis, and destruction of paraventricular nucleus failed to decrease the content of DYN-LI in the spinal cord (Millan et al., 1984).

Corticotropin-releasing hormone

Distribution and fibre projection

The distribution of corticotropin-releasing hormone (CRF)-like immunoreactive (CRF-LIr) structures in the brain is reviewed by Swanson et al. (1983), but little is known about the detailed localization of CRF-LIr structures in the spinal cord. However, it is likely that numerous CRF-LIr fibres exist in the spinal cord, because injection of the retrograde tracer, true blue, into the spinal cord resulted in the labeling of numerous CRF-LIr neurons in the "Barrington" nucleus (Vincent and Satoh, 1984; see Tohyama et al., 1978 for nomenclature) which is related to the micturition reflex. Thus, it is possible that CRF in the spinal cord may be present in the region related to autonomic function and be involved with the micturition reflex, although the involvement of CRF in other spinal functions remains to be established.

Thyrotropin-releasing hormone

Distribution and fibre connections

In the spinal cord, TRH-like immunoreactive (TRH-LIr) fibres are numerous in the ventral horn, and these fibres are supplied by TRH-LIr neurons located in the caudal raphe nuclei (Johansson et al., 1980; Gilbert et al., 1982). The fibre connections are summarized in Fig. 26.

(See also section on substance P.)

Bombesin/gastrin-releasing peptide

Distribution and fibre connections

In the posterior horn, bombesin/gastrin-releasing peptide (BN-GRP)-like immunoreactive (BN-GRP-LIr) fibres were found to form a dense plexus, particularly in laminae I and II (Panula et al., 1982). In addition, the presence of BN-GRP-LIr structures in the small sized cells of the dorsal ganglion was reported and dorsal rhizotomy resulted in a decrease in immunoreactivity in the dorsal horn (Panula et al., 1982, 1983), suggesting that BN-GRP is also one of the components of the primary sensory afferents.

General discussion

Distribution of neuropeptides in the spinal cord and fibre connections

The distribution of specific neuropeptides in the spinal cord has been described. The most characteristic feature of their distribution is the very high concentration of immunoreactive structures in the dorsal horn. SRIF-, SP-, CGRP-, ENK-, NT-, VIP-, CCK-, NPY- and BN-GRP-like immunoreactivities are particularly concentrated in the dorsal horn, suggesting that these peptides are closely involved with sensory functions. However, it should be stressed that it is still obscure how these peptides may be involved in the transmission of sensory information through the substantia gelatinosa.

Since SRIF-, SP-, CGRP-, and VIP-LIr neurons are seen in the dorsal root ganglia, it is certain that many of the immunoreactive fibres seen in the dorsal horn are supplied by these cells located in the ganglia. As some ENK- and BN-GRP-LIr neurons are also detected in the dorsal root ganglia (probably also CCK), it is likely that these peptides may also contribute to the spinal function as components of the primary sensory afferents. Among the peptidergic neurons found in the dorsal ganglia, the coexistence of CGRP-LI and SP-LI in single cells is clearly established (Lee et al., 1985). SP-LI is concentrated particularly in small to medium sized cells most of which also contain CGRP-LI. On the other hand, although CGRP-LI is localized both in the small, medium sized and large sized cells, most of the large sized CGRP-LIr cells lack SP-LI. In addition, a few SP-LIr cells without CGRP-LI are also detected. Thus, the composition of SP-LIr and CGRP-LIr cells is very complicated and different subpopulations may play different roles. Apart from the numerous large sized CGRP-LIr and a few large sized SP-LIr cells, peptides have been shown to be localized exclusively to small and medium sized cells. So far no identified neuromodulator or neurotransmitter substance has been localized to the giant sized cells (>40 μm in diameter). Tuchscherer and Seybold (1985) denied the possibility of the coexistence of SP and CCK in single cells of the dorsal root ganglion. In contrast, Leah et al. (1985) reported the presence of dorsal root ganglion neurons which contain SP, CCK, SRIF and VIP in single cells. Since dorsal root ganglion cells are composed of a very heterogeneous population in terms of size, neuropeptide content and coexistence, one of the first steps in elucidating the function of these various neuropeptides in the dorsal horn is to try to elucidate what sensory information individual peptidergic cells convey.

Although the primary sensory afferent system is complicated, the chemical neuroanatomy of the dorsal horn is even more so. Peptidergic fibres from the primary sensory afferents are abundant in the dorsal horn, but in addition to the peptidergic afferents a number of intrinsic peptidergic cells containing peptides such as SRIF, SP, NT, VIP, CCK, ENK and APP cells are also abundant. The processes of most of these cells also give rise to fibres in the dorsal horn, particularly in laminae I and II, where the primary sensory afferent peptidergic systems also terminate. Moreover, other non-peptidergic neuronal elements are also abundantly represented in the dorsal horn. To elucidate the involvement of these ubiquitous peptides in the sensory system, a detailed analysis of the local circuit among the neuronal elements mentioned above is needed. For this purpose, the electron microscopic double-labeling method developed by Shiosaka and

Tohyama (1984) seems to be a powerful tool (see Shiosaka and Tohyama, this volume).

In addition, peptidergic fibres whose cell bodies are not present in the dorsal root ganglia and dorsal horn have also been found. Immunoreactive fibres derived from pro-opiomelanocortin, OT, VP and TRH belong to this category. Furthermore, some of the SP-, ENK-, SRIF-, and CCK-LIr fibres in the spinal cord are supplied by supraspinal immunoreactive neurons, though their precise terminal fields are unknown. In addition to supraspinal peptidergic projections to the spinal cord, monoamine systems also project from brain stem to the spinal cord (see Ljungdahl and Bjorklund, 1983, and Steinbusch and Nieuwenhuys, 1983 for a review) and may also include histamine (Watanabe et al., 1984). The third stage for analyzing the dorsal horn sensory systems is to explore the types of neurons, including the neurotransmitter status of these supraspinal systems and their terminations. For this purpose, the electron-microscopic double-labeling method is also useful.

As with the dorsal horn, the lateral horn and lamina X are also enriched in neuropeptides and monoamines. Recently ENK-LI has been found to be localized in the parasympathetic preganglion cells, suggesting that ENK-LI coexists with ACH in these cells and plays some role in visceromotor functions. Another characteristic profile of the lateral horn and lamina X is the "grid-like" fibre band passing from the lateral horn to the central canal (Figs. 1–4; see also Krunkoff et al., 1985a,b). This is most conspicuous at the thoracic level. This can be readily visualized by parasagittal sections. The functional significance of this grid is so far obscure.

Although much attention has been paid to the functions of the peptides in the sensory system, numerous peptidergic fibres are also found in the ventral horn. For example, SP-, ENK- and TRH-LIr fibres are abundant around the motor neurons, suggesting that these peptidergic fibres are also involved in influencing motor function. In addition to these peptidergic fibres, numerous NA and serotonin fibres can be detected in the ventral horn. The most interesting current observation on peptides in the ventral horn is that CGRP-LI coexists with ACH in motor neurons. Recent pharmacological experiments have shown that CGRP may enhance the contraction of striated muscle induced by ACH (Takami et al., 1985a).

Peptidergic fibre connections can be approximately divided into five groups: projections from the supraspinal area, projections to the supraspinal area, intrinsic spinal system, peripheral afferent system and peripheral efferent system. The pattern of peptides is different for each of these and the functional correlates of this remain to be determined.

Notes added in proof

(1) Recent immunocytochemical and radioimmunoassay studies (Rokaeus et al., 1984; Ch'ng et al., 1985) have demonstrated the presence of galanin-like immunoreactive (GR-LIr) neurons and fibers in the spinal cord of mammals. These studies suggest that there will be at least two GR-LIr pathways, one from the dorsal root ganglia to the substantia gelatinosa (primary afferent sensory fibers) and an ascending projection from the sacral cord.

(2) In addition to substance P, two further novel tachykinins have been identified in the mammalian nervous system. These have been termed neurokinin A (NKA) (or neuromedin L, or substance K) and neurokinin B (NKB) (or neuromedin B) (Kimura et al., 1983; Nawa et al., 1983; and Kangawa et al., 1983). (By convention the original names neurokinin A and B, have precedence over the other proposed names for these peptides). Nawa et al. (1983) identified two precursors for the tachykinins in bovine striatum, they termed these α and β-preprotachykinin. The α-preprotachykinin contains only substance P, whilst the β-preprotachykinin contains both substance P and neurokinin A. A precursor for neurokinin B has not yet been identified. The distribution of neurokinin A and substance P in the brain have been shown to be similar (Takano et al., 1986) and this is not suprising as both tachykinins originate from one precursor (β-preprotachykinin).

Acknowledgements

Much of the original work described here was performed in collaboration with Dr. E. Senba, S. Shiosaka, Y. Kawai, M. Yamano, H. Kiyama, K. Shinoda, K. Takami, S. Shimada, S. Yoshida, Y. Tatehata, Y. Lee, Y. Yamazoe and K. Umegaki (Department of Neuroanatomy, Institute of Higher Nervous Activity, Osaka University Medical School), Drs. K. Fuji and Y. Ueda (Department of Anaesthesiology, Osaka Dental School), Drs. S. Inagaki and Y. Kubota (Third Department of Internal Medicine, Hiroshima University School of Medicine), Dr. P. C. Emson (MRC, Cambridge) and Dr. I. MacIntyre (University of London).

References

Amara, S. G., Jonas, V., Rosenfeld, M. G., Ong, E. S. and Evans, R. M. (1982) Alternative RNA processing in calcitonin gene expression generates mRNA encoding different polypeptide products. *Nature*, 298: 240–244.

Anand, P., Gibson, S. J., McGregor, G. P., Blank, M. A., Ghatei, M. A., Bacarese-Hamilton, A. J., Polak, J. M. and Bloom, S. R. (1983) A VIP-containing system concentrated in the lumbosacral region of human spinal cord. *Nature*, 305: 143–145.

Armstrong, W. E., Warach, S., Hatton, G. I. and McNeill, T. H. (1980) Subnuclei in the rat hypothalamic paraventricular nucleus: a cytoarchitectural, horseradish peroxidase and immunocytochemical study. *Neuroscience*, 5: 1931–1958.

Aronin, N., DiFiglia, M., Liotta, A. S. and Martin, J. B. (1981) Ultrastructural localization and biochemical features of immunoreactive Leu-enkephalin in monkey dorsal horn. *J. Neurosci.*, 1: 561–577.

Barber, R. P., Vaughn, J. E., Slemmon, J. R., Salvaterra, P. M., Roberts, E. and Leeman, S. E. (1979) The origin, distribution, and synaptic relationships of substance P axons in rat spinal cord. *J. Comp. Neurol.*, 184: 331–351.

Basbaum, A. I. and Glazer, E. J. (1983) Immunoreactive vasoactive intestinal polypeptide is concentrated in the sacral spinal cord: a possible marker for pelvic visceral afferent fibers. *Somatosensory Res.*, 1: 69–82.

Basbaum, A. I., Glazer, E. J., Steinbusch, H. and Verhofstad, A. (1980) Serotonin and enkephalin coexist in neurons involved in opiate and stimulation produced analgesia. *Neurosci. Abstr.*, 6: 540.

Blomquist, A., Flink, R., Bowsher, S., Griph, S. and Westman, J. (1978) Tectal and thalamic projections of dorsal column and lateral cervical nuclei: A quantitative study in cat. *Brain Res.*, 141: 335–341.

Boivie, J. (1970) Termination of the cervicothalamic tract in the cat. An experimental study with silver impregnation methods. *Brain Res.*, 19: 333–360.

Bowker, R. M., Steinbusch, H. W. M. and Coutler, J. D. (1981) Serotonergic and peptidergic projections to the spinal cord demonstrated by a combined retrograde HRP histochemical and immunocytochemical staining method. *Brain Res.*, 211: 412–417.

Buijs, R. M. (1978) Intra- and extra-hypothalamic vasopressin and oxytocin pathways in the rat. *Cell Tissue Res.*, 192: 423–434.

Burnweit, C. and Forsmann, W. G. (1979) Somatostatinergic nerves in the cervical cord of the monkey. *Cell Tissue Res.*, 200: 83–90.

Chan-Palay, V. and Palay, S. L. (1977) Immunocytochemical identification of substance P cells and their processes in rat sensory ganglia and their terminals in the spinal cord: Light microscopic studies. *Proc. Natl. Acad. Sci. U.S.A.*, 74: 3597–3601.

Ch'ng, J. L. C., Christofides, N. D., Anand, P., Gibson, S. J., Allen, Y. S., Su, H. C., Tatemoto, K., Morrison, J. F. B., Polak, J. M. and Bloom, S. R. (1985) Distribution of galanin immunoreactivity in the central nervous system and the responses of galanin-containing neuronal pathways to injury. *Neuroscience*, 16: 343–354.

Conrath-Verrier, M., Dietl, M. and Traum, G. (1984) Cholecystokinin-like immunoreactivity in the dorsal horn of the spinal cord of the rat: A light and electron microscopic study. *Neuroscience*, 13: 871–885.

Craig, A. D., Jr. and Burton, H. (1979) The lateral cervical nucleus in the cat: Autonomic organization of cervicothalamic neurons. *J. Comp. Neurol.*, 185: 329–346.

Cuello, A. C. and Kanazawa, I. (1978) The distribution of substance P immunoreactive fibers in the rat central nervous system. *J. Comp. Neurol.*, 178: 129–156.

Cuello, A. C., Polak, J. M. and Pearse, A. G. E. (1976) Substance P: A naturally occurring transmitter in human spinal cord. *Lancet*, 2: 1054–1056.

Dalsgaard, C.-J., Hökfelt, T., Johansson, O. and Elde, R. (1981) Somatostatin immunoreactive cell bodies in the dorsal horn and the parasympathetic intermediolateral nucleus of the rat spinal cord. *Neurosci. Lett.*, 27: 335–339.

Dalsgaard, C.-J., Hökfelt, T., Elfvin, L.-G. and Terenius, L. (1982) Enkephalin-containing sympathetic preganglionic neurons projecting to the inferior mesenteric ganglion: evidence from combined retrograde tracing and immunohistochemistry. *Neuroscience*, 7: 2039–2050.

De Lanerolle, N. C. and LaMotte, C. C. (1983) Ultrastructure of chemically defined neuron systems in the dorsal horn of the monkey. I. Substance P immunoreactivity. *Brain Res.*, 274: 31–49.

DiFiglia, M., Aronin, N. and Leeman, S. E. (1982) Light microscopic and ultrastructural localization of immunoreactive substance P in the dorsal horn of monkey spinal cord. *Neuroscience*, 7: 1127–1139.

DiTirro, F. J., Ho, R. H. and Martin, G. F. (1981) Immunohistochemical localization of substance P, somatostatin, and methionine-enkephalin in the spinal cord and dorsal root ganglia of the North American opossum, *Didelphis virginiana*. *J. Comp. Neurol.*, 198: 351–363.

Elde, R., Hökfelt, T., Johansson, O. and Terenius, L. (1976) Immunohistochemical studies using antibodies to leucine-enkephalin: Initial observations on the nervous system of the rat. *Neuroscience,* 1: 349–350.

Elde, R., Hökfelt, T., Johansson, M., Schultzberg, M., Efendić, S. and Luft, R. (1978) Cellular localization of somatostatin. *Metabolism,* 27: 1151–1159.

Emson, P. C., Gilbert, R. F. T., Loren, I., Fahrenkrug, J., Sundler, F. and Schaffalitzky de Muckadell, O. B. (1979) Development of vasoactive intestinal polypeptide (VIP)-containing neurons in the rat brain. *Brain Res.,* 177: 437–444.

Everitt, B. J., Hökfelt, T., Terenius, L., Tatemoto, K., Mutt, V. and Goldstein, M. (1984) Differential co-existence of neuropeptide Y (NPY)-like immunoreactivity with catecholamine in the central nervous system of the rat. *Neuroscience,* 11: 443–462.

Finley, J. C. W., Maderdrut, J. L. and Petrusz, P. (1981a) The immunocytochemical localization of enkephalin in the central nervous system of the rat. *J. Comp. Neurol.,* 198: 541–565.

Finley, J. C. W., Maderdrut, J. L., Roger, L. J. and Petrusz, P. (1981b) The immunocytochemical localization of somatostatin-containing neurons in the rat central nervous system. *Neuroscience,* 6: 2173–2192.

Forsmann, W. G. (1978) A new somatostatinergic system in the mammalian spinal cord. *Neurosci. Lett.,* 10: 293–297.

Fuji, K., Senba, E., Ueda, Y. and Tohyama, M. (1983) Vasoactive intestinal polypeptide (VIP)-containing neurons in the spinal cord of the rat and their projections. *Neurosci. Lett.,* 37: 51–55.

Fuji, K., Senba, E., Tohyama, M., Fujii, S., Ueda, Y. and Wu, J.-W. (1985) Distribution, ontogeny and fiber connections of cholecystokinin-8, vasoactive intestinal polypeptide and γ-aminobutyrate-containing neuron systems in the rat spinal cord: An immunohistochemical analysis. *Neuroscience,* 14: 881–896.

Gibson, S. J., Polak, J. M., Bloom, S. R. and Wall, P. D. (1981) The distribution of nine peptides in rat spinal cord with special emphasis on the substantia gelatinosa and on the area around the central canal (Lamina X). *J. Comp. Neurol.,* 201: 65–79.

Gibson, S. J., Polak, J. M., Allen, J. M., Adrian, T. E., Kelly, J. S. and Bloom, S. R. (1984a) The distribution and origin of a novel brain peptide, neuropeptide Y, in the spinal cord of several mammals. *J. Comp. Neurol.,* 227: 78–91.

Gibson, S. J., Polak, J. M., Bloom, S. R., Sabate, I. M., Mulderry, P. M., Ghatei, M. A., McGregor, G. P., Morrison, J. F. B., Kellet, J. S., Evans, R. M. and Rosenfeld, M. G. (1984b) Calcitonin gene-related peptide immunoreactivity in the spinal cord of man and of eight other species. *J. Neurosci.,* 4: 3101–3111.

Giesler, G. J., Menetrey, D. and Basbaum, A. I. (1979) Differential origins of spinothalamic tract projections to medial and lateral thalamus in the rat. *J. Comp. Neurol.,* 184: 107–126.

Gilbert, R. F. T., Emson, P. C., Hunt, S. P., Bennett, G. W., Marsden, C. A., Sandberg, B. E. B., Steinbusch, H. W. M. and Verhofstad, A. A. J. (1982) The effects of monoamine neurotoxins on peptides in the rat spinal cord. *Neuroscience,* 7: 69–87.

Glazer, E. J. and Basbaum, A. I. (1980) Leucine enkephalin: localization in and axoplasmic transport by sacral parasympathetic preganglionic neurons. *Science,* 208: 1479–1481.

Glazer, E. J. and Basbaum, A. I. (1981) Immunohistochemical localization of leucine-enkephalin in the spinal cord of the cat: enkephalin-containing marginal neurons and pain modulation. *J. Comp. Neurol.,* 196: 377–389.

Go, V. L. W. and Yaksh, T. L. (1980) Vasoactive intestinal peptide (VIP) and cholecystokinin (CCK)-8 in cat spinal cord and dorsal root ganglion: release from cord by peripheral nerve stimulation. *Regul. Peptides,* Suppl., 1S: 43.

Gwyn, D. G. and Waldron, H. A. (1968) A nucleus in the dorsolateral funiculus of the spinal cord of the rat. *Brain Res.,* 10: 342–351.

Haber, S. and Elde, R. (1982) The distribution of enkephalin immunoreactive fibers and terminals in the monkey central nervous system: An immunohistochemical study. *Neuroscience,* 7: 1049–1095.

Hökfelt, T., Elde, R., Johansson, O., Luft, R. and Arimura, A. (1975a) Immunohistochemical evidence for the presence of somatostatin, a powerful inhibitory peptide, in some primary sensory neurons. *Neurosci. Lett.,* 1: 231–235.

Hökfelt, T., Kellerth, J.-O., Nilsson, G. and Pernow, B. (1975b) Substance P: Localization in the central nervous system and in some sensory neurons. *Science,* 190: 889–890.

Hökfelt, T., Kellerth, J.-O., Nilsson, G. and Pernow, B. (1975c) Experimental immunohistochemical studies on the localization and distribution of substance P in cat primary sensory neurons. *Brain Res.,* 100: 235–252.

Hökfelt, T., Elde, R., Johansson, O., Luft, R., Nilsson, G. and Arimura, A. (1976) Immunohistochemical evidence for separate populations of somatostatin containing and substance P containing primary afferent neurons in the rat. *Neuroscience,* 1: 131–136.

Hökfelt, T., Elde, R., Johansson, O., Terenius, L. and Stein, L. (1977a) The distribution of enkephalin-immunoreactive cell bodies in the rat central nervous system. *Neurosci. Lett.,* 5: 25–31.

Hökfelt, T., Ljungdahl, A., Terenius, L., Elde, R. and Nilsson, G. (1977b) Immunohistochemical analysis of peptide pathways possibly related to pain and analgesia: Enkephalin and substance P. *Proc. Natl. Acad. Sci. U.S.A.,* 74: 3081–3085.

Hökfelt, T., Terenius, L., Kuypers, H. G. J. M. and Dann, O. (1979) Evidence for enkephalin immunoreactive neurons in the medulla oblongata projecting to the spinal cord. *Neurosci. Lett.,* 14: 55–60.

Hökfelt, T., Johansson, O., Ljungdahl, A., Lundberg, J. M. and Schultzberg, M. (1981) Peptidergic neurons. *Nature,* 284: 515–521.

Honda, C. N., Rethelyi, M. and Petrusz, P. (1983) Preferential immunohistochemical localization of vasoactive intestinal polypeptide (VIP) in the sacral spinal cord of the cat: light and electron microscopic observation. *J. Neurosci.*, 3: 2183–2196.

Hunt, S. P. and Lovik, T. A. (1982) The distribution of 5-HT, Met enkephalin and β-lipotropin like immunoreactivity in neuronal perikarya in the cat brain stem. *Neurosci. Lett.*, 30: 139–145.

Hunt, S. P., Kelly, J. S. and Emson, P. C. (1980) The electron microscopic localization of methionine-enkephalin within the superficial layers (I and II) of the spinal cord. *Neuroscience*, 5: 1871–1890.

Hunt, S. P., Emson, P. C., Gilbert, R., Goldstein, M. and Kimmel, J. R. (1981a) Presence of avian pancreatic polypeptide-like immunoreactivity in catecholamine and methionine-enkephalin containing neurons within the central nervous system. *Neurosci. Lett.*, 21: 125–130.

Hunt, S. P., Kelly, J. S., Emson, P. C., Kimmel, J. R., Miller, R. and Wu, J.-Y. (1981b) An immunohistochemical study of neuronal populations containing neuropeptidergic or GABA within the superficial layers of the rat dorsal horn. *Neuroscience*, 6: 1883–1898.

Inagaki, S., Senba, E., Shiosaka, S., Takagi, H., Kawai, Y., Takatsuki, K., Sakanaka, M., Matsuzaki, T. and Tohyama, M. (1981a) Regional distribution of substance P-like immunoreactivity in the frog brain and spinal cord: Immunohistochemical analysis. *J. Comp. Neurol.*, 201: 243–254.

Inagaki, S., Shiosaka, S., Takatsuki, K., Sakanaka, M., Takagi, H., Senba, E., Matsuzaki, T. and Tohyama, M. (1981b) Distribution of somatostatin in the frog brain, *Rana catesbeiana*, in relation to location of catecholamine-containing neuron system. *J. Comp. Neurol.*, 202: 89–101.

Inagaki, S., Kawai, Y., Matsuzaki, T., Shiosaka, S. and Tohyama, M. (1983) Precise terminal fields of the descending somatostatinergic neuron system from the amygdaloid complex of the rat. *J. Hirnforsch.*, 24: 345–356.

Jancso, G., Hökfelt, T., Lundberg, J. M., Kiraly, E., Lalasz, N., Nilsson, G., Terenius, L., Rehfeld, J., Steinbusch, H., Verhofstad, A., Elde, R., Said, S. and Brown, M. (1981) Immunohistochemical studies on the effect of capsaisin on peptide and monoamine neurons using antisera to substance P, gastrin/CCK, somatostatin, VIP, enkephalin, neurotensin and 5-hydroxytryptamine. *J. Neurocytol.*, 10: 963–980.

Jessell, T. M. and Iversen, L. L. (1977) Opiate analgesics inhibit substance P release from rat trigeminal nucleus. *Nature*, 268: 549–551.

Johansson, O., Hökfelt, T., Elde, R., Schultzberg, M. and Terenius, L. (1978) Immunohistochemical distribution of enkephalin neurons. In *Advances in Biochemical Psychopharmacology, Vol. 18*, Raven Press, New York, pp. 51–70.

Johansson, O., Hökfelt, T., Pernow, B., Jeffcoate, S. L., White, N., Steinbusch, H. W. M., Verhofstad, A. A. J., Emson, P. C. and Spindle, E. (1981) Immunohistochemical support for three putative transmitters in one neuron: Coexistence of 5-hydroxytryptamine, substance P- and thyrotropin releasing hormone-like immunoreactivity in medullary neurons projecting to the spinal cord. *Neuroscience*, 6: 1857–1882.

Kangawa, K. Minamino, N., Fukuda, A. and Matsuo, H. (1983) Neurokinin K: a novel mammalian tachykinin identified in porcine spinal cord. *Biochem. Biophys. Res. Commun.*, 114: 533–540.

Kawai, Y., Inagaki, S., Shiosaka, S., Senba, E., Hara, Y., Sakanaka, M., Takatsuki, K. and Tohyama, M. (1982) Long descending projection from amygdaloid somatostatin-containing cells to the lower brain stem. *Brain Res.*, 239: 603–607.

Kawai, Y., Inagaki, S., Shiosaka, S., Shibasaki, T., Ling, N., Tohyama, M. and Shiotani, Y. (1984) The distribution and projection of β-melanocyte stimulating hormone in the rat brain: An immunohistochemical analysis. *Brain Res.*, 297: 21–32.

Kawai, Y., Takami, K., Shiosaka, S., Emson, P. C., Hillyard, C. J., Girgis, S., MacIntyre, I. and Tohyama, M. (1985) Topographic localization of calcitonin gene-related peptide in the rat brain: An immunohistochemical analysis. *Neuroscience* (in press).

Kawatani, M., Lowe, I. P., Nadelhaft, I., Morgan, C. and De Groat, W. C. (1983) Vasoactive intestinal polypeptide in visceral afferent pathways to the sacral spinal cord of the cat. *Neurosci. Lett.*, 42: 311–316.

Kimura, S., Okada, M., Sugita, Y., Kanazawa, I. and Munekata, E. (1983) Novel neuropeptides, neurokinin α and β, isolated from porcine spinal cord. *Proc. Acad. Ser.*, B59: 101.

Kiyama, H., Katayama, Y., Hillyard, C. J., Girgis, S., MacIntyre, I., Emson, P. C. and Tohyama, M. (1985) Occurrence of calcitonin gene-related peptide in the chicken amacrine cells. *Brain Res.*, 327: 367–369.

Köhler, C., Haglund, L. and Swanson, L. W. (1984) A diffuse α-MSH immunoreactive projection to the hippocampus and spinal cord from individual neurons in the lateral hypothalamic area and zona incerta. *J. Comp. Neurol.*, 223: 501–514.

Krisch, B. (1981) Somatostatin-immunoreactive fiber projections into the brain stem and the spinal cord of the rat. *Cell Tissue Res.*, 217: 531–552.

Krunkoff, T. L., Ciriello, J. and Calaresu, F. R. (1985a) Segmental distribution of peptide-like immunoreactivity in cell bodies of the thoracolumbar sympathetic nuclei of the cat. *J. Comp. Neurol.*, 240: 90–102.

Krunkoff, T. L., Ciriello, J. and Calaresu, F. R. (1985b) Segmental distribution of peptide- and 5-HT-like immunoreactivity in nerve terminals and fibers of the thoracolumbar sympathetic nuclei of the cat. *J. Comp. Neurol.*, 240: 103–116.

Kubota, Y., Takagi, H., Shiosaka, S., Tohyama, M. and Shiotani, Y. (1983) Somatostatin-like immunoreactive neurons and axon terminals in the substantia gelatinosa of rat spinal cord (L2-3). *Acta Histochem. Cytochem. Abstr.*, 16: 631.

LaMotte, C. C. and De Lanerolle, N. C. (1983) Ultrastructure of chemically defined neuron system in the dorsal horn of the monkey. II. Methionine-enkephalin immunoreactivity. *Brain Res.*, 244: 51–63.

Lang, R. E., Heil, J., Ganten, D., Hermann, K., Rascher, W.

and Unger, Th. (1983) Effect of lesions in the paraventricular nucleus on vasopressin and oxytocin contents in brainstem and spinal cord. *Brain Res.*, 260: 326–329.

Larsson, L.-I. and Rehfeld, J. F. (1979) Localization and molecular heterogeneity of cholecystokinin in the central and peripheral nervous system. *Brain Res.*, 165: 201–218.

Lavalley, A. L. and Ho, R. H. (1983) Substance P, somatostatin, and methionine enkephalin immunoreactive elements in the spinal cord of the domestic fowl, *Gallus domesticus*. *J. Comp. Neurol.*, 213: 406–413.

Leah, J. D., Cameron, A. A., Kelly, W. L. and Snow, P. J. (1985) Coexistence of peptide immunoreactivity in sensory neurons of the cat. *Neuroscience*, 16: 683–690.

Lee, Y., Kawai, Y., Shiosaka, S., Takami, K., Kiyama, H., Hillyard, C. J., Girgis, S., MacIntyre, I., Emson, P. C. and Tohyama, M. (1985a) Coexistence of calcitonin gene-related peptide and substance P-like peptide in single cells of the trigeminal ganglion of the rat: immunohistochemical analysis. *Brain Res.*, 330: 194–196.

Lee, Y., Takami, K., Kawai, Y., Girgis, S., Hillyard, C. J., MacIntyre, I., Emson, P. C. and Tohyama, M. (1985b) Distribution of calcitonin gene-related peptide in the rat peripheral nervous system with special reference to its coexistence with substance P. *Neuroscience*, 15: 1227–1237.

Lindvall, O. and Björklund, A. (1983) Dopamine- and norepinephrine-containing neuron systems: Their anatomy in the rat brain. In P. C. Emson (Ed.), *Chemical Neuroanatomy*, Raven Press, New York, pp. 229–256.

Ljungdahl, A., Hökfelt, T. and Nilsson, G. (1978) Distribution of substance P-like immunoreactivity in the central nervous system of the ray. I. Cell bodies and nerve terminals. *Neuroscience*, 3: 861–943.

Lorént, I., Alumets, R., Hakanson, R. and Sundler, F. (1979) Distribution of gastrin and CCK-like peptides. *Histochemistry*, 59: 249–257.

Lorez, H. P. and Kemali, M. (1981) Substance P-, Met-enkephalin and somatostatin-like immunoreactivity distribution in the frog spinal cord. *Neurosci. Lett.*, 26: 119–124.

Lundberg, J. M., Hökfelt, T., Änggård, A., Uvnäs-Wallenstein, K., Brimijoin, S., Brodin, E. and Fahrenkrug, J. (1980) Peripheral peptide neurons: distribution, axonal transport and some aspects on possible functions. In E. Costa and M. Trabucchi (Eds.), *Neuronal Peptides and Neuronal Communication*, Raven Press, New York, pp. 25–36.

Macdonald, R. L. and Nelson, P. G. (1978) Specific-opiate induced depression of transmitter release from dorsal root ganglion cells in culture. *Science*, 199: 1449–1451.

Manderdrut, J. L., Yaksh, T. L., Petrusz, P. and Go, V. L. W. (1982) Origin and distribution of cholecystokinin containing nerve terminals in the lumbar dorsal horn and nucleus caudalis of the cat. *Brain Res.*, 243: 363–369.

Mantyh, P. W. and Hunt, S. P. (1984) Evidence for cholecystokinin-like immunoreactive neurons in the rat medulla oblongata which project to the spinal cord. *Brain Res.*, 291: 49–54.

Micevych, P. E. and Elde, R. P. (1982) Neurons containing α-melanocyte stimulating hormone and β-endorphin immunoreactivity in the cat hypothalamus. *Peptides*, 3: 655–662.

Millan, M. J., Millan, M. H., Czlonkowski, A. and Herez, A. (1984) Vasopressin and oxytocin in the rat spinal cord: Distribution and origins in comparison to [Met] enkephalin, dynorphin and related opioids and their irresponsiveness to stimuli modulating neurohypophyseal secretion. *Neuroscience*, 13: 179–187.

Morin, F. and Catalano, J. F. (1955) Central connections of a cervical nucleus (nucleus cervicalis lateralis of the cat). *J. Comp. Neurol.*, 103: 17–32.

Naik, D. R., Sar, M. and Stumpf, W. E. (1981) Immunohistochemical localization of enkephalin in the central nervous system and pituitary of the lizard, *Anolis carolonensis*. *J. Comp. Neurol.*, 198: 583–601.

Nawa, H., Hirose, T., Takahashi, H., Inayama, S. and Nakanishi, S. (1900) Nucleotide sequence of cloned cDNAs for two types of bovine brain substance P precursor. *Nature (London)*, 306: 32-36.

Nilsson, G., Hökfelt, T. and Pernow, B. (1974) Distribution of substance P-like immunoreactivity in the rat central nervous system as revealed by immunohistochemistry. *Med. Biol.*, 52: 424–427.

Ninkovic, M., Hunt, S. P. and Kelly, J. S. (1981) Effect of dorsal rhizotomy on the autoradiographic distribution of opiate and neurotensin receptors and neurotensin-like immunoreactivity within the rat spinal cord. *Brain Res.*, 230: 111–119.

Osamura, R. Y., Komatsu, N., Watanabe, K., Nakai, Y., Tanaka, I. and Imura, H. (1982) Immunohistochemical and immunocytochemical localization of γ-melanocyte stimulating hormone (γ-MSH)-like immunoreactivity in human and rat hypothalamus. *Peptides*, 3: 781–787.

Panula, P., Yang, T.-Y. H. and Costa, E. (1982) Neuronal location of bombesin-like immunoreactivity in the rat central nervous system. *Regul. Peptides*, 4: 275–283.

Panula, P., Hadjiconstantinou, M., Yang, H.-Y. T. and Costa, E. (1983) Immunohistochemical localization of bombesin/gastrin-releasing peptide and substance P in the primary sensory neurons. *J. Neurosci.*, 3: 2021–2029.

Pickel, V. M., Reis, D. J. and Leeman, S. E. (1977) Ultrastructural localization of substance P in neurons of rat spinal cord. *Brain Res.*, 122: 534–540.

Preistley, J. V., Bramwell, S., Butcher, L. L. and Cuello, A. C. (1982) Effect of capsaisin on neuropeptides in areas of termination of primary sensory neurons. *Neurochem. Int.*, 4: 57–65.

Rexed, B. and Strom, G. (1952) Afferent neuron connections of the lateral cervical nucleus. *Acta Physiol. Scand.*, 25: 219–229.

Rokaeus, A., Melander, T., Hökfelt, T., Lundberg, J. M., Tatemoto, K., Carlquist, M. and Mutt, V. (1984) A galanin-like peptide in the central nervous system and intestine of the rat.

Rosenfeld, M. G., Mermod, J. J., Amara, S. G., Swanson, L. W., Sawchenko, P. E., Rivier, J., Vale, W. W. and Evans, R. M. (1983) Production of novel neuropeptide encoded by the calcitonin gene via tissue specific RNA processing. *Nature*, 304: 129–135.

Sar, M., Stumpf, W. E., Miller, R. J., Chang, K. J. and Cuatrecasas, P. (1978) Immunohistochemical localization of enkephalin in rat brain and spinal cord. *J. Comp. Neurol.*, 96: 415–495.

Sasek, C. A., Seybold, V. S. and Elde, R. P. (1984) The immunohistochemical localization of nine peptides in the sacral parasympathetic nucleus and the dorsal gray commissure in rat spinal cord. *Neuroscience*, 12: 855–873.

Schroder, H. D. (1983) Localization of cholecystokinin like immunoreactivity in the rat spinal cord, with particular reference to the autonomic innervation of the pelvic organs. *J. Comp. Neurol.*, 217: 176–186.

Senba, E., Shiosaka, S., Inagaki, S., Takagi, H., Takatsuki, K., Sakanaka, M., Kawai, Y., Tohyama, M. and Shiotani, Y. (1981) Peptidergic neuron system in the rat spinal cord. I. Experimental immunohistochemical study. *Neurosci. Lett. Abstr.*, 6: S77.

Senba, E., Shiosaka, S., Hara, Y., Inagaki, S., Sakanaka, M., Takatsuki, K., Kawai, Y. and Tohyama, M. (1982) Ontogeny of the peptidergic system in the rat spinal cord: Immunohistochemical analysis. *J. Comp. Neurol.*, 208: 54–66.

Shimada, S., Shiosaka, S., Takami, K., Yamano, M. and Tohyama, M. (1985) Somatostatinergic neurons in the insular cortex project to the spinal cord: combined retrograde axonal transport and immunohistochemical study. *Brain Res.*, 326: 197–200.

Shiosaka, S. and Tohyama, M. (1984) Evidence for an α-MSH-ergic hippocampal commissural connection in the rat, revealed by a double-labeling technique. *Neurosci. Lett.*, 49: 213–216.

Shiosaka, S., Takatsuki, K., Inagaki, S., Sakanaka, M., Takagi, H., Senba, E., Matsuzaki, T. and Tohyama, M. (1981) Topographic atlas of somatostatin-containing neuron system in the avian brain in relation to catecholamine-containing neuron system. II. Mesencephalon, rhombencephalon, and spinal cord. *J. Comp. Neurol.*, 202: 115–124.

Shiosaka, S., Shibasaki, T. and Tohyama, M. (1984) Bilateral α-melanocyte stimulating hormonergic fiber system from zona incerta to cerebral cortex: combined retrograde axonal transport and immunocytochemical study. *Brain Res.*, 309: 350–353.

Shiosaka, S., Kawai, Y., Shibasaki, T. and Tohyama, M. (1985a) The descending alpha-melanocyte stimulating hormone (α-MSH)ergic projections from the zona incerta and lateral hypothalamic area to the inferior colliculus and spinal cord. *Brain Res.*, 338: 371–375.

Shiosaka, S., Shimada, S. and Tohyama, M. (1985b) Sensitive double-labeling technique of retrograde biotinized tracer (biotin-WGA) and immunocytochemistry: Light and electron microscopic analysis. *J. Neurosci. Res. Abstr.* (in press).

Simantov, R. M., Kuhar, M., Uhl, G. and Snyder, S. H. (1977) Opioid enkephalin: Immunohistochemical mapping in rat central nervous system. *Proc. Natl. Acad. Sci. U.S.A.*, 74: 2167–2171.

Skirboll, L., Hökfelt, T., Dockray, G., Rehfeld, J., Brownstein, M. and Cuello, A. C. (1983) Evidence of periaqueductal cholecystokinin-substance P neurons projecting to the spinal cord. *J. Neurosci.*, 3: 1151–1158.

Sofroniew, M. V. and Weindel, A. (1981) Central nervous system distribution of vasopressin, oxytocin and neurophysin. In J. L. Martinez, R. A. Jenson, R. B. Messing, H. Rigter and J. L. McGaugh (Eds.), *Endogenous Peptides and Learning and Memory Processes*, Academic Press, New York, pp. 377–389.

Steinbusch, H. W. and Nieuwenhuys, R. (1983) The raphe nuclei of the rat brainstem: A cytoarchitectonic and immunohistochemical study. In P. C. Emson (Ed.), *Chemical Neuroanatomy*, Raven Press, New York, pp. 131–208.

Swanson, L. W. and McKellar, S. (1979) The distribution of oxytocin- and neurophysin-stained fibres in the spinal cord of the rat and monkey. *J. Comp. Neurol.*, 188: 87–106.

Swanson, L. W. and Sawchenko, P. E. (1980) Paraventricular nucleus: a site for the integration of neuroendocrine and autonomic mechanisms. *Neuroendocrinology*, 31: 410–417.

Swanson, L. W., Sawchenko, P. E., Rivier, J. and Vale, W. W. (1983) Organization of ovine corticotropin-releasing factor immunoreactive cells and fibers in the rat brain. An immunohistochemical study. *Neuroendocrinology*, 36: 165–186.

Takahashi, T. and Otsuka, M. (1975) Regional distribution of substance P in the spinal cord and nerve root of the cat and the effect of dorsal root section. *Brain Res.*, 87: 1–11.

Takami, K., Kawai, Y., Vehida, H., Tohyama, M., Shiotani, Y., Yoshida, H., Emson, P. C., Girgis, S., Hillyard, C. J. and MacIntyre, I. (1985a) Effect of calcitonin gene-related peptide on contraction of striated muscles in the mouse. *Neurosci. Lett.*, 60: 227–231.

Takami, K., Kawai, Y., Shiosaka, S., Lee, Y., Girgis, S., Hillyard, C. J., MacIntyre, I., Emson, P. C. and Tohyama, M. (1985b) Immunohistochemical evidence for the coexistence of calcitonin gene-related peptide- and choline acetyltransferase--like immunoreactivity in neurons of the rat hypoglossal, facial and ambiguous nuclei. *Brain Res.*, 328: 386–389.

Takano, Y., Nagashima, A., Masui, H., Kuromizu, K. and Kamiya, H. (1986) Distribution of substance K (neurokinin A) in the brain and peripheral tissues of rats. *Brain Res.* (in press).

Tohyama, M., Satoh, K., Sakumoto, T., Kimoto, Y., Takahashi, Y., Yamamoto, K. and Itakura, T. (1978) Organization and projection of the neurons in the dorsal tegmental area of the rat. *J. Hirnforsch.*, 19: 165–176.

Tuchscherer, M. M. and Seybold, V. S. (1985) Immunohistochemical studies of substance P, cholecystokinin-octapeptide and somatostatin in dorsal root ganglia of the rat. *Neuroscience*, 14: 593–605.

Uhl, G. R., Kuhar, M. J. and Snyder, S. H. (1977) Neurotensin: Immunohistochemical localization in rat central nervous system. *Proc. Natl. Acad. Sci. U.S.A.*, 74: 4059–4063.

Umegaki, K., Shiosaka, S., Kawai, Y., Shinoda, K., Yagura, A., Shibasaki, T., Ling, N. and Tohyama, M. (1983) The distribution of α-melanocyte stimulating hormone (α-MSH) in the central nervous system of the rat. An immunohistochemical study. I. Forebrain and upper brain stem. *Cell. Mol. Biol.*, 29: 377–386.

Vanderhaeghen, J. J., Deschepper, C., Lotstra, F., Vierendeels, G. and Schoenen, J. (1982) Immunohistochemical evidence for cholecystokinin-like peptides in neuronal cell bodies of the rat spinal cord. *Cell. Tissue Res.*, 223: 463–467.

Vincent, S. R. and Satoh, K. (1984) Corticotropin-releasing factor (CRF) immunoreactivity in the dorsolateral pontine tegmentum: further studies on the micturition reflex system. *Brain Res.*, 308: 387–391.

Watanabe, T., Taguchi, Y., Shiosaka, S., Tanaka, J., Kubota, H., Terano, Y., Tohyama, M. and Wada, H. (1984) Distribution of the histaminergic neuron system in the central nervous system of rats: a fluorescent immunohistochemical analysis with histidine decarboxylase as a marker. *Brain Res.*, 295: 13–25.

Watson, S. J. and Akil, H. (1980) α-MSH in rat brain: occurrence within and outside of β-endorphin neurons. *Brain Res.*, 182: 217–223.

Watson, S. J., Richard, C. W. III and Barchas, J. D. (1978) Adrenocorticotropin in the rat brain: immunocytochemical localization in cells and axons. *Science*, 200: 1180–1182.

Yaksh, T. L., Abay, E. O. and Go, V. L. W. (1982) Studies on the location and release of cholecystokinin and vasoactive intestinal peptide in rat and cat spinal cord. *Brain Res.*, 242: 279–290.

Yamano, M., Bai, F. L., Tohyama, M. and Shiotani, Y. (1985) Ultrastructural evidence of direct contact of catecholamine terminals with oxytocin-containing neurons in the parvocellular portion of the rat hypothalamic paraventricular nucleus. *Brain Res.*, 336: 176–179.

Yamazoe, M., Shiosaka, S., Yagura, A., Kawai, Y., Shibasaki, T., Ling, N. and Tohyama, M. (1984) The distribution of α-melanocyte stimulating hormone (α-MSH) in the central nervous system of the rat: An immunohistochemical study. II. Lower brain stem. *Peptides*, 5: 721–727.

Yamazoe, M., Shiosaka, S., Emson, P. C. and Tohyama, M. (1985) Distribution of neuropeptide Y in the lower brainstem: an immunohistochemical analysis. *Brain Res.*, 335: 109–120.

Zimmerman, E. A., Liotta, A. and Krieger, D. T. (1978) β-Lipotropin in brain: localization in hypothalamic neurons by immunoperoxidase technique. *Cell Tissue Res.*, 186: 393–398.

P. C. Emson, M. N. Rossor and M. Tohyama (Eds.),
Progress in Brain Research, Vol. 66.

CHAPTER 11

Neurochemistry and neural circuitry in the dorsal horn

M. A. Ruda, G. J. Bennett and R. Dubner

Neurobiology and Anesthesiology Branch, National Institute of Dental Research, National Institutes of Health, Bethesda, MD 20892, U.S.A.

Introduction

We attempt here to summarize the wealth of immunocytochemical data about the neurochemicals (the classical neurotransmitters and the neuropeptides) that are found in the mammalian spinal cord. Our review concentrates on those regions of the spinal cord that are likely to be concerned with somatosensation rather than movement or autonomic function. Relay nuclei in the central nervous system characteristically include three major functional components: (1) the terminals of primary afferent neurons that relay signals *from* distant sites; (2) intrinsic neurons which include output neurons that transmit signals *to* distant sites and local circuit neurons that modify sensory transmission in a restricted region; and (3) the axon terminals of extrinsic neurons arising from distant sites. The spinal dorsal horn is no exception to this common organization. The discussion that follows treats each of these components separately. In addition, we have summarized the progress which has been made towards integrating the immunocytochemical data with the emerging picture of spinal circuitry derived from anatomical and electrophysiological work (for reviews see Willis and Coggeshall, 1978; Dubner and Bennett, 1983).

The use of immunocytochemical techniques in combination with retrograde tracers or intracellular markers results in the labelling of two neural elements simultaneously or in the identification of the neurochemical contained in a known morphological component. When these methods are used in conjunction with electrophysiological techniques, knowledge about the functional aspects of the system is further enhanced. These methods provide information about neural connectivity at the level of the light and electron microscope. At the ultrastructural level, immunocytochemistry can readily identify the neurochemical contained in a neural process whereas earlier studies relied on inconclusive morphological criteria. Technical progress has also resulted in multiple-labelling experiments in which two or more substances are identified in a single neuron or multiple neurochemically identified afferents are shown contacting the same neuron. The increasing use of methodologies that combine two or more labels in a single experiment promises to elucidate the details of spinal circuitry to an unprecedented degree.

It is well known that the immunocytochemical method does not unequivocally identify the substance that is labelled (Petrusz et al., 1976, 1977). Thus it should be noted that any given substance detected immunocytochemically is identified only as, for example, "enkephalin-like immunoreactivity" (ENK-LI). Such qualifications are to be inferred when we have omitted them in the discussion that follows.

Somatosensory primary afferents

The identification of the regions of termination of peptide-containing primary afferents is inferred from the depletion of immunostaining that follows surgical deafferentation or capsaicin poisoning. Capsaicin-induced deafferentation has several shortcomings. Its relative specificity for unmyelinated afferents is not a predictable function of dose (Nagy et al., 1981). Perhaps more importantly, capsaicin-induced deafferentation is generally produced by exposing neonates and then examining the animals as adults. Because the nervous system in such a case has developed in a deafferented state, one must be concerned about comparing it to a normally developed case. Acute, surgical deafferentation is preferable but also not without problems. As with capsaicin, the nervous system can be assumed to respond to the insult in a dynamic way, as many investigations have already shown (e.g., Tessler et al., 1981; McGregor et al., 1984). These dynamic responses are generally undetectable until weeks after deafferentation, suggesting that short-term changes are solely the result of the intended effects. Yet it is not known how rapidly substances within severed afferents are removed. Finally, deafferentation-induced depletion may be obscured by the remaining presence of axons of non-primary afferent neurons containing the same substance. This seems to be true, for example, in the case of primary afferent somatostatin. Double-label studies will be needed to localize unequivocally immunocytochemically defined primary afferents (and many kinds of descending axons as well).

Substance P

Substance P[a] (SP)-containing axons of diverse origin are found throughout the spinal grey matter, but with very pronounced regional differences in density (Hökfelt et al., 1975c,d, 1976; Cuello et al., 1978; Ljungdahl et al., 1978; Gibson et al., 1981; DeLanerolle and LaMotte, 1982, 1983; Charnay et al., 1983). SP staining is densest in laminae I–IIa and Lissauer's tract (Fig. 1). Many of the varicose SP-like immunoreactive (SP-LIr) fibres in laminae I and IIa have a conspicuous rostrocaudal orientation. Less dense staining is seen in the reticulated region at the neck of the dorsal horn (largely lamina V). In sagittal sections through this region of the cat's spinal cord, SP-LIr fibres form dense, regularly spaced plexi separated by areas of less dense SP--LIr staining (Fig. 1C). Regularly spaced bundles of dorsoventrally oriented SP-LIr axons traverse laminae IIb–IV (Fig. 1A, C). It has been suggested that these bundles terminate in lamina IV of the rat (Nagy et al., 1981). However, our observations in the cat suggest that the bundles may begin to disaggregate in lamina IV but terminate and create the plexi in lamina V (Fig. 1A, C). The area around the central canal (lamina X) stains densely. The adjacent neuropil (largely the dorsomedial part of lamina VII) is stained only slightly less densely. Because the peptidergic staining in these regions melds imperceptibly, we will refer to them collectively as the central canal region. In some sections, a narrow stripe of SP-LIr staining lines the medial wall of the dorsal horn from laminae I–IIa to the central canal region. Varicose SP-LIr fibres form a relatively modest plexus in laminae IIb–IV. In the dorsolateral funiculus, SP-LIr fibres form an arc extending from the apex of the dorsal horn to the lateral reticulated region[b].

The SP-containing cells in the spinal and trige-

[a] Recent studies (Kimura et al., 1983; Minamino et al., 1984) show that SP has the same N-terminal tripeptide as two other substances: neuromedin L (= substance K, neurokinin α) and neuromedin K (= neurokinin β). There is evidence that SP and neuromedin L are coded by the same gene (Nawa et al., 1983). SP-neuromedin L coexistence is thus possible. It is not known whether SP antisera crossreact with the neuromedins.

[b] The neurons in the dorsolateral funiculus are part of a very loosely aggregated cell column that extends along the entire length of the spinal cord. Recent evidence indicates that this cell group, the lateral spinal nucleus (LSN), is distinct from the lateral cervical nucleus. LSN neurons are innervated by a variety of neurochemically identified terminals, including some that contain SP-LI. The source and significance of the LSN's innervation is discussed in detail by Giesler and Elde (1985) and by Bresnahan et al. (1984).

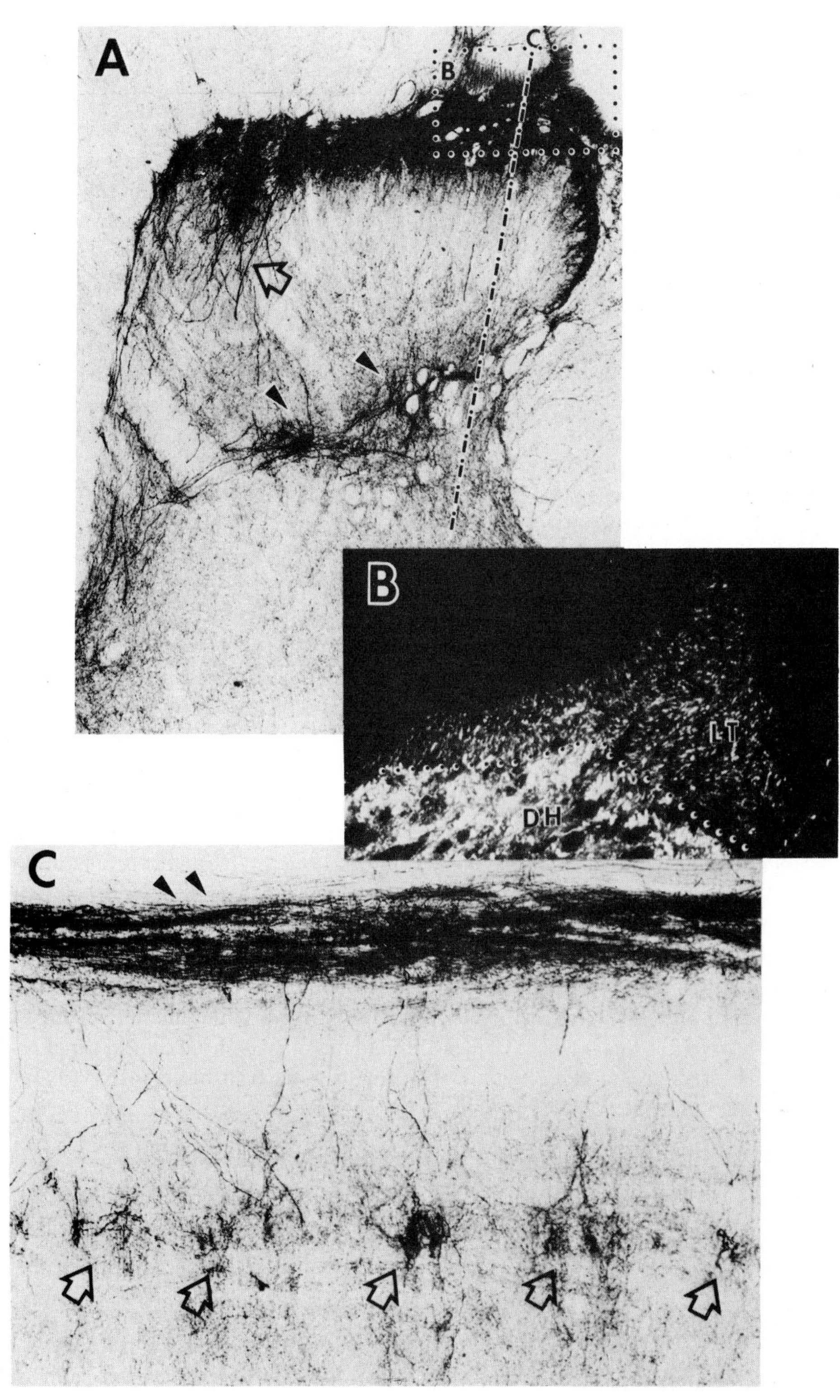

Fig. 1. SP-LIr profiles in cat lumbar spinal cord. (A) Transverse section demonstrating dense peroxidase–anti-peroxidase staining in laminae I and II. Bundles of SP-LIr axons (open arrow) traverse laminae III and IV. A second dense focus of SP-LIr staining is found in lamina V (arrowheads). Magnification: × 54. (B) Immunofluorescence staining of SP axons in Lissauer's tract (LT). The SP-LIr axons appear as punctate foci in the transverse plane because of their rostrocaudal orientation. The axons are distributed throughout Lissauer's tract. Area of dorsal horn (DH) in B is indicated in the box in A. (C) Sagittal section through the dorsal horn at the approximate position indicated by the dashed line in A. In lamina V, SP-LIr profiles form discrete, dense foci (open arrows), separated by less densely stained patches of neuropil. The discontinuous nature of SP-LI in lamina V is in contrast to the continuous band of SP-LI found in the superficial laminae (arrowheads). Magnification: × 68.

minal ganglia are of the small type (Hökfelt et al., 1975c,d, 1976; Lundberg et al., 1978; DelFiacco and Cuello, 1980; Senba et al., 1982). Recent reports suggest that in some segments 35–50% of the rat's small dorsal root ganglion cells contain SP-LI (Lehtosalo et al., 1984; Tuchscherer and Seybold, 1985). Most SP-LIr cells have unmyelinated axons as demonstrated by electron microscopy and by the concomitant loss of unmyelinated dorsal root axons and intraspinal SP-LI that follows capsaicin poisoning (Nagy et al., 1980, 1981). It is of interest, however, that capsaicin spares a small patch of SP--LIr terminals in laminae I–IIa in the rat (Jancso et al., 1981; Schultzberg et al., 1982). Because this patch may be of primary afferent origin (it has not been noted to survive surgical deafferentation), it is possible that there is a subpopulation of myelinated (i.e., capsaicin-resistant) SP primary afferents.

Deafferentation produces a pronounced, but subtotal, depletion of SP-LI in Lissauer's tract and laminae I–IIa (Hökfelt et al., 1975c,d; Cuello et al., 1978; Barber et al., 1979; DelFiacco and Cuello, 1980). The rostrocaudally oriented varicose fibres in laminae I–IIa are especially affected (Nagy et al., 1981). Tessler et al. (1980, 1981) have noted that the SP-LI in the normal dorsal horn is a mixture of coarse granules and fine, punctate granules, suggesting that there are two populations of SP-LIr fibres. The latter type is believed to represent primary afferents because it disappears after deafferentation. Tessler and his colleagues have also noted a more subtle depletion of the fine, punctate type of SP-LI in the reticulated region of lamina V. In addition, a subtle depletion of SP-LI in lamina V of the medullary dorsal horn has been noted after destruction of the trigeminal ganglion (Cuello et al., 1978). Inspection of published photomicrographs from other laboratories (Hökfelt et al., 1975c,d; Barber et al., 1979) also suggests a depletion of SP--LI in lamina V after deafferentation. Furthermore, the dorsoventrally oriented SP-LIr bundles that we believe to terminate in the lamina V plexi are absent after neonatal exposure to capsaicin (Nagy et al., 1981). An unequivocal demonstration of SP primary afferent input to lamina V would be of great import, because it would suggest that the large population of spinothalamic neurons in this region may receive a monosynaptic input from unmyelinated nociceptors.

Inspection of published photomicrographs (Hökfelt et al., 1975c,d; Tessler et al., 1980) suggests that deafferentation may deplete SP-LI in the stripe that lines the medial wall of the dorsal horn. The sparse plexus of varicose SP-LIr fibres in laminae IIb–IV is depleted by capsaicin (Nagy et al., 1981). It is important to note that some SP-LI remains in each of the regions of depletion. Thus the terminal fields of SP-containing primary afferents in these regions must overlap with the terminal fields of SP-containing intrinsic and descending neurons.

Cholecystokinin

Although the presence of authentic CCK in the spinal cord is not in doubt, the identity of the substance(s) labelled by CCK antisera is undetermined[c]. As recently shown by Schultzberg and her colleagues (1982), and confirmed by Marley et al. (1982), deafferentation produced by neonatal exposure to capsaicin causes a near total depletion of immunocytochemically labelled CCK from the superficial laminae, but no significant reduction in radioimmunoassayable CCK.

The distribution of CCK-like immunoreactivity (CCK-LI) in laminae I–V and the central canal region is very similar to the distribution of SP-LI and

[c] As in the rest of the CNS, intraspinal CCK is believed to be largely in the form of the C-terminal octapeptide (CCK-8) fragment of the full CCK molecule (CCK-33). There are reports, however, of considerable variability in regards to the size of the CNS-active C-terminal fragments present in different species (Rehfeld, 1978; Straus and Yalow, 1979). In addition, Straus and Yalow (1979) have argued that large forms of CCK are more common than generally observed because of the rapid enzymatic degradation that is unchecked in most extraction procedures. It should be noted also that gastrin and CCK have the same C-terminal pentapeptide. Gastrin, however, is absent from nearly all regions of the CNS (Dockray et al., 1981a; Marley et al., 1982).

deafferentation depletes both from the same regions[d] (Larsson and Rehfeld, 1979; Gibson et al., 1981; Jancso et al., 1981; Dalsgaard et al., 1982; Marley et al., 1982; Schrøder, 1983; Conrath-Verrier et al., 1984). An early report (Dalsgaard et al., 1982) suggested that CCK-LI and SP-LI always occurred together in primary afferent somata. However, a recent study (Tuchscherer and Seybold, 1985) indicates that although CCK/SP coexistence is very common, relatively small populations of primary afferent somata contain either CCK-LI or SP-LI without the other peptide.

Gibson et al. (1981) have noted that CCK-LI in the rat consists of two kinds of varicose fibres. One kind appears as fine (< 0.5 μm) fibres with small varicosities. These fibres are found especially in laminae I–IIa, where they have a clear rostrocaudal orientation, and in the central canal region. The second kind, most abundant in the ventral horn, has a larger axon (> 1 μm) and carries larger varicosities. They also occur in laminae I–IIa with a predominantly dorsoventral orientation. In the central canal region, plexi composed of both fibre types surround single large neurons; these plexi form a regularly spaced row along the spinal cord's long axis. Gibson et al.'s description of two populations of CCK-LIr fibres recalls Tessler et al.'s (1980, 1981) description of two populations of SP-LIr fibres. Given the high degree of coexistence of SP-LI and CCK-LI in primary afferents, we may conclude that the small CCK-LIr fibres are of primary afferent origin. Their presence in the central canal region suggests that this region may be innervated by CCK/SP primary afferents with small-diameter axons[e]. The absence of reports of deafferentation-induced depletion of SP-LI or CCK-LI from the central canal region contradicts this hypothesis, but one can easily imagine how the presence of numerous CCK and SP non-primary afferent axons might mask the depletion of a relatively small number of primary afferents.

[d] A small patch of capsaicin-resistant CCK terminals resembling the patch of capsaicin-resistant SP terminals is also present in laminae I–IIa of the rat. It may be that these capsaicin-resistant terminals originate from the occasional, relatively large CCK dorsal root ganglia cells noted by Lundberg et al. (1978).

[e] Input to this region from small, myelinated nociceptors has been documented unequivocally (Light and Perl, 1979), but there is no anatomical evidence regarding unmyelinated primary afferent input to the central canal region.

Somatostatin

All investigators agree that the highest concentration of somatostatin (SRIF)-like immunoreactivity (SRIF-LI) is in the superficial laminae, but there is some confusion regarding the exact position of the immunoreactivity with respect to the superficial laminar borders (reviewed by Schrøder, 1984). The summary given here is intended to be a reasonable synthesis[f] (Hökfelt et al., 1975a; Forssmann, 1978; Burnweit and Forssmann, 1979; Dalsgaard et al., 1981; Hunt et al., 1981b; Johansson et al., 1984; Schrøder, 1984).

The densest concentration of SRIF-LI corresponds very closely to lamina IIa, but the possibility that this heavily stained band overlaps into the inner part of lamina I cannot be excluded. Lamina I (perhaps only its superficial part) and lamina IIb are less densely stained. The band of SRIF-LI in lamina IIb may overlap slightly into the dorsal-most part of lamina III. SRIF-LIr varicosities in the superficial laminae are sometimes very large. Electron microscopy shows that such a varicosity can be presynaptic to multiple targets (Fig. 2). A

[f] We cannot exclude the possibility that the differing accounts arise from the use of different antisera with specificities for a possibly large number of molecular forms of SRIF (Morrison et al., 1982). Inspection of the published material and our own experience with immunocytochemistry, however, lead us to proceed on the premise that the discrepancies arise from methodological differences. Differentiation of adjacent areas of slightly different darkness will depend on how the final image is developed. If the image is underdeveloped, only the darkest areas will show; if overdeveloped, different areas will seem to be equally dark. Development of the final image will vary with the concentrations and incubation times used for the various reagents and with the choice of label.

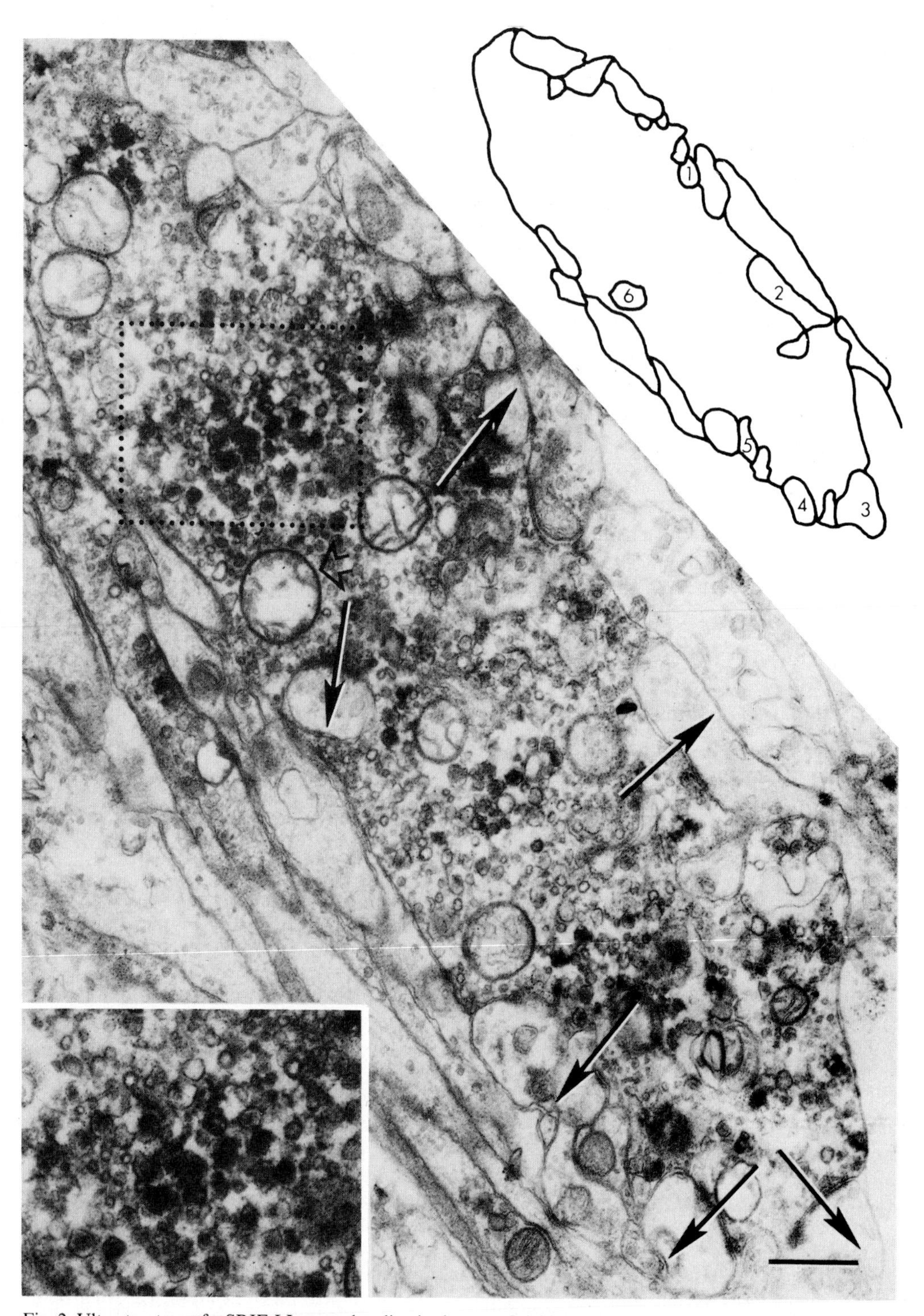

Fig. 2. Ultrastructure of a SRIF-LIr axonal ending in the superficial laminae of cat lumbar dorsal horn. The ending is large, extending more than 7 μm in length. In this single ultrathin section, 25 different profiles, six of which receive asymmetrical synapses (arrows), are impressed into its surface (see drawing). The peroxidase–anti-peroxidase reaction product is mainly associated with a subpopulation of dense core vesicles (box and insert). The mitochondria of SRIF-LIr axonal endings (open arrow) typically exhibit a light matrix in the inner mitochondrial compartment and irregular cristae. Magnification: ×26 730. Scale bar represents 0.5 μm.

sparse innervation from varicose SRIF-LIr fibres is seen in the rest of lamina III, in lamina IV, in the reticulated region of lamina V, and in the central canal region. A narrow stripe of SRIF-LI has been noted to line the medial wall of the dorsal horn (Senba et al., 1982). Many SRIF-LIr axons are present in the dorsolateral funiculus.

SRIF-LI is present in 10–15% of the rat's dorsal root ganglion cells (Hökfelt et al., 1975a, 1976; Senba et al., 1982; Tuchscherer and Seybold, 1985). It has been shown that the population of SRIF-LIr primary afferent neurons is separate from the populations of similarly small cells that contain SP-LI and the enzyme, fluoride-resistant acid phosphatase (Hökfelt et al., 1976; Nagy and Hunt, 1981). The absence of SRIF-LI in the dorsal root ganglion cells of rats treated neonatally with capsaicin suggests that SRIF primary afferent axons are unmyelinated (Jansco et al., 1981; Nagy et al., 1981).

Very little is known about the regions in which SRIF primary afferents terminate because deafferentation produces hardly any change in the immunocytochemical picture (Jancso et al., 1981; Schrøder, 1984). Radioimmunoassays, however, document that both surgical and capsaicin-induced deafferentation produces a small (ca. 20%) decrease in intraspinal SRIF-LI (Nagy et al., 1981; Stine et al., 1982).

Vasoactive intestinal polypeptide

Radioimmunoassays show that the concentration of vasoactive intestinal polypeptide (VIP)-like immunoreactivity (VIP-LI) is very much greater in lumbosacral segments than in cervicothoracic segments. This has been confirmed in every mammal studied: mouse, rat, guinea pig, cat, horse, marmoset, and human (Anand et al., 1983; Gibson et al., 1984a). The segmental differences are greatest in the cat. Immunocytochemically, no VIP-LI is seen in the medullary dorsal horn[g] and only a very few VIP-LIr fibres are found in the cervical and thoracic spinal cord (Kawatani et al., 1983; M. A. Ruda, unpublished observations). Very dense VIP-LIr staining is seen only within, and adjacent to, the sacral segments (Basbaum and Glazer, 1983; Honda et al., 1983; Kawatani et al., 1983).

In sacral segments, Lissauer's tract contains long, rostrocaudally oriented VIP-LIr axons (Figs. 3A, 4A). Beneath Lissauer's tract in the curving part of lamina I, a densely stained arc of VIP-LIr axons (originating from Lissauer's tract) sweeps into the neck of the dorsal horn (Fig. 3A; Basbaum and Glazer, 1983; Honda et al., 1983; Kawatani et al., 1983). Light microscopically, many of the immunoreactive profiles in Lissauer's tract and the subjacent part of lamina I appear to be quite thick (Figs. 3A, 4C). However, inspection of 1-μm Epon sections and electronmicrographs (Fig. 4) reveals that these thick processes are actually fascicles composed exclusively of unmyelinated axons, only some of which contain VIP-LI.

VIP-LI is found in a dense band in lamina I (Fig. 3A). This band is thickest in the lateral part of the lamina that curls around the dorsal horn and distinctly thinner in the dorsal part of the lamina. Occasionally, a varicose VIP-LIr fibre from lamina I dips into lamina IIa. Long strands of varicose VIP-LIr fibres within lamina I are seen coursing rostrocaudally or mediolaterally. These axons are generally fine (< 0.5 μm in the rat) and carry a mixture of small and large varicosities (sometimes > 2 μm; Gibson et al., 1981) that issue axodendritic synapses within lamina I (Fig. 3B).

The reticulated region at the neck of the dorsal horn contains dense, regularly spaced plexi of VIP-LIr terminals that resemble those that stain for SP (Fig. 3A). These plexi are generated by the axons that leave Lissauer's tract and form the dorsolateral arc of VIP-LIr fibres (Basbaum and Glazer, 1983; Honda et al., 1983; Kawatani et al., 1983). Some of these fibres continue medially to contribute to a sparse VIP-LIr plexus in medial lamina V.

The central canal region (especially dorsal to the canal) is innervated by a modest number of varicose VIP-LIr fibres. The density of this innervation is

[g] The nucleus of the solitary tract, however, stains densely for VIP-LI (Lundberg et al., 1978).

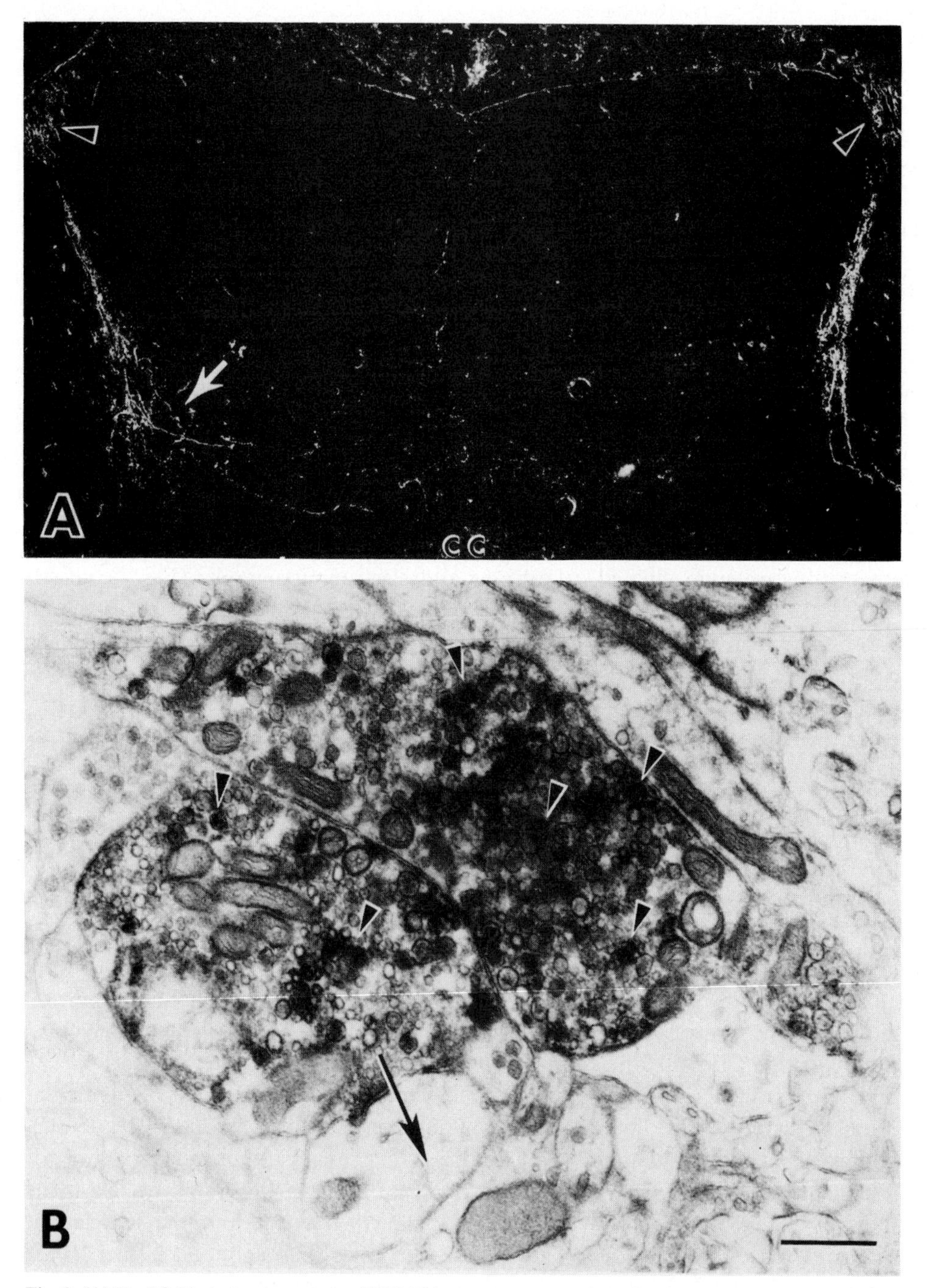

Fig. 3. (A) Darkfield photomicrograph of VIP-LI in a transverse section of cat sacral spinal cord. Bundles of VIP-LIr axons are found in Lissauer's tract (arrowheads). Most VIP-LIr axons traverse lamina I in the lateral part of the dorsal horn to terminate in the area of lamina V (arrow). Some VIP-LIr axons appear to cross the midline above the central canal (CC). Few VIP-LIr axons are found in the dorsal part of lamina I. Magnification: ×60. (B) Ultrastructure of VIP-LIr varicosities in lateral lamina I; two VIP-LIr varicosities are seen side-by-side. In this section, one varicosity synapses (arrow) on an adjacent dendrite. VIP-LIr varicosities contain numerous dense core vesicles in addition to round, agranular vesicles. The peroxidase–anti-peroxidase reaction product is associated with only a subpopulation of the dense core vesicles (arrowheads). Magnification: ×26 350. Scale bar represents 0.5 μm.

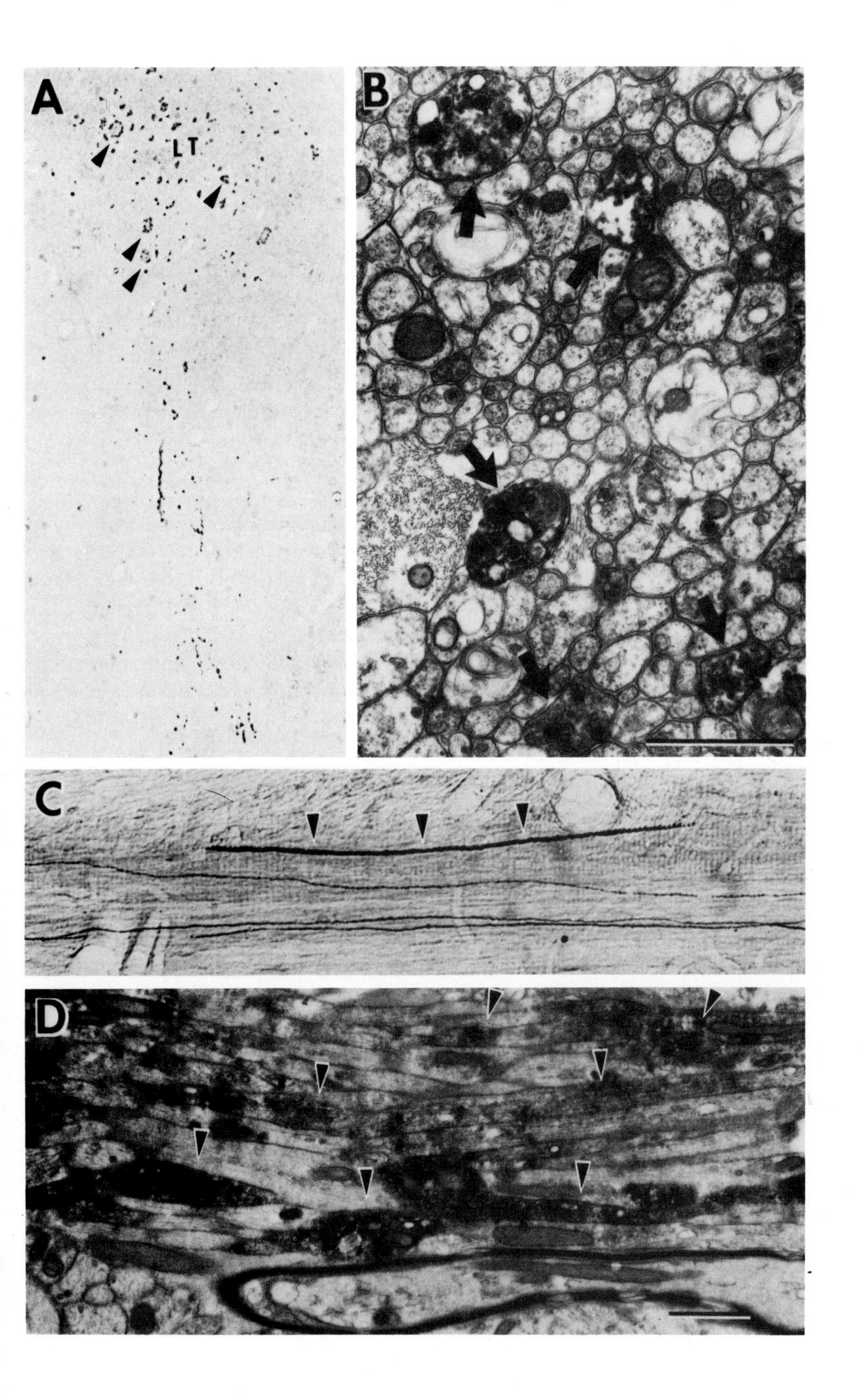
A
LT
B
C
D

reported to be approximately equal in all segments (Kawatani et al., 1983). In sacral segments, these fibres appear to have two origins. One is a continuation of some of the fibres that sweep into the neck of the dorsal horn (some of these fibres continue to the opposite side via the dorsal commissure; Basbaum and Glazer, 1983). The second origin is from fibres that travel across lamina I and then descend along the medial edge of the dorsal horn.

VIP-LI has been visualized in primary afferent somata (Lundberg et al., 1978; Hökfelt et al., 1982; Kawatani et al., 1983). VIP-LI is depleted in lamina I after neonatal exposure to capsaicin (Jancso et al., 1981) and ultrastructural studies show unmyelinated VIP-LIr axons in Lissauer's tract (Fig. 4B, D) and in the dorsal roots (Honda et al., 1983), but the possibility that a subset of VIP primary afferents are myelinated cannot be excluded. There is conclusive evidence that VIP-LI is especially common in a subpopulation of sacral primary afferents that innervate the pelvic viscera (Morgan et al., 1981; Basbaum and Glazer, 1983; Honda et al., 1983; Kawatani et al., 1983). The presence of a few VIP-LIr axons in Lissauer's tract at all segmental levels (Kawatani et al., 1983) and Lundberg et al.'s (1978) observation of VIP-LIr axons in the vagus nerve suggest that VIP-LI is present in other visceral afferents.

Deafferentation produces a marked, but subtotal, reduction of VIP-LI from Lissauer's tract, lamina I, lamina V, and the arc of fibres in the dorsolateral white matter (Jancso et al., 1981; Kawatani et al., 1983; Gibson et al., 1984a). The fate of the VIP-LIr fibres in the central canal region is controversial; Kawatani et al. (1983) report them to be unaffected, but Gibson et al. (1984a) have noted a decrease.

Fluoride-resistant acid phosphatase

Fluoride-resistant acid phosphatase (FRAP) is an enzyme (or family of enzymes; Dodd et al., 1983) that is found only in rodents. Its intraspinal distribution is restricted to Lissauer's tract and lamina II. In lamina II it forms a dense band that spans the laminae IIa–IIb border but fills only a fraction of the full thickness of either sublamina (Coimbra et al., 1974; Nagy and Hunt, 1981; Nagy et al., 1981).

The dorsal root ganglion cells that contain FRAP are small and, based on their sensitivity to capsaicin, issue unmyelinated axons (Nagy and Hunt, 1981; Nagy et al., 1981). It has been shown that no more than a very few FRAP primary afferents contain SRIF-LI or SP/CCK-LI (Nagy and Hunt, 1981; Dodd et al., 1983). The location of the FRAP band is so clearly separate from the region of VIP-LI that one can also assume that FRAP and VIP-LI do not coexist.

Dorsal rhizotomy and neonatal exposure to capsaicin deplete FRAP from Lissauer's tract. The band of FRAP in the middle of lamina II is severely depleted, but a small amount remains in the dorsalmost part of the band in lamina IIa (Coimbra et al., 1974; Nagy et al., 1981; Dodd et al., 1983).

There is preliminary evidence supporting the hypothesis that FRAP is in primary afferents that use adenosine triphosphate as their neurotransmitter

Fig. 4. Immunocytochemically stained VIP axons in the cat sacral spinal cord. (A) Phase contrast photomicrograph of a 1 μm transverse section through the lateral part of the dorsal horn. In Lissauer's tract (LT), large bundles of axons (arrowheads) appear to contain some VIP-LIr axons. Magnification: × 220. (B) At the ultrastructural level, the bundles indicated in (A) are found to be composed exclusively of unmyelinated axons. Some axons in the bundle contain VIP-LI (arrows) but many do not, suggesting that they contain a different neurochemical. Magnification: × 25 200. Scale bar represents 1 μm. (C) Sagittal section through Lissauer's tract. Both large, immunoreactive profiles (arrowheads) and thinner, varicose strands are immunocytochemically labelled for VIP. Magnification: × 1150. (D) Electron-microscope analysis of the large immunoreactive profiles in Lissauer's tract reveals that they correspond to the bundles of unmyelinated axons which contain a mixed neurochemical population seen in transverse sections (A and B). Some of the unmyelinated axons display VIP-LI (arrowheads). These observations demonstrate that VIP afferents in Lissauer's tract are unmyelinated axons which travel in a neurochemically diverse bundle composed exclusively of unmyelinated axons. Magnification: × 13 600. Scale bar represents 1 μm.

(Dodd et al., 1983; Jahr and Jessell, 1983; Yoshioka and Jessell, 1984).

Other identified neurochemicals in primary afferents

Antisera raised against bombesin (BOM) stain several regions of the spinal grey matter (Panula et al., 1982, 1983; Fuxe et al., 1983). However, there is no evidence that BOM itself is present in the mammalian CNS (Moody and Pert, 1979). Column chromatographic analyses of BOM-like immunoreactivity in neural tissue reveal two peaks corresponding to substances larger and smaller than BOM. The larger substance co-elutes with the complete gastrin-releasing peptide (GRP) molecule (GRP_{1-27}; McDonald et al., 1979; Panula et al., 1983). The smaller substance may correspond to a C-terminal fragment of GRP (GRP_{14-27}; Yanaihara et al., 1981; Panula et al., 1983). The crossreactivity is explained by the observation that GRP and BOM have homologous C-terminal heptapeptides[h] (McDonald et al., 1979). We will refer to the immunocytochemically demonstrated substance(s) as GRP-like immunoreactivity (GRP-LI). Dense concentrations of GRP-LI are seen in laminae I–II, Lissauer's tract, and the adjacent part of the dorsolateral funiculus. Inspection of the published photomicrographs suggests that lamina IIa is more heavily stained than lamina IIb and that a thin stripe of GRP-LI lines the medial wall of the dorsal horn. Scattered fibres have been reported in laminae III–V, the central canal region, and the ventral horn (Panula et al., 1982, 1983; Fuxe et al., 1983). About 5% of the rat's primary afferent somata contain GRP-LI. The cells are of the small type. Fuxe et al. (1983) detected SP-LI in GRP-LIr somata, but this result has not been confirmed (Panula et al., 1983). Dorsal rhizotomies cause a marked, but subtotal, depletion of GRP-LI in laminae I–III but no detectable change in the deeper laminae (Panula et al., 1982; Massari et al., 1983).

Gibson et al. (1984b) have examined the intraspinal distribution of calcitonin gene-related peptide (CGRP) in several mammals, including human. Approximately one-third of the CGRP-LIr somata were also shown to contain SP-LI; conversely, every SP-containing cell contained CGRP-LI. Thus, if CGRP and CCK antisera are recognizing different substances, then some SP primary afferent somata may contain three different peptides (perhaps four, if the presence of GRP is confirmed). CGRP-LIr fibres are concentrated in Lissauer's tract and laminae I–II. Slightly less dense concentrations are seen in the central canal region, lateral lamina V, the dorsal part of lamina III, and the dorsolateral white matter. A relatively sparse fibre density is seen throughout the deep parts of lamina III, all of lamina IV, and the ventral horn. The thoracic sympathetic nucleus and sacral parasympathetic nucleus are also innervated. Dorsal rhizotomies in rats and cats deplete CGRP-LI from Lissauer's tract, laminae I–III, V, and the central canal region. The ventral horn is unaffected.

Primary afferent somata that are stained with antisera raised against Leu-enkephalin (LEU-ENK) have recently been reported in the rat, cat and monkey (Senba et al., 1982; Kawatani et al., 1984; Ropollo et al., 1984). Nothing is known of the intraspinal terminations of these afferents since deafferentation produces no detectable change in the quantity or distribution of ENK as assayed immunocytochemically. A few dynorphin (DYN) dorsal root ganglion cells have been seen in the cat (Kawatani et al., 1984). Very recent reports suggest that DYN-like immunoreactivity (DYN-LI) is found in sacral primary afferents from the pelvic viscera and that these afferents terminate in laminae I and V (Basbaum, 1985). Itoga et al. (1980) have reported β-endorphin-like immunoreactivity in a small number of cat primary afferent cell bodies, but we know of no replication of this observation.

The neurochemical labels described above are all found in the small- or medium-size cells of the dorsal root ganglion. There is very little data concern-

[h] The C-terminal dipeptides of BOM, GRP, and SP are identical, but radioimmunoassay and immunocytochemical competition tests have consistently shown that GRP/BOM antisera have negligible crossreactivity with SP.

ing the neurochemistry of the ganglion's large cells that issue large, myelinated axons. It has been proposed that at least some of the somatosensory afferents with large, myelinated axons utilize amino acids (L-glutamate, L-aspartate, or L-cysteine sulphinate) as neurotransmitters. The evidence for this hypothesis is reviewed elsewhere (Salt and Hill, 1983). An enzyme, carbonic anhydrase (CA), has been detected in a subpopulation of large primary afferent somata and large, myelinated dorsal root axons in the rat (Wong et al., 1983; Riley et al., 1984). The presence of CA in axons in cutaneous branches of the sciatic nerve and in the dorsal column nuclei suggests that CA may be present in some somatosensory afferents. The presence of α-melanocyte stimulating hormone (α-MSH)-like immunoreactivity (α-MSH-LI) in cells and myelinated axons within the rat's dorsal root ganglion has been reported (Swaab and Fisser, 1977). The detection of a few α-MSH-LIr fibres in laminae III–V of the medullary dorsal horn (Yamazoe et al., 1984) supports this finding and suggests that at least some of the MSH-containing primary afferents may be Aβ low-threshold mechanoreceptors.

Conclusions

Nine substances have been detected in small dorsal root ganglion (DRG) cells: SP, CCK, SRIF, VIP, FRAP, GRP, CGRP, ENK, and DYN. These DRG cells can be divided into at least eleven subsets on the basis of their neurochemistry. Five of the eleven subsets contain a single identified substance that is not known to coexist to any appreciable extent with any other substance in DRG cells: SRIF, VIP, FRAP, ENK, and DYN. Three subsets contain substances (CCK, CGRP and GRP) that are found singly in some cells and coexisting in others. SP may be present in three subsets with different patterns of coexistence: SP/CGRP, SP/CCK/CGRP, and SP/CCK/CGRP/GRP.

Single tracer studies utilizing intracellular, retrograde, and anterograde horseradish peroxidase (HRP) methodologies have revealed a differential distribution of primary afferents in the dorsal horn. Nociceptive, finely myelinated (Aδ) and unmyelinated (C) afferents terminate in laminae I–IIa and lamina V. Low-threshold mechanoreceptive Aδ and C afferents appear to be localized to laminae IIb and III (Light and Perl, 1979; Bennett et al., 1980), whereas different functional types of large, myelinated (Aβ), mechanoreceptive afferents terminate in laminae III, IV and V (Brown et al., 1977). Do any of the neurochemically defined subsets correspond uniquely to functionally defined categories of primary afferents? There is not enough evidence to give a definitive answer, but the data currently available suggest that the answer is probably "No". We have so little information about the primary afferents that contain SRIF-, ENK-, and DYN-like immunoreactivities that the question cannot even be addressed. Primary afferent FRAP is found only in a band in the middle of lamina II; no known, functionally characterized group of primary afferents has this terminal distribution. Deafferentation depletes CGRP-LI and GRP-LI from laminae I–III. No single functional category of primary afferents is known (or suspected) to terminate in laminae I–II *and* III. VIP-LI appears to be confined to visceral afferents, but these are not necessarily functionally homogeneous (Cervero, 1984). It has been noted (Basbaum and Glazer, 1983) that the laminar distribution of VIP-containing pelvic visceral afferents is identical to that of cutaneous Aδ high-threshold mechanoreceptors. They suggest that visceral high-threshold mechanoreceptors may be uniquely marked by VIP. But, (1) many VIP primary afferents are probably unmyelinated (Fig. 4) and, (2) *both* Aδ and C afferents from the viscera are distributed like Aδ high-threshold mechanoreceptors (Morgan et al., 1981; Neuhuber, 1982; Cervero and Connell, 1984a,b). Finally, it has been suggested on many occasions that unmyelinated polymodal nociceptors contain SP. The laminar distribution of primary afferent SP-LI is consistent with the hypothesis. However, the probable presence of primary afferent SP in laminae IIb–IV (Nagy et al., 1981) suggests that non-nociceptive afferents also contain SP. The effects of capsaicin on nociception (especially chemogenic pain) and neurogenic inflammation are also consistent with the hypothesis, but the capsaicin data are essentially circumstantial (Dubner and Bennett, 1983; Fitzgerald, 1983).

Primary afferents with Aδ and C axons are classically described as travelling in Lissauer's tract and terminating in laminae I–II. Recent work indicates that this description is incomplete. First, Light and Perl (1979) have demonstrated that Aδ nociceptors in cats and monkeys issue collaterals that terminate not only in lamina I and lamina II, but also in lamina V and the central canal region. Second, Morgan and his colleagues (1981) have shown that cat primary afferents from the pelvic viscera (which are nearly all Aδ or C axons) terminate in laminae I–IIa, V, and the central canal region. Thoracic visceral afferents have also been shown to terminate in laminae I and V, but terminals in the central canal region have not been detected (Neuhuber, 1982; Cervero and Connell, 1984a,b). The deep regions of Aδ and C fibre termination are reached by several pathways (Morgan et al., 1981). One, designated the lateral collateral pathway, arises from Lissauer's tract, forms an arc along the lateral aspect of the dorsal horn in lamina I, and sweeps ventromedially into lamina V. The pathway issues terminals in the reticulated part of lamina V and many of its axons continue medially and issue synapses through medial lamina V and the central canal region. A smaller, medial collateral pathway also arises from Lissauer's tract; it travels across the top of the dorsal horn, descends along the dorsal horn's medial wall, and terminates largely in the central canal region. In addition, scattered afferents travel in the dorsolateral funiculus; some of these course ventromedially and join the lateral collateral pathway.

Immunocytochemical data suggest that these previously unrecognized pathways and terminations for Aδ and C afferents may be characteristic of afferents from both the skin and viscera at all spinal levels. Antisera directed against SP, CCK, VIP, ENK, SRIF, GRP, and CGRP reveal lateral and medial collateral pathway-like fibre systems in various segments of several species and there is evidence that at least some of these immunolabelled pathways are depleted by deafferentation.

The presumed target neurons of the lateral and medial collateral pathway afferents in lamina V and the central canal region are known to be excited by nociceptors (Willis and Coggeshall, 1978; Nahin et al., 1983), but there is no evidence that they are excited by thermoreceptors or Aδ/C low-threshold mechanoreceptors. It is thus probable that these afferents are mostly nociceptors. The import of this hypothesis is that the spinothalamic, spinomesencephalic and spinoreticular projection neurons that lie in lamina V and the central canal region may receive a direct Aδ and C nociceptor input independent of the input to laminae I–II and independent of the presumed modulatory effects of the circuitry of the substantia gelatinosa.

Intrinsic spinal neurons

The immunocytochemical detection of peptidergic intrinsic spinal neurons usually requires the use of colchicine to increase the intraneuronal concentration of antigen. But even with the use of colchicine it is rare to see more than a small part of a neuron's proximal dendritic arbor. Thus it is often very difficult to identify neurons immunocytochemically that have been defined morphologically by other methods (e.g., Golgi and intracellular HRP staining). Moreover, it is never possible to trace an axon of a neuron which has been immunocytochemically visualized. Thus we cannot identify directly the cells of origin of the immunolabelled axon terminals arising from intrinsic neurons. Finally, it should be noted that the efficacy of colchicine is variable and, thus, negative findings are very uncertain.

Substance P

SP-LI has been detected in neurons in laminae I–V, the central canal region, and the lateral spinal nucleus (Hökfelt et al., 1977b; Ljungdahl et al., 1978; DelFiacco and Cuello, 1980; Gibson et al., 1981; Hunt et al., 1981b; Nagy et al., 1981; Tessler et al., 1981; Senba et al., 1982). There are very few data about the morphology of these neurons.

The SP-LIr neurons in lamina I are described as being large with prominent, rostrocaudally oriented dendrites. Small SP-LIr somata (ca. 15 μm) are present throughout lamina II but they appear to be most numerous in lamina IIb. The SP-LIr neurons in lamina III are generally small (< 10 μm) with perikarya that are barely larger than their nuclei. Some SP-LIr neurons in laminae III and IV have apical dendrites that penetrate the overlying substantia gelatinosa (Hunt et al., 1981b; Nagy et al., 1981).

Tessler et al. (1980, 1981) have shown that SP-LIr fibres of non-primary afferent origin (i.e., those that have large varicosities and survive deafferentation) are distributed to the same regions as primary afferent SP. About one month after deafferentation, there is an increase of SP-LI in these regions. This SP-LI must originate from intrinsic spinal neurons because it is unaltered in animals with contralateral dorsal rhizotomies, unilateral ganglionectomies, spinal transections, or isolated hemisegments (i.e., fore and aft hemisections plus rhizotomies and a midline myelotomy).

Cholecystokinin

CCK-like immunoreactivity has been seen in neurons in laminae I–V and the central canal region in the rat (Gibson et al., 1981; Vanderhaeghen et al., 1982; Schrøder, 1983). Vanderhaeghen et al. (1982) noted that the CCK-LIr cells in the central canal region have average diameters of about 30 μm and form a "discontinuous column" that can be detected only in lumbosacral segments.

Somatostatin

Neurons that contain SRIF-LI have been located in laminae I–V, the central canal region, the lateral spinal nucleus, the sacral parasympathetic nucleus, and Onuf's nucleus (Forssmann, 1978; Burnweit and Forssmann, 1979; Dalsgaard et al., 1981; Hunt et al., 1981b; Senba et al., 1982; Johansson et al., 1984; Schrøder, 1984). The morphology of SRIF-LIr neurons has received little study.

SRIF-LIr neurons in lamina II are small (10–12 μm) and concentrated in the deeper part of the lamina (IIb). It is noteworthy that the efferent targets (and hence the identities) of the SRIF-LIr neurons in the parasympathetic and Onuf's nuclei have been confirmed in a double-labelling study (Schrøder, 1984).

Vasoactive intestinal polypeptide

VIP-LIr neurons have been seen in laminae II–IV, the central canal region, the lateral spinal nucleus, and the sacral parasympathetic nucleus (Fuji et al., 1983; Gibson et al., 1984a). The VIP-LIr neurons in laminae II–III are small, bipolar, and concentrated in the lateral part of the laminae (Gibson et al., 1984a). The cells in the central canal region are also small (Fuji et al., 1983). Nothing else is known about the morphology of these cells.

McGregor et al. (1984) have described a remarkable change in the laminar distribution of varicose VIP-LIr fibres following damage to a peripheral nerve. Laminae II–III, which are essentially devoid of VIP-LIr fibres in the normal state, develop a significant VIP innervation (heaviest in lamina II and sparse in lamina III) about 2 weeks after peripheral nerve transection. In addition, the normally dense innervation in lamina I is intensified. This effect is seen only in the ipsilateral dorsal horn. The source of this emergent VIP-LI is unknown, but it is tempting to speculate that it comes from the VIP-LIr neurons in laminae II and III[i].

Neurotensin

Spinal neurons that contain neurotensin (NT)-like

[i] McGregor et al. (1984) have also demonstrated that a peptide designated PHI (**p**eptide with an N-terminal **h**istidine and a C-terminal **i**soleucine) is present in the dorsal horn. Its normal distribution and reaction to peripheral nerve damage appear to be identical to VIP. It is probable that PHI and VIP are coexistent because they are derived from the same precursor (Itoh et al., 1983; Christofides et al., 1982).

immunoreactivity (NT-LI) have been found in laminae I–III (Hunt et al., 1981b; Seybold and Elde, 1982; DiFiglia et al., 1984). They are most numerous within the deepest part of lamina IIb and the superficial part of lamina III. The cells in both locations have small perikarya (15 μm) but, as shown by Seybold and Elde (1982), their proximal dendritic arbors differ. The lamina IIb NT-LIr cells have rostrocaudally oriented dendritic arbors and recurrent branches; they resemble lamina IIb islet cells (see Conclusions of this section). The cells in lamina III have proximal dendritic arbors with no clearly preferred orientation and they often occur together in clusters. The transition from one type to the other is fairly abrupt with the islet cell found within the deep, dense band of NT-LIr staining (lamina IIb) and the other cell type appearing beneath the band (lamina III). A smaller number of NT-LIr cells is found in lamina IIa. These are small, spindle-shaped, and have rostrocaudally oriented dendritic arbors with recurrent branches; they resemble lamina IIa islet cells. A few small, bipolar NT-LIr cells have been seen in lamina I (Hunt et al., 1981b).

There is no evidence for the existence of NT-LI in primary afferent somata. Intraspinal NT-LI is thus presumed to arise from intrinsic spinal neurons and, perhaps, from descending axons.

NT-LIr axons and varicosities are concentrated in laminae I–II (Uhl et al., 1979b; Gibson et al., 1981; Hunt et al., 1981b; Seybold and Elde, 1982). The densest NT-LIr staining is seen as a band, in the deep part of lamina II. It is not clear whether this band occupies the full thickness of lamina IIb or just its deepest part. It is also unclear whether the band's inner border corresponds to the inner border of lamina IIb or whether there is some slight overlap into lamina III. In favourable preparations (Gibson et al., 1981; Ninkovic et al., 1981; Seybold and Elde, 1982), a second band of NT-LI is seen more dorsally. Gibson et al. (1981) place this band in the inner half of lamina I, but others (Ninkovic et al., 1981; Seybold and Elde, 1982) have it occupying laminae I and IIa. A relatively modest plexus of NT-LIr fibres has been detected in the lateral, reticulated region of lamina V and the corresponding region of the medullary dorsal horn (Uhl et al., 1979; Gibson et al., 1981). In the spinal cord this region appears to be supplied by an arc of NT-LIr fibres in the adjacent dorsolateral funiculus. A very few fibres have been observed in laminae III (particularly its dorsal part), IV, and the central canal region (Gibson et al., 1981; DiFiglia et al., 1984). NT-LI has not been visualized immunocytochemically in the ventral horn, but radioimmunoassay studies suggest that it is present (Kataoka et al., 1979).

Enkephalin

We will make no distinction here between immunocytochemical data obtained with antisera directed against Met- and Leu-enkephalin (MET-ENK and LEU-ENK). While a distinction between the two is clearly desirable, it seems unwarranted at present for the following reasons. In radioimmunoassays, MET-ENK and LEU-ENK antisera often show appreciable (10–30%) crossreactivity for the other pentapeptide and their crossreactivity to other opioid peptides with homologous sequences (e.g., MET-ENK-^{6}Arg-^{7}Phe, DYN) is generally unknown. Even less is known about their crossreactivities in fixed tissue. Finally, no obvious differences have been noted between the intraspinal staining patterns produced by MET-ENK and LEU-ENK antisera (Simantov et al., 1977; Hunt et al., 1981b).

Deafferentation and spinal transection produce no detectable decrease in the amount or distribution of ENK-LI in the dorsal horn (DelFiacco and Cuello, 1980; Jancso et al., 1981; Naftchi et al., 1981; Sumal et al., 1982; Ruda et al., 1983). Thus, most of the ENK-LI in the dorsal horn is believed to come from intrinsic spinal neurons. ENK-LIr somata have been found in laminae I–V and the central canal region (Hökfelt et al., 1977a; Uhl et al., 1979a; DelFiacco and Cuello, 1980; Aronin et al., 1981; Glazer and Basbaum, 1981; Hunt et al., 1981b; Bennett et al., 1982; Sumal et al., 1982; Conrath-Verrier et al., 1983; Hunt, 1983). In lamina

I the ENK-LIr cells have dendritic trees that are typical of this region. Both large (Waldeyer type) and small cells are stained. In lamina II the ENK-LIr somata are most numerous along the laminae I–IIa border and in lamina IIb. The ENK-LIr cells that lie along the laminae I–IIa border have been identified as stalked cells (see Conclusions of this section; Glazer and Basbaum, 1981; Bennett et al., 1982; Hunt, 1983). The ENK-LIr cells in lamina IIb have the light and electron microscopic character-

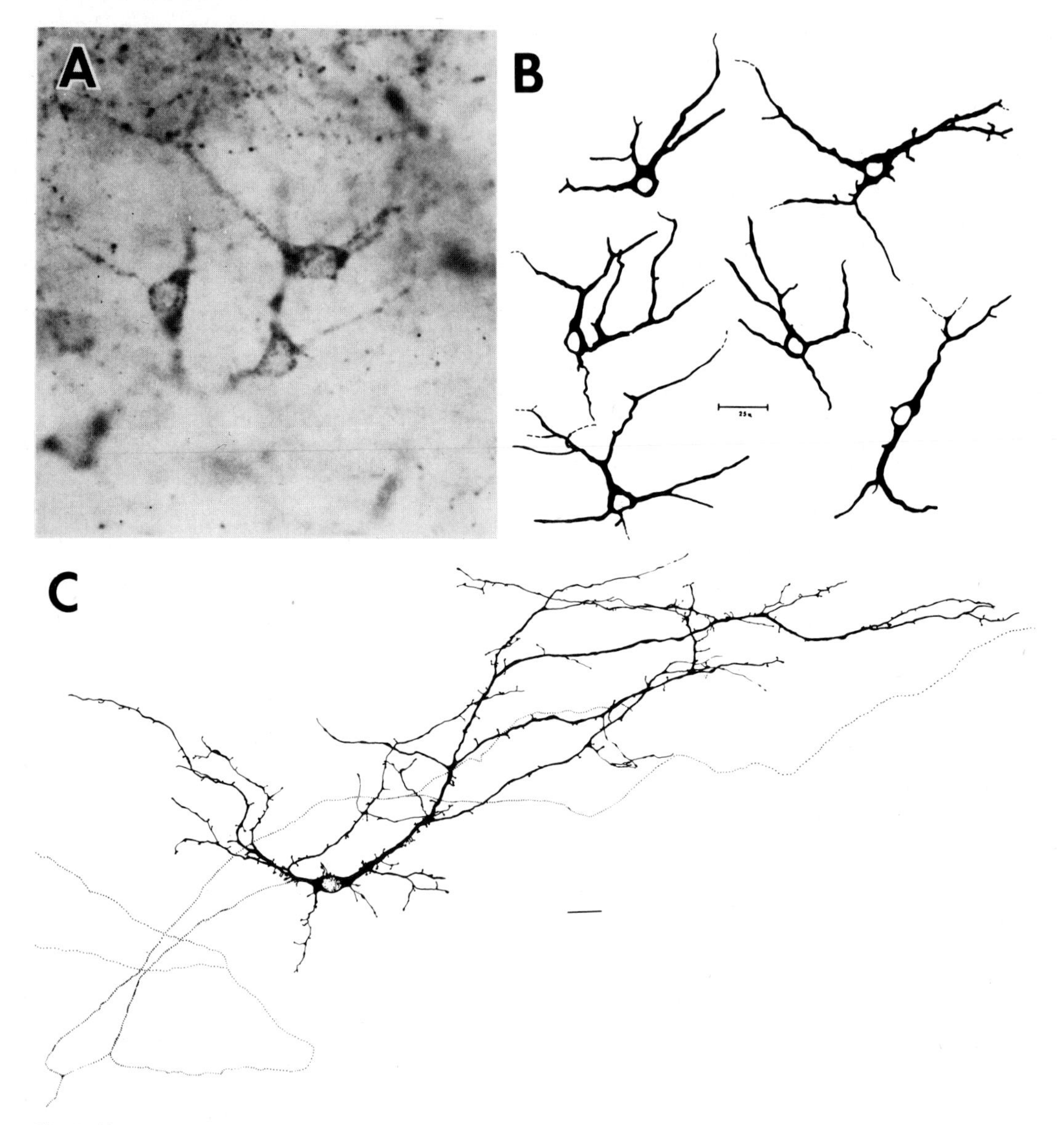

Fig. 5. (A) Photomicrograph in the sagittal plane of three ENK-LIr neurons in lumbar lamina III from a colchicine-pretreated cat. The varicose, ENK-LIr fibres at the top of the picture are in lamina IIb. (B) Camera lucida drawings of lamina III ENK-LIr cells. Scale bar represents 25 μm. (C) Camera lucida reconstruction of a lamina III interneuron that was intracellularly stained with HRP. Note the morphological similarities with the ENK neurons in A and B. The cell's nucleus is stippled. The axon (dotted line) courses through laminae III and IV and emits numerous terminal and en passant varicosities (not shown). Scale bar represents 10 μm. Sagittal section.

istics of lamina IIb islet cells (Glazer and Basbaum, 1981; Hunt et al., 1981b; Bennett et al., 1982; Hunt, 1983). The ENK-LIr neurons in lamina III have small perikarya and usually issue 2–3 primary dendrites. Their dendritic arbors are oriented primarily dorsally (Fig. 5; Bennett et al., 1981b; Glazer and Basbaum, 1981). The ENK-LIr somata in laminae IV and V are larger (ca. 20–30 μm) and multipolar (Fig. 6; DelFiacco and Cuello, 1980; Glazer and Basbaum, 1981). In the central canal region the ENK-LIr neurons have small perikarya (Conrath-Verrier et al., 1983).

The greatest concentration of ENK-LI is found in laminae I and II. In the cat, the ENK-LI concentration in lamina IIa is distinctly less than that in laminae I and IIb (Glazer and Basbaum, 1981; Bennett et al., 1982; Conrath-Verrier et al., 1983). It is not clear whether this banding pattern is present in other species. Hunt et al. (1981b), for example, report that in the rat ENK-LI is denser in lamina IIa than in laminae I or IIb, but Bresnahan et al. (1984) report that the pattern in the rat is similar to that found in the cat (lamina IIa lighter than laminae I and IIb). The ENK-LIr terminals in the monkey's dorsal horn appear to be equally dense in laminae I and IIa and distinctly lighter in lamina IIb (LaMotte and DeLanerolle, 1983a). However, ENK-LI in the human dorsal horn appears to be most dense in lamina IIb (DeLanerolle and LaMotte, 1982).

The concentration of ENK-LI in the central canal region is less than that in the superficial laminae. The lateral part of lamina V contains a moderate amount of ENK-LI. A thin stripe of ENK-LIr fibres lining the medial wall of the dorsal horn has been noted (Gibson et al., 1981; Senba et al., 1982). An arc of ENK-LIr fibres is present in the dorsolateral funiculus. Laminae III–IV and the ventral horn contain relatively little ENK, but varicose ENK fibres are still quite common.

Neuropeptide Y and the pancreatic polypeptides

It is now reasonably certain that previous demonstrations of neuronal structures with antisera raised against avian or bovine pancreatic polypeptides were actually demonstrating the presence of neuropeptide Y (NPY) or NPY-related peptides (Allen et al., 1983; Lundberg et al., 1984). The crossreactivities are not unexpected given the extensive se-

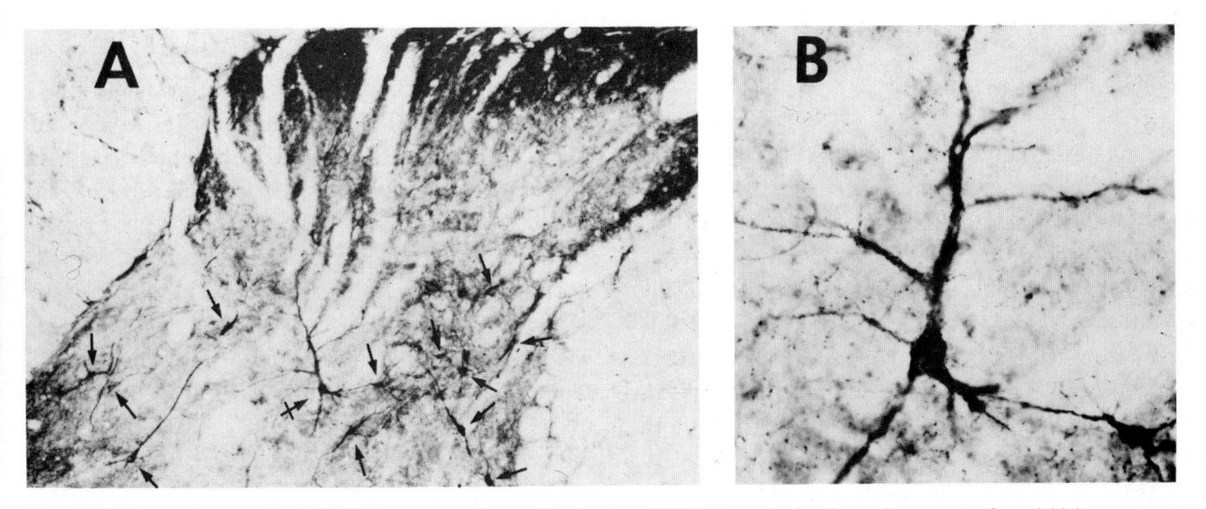

Fig. 6. (A) Photomicrograph of ENK-LIr neurons (arrows) in laminae V–VII from the lumbar enlargement of a colchicine-pretreated cat. Thirteen ENK-LIr cells are visible in this 50-μm thick section stained with the double-bridge peroxidase–anti-peroxidase method with nickel-intensified diaminobenzidine histochemistry. The ENK-LIr cells in laminae I–II are obscured by the dense reaction product while the small ENK-LIr cells in lamina III are just barely visible at this magnification. (B) Higher magnification (with a slightly different orientation) of the ENK-LIr neuron in A marked by the crossed arrow.

quence homologies of NPY and the pancreatic polypeptides (Tatemoto, 1982; Tatemoto et al., 1982). In the summary that follows, we will refer to the immunocytochemically demonstrated substance as NPY, regardless of whether the antisera were raised against NPY or avian or bovine pancreatic polypeptides.

Intrinsic spinal neurons that contain NPY-like immunoreactivity (NPY-LI) have been reported in laminae I–III, V, the central canal region, and the sacral parasympathetic nucleus (Lundberg et al., 1980; Hunt et al., 1981a,b; Olschowka et al., 1981; Allen et al., 1983). NPY-LIr cells are most common in lamina I and the region composed of lamina IIb and the dorsal part of lamina III. Only a few cells are seen in lamina IIa. The NPY-LI cells in the sacral parasympathetic nucleus contain coexistent MET-ENK-LI (Hunt et al., 1981a).

It is possible that all intraspinal NPY-LI originates from intrinsic spinal neurons because dorsal rhizotomies and spinal hemisections do not produce any noticeable depletion (Hunt et al., 1981a,b; Gibson et al., 1984c). Varicose NPY-LIr fibres are particularly numerous in laminae I–II. NPY-LIr terminals are also seen in laminae III–V and the central canal region (Hunt et al., 1981a,b; Gibson et al., 1984c).

Gamma-aminobutyric acid

Gamma-aminobutyric acid (GABA) is demonstrated immunocytochemically by antisera raised against its synthesizing enzyme, glutamic acid decarboxylase (GAD). GAD-like immunoreactive (GAD-LIr) terminals are found in every part of the spinal gray matter, but with marked regional differences in density (McLaughlin et al., 1975; Hunt et al., 1981b). Particularly dense concentrations appear as two very thin bands in the superficial laminae. The outer-most band corresponds closely to lamina I, although it is most pronounced in the medial half of the dorsal horn. The deeper band corresponds approximately to the laminae IIb–III border zone. Slightly less dense fields of GAD-LIr terminals fill the rest of laminae II and III. A relatively lower concentration of GAD-LI fills lamina IV, the central canal region, and the medial border of the dorsal horn. A sparse innervation covers lamina V and the ventral horn. There is, at present, no reason to suspect that GABA is contained within primary afferents or descending axons. Thus, all intraspinal GABA is assumed to come from intrinsic spinal neurons.

GAD-LIr cell bodies are found in the vicinity of the lateral spinal nucleus and everywhere in the spinal grey matter except lamina IX (Barber et al., 1982). The somata in the superficial laminae are generally small (<15 μm) although occasional medium-size (20–30 μm) somata are also found (Ribiero-da-Silva and Coimbra, 1980; Hunt et al., 1981b; Barber et al., 1982).

Acetylcholine

The localization of acetylcholine (ACH) is inferred from the localization of the enzymes that catalyze its synthesis and degradation, choline acetyltransferase (ChAT) and acetylcholinesterase (ACHE), respectively. Exceptions to the expected colocalization of these two enzymes have been reported in the spinal cord. These exceptions are unexplained.

Both ChAT and ACHE terminal fields are present throughout the spinal grey matter. Their presence in the ventral horn, of course, is expected, but the dorsal horn also contains dense terminal fields (Navaratnam and Lewis, 1970; Schrøder, 1977, 1983; Kimura et al., 1981; Barber et al., 1984). ChAT and ACHE within the superficial dorsal horn have different distributions. The distribution of ACHE seems to be most dense in lamina IIa, with laminae I and IIb markedly less dense and lamina III only moderately less dense (Navaratnam and Lewis, 1970; Schrøder, 1977, 1983; but compare with Silver and Wolstencroft, 1971). ChAT, however, is very scarce in lamina IIa. Its densest distribution forms a narrow band across the laminae IIb–III border zone. The rest of lamina III contains a moderately reduced concentration. In lamina I the concentration is even less dense but still noticeably denser than that seen in superficial lamina II (Barber et al., 1984).

There is no reason to suspect that motorneuron axon collaterals terminate in the superficial laminae. A contribution from preganglionic autonomic neuron axon collaterals, however, is a theoretical possibility. Neither ACHE nor ChAT has been shown to be in primary afferent somata or dorsal root axons. Thus it is not unreasonable to suppose that cholinergic spinal neurons generate the terminal fields in the superficial laminae.

Both ACHE and ChAT have been located within dorsal horn neurons (Marchand and Barbeau, 1982; Barber et al., 1984). Their laminar origins are overlapping but not completely identical. Laminae I and II and the lateral spinal nucleus contain ACHE neurons, but no ChAT neurons. Both ACHE- and ChAT-containing neurons are present in laminae III, IV and lateral V. The lamina III ChAT neurons have small somata ($<15\ \mu m$) and issue 2–3 primary dendrites that travel predominately in the dorsal direction. ChAT (but not ACHE) neurons are also occasionally seen within the dorsal columns and lying along the medial wall of the dorsal horn (Barber et al., 1984).

Other neurochemicals in intrinsic spinal neurons[j]

Jaeger et al. (1983) have shown that aromatic-L-amino decarboxylase (AADC)-like immunoreactivity (AADC-LI) is present in numerous small neurons that surround the central canal. Most of them have at least one process that passes through the ependymal layer and extends into the canal's lumen. The identity of the substance that is synthesized by AADC in these neurons is unknown; tyramine, tryptamine, and phenethylamine are possible candidates. The AADC-LIr neurons are probably not monoaminergic because they do not contain serotonin, tyrosine hydroxylase, or phenylethanolamine-*N*-methyltransferase.

Thyrotropin releasing hormone (TRH)-like immunoreactive (TRH-LIr) fibres are seen in the central canal region, the thoracic sympathetic nucleus, and in the ventral horn surrounding motorneurons (Hökfelt et al., 1975b; Johansson and Hökfelt, 1980; Gibson et al., 1981). Until just recently, the only suspected source of intraspinal TRH-LI was from descending axons (Johansson et al., 1981; Gilbert et al., 1982). However, Coffield et al. (1984) have now demonstrated the presence of TRH-LIr neurons in the dorsal horn of the mouse. The TRH-LIr cells have small perikarya ($<15\ \mu m$) and are found in the laminae IIb–III border region.

Corticotropin-releasing factor (CRF)-like immunoreactive (CRF-LIr) somata have been visualized only after the peptide concentration has been increased in response to hypophysectomy. The CRF-LIr cells are located in lamina V, the thoracic intermediolateral column, and the central canal region (Merchenthaler et al., 1983). CRF-LI accumulates caudal to a spinal transection, indicating an ascending CRF projection (Merchenthaler et al., 1983). The presence of CRF-LIr neurons in the paraventricular nucleus and locus coeruleus, however, suggests that there may also be descending CRF systems (Olschowska et al., 1982; Cummings et al., 1983; Paull and Gibbs, 1983; Swanson et al., 1983). CRF-LIr fibres are present in the dorsolateral funiculus, lamina I, lateral lamina V, the thoracic sympathetic nucleus, and the central canal region (Olschowka et al., 1982; Cummings et al., 1983; Merchenthaler et al., 1983; Swanson et al., 1983). The fibres in lamina I have a pronounced rostrocaudal orientation. The fibres in lamina V and the intermediolateral column travel across the neck of the dorsal horn and pass through the central canal region to the other side.

[j] For the sake of completeness, we note here two other substances that have been detected in the spinal cord. First, antisera raised against FMRFamide reveal axon terminal-like processes in laminae I–II, the central canal region, and the ventral horn (Williams and Dockray, 1983; Chronwall et al., 1984). The identity of the FMRFamide-like immunoreactivity is uncertain (Dockray et al., 1981b, 1983; Dockray and Williams, 1983). A few FMRFamide-containing cells have been seen in the central canal region of some spinal segments (Sasek et al., 1984). Second, Gibson et al. (1981) describe adrenocorticotropic hormone immunoreactive fibres coursing in the dorsolateral funiculus and very sparse terminal-like processes in laminae I–III and the central canal region.

Panula et al. (1982) have reported GRP-LI cell bodies in the "substantia gelatinosa" (lamina II?) of the rat's medullary dorsal horn. The antisera that they used were raised against protein-conjugated [Glu[7]]BOM_{6-14}. The C-terminal heptapeptide of this immunogen is shared by BOM and GRP.

CGRP-LI is present in large, ventral horn neurons and in neurons in the sacral parasympathetic nucleus. If, as seems likely, the ventral horn neurons are motorneurons, then CGRP-LI in motorneuron axon collaterals may contribute to the CGRP-LIr fibre plexus in the ventral horn. No CGRP-LIr neurons have been detected in the dorsal horn, despite colchicine pretreatment (Gibson et al., 1984b).

Dynorphin-containing spinal neurons have recently been described (Basbaum, 1985). They are very numerous in laminae I and V. A few are found in laminae IIa, VI and VII.

The sequence of 29 amino acids in the recently discovered substance galanin (GAL) has very little resemblance to the sequences of other known neuropeptides (Tatemoto et al., 1983). In the rat, GAL-like immunoreactive (GAL-LIr) cell bodies and fibre systems are present in laminae I and II of the spinal and medullary dorsal horns (Melander et al., 1984).

Antisera raised against angiotensin II (AII) reveal fibre systems in the spinal and medullary dorsal horns. AII-like immunoreactivity (AII-LI) is most dense in laminae I, II, V, and the central canal region. A few fibres are present in laminae III–IV and the ventral horn. AII-LIr fibres are also present in the thoracic sympathetic nucleus (Fuxe et al., 1976; DeLanerolle and Coen, 1984). Changaris et al. (1978) report that AII-LIr *cell bodies* are present in Lissauer's tract, the spinal trigeminal tract, and (occasionally) in the dorsal columns.

Electronmicrographs show FRAP within the dendritic profiles that surround the central (primary afferent) terminals of synaptic glomeruli in lamina II of the rat (Coimbra et al., 1974). This suggests that FRAP may be present within intrinsic spinal neurons. These neurons may account for the small quantity of FRAP in lamina IIa that survives capsaicin-induced deafferentation (Nagy et al., 1981).

Conclusions

There is evidence for the existence of at least 18 substances in intrinsic spinal neurons. The distributions of neurons that contain AADC, GRP, and FRAP are very limited, suggesting that these neurochemicals may occur in only a single type of neuron. The distributions of neurons containing TRH-, GAL-, CRF-, CGRP-, and AII-like immunoreactivities are also limited and these substances may mark only one or two kinds of neurons. The remaining substances (SP-, CCK-, SRIF-, VIP-, NT-, ENK-, DYN-, NPY-, ACH-like immunoreactivities, and GABA) are almost certainly present in at least several different kinds of neurons. With the exception of the lateral spinal nucleus, there is no evidence that any of these substances are present in somatosensory long projection pathways. It is noteworthy that there are no demonstrations of coexistence in spinal neurons of presumed somatosensory function; coexistence has been shown, however, in autonomic preganglionic neurons and the motorneurons of Onuf's nucleus (Hunt et al., 1981a; Schrøder, 1984).

Ten substances (SP-, CCK-, SRIF-, VIP-, NT-, ENK-, NPY-, DYN-, and GAL-like immunoreactivities, ACH and GABA) have been found in *lamina I* neurons. This lamina contains a mixture of local circuit and projection neurons. Physiologically, lamina I contains three general kinds of neurons: nociceptive-specific neurons (NS), wide-dynamic-range neurons (WDR) which respond to both nociceptors and low-threshold mechanoreceptors, and thermoreceptive neurons which respond to warming or cooling. It is not yet possible to assign a neurochemical label to any of these physiologically or anatomically defined categories. There is evidence, however, that lamina I spinothalamic neurons do not contain ENK-LI (Basbaum, 1982).

Lamina II contains neurons with 14 identified substances (SP-, CCK-, SRIF-, VIP-, NT-, ENK-, NPY-, DYN-, GAL-, GRP-, TRH-like immunoreactivities, ACH, GABA and, perhaps, FRAP). The physiological responses of lamina II neurons vary with position within the lamina. The cells in lamina IIa are either NS or WDR cells and those in lamina IIb respond only to innocuous mechanical stimuli (for review see Dubner and Bennett, 1983). Several morphological types of lamina II cells are known, but only two, stalked cells and islet cells, are believed to be especially numerous (Gobel, 1975, 1978; Bennett et al., 1979, 1980; Gobel et al., 1980). Stalked cells have their somata along the laminae I–IIa border, dendritic arbors that descend and fan-out through lamina II (and sometimes III), and axons that arborize profusely in lamina I. Islet cell somata lie in the middle of their rostrocaudally elongated dendritic arbors which are largely confined to either lamina IIa or IIb. Their axonal arbors are especially profuse and largely confined to the same sublamina as their dendritic trees. There is good evidence that some stalked cells and some lamina IIb islet cells contain ENK-LI (Hunt et al., 1980, 1981b; Glazer and Basbaum, 1981; Bennett et al., 1982; Hunt, 1983). The pictures of Seybold and Elde (1982) leave little doubt that some lamina IIb islet cells contain NT-LI. There is also good evidence that some stalked cells and islet cells are GABAergic (Hunt et al., 1981b; Barber et al., 1982). In addition, it is probable that some TRH-LIr cells are lamina IIb islet cells (Coffield et al., 1984).

Ten substances (SP-, CCK-, SRIF-, VIP-, NT-, ENK-, NPY-, TRH-like immunoreactivities, ACH and GABA) are found in neurons in the nucleus proprius *(laminae III–IV)*. Lamina III contains a population of small interneurons whose somata have average diameters of 10–15 μm. Their dendritic arbors are generated by 2–3 primary dendrites that branch infrequently. The arbors extend slightly further rostrocaudally than mediolaterally, have a predominantly dorsal orientation, and are largely contained with lamina III, although sometimes terminal branches pass into lamina IIb. The axons issue terminal-bearing collaterals that, to date, have been found in laminae III–V. Intracellular HRP studies show that cells with these morphological characteristics (Fig. 5C) respond only to stimulation of Aβ low-threshold mechanoreceptors (Bennett et al., 1981b, and unpublished observations; Maxwell et al., 1983a). This morphologically and physiologically defined population of lamina III neurons appears to be neurochemically heterogeneous. Antisera to ENK, SP, NT, TRH, and ChAT reveal numerous lamina III neurons with morphological features that resemble those seen with the intracellular HRP method (Fig. 5; Bennett et al., 1981b; Glazer and Basbaum, 1981; Hunt et al., 1981b; Seybold and Elde, 1982; Barber et al., 1984; Coffield et al., 1984). The function of these lamina III neurons is unknown, but preliminary data suggest that they are inhibitory local circuit neurons involved in the generation of IPSPs triggered by Aβ afferent volleys (G. J. Bennett et al., unpublished observations). The nucleus proprius contains a distinctive population of neurons with prominent apical dendrites that penetrate and branch within laminae I–II and, sometimes, Lissauer's tract (Gobel et al., 1982; Maxwell et al., 1983a; G. J. Bennett et al., unpublished observations). It is probable that these cells correspond to the "antennae" cells described by Rethelyi and Szentagothai (1973). There is very good evidence that at least some of these cells contain SP and receive input from SP and serotonin afferents (Hunt et al., 1981b; Hoffert et al., 1982). The function of these neurons is not known. The nucleus proprius contains many somatosensory projection neurons (chiefly spinocervical tract and dorsal column postsynaptic tract neurons). There is no evidence that any of the substances immunocytochemically identified in the nucleus proprius are contained within these neurons.

Lamina V neurons have been identified as containing at least nine substances (SP-, CCK-, SRIF-, ENK-, NPY-, CRF-, DYN-like immunoreactivities, ACH and GABA). We know practically nothing about the morphology of these neurons. Large populations of somatosensory projection neurons lie in lamina V, but we know of no data that identify any of these projection neurons neurochemically.

The neurons of the *central canal region* contain

at least ten identified substances (SP-, CCK-, SRIF-, VIP-, ENK-, NPY-, AADC-, CRF-, DYN-like immunoreactivities and GABA). We know very little about this region of the spinal cord. It is known, however, that the region receives an input from myelinated nociceptors (Light and Perl, 1979) and that many of its cells project to the medullary reticular formation (Nahin et al., 1983).

Descending afferent input

Descending inputs have been shown to modulate sensory, motor and autonomic systems at the spinal level. Several cell groups in the brain stem as well as the hypothalamus and cortex contribute afferent axons to the spinal cord. Recent methodological advances, especially the technique of immunocytochemistry, have helped identify the neurotransmitters of some of these descending axons. Both monoaminergic and peptidergic descending axons originate from the brain stem. Some brain stem neurons which contain a monoamine neurotransmitter and project to the spinal cord have also been identified as containing a coexistent peptide neurotransmitter. The functional role of axons with coexistent neurotransmitters in spinal cord synaptic transmission is unclear.

This section will discuss the neurotransmitters of descending afferents to the spinal cord, their origin at supraspinal levels and termination in specific spinal cord laminae.

Serotonin

Serotonin (5-HT) is an important neurotransmitter involved in the descending control of nociception. Iontophoresis of 5-HT onto dorsal horn neurons usually leads to inhibition of the neurons' evoked responses, although excitation of some neurons has been observed in the deeper layers of the dorsal horn (Randic and Yu, 1976; Belcher et al., 1978; Headley et al., 1978; Jordan et al., 1979; Willcockson et al., 1984a). Some of the effects of 5-HT on dorsal horn neurons can be antagonized by methysergide, a 5-HT antagonist (Headley et al., 1978; Griersmith and Duggan, 1980). The effects of 5-HT in the dorsal horn are mimicked by nucleus raphe magnus stimulation. The usual effect of n. raphe magnus stimulation is inhibition (Beall et al., 1976; Fields et al., 1977; Belcher et al., 1978; Yezierski et al., 1982). However, some excitatory effects of n. raphe magnus stimulation have been seen in the deeper layers of the dorsal horn (Belcher et al., 1978; Yezierski et al., 1982). Although it is clear that the 5-HT neurons in the n. raphe magnus play a role in descending control, the non-5-HT-containing neurons in the n. raphe magnus are also likely to be involved (Bowker et al., 1981a,b). 5-HT inputs to the dorsal horn and n. raphe magnus afferents play a role in both opiate- and stimulation-produced analgesia, although neither appears to be solely responsible or essential (for review see Dubner and Bennett, 1983).

Origin of spinal serotonin axons

Essentially all of the 5-HT in the spinal cord originates from supraspinal sites. Four weeks following transection of the thoracic spinal cord in cats, only an occasional immunocytochemically stained 5-HT axon can be found in the lumbar cord. Eight weeks after transection, no 5-HT-like immunoreactivity (5-HT-LI) can be demonstrated (Fig. 7; Ruda et al., 1983). In the monkey, following treatment with L-tryptophan, a few neurons in the spinal cord appear to show 5-HT-LI (LaMotte et al., 1982). The role of these neurons in spinal cord 5-HT mechanisms is unknown.

Cell groups which contain 5-HT neurons are mainly localized in the brain stem (Dahlström and Fuxe, 1964; Steinbusch, 1981; Wiklund et al., 1981; Takeuchi, 1982a,b; Felton and Sladek, 1983; Jacobs et al., 1984). Studies by Bowker and his colleagues (1981a,b), in which spinally projecting 5-HT-LIr neurons were double-labelled by immunocytochemistry and retrograde HRP, demonstrate that 5-HT-LIr neurons in the B1–3, B7 and B9 cell groups project to the spinal cord. Some 5-HT-LIr cell groups project to all levels of the spinal cord while others, such as those in the rostral midbrain reticular formation, are more restricted and appear

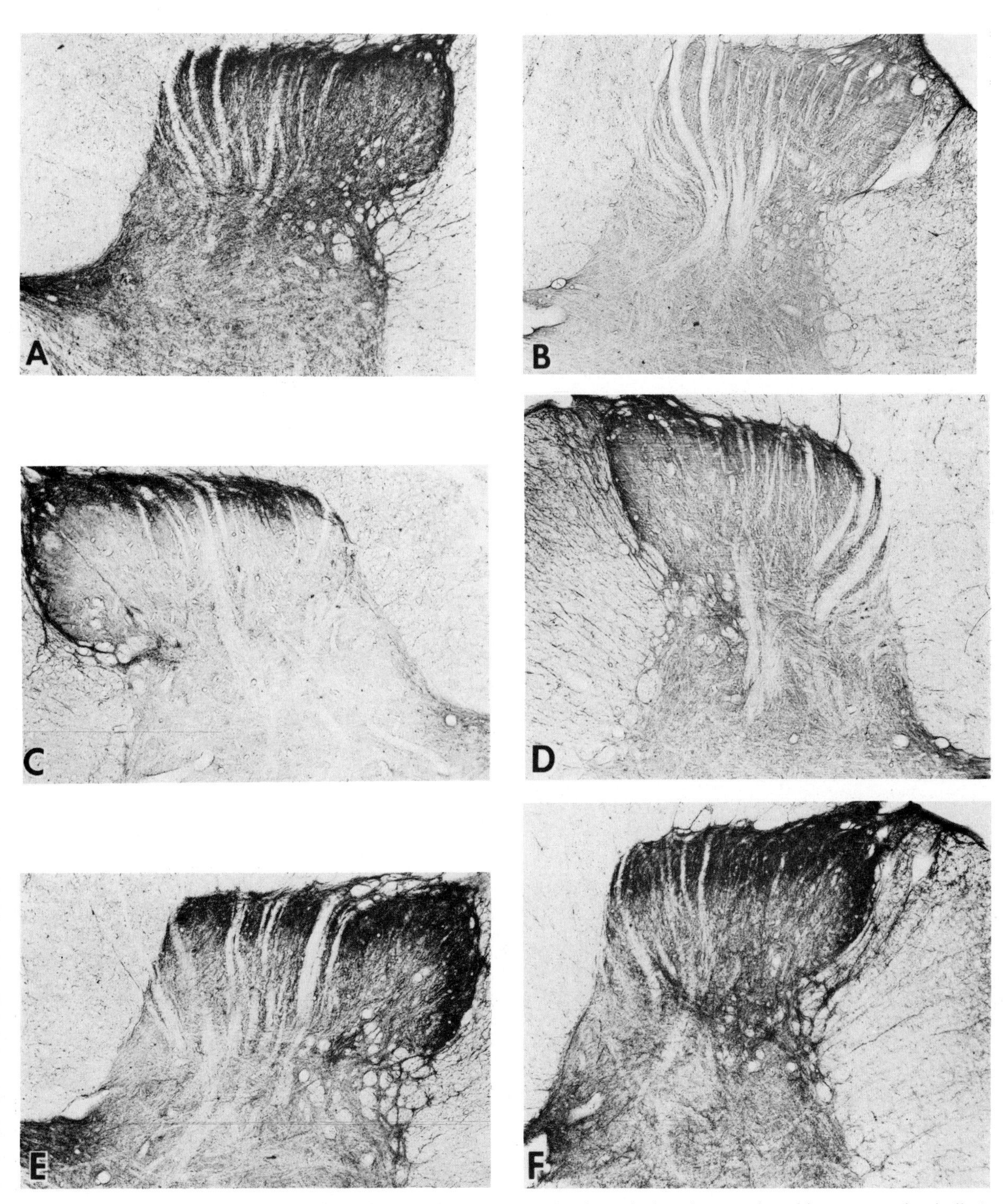

Fig. 7. The contribution of descending 5-HT, SP and ENK axons to cat lumbar spinal cord was evaluated immunocytochemically 8 weeks after transection of the thoracic spinal cord. A, C, and E are views of the cervical spinal cord processed for 5-HT, SP and ENK, respectively. B, D, and F represent 5-HT-, SP- and ENK-staining in the lumbar spinal cord of the same cat following transection. All 5-HT-LI has disappeared from the lumbar spinal cord 8 weeks following transection. There is perhaps a slight decrease in SP-LI in the dorsal horn while there is no noticeable decrease in ENK (maybe even a slight increase). Magnification: ×50.

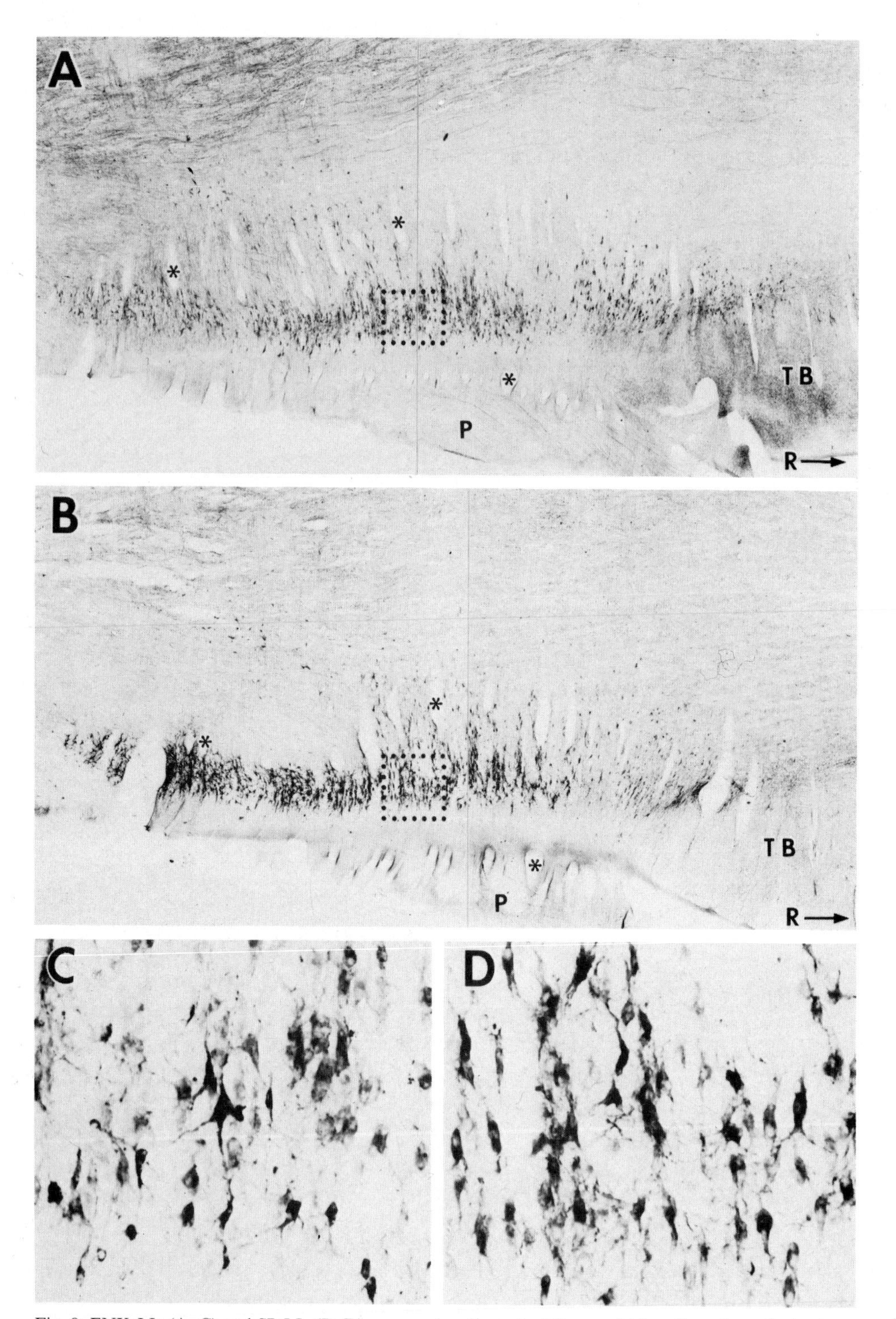

Fig. 8. ENK-LIr (A, C) and SP-LIr (B, D) neurons in adjacent midline sagittal sections from the brain stem of a cat pretreated with colchicine. Three corresponding blood vessels (asterisks) are marked in A and B for orientation. Neurons containing ENK-LI or SP-LI are distributed similarly over several millimeters rostrocaudally. The neurons lie above the pyramidal tract (P) and extend rostrally (R) to the level of the trapezoid body (TB). The areas outlined in A and B are shown at higher magnification in C and D, respectively. In the brain stem raphe nuclei, ENK-LI and SP-LI are found mainly in cell bodies and dendrites. A few varicose processes are also visible. The density of peroxidase–anti-peroxidase reaction product varies in individual neurons. Magnification in A and B: × 14, and C and D: × 113.

to project only to the cervical spinal cord (Bowker et al., 1981b).

The laminar termination of the different 5-HT-LIr cell groups is not clearly defined. However, evidence is accumulating to suggest that 5-HT neurons in the nucleus raphe magnus project to the dorsal horn via the dorsolateral funiculus (DLF) while nucleus raphe pallidus and oralis neurons terminate mainly in the ventral horn (Basbaum et al., 1978; Leichnetz et al., 1978; Martin et al., 1978; Watkins et al., 1980; Bowker et al., 1982b; Holstege and Kuypers, 1982; Hylden et al., 1985).

Funicular trajectory of serotonin axons

The location of 5-HT axons in the white matter of the spinal cord was first observed in the rat using the histofluorescence method (Dahlström and Fuxe, 1965). Recent studies have immunocytochemically localized 5-HT fibre tracts in the monkey (Kojima et al., 1983) and cat (Hylden et al., 1985). The 5-HT axons are found at the lumbar level throughout the white matter with two areas of dense concentration. A well-defined wedge-shaped fibre bundle occurs in the DLF. Axons from this area probably terminate in the dorsal horn. The other bundle of 5-HT axons occurs in the ventral and ventromedial white matter. Ventral horn 5-HT terminals probably originate from this fibre bundle. 5-HT axons in the white matter generally travel close to the pial surface of the spinal cord. The presence of two distinct 5-HT fibre bundles in the white matter suggests a separation in the funicular trajectory of 5-HT input to sensory and motor systems.

Serotonin axonal endings

5-HT axonal endings in the spinal cord were first identified using the Falk–Hillarp histofluorescence technique (Dahlström and Fuxe, 1965). More recently, using the immunocytochemical method, 5-HT-LIr axons have been examined by numerous investigators in the rat, cat, dog and monkey (Steinbusch, 1981; Ruda et al., 1982; Kojima et al., 1982, 1983; LaMotte and DeLanerolle, 1983b; Light et al., 1983; Maxwell et al., 1983b). 5-HT axonal endings are found in all laminae of the spinal cord. In the dorsal horn (Fig. 7) 5-HT-LIr axons are densest in laminae I, IIa and V while lamina IIb contains the fewest. Another concentration of 5-HT-LIr axons occurs near the central canal. In the ventral horn, 5-HT-LIr axons are densest in the motorneuron pools of lamina IX. At the level of the thoracic spinal cord, the densest concentration of 5-HT-LIr axons of any spinal location is found in the intermediolateral cell column. The primary orientation of 5-HT-LIr axons varies in the different laminae. In the dorsal horn, especially laminae I and II, the 5-HT-LIr axons can be followed for several hundred microns as they travel rostrocaudally. In the motorneuron pools of the ventral horn, the 5-HT-LIr axons have a more oblique trajectory. The size and shape of varicosities along a 5-HT immunoreactive strand are variable (range of 0.8–3.0 μm). Axonal strands with many large, thick varicosities are more typically encountered in the ventral horn than in the dorsal horn.

Coexistence of serotonin and peptides

5-HT-, SP-, ENK- and TRH-like immunoreactive neurons are found in large numbers in the same brain stem cell groups (see Fig. 8 for example). A subpopulation of these brain stem neurons appears to contain a coexistent monoamine and peptide. Some of these neurons may project to the spinal cord. It is also possible that some peptidergic raphe-spinal neurons do not contain a monoamine. The laminar termination of descending peptide and coexistent neurotransmitters is a key question in the neurochemical organization of the spinal cord since the same peptide (i.e., SP) may originate from either descending axons, intrinsic neurons or primary afferent axons. Identification of the cells of origin of the peptide would help clarify its functional role in dorsal horn neural circuits. Double-label experiments employing immunocytochemistry and retrograde transport demonstrate that some ENK-, SP- and TRH-like immunoreactive neurons in the raphe complex project to the spinal cord (Hökfelt et al., 1979b; Bowker et al., 1982a). Using the 5-HT neurotoxins 5,6- or 5,7-dihydroxytryptamine, a loss

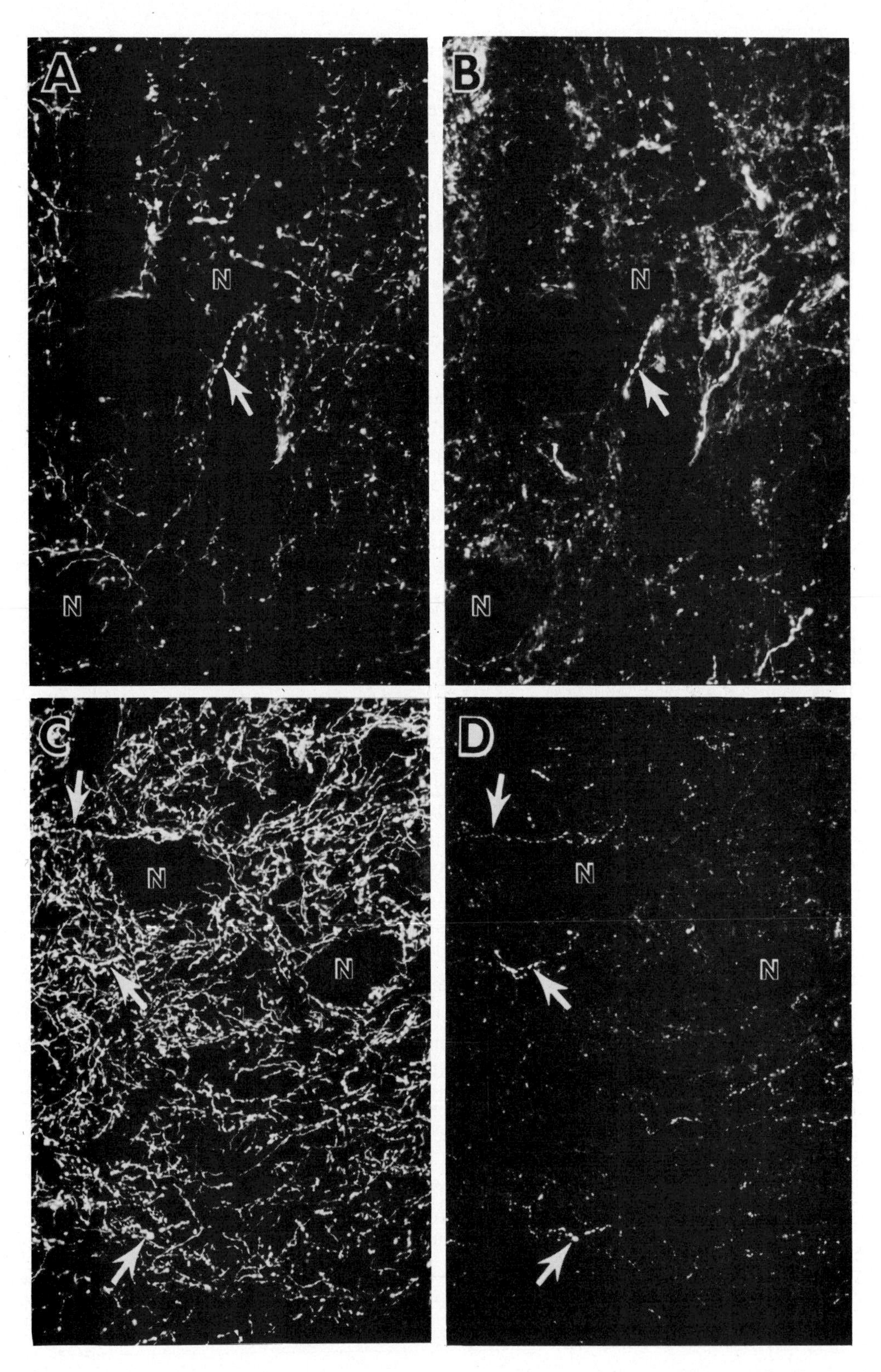
A
N
N
B
N
N
C
N
N
D
N
N

of 5-HT-, SP- and TRH-LI in the spinal cord has been shown immunocytochemically and by radioimmunoassay (Hökfelt et al., 1978; Björklund et al., 1979; Singer et al., 1979; Johansson et al., 1981; Gilbert et al., 1982). A total loss of TRH-LI in the ventral horn was observed (Johansson et al., 1981; Gilbert et al., 1982). Almost all of the ventral horn SP-LI also disappeared while only a 25% decrease of SP-LI was found in the dorsal horn (Gilbert et al., 1982). These observations suggest that brain stem 5-HT neurons which contain either SP- or TRH-LI distribute preferentially to the ventral horn rather than the dorsal horn and are thus likely to be involved in motor function. These observations are supported by immunocytochemical localization of peptides following transection of the spinal cord. Spinal cord transection results in an almost complete loss of SP-LI in the ventral horn (Hökfelt et al., 1977b; Kanazawa et al., 1979; M. A. Ruda and G. J. Bennett, unpublished observations). In contrast, only a slight reduction of SP-LI in the dorsal horn can be observed (Fig. 7C, D). ENK immunocytochemical staining following spinal transection exhibits no detectable diminution in the dorsal horn (Fig. 7E, F) while some reduction is apparent around the motorneurons in lamina IX (M. A. Ruda and G. J. Bennett, unpublished observations). Descending ENK inputs to the spinal cord thus represent only a small part of the ENK-LIr staining one observes and it is more likely to be related to the modulation of motor than sensory functions.

The course of descending peptidergic axons in the spinal white matter has been difficult to determine. Tissue sections, in which 5-HT axons in the DLF and ventral funiculi are well stained, display few if any immunoreactive SP or ENK axons within the white matter (M. A. Ruda, unpublished observations). However, following knife cuts in the medulla or cervical spinal cord, Pickel et al. (1983) reported some ENK-LIr and SP-LIr axons in the ventrolateral funiculus extending into the lateral funiculus. Immunofluorescence studies of double-stained 5-HT and SP axons do find occasional 5-HT-LIr axons in the DLF and ventral funiculi which also display SP-LI (T. Tashiro and M. A. Ruda, unpublished observations).

Recent experiments, using double immunocytochemical staining of individual axons, directly demonstrate coexistent 5-HT and peptide axon terminals in the spinal cord (Fig. 9; Wessendorf and Elde, 1984; Tashiro and Ruda, 1985). Most varicose axonal strands which contain 5-HT-LI and SP-LI are found in the ventral horn in association with motorneurons (Fig. 9C, D). However, some coexistent 5-HT- and SP-LIr axons are found in all laminae of the spinal cord. In the dorsal horn, they represent only a small percentage of either 5-HT- or SP-LIr axons (Tashiro and Ruda, 1985). Based on these observations, it appears that the brain stem raphe neurons which contain 5-HT with coexistent SP-LI or ENK-LI or TRH-LI are involved mainly in motor rather than sensory function.

Noradrenaline

Noradrenaline (NR) appears to be involved in analgesic mechanisms since the responses of dorsal horn neurons to noxious stimulation can be inhibited by iontophoretic application of NA (Belcher et al., 1978; Headley et al., 1978). In addition, this aminergic neurotransmitter appears to be critical

Fig. 9. Coexistence of 5-HT-LI (A, C) and SP-LI (B, D) in the same axons in laminae V (A, B) and IX (C, D) of the cat lumbar spinal cord. Antisera to 5-HT and SP were raised in different species and identified in single tissue sections using either a fluorescein-isothiocyanate or TRITC (rhodamine) conjugated secondary antiserum specific for the species in which the primary antiserum was raised. A and C were photographed with a fluorescent filter combination specific for rhodamine fluorescence. B and D were photographed with a narrow band fluorescein-isothiocyanate filter combination which prevents rhodamine fluorescence from being visible. Some axons double-labelled for 5-HT-LI and SP-LI are indicated by arrows. The same neuronal cell bodies (N) in A and B, and C and D, are marked for orientation. The density of axons with coexistent 5-HT-LI and SP-LI is substantially greater in lamina IX than in lamina V. (Illustrations from T. Tashiro and M. A. Ruda, unpublished observations.)

for opiate-induced analgesia. Alpha-adrenergic blockers (phentolamine or phenoxybenzamine), when given intrathecally, attenuate the analgesia produced by morphine (Proudfit and Hammond, 1981) or by electrical stimulation of the magnocellular tegmental field (Yaksh et al., 1981).

NA effects in the spinal cord are not limited to pain mechanisms and sensory function. The dense NA fibre plexi in the thoracic intermediolateral cell column and sacral preganglionic parasympathetic cell column suggest involvement in autonomic functions. In the ventral horn, a dense network of NA axons occurs in the motorneuron cell groups, indicating a role for NA in movement.

Catecholamine terminals in the spinal cord appear to arise exclusively from supraspinal sources (Dahlström and Fuxe, 1965; Glazer and Ross, 1980). In the brain stem, neurons in the A5, A6 and A7 groups can be labelled by retrograde transport of antibodies to dopamine-β-hydroxylase (DBH), an enzyme involved in the synthesis of NA (Westlund et al., 1981), or double-labelled with retrograde HRP and DBH-like immunoreactivity (Westlund and Coulter, 1980; Westlund et al., 1982). It is likely that these three cell groups represent the major sources of spinal cord NA, although which cell groups contribute the majority of spinal afferents is unclear (Stevens et al., 1982). The catecholamine neurons in the A1 and A2 cell groups were once thought to project to the spinal cord. However, recent studies have shown that it is the noncatecholaminergic neurons in these cell groups that project to the spinal cord (McKellar and Loewy, 1982).

Coexistent neurotransmitters have been identified in some catecholamine cell groups. In cat locus coeruleus, neurons identified as catecholaminergic by the presence of tyrosine hydroxylase (TH)-like immunoreactivity (TH-LI) were found in double-label experiments to be immunoreactive for ENK (Charnay et al., 1982). In rat, avian pancreatic polypeptide (NPY) was localized to catecholamine-containing neurons (Hunt et al., 1981a). It is not known if these coexistent catecholamine- and peptide-containing neurons contribute afferents to the spinal cord.

The location of NA axons in the spinal cord has been studied with the histofluorescence technique (Dahlström and Fuxe, 1965) and immunocytochemistry (Glazer and Ross, 1980; Westlund and Coulter, 1980) using antisera to DBH. NA axons descend in the dorsal and ventral part of the lateral funiculus (Dahlström and Fuxe, 1965). NA axons can be localized to all laminae of the spinal cord. The most dense terminal labelling is found in the ventral horn, especially in association with motorneurons. A particularly dense plexus of NA terminals occurs in the intermediolateral cell column at thoracic levels of the spinal cord. In the sacral spinal cord, the preganglionic parasympathetic cell column is also heavily innervated. In the dorsal horn, NA axon terminals are particularly prominent in lamina I, the outer part of lamina II and lamina V.

Anterograde tracing experiments give some indication of the cell groups whose NA neurons project to the different spinal cord laminae (Westlund and Coulter, 1980). The A6 and A7 cell groups project most densely to the ventral horn, especially to the motorneuron groups. The A6 and A7 cell groups also contribute afferents to laminae I and IIa, IV–VI and X. Neurons within A6 (n. locus coeruleus) have a dense projection to the parasympathetic preganglionic cell column in the sacral spinal cord. Neurons within A7 (n. subcoeruleus and the medial parabrachial nucleus) contribute axons to the intermediolateral cell column at thoracic spinal cord levels.

Dopamine

Dopamine (DA)-containing neurons in the A11 cell group of the diencephalon have been shown to descend to the spinal cord (Björklund and Skagerberg, 1979; Hökfelt et al., 1979a). The terminals of DA axons are found mainly in the dorsal horn and lamina X at all levels of the spinal cord (Skagerberg et al., 1982). Some DA axons are found in the ventral horn. The intermediolateral cell column of the thoracic spinal cord exhibits a plexus of DA fluorescence clearly denser than that found in any other part of the spinal cord. This terminal site suggests an important involvement of DA in autonomic regulatory processes. The role of DA axons in the dorsal horn is unclear, but their presence suggests an involvement in sensory processing. Iontophoresis of DA has been shown to inhibit the responses of spinothalamic tract neurons (Willcockson et al., 1984a). However, DA axons represent only a minor component of descending catecholaminergic modulation since the NA innervation of the dorsal horn has been shown to be approximately 10-fold greater than the DA innervation (Skagerberg et al., 1982).

Adrenaline

The contribution of adrenaline-containing axons to the spinal cord has received limited study. Adrenaline-containing neurons in the C1 cell group of the medulla have been shown to project to the spinal cord at the thoracic level (Ross et al., 1981). Adrenaline neurons were characterized as those which contain the enzyme phenylethanolamine-*N*-methyltransferase (PNMT), which converts NA to adrenaline. In the spinal cord, PNMT-like immunoreactive axons have only been observed in the intermediolateral cell column of the thoracic cord (Hökfelt et al., 1974). Adrenaline at the spinal level thus appears to be primarily associated with autonomic function.

Hypothalamospinal afferents

A direct hypothalamospinal projection, arising largely from the paraventricular nucleus (PVN), travels to all levels of the spinal cord (Kuypers and Maisky, 1975; Saper et al., 1976; Sawchenko and Swanson, 1982). PVN neurons that contain vasopressin (VP)-like immunoreactivity (VP-LI), oxytocin (OT)-like immunoreactivity (OT-LI) and their respective neurophysins have been observed to send afferents to the spinal cord. A very small number of the MET-ENK and SRIF-LIr neurons in the PVN also appear to project to the spinal cord (Sawchenko and Swanson, 1982).

Intraspinal OT-LI and VP-LI originate largely, if not entirely, from the PVN (Lang et al., 1983; Millan et al., 1984). Their distributions appear similar, although OT-LI is present in greater amounts. Axons containing OT-LI and VP-LI descend in the DLF to terminate in the superficial dorsal horn and the intermediolateral cell column at thoracic and sacral levels. Terminals in lamina X may originate from the DLF bundle or from axons descending through the central canal region (Swanson and McKellar, 1979). The fibres in lamina I have a pronounced rostrocaudal orientation and issue side-branches to lamina II. The input to lamina II has been described as patchy (Gibson et al., 1981) and, like the input to lamina III, it is very sparse. There are some axons in lateral lamina V and the homologous area in the medullary dorsal horn (Morris et al., 1980; Gibson et al., 1981).

OT-LIr and VP-LIr afferents to the spinal cord are few in number when compared with other identified afferents. They are probably related to sympathetic and parasympathetic preganglionic neurons, indicating a direct hypothalamic influence on autonomic function. Their inputs to laminae I and II suggest that the hypothalamus may influence somatosensory function.

Corticospinal afferents

The location of corticospinal afferents in the spinal cord has been examined by numerous investigators using a variety of techniques. A recent study by Cheema et al. (1984) is the first to demonstrate the presence of a small contingent of corticospinal terminations in laminae I and II. Corticospinal tract afferents from somatosensory cortex terminate in laminae I–VII while fibres from motor cortical areas terminate in VI–IX. Corticospinal tract axons descend in the dorsolateral funiculus and terminate mainly on the contralateral side. At the ultrastructural level, the terminals of corticospinal tract axons in laminae I and II are small, dome-shaped, and contain agranular spherical or pleomorphic vesicles and an occasional dense core vesicle. Symmetrical and asymmetrical synapses are mainly axodendritic and never axoaxonic. Corticospinal fibres in laminae I and II thus exert a direct postsynaptic control on dorsal horn neurons. The neurotransmitter employed by corticospinal axons is unknown.

Conclusions

Brain stem neurons provide a neurochemically diverse afferent input to the spinal cord which includes nine different substances (5-HT, NA, DA, adrenaline and ENK-, SP-, TRH-, OT- and VP-like immunoreactivities). The monoaminergic cell groups contribute inputs to spinal cord laminae involved in sensory, motor and autonomic function. A dense plexus of 5-HT axons is found in all laminae of the spinal cord, especially around the motorneurons of laminae IX and the autonomic nuclei in the thoracic and sacral spinal segments. NA axons are fewer in number although similarly distributed. Some DA and adrenaline axons are likewise present. Essentially all of the monoaminergic axons in the spinal cord descend from brain stem neurons via the dorsolateral and ventral funiculi.

Coexistence of 5-HT and a peptide (i.e., SP, TRH, ENK) in neurons which contribute afferents to the spinal cord has been observed. The majority of these afferents appear to terminate in the ventral horn, especially in association with motorneurons.

Little ultrastructural information on descending, neurochemically defined afferents is available except in the case of 5-HT. 5-HT axons in the dorsal horn commonly form symmetrical axodendritic synapses on the dendrites of intrinsic neurons. Little anatomical evidence for axoaxonic relationships has been observed. 5-HT in the dorsal horn thus acts mainly through synapses on dorsal horn neurons.

In the dorsal horn, both 5-HT and NA act predominantly through inhibition of intrinsic neurons. Each monoamine is involved to a variable degree in both opiate and non-opiate analgesic mechanisms.

Ultrastructure of neurochemically identified elements in the dorsal horn

Few of the neurochemically identified elements in the dorsal horn have been examined at the ultrastructural level. Most studies have focused on 5-HT, ENK and SP, while a few have looked at GABA, NT, SRIF and VIP. This section will describe the morphology and synaptic connectivity of these elements in the superficial laminae of the dorsal horn.

Enkephalin

The ultrastructure of ENK-LIr profiles in the superficial laminae has been examined in the rat (Hunt et al., 1980; Sumal et al., 1982), cat (Glazer and Basbaum, 1983; Ruda et al., 1984) and monkey (Aronin et al., 1981; LaMotte and DeLanerolle, 1983a; Ruda et al., 1984). In each species, ENK-LIr axons synapse primarily on dendrites and occasionally on cell somata. Most ENK-LIr synapses are asymmetrical, although symmetrical synapses are also present. The synaptic specializations are often small (0.1–0.3 μm; Glazer and Basbaum, 1982). ENK-LIr boutons are dome-shaped, vary in length from less than 1 μm to greater than 3 μm, and contain a mixture of spherical clear vesicles and a few dense core vesicles. There is little morphological evidence for ENK involvement in axoaxonic interactions (in which the ENK axon is presynaptic). Double-label studies in which ENK-LIr terminals and primary afferent terminals were identified, provide no evidence for synaptic interaction (Hunt et al., 1980; Sumal et al., 1983). In single label studies, ENK-LIr profiles presynaptic to a central terminal (of presumed primary afferent origin) have been observed only rarely (Aronin et al., 1981) or not at all (Glazer and Basbaum, 1983). There is thus little evidence for a morphological correlate of presynaptic opioid modulation of primary afferent axons. However, two other types of ENK relationships with possible central terminals of primary axons have been described. First, vesicle-containing ENK-LIr dendrites have been observed postsynaptic to large central terminals (Fig. 10; Bennett et al., 1982; Glazer and Basbaum, 1983), although, even after serial sectioning, the vesicle-containing ENK-LIr dendrite was not observed to synapse on the central terminal. The second type of interaction is characterized by a labelled (degenerating) primary afferent terminal synapsing on an unlabelled dendrite which, in turn, received a synapse from a vesicle-containing ENK-LIr profile (Sumal et al., 1982). These observations support a postsynaptic role for ENK modulation that is in close proximity to the primary afferent terminal itself.

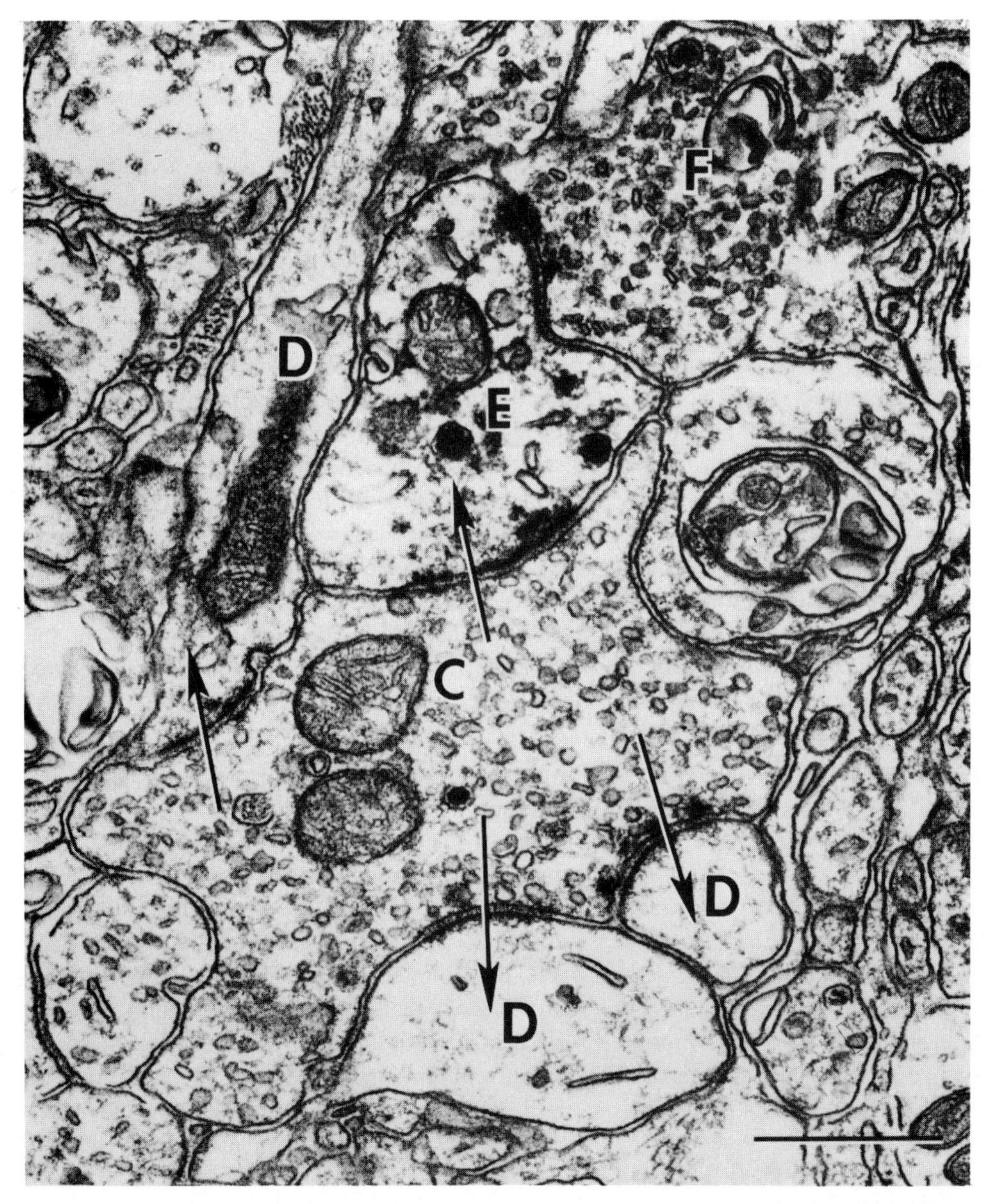

Fig. 10. Ultrastructural analysis of the synaptic relationships of an ENK-LIr dendrite in the superficial laminae of cat lumbar dorsal horn. The ENK-LIr dendrite (E) contains a few dense core vesicles and agranular reticulum which are associated with the peroxidase–anti-peroxidase reaction product. The ENK-LIr dendrite receives an asymmetrical synapse (arrow) from a central ending (C) which also synapses (arrows) on three unlabelled dendrites (D). The central ending resembles those which are of primary afferent origin. The ENK-LIr dendrite is also postsynaptic to an axonal ending (F) which contains flattened vesicles. Magnification: × 52 650. Scale bar represents 0.5 μm.

Substance P

The ultrastructure of SP-LIr axons has been investigated in laminae I and II of the dorsal horn of the rat (Barber et al., 1979; Priestley et al., 1982), cat (Ruda, 1985) and monkey (DeLanerolle and LaMotte, 1983). Only unmyelinated SP-LIr axons (ca. 0.2 μm in diameter) are encountered. The SP-LIr axonal varicosities display different morphological characteristics. Many SP-LIr axons are dome-shaped and form a single synapse. Others are the central endings of glomeruli and are probably of primary afferent origin. The central endings are surrounded by several different neuronal profiles which are impressed into their surface, imparting a characteristic scalloped shape. Synaptic contacts are typically asymmetrical. Both the dome-shaped and scalloped-shaped SP-LIr varicosities contain a mixture of oval agranular vesicles and dense core vesicles, but the number of dense core vesicles is generally greater in the central ending. SP-LIr axon terminals have not been found presynaptic to other axons. However, there is some evidence (Priestley et al., 1982; DeLanerolle and LaMotte, 1983) that scalloped-shaped SP-LIr terminals are postsynaptic to what are probably vesicle-containing dendrites. These observations emphasize a role for SP in post-synaptic interactions in the superficial dorsal horn.

Other peptide-containing dorsal horn axons

SRIF-LIr axon terminals in lamina II of the dorsal horn are typically large, often greater than 5 μm in length (Fig. 2). Numerous dendrites are impressed into the terminal's perimeter and receive asymmetrical synapses from the SRIF-LIr axon. In addition to the round agranular vesicles, numerous dense core vesicles are also present. A subpopulation of the agranular and dense core vesicles are associated with the immunocytochemical reaction product. These large SRIF-LIr central endings resemble the central terminals of primary afferent axons.

The ultrastructure of VIP-LIr axons has been examined in the cat (Fig. 3B; Honda et al., 1983). Only unmyelinated axons in the dorsal horn and Lissauer's tract contain VIP-LI. VIP-LIr axon terminals present a somewhat unique morphology both with respect to size (often measuring 4–5 μm in length) and the large number of dense core vesicles which they contain. Only a small subpopulation of the numerous dense core vesicles are associated with the immunocytochemical reaction product, suggesting that the other dense core vesicles contain other neurochemicals. Synaptic sites are encountered on only a small portion of the varicosity's perimeter. The synaptic contacts are usually asymmetrical and typically occur on a dendritic shaft.

The ultrastructure of NT-LIr axonal endings has been examined in monkey (DiFiglia et al., 1984). They originate mainly from unmyelinated axons, although occasional thinly myelinated axons are also observed. The dome-shaped NT-LIr axonal endings vary in length from 1 to 3 μm and contain numerous round clear vesicles and a few large granular vesicles. Asymmetrical synaptic contacts, some with a prominent subsynaptic density, occur on cell somata and dendrites in laminae I and II of the dorsal horn. In lamina II, some NT-LIr endings resemble the central endings of glomeruli.

Monoamines: serotonin and noradrenaline

Studies of the ultrastructure of 5-HT-LIr axon terminals have focused on the rat (Maxwell et al., 1983b), cat (Ruda et al., 1982; Light et al., 1983) and monkey (LaMotte and DeLanerolle, 1983b). 5-HT-LIr axonal endings are generally dome-shaped and typically form a single synapse with an adjacent neuronal profile. The 5-HT-LIr axonal endings contain mainly round or oval agranular vesicles. Some endings contain pleomorphic agranular vesicles and some of these are highly flattened. A variable number of dense core vesicles are also present. The immunocytochemical reaction product is associated with both the agranular and dense core vesicles. Most 5-HT-LIr synapses occur on small calibre (ca. 1 μm) dendritic shafts, although synapses on somata and dendritic spines are also found. The synaptic specialization is small and most frequently of the symmetrical type. Although non-synaptic mechanisms of action for 5-HT have been proposed (Maxwell et al., 1983b), the frequency of ultrastructurally identified, conventional synaptic junctions between a presynaptic 5-HT-LIr axonal ending and a postsynaptic dendritic profile (Ruda et al., 1982; LaMotte and DeLanerolle, 1983b; Light et al., 1983) argues for a predominant synaptic mode of action.

Due in part to technical limitations, NA axons in the dorsal horn have received little attention at the ultrastructural level (Senba et al., 1981; Satoh et al., 1982). Studies have relied on the potassium permanganate fixation technique to mark NA axons. NA axons most commonly contact small calibre dendritic shafts. In one study in which primary afferent axons were marked by degeneration, no NA terminals were found in contact with the degenerating axons (Satoh et al., 1982). These findings suggest that NA afferents act mainly through effects on dorsal horn neurons.

Gamma-aminobutyric acid

The ultrastructure of GABA profiles has been examined in the rat using antisera to GAD (McLaughlin et al., 1975; Barber et al., 1978). GAD-LIr profiles are presynaptic to dendrites, cell somata and axon terminals in laminae I and II. They can be divided into two groups based on vesicle morphology. One type contains round synaptic vesicles and makes asymmetrical synaptic contacts. The other, more numerous type, contains pleomorphic or flattened vesicles and makes symmetrical synaptic contacts. Following dorsal rhizotomy, GAD-LIr terminals are found presynaptic to degenerating primary afferent terminals. These observations suggest a role for GABA in both postsynaptic and presynaptic modulation of dorsal horn neuronal circuits.

Conclusions

The immunocytochemical technique has provided an opportunity to examine the ultrastructure of neural elements which contain an identified neurochemical. These studies indicate that morphological criteria are not sufficient to identify the neurochemical contained within an axon since axons containing the same neurochemicals are often morphologically heterogeneous, whereas axons containing different neurochemicals may have the same morphology. For example, virtually all the neurochemically identified axons in the superficial laminae have a subpopulation of axon terminals which are dome-shaped, contain mainly oval agranular vesicles, an occasional dense core vesicle and form axodendritic synapses. On the other hand, axonal endings, such as those which contain SP-LI, may vary from small dome-shaped terminals to large central terminals within glomeruli.

All the neurochemically identified axonal endings examined to date form membrane specializations which conform to the conventional morphological criteria of a synapse. Thus, modulation of the output of information from the dorsal horn occurs in large part through synaptic mechanisms.

Little morphological evidence is available for presynaptic modulation of primary afferent axons. GABA is the only neurochemical for which a presynaptic role has been observed. Of the peptides, only ENK profiles are associated with primary afferent terminals. However, in this instance it appears that ENK vesicle-containing dendrites (probably from the ENK IIb islet cell) are postsynaptic to scalloped-shaped central terminals that are thought to be of primary afferent origin.

SP and SRIF have been identified as the central terminals within glomeruli. Since most central endings probably originate from primary afferent axons, this observation suggests a role for SP and SRIF in primary afferent input to the dorsal horn. However, we cannot conclude that all primary afferent axons form only central terminals. VIP-LI in lamina I almost certainly is primary afferent in origin and its axon terminals are large, bulbous, dome-shaped varicosities rather than scalloped-shaped central terminals.

Neurochemically identified components of dorsal horn circuitry

Multiple-labelling techniques allow one to examine the direct interaction of neurochemically defined axon terminals with projection neurons and local circuit neurons in the spinal cord dorsal horn. For example, in our laboratory we have combined the retrograde HRP method with immunocytochemistry. In this technique, HRP is injected into the site of axonal termination of the dorsal horn neurons (e.g., thalamus, brain stem), transported back to the soma, and then visualized by histochemical methods. Tissue sections containing the soma and proximal dendrites (distal dendrites are not labelled by this method) of projection neurons are subsequently immunocytochemically labelled to identify axonal varicosities. Sites of apposition between the retrogradely labelled cells and immunoreactive axonal varicosities can be visualized at the light microscope level and synaptic relationships can be verified at the electron microscope level. We have also combined the intracellular HRP method with immunocytochemistry. The intracellular HRP method results in a Golgi-like appearance of the neuron's dendritic arbor and axon and permits an analysis of the distribution of immunoreactive axonal terminals on the soma and dendrites. In the following discussion of neurochemically defined spinal circuitry, we will focus our attention mainly on SP, ENK and 5-HT since their roles in dorsal horn circuitry have been examined in depth.

Substance P

Microiontophoresis of SP excites spinothalamic tract neurons (Willcockson et al., 1984b) and other, unidentified dorsal horn neurons (Henry, 1976; Randic and Miletic, 1977). High intensity peripheral nerve stimulation releases SP-LI from the spinal cord (Yaksh et al., 1980). SP-LIr contacts have been found on lamina I projection neurons (Priestley and Cuello, 1983). However, it cannot be assumed that these contacts originate from primary afferents since SP-LI is found in some intrinsic neurons and in descending axons. Electron microscope data indicate that many SP-LIr terminals in monkey and cat are scalloped-shaped central terminals of glomeruli, suggesting that they are the terminal boutons of primary afferents (DiFiglia et al., 1982; DeLanerolle and LaMotte, 1983; Ruda, 1985). Since dense SP-LI staining occurs in dendritic spines in the superficial laminae (Hoffert et al., 1982) and many synaptic contacts in primary afferent glomeruli occur on spines, it is likely that SP-LIr contacts on the spines of lamina I neurons are primary afferent in origin.

SP-LIr contacts have also been reported on lamina I local circuit neurons, lamina II stalked cells and lamina IV neurons (Hoffert et al., 1982 and unpublished observations). Fig. 11 is a camera lucida drawing of a lamina II stalked cell intracellularly filled with HRP. The tissue section was subsequently labelled immunocytochemically for SP. Most of the SP-LIr contacts occur on dendritic spines.

Two lines of evidence suggest that SP primary afferents directly contact ENK-LIr lamina I and II local circuit neurons. First, in lamina II, ENK-LIr dendritic spines are postsynaptic to scalloped-shaped central endings that resemble the central terminals of SP-containing primary afferents (Glazer and Basbaum, 1982; LaMotte and DeLanerolle, 1983a). Second, all local circuit neurons in laminae I and IIa appear to receive SP-LIr contacts (Hoffert et al., 1982); a subpopulation of them contains ENK-LI (see below). SP-LIr primary afferents also may synapse on lamina III SP-LIr local circuit neurons (Hunt et al., 1981b). Table I summarizes some of the established connections between SP and identified dorsal horn neurons.

Although SP excites dorsal horn neurons, it does not appear to act as a conventional neurotransmitter that changes ionic conductance with resulting membrane depolarization. Manipulations that reduce SP-LI levels in the dorsal horn, such as peripheral nerve section or capsaicin treatment, do not eliminate dorsal horn activity evoked by peripheral nerve electrical stimulation (Wall et al., 1981). SP appears to mediate a slow depolarization of dorsal

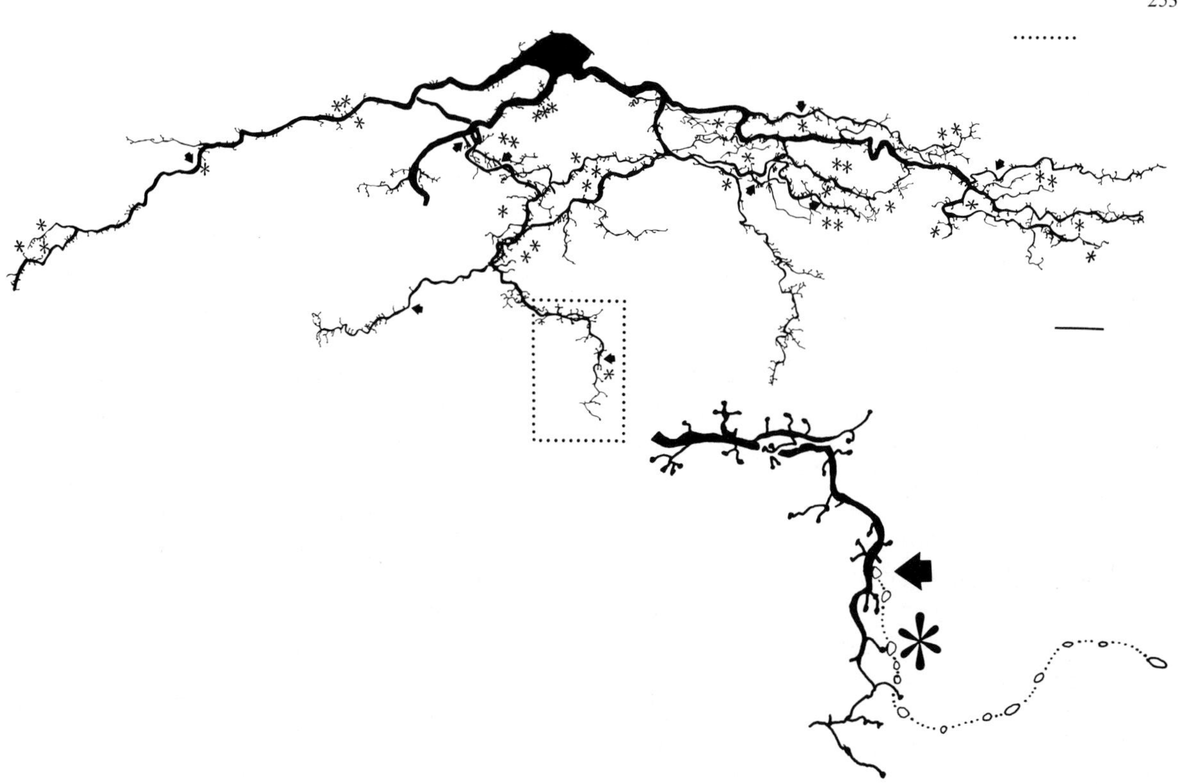

Fig. 11. Camera lucida drawing of a lamina II stalked cell labelled with the intracellular HRP method. The tissue section was subsequently stained with SP antisera. Arrows represent sites of SP-LIr contacts on dendritic shafts; asterisks denote sites of SP-LIr contacts on spines. Note the predominance of contacts on spines. Inset is an enlargement of the area outlined in dots and shows a SP-LIr axon with its varicosities contacting a spine and the dendritic shaft. Dotted horizontal line represents the laminae I–II border (from M. J. Hoffert et al., unpublished observations). Scale bar represents 10 μm.

TABLE I

Neurochemicals innervating identified dorsal horn neurons

Histochemically characterized afferents	Projection neurons			Physiologically and morphologically characterized neurons				Histochemically characterized neurons	
	LI TPN	LV TPN	DCPS	LI cells	LII stalked cells	LII islet cells	LIV cells	ENK LI & II neurons	SP LIII neurons
5-HT	C[a], M[a,b]	C[a], M[a,b]	(C, M)[c]	C[d,e,h]	C[d,e]	C[d,e,*]	C[h]	C[f]	–
SP	R[g]	–	–	C[h]	–	–	C[h]	–	R[m]
ENK	(C, M)[i]	C[i,j], M[i]	C[n]	–	–	–	–	C[k], M[l]	–

TPN, thalamic projection neurons; DCPS, dorsal column postsynaptic neurons; LI, lamina I; LII, lamina II; LV, lamina V; C, cat; M, monkey; R, rat.

* Islet cells receive significantly fewer contacts than either marginal or stalked cells.

[a] Ruda et al. (1983a); [b] Ruda and Coffield (1983); [c] Nishikawa et al. (1984); [d] Hoffert et al. (1983); [e] Miletic et al. (1984); [f] Glazer and Basbaum (1984b); [g] Priestley and Cuello (1983); [h] Hoffert et al. (1983); [i] Ruda et al. (1984); [j] Ruda (1981); [k] Glazer and Basbaum (1984a); [l] Aronin et al. (1981); [m] Hunt et al. (1981b); [n] unpublished observations.

Adapted from Ruda (1985).

horn neurons (Murase and Randic, 1984; Urban and Randic, 1984) and presumably enhances the effectiveness of a conventional but unknown chemical mediator.

Enkephalin

Retrogradely labelled trigeminothalamic and spinothalamic tract neurons receive synaptic contacts from ENK-LIr axon terminals in cat and monkey in laminae I and V (Ruda, 1982; Ruda et al., 1984; Ruda, 1985). ENK-LIr contacts are found on approximately one-third of lamina I thalamic projection neurons. In lamina V, more than 50% of the thalamic projection neurons receive ENK-LIr contacts. The large multipolar cells often have the greatest density of contacts with bipolar cells exhibiting less ENK-LIr innervation. In laminae I and V, some multipolar and bipolar neurons do not have ENK-LIr contacts, indicating that these morphological subtypes are heterogeneous with respect to their neurochemical input. Thalamic projection neurons in other laminae (VI, VII, VIII and X) receive few ENK-LIr contacts. This finding suggests that if such projection neurons receive an ENK-LIr input, then it must occur on the more distal dendrites not labelled by the retrograde HRP method.

Ultrastructural analysis has confirmed that the ENK-LIr contacts viewed at the light microscopic level are sites of synaptic interaction (Ruda, 1982, 1985; Ruda et al., 1984). Thalamic projection neurons which receive an ENK-LIr input proximally are also often contacted by unlabelled varicosities with similar morphological characteristics. Thus, many thalamic projection neurons receive input from multiple neurotransmitters whose identity cannot be ascertained by morphological criteria alone.

Dorsal column postsynaptic (DCPS) neurons (Bennett et al., 1983), whose axons terminate mainly in the dorsal column nuclei, have also been shown to receive ENK-LIr contacts on their somata and proximal dendrites using the combined techniques of retrograde HRP and immunocytochemistry (G. J. Bennett et al., unpublished observations). The DCPS projection system includes low-threshold mechanoreceptive neurons and wide dynamic-range neurons in about equal numbers (Lu et al., 1983), suggesting that opioid modulation involves non-nociceptive as well as nociceptive long projection output pathways from the dorsal horn.

The demonstration of direct ENK-LIr contacts suggests that opioid peptides act via postsynaptic mechanisms on dorsal horn neurons. This evidence is supported by iontophoresis studies in which glutamate-evoked excitation of dorsal horn neurons is antagonized by ENK (Zieglgänsberger and Tulloch, 1979; Willcockson et al., 1984b). ENK has no detectable effect on the resting membrane potential or membrane resistance of spinal neurons studied in vivo or in tissue culture (Barker et al., 1978; Zieglgänsberger and Tulloch, 1979) although it produces membrane hyperpolarization of dorsal horn neurons in tissue slice preparations (Werz and MacDonald, 1983; Yoshimura and North, 1983). Thus, ENK in the spinal cord may act either as a conventional neurotransmitter and produce changes in ionic conductance or exert a neuromodulating effect and alter the effectiveness of other chemical mediators.

Ultrastructural studies indicate that the vast majority of ENK-LIr terminals are presynaptic to dendrites in the dorsal horn (Hunt et al., 1980; Aronin et al., 1981; Sumal et al., 1982; Arluison et al., 1983; Glazer and Basbaum, 1983; LaMotte and DeLanerolle, 1983a). The ENK-LIr terminals are associated with synaptic specializations, supporting the idea that ENK is released at these sites and binds to opiate receptors on the postsynaptic dendritic membrane.

Opiate receptor studies and electrophysiological studies provide evidence for a possible presynaptic site of action of ENK on primary afferents, including those containing SP-LI (for review see Ruda et al., 1984; Basbaum, 1985). However, such a hypothesis would predict that axoaxonic synapses are present between primary afferent fibres and ENK-LIr profiles at the electron-microscope level. Unequivocal examples of such a synaptic relationship have been intensively searched for by many investigators, but none have been found (Hunt et al., 1980; Aronin et al., 1981; Sumal et al., 1982; Arluison et al., 1983; Glazer and Basbaum, 1983; LaMotte and DeLanerolle, 1983a). If ENK synapses on primary afferent axons do not exist, then it is necessary to postulate that ENK acts presynaptically by binding to opiate receptors located remote from its site of release, a situation known to occur with peripheral peptidergic synapses (Jan and Jan, 1982). It should be noted however, that scalloped-shaped central terminals, resembling SP terminals, synapse on dorsal horn dendrites that also are postsynaptic to ENK-LIr axons (Glazer and Basbaum, 1983). Such a feedforward inhibitory pathway can readily account for the modulatory effects of opioid peptides on dorsal horn input and output pathways.

What is the source of the ENK input onto long projection output pathways in the dorsal horn? As already discussed, some ENK-LI appears to be localized in primary afferent and descending axons. However, studies of ENK-LI following spinal transection or rhizotomy reveal only negligible changes in ENK-LI, suggesting that most of the ENK-LI in the dorsal horn originates from intrinsic neurons. ENK-LIr neurons are found in laminae I, II, III, IV/V, X and the intermediate gray (see above).

ENK-LIr lamina I neurons are not thalamic projection neurons since studies using a retrograde tracer in combination with ENK immunocytochemistry revealed no double labelling of thalamic projection neurons (Basbaum, 1982). ENK-LIr lamina I neurons with local axonal arbors are reasonable candidates for a source of ENK input to the projection neurons located in the same lamina. Nociceptive lamina I neurons with local axon collaterals have been described (Bennett et al., 1981a; J. L. K. Hylden, H. Hayashi and G. J. Bennett, unpublished observations) and some may contain ENK.

Evidence already has been presented that some stalked cells and some lamina IIb islet cells are ENK-LIr neurons. Since the axons of stalked cells arborize in lamina I, they are a second reasonable source of ENK input to lamina I thalamic projection neurons. Islet cell axons and dendrites are confined to lamina II and cannot directly influence lamina I activity. However, islet cells have vesicle-containing dendrites that are presynaptic to stalked cell dendrites and thereby can indirectly influence the output of lamina I thalamic projection neurons. We have proposed (Dubner et al., 1984) that ENK-LIr stalked cells are inhibitory local circuit neurons that suppress the activity of lamina I projection neurons. Other stalked cells that are not ENK-LIr may be excitatory local circuit neurons (Gobel, 1978).

ENK-LIr local circuit neurons also are present in other dorsal horn laminae. Lamina III contains small cells that receive only low-threshold mechanoreceptive input and have extensive axonal arbors in laminae III–IV (Bennett et al., 1981b; Maxwell, 1983a). Some of these lamina III cells are probably ENK-LIr (Fig. 5; Bennett et al., 1981b) and are ideally located to provide ENK input to projection neurons with dendrites in laminae III–IV.

Little is known about the ENK-LIr local circuit neurons in lamina V (Fig. 6), but their presence suggests that they are logical candidates to provide the ENK input to lamina V thalamic projection neurons.

Serotonin

Essentially all of the 5-HT found in the dorsal horn originates from descending brain stem neurons since chronic spinal cord transection results in complete depletion of 5-HT-LI below the lesion (see above). Sites of interaction between descending 5-HT axons and long projection neurons have been identified in two different double-label experiments. In one study, physiological properties were correlated with cell morphology and 5-HT input by combining 5-HT immunocytochemistry with the intracellular HRP method (Hoffert et al., 1983; Miletic et al., 1984). All intracellularly filled lamina I neurons were physiologically characterized as nociceptive neurons and all were inhibited by n. raphe magnus stimulation (Miletic et al., 1984). They invariably exhibited many 5-HT-LIr axonal contacts on their dendrites (average of 74 contacts per cell, $n = 12$). These contacts were mainly on dendritic shafts with few contacts on spines. This is in contrast to SP-LIr contacts, which were mainly on dendritic spines of lamina I spiny neurons (see above). In a similar study, lamina I spinomesencephalic neurons projecting to the parabrachial area of the midbrain (Hylden et al., 1985) exhibited a varying number of 5-HT-LIr contacts (J. L. K. Hylden et al., unpublished observations). Some were densely innervated, whereas others had few contacts.

A second double-label study was performed to determine whether 5-HT-LIr contacts were present on thalamic projection neurons (Ruda and Coffield, 1983). In this experiment, the retrograde HRP method was combined with 5-HT immunocytochemistry. Retrogradely filled thalamic projection neurons in laminae I and V had a variable number of contacts on their somata and proximal dendrites. More than 50% of the retrogradely labelled thalamic projection neurons received contacts from 5-HT-LIr axonal endings. These contacts were generally less dense than ENK contacts on thalamic projection neurons, with most occurring on the proximal dendrites rather than on the cell soma. 5-HT-LIr contacts on thalamic projection neurons were shown to be sites of synaptic interaction at the ultrastructural level. They formed symmetrical synaptic contacts in contrast to the predominantly asymmetrical ENK-LIr contacts on the same class of cells.

The same retrograde HRP and 5-HT immunocytochemistry double-labelling method has been used to provide direct evidence of 5-HT contacts on DCPS neurons (Nishikawa et al., 1983). Electron-microscope analysis has verified that these 5-HT-LIr varicosities are sites of synaptic interaction. Almost all of the DCPS neurons examined in cat and monkey at the light-microscope level had 5-HT-LIr axonal varicosities in contact with their somata and proximal dendrites. It appears that 5-HT modulates the output of both non-nociceptive and nociceptive neurons because the DCPS population is known to contain about equal numbers of both kinds of neurons (Lu et al., 1983).

These observations of direct postsynaptic contacts of 5-HT-LIr varicosities on long projection neurons in the dorsal horn are in agreement with microiontophoretic studies in which 5-HT has been shown to inhibit the glutamate-evoked excitation of cat DCPS neurons (Nishikawa et al., 1983) and monkey spinothalamic tract neurons (Willcockson et al., 1984a). It appears that four different long projection output systems (lamina I and V thalamic projection neurons, lamina I spinomesencephalic neurons, and DCPS neurons) receive direct postsynaptic inhibitory input from descending 5-HT pathways. Such pathways are of general interest since few long inhibitory pathways without intervening local circuit neurons have been described in the vertebrate central nervous system.

Numerous 5-HT contacts are found on local circuit neurons in the dorsal horn. Using the intracellular HRP method and 5-HT immunocytochemistry, it has been observed that stalked cells are inhibited by n. raphe magnus stimulation and exhibit many 5-HT-LIr contacts at the light-microscope level (Miletic et al., 1984). In contrast, nucleus raphe magnus stimulation had no effect on lamina IIa or IIb islet cells and these cells had few 5-HT-LIr contacts. Thus, direct descending 5-HT postsynaptic inhibition modifies the output of lamina II local circuit neurons differentially; stalked cells that can directly modulate the output of lamina I projection neurons are heavily innervated whereas islet cells whose axons are confined to lamina II have few 5-HT contacts.

Recent findings using double-label methods at the ultrastructural level indicate that ENK-LIr neurons receive descending 5-HT input (Glazer and Basbaum, 1984). Such findings are consistent with the above observation that stalked cells, some of which contain ENK-LI, receive 5-HT input and suggest that lamina I output pathways may be indirectly influenced by this 5-HT/ENK pathway.

Serotonin contacts have been observed on one other identified dorsal horn neuron: lamina IV low-threshold mechanoreceptive neurons with apical dendrites entering lamina II (Hoffert et al., 1982).

This cell type receives 5-HT input on its distal dendritic arbor in lamina II.

The 5-HT connections with identified dorsal horn neurons are summarized in Table I.

Neurochemically identified intrinsic dorsal horn circuitry

With the exception of lamina I local circuit neurons and stalked and islet cells in lamina II, very little is known about the morphology, physiology and connections of other local circuit neurons in the dorsal horn. Lamina I local circuit neurons may receive synaptic input from stalked cells whose axons arborize in lamina I. Stalked cells receive input from islet cells in lamina IIa and IIb. These connections probably are quite complex since each of these morphological types of local circuit neurons is known to contain a variety of neurochemicals. Lamina I local circuit neurons may contain either GABA, ENK or DYN. Stalked cells are reported to contain either ENK or GABA or some unidentified excitatory neurotransmitters. Islet cells may contain either NT, ENK or GABA.

Which types of connections between neurochemically identified local circuit neurons in laminae I and II can be predicted based on the above evidence? Stalked cells that release excitatory neurotransmitters may synapse on lamina I inhibitory local circuit neurons containing either GABA or an opioid peptide. Stalked cells and lamina I neurons that release a presumed inhibitory neurotransmitter such as GABA may provide input to ENK-containing neurons in laminae I and II.

There is evidence suggesting that the islet cell-stalked cell circuitry involves NT, ENK and GABA. NT has excitatory effects in microiontophoretic studies, yet it has antinociceptive effects in behavioral studies (Yaksh et al., 1982). These antinociceptive effects are reversed by naloxone. Such findings suggest that NT-LIr islet cells are part of an opioid pathway involving ENK-LIr stalked cells that act as inhibitory local circuit neurons. ENK- or GABA-containing islet cells may provide inhibitory input to excitatory stalked cells and suppress the transmission of nociceptive excitatory input from lamina II to lamina I. Finally, it is possible that ENK–ENK connections (Aronin et al., 1981; Glazer and Basbaum, 1983) involve islet cell–stalked cell interactions.

New principles of neural connectivity

The functional diversity of individual neurochemicals has led to a re-evaluation of how chemical messengers act in the central nervous system. Some neurochemicals such as GABA and glutamate act as conventional inhibitory and excitatory neurotransmitters. They produce changes in ionic conductance with resulting membrane hyperpolarization or depolarization. On the other hand, many neuropeptides do not produce changes in membrane ionic permeability or polarization but only enhance or reduce the effectiveness of a conventional transmitter. Some neurotransmitters produce different effects even in the same region. For example, it has been mentioned that ENK in the dorsal horn may alter specific ionic conductances like a conventional neurotransmitter or exert a modulatory role without any direct membrane effects. This multiplicity of actions by a neurotransmitter has led to new principles of neural connectivity: (1) a neuron may receive input from many neurotransmitters; (2) each neurotransmitter may have multiple actions in a given region; and (3) multiple neurotransmitters may exist in a single neuron.

Fig. 12 is a diagram illustrating many of the identified connections between primary afferent neurons, dorsal horn neurons and descending axons. New principles of neurochemical transmission appear to apply here. Thalamic projection neurons in laminae I and V can receive ENK, 5-HT, SP and possibly NA input (not shown). Other unidentified neurotransmitters may also synapse on these long projection output neurons since each labelled transmitter represents only a small part of the total synaptic input to some of these cells (Ruda et al., 1984; Ruda, 1985). Thus, a single neuron appears to receive input from many neurotransmitters. The principle that a single transmitter can arise from mul-

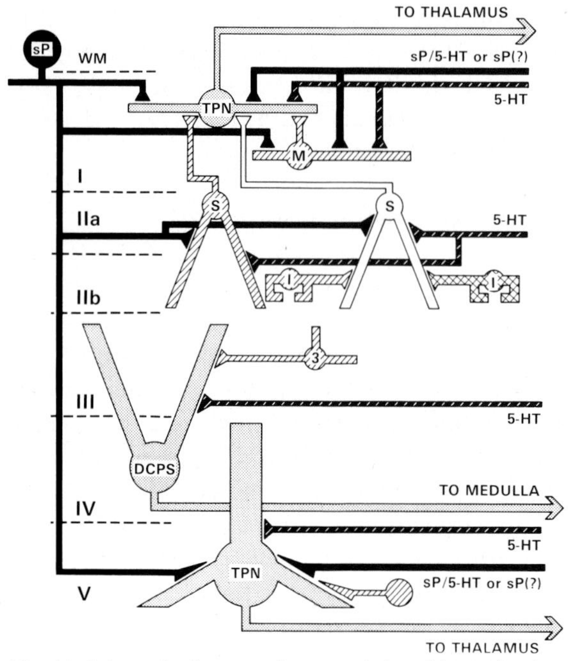

Fig. 12. Schematic diagram of proposed dorsal horn circuitry. Lamina borders and white matter (WM) are represented at the left. Primary afferent axons enter at the left and descending, extrinsic axons enter at the right. TPN, thalamic projection neuron; M, lamina I local circuit neuron; S, stalked cell; I, islet cell; 3, lamina III local circuit neuron; sP, substance P; 5-HT, serotonin; sP/5-HT or sP: descending axons with co-existent 5-HT and SP or SP only. Striped cells are ENK-containing local circuit neurons. Cross-hatched cell is a GABA-containing local circuit neuron. See text for details.

tiple sites and may have multiple actions in a given region is demonstrated by SP. It is present in primary afferent terminals and descending axons and both inputs may occur on the same cell (Fig. 12). Another new principle of neurochemical diversity is illustrated in this circuit diagram. The same morphological subtype of neuron may contain different neurotransmitters and provide excitatory or inhibitory input. Different stalked cells contain ENK or GABA, or may contain an unknown excitatory transmitter. Different islet cells contain ENK, GABA, or NT.

What functional roles can we attribute to this proposed neural circuitry? Fig. 12 shows two candidate ENK-containing local circuit neurons that could provide direct input to lamina I thalamic projection neurons — lamina I neurons and lamina II stalked cells. This is a nociceptive circuit (excluding thermoreceptive neurons) since all lamina I neurons and stalked cells respond maximally to noxious stimuli (Willis and Coggeshall, 1978; Bennett et al., 1979, 1980, 1981a; Miletic et al., 1984). Modulation of this long projection output system by ENK-containing neurons provides a neural circuit for naloxone-sensitive analgesia or naloxone-sensitive inhibition of lamina I thalamic projection neurons produced by intense transcutaneous electrical stimulation (Sjölund and Eriksson, 1979; Chung et al., 1984). GABA-containing local circuit neurons in laminae I and II may produce similar segmental inhibitory effects.

Non-nociceptive modulation of nociceptive circuits could occur via low-threshold mechanoreceptive, lamina IIb islet cells that are ENK- or GABA-containing, or lamina III ENK-containing local circuit neurons. As shown in Fig. 12, lamina IIb islet cells would modulate lamina I activity via dendrodendritic synapses on stalked cells. The lamina III local circuit neurons can provide ENK input to the soma and proximal dendrites of lamina III–IV DCPS neurons and possibly to the apical dendrites of lamina V thalamic projection neurons. Such circuits would provide a morphological basis for opioid modulation of dorsal horn neuronal activity that is evoked by innocuous stimulation.

SP produces slow depolarization and may augment noxious signals by enhancing the effectiveness of the conventional neurotransmitter. The input–output system of primary afferent neurons and long projection output systems involves highly secure synaptic transmission. The conventional type of excitatory transmitter in this pathway is not known, although there is some evidence that amino acids may play a role (Salt and Hill, 1983).

Descending 5-HT- and NA-containing axons distribute to widespread regions of the medullary and spinal dorsal horns. These axons give off an enormous number of boutons en passant along their paths through the spinal cord, yet they arise from a relatively small number of neurons in the brain stem (Bowker et al., 1981b; Westlund et al., 1982). The multi-targeted effects of these 5-HT- and NA-containing neurons suggest that they provide a global enhancement or suppression that enables dorsal horn neurons to respond more effectively to incoming sensory information. Such mechanisms probably form the neural basis of attentional mechanisms and the ability of animals to extract behaviorally relevant information from the environment (Dubner, 1985).

Acknowledgements

We thank J. L. K. Hylden and W. Maixner for reviewing the manuscript and E. Welty for secretarial services. We are also indebted to A. I. Basbaum, J. Coffield and V. Miletic, G. J. Giesler, Jr. and T. M. Jessell for pre-publication copies of their work.

References

Allen, Y. S., Adrian, T. E., Allen, J. M., Tatemoto, K., Crow, T. J., Bloom, S. R. and Polak, J. M. (1983) Neuropeptide Y distribution in rat brain. *Science,* 221: 877–879.

Anand, P., Gibson, S. J., McGregor, G. P., Blank, M. A., Ghatei, M. A., Bacarese-Hamilton, A. J., Polak, J. M. and Bloom, S. R. (1983) A VIP-containing system concentrated in the lumbosacral region of human spinal cord. *Nature,* 305: 143–145.

Arluison, M., Conrath-Verrier, M., Tauc, M., Mailly, P., De La Manche, I. S., Dietl, M., Cesselin, F., Bourgoin, S. and Hamon, M. (1983) Met-enkephalin-like immunoreactivity in rat forebrain and spinal cord using hydrogen peroxide and Triton X-100. Ultrastructural study. *Brain Res. Bull.,* 11: 573–586.

Aronin, N., DiFiglia, M., Liotta, A. S. and Martin, J. B. (1981) Ultrastructural localization and biochemical features of immunoreactive leu-enkephalin in monkey dorsal horn. *J. Neurosci.,* 1: 561–577.

Barber, R. P., Vaughn, J. E., Saito, K., McLaughlin, B. J. and Roberts, E. (1978) GABAergic terminals are presynaptic to primary afferent terminals in the substantia gelatinosa of the rat spinal cord. *Brain Res.,* 141: 35–55.

Barber, R. P., Vaughn, J. E., Slemmon, J. R., Salvaterra, P. M., Roberts, E. and Leeman, S. E. (1979) The origin, distribution and synaptic relationships of substance P axons in rat spinal cord. *J. Comp. Neurol.,* 184: 331–352.

Barber, R. P., Vaughn, J. E. and Roberts, E. (1982) The cytoarchitecture of GABAergic neurons in rat spinal cord. *Brain Res.,* 238: 305–328.

Barber, R. P., Phelps, P. E., Houser, C. R., Crawford, G. D., Salvaterra, P. M. and Vaughn, J. E. (1984) The morphology and distribution of neurons containing choline acetyltransferase in the adult rat spinal cord: An immunocytochemical study. *J. Comp. Neurol.,* 229: 329–346.

Barker, J. L., Neale, J. H., Smith, Jr., T. G. and McDonald, R. L. (1978) Opiate peptide modulation of amino acid responses suggests novel form of neuronal communication. *Science,* 199: 1451–1453.

Basbaum, A. I. (1982) Anatomical substrates for the descending control of nociception. In B. Sjölund and A. Björklund (Eds.), *Brain Stem Control of Spinal Mechanisms,* Elsevier, Amsterdam, pp. 119–133.

Basbaum, A. I. (1985) A functional analysis of the cytochemistry of the spinal dorsal horn. In *Proceedings of the IV World Congress on Pain,* Raven Press, New York (in press).

Basbaum, A. I. and Glazer, E. J. (1983) Immunoreactive vasoactive intestinal polypeptide is concentrated in the sacral spinal cord: A possible marker for pelvic visceral afferent fibers. *Somatosensory Res.,* 1: 69–82.

Basbaum, A. I., Clanton, C. H. and Fields, H. L. (1978) Three bulbospinal pathways from the rostral medulla of the cat: An autoradiographic study of pain modulating systems. *J. Comp. Neurol.,* 178: 209–224.

Beall, J. E., Martin, R. F., Applebaum, A. E. and Willis, W. D. (1976) Inhibition of primate spinothalamic tract neurons by stimulation in the region of the nucleus raphe magnus. *Brain Res.,* 114: 328–333.

Belcher, G., Ryall, R. W. and Schaffner, R. (1978) The differential effects of 5-hydroxytryptamine, noradrenaline and raphe stimulation on nociceptive and non-nociceptive dorsal horn interneurones in the cat. *Brain Res.,* 151: 307–321.

Bennett, G. J., Hayashi, H., Abdelmoumene, M. and Dubner, R. (1979) Physiological properties of stalked cells of the substantia gelatinosa intracellularly stained with horseradish peroxidase. *Brain Res.,* 164: 285–289.

Bennett, G. J., Abdelmoumene, M., Hayashi, H. and Dubner, R. (1980) Physiology and morphology of substantia gelatinosa neurons intracellularly stained with horseradish peroxidase. *J. Comp. Neurol.,* 194: 809–827.

Bennett, G. J., Abdelmoumene, M., Hayashi, H., Hoffert, M. J. and Dubner, R. (1981a) Spinal cord layer I neurons with axon collaterals that generate local arbors. *Brain Res.,* 209: 421–426.

Bennett, G. J., Abdelmoumene, M., Hayashi, H., Hoffert, M. J., Ruda, M. A. and Dubner, R. (1981b) Physiology, morphology and immunocytology of dorsal horn layer III neurons. *Pain, Suppl. 1,* S240.

Bennett, G. J., Ruda, M. A., Gobel, S. and Dubner, R. (1982) Enkephalin immunoreactive stalked cells and lamina IIb islet cells in cat substantia gelatinosa. *Brain Res.,* 240: 162–166.

Bennett, G. J., Seltzer, Z., Lu, G.-W., Nishikawa, N. and Dubner, R. (1983) The cells of origin of the dorsal column postsynaptic projection in the lumbosacral enlargements of cats and monkeys. *Somatosensory Res.,* 1: 131–149.

Björklund, A. and Skagerberg, G. (1979) Evidence for a major spinal cord projection from the diencephalic A11 dopamine cell group in the rat using transmitter-specific fluorescent retrograde tracing. *Brain Res.*, 177: 170–175.

Björklund, A. J., Emson, P. C., Gilbert, R. F. T. and Skagerberg, G. (1979) Further evidence for the possible coexistence of 5-hydroxytryptamine and substance P in medullary raphe neurons of rat brain. *Br. J. Pharmacol.*, 66: 112–113.

Bowker, R. M., Steinbusch, H. W. M. and Coulter, J. D. (1981a) Serotonergic and peptidergic projections to the spinal cord demonstrated by a combined retrograde HRP histochemical and immunocytochemical staining method. *Brain Res.*, 211: 412–417.

Bowker, R. M., Westlund, K. N. and Coulter, J. D. (1981b) Origins of serotonergic projections to the spinal cord in rat: An immunocytochemical-retrograde transport study. *Brain Res.*, 226: 187–199.

Bowker, R. M., Westlund, K. N., Sullivan, M. C., Wilber, J. F. and Coulter, J. D. (1982a) Transmitters of the raphe-spinal complex: Immunocytochemical studies. *Peptides*, 3: 291–298.

Bowker, R. M., Westlund, K. N., Sullivan, M. C. and Coulter, J. D. (1982b) Organization of descending serotonergic projections to the spinal cord. *Prog. Brain Res.*, 57: 239–265.

Bresnahan, J. C., Ho, R. H. and Beattie, M. S. (1984) A comparison of the ultrastructure of substance-P and enkephalin immunoreactive elements in the nucleus of the dorsolateral funiculus and laminae I and II of the rat spinal cord. *J. Comp. Neurol.*, 229: 497–511.

Brown, A. G., Rose, P. K. and Snow, P. J. (1977) The morphology of hair follicle afferent fibre collaterals in the spinal cord of the cat. *J. Physiol. (London)*, 272: 779–797.

Burnweit, C. and Forssmann, W. G. (1979) Somatostatinergic nerves in the cervical spinal cord of the monkey. *Cell Tiss. Res.*, 200: 83–90.

Cervero, F. (1984) Functional properties and central actions of visceral nociceptors. In W. Hamann and A. Iggo (Eds.), *Sensory Receptor Mechanisms*, World Scientific Publ. Co., Singapore, pp. 275–282.

Cervero, F. and Connell, L. A. (1984a) Distribution of somatic and visceral primary afferent fibres within the thoracic spinal cord of the cat. *J. Comp. Neurol.*, 230: 88–98.

Cervero, F. and Connell, L. A. (1984b) Fine afferent fibres from viscera do not terminate in the substantia gelatinosa of the thoracic spinal cord. *Brain Res.*, 294: 370–374.

Changaris, D. G., Keil, L. C. and Severs, W. B. (1978) Angiotensin II immunohistochemistry of the rat brain. *Neuroendocrinology*, 25: 257–274.

Charnay, Y., Leger, L., Dray, F., Berod, A., Jouvet, M., Pujol, J. F. and Dubois, P. M. (1982) Evidence for the presence of enkephalin in catecholaminergic neurones of cat locus coeruleus. *Neurosci. Lett.*, 30: 147–151.

Charnay, Y., Paulin, C., Chayvialle, J.-A. and Dubois, P. M. (1983) Distribution of substance P-like immunoreactivity in the spinal cord and dorsal root ganglia of the human foetus and infant. *Neuroscience*, 10: 41–55.

Cheema, S. S., Rustioni, A. and Whitsel, B. L. (1984) Light and electron microscopic evidence for a direct corticospinal projection to superficial laminae of the dorsal horn in cats and monkeys. *J. Comp. Neurol.*, 225: 276–280.

Christofides, N. D., Yiangou, Y., Blank, M. A., Tatemoto, K., Polak, J. M. and Bloom, S. R. (1982) Are peptide histidine isoleucine and vasointestinal peptide co-synthesized in the same prohormone? *Lancet*, ii: 1398.

Chronwall, B. M., Olschowska, J. A. and O'Donohue, T. L. (1984) Histochemical localization of FMRFamide-like immunoreactivity in the rat brain. *Peptides*, 5: 569–581.

Chung, J. M., Fang, Z. R., Hori, Y., Lee, K. H. and Willis, W. D. (1984) Prolonged inhibition of primate spinothalamic tract cells by peripheral nerve stimulation. *Pain*, 19: 259–275.

Coffield, J. A., Zimmerman, E. M., Hoffert, M. J., Miletic, V. and Brooks, B. R. (1984) Immunocytochemical localization of TRH-LI in dorsal horn neurons of the mouse spinal cord. *Soc. Neurosci. Abstr.*, 10: 488.

Coimbra, A., Sodre-Borges, B. P. and Magalhaes, M. M. (1974) The substantia gelatinosa Rolandi of the rat. Fine structure, cytochemistry (acid phosphatase) and changes after dorsal root section. *J. Neurocytol.*, 3: 199–217.

Conrath-Verrier, M., Dietl, M., Arluison, M., Cesselin, F., Bourgoin, S. and Hammon, M. (1983) Localization of Met-enkephalin-like immunoreactivity within pain-related nuclei of cervical spinal cord, brainstem and midbrain in the cat. *Brain Res. Bull.*, 11: 587–604.

Conrath-Verrier, M., Dietl, M. and Tramu, G. (1984) Cholecystokinin-like immunoreactivity in the dorsal horn of the spinal cord of the rat: a light and electron microscopic study. *Neuroscience*, 13: 871–886.

Cuello, A. C., Del Fiacco, M. and Paxinos, G. (1978) The central and peripheral ends of the substance P-containing sensory neurones in the rat trigeminal system. *Brain Res.*, 152: 499–509.

Cummings, S., Elde, R., Ells, J. and Lindall, A. (1983) Corticotropin-releasing factor immunoreactivity is widely distributed within the central nervous system of the rat: An immunohistochemical study. *J. Neurosci.*, 3: 1355–1368.

Dahlström, A. and Fuxe, K. (1964) Evidence for the existence of monoamine-containing neurons in the central nervous system. I. Demonstration of monoamines in the cell bodies of brain stem neurons. *Acta Physiol. Scand.*, 62 Suppl. 232: 1–55.

Dahlström, A. and Fuxe, K. (1965) Evidence for the existence of monoamine-containing neurons in the central nervous system. II. Experimentally induced changes in the intraneuronal amine levels of bulbospinal neuron systems. *Acta Physiol. Scand.*, 64 Suppl. 247: 7–36.

Dalsgaard, C.-J., Hökfelt, T., Johansson, O. and Elde, R. (1981) Somatostatin immunoreactive cell bodies in the dorsal horn and the parasympathetic intermediolateral nucleus of the rat spinal cord. *Neurosci. Lett.*, 27: 335–339.

Dalsgaard, C.-J., Vincet, S. R., Hökfelt, T., Lundberg, J. M., Dahlström, A., Schultzberg, M., Dockray, G. J. and Cuello, A. C. (1982) Coexistence of cholecystokinin- and substance P-like peptides in neurons of the dorsal root ganglia of the rat. *Neurosci. Lett.*, 33: 159–163.

DeLanerolle, N. C. and Coen, C. W. (1984) Differential patterns of peptidergic immunoreactivity in the human spinal cord. *Soc. Neurosci. Abstr.*, 10: 435.

DeLanerolle, N. C. and LaMotte, C. C. (1982) The human spinal cord: substance P and methionine-enkephalin immunoreactivity. *J. Neurosci.*, 2: 1369–1386.

DeLanerolle, N. C. and LaMotte, C. C. (1983) Ultrastructure of chemically defined neuron systems in the dorsal horn of the monkey. I. Substance P immunoreactivity. *Brain Res.*, 274: 31–49.

DelFiacco, M. and Cuello, A. C. (1980) Substance P- and enkephalin-containing neurones in the rat trigeminal system. *Neuroscience*, 5: 803–815.

DiFiglia, M., Aronin, N. and Leeman, S. E. (1982) Light microscopic and ultrastructural localization of immunoreactive substance P in the dorsal horn of monkey spinal cord. *Neuroscience*, 7: 1127–1139.

DiFiglia, M., Aronin, N. and Leeman, S. E. (1984) Ultrastructural localization of immunoreactive neurotensin in the monkey superficial dorsal horn. *J. Comp. Neurol.*, 225: 1–12.

Dockray, G. J. and Williams, R. G. (1983) FMRF-amide-like immunoreactivity in rat brain: development of a radioimmunoassay and its application in studies of distribution and chromatographic properties. *Brain Res.*, 266: 295–303.

Dockray, G. J., Gregory, R. A., Tracy, H. J. and Zhu, W.-Y. (1981a) Transport of cholecystokinin-octapeptide-like immunoreactivity toward the gut in afferent vagal fibres in cat and dog. *J. Physiol. (London)*, 314: 501–511.

Dockray, G. J., Vaillant, C. and Williams, R. G. (1981b) New vertebrate brain-gut peptide related to a molluscan neuropeptide and an opioid peptide. *Nature*, 293: 656–657.

Dockray, G. J., Reeve, Jr., J. R., Shively, J., Gayton, R. J. and Barnard, C. S. (1983) A novel pentapeptide from chicken brain identified by antibodies to FMRFamide. *Nature*, 305: 328–330.

Dodd, J., Jahr, C. E., Hamilton, P. N., Heath, M. J. S., Matthew, W. D. and Jessell, T. M. (1983) Cytochemical and physiological properties of sensory and dorsal horn neurons that transmit cutaneous sensation. *Cold Spring Harbor Symp. Quant. Biol.*, 48: 685–695.

Dubner, R. (1985) Specialization in nociceptive pathways: sensory-discrimination, sensory modulation and neuronal connectivity. In H. L. Fields, R. Dubner and F. Cervero (Eds.), *Advances in Pain Research and Therapy*, Vol. 9, Raven Press, New York, pp. 111–137.

Dubner, R. and Bennett, G. J. (1983) Spinal and trigeminal mechanisms of nociception. *Annu. Rev. Neurosci.*, 6: 381–418.

Dubner, R., Ruda, M. A., Miletic, V., Hoffert, M. J., Bennett, G. J., Nishikawa, N. and Coffield, J. (1984) Neural circuitry mediating nociception in the medullary and spinal dorsal horns. In L. Kruger and J. C. Liebeskind (Eds.), *Advances in Pain Research and Therapy, Vol. 6*, Raven Press, New York, pp. 151–166.

Felton, D. L. and Sladek, J. R. (1983) Monoamine distribution in primate brain. V. Monoaminergic nuclei: Anatomy, pathways and local organization. *Brain Res. Bull.*, 10: 171–284.

Fields, H. L., Basbaum, A. I., Clanton, C. H. and Anderson, S. D. (1977) Nucleus raphe magnus inhibition of spinal cord dorsal horn neurons. *Brain Res.*, 126: 441–453.

Fitzgerald, M. (1983) Capsaicin and sensory neurones – a review. *Pain*, 15: 109–130.

Forssmann, W. G. (1978) A new somatostatinergic system in the mammalian spinal cord. *Neurosci. Lett.*, 10: 293–297.

Fuji, K., Senba, E., Ueda, Y. and Tohyama, M. (1983) Vasoactive intestinal polypeptide (VIP)-containing neurons in the spinal cord of the rat and their projections. *Neurosci. Lett.*, 37: 51–55.

Fuxe, K., Ganten, D., Hökfelt, T. and Bolme, P. (1976) Immunohistochemical evidence for the existence of angiotensin II-containing nerve terminals in the brain and spinal cord in the rat. *Neurosci. Lett.*, 2: 229–234.

Fuxe, K., Agnati, L. F., McDonald, T., Locatelli, V., Hökfelt, T., Dalsgaard, C.-J., Battistini, N., Yanaihara, N., Mutt, V. and Cuello, A. C. (1983) Immunohistochemical indications of gastrin releasing peptide – bombesin-like immunoreactivity in the nervous system of the rat. Codistribution with substance P-like immunoreactive nerve terminal systems and coexistence with substance P-like immunoreactivity in dorsal root ganglion cell bodies. *Neurosci. Lett.*, 37: 17–22.

Gibson, S. J., Polak, J. M., Bloom, S. R. and Wall, P. D. (1981) The distribution of nine peptides in rat spinal cord with special emphasis on the substantia gelatinosa and on the area around the central canal (lamina X). *J. Comp. Neurol.*, 201: 65–79.

Gibson, S. J., Polak, J. M., Anand, P., Blank, M. A., Morrison, J. F. B., Kelly, J. S. and Bloom, S. R. (1984a) The distribution and origin of VIP in the spinal cord of six mammalian species. *Peptides*, 5: 201–207.

Gibson, S. J., Polak, J. M., Bloom, S. R., Sabata, I. M., Mulderry, P. M., Ghatel, M. A., McGregor, G. P., Morrison, J. F. B., Kelly, J. S., Evans, R. M. and Rosenfield, M. G. (1984b) Calcitonin gene-related peptide immunoreactivity in the spinal cord of man and of eight other species. *J. Neurosci.*, 4: 3101–3111.

Gibson, S. J., Polak, J. M., Allen, T. E., Adrian, T. E., Kelly, J. S. and Bloom, S. R. (1984c) The distribution and origin of a novel brain peptide, neuropeptide Y, in the spinal cord of several mammals. *J. Comp. Neurol.*, 227: 78–91.

Giesler, Jr., G. J. and Elde, R. (1985) Immunocytochemical studies of the content of axons and terminals in the lateral cervical nucleus and the lateral spinal nucleus. *J. Neurosci.* (in press).

Gilbert, R. F. T., Emson, P. C., Hunt, S. P., Bennett, G. W., Marsden, C. A., Sandberg, B. E. B., Steinbusch, H. W. M.

and Verhofstad, A. A. J. (1982) The effects of monoamine neurotoxins on peptides in the rat spinal cord. *Neuroscience,* 7: 69–87.

Glazer, E. J. and Basbaum, A. I. (1981) Immunohistochemical localization of leucine-enkephalin in the spinal cord of the cat: Enkephalin-containing marginal neurons and pain modulation. *J. Comp. Neurol.,* 196: 377–389.

Glazer, E. J. and Basbaum, A. I. (1983) Opiate neurons and pain modulation: an ultrastructural analysis of enkephalin in cat superficial dorsal horn. *Neuroscience,* 10: 357–376.

Glazer, E. J. and Basbaum, A. I. (1984) Axons which take up [^{3}H]serotonin are presynaptic to enkephalin immunoreactive neurons in cat dorsal horn. *Brain Res.,* 298: 386–391.

Glazer, E. J. and Ross, L. L. (1980) Localization of noradrenergic terminals in sympathetic preganglionic nuclei of the rat: Demonstration by immunocytochemical localization of dopamine-β-hydroxylase. *Brain Res.,* 185: 39–49.

Gobel, S. (1975) Golgi studies of the substantia gelatinosa neurons in the spinal trigeminal nucleus. *J. Comp. Neurol.,* 162: 397–415.

Gobel, S. (1978) Golgi studies of the neurons in layer II of the dorsal horn of the medulla (trigeminal nucleus caudalis). *J. Comp. Neurol.,* 180: 395–414.

Gobel, S., Falls, W. M., Bennett, G. J., Abdelmoumene, M., Hayashi, H. and Humphrey, E. (1980) An EM analysis of the synaptic connections of horseradish peroxidase-filled stalked cells and islet cells in the substantia gelatinosa of adult cat spinal cord. *J. Comp. Neurol.,* 194: 781–807.

Gobel, S., Bennett, G. J., Allen, B., Humphrey, E., Seltzer, Z., Abdelmoumene, M., Hayashi, H. and Hoffert, M. J. (1982) Synaptic connectivity of substantia gelatinosa neurons with reference to potential termination sites of descending axons. In B. Sjölund and A. Björklund (Eds.), *Brain Stem Control of Spinal Mechanisms,* Elsevier, Amsterdam, pp. 135–158.

Griersmith, B. T. and Duggan, A. W. (1980) Prolonged depression of spinal transmission of nociceptive information by 5-HT administered in the substantia gelatinosa: antagonism by methysergide. *Brain Res.,* 187: 231–236.

Headley, P. M., Duggan, A. W. and Griersmith, B. T. (1978) Selective reduction by noradrenaline and 5-hydroxytryptamine of nociceptive responses of cat dorsal horn neurones. *Brain Res.,* 145: 185–189.

Henry, J. L. (1976) Effects of substance P on functionally identified units in cat spinal cord. *Brain Res.,* 114: 439–451.

Hoffert, M. J., Miletic, V., Ruda, M. A. and Dubner, R. (1982) A comparison of substance P and serotonin axonal contacts on identified neurons in cat spinal dorsal horn. *Soc. Neurosci. Abstr.,* 8: 805.

Hoffert, M. J., Miletic, V., Ruda, M. A. and Dubner, R. (1983) Immunocytochemical identification of serotonin axonal contacts on characterized neurons in laminae I and II of the cat dorsal horn. *Brain Res.,* 267: 361–364.

Hökfelt, T., Fuxe, K., Goldstein, M. and Johansson, O. (1974) Immunohistochemical evidence for the existence of adrenaline neurons in the rat brain. *Brain Res.,* 66: 235–251.

Hökfelt, T., Efendic, S., Hellerström, C., Johansson, O., Luft, R. and Arimura, A. (1975a) Cellular localization of somatostatin in endocrine-like cells and neurons of the rat with special references to the A_1-cells of the pancreatic islets and to the hypothalamus. *Acta Endocrinol.,* 80 (Suppl. 200): 5–41.

Hökfelt, T., Fuxe, K., Johansson, O., Jeffcoate, S. and White, N. (1975b) Thyrotropin releasing hormone (TRH)-containing nerve terminals in certain brain stem nuclei and in the spinal cord. *Neurosci. Lett.,* 1: 133–139.

Hökfelt, T., Kellerth, J. O., Nilsson, G. and Pernow, B. (1975c) Substance P: localization in the central nervous system and in some primary sensory neurons. *Science,* 190: 889–890.

Hökfelt, T., Kellerth, J. O., Nilsson, G. and Pernow, B. (1975d) Experimental immunohistochemical studies on the localization and distribution of substance P in the cat primary sensory neurons. *Brain Res.,* 100: 235–252.

Hökfelt, T., Elde, R., Johansson, O., Luft, R., Nilsson, G. and Arimura, A. (1976) Immunohistochemical evidence for separate population of somatostatin-containing and substance P-containing primary afferent neurons in the rat. *Neuroscience,* 1: 131–136.

Hökfelt, T., Elde, R., Johansson, O., Terenius, L. and Stein, L. (1977a) The distribution of enkephalin-immunoreactive cell bodies in the rat central nervous system. *Neurosci. Lett.,* 5: 25–31.

Hökfelt, T., Ljungdahl, A., Terenius, L., Elde, R. and Nilsson, G. (1977b) Immunohistochemical analysis of peptide pathways possibly related to pain and analgesia: enkephalin and substance P. *Proc. Natl. Acad. Sci. U.S.A.,* 74: 3081–3085.

Hökfelt, T., Ljungdahl, A., Steinbusch, H., Verhofstad, A. N., Nilsson, G., Brodin, E., Pernow, B. and Goldstein, M. (1978) Immunohistochemical evidence of substance P-like immunoreactivity in some 5-hydroxytryptamine containing neurons in the rat central nervous system. *Neuroscience,* 3: 517–538.

Hökfelt, T., Phillipson, O. and Goldstein, M. (1979a) Evidence for a dopaminergic pathway in the rat descending from the A11 cell group to the spinal cord. *Acta Physiol. Scand.,* 107: 393–395.

Hökfelt, T., Terenius, L., Kuypers, H. G. J. M. and Dann, O. (1979b) Evidence for enkephalin immunoreactive neurons in the medulla oblongata projecting to the spinal cord. *Neurosci. Lett.,* 14: 55–60.

Hökfelt, T., Schultzberg, M., Lundberg, J. M., Fuxe, K., Mutt, V., Fahrenkrug, J. and Said, S. I. (1982) Distribution of vasoactive intestinal polypeptide in the central and peripheral nervous systems as revealed by immunocytochemistry. In S. I. Said (Ed.), *Vasoactive Intestinal Polypeptide,* Raven Press, New York, pp. 65–90.

Holstege, G. and Kuypers, H. G. J. M. (1982) The anatomy of brain stem pathways to the spinal cord in the cat. A labeled amino acid tracing study. *Prog. Brain Res.,* 57: 145–175.

Honda, C. N., Rethelyi, M. and Petruz, P. (1983) Preferential immunohistochemical localization of vasoactive intestinal polypeptide (VIP) in the sacral spinal cord of the cat: light and electron microscopic observations. *J. Neurosci.*, 3: 2183–2196.

Hunt, S. P. (1983) Cytochemistry of the spinal cord. In P. C. Emson (Ed.), *Chemical Neuroanatomy*, Raven Press, New York, pp. 53–84.

Hunt, S. P., Kelly, J. S. and Emson, P. C. (1980) The electron microscopic localization of methionine-enkephalin within the superficial layers (I and II) of the spinal cord. *Neuroscience*, 5: 1871–1890.

Hunt, S. P., Emson, P. C., Gilbert, R., Goldstein, M. and Kimmel, J. R. (1981a) Presence of avian pancreatic polypeptide-like immunoreactivity in catecholamine and methionine-enkephalin containing neurons within the central nervous system. *Neurosci. Lett.*, 21: 125–130.

Hunt, S. P., Kelly, J. S., Emson, P. C., Kimmel, J. R., Miller, R. J. and Wu, J.-Y. (1981b) An immunohistochemical study of neuronal populations containing neuropeptides or gamma--aminobutyrate within the superficial layers of the rat dorsal horn. *Neuroscience*, 6: 1883–1898.

Hylden, J. L. K., Hayashi, H., Bennett, G. J. and Dubner, R. (1985) Spinal lamina I neurons projecting to the parabrachial area of the cat midbrain. *Brain Res.* (in press).

Hylden, J. L. K., Ruda, M. A., Hayashi, H. and Dubner, R. (1985) Descending serotonergic fibers in the dorsolateral and ventral funiculi of cat spinal cord. *Neurosci. Lett.*, 62: 299–304.

Itoga, E., Kito, S., Kishida, T., Yanaihara, N., Ogawa, N. and Wakabayashi, I. (1980) Ultrastructural localization of neuropeptides in the rat primary sensory neurones. *Acta Histochem. Cytochem.*, 13: 407–420.

Itoh, N., Obata, K., Yanaihara, N. and Okamoto, M. (1983) Human preprovasoactive intestinal polypeptide contains a novel PHI-27-like peptide, PHM-27. *Nature*, 304: 547–549.

Jacobs, B. L., Gannon, P. J. and Azmitia, E. C. (1984) Altas of serotonergic cell bodies in the cat brainstem: An immunocytochemical analysis. *Brain Res. Bull.*, 13: 1–31.

Jaeger, C. B., Teitelman, G., Joh, T. H., Albert, V. R., Park, D. H. and Reiss, D. J. (1983) Some neurons of the rat central nervous system contain aromatic-L-amino decarboxylase but not monoamines. *Science*, 219: 1233–1235.

Jahr, C. E. and Jessell, T. M. (1983) ATP excites a subpopulation of rat dorsal horn neurones. *Nature*, 304: 730–733.

Jan, L. Y. and Jan, Y. N. (1982) Peptidergic transmission in sympathetic ganglia of the frog. *J. Physiol. (London)*, 327: 219–246.

Jansco, G., Hökfelt, T., Lundberg, J. M., Kiraly, E., Halasz, N., Nilsson, G., Terenius, L., Rehfeld, J., Steinbusch, H., Verhofstad, A., Elde, R., Said, S. and Brown, M. (1981) Immunohistochemical studies on the effect of capsaicin on spinal and medullary peptide and monamine neurons using antisera to substance P, gastrin/CCK, somatostatin, VIP, enkephalin, neurotensin and 5-hydroxytryptamine. *J. Neurocytol.*, 10: 963–980.

Johansson, O. and Hökfelt, T. (1980) Thyrotropin releasing hormone, somatostatin, and enkephalin: Distribution studies using immunohistochemical techniques. *J. Histochem. Cytochem.*, 28: 364–366.

Johansson, O., Hökfelt, T., Pernow, B., Jeffcoate, S. L., White, N., Steinbusch, H. W. M., Verhofstad, A. A. J., Emson, P. C. and Spindel, E. (1981) Immunohistochemical support for three putative transmitters in one neuron: Coexistence of 5-hydroxytryptamine, substance P- and thyrotropin releasing hormone-like immunoreactivity medullary neurons projecting to the spinal cord. *Neuroscience*, 6: 1857–1881.

Johansson, O., Hökfelt, T. and Elde, R. P. (1984) Immunohistochemical distribution of somatostatin-like immunoreactivity in the central nervous system of the adult rat. *Neuroscience*, 13: 265–339.

Jordan, L. M., Kenshalo, Jr., D. R., Martin, R. F., Haber, L. H. and Willis, W. D. (1979) Two populations of spinothalamic tract neurons with opposite responses to 5-hydroxytryptamine. *Brain Res.*, 164: 342–346.

Kanazawa, I., Sutoo, D., Oshima, I. and Saito, S. (1979) Effect of transection on choline acetyltransferase thyrotropin releasing hormone and substance P in the cat cervical spinal cord. *Neurosci. Lett.*, 13: 325–330.

Kataoka, K., Mizuno, N. and Frohman, L. A. (1979) Regional distribution of immunoreactive neurotensin in monkey brain. *Brain Res. Bull.*, 4: 57–60.

Kawatani, M., Lowe, I. P., Nadelhaft, I., Morgan, C. and De-Groat, W. C. (1983) Vasoactive intestinal polypeptide in visceral afferent pathways to the sacral spinal cord of the cat. *Neurosci. Lett.*, 42: 311–316.

Kawatani, M., Nagel, J., Houston, M. B., Eskay, R., Lowe, I. P. and DeGroat, W. C. (1984) Identification of leucine-enkephalin and other neuropeptides in pelvic and pudendal afferent pathways to the spinal cord of the cat. *Soc. Neurosci. Abstr.*, 10: 589.

Kimura, H., McGeer, P. L., Peng, J. H. and McGeer, E. G. (1981) The central cholinergic system studied by choline acetyltransferase immunohistochemistry in the cat. *J. Comp. Neurol.*, 200: 151–202.

Kimura, S., Okada, M., Sugita, Y., Kanazawa, I. and Munekata, E. (1983) Novel neuropeptides, neurokinin α and β, isolated from porcine spinal cord. *Proc. Jpn. Acad.*, 59 (ser. B): 101–104.

Kojima, M., Takeuchi, Y., Goto, M. and Sano, Y. (1982) Immunohistochemical study on the distribution of serotonin fibers in the spinal cord of the dog. *Cell Tissue Res.*, 226: 477–491.

Kojima, M., Takeuchi, Y., Goto, M. and Sano, Y. (1983) Immunohistochemical localization of serotonin fibers and terminals in the spinal cord of the monkey *(Macca fuscata)*. *Cell Tissue Res.*, 229: 23–36.

Kuypers, H. G. J. M. and Maisky, V. A. (1975) Retrograde axonal transport of horseradish peroxidase from spinal cord to brain stem cell groups in the cat. *Neurosci. Lett.*, 1: 9–14.

LaMotte, C. C. and DeLanerolle, N. C. (1983a) Ultrastructure of chemically defined neuron systems in the dorsal horn of the monkey. II. Methionine-enkephalin immunoreactivity. *Brain Res.*, 274: 51–63.

LaMotte, C. C. and DeLanerolle, N. C. (1983b) Ultrastructure of chemically defined neuron systems in the dorsal horn of the monkey. III. Serotonin immunoreactivity. *Brain Res.*, 274: 65–77.

LaMotte, C. C., Johns, D. R. and DeLanerolle, N. C. (1982) Immunohistochemical evidence of indolamine neurons in monkey spinal cord. *J. Comp. Neurol.*, 206: 359–370.

Lang, R. E., Heil, J., Ganten, D., Hermann, K., Rascher, W. and Unger, Th. (1983) Effects of lesions in the paraventricular nucleus of the hypothalamus on vasopressin and oxytocin contents in brainstem and spinal cord of rat. *Brain Res.*, 260: 326–329.

Larsson, L. I. and Rehfeld, J. F. (1979) Localization and molecular heterogeneity of cholecystokinin in the central and peripheral nervous system. *Brain Res.*, 165: 201–218.

Lehtosalo, J. I., Uusitalo, H., Stjernschantz, J. and Palkama, A. (1984) Substance P-like immunoreactivity in the trigeminal ganglion: A fluorescence, light and electron microscopic study. *Histochemistry*, 80: 421–427.

Leichnetz, G. R., Watkins, L., Griffin, G., Murfin, R. and Mayer, D. J. (1978) The projection from nucleus raphe magnus and other brain stem nuclei to the spinal cord in the rat: a study using the HRP blue-reaction. *Neurosci. Lett.*, 8: 119–124.

Light, A. R. and Perl, E. R. (1979) Spinal termination of functionally identified primary afferent neurons with slowly conducting myelinated fibers. *J. Comp. Neurol.*, 186: 133–150.

Light, A. R., Kavookjian, A. M. and Petrusz, P. (1983) The ultrastructure and synaptic connections of serotonin-immunoreactive terminals in spinal laminae I and II. *Somatosensory Res.*, 1: 33–50.

Ljungdahl, A., Hökfelt, T. and Nilsson, G. (1978) Distribution of substance P-immunoreactivity in the central nervous system of the rat. I. Cell bodies and nerve terminals. *Neuroscience*, 3: 861–943.

Lu, G.-W., Bennett, G. J., Nishikawa, N., Hoffert, M. J. and Dubner, R. (1983) Extra- and intracellular recordings from dorsal column postsynaptic spinomedullary neurons in the cat. *Exp. Neurol.*, 82: 456–477.

Lundberg, J. M., Hökfelt, T., Nilsson, G., Terenius, L., Rehfeld, J., Elde, R. and Said, S. (1978) Peptide neurons in the vagus, splanchnic and sciatic nerves. *Acta Physiol. Scand.*, 104: 499–501.

Lundberg, J. M., Hökfelt, T., Anggard, A., Kimmel, J., Goldstein, M. and Markey, K. (1980) Co-existence of avian pancreatic polypeptide (APPP) immunoreactive substance and catecholamines in some peripheral and central neurons. *Acta Physiol. Scand.*, 110: 107–109.

Lundberg, J. M., Terenius, L., Hökfelt, T. and Tatemoto, K. (1984) Comparative immunohistochemical and biochemical analysis of pancreatic polypeptide-like peptides with special reference to presence of neuropeptide Y in central and peripheral neurons. *J. Neurosci.*, 4: 2376–2386.

Marchand, R. and Barbeau, H. (1982) Vertically oriented alternating acetylcholinesterase rich and poor territories in laminae VI, VII, VIII of the lumbosacral cord of the rat. *Neuroscience*, 7: 1197–1202.

Marley, P. D., Nagy, J. I., Emson, P. C. and Rehfeld, J. F. (1982) Cholecystokinin in the rat spinal cord: distribution and lack of effect of neonatal capsaicin treatment and rhizotomy. *Brain Res.*, 238: 494–498.

Martin, R. F., Jordan, L. M. and Willis, W. D. (1978) Differential projections of cat medullary raphe neurons demonstrated by retrograde labeling following spinal cord lesions. *J. Comp. Neurol.*, 182: 77–88.

Massari, V. J., Tizabi, Y., Park, C. H., Moody, T. W., Helke, C. J. and O'Donohue, T. L. (1983) Distribution and origin of bombesin, substance P and somatostatin in cat spinal cord. *Peptides*, 4: 673–681.

Maxwell, D. J., Fyffe, R. E. W. and Rethelyi, M. (1983a) Morphological properties of physiologically characterized lamina III neurones in the cat spinal cord. *Neuroscience*, 10: 1–22.

Maxwell, D. J., Leranth, C. and Verhofstad, A. A. J. (1983b) Fine structure of serotonin-containing axons in the marginal zone of the rat spinal cord. *Brain Res.*, 266: 253–259.

McDonald, T. J., Jörnvall, H., Nilsson, G., Vagne, M., Ghatei, M., Bloom, S. R. and Mutt, V. (1979) Characterization of a gastrin releasing peptide from porcine non-antral gastric tissue. *Biochem. Biophys. Res. Comm.*, 90: 227–233.

McGregor, G. P., Gibson, S. J., Sabate, I. M., Blank, M. A., Christofides, N. D., Wall, P. D., Polak, J. M. and Bloom, S. R. (1984) Effect of peripheral nerve section and nerve crush on spinal cord neuropeptides in the rat; increased VIP and PHI in the dorsal horn. *Neuroscience*, 13: 207–216.

McKellar, S. and Loewy, A. D. (1982) Efferent projections of the A1 catecholamine cell group in the rat: An autoradiographic study. *Brain Res.*, 241: 11–29.

McLaughlin, B. J., Barber, R., Saito, K., Roberts, E. and Wu, J. Y. (1975) Immunocytochemical localization of glutamate decarboxylase in rat spinal cord. *J. Comp. Neurol.*, 164: 305–322.

Melander, T., Hökfelt, T., Rökaeus, A., Tatemoto, K. and Mutt, V. (1984) Galanin immunoreactive neurons in the central and peripheral nervous system. *Soc. Neurosci. Abstr.*, 10: 694.

Merchenthaler, I., Hynes, M. A., Vigh, S., Shally, A. V. and Petrusz, P. (1983) Immunocytochemical localization of corticotropin releasing factor (CRF) in the rat spinal cord. *Brain Res.*, 275: 373–377.

Miletic, V., Hoffert, M. J., Ruda, M. A., Dubner, R. and Shigenaga, Y. (1984) Serotonergic axonal contacts on identified cat spinal dorsal horn neurons and their correlation with nucleus raphe magnus stimulation. *J. Comp. Neurol.*, 228: 129–141.

Millan, M. J., Millan, M. H., Czlonkowski, A. and Herz, A. (1984) Vasopressin and oxytocin in the rat spinal cord: distribution and origins in comparison to [Met]enkephalin, dynorphin and related opioids and their irresponsiveness to stimuli modulating neurohypophyseal secretion. *Neuroscience,* 13: 179–188.

Minamino, N., Kangawa, K., Fukuda, A. and Matsuo, H. (1984) Neuromedin L: A novel mammalian tachykinin identified in porcine spinal cord. *Neuropeptides,* 4: 157–166.

Moody, T. W. and Pert, C. B. (1979) Bombesin-like peptides in rat brain: quantitation and biochemical characterization. *Biochem. Biophys. Res. Comm.,* 90: 7–14.

Morgan, C., Nadelhaft, I. and DeGroat, W. C. (1981) The distribution of visceral primary afferents from the pelvic nerve to Lissauer's tract and the spinal gray matter and its relationship to the sacral parasympathetic nucleus. *J. Comp. Neurol.,* 201: 415–440.

Morris, R., Salt, T. E., Sofroniew, M. V. and Hill, R. G. (1980) Actions of microiontophoretically applied oxytocin, and immunohistochemical localization of oxytocin, vasopressin and neurophysin in the rat caudal medulla. *Neurosci. Lett.,* 18: 163–168.

Morrison, J. H., Benoit, R., Magistretti, P. J., Ling, N. and Bloom, F. E. (1982) Immunohistochemical distribution of pro-somatostatin-related peptides in hippocampus. *Neurosci. Lett.,* 34: 137–142.

Murase, K. and Randic, M. (1984) Action of substance P on rat spinal dorsal horn neurones. *J. Physiol. (London),* 346: 203–217.

Naftchi, N. E., Abrahams, S. J., St. Paul, H. M. and Vacca, L. L. (1981) Substance P and leucine-enkephalin changes after chordotomy and morphine treatment. *Peptides,* 2: 61–70.

Nagy, J. I. and Hunt, S. P. (1981) Fluoride-resistant acid phosphatase-containing neurones in dorsal root ganglia are separate from those containing substance P or somatostatin. *Neuroscience,* 7: 89–97.

Nagy, J. I., Vincent, S. R., Staines, W. A., Fibiger, H. C., Reisine, T. D. and Yamamura, H. I. (1980) Neurotoxic action of capsaicin on spinal substance P neurons. *Brain Res.,* 186: 435–444.

Nagy, J. I., Hunt, S. P., Iversen, L. L. and Emson, P. C. (1981) Biochemical and anatomical observations on the degeneration of peptide-containing primary afferent neurons after neonatal capsaicin. *Neuroscience,* 6: 1923–1934.

Nahin, R. L., Madsen, A. M. and Giesler, Jr., G. J. (1983) Anatomical and physiological studies of the gray matter surrounding the spinal cord central canal. *J. Comp. Neurol.,* 220: 321–335.

Navaratnam, V. and Lewis, P. R. (1970) Cholinesterase-containing neurons in the spinal cord of the rat. *Brain Res.,* 18: 411–425.

Nawa, H., Hirose, T., Takashima, H., Inayama, S. and Nakanishi, S. (1983) Nucleotide sequence of cloned cDNAs for two types of bovine brain substance P precursor. *Nature,* 306: 32–36.

Neuhuber, W. (1982) The central projections of visceral primary afferent neurons of the inferior mesenteric plexus and hypogastric nerve and the location of the related sensory and preganglionic sympathetic cell bodies in the rat. *Anat. Embryol.,* 164: 413–425.

Ninkovic, M., Hunt, S. P. and Kelly, J. S. (1981) Effect of dorsal rhizotomy on the autoradiographic distribution of opiate and neurotensin receptors and neurotensin-like immunoreactivity within the rat spinal cord. *Brain Res.,* 230: 111–119.

Nishikawa, N., Bennett, G. J., Ruda, M. A., Lu, G.-W. and Dubner, R. (1983) Immunocytochemical evidence for a serotonergic innervation of dorsal column postsynaptic neurons in cat and monkey. *Neuroscience,* 10: 1333–1340.

Olschowka, J. A., O'Donohue, T. L. and Jacobowitz, D. M. (1981) The distribution of bovine pancreatic polypeptide-like immunoreactive neurons in rat brain. *Peptides,* 2: 309–331.

Olschowka, J. A., O'Donohue, T. L., Mueller, G. P. and Jacobowitz, D. M. (1982) Hypothalamic and extrahypothalamic distribution of CRF-like immunoreactive neurons in the rat brain. *Neuroendocrinology,* 35: 305–308.

Panula, P., Yang, H.-Y. T. and Costa, E. (1982) Neuronal location of the bombesin-like immunoreactivity in the central nervous system of the rat. *Regul. Peptides,* 4: 275–283.

Panula, P., Hadjiconstantinou, M., Yang, H.-Y. T. and Costa, E. (1983) Immunohistochemical localization of bombesin/gastrin-releasing peptide and substance P in primary sensory neurons. *J. Neurosci.,* 3: 2021–2029.

Paull, W. K. and Gibbs, F. P. (1983) The corticotropin releasing factor (CRF) neurosecretory system in intact, adrenalectomized and adrenalectomized-dexamethasone treated rats: An immunocytochemical analysis. *Histochemistry,* 78: 303–316.

Petrusz, P., Sar, M., Ordronneau, P. and DiMeo, P. (1976) Specificity in immunocytochemical staining. *J. Histochem. Cytochem.,* 24: 1110–1115.

Petrusz, P., Sar, M., Ordronneau, P. and DiMeo, P. (1977) Reply to the letter of Swaab et al. Can specificity ever be proved in immunocytochemical staining? *J. Histochem. Cytochem.,* 25: 390–391.

Pickel, V. M., Miller, R., Chan, J. and Sumal, K. K. (1983) Substance P and enkephalin in transected axons of medulla and spinal cord. *Regul. Peptides,* 6: 121–135.

Priestley, J. V. and Cuello, A. C. (1983) Substance P immunoreactive terminals in the spinal trigeminal nucleus synapse with lamina I neurons projecting to the thalamus. In P. Skraborek and D. Powell (Eds.), *Substance P,* Book Press, Dublin, pp. 251–252.

Priestley, J. V., Somogyi, P. and Cuello, A. C. (1982) Immunocytochemical localization of substance P in the spinal trigeminal nucleus of the rat: A light and electron microscopic study. *J. Comp. Neurol.,* 211: 31–49.

Proudfit, H. K. and Hammond, D. L. (1981) Alterations in nociceptive threshold and morphine-induced analgesia produced by intrathecally administered amine antagonists. *Brain Res.,* 218: 393–399.

Randic, M. and Miletic, V. (1977) Effect of substance P in cat dorsal horn neurones activated by noxious stimuli. *Brain Res.*, 128: 164–169.

Randic, M. and Yu, H. H. (1976) Effects of 5-hydroxytryptamine and bradykinin in cat dorsal horn neurones activated by noxious stimuli. *Brain Res.*, 111: 197–203.

Rehfeld, J. F. (1978) Immunochemical studies on cholecystokinin. II. Distribution and molecular heterogeneity in the central nervous system and small intestine of man and hog. *J. Biol. Chem.*, 253: 4022–4030.

Rethelyi, M. and Szentagothai, J. (1973) Distribution and connections of afferent fibres in the spinal cord. In A. Iggo (Ed.), *Handbook of Sensory Physiology: Somatosensory System, Vol. 2,* Springer-Verlag, New York, pp. 207–252.

Ribeiro-da-Silva, A. and Coimbra, A. (1980) Neuronal uptake of [^{3}H]GABA and [^{3}H]glycine in laminae I–III (substantia gelatinosa Rolandi) of the rat spinal cord. An autoradiographic study. *Brain Res.*, 188: 449–464.

Riley, D. A., Ellis, S. and Bain, J. L. W. (1984) Ultrastructural cytochemical localization of carbonic anhydrase activity in rat peripheral sensory and motor nerves, dorsal root ganglia and dorsal column nuclei. *Neuroscience,* 13: 189–206.

Roppolo, J. R., Lowe, I. P. and DeGroat, W. C. (1984) Immunohistochemical identification of leucine enkephalin in dorsal root ganglion cells of the rhesus monkey. *Soc. Neurosci. Abstr.*, 10: 993.

Ross, C. A., Armstrong, D. M., Ruggiero, D. A., Pickel, V. M., Joh, T. H. and Reis, D. J. (1981) Adrenaline neurons in the rostral ventrolateral medulla innervate thoracic spinal cord: A combined immunocytochemical and retrograde transport demonstration. *Neurosci. Lett.*, 25: 257–262.

Ruda, M. A. (1982) Opiates and pain pathways: Demonstration of enkephalin synapses on dorsal horn projection neurons. *Science,* 215: 1523–1525.

Ruda, M. A. (1985) The pattern and place of nociceptive modulation in the dorsal horn: A discussion of the anatomically characterized neural circuitry of enkephalin, serotonin and substance P. In T. L. Yaksh (Ed.), *Functional Organization of Spinal Afferent Processing,* Plenum Press, New York (in press).

Ruda, M. A. and Coffield, J. (1983) Light and ultrastructural immunocytochemical localization of serotonin synapses on primate spinothalamic tract neurons. *Soc. Neurosci. Abstr.*, 9: 1.

Ruda, M. A., Coffield, J. and Steinbusch, H. W. M. (1982) Immunocytochemical analysis of serotonergic axons in laminae I and II of the lumbar spinal cord of the cat. *J. Neurosci.*, 2: 1660–1671.

Ruda, M. A., Coffield, J., Bennett, G. J. and Dubner, R. (1983) Role of serotonin (5-HT) and enkephalin (ENK) in trigeminal and spinal pain pathways. *J. Dent. Res.*, 62: 691.

Ruda, M. A., Coffield, J. and Dubner, R. (1984) Demonstration of postsynaptic opioid modulation of thalamic projection neurons by the combined techniques of retrograde horseradish peroxidase and enkephalin immunocytochemistry. *J. Neurosci.*, 4: 2117–2132.

Salt, T. E. and Hill, R. G. (1983) Neurotransmitter candidates of somatosensory primary afferent fibres. *Neuroscience,* 10: 1083–1103.

Saper, C. B., Loewy, A. D., Swanson, L. W. and Cowan, W. M. (1976) Direct hypothalamo-autonomic connections. *Brain Res.*, 117: 305–312.

Sasek, C. A., Seybold, V. S. and Elde, R. P. (1984) The immunohistochemical localization of nine peptides in the sacral parasympathetic nucleus and the dorsal gray commissure in rat spinal cord. *Neuroscience,* 12: 855–874.

Satoh, K., Kashiba, A., Kimura, H. and Maeda, T. (1982) Noradrenergic axon terminals in the substantia gelatinosa of the rat spinal cord. An electron-microscopic study using glyoxylic acid-potassium permanganate fixation. *Cell Tissue Res.*, 222: 359–378.

Sawchenko, P. E. and Swanson, L. W. (1982) Immunohistochemical identification of neurons in the paraventricular nucleus of the hypothalamus that project to the medulla or to the spinal cord in the rat. *J. Comp. Neurol.*, 205: 260–272.

Schrøder, H. D. (1977) Sulfide silver architectonics of the rat, cat, and guinea pig spinal cord. A light microscopic study with Timm's method for demonstration of heavy metals. *Anat. Embryol.*, 150: 251–267.

Schrøder, H. D. (1983) Localization of cholecystokinin like immunoreactivity in the rat spinal cord, with particular reference to the autonomic innervation of the pelvic organs. *J. Comp. Neurol.*, 217: 176–186.

Schrøder, H. D. (1984) Somatostatin in the caudal spinal cord: an immunohistochemical study of the spinal centers involved in the innervation of pelvic organs. *J. Comp. Neurol.*, 223: 400–414.

Schultzberg, M., Dockray, G. J. and Williams, R. G. (1982) Capsaicin depletes CCK-like immunoreactivity detected by immunohistochemistry, but not that measured by radioimmunoassay in rat dorsal spinal cord. *Brain Res.*, 235: 198–204.

Senba, E., Tohyama, M., Shiosaka, S., Takagi, H., Sakanaka, M., Matsuzaki, T., Takahashi, Y. and Shimizu, N. (1981) Experimental and morphological studies of the noradrenaline innervations in the nucleus tractus spinalis nervi trigemini of the rat with special reference to their fine structures. *Brain Res.*, 206: 39–50.

Senba, E., Shiosaka, S., Hara, Y., Inagaki, S., Sakanaka, M., Takatsuki, K., Kawai, Y. and Tohyama, M. (1982) Ontogeny of the peptidergic system in the rat spinal cord: immunohistochemical analysis. *J. Comp. Neurol.*, 208: 54–66.

Seybold, V. S. and Elde, R. P. (1982) Neurotensin immunoreactivity in the superficial laminae of the dorsal horn of the rat: I. Light microscopic studies of cell bodies and proximal dendrites. *J. Comp. Neurol.*, 205: 89–100.

Silver, A. and Wolstencroft, J. H. (1971) The distribution of cholinesterase in relation to the structure of the spinal cord in the cat. *Brain Res.*, 34: 205–227.

Simantov, R., Kuhar, M. J., Uhl, G. R. and Snyder, S. H. (1977) Opioid peptide enkephalin: immunohistochemical mapping in rat central nervous system. *Proc. Natl. Acad. Sci. U.S.A.*, 74: 2167–2171.

Singer, E., Sperk, G., Placheta, P. and Leeman, S. E. (1979) Reduction of substance P levels in the ventral cervical spinal cord of the rat after cisternal 5,7 dihydroxytryptamine injection. *Brain Res.*, 174: 362–365.

Sjölund, B. H. and Eriksson, M. B. E. (1979) The influence of naloxone on analgesia produced by peripheral conditioning stimulation. *Brain Res.*, 178: 295–302.

Skagerberg, G., Björklund, A., Lindvall, O. and Schmidt, R. H. (1982) Origin and termination of the diencephalo-spinal dopamine system in the rat. *Brain Res. Bull.*, 9: 237–244.

Steinbusch, H. W. M. (1981) Distribution of serotonin-immunoreactivity in the central nervous system of the rat – Cell bodies and terminals. *Neuroscience*, 6: 557–618.

Stevens, R. T., Hodge, Jr., C. J. and Apkarian, A. V. (1982) Kölliker-Fuse nucleus: the principal source of pontine catecholaminergic cells projecting to the lumbar spinal cord of cat. *Brain Res.*, 239: 589–594.

Stine, S. M., Yang, H.-Y. and Costa, E. (1982) Evidence for ascending and descending intraspinal as well as primary sensory somatostatin projections in the rat spinal cord. *J. Neurochem.*, 38: 1144–1150.

Straus, E. and Yalow, R. S. (1979) Gastrointestinal peptides in the brain. *Fed. Proc.*, 38: 2320–2324.

Sumal, K. K., Pickel, V. M., Miller, R. J. and Reis, D. J. (1982) Enkephalin-containing neurons in substantia gelatinosa of spinal trigeminal complex: ultrastructure and synaptic interaction with primary sensory afferents. *Brain Res.*, 248: 223–236.

Swaab, D. F. and Fisser, B. (1977) Immunocytochemical localization of α-melanocyte stimulating hormone (αMSH)-like compounds in the rat nervous system. *Neurosci. Lett.*, 7: 313–317.

Swanson, L. W. and McKellar, S. (1979) The distribution of oxytocin and neurophysin stained fibers in the spinal cord of the rat and monkey. *J. Comp. Neurol.*, 188: 87–106.

Swanson, L. W., Sawchenko, P. E., Rivier, J. and Vale, W. W. (1983) Organization of ovine corticotropin-releasing factor immunoreactive cells and fibers in the rat brain: an immunohistochemical study. *Neuroendocrinology*, 36: 165–186.

Takeuchi, Y., Kimura, H., Matsuura, T. and Sano, Y. (1982a) Immunohistochemical demonstration of the organization of serotonin neurons in the brain of the monkey. *Acta Anat.*, 114: 106–124.

Takeuchi, Y., Kimura, H. and Sano, Y. (1982b) Immunohistochemical demonstration of the distribution of serotonin neurons in the brain stem of the rat and cat. *Cell Tissue Res.*, 224: 247–267.

Tashiro, T. and Ruda, M. A. (1985) Immunocytochemical identification of axons containing coexistent serotonin and substance P in the dorsal horn and lamina X of the cat lumbar spinal cord. *Soc. Neurosci. Abstr.*, 11.

Tatemoto, K. (1982) Neuropeptide Y: the complete amino acid sequence of the brain peptide. *Proc. Natl. Acad. Sci. U.S.A.*, 79: 5485–5489.

Tatemoto, K., Carlquist, M. and Mutt, V. (1982) Neuropeptide Y – a novel brain peptide with structural similarities to peptide YY and pancreatic polypeptides. *Nature*, 269: 659–660.

Tatemoto, K., Rökaeus, A., Jörnvall, H., McDonald, T. J. and Mutt, V. (1983) Galanin – A novel biologically active peptide from porcine intestine. *FEBS Lett.*, 164: 124–128.

Tessler, A., Glazer, E., Artmyshyn, R., Murray, M. and Goldberger, M. E. (1980) Recovery of substance P in the cat spinal cord after unilateral lumbosacral deafferentation. *Brain Res.*, 191: 459–470.

Tessler, A., Himes, B. T., Artmyshyn, R., Murray, M. and Goldberger, M. E. (1981) Spinal neurons mediate return of substance P following deafferentation of cat spinal cord. *Brain Res.*, 230: 263–281.

Tuchscherer, M. M. and Seybold, V. S. (1985) Immunohistochemical studies of substance P, cholecystokinin-octapeptide and somatostatin in dorsal root ganglia of the rat. *Neuroscience*, 14: 593–605.

Uhl, G. R., Goodman, R. R., Kuhar, M. J., Children, S. R. and Snyder, S. H. (1979a) Immunohistochemical mapping of enkephalin containing cell bodies, fibers and nerve terminals in the brain stem of the rat. *Brain Res.*, 166: 75–94.

Uhl, G. R., Goodman, R. R. and Snyder, S. H. (1979b) Neurotensin-containing cell bodies and nerve terminals in the brain stem of the rat: immunohistochemical mapping. *Brain Res.*, 167: 77–92.

Urban, L. and Randic, M. (1984) Slow excitatory transmission in rat dorsal horn: possible mediation by peptides. *Brain Res.*, 290: 336–341.

Vanderhaeghen, J. J., Deschepper, C., Lotstra, F., Vierendeels, G. and Schoenen, J. (1982) Immunohistochemical evidence for cholecystokinin-like peptides in neuronal cell bodies of the rat spinal cord. *Cell Tiss. Res.*, 223: 463–467.

Wall, P. D., Fitzgerald, M. and Gibson, S. J. (1981) The response of rat spinal cord cells to unmyelinated afferents after peripheral nerve section and after changes in substance P levels. *Neuroscience*, 6: 2205–2215.

Watkins, L. R., Griffin, G., Leichnetz, G. R. and Mayer, D. J. (1980) The somatotropic organization of the nucleus raphe magnus and surrounding brain stem structures as revealed by HRP slow-releasing gels. *Brain Res.*, 181: 1–15.

Werz, M. A. and MacDonald, R. L. (1983) Opioid peptides selective for mu- and delta-opiate receptors reduce calcium-dependent action potential duration by increasing potassium conductance. *Neurosci. Lett.*, 42: 173–178.

Wessendorf, M. W. and Elde, R. P. (1984) Distribution of spinal fibers and terminals in which serotonin (5-HT) and substance P (SP) like immunoreactivities coexist. *Soc. Neurosci. Abstr.*, 10: 696.

Westlund, K. N. and Coulter, J. D. (1980) Descending projections of the locus coeruleus and subcoeruleus/medial parabrachial nuclei in monkey: Axonal transport studies and dopamine-β-hydroxylase immunocytochemistry. *Brain Res. Rev.,* 2: 235–264.

Westlund, K. N., Bowker, R. M., Ziegler, M. G. and Coulter, J. D. (1981) Origins of spinal noradrenergic pathways demonstrated by retrograde transport of antibody to dopamine-β-hydroxylase. *Neurosci. Lett.,* 25: 243–249.

Westlund, K. N., Bowker, R. M., Ziegler, M. G. and Coulter, J. D. (1982) Descending noradrenergic projections and their spinal terminations. *Prog. Brain Res.,* 57: 219–238.

Wiklund, L., Leger, L. and Persson, M. (1981) Monoamine cell distribution in the cat brain stem: A fluorescence histochemical study with quantification of indolaminergic and locus coeruleus cell groups. *J. Comp. Neurol.,* 203: 613–647.

Willcockson, W. S., Chung, J. M., Hori, Y., Lee, K. H. and Willis, W. D. (1984a) Effects of iontophoretically released amino acids and amines on primate spinothalamic tract cells. *J. Neurosci.,* 4: 732–740.

Willcockson, W. S., Chung, J. M., Hori, Y., Lee, K. H. and Willis, W. D. (1984b) Effects of iontophoretically released peptides on primate spinothalamic tract cells. *J. Neurosci.,* 4: 741–750.

Williams, R. G. and Dockray, G. J. (1983) Immunohistochemical studies of FMRF-amide-like immunoreactivity in rat brain. *Brain Res.,* 276: 213–229.

Willis, W. D. and Coggeshall, R. E. (1978) *Sensory Mechanisms of the Spinal Cord,* Plenum Press, New York, pp. 1–485.

Wong, V., Barrett, C. P., Donati, E. J., Eng, L. F. and Guth, L. (1983) Carbonic anhydrase activity in first-order sensory neurons of the rat. *J. Histochem. Cytochem.,* 31: 293–300.

Yaksh, T. L., Jessell, T. M., Gamse, R., Mudge, A. W. and Leeman, S. E. (1980) Intrathecal morphine inhibits substance P release from mammalian spinal cord in vivo. *Nature,* 286: 155–157.

Yaksh, T. L., Hammond, D. L. and Tyce, G. M. (1981) Functional aspects of bulbospinal monoaminergic projections in modulating processing of somatosensory information. *Fed. Proc.,* 40: 2786–2794.

Yaksh, T. L., Schmauss, C., Micevych, P. E., Abay, E. O. and Go, V. L. W. (1982) Pharmacological studies on the application, disposition, and release of neurotensin in the spinal cord. *Ann. N.Y. Acad. Sci.,* 400: 228–243.

Yamazoe, M., Shiosaka, S., Yagura, A., Kawai, Y., Shibasaki, T., Ling, N. and Tohyama, M. (1984) The distribution of α-melanocyte stimulating hormone (αMSH) in the central nervous system of the rat: an immunohistochemical study. II. Lower brain stem. *Peptides,* 5: 721–727.

Yanaihara, N., Yanaihara, C., Mochizuki, T., Iwahara, K., Fujita, T. and Iwanaga, T. (1981) Immunoreactive GRP. *Peptides,* 2 (Suppl. 2): 185–191.

Yezierski, R. P., Wilcox, T. K. and Willis, W. D. (1982) The effects of serotonin antagonists on the inhibition of primate spinothalamic tract cells produced by stimulation in the nucleus raphe magnus or periaqueductal gray. *J. Pharmacol. Exp. Ther.,* 220: 266–277.

Yoshimura, M. and North, R. A. (1983) Substantia gelatinosa neurones hyperpolarized *in vitro* by enkephalin. *Nature,* 305: 529–530.

Yoshioka, K. and Jessell, T. M. (1984) ATP release from the dorsal horn of rat spinal cord. *Soc. Neurosci. Abstr.,* 10: 993.

Zieglgänsberger, W. and Tulloch, I. F. (1979) The effects of methionine- and leucine-enkephalin on spinal neurones of the cat. *Brain Res.,* 167: 53–64.

P. C. Emson, M. N. Rossor and M. Tohyama (Eds.),
Progress in Brain Research, Vol. 66.

CHAPTER 12

Peptides in the peripheral nervous system

Shinobu Inagaki and Shozo Kito

Third Department of Internal Medicine, Hiroshima University Medical School, 1-2-3 Kasumi, Minamiku, Hiroshima, 734 Japan

Introduction

Until recently only the classical neurotransmitters catecholamine (CA) and acetylcholine (ACH) were found in the peripheral nervous system. However, recent immunological studies have demonstrated a great variety of neurotransmitter or neuromodulator candidates, particularly peptides, in the peripheral, as well as in the central, nervous system. In this chapter, we will first describe the general morphological profiles of these various bioactive peptides in the peripheral nervous system, and then, their proposed roles in several specific organs. In addition to the peptides, the distribution of serotonin (5-HT) will be described briefly.

Morphological observations on the distribution of peptides in the peripheral nervous system

Sensory ganglia

Primary sensory neurons have been shown by immunohistochemistry to contain a variety of peptides including substance P (Hökfelt et al., 1975c, 1976; Cuello et al., 1978), a calcitonin gene-related polypeptide (Rosenfeld et al., 1983; Lee et al., 1985a), enkephalin (Senba et al., 1982), somatostatin (Hökfelt et al., 1975a, 1976), vasoactive intestinal polypeptide (Lundberg et al., 1978; Fuji et al., 1985) and cholecystokinin (Lundberg et al., 1978; Fuji et al., 1985). Fig. 5 shows the distribution of these peptides in the dorsal root ganglion and spinal cord.

Substance P/calcitonin gene-related peptide

Substance P (SP)-like immunoreactivity (SP-LI) is observed in a small number of fibres, and in small- to medium-size (15–30 μm) sensory neuronal cell bodies following colchicine treatment in the cat (Hökfelt et al., 1975c) and rat (Cuello et al., 1978; Senba et al., 1982; Lee et al., 1985a) (see Figs. 1A, 2A, 3A). SP is located in the primary sensory afferent neuron system and is believed to act as a neurotransmitter or neuromodulator (Otsuka and Konishi, 1976). Ontogenetic immunohistochemical (Senba et al., 1982) and radioimmunoassay (Kessler and Black, 1981) studies have demonstrated that SP-LIr cells appear on gestational day 17 of rat and gradually increase in number. Five to ten percent of the ganglion cells are SP-positive at birth and reach their maximum number on postnatal days 5–7. About 20% of the ganglion cells are SP-positive. In the adult rat, SP-LI can be detected in colchicine-treated ganglia but is not readily detected in the untreated ganglia. Most of the SP-LI cells are small (20–30 μm in diameter) (Figs. 1A, 2A, 3A). In the dorsal roots a few SP-LI fibres are observed on gestational day 20 and they increase in number as the rat grows (Fig. 2A). Numerous fibres can be seen in the peripheral branch of the spinal ganglion (Fig. 2A). No obvious difference between the cervical, thoracic or lumbar ganglia is found.

Calcitonin gene-related peptide (CGRP) is composed of 37 amino acids, whose presence has been demonstrated in nervous tissue by recombinant DNA and molecular biological techniques (Amara et al., 1982). The presence of neurons with CGRP-

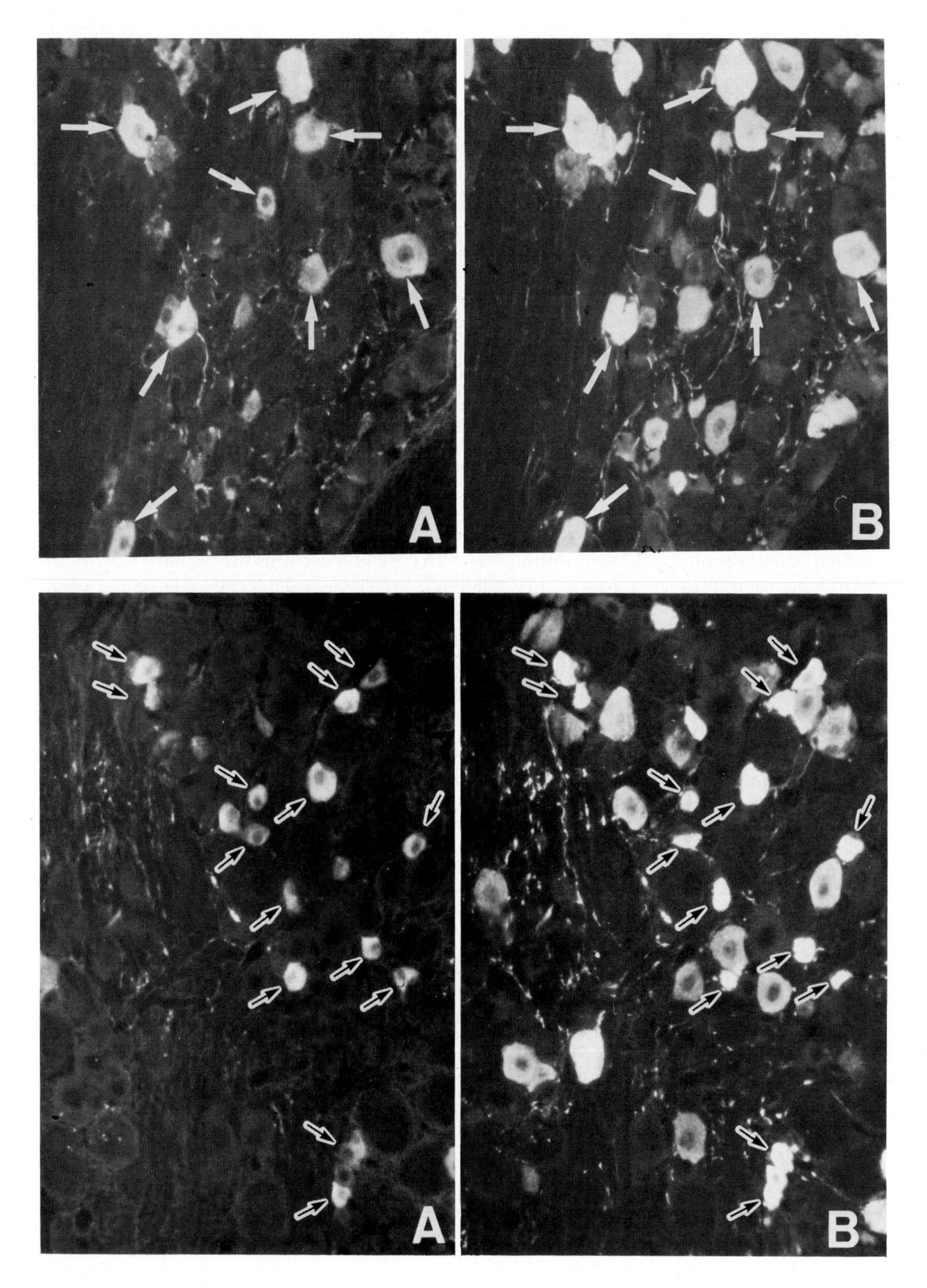
A
B
A
B

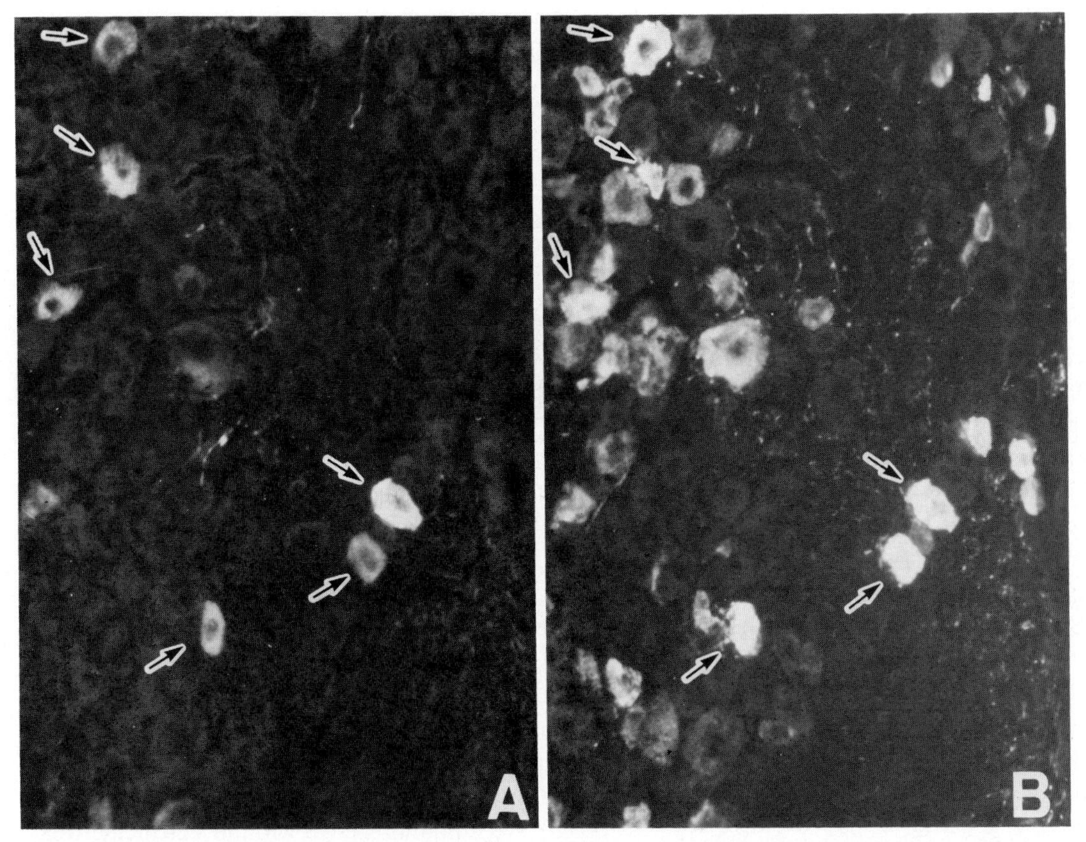

Fig. 3. Fluorescent photomicrographs of a frozen section of the nodosal ganglion of colchicine-treated rat. SP-LIr structures show fluorescein isothiocyanate green fluorescence (A) and CGRP-LIr structures show red fluorescence of Texas red (B) in the same section. Arrows show double-labeled ganglion cells with SP and CGRP antibodies. They are small- to medium-size ganglion cells. Large cells with CGRP-LI (B) but without SP-LI are seen in the nodosal ganglion as found in the trigeminal and dorsal root ganglia. ×200. (Adapted from Lee et al., 1985c, with permission.)

-like immunoreactivity (CGRP-LI) was reported in the peripheral nervous system by Rosenfeld et al. (1983) in the trigeminal ganglion and in fibres of the tongue.

Recent studies using consecutive sections and double staining techniques which are described in detail in this volume in the chapter by Shiosaka and Tohyama have demonstrated that both CGRP-LI

Fig. 1. Fluorescent photomicrographs showing the distribution of SP (A) and CGRP (B) immunoreactivities in the trigeminal ganglion of the colchicine-treated rat in consecutive sections. Arrows indicate SP-LIr neurons (A) which simultaneously contain CGRP-LI (B). Note that CGRP-LIr neurons are more numerous than SP-LIr neurons. ×200. (From Lee et al., 1985a, with permission.)

Fig. 2. Immunofluorescent photomicrographs of a frozen section of the rat dorsal root ganglion with colchicine treatment. SP-LI was visualized by fluorescein isothiocyanate (FITC) green fluorescence following incubation with monoclonal antibody to SP followed by FITC-labeled anti rat IgG (A). CGRP-LI was revealed by Texas red fluorescence following the incubations first with anti-CGRP rabbit serum and then with Texas red-labeled anti-rabbit IgG (B). In the dorsal root ganglion, many small- to medium-sized CGRP--LIr cells are positive for SP in the same section. Arrows indicate double-labeled ganglion cells with SP (A) and CGRP (B). They are small- to medium-size ganglion cells. As found in the trigeminal ganglion (Fig. 1), some of the large-size ganglion cells are CGRP-LIr (B) but lack SP-LI in the dorsal root ganglion. ×200. (Adapted from Lee et al., 1985c, with permission.)

and SP-LI coexist in single cells of the trigeminal ganglion (Gibson et al., 1984; Lee et al., 1985a, 1985b). The distribution patterns of CGRP-LI and SP-LI are in general similar in the trigeminal (Fig. 1), dorsal root (Fig. 2) and nodosal ganglia (Fig. 3), though several differences are seen. In the dorsal root ganglion, about 40% of the ganglion cells have CGRP-LI (Fig. 2B), while only about 20% of these cells show SP-LI (Fig. 2A) (Lee et al., 1985c). Forty percent of the CGRP-LI cells are large (30–45 μm in diameter), 30% are medium (20–30 μm in diameter), 30% are small (less than 20 μm in diameter). Four percent of the SP-LIr cells are large, 48% are medium, and 48% are small.

Double-immunostaining experiments show that the majority of the SP-LIr cells are also immunoreactive for CGRP (Figs. 1–3), although the majority of large cells with CGRP-LI lack SP-LI. A few cells with SP-LI but without CGRP-LI are also detected. In the trigeminal ganglion, a similar distribution pattern of the SP- and CGRP-positive cells is observed (Lee et al., 1985a) (Fig. 1). However, in the nodosal ganglion, the percentage of CGRP-LIr cells lacking SP-LI seems to be greater (Fig. 3) (Lee et al., 1985c).

The location of CGRP-LIr fibres in the trigeminal and dorsal root ganglia (Figs. 1B, 2B) is indistinguishable from that of the SP-LIr fibres (Figs. 1A, 2A) although the CGRP-LIr fibres are more numerous. However, in the nodosal ganglion, CGRP-LIr fibres which run in a different pattern from the SP-LIr fibres can be seen in much greater abundance together with other CGRP-LIr fibres which are arranged in the same way as the SP-LIr fibres. The presence of many large sized cells with CGRP in the primary sensory ganglia suggests that CGRP is closely related and SP is partly related to thermo- or mechanoreception in addition to any involvement in nociception.

Somatostatin

Ten to 15% of the spinal ganglion cells are somatostatin (SRIF)-immunoreactive (SRIF-LIr) in the colchicine-treated rats (Hökfelt et al., 1976; Senba et al., 1982) (Fig. 4). They are small- to medium-sized cells (15–30 μm). SRIF-LIr cells appear on gestational day 15 before the establishment of normal synaptic transmission. They gradually increase in immunoreactivity and 6–9% of the dorsal root ganglion cells are SRIF-positive at birth. On the 7–10th postnatal day, about 15% of the ganglion cells are immunoreactive without colchicine treatment. SRIF-LIr fibres are first detected in the dorsal root ganglion on gestational day 21. No difference concerning the distribution and ontogeny of SRIF-LIr cells was noted between the cervical and thoracic ganglia.

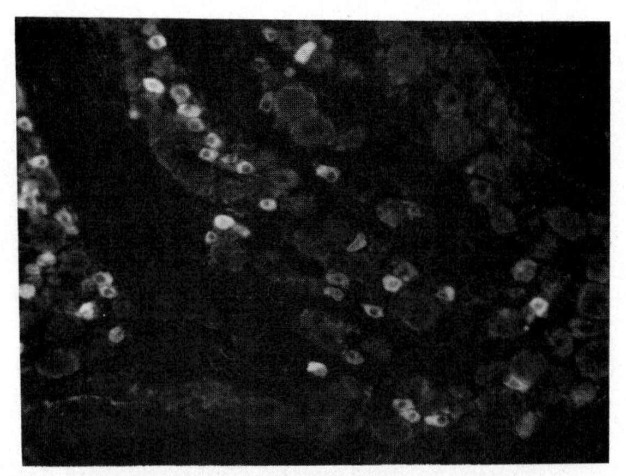

Fig. 4. Fluorescent photomicrographs showing SRIF-LI in the dorsal root ganglion of adult rat. SRIF-LIr ganglion cells are small to medium ganglion cells in the primary sensory ganglia. × 123. (From Senba et al., 1982, with permission.)

The fact that the SRIF-LIr fibres located in the dorsal horn maintain their immunoreactivity even in the adult rats plus the fact that they are supplied by SRIF-LIr cells found in the dorsal root ganglia suggest that SRIF might function as a neurotransmitter in the primary sensory system.

Enkephalin

In the dorsal root ganglion, a few enkephalin (ENK)-positive cells first appear in neonatal rats without colchicine treatment (Senba et al., 1982). Only a few ENK-like immunoreactive (ENK-LIr) cells are found in the colchicine-treated ganglion of adult rats. This group of cells is small to medium in size, about 15–30 μm. A few ENK-LIr fibres are

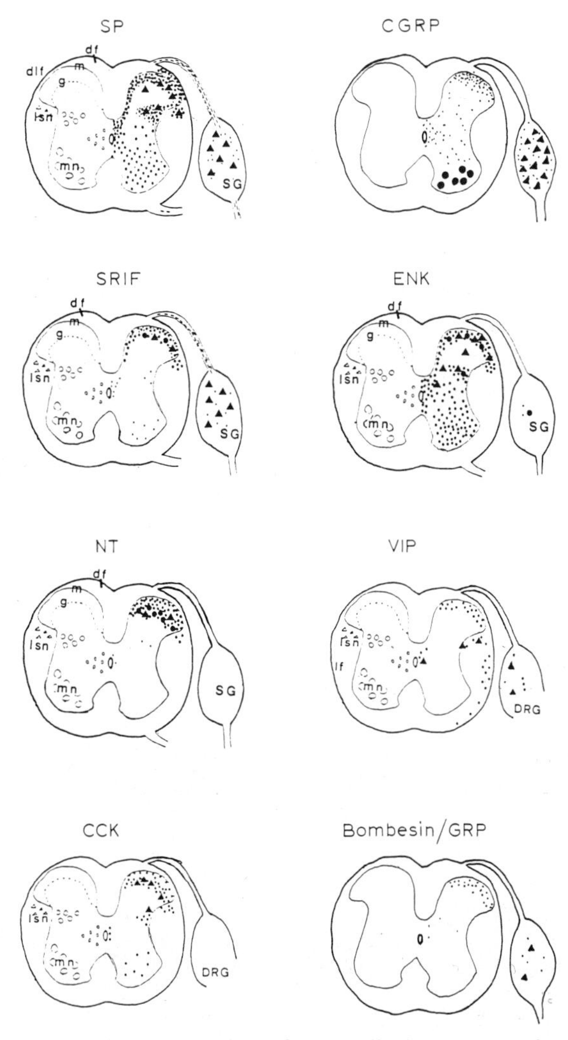

Fig. 5. Schematic drawings of the distribution patterns of various peptides in the dorsal root ganglion and spinal cord of the rat at the thoracic level. (●) immunoreactive perikarya; (·) immunoreactive fibres; (▲) immunoreactive perikarya detected following pretreatment with colchicine. Abbreviations: df, dorsal funiculus; DRG, dorsal root ganglion; lsn, a small nucleus of dorsal lateral funiculus; m, marginal layer; mn, motor neuron; SG, spinal ganglion. (Adapted and reconstructed from Senba et al., 1982, and Fuji et al., 1985.)

seen in the dorsal root. These data suggest that ENK is also involved in the primary sensory system, though the majority of the ENK-LIr fibres in the dorsal horn of the spinal cord are intrinsic. No remarkable difference concerning the location and ontogeny of ENK-LIr structures has been identified between the cervical and thoracic ganglia.

Vasoactive intestinal polypeptide

Vasoactive intestinal polypeptide (VIP)-like immunoreactivity (VIP-LI) has been demonstrated in the sacral sensory ganglion cells in a limited number of colchicine-treated cats (Lundberg et al., 1978), but no VIP-LIr neural cells were found in the untreated, cervical and thoracic sensory ganglia of rat (Fuji et al., 1985). Immunostaining was found predominantly in small- and medium-sized cells and VIP-LIr axons observed in close proximity to ganglion cells and coiled axonal profiles. Unilateral deafferentation of the sacral cord (S_1 to S_3) by either dorsal rhizotomy or ganglionectomy resulted in a nearly complete loss of VIP-like immunostaining in the ipsilateral dorsal horn (Honda et al., 1983). Thus VIP may be contained preferentially within pelvic visceral afferent fibres confined mostly to sacral segments. Ontogenetic development has also been demonstrated by Fuji et al. (1985). VIP-LIr fibres were first detected on postnatal day 1, after which they slightly increased in number. No immunoreactive neurons are observed in rat spinal ganglia at the cervical level throughout ontogenesis.

Cholecystokinin

The spinal ganglia also contain scattered cholecystokinin (CCK)-like immunoreactive (CCK-LIr) nerves in the guinea pig, which occur in the vicinity of ganglion cell bodies (Larsson et al., 1979). However no CCK-LIr structures have been observed in the rat during ontogenesis (Fuji et al., 1985). No evidence for the occurrence of CCK-LIr cell somata was obtained in guinea pig (Larsson et al., 1979) nor rat (Fuji et al., 1985).

Bombesin/gastrin-releasing peptide

About 5% of the spinal sensory ganglion cells exhibit bombesin/gastrin-releasing peptide (GRP)-like immunoreactivity (GRP-LI) (Panula et al., 1983). When the consecutive sections are stained, different cells exhibit SP and bombesin/GRP-LI, though no difference is found in the morphology of the small and medium size cells that contain the two peptides. The large cells in the ganglia do not exhibit bombesin/GRP-LI. Less numerous fibres are found in the ganglia than those exhibiting SP-LI.

Neurotensin

No neurotensin (NT)-positive structures have been found in the spinal sensory ganglion at the cervical level in the various ontogenic stages (Senba et al., 1982).

Neuropeptide Y

So far no neuropeptide Y (NPY)-like immunoreactive (NPY-LIr) ganglion cells have been found in the primary sensory ganglia (Lundberg et al., 1982; Lee et al., 1985b).

Motor systems innervating striated muscle

Classically, ACH is the recognized transmitter in motor neurons. In addition, however, recent reports have shown that CGRP-LI coexists with ACH in the somatomotor and branchiomotor neurons (Takami et al., 1985a).

Calcitonin gene-related polypeptide

Hypoglossal, facial and ambiguus nuclei contain numerous CGRP-LIr and choline acetyltransferase (ChAT)-like immunoreactive (ChAT-LIr) neurons (Takami et al., 1985a) (Figs. 6, 7). Simultaneous immunostaining for CGRP showed CGRP-LI in about 50–70% of the ChAT-LIr neurons in the hypoglossal nucleus, in about 80% of the ChAT-LIr cells in the facial nucleus (Fig. 6) and in about 90% of the ChAT-LIr cells in the ambiguus nucleus (Fig. 7). Numerous CGRP-LIr fibres are found in the terminal profiles of neuromuscular junctions of striated muscle, such as in the tongue (Figs. 8, 9A),

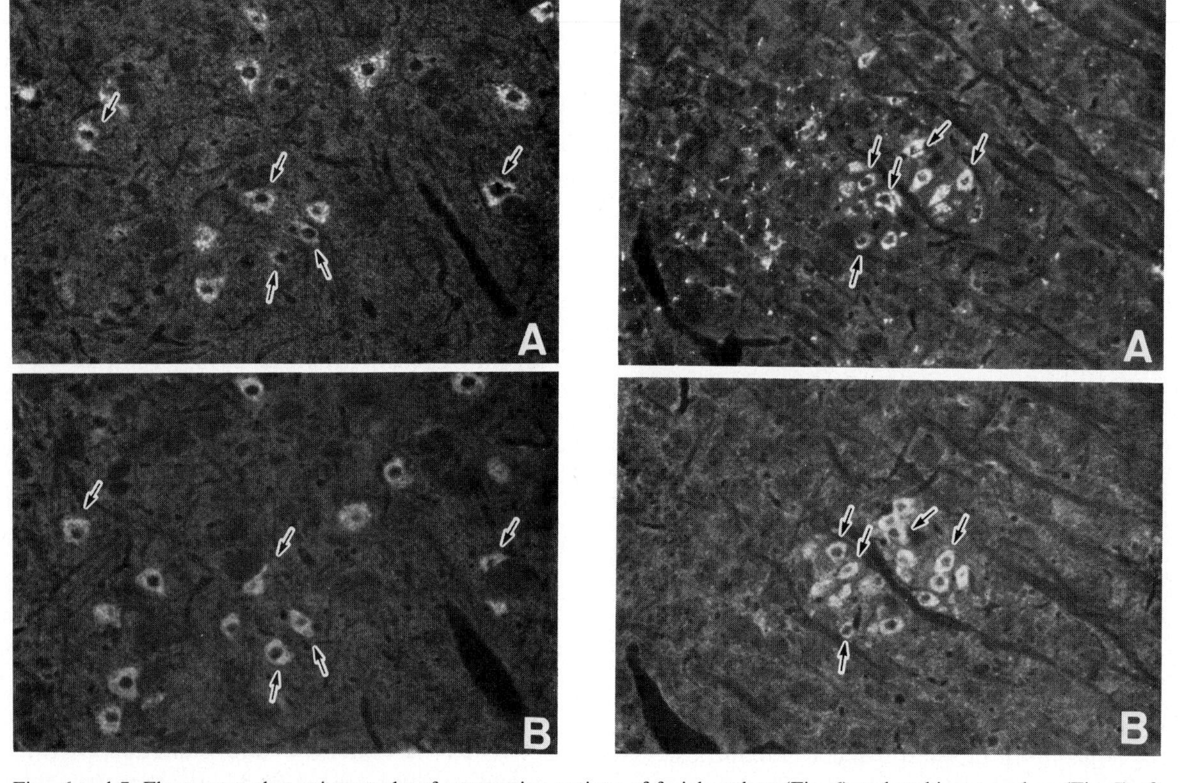

Figs. 6 and 7. Fluorescent photomicrographs of consecutive sections of facial nucleus (Fig. 6) and ambiguus nucleus (Fig. 7) after incubation with CGRP antiserum (A) and ChAT antibody (B). Most of the CGRP-LIr cells are simultaneously ChAT-positive (arrows). Frontal sections. × 160. (From Takami et al., 1985, with permission.)

Fig. 8. Fluorescent photomicrograph showing CGRP-LIr fibres that terminate at the striated muscles of the tongue. Arborization of CGRP-LIr terminals shows a neuromuscular junction profile. × 560. (From Takami et al., 1985, with permission.)

which also stain for acetylcholinesterase (Fig. 9B). Therefore, the evidence suggests that CGRP and ACH are colocalized in neurons of several motor systems which terminate in striated muscles to form motor end plates.

Using a phrenic nerve-diaphragm preparation, CGRP application has also been shown to enhance muscle contraction both during stimulation of the nerve fibres and during direct stimulation of the muscle (Takami et al., 1985b). This effect appears to be mediated via a CGRP receptor, rather than by the ACH receptor as it is not blocked by cholinergic antagonists.

Preganglionic parasympathetic system

ACH has been established as the transmitter of pre- and post-ganglionic parasympathetic fibres but recent reports have shown that several peptides coexist in this system (Hunt et al., 1981; Dalsgaard et al., 1983).

Enkephalin

ENK-LIr neural perikarya are scattered throughout the intermediate gray matter from the lateral edge to the central canal in the sacral cord from S_2 to S_3 in colchicine-treated cats (Glazer et

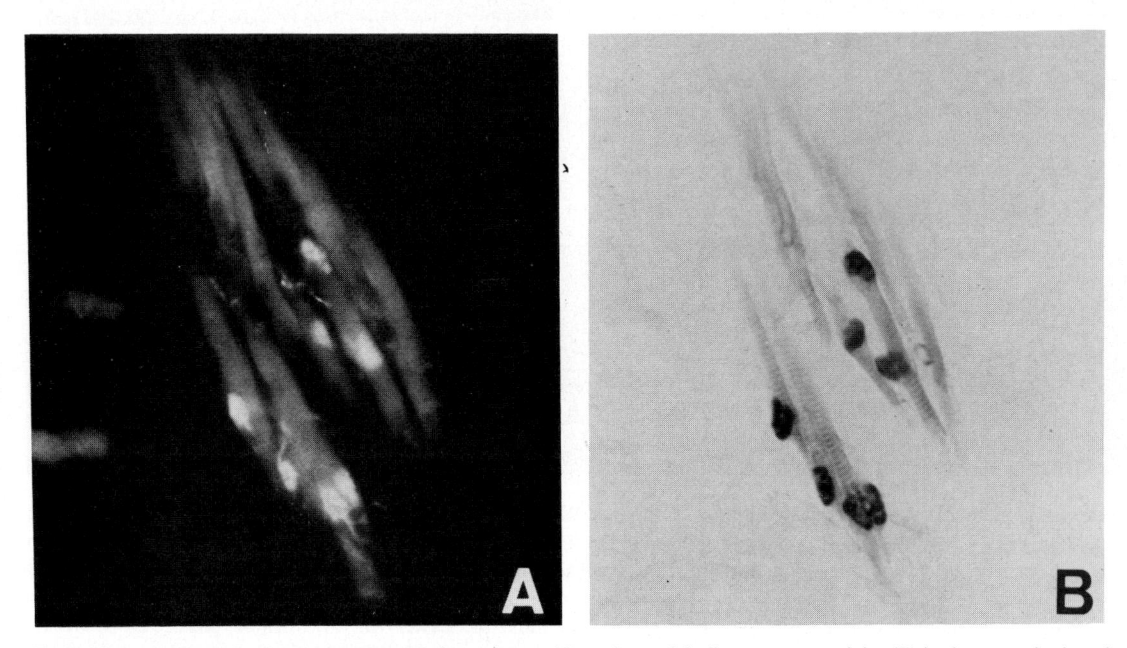

Fig. 9. Photomicrographs showing CGRP-LIr fluorescence (A) and acetylcholinesterase reactivity (B) in the muscular junctions of the striated muscle of the tongue on the same sections. × 240. (From Takami et al., 1985, with permission.)

al., 1978). The distribution and morphology of these cells in the cat are strikingly similar to those of the sacral parasympathetic preganglionic neurons which innervate the smooth musculature of the pelvic viscera. They have somata 30–60 μm in diameter and the dendrites are oriented dorsoventrally or mediolaterally. No ENK-LIr perikarya are found in the motor column of Onuf which innervates striated pelvic sphincter muscles. Moreover, ligation of the sacral ventral roots produced a buildup of ENK-LI in the S_2 ventral root proximal to the ligature and appearance of ENK-LIr perikarya in the intermediate gray matter of the S_2 sacral cord. No ENK-LIr products were seen in axonal processes of the distal S_2 root or anywhere in the S_1 or S_3 ventral root (Glazer et al., 1978). Therefore, parasympathetic preganglionic neurons contain and transport ENK-LIr compounds to the peripheral pelvic viscera.

The presence of ENK-LI has also been demon-

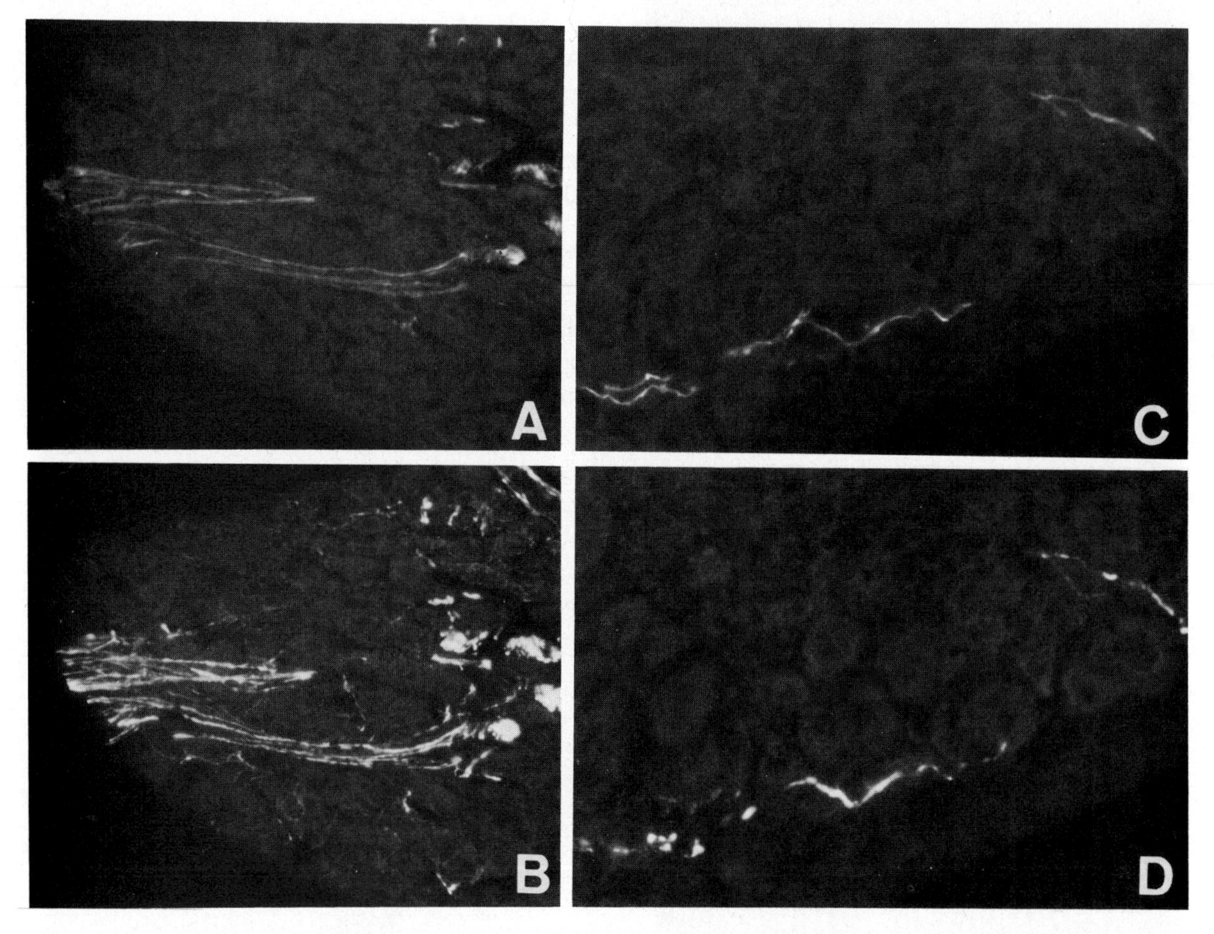

Fig. 10. Fluorescent photomicrographs showing SP-LI (A, C) and CGRP-LI (B, D) in the coeliac ganglion (A, B) and superior cervical ganglion (C, D). No SP-LIr or CGRP-LIr cell bodies are seen in the coeliac or superior cervical ganglia. The distribution patterns of labeled fibres differ between the two sympathetic ganglia. In the coeliac ganglion, thick fibre bundles which are immunoreactive for both SP and CGRP are seen. However, such thick-fibre bundles are not found in the superior cervical ganglion. Arrows indicate the transverse sections of the immunoreactive thick-fibre bundles. Note that numerous CGRP-LIr varicose fibres lacking SP-LI are seen in the coeliac ganglion (A, B), although this phenomenon is less prominent in the superior cervical ganglion (C, D). A, B, × 170; C, D, × 420. (From Lee et al., 1985c, with permission.)

strated in the vagus and splanchnic nerves which are composed of parasympathetic preganglionic and sensory nerves (Lundberg et al., 1978).

Neuropeptide Y

Since avian pancreatic polypeptide (APP)-like immunoreactivity (APP-LI) is thought to be the same as NPY-LI, APP-LIr structures will be described under NPY.

Intense APP-LI has been found in cell bodies of sacral parasympathetic systems (Hunt et al., 1981). These APP-LIr sacral parasympathetic preganglionic cells also contain ENK-LI.

Preganglionic sympathetic system

No information on the localization of the peptides in this system is available at present.

Postganglionic sympathetic ganglia

Noradrenaline (NA) is established as the neurotransmitter in postganglionic sympathetic fibres, but recently several neuropeptides have also been demonstrated in this system.

Substance P

Fine networks of SP-LIr fibres are found in the prevertebral sympathetic ganglia such as the coeliac and inferior mesenteric ganglia of rat, cat and guinea pig (Hökfelt et al., 1977b; Lee et al., 1985c). SP-LIr fibres are in general thin, smooth, and varicose (Fig. 10A, C). These SP-LIr fibres were demonstrated to be the peripheral branches of SP neurons in the primary sensory ganglia by denervating the inferior mesenteric ganglia in guinea pig (Konishi et al., 1979; Dalsgaard et al., 1983). Most of the SP-LIr fibres which are the peripheral branches of primary sensory ganglia seem to pass through the sympathetic ganglia and innervate various peripheral organs (Dalsgaard et al., 1983). In the stellate and superior cervical ganglia, only a few SP-LIr fibres are observed (Fig. 10C).

Calcitonin gene-related polypeptide

In the superior and coeliac ganglia of rat, no cells with CGRP-LI have been seen either with or without colchicine treatment (Lee et al., 1985c) (Fig. 10B, D). A network of CGRP-LIr fibres is found in these sympathetic ganglia. In the coeliac ganglion more numerous CGRP-LIr fibres are observed (Fig. 10B) than in the superior cervical ganglion (Fig. 10D). Two types of fibres, non-varicose and varicose, are found. The distribution pattern of CGRP-LIr fibres (Fig. 10B, D) is almost identical to that of the SP-LIr fibres (Fig. 10A, C) in the same double-stained sections (Fig. 10). However, more varicose CGRP-LIr fibres occur in the coeliac ganglion than SP-LIr fibres (Fig. 10).

Enkephalin

In the guinea pig, a very dense network of ENK-LIr fibres is found in the inferior mesenteric ganglion, while a less dense network of ENK-LIr fibres is seen in the coeliac ganglion, and only a few fibres in the superior cervical ganglion. Several small cells are also ENK-LIr in this sympathetic ganglion (Schultzberg et al., 1979).

In the rat, the inferior and coeliac ganglia contain a medium dense network of ENK-LIr fibres, while irregularly distributed fibres are present in the superior cervical ganglion where several ganglion cells are also ENK-positive (Schultzberg et al., 1979).

Vasoactive intestinal polypeptide

A few VIP-LIr cells are observed in the inferior mesenteric ganglion. Colchicine treatment enhances VIP-immunostaining in these cells with a decrease in the number of positive fibres. These VIP-LIr neurons are probably local circuit neurons or project to further sympathetic ganglia since VIP-LIr fibres were observed to decrease markedly but a few VIP-LIr fibres to remain in totally denervated ganglia (Dalsgaard et al., 1983).

A dense but uneven network of VIP-LIr fibres is observed over the inferior mesenteric and coeliac ganglia of rat and guinea pig (Hökfelt et al., 1977a). Some regions are densely innervated with VIP-LIr fibres surrounding the ganglion cells, whereas others are almost devoid of VIP-LIr fibres. Dalsgaard et al. (1983) by using various denervation techniques suggested that VIP-LIr fibres found in the inferior mesenteric ganglion arise in the perikarya of the colon. VIP-LIr fibres are more sparse in the superior cervical ganglion than in the coeliac and inferior mesenteric ganglia.

Somatostatin

In the coeliac and inferior mesenteric ganglia, a large number of ganglion cells contain SRIF-LI and NA (Hökfelt et al., 1977a). In the rat, these sympathetic ganglia contain weak SRIF immunostaining with a similar distribution pattern to that found in the guinea pig, though SRIF-LIr ganglion cells are less numerous in the rat than in the guinea pig. However, there is an inconsistency in that SRIF-LIr ganglionic neurons in the inferior mesenteric ganglion are supposed to project to the gut, as NA neurons do, while denervation of the gut did not result in a change in the network of SRIF-LIr fibres. Thus SRIF in gut may arise in the SRIF-LIr perikarya in the gut but not in those in inferior mesenteric ganglion, though the site where the SRIF-LIr ganglion cells located in the inferior mesenteric ganglion project to has not been elucidated. In the superior cervical ganglion, only a few SRIF--LIr cells are detected.

Neuropeptide Y

As described above, pancreatic polypeptide immunoreactivities such as those to avian pancreatic polypeptide (APP) and bovine pancreatic polypeptide (BPP) appear to be due to cross-reactivity with NPY, which is the PP-like peptide found in the mammalian nervous system (Tatemoto et al., 1982).

APP- and NPY-like immunoreactivities were observed in the majority of the principal ganglionic cells in the superior cervical, stellate and coeliac ganglia of the rat (Lee et al., 1985b) and cat (Lundberg et al., 1982). These NPY-LIr cells were also positive for tyrosine hydroxylase (TH) in consecutive sections (Fig. 11) (Lee et al., 1985b), though NPY-LIr neurons were less numerous in number than TH-LIr perikarya (Fig. 11), indicating the coexistence of NPY and CA in sympathetic ganglia. It is likely that NPY cells in coeliac ganglia innervate the myenteric plexus (Lee et al., 1985b). However, no NPY-LIr cells are found in the sphenopalatine ganglion which belongs to the parasympathetic system (Lundberg et al., 1982) nor in the primary sensory ganglia (Lee et al., 1985b).

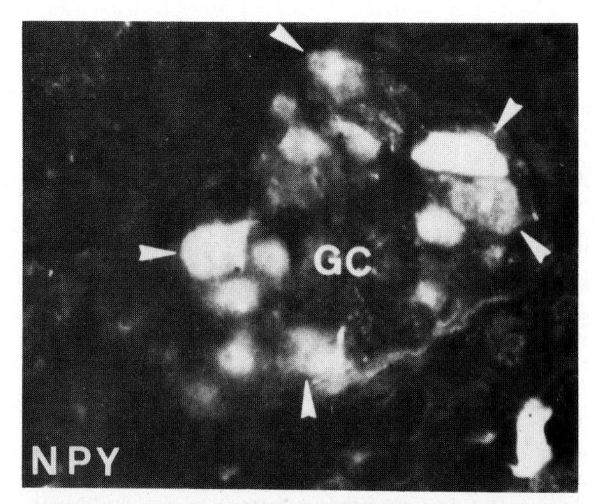

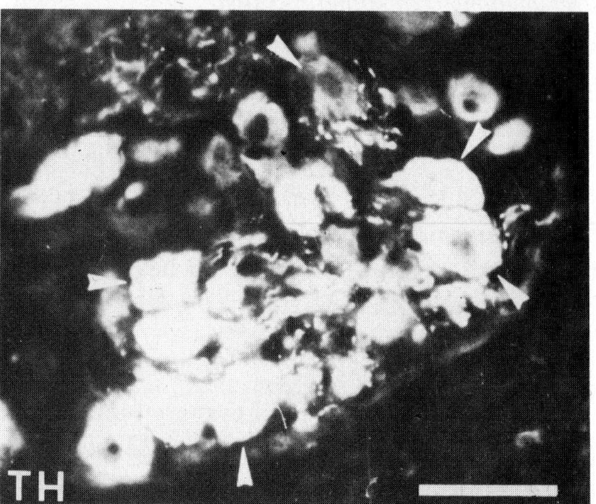

Fig. 11. Fluorescent photomicrographs showing NPY-LIr (NPY) and TH-LIr (TH) neurons in the coeliac ganglion (GC) in the consecutive sections. Arrowheads indicate the neurons which contain both NPY- and TH-LI. Note that TH-LIr neurons outnumber the NPY-LIr neurons in this ganglion. Bar = 50 μm. (From Lee et al., 1985b, with permission.)

Cholecystokinin

A dense network of CCK-LIr fibres is found around the ganglion cells of the inferior mesenteric ganglion of guinea pig (Larsson et al., 1979), which disappears with total denervation. However, when the lumbar splanchnic, hypogastric and intermesenteric nerves were cut and the colonic nerves were left intact, a dense network of CCK-LIr fibres was still seen in the operated ganglion with a similar pattern to that found in the intact ganglion. Thus most of the CCK-LIr fibres in the inferior mesenteric ganglion seem to arise from perikarya of the colon (Dalsgaard et al., 1983).

Bombesin/gastrin-releasing peptide

Very dense but weakly stained bombesin-like immunoreactive fibres are observed in the inferior mesenteric ganglion of the guinea pig. When the lumbar splanchnic and intermesenteric nerves were cut together with the hypogastric nerves, leaving the colonic nerves intact, a moderately dense network of bombesin-like immunoreactive fibres was still observed (Dalsgaard et al., 1983), while the adjacent sections incubated with SP antiserum showed no SP-LIr fibres. However, when the supply from the hypogastric nerves was cut, no bombesin-like immunoreactive fibres were seen. In addition, ligation of the colonic nerve caused an accumulation of bombesin-like immunoreactivity distal to the ligation. In the other nerves no bombesin-like immunostaining was observed. These data (Dalsgaard et al., 1983) suggest that the bombesin-like immunoreactive fibres in the inferior mesenteric ganglion reach the ganglion exclusively via the colonic nerves from the distal colon to the inferior mesenteric ganglion.

Neurotensin

So far no or few NT-LIr structures have been found in the sympathetic ganglia of rat and guinea pig, but a few NT-LIr fibres were reported to be found in the sympathetic ganglia of cat (Schultzberg et al., 1984).

Serotonin

Numerous small cells (10–15 μm in diameter) are distributed in the superior cervical ganglion of the rat (Verhofstad et al., 1981) and guinea pig (T. Matsuyama et al., in preparation). 5-HT-LIr cells possess processes in rat (probably SIF cells), but CA-LIr cells are rounded without processes (Verhofstad et al., 1981). Two types of cells have been identified in the guinea pig; one is always solitary whilst the other forms cell clusters of 2–5 cells (T. Matsuyama et al., in preparation). 5-HT-LIr fibres are also observed to run randomly among the superior cervical ganglion, some of which show a close relationship with blood vessels. Species differences are seen in the 5-HT system of the superior cervical ganglia of rats and guinea pigs. In the rat, 5-HT-LIr cell bodies are largely localized in the rostral and caudal portions of this ganglion and in general form cell clusters (referred to as small, intense fluorescence (SIF) cells), while in the guinea pig 5-HT-LI is found in both cell clusters and evenly distributed solitary cells in the superior cervical ganglion (T. Matsuyama et al., in preparation).

Vasopressin

Recently, a vasopressin (VP)-like immunoreactivity (VP-LI) was reported in the principal sympathetic ganglion cells and in the nerve fibres innervating peripheral tissues (Hanley et al., 1984).

Parasympathetic ganglia

Parasympathetic ganglia will be described in greater detail in the section dealing with specific organs.

Substance P

No SP-LIr cells have been found in the parasympathetic pterygopalatine ganglion of the rat (Lee et al., 1985c). Scattered SP-LIr fibres are observed in this ganglion (Fig. 12).

Calcitonin gene-related peptide

In the pterygopalatine ganglion, a few CGRP-LIr cells are detected (Fig. 12) (Lee et al., 1985c). These cells are medium in size (20–30 μm in diameter). Scattered CGRP-LIr fibres are observed with an identical location to SP-LIr fibres on the same sections (Fig. 12).

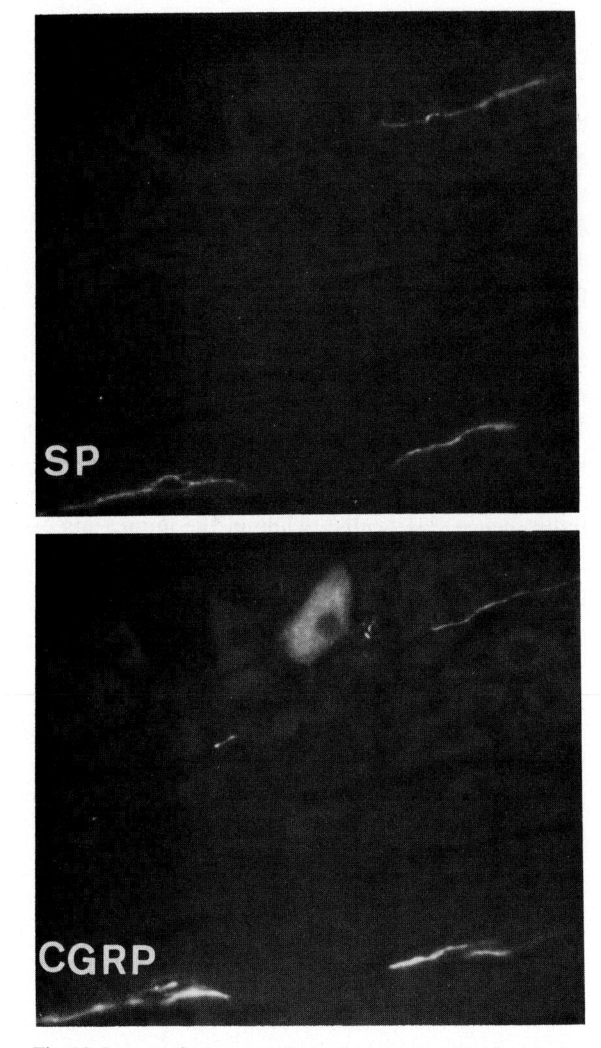

Fig. 12. Immunofluorescent photomicrographs of the frozen sections of pterygopalatine ganglion which are double stained with SP-antiserum (Texas red fluorescence) and CGRP (fluorescein isothiocyanate). No SP-LIr perikarya are found in this ganglion but a few CGRP-LIr cell bodies are seen. The running patterns of CGRP-LIr and SP-LIr fibres are mostly identical. × 190. (From Lee et al., 1985c, with permission.)

Vasoactive intestinal polypeptide

Most neurons in the local submandibular ganglia and almost all neurons in the sphenopalatine ganglion of the cat are immunoreactive for VIP and rich in the acetylcholinesterase (Lundberg et al., 1979b). This was also observed for local intramural ganglia of the tongue, bronchi and pancreas (Lundberg et al., 1979b). In the parasympathetic pelvic ganglion of the cat, the number of VIP-LIr cells is variable (Lundberg et al., 1979b). The majority are acetylcholinesterase-positive. VIP-LIr cell bodies are also found in the small intramural ganglia of the intestine. Some ganglia, particularly those found in the lamina propria, contain many VIP-LIr cells, but only a few VIP-LIr cells are observed outside of the muscle layers.

Peptidergic innervation of various organs

Cerebral arteries

Overall distribution patterns of SP-LIr and VIP-LIr fibres in the cerebral arteries are shown in the Figs. 14 and 16. Innervation of cerebral arteries by SP, CGRP, VIP, NPY and 5-HT is schematically shown in Fig. 19. The relationship between aminergic terminals and peptidergic terminals such as SP, VIP and NPY in the anterior cerebral artery is also shown schematically in Fig. 21.

Substance P

Overall distribution of SP-LIr fibres in the cerebral arteries of the guinea pig has been studied using frozen and whole-mount sections (Furness et al., 1982; Matsuyama et al., 1983a; Yamamoto et al., 1983).

In the wall of the cerebral arteries two types of fibres are identified. One forms a SP-LIr fibre bundle and the other a meshwork of varicose fibres (Fig. 13A). The SP-LIr fibre bundle is composed of several SP-LIr fibres, the number varying in relation to the diameter of the vessels, and is located in the periadventitial nerve bundles (Matsuyama et al., 1983b; Yamamoto et al., 1983). SP-LIr fibres in the periadventitial nerve bundle run mostly longitudinally in the wall of the vessels (Fig. 13A). Fig. 15 schematically shows the overall distribution in the cerebral arteries of the guinea pig found using whole-mounts.

Unilateral destruction of the trigeminal ganglion resulted in: (1) disappearance of SP-LIr fibres from the ipsilateral distal part of the internal carotid ar-

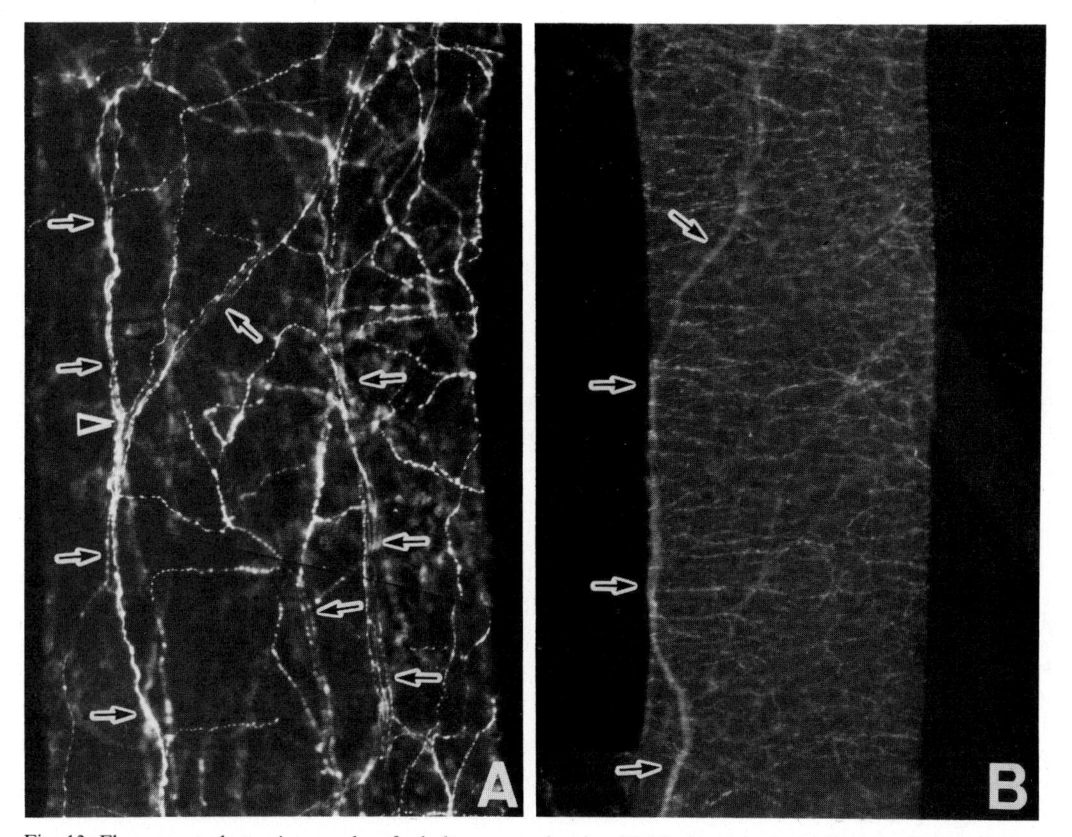

Fig. 13. Fluorescent photomicrographs of whole-mounts showing SP-LIr fibres in the wall of the internal carotid artery inside the circle of Willis of the guinea pig (A) and VIP-LIr fibres in the wall of the anterior cerebral artery of the rat (B). In the wall, some periadventitial nerve bundles contain SP-LI (indicated by arrows) (A). In a rostral direction the periadventitial nerve bundles containing SP-LIr fibres dissociate into thinner ones (branching indicated by arrowhead). Numerous spirally running VIP-LIr fibres give a dense grid-like appearance on the wall of cerebral arteries (B). Arrows indicate periadventitial nerve containing VIP-LIr fibres. A, × 120; B, × 60. (From Matsuyama et al., 1983b; Yamamoto et al., 1983.)

tery, posterior communicating artery, anterior cerebral artery within the circle of Willis, and the middle cerebral artery; (2) a marked decrease of SP-LIr fibres in the wall of the azygos anterior cerebral artery, anterior cerebral artery outside the circle of Willis, and its branches, compared with the normal high density of the SP-LIr fibre meshwork in unoperated sections; and (3) a slight decrease of SP-LIr fibres in the wall of the basilar artery occurring only in the wall of the rostral one-third of the artery (Yamamoto et al., 1983). In the caudal direction these changes become less conspicuous and no significant changes are seen in the caudal two-thirds of the basilar artery, nor in the wall of the vertebral artery after unilateral destruction of the trigeminal ganglion. Bilateral destruction of the trigeminal ganglia resulted in the disappearance of SP-LIr fibres from all the walls of the circle of Willis and its branches. The alterations in SP-LIr fibres in the basilar artery and its branches are fundamentally similar to those seen following unilateral destruction of the trigeminal ganglion, though the changes are more conspicuous. No changes in the SP-LIr fibres in the wall of the vertebral artery were found after the bilateral destruction of the trigeminal ganglion (Yamamoto et al., 1983). By contrast, bilateral excisions of the superior cervical ganglia failed to reduce the number of SP-LIr fibres in the wall

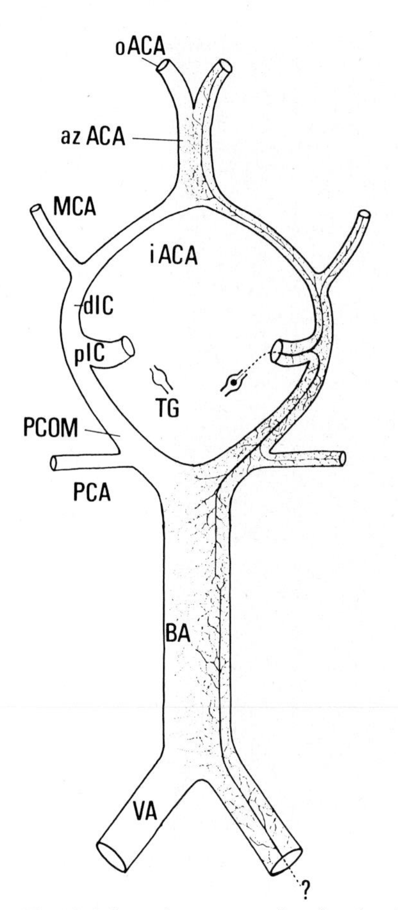

Fig. 14. Schematic representation showing the innervation of SP-LIr fibres in the wall of the cerebral arteries of the guinea pig. Thick line indicates periadventitial nerve bundles containing SP-LI and thin SP-LIr fibre meshwork, respectively. Abbreviations: oACA, anterior cerebral artery (outside of the circle of Willis); az ACA, azygos anterior cerebral artery; iACA, anterior cerebral artery (inside the circle of Willis); MCA, mid cerebral artery; dIC, distal part of internal carotid artery; pIC, proximal part of internal carotid artery; PCOM, posterior communicating artery; PCA, posterior cerebral artery; BA, basilar artery; VA, vertebral artery; TG, trigeminal ganglion. (From Yamamoto et al., 1983.)

of the cerebral arteries. Therefore, it is suggested that: (1) SP-LIr fibres which are located in the circle of Willis and their branches originate exclusively from SP-LIr cells in the trigeminal ganglion; (2) SP-LIr fibres of the rostral one-third of the basilar artery originate partly from trigeminal SP-LIr cells; and (3) SP-LIr fibres in the caudal two-thirds of the basilar artery originate exclusively from other SP-LIr cells, apart from the trigeminal ganglion. This dual innervation was also confirmed experimentally in the cerebral arteries of the gerbil and guinea pig by Matsuyama et al. (1984a). SP-LIr trigeminal ganglion cells mainly innervate the carotid vascular system and other non-trigeminal SP-LIr cells mainly innervate the vertebrobasilar system. Matsuyama et al. (1985) also examined the fine structure of SP-LIr terminals along the cerebral blood vessels. In the SP-LIr terminals, immunoreactivity was localised only over large granular vesicles (100–120 nm in diameter) or found associated with the surface membranes of cell organelles. About 5% of the fibres in the anterior cerebral arteries are SP-LIr. In the periadventitial nerve bundles, SP-LIr fibres can be identified along with non-immunoreactive fibres, both being enclosed by the cytoplasm of the Schwann cells similar to that seen for the VIP-LIr fibres. SP-LIr fibres also leave the periadventitial nerve to form the fibre plexus on the walls of the cerebral arteries. Two types of SP-LIr fibres are observed. One is enclosed by the cytoplasm of the Schwann cells together with non-immunoreactive fibres (Fig. 15A, C) and the other is a solitary SP-LIr fibre (Fig. 15B). The former is close to and opposite the basal lamina together with more than one non-immunoreactive fibre, and is enclosed by the cytoplasm of the Schwann cells. On the other hand,

Fig. 15. Immunoelectron photomicrographs showing SP-LIr nerve terminals (indicated by arrows) in the anterior cerebral artery of the rat without (A and B) and with 5-OHDA treatment. (A) SP-LIr terminal in the periadventitial nerve bundles. In this terminal, SP--LI occurs within some large vesicles (arrowheads) as well as in the axoplasm. A non-immunoreactive fibre is also found in the nerve bundle. (B) A solitary SP-LIr terminal (arrow) is shown. Note that this terminal is close to the basal lamina and smooth muscle (SM) with no ensheathment by the Schwann cell cytoplasm. (C) SP-LIr (indicated by arrow), CA (indicated by asterisk), and other (neither SP-LIr nor CA) terminals are close to one another, and directly apposed to the basal lamina and smooth muscle (SM). These terminals are enclosed by the Schwann cell cytoplasm, although no axo-axonic contact is present. SP-LIr fibres included in the terminal cluster, as shown in this figure, are seen less frequently than solitary SP-LIr terminals. Bar = 1 μm. (Adapted from Matsuyama et al., 1985, with permission.)

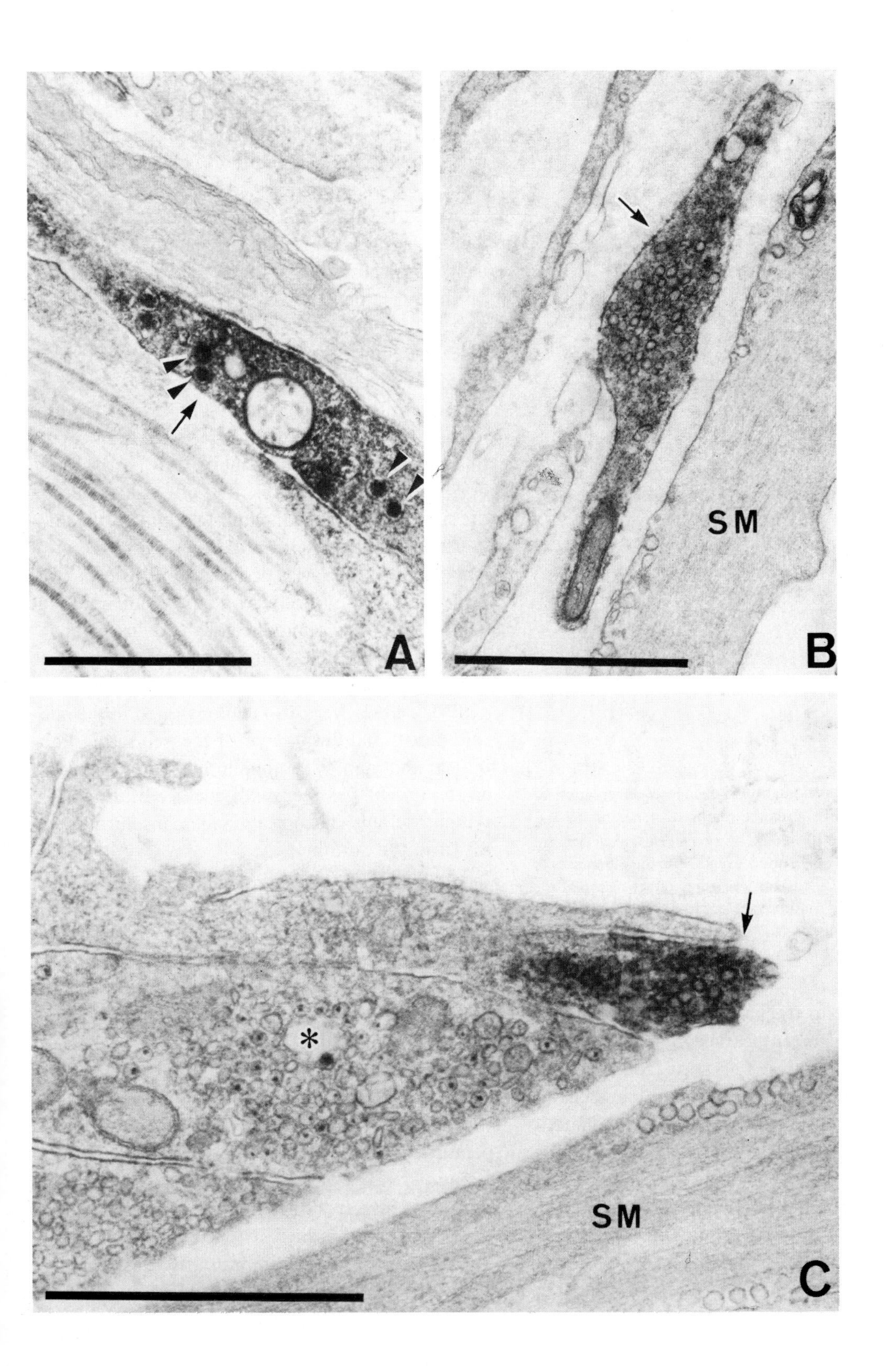
A
SM
B
SM
C

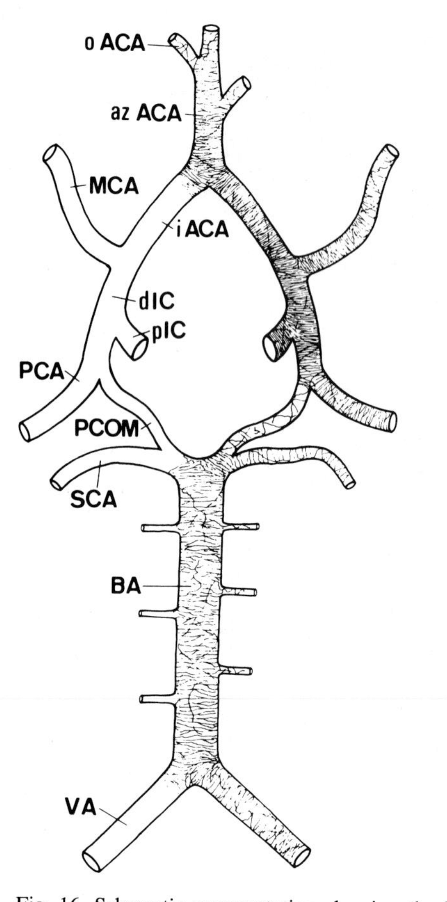

Fig. 16. Schematic representation showing the innervation of VIP-LIr fibres on the wall of the cerebral arteries of the rat. Thick line indicates VIP-containing periadventitial nerve and thin VIP-LIr fibre network, respectively. Note the differences in the distribution of VIP-LIr fibres according to the region of the cerebral arteries. Abbreviations: see legend to Fig. 14. (Matsuyama et al., 1983b, with permission.)

the latter is close to the basal lamina and smooth muscle; they are not ensheathed by the Schwann cell cytoplasm.

By using the combination of immunocytochemistry and false transmitter histochemistry, the coexistence of SP-LI and CA and their topographic interrelationship were examined ultrastructurally (Matsuyama et al., 1985) (Fig. 15C). 5-Hydroxydopamine (5-OHDA) was used as a marker for CA terminals. 5-OHDA-labeled CA fibres are mostly detected extending from the periadventitial nerve and basal lamina. In the nerve fibre clusters which are close to the basal lamina and enclosed by the cytoplasm of Schwann cells, CA terminals are located next to the SP-LIr fibres (Fig. 15C), though no axo-axonic contact could be identified between them. No SP-LIr terminals which simultaneously contain amines are seen in the periadventitial, adventitial or lamina propria of the cerebral arteries of the guinea pig (Fig. 15C). No double-labeled terminals for both SP antiserum and 5-OHDA were detected, indicating that these two bioactive substances are localized in different neuron systems. In support of this finding, the origin of SP-LIr fibres in the walls of the cerebral artery has been shown to lie in the trigeminal ganglion and partly in the dorsal root ganglion, while that of CA fibres lies in the superior cervical ganglion. In addition, the finding that bilateral excision of the superior cervical ganglia failed to cause a decrease of SP-LI was confirmed by immunohistochemistry (Yamamoto et al., 1983) and radioimmunoassay (Edvinsson et al., 1983b). The possibility of coexistence with VIP is discussed below.

SP-LIr fibres were also demonstrated in the choroid plexus and dura mater of the guinea pig, rabbit, cat and man by radioimmunoassay (Edvinsson et al., 1983b). As seen with the cerebral arteries, sympathectomy did not alter the concentration of SP-LI in the dura mater.

The high density of SP-LIr fibres in the cerebral arteries and dura mater provides support for a functional role of perivascular SP within the cranial circulation, though it is not clear why peripheral branches of primary sensory ganglion cells contain SP.

Calcitonin gene-related polypeptide

Recently, Wanaka et al. have examined the distribution and origins of the fibres containing this peptide (A. Wanaka et al., in preparation). The distribution pattern of CGRP-LIr fibres is very similar to that of SP-LIr fibres and destruction of the trigeminal ganglion, which is also the major origin of SP-LIr fibres in the cerebral arteries, caused a marked reduction of these fibres. In addition, in the trigeminal ganglion, most of the SP-LIr cells contain this peptide. These findings indicate that most of the CGRP-containing fibres are likely to be found in the adventitia of the arteries and veins of the cardiovascular system of the rat (Mulderry et al., 1985).

Vasoactive intestinal polypeptide

VIP-LIr fibres run along the lateral surface of the coat of the adventitia and often turn medially to reach the area close to the muscular layer (Larsson et al., 1976a; Edvinsson et al., 1980; Matsuyama et al., 1983a,b). Recently, the overall distribution of VIP-LIr structures in the walls of the cerebral arteries using whole-mounted sections has been demonstrated (Matsuyama et al., 1983a,b).

VIP-LIr fibres run spirally on the walls of the cerebral arteries. On the walls of the large arteries, such as the vertebral artery, basilar artery, internal carotid artery, within or without the circle of Willis, posterior and anterior communicating arteries, proximal parts of anterior, mid and posterior cerebral arteries, these fibres are richly distributed and show a dense grid-like appearance (Figs. 13B, 16). The highest density is found on the wall of the anterior cerebral artery (Fig. 13B), internal carotid artery and anterior communicating artery, while the lowest density is on the posterior communicating artery. On the other hand, on the walls of the branches of these arteries or along distal parts of the anterior, mid and posterior cerebral arteries, the number of VIP-LIr fibres decreases markedly (Matsuyama et al., 1983a,b). Fig. 16 shows the overall distribution of VIP-LIr fibres in the cerebral arteries of the guinea pig.

Excision of the superior cervical ganglion does not produce any change in VIP-LIr fibres on the wall of any cerebral arteries (Matsuyama et al., 1983b). In addition, injection of capsaicin has no effect on the number of VIP-LIr fibres on the wall of the cerebral arteries, though this injection causes a marked disappearance of SP-LIr fibres from these arteries. The possibility of the coexistence of CA and VIP seems to be excluded, since there is no decrease of VIP-LIr fibres in the cerebral arteries after excision of the bilateral superior cervical ganglia. In addition, the possibility of coexistence of SP and VIP also seems to be excluded, because capsaicin treatment failed to decrease the number of the VIP-LIr fibres on the wall of the cerebral arteries.

Ultrastructural studies show that these VIP-LIr endproducts occur in the terminals which contain a number of small clear vesicles (about 50 nm in diameter) but lack large granular vesicles (Matsuyama et al., 1985) (Fig. 17). In the periadventitial nerve bundle, VIP-LIr fibres are surrounded by the cytoplasm of Schwann cells together with non-immunoreactive fibres (Fig. 17A). At the internal surface of the adventitia, as shown in Fig. 17B, C, VIP-LIr fibres together with more than one non-immunoreactive fibre are seen close (100 nm) to the basal lamina and to the smooth muscle (Fig. 17B, C). VIP-LIr fibres are always located together with non-immunoreactive fibres and no solitary VIP-LIr fibres are detected (see Fig. 21).

An ultrastructural experiment combined with 5-OHDA (Matsuyama et al., 1985), has demonstrated that no VIP-LIr terminals and fibres contain CA-containing small granular vesicles in the anterior cerebral artery. No profiles suggesting axo-axonic contact between VIP-LIr terminals and aminergic terminals are seen, though aminergic terminals are always found next to the VIP-LIr terminals (Fig. 17C). This is shown schematically in Fig. 21.

Furthermore, the fact that the distribution pattern of the VIP-LIr fibres on the wall of the cerebral artery (grid-like appearance) (Figs. 13B, 16) differs from that of the other putative transmitters suggests that VIP may be localized in terminals separate from those of CA, SP and ACH (see Figs. 13, 14, 16, 18).

Neuropeptide Y

In the cerebral arteries, abundant NPY-LIr fibres are found with a very similar distribution pattern to that of adrenergic fibres (see Fig. 18). NPY coexists with NA in the superior cervical ganglion cells (Lundberg et al., 1982) which innervate cerebral arteries (Matsuyama et al., 1985) (see Fig. 19).

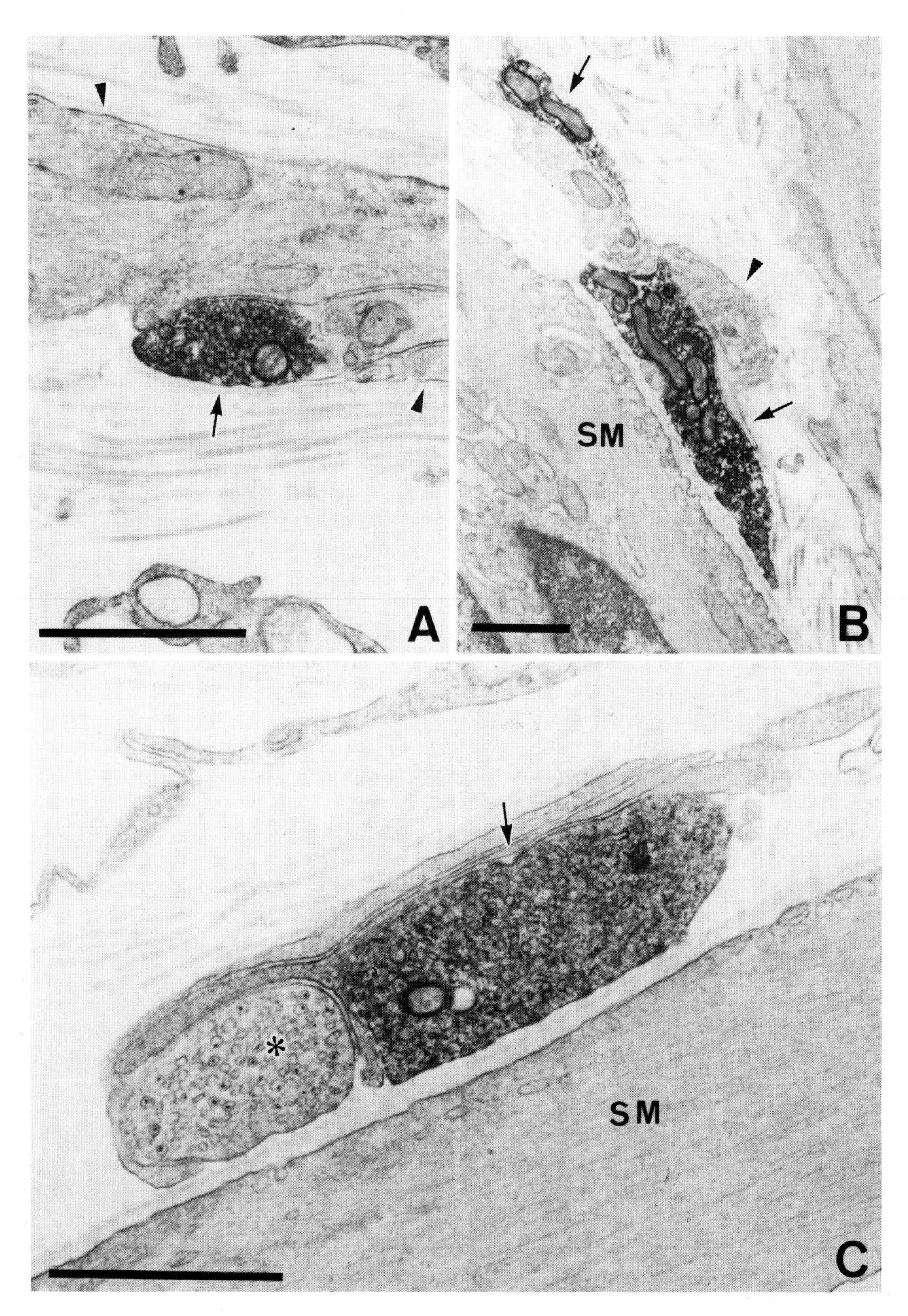
A
SM
B
SM
C

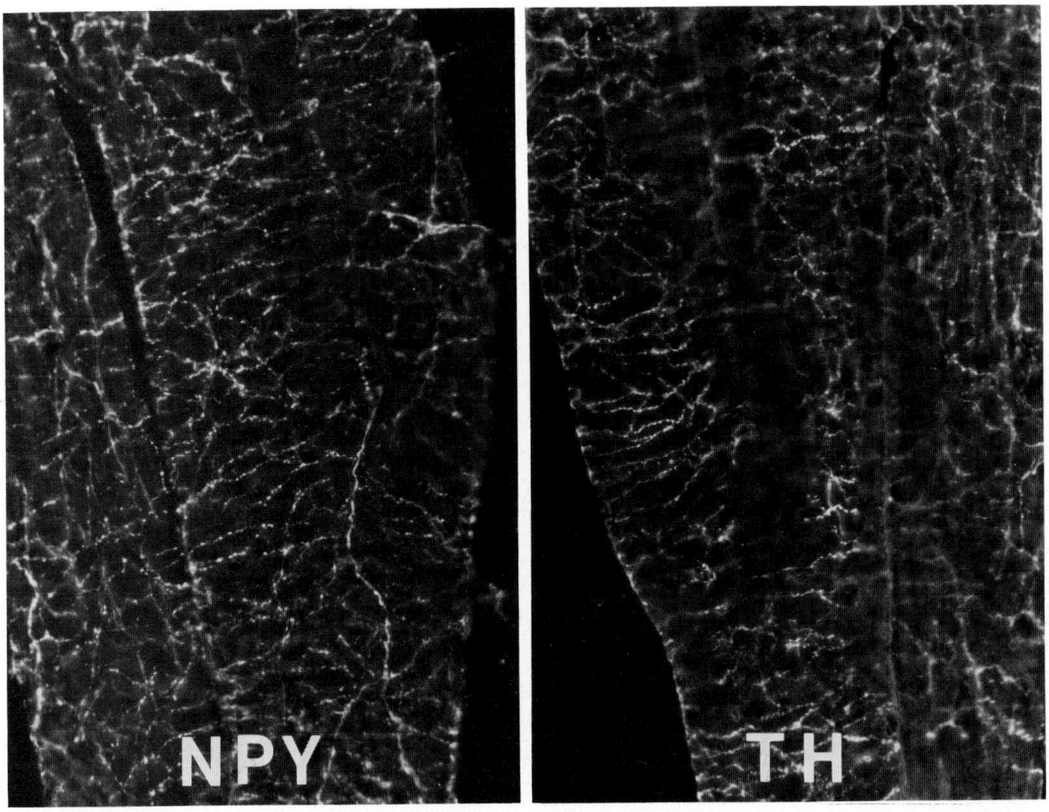

Fig. 18. Fluorescent photomicrographs showing NPY-LIr and TH-LIr fibres in the left and right anterior cerebral arteries of the same rat. NPY-LIr fibres show a distribution type mixed with grid-like and mesh-like patterns. For example, as described above, SP-LIr fibres exhibit a meshwork, and VIP-LIr fibres show grid-like network. Moreover, the distribution patterns of NPY-LIr and TH-LIr fibres in the cerebral arteries are very similar to each other. × 120. (From T. Matsuyama et al., unpublished data.)

Following the bilateral excision of the superior cervical ganglia or chemical sympathectomy by injection of 6-OHDA, there is disappearance of NPY-LIr and TH-LIr fibres on the cerebral blood vessels, while VIP-LIr and SP-LIr fibres are numerous.

A recent study on the ultrastructure of peptidergic and catecholaminergic fibres in the cerebral artery has demonstrated that NPY-LI coexists with CA in nerve terminals in the anterior cerebral artery (Matsuyama et al., 1985) (Fig. 20). Ca. 36% of fibres found in the wall of the anterior cerebral artery were found to contain NPY-LI. The profile of these fibres is very similar to that of the VIP-LIr fibres, in that they are enclosed by the cytoplasm of

---

Fig. 17. Immunoelectron photomicrographs showing VIP-LIr terminals (indicated by arrows) in the anterior cerebral artery of the rat without (A, B) and with (C) 5-OHDA treatment. (A) VIP-LIr terminals are seen in the periadventitial nerve bundle together with non--immunoreactive terminals (arrowheads). This terminal proceeds medially (bottom of the figures). (B) Two VIP-LIr terminals (indicated by arrows) are visible. One is directly apposed to the smooth muscle cells (SM), though the basal lamina is in between. Seven non-immunoreactive terminals (arrowheads) are seen next to the VIP-LIr terminals, both ensheathed by Schwann cells. (C) VIP-LIr (indicated by arrow) and CA (asterisk) terminals form a nerve cluster and are directly apposed to the SM. It should be noted that no axo-axonic contact is present. Bars = 1 μm. (From Matsuyama et al., 1985, with permission.)

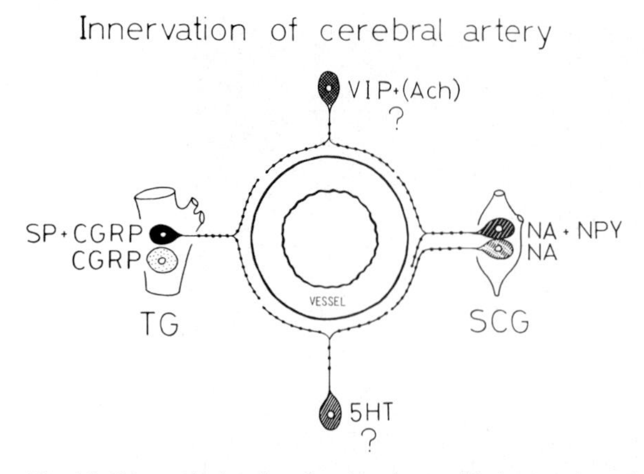

Fig. 19. Schematic drawing showing the peptide innervation of cerebral arteries. SCG, superior cervical ganglion; TG, trigeminal ganglion. (From T. Matsuyama et al., unpublished data.)

Schwann cells together with non-immunoreactive fibres (Fig. 20A), and lie close (100 nm) to the basal lamina and to the smooth muscle cells (Fig. 20B).

NPY-LIr terminals simultaneously labeled with 5-OHDA (see Fig. 20) are found in large numbers. Thus, the majority of NPY-LIr fibres (95–99%) are catecholaminergic. On the other hand, catecholaminergic terminals do not always exhibit NPY-LI. About 50% of the fibres in the adventitia are found to be catecholaminergic but 30% of the CA terminals are devoid of NPY-like immunoprecipitates (Matsuyama et al., 1985) (see Fig. 21).

Serotonin

The presence (Iijima and Wasano, 1980; Griffith et al., 1982) and overall distribution of 5-HT-LIr structures have been demonstrated in the wall of the cerebral arteries of the guinea pig by using whole-mount sections (T. Matsuyama et al., in preparation).

Dense networks of 5-HT-LIr fibres at varying densities are found in the wall of the cerebral arteries of guinea pig. The highest density is found in the large cerebral artery such as the basilar artery and the arteries of the circle of Willis. Their number and density decrease in distal areas. 5-HT-LIr fibres form mesh-like plexi in these arteries, and, circularly running fibres are more common in the arteries of the circle of Willis. In distal areas, these fibres decrease slightly in number. At the proximal part of the branching arteries of the basilar artery and circle of Willis, numerous 5-HT-LIr fibres run circumferentially, resulting in a grid-like appearance. However, this configuration progressively develops into a mesh-like plexus in a distal direction. The number of 5-HT-LIr fibres continues to decrease in distal areas, though they are still present up to the arterioles.

Two different findings have been obtained concerning their origins. Bilateral excision of the superior cervical ganglia resulted in the disappearance of 5-HT-LIr fibres in the cerebral artery, while intracerebroventricular injection of 5,7-dihydroxytryptamine did not change the density of these 5-HT fibres, though simultaneously, a marked decrease of 5-HT-LIr fibres was observed (T. Matsuyama et al., in preparation). These findings suggest that 5-HT-LIr cell bodies in the superior cervical ganglion but not in the brain project to the cerebral artery in the guinea pig. On the other hand, it was earlier speculated in the rat that 5-HT-LIr fibres running along the cerebral artery originate from central 5-HT-LIr neurons, and that 5-HT-LIr neurons located in the superior cervical ganglion probably do not contribute to these innervations (Edvinsson et al., 1983a). This discrepancy in the origin of 5-HT-LIr fibres in the cerebral artery may be attributed to a species difference.

Glucagon

The presence of immunoreactive glucagon was demonstrated immunohistochemically in the smooth muscle of the cerebral arteries of rats (Fig. 22) (Matsuyama et al., 1984b). This result together with previous radioimmunoassay and immunohistochemical studies (Tanaka et al., 1983), suggests that smooth muscle cells of the blood vessels may be one of the extrapancreatic sources of immunoreactive glucagon in the plasma.

Gastro-intestinal tract

The distribution of various peptides such as SP,

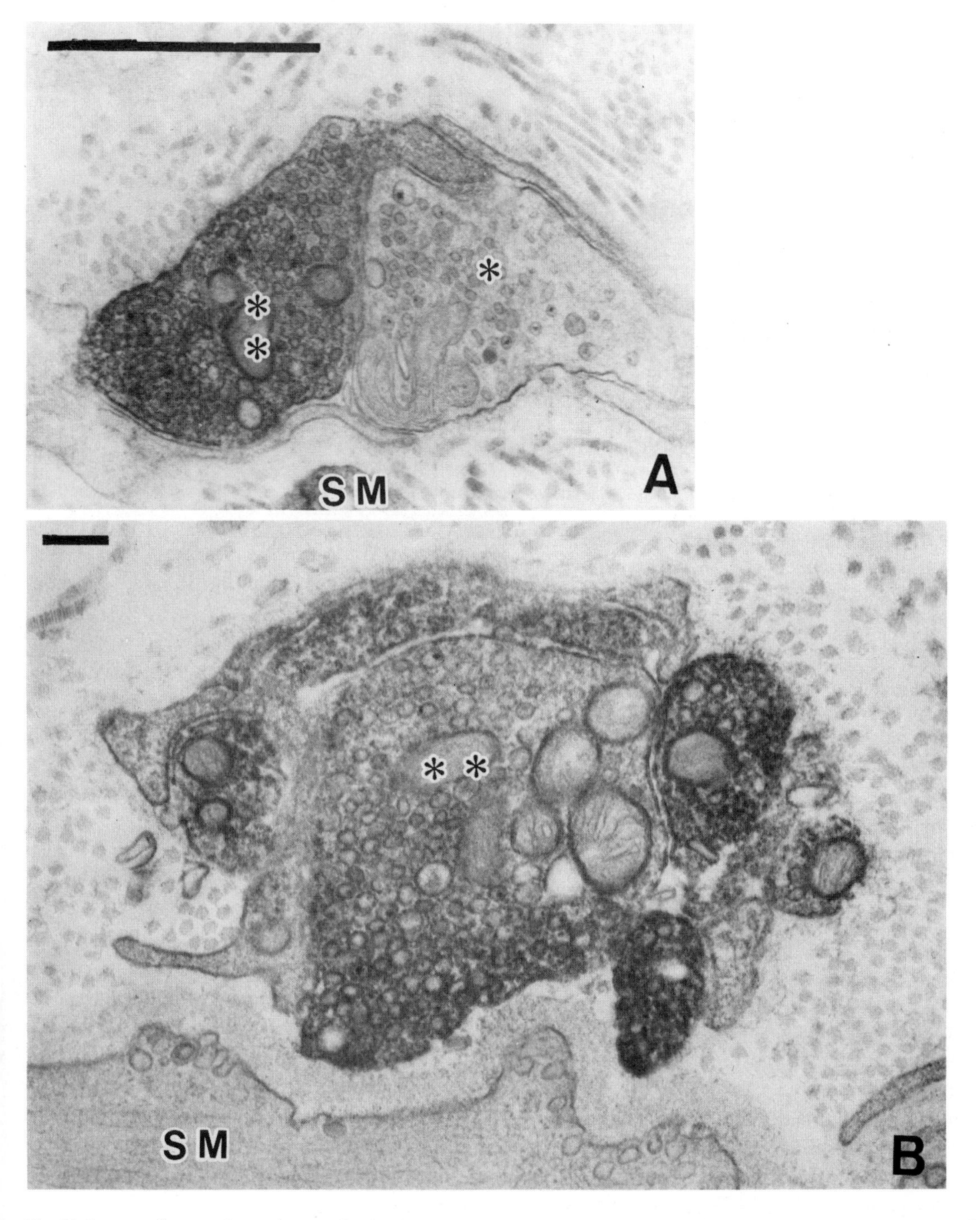

Fig. 20. Immunoelectron photomicrographs showing NPY-LIr terminals in the anterior cerebral artery of a rat treated with 5-OHDA. (A) Two CA terminals, which have both small and large granulated vesicles, are present in the Schwann cell. One (**) is NPY-LIr and the other (*) not. (B) NPY-LIr terminals contain small granulated vesicles, indicating coexistence of NPY and CA. These data suggest that there are two types of CA terminals in the cerebral arteries, one (major) simultaneously contains NPY-LI and the other (minor) does not. Bars = 1 μm (A) and 0.1 μm (B). SM, smooth muscle. (Adapted from Matsuyama et al., 1985, with permission.)

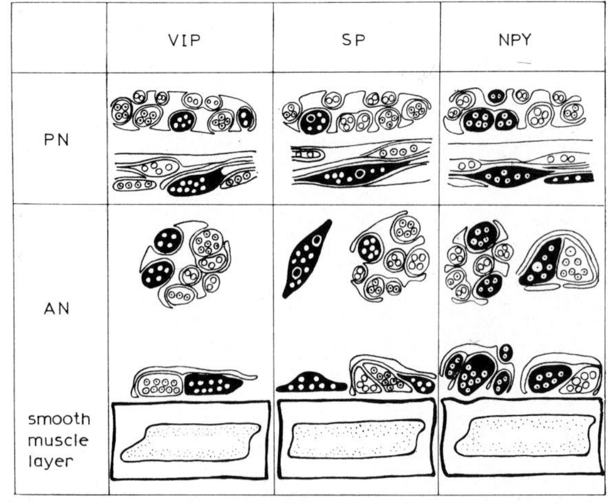

Fig. 21. Schematic representation of the relationships between VIP-LIr and CA terminals, SP-LIr and CA terminals, and NPY-LIr and CA terminals in the wall of the rat anterior cerebral artery. Terminals which are immunoreactive for VIP, SP, or NPY are shaded, while non-immunoreactive terminals are not shaded. CA terminals are those containing some open circles with a small dot (⊙). In the SP-LIr terminals immunoreactivity is also visible in the large vesicles, represented by closed circles (●). Non-labeled small clear vesicles are indicated by open circles (○). PN, periadventitial nerve bundle; AN, adventitial nerve. (From Matsuyama et al., 1985, with permission.)

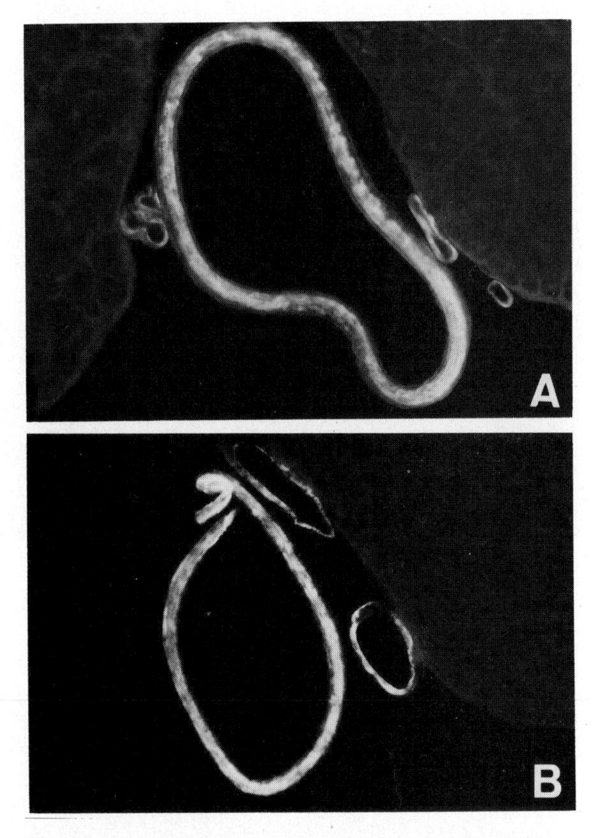

Fig. 22. Fluorescent photomicrographs showing immunoreactive glucagon in the cerebral arteries. (A) Basilar artery; (B) anterior cerebral arteries. Note that immunoreactive glucagon is located in the cytoplasm of the smooth muscle cells. × 170. (From Matsuyama et al., 1984b, with permission.)

VIP, ENK, SRIF, CCK and NT has already been reviewed in detail for the small intestine of guinea pig (Schultzberg, 1984). Thus in this report, the peptidergic system will be described with special reference to the stomach in comparison with that of the intestine or oesophagus.

Substance P

The stomach contains a very high concentration of SP-LI in the myenteric plexus in the guinea pig and rat (Schultzberg et al., 1980; Minagawa et al., 1984). In this section, the distribution and origin of SP-LIr structures in the rat will be described.

SP-LI cell bodies are observed in the myenteric plexus of the stomach (Fig. 24C) and submucosal and myenteric plexus of the small and large intestines except the rectum and jejunum. The highest density of SP-LIr fibres is found in the myenteric plexus and a small or moderate number of fibres in the mucous and submucous layers and in the longitudinal muscle layer, though there are some exceptions. The guinea pig stomach represents a special case in that the myenteric plexus and the circular muscle layer only contain a few fibres (Schultzberg et al., 1980), but the rat stomach shows a high density of SP-LIr fibres in the myenteric plexus (Fig. 23) and moderate densities in the submucous (Fig. 24A) and muscle layers (Fig. 24B) (Minagawa et al., 1984).

The overall distribution and origins of SP-LIr structures in the rat stomach have been investigated using whole-mount sections prepared from animals subjected to surgery (Minagawa et al., 1984). SP-

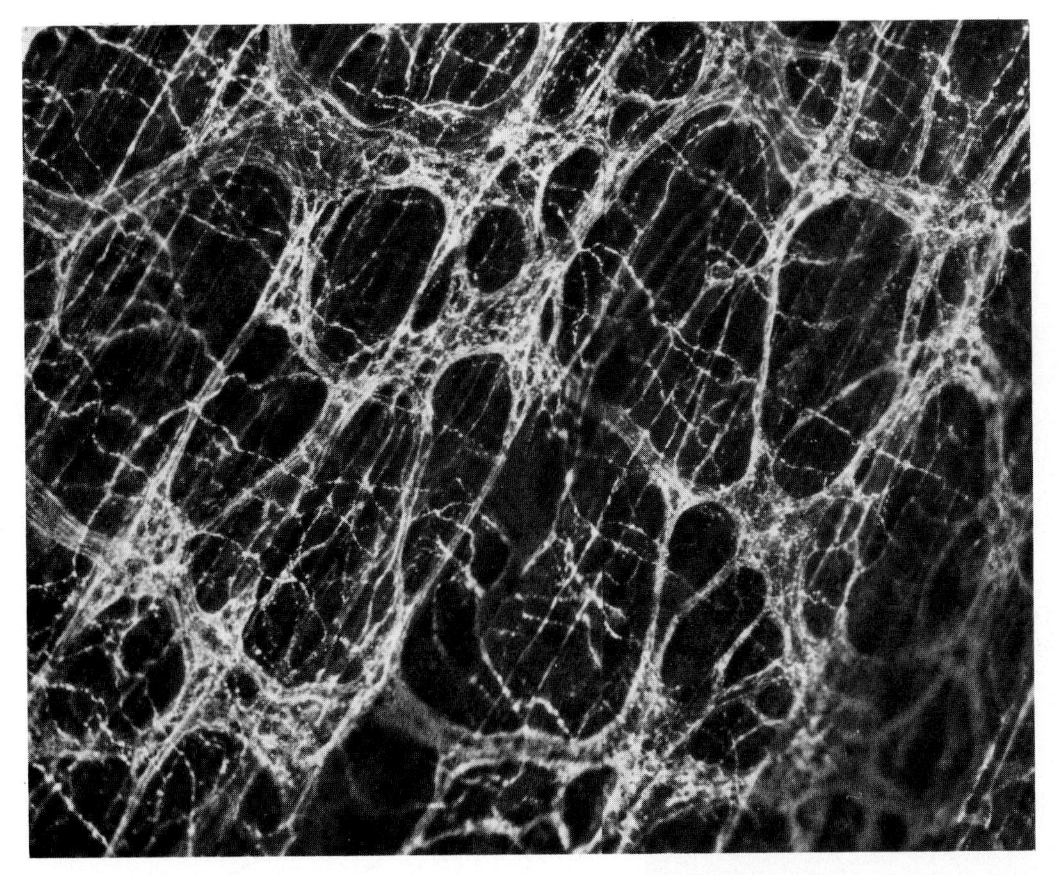

Fig. 23. Fluorescent photomicrograph of whole-mount preparations showing SP-LIr fibre meshwork in the myenteric plexus of the rat stomach. Numerous SP-LIr fibres are seen both in the myenteric ganglia and internodal strands. × 120. (From Minagawa et al., 1984, with permission.)

LIr fibres enter the stomach at the lesser curvature and form a compact fibre bundle that extends to the greater curvature and dissociates into minor bundles in several directions. These bundles project via many collaterals to the muscle layer and form a dense fibre meshwork in the myenteric plexus; some run along the blood vessels (Fig. 24A). Meshworks of SP-LIr fibres are observed in the circular muscle layer (Fig. 24B), the myenteric plexus (Fig. 23) and in the longitudinal muscle layer. The SP-LIr fibres in the muscle fibres run parallel to the muscles and are distributed evenly throughout the entire stomach, although more are seen in the circular muscle layer than in the longitudinal. The large numbers of SP-LIr fibres detected in the myenteric plexus are located in the ganglia internodal strands (Fig. 23). Although these fibre meshworks are observed throughout the stomach, the highest concentration is seen in the greater curvature, with those closer to the lesser curvature decreasing in number (see Fig. 25).

Identification of SP-LIr cells in the normal stomach is difficult due to a high density of SP-LIr fibres (see Fig. 23) and immunoreactive cells in the myenteric ganglia are clearly visible after colchicine treatment (Fig. 24C); these are moderately concentrated in the area near the greater curvature. Similar results have been obtained in animals in which either the greater splanchnic nerves were transected or dorsal spinal ganglionectomy was performed (Min-

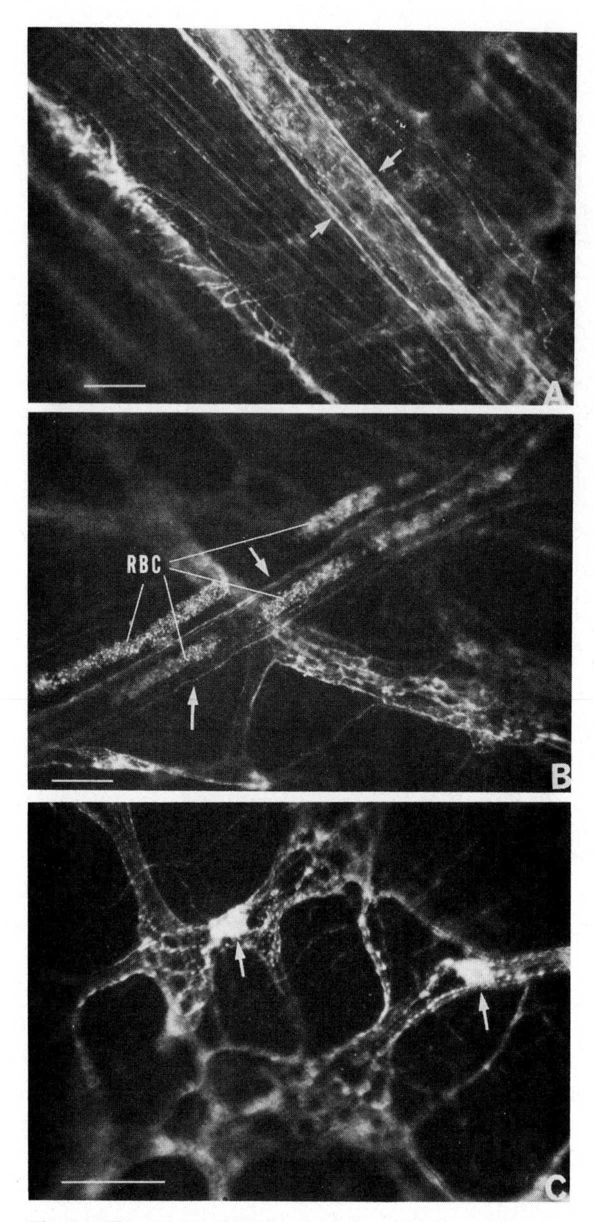

Fig. 24. Fluorescent photomicrographs of whole-mount preparations showing SP-LIr fibres (arrows) running along the blood vessels (A, B) and SP-LIr perikarya in the myenteric ganglia (arrows) of colchicine-treated rat (C). (A) Circular muscle layer; (B) myenteric plexus. Bars = 100 μm. RBC, red blood cells. (From Minagawa et al., 1984, with permission.)

agawa et al., 1984). Fig. 26 schematically shows the distribution of SP-LIr structures in the myenteric plexus of the rat stomach.

After bilateral vagotomy performed in the neck,

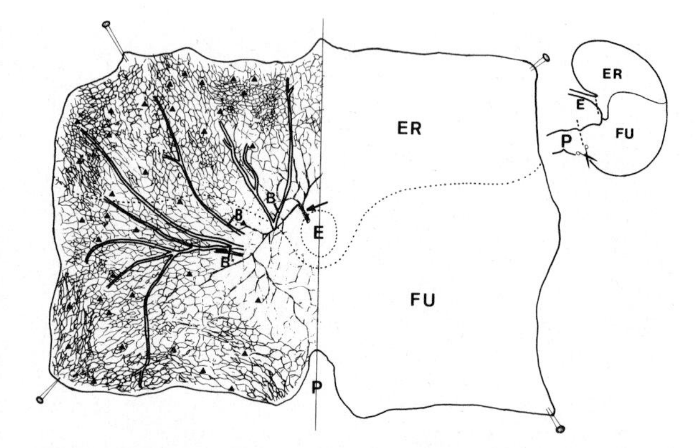

Fig. 25. Schematic drawing showing the distribution of SP-LIr structures in the myenteric plexus of the stomach. SP-LIr fibres enter the stomach forming a dense fibre bundle (arrow). This fibre bundle divides into several thinner segments to form a dense plexus in the myenteric plexus. More SP-LIr fibres are seen in the greater curvature than in lesser curvature. Triangles indicate SP-LIr perikarya. Note the similar distribution of these cells to those of the SP-LIr fibres in the stomach. B, blood vessels; E, oesophagus; ER, oesophageal region; FU, fundus; P, pylorus. (From Minagawa et al., 1984, with permission.)

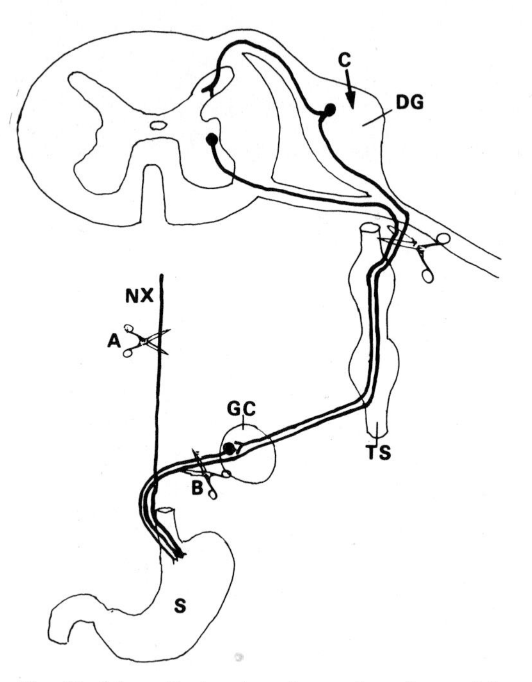

Fig. 26. Schematic drawing of operation sites used in peptide studies of the gut. Vagotomy at the level of the neck (A), neurotomy of the greater splanchnic nerve proximal to the junction with the vagal nerve (NX) (B), and dorsal spinal ganglionectomy (C) from T_5 to L_2. DG, dorsal spinal ganglion; GC, coeliac ganglion; S, stomach; TS, sympathetic trunks. (From Minagawa et al., 1984, with permission.)

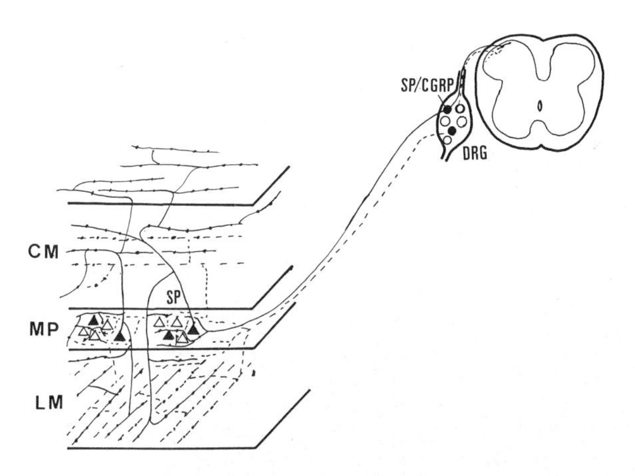

Fig. 27. Schematic representation showing the innervation of VIP-LIr structures in the stomach. SP-LIr fibres in the myenteric plexus (MP) originate mainly from the SP-LIr (●) cells in the dorsal spinal ganglia (DRG) and secondarily from intrinsic SP-LIr nerve cells (▲) in the MP. The SP-LIr fibres located in muscle layers (LM), originate from SP-LIr cells in the myenteric ganglia. CM, circular muscle layer. ○, △, non-immunoreactive neurons.

no conspicuous changes in SP-LIr fibres are seen in the muscle layer of the stomach (see Fig. 26). However, bilateral transection of the greater splanchnic nerve proximal to the junction with the vagal nerve results in a remarkable decrease of SP-LIr fibres in the ipsilateral myenteric plexus of the stomach, although the number of fibres in the muscle layer is still considerable. Similarly, bilateral dorsal spinal ganglionectomy from T_5 to L_2 results in a marked reduction of SP-LIr fibres in the myenteric plexus, whereas no remarkable changes are seen in the muscle layer. Following these operations, with the exception of vagotomy, numerous SP-LIr cells are seen in the myenteric plexus of the stomach. Occasionally, the processes of these cells can be traced to the longitudinal muscle layer, circular muscle layer, or mucosal layer. Fig. 27 schematically shows the innervation of the SP-LIr structures in the rat stomach.

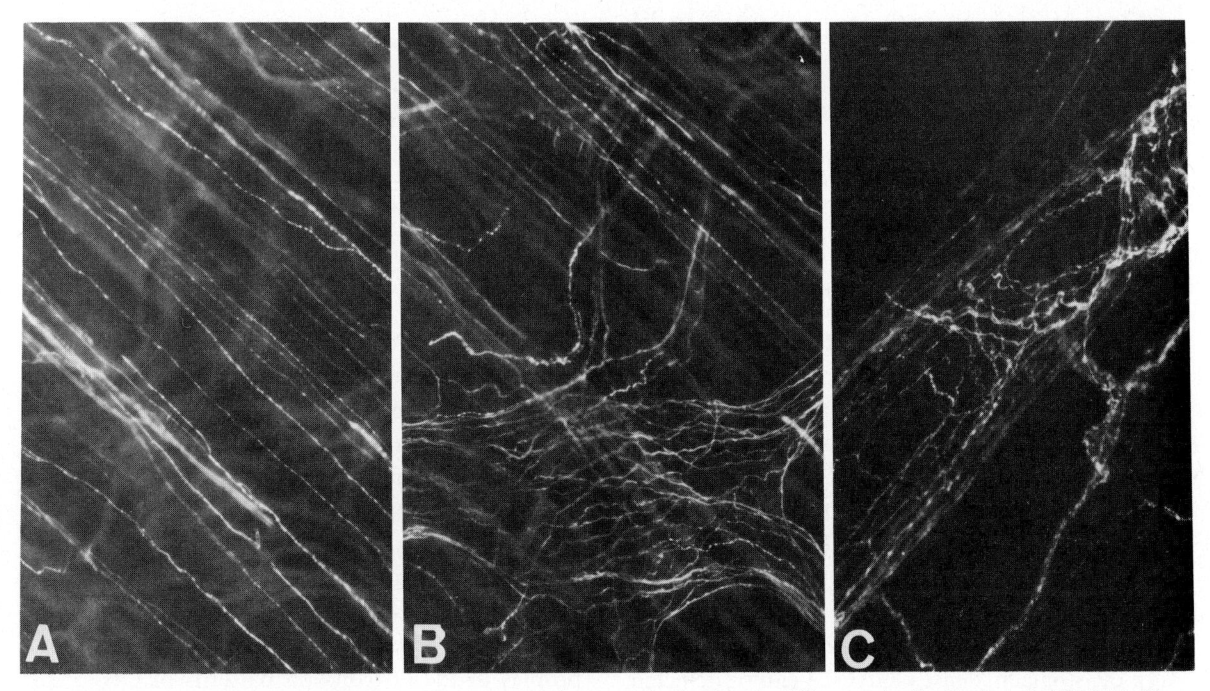

Fig. 28. Fluorescent photomicrographs of a flat-mount of the stomach showing the differences in distribution of CGRP-LIr fibres according to layer. A and B are the same areas photographed with a different focus. A shows CGRP-LIr fibres in the circular muscles; B shows a dense CGRP-LIr fibre meshwork in the myenteric plexus. Numerous CGRP-LIr fibres are seen in both the myenteric ganglion and the internodal strands. C shows a group of CGRP-LIr fibres running along the blood vessels in the circular muscle layer. (A, B) × 100; (C) × 65. (From Lee et al., 1985c, with permission.)

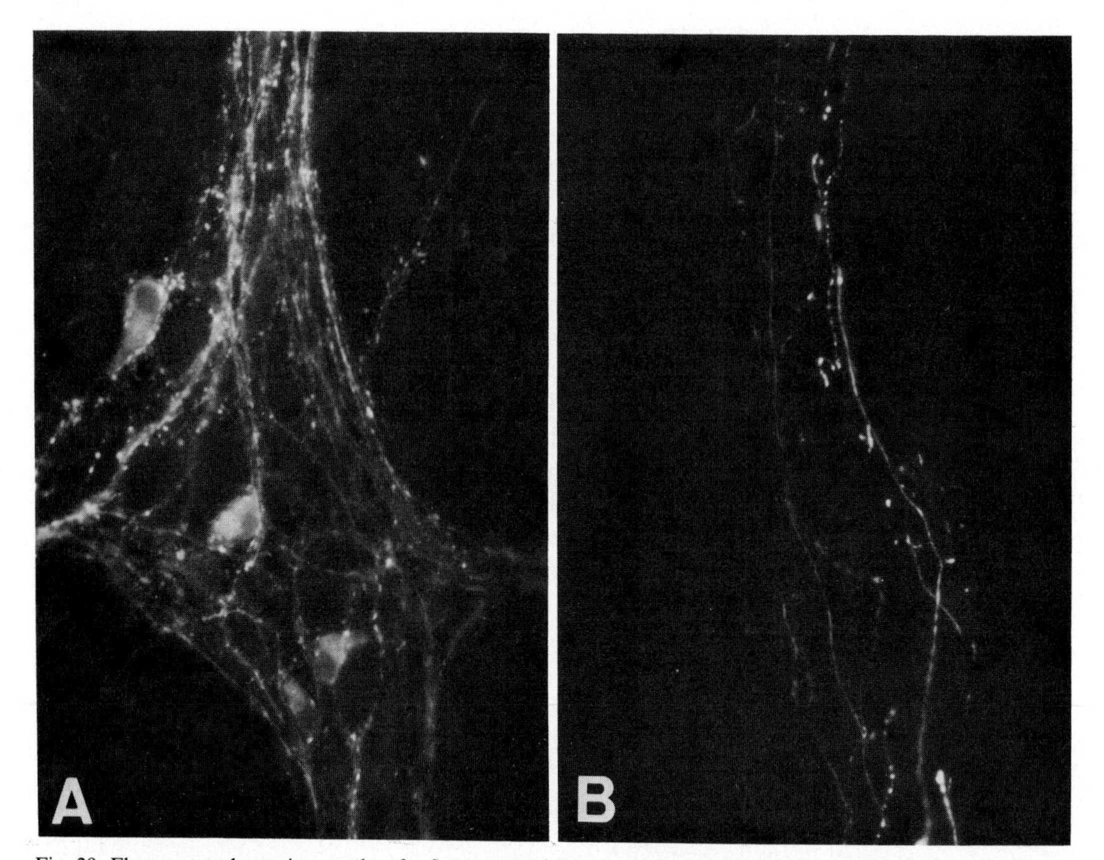

Fig. 29. Fluorescent photomicrographs of a flat-mount of the rat stomach following colchicine treatment. (A) SP-LIr fluorescence; (B) CGRP-LIr fluorescence. Colchicine treatment resulted in the decrease of immunoreactive fibres and the visualization of SP-LIr neurons (A), but failed to demonstrate CGRP-LIr perikarya (B) in the same section. ×200. (Adapted from Lee et al., 1985c, with permission.)

These findings indicate that SP-LIr fibres in the gastrointestinal tract probably originate from both intrinsic and extrinsic sources, including primary sensory ganglion cells (Hayashi et al., 1982; Minagawa et al., 1984). Thus, the SP-LIr cells in the dorsal spinal ganglion may terminate on SP-LIr cells in the myenteric ganglia which, in turn, send projections to the muscle layers and submucous layer.

Calcitonin gene-related polypeptide

In the stomach, dense CGRP-LI can be detected (Fig. 28) (Lee et al., 1985c). Although there are numerous CGRP-LIr fibres, no reactive cells are detected even with colchicine treatment (Fig. 29B). SP-LIr cell bodies are observed in the myenteric plexus following colchicine injection (Fig. 29A), but no CGRP-LIr cells are found in the same section (Fig. 29B). There are many CGRP-LIr fibres in the myenteric plexus (Fig. 28B) and a moderate number in the submucosal plexus, in the circular (Fig. 28A) and longitudinal muscle layers, and in the mucosa. The lamina propria contains fibres extending superficially to approach the epithelium. Dense CGRP-LIr fibres run along the blood vessels with mesh-like profiles (Fig. 29B). Three-dimensional profiles of the distribution patterns of CGRP-LIr structures are identified on the flat-mounted preparation. Fig. 28A, B shows the distribution pattern of CGRP-LIr fibres in different layers of the same region. In the muscle layer, CGRP-LIr fibres run in the same direction as the muscle (Fig. 28A). Some of these fibres are closely associated with blood vessels. In the myenteric plexus, numerous CGRP-LIr fibres are seen both in the myenteric ganglia and the internodal strands (Fig. 28B). The distribution pattern of CGRP-LIr fibres is identical to that of SP-LIr fibres, although CGRP-LIr fibres with a varicose appearance are more numerous than SP-LIr fibres with SP-LIr varicose fibres.

Enkephalin

ENK-LIr perikarya are found in most parts of the gastrointestinal tract in moderate numbers in the guinea pig and in small numbers in the rat (Elde et al., 1976; Polak et al., 1977; Alumets et al., 1978; Linnoila et al., 1978; Jessen et al., 1980; Schulzberg et al., 1980; Furness et al., 1983). In the guinea pig, large numbers are seen in the region of the cardia of the stomach and in the intestine. In the myenteric plexus ENK-LIr cells are observed in all parts of the gastrointestinal tract except for the jejunum of the guinea pig. In the submucous plexus ENK-LIr cells are only seen in the duodenum and colon of the rat and cecum of the guinea pig, and the number of cells is always small.

In the guinea pig, ENK-LIr fibres are found in large numbers in the myenteric plexus, circular muscle layer, lamina muscularis mucosae and the longitudinal muscle layer of the stomach and cecum. The oesophagus is an exception with only a small number of fibres present in these layers. In the rat the longitudinal muscle layer has a moderate density of ENK-LIr fibres in the stomach whereas in the small intestine this layer has ENK-LIr fibres.

Experiments using microsurgical lesions of the small intestine of guinea pig have demonstrated that the ENK neurons are intrinsic to the intestine (Furness et al., 1983).

Vasoactive intestinal polypeptide

VIP-LIr structures are observed in large numbers and with a characteristic distribution in all parts of the gastrointestinal tract (Larsson et al., 1976b; Alumets et al., 1979; Costa et al., 1980a; Schultzberg et al., 1980; Inoue et al., 1984). In the intestine, the VIP-LIr cell bodies are more numerous in the submucous than in the myenteric plexus, whereas in the stomach there seem to be more VIP-LIr cells in the myenteric plexus. In addition, single or small groups of VIP-LIr cells are observed in the lamina propria, mostly close to the lamina muscularis and sometimes more superficial. No VIP-LIr cells are observed in the submucous plexus of the oesophagus (Schulzberg, 1980).

VIP-LIr fibres are found in large or moderate numbers in the myenteric plexus, around the ganglion cells of the submucous plexus, in the circular muscle layer and the mucosa. In the lamina propria the fibres form an extensively ramifying network up to the surface, sometimes coming close to the epithelium. As with the SP-LIr fibres only a few VIP-LIr fibres are seen in the longitudinal muscle layer except the stomach, where in both species there is a moderate or even a high density. VIP-LIr fibres are observed around the blood vessels, particularly in the submucous layers of the rat stomach and cecum.

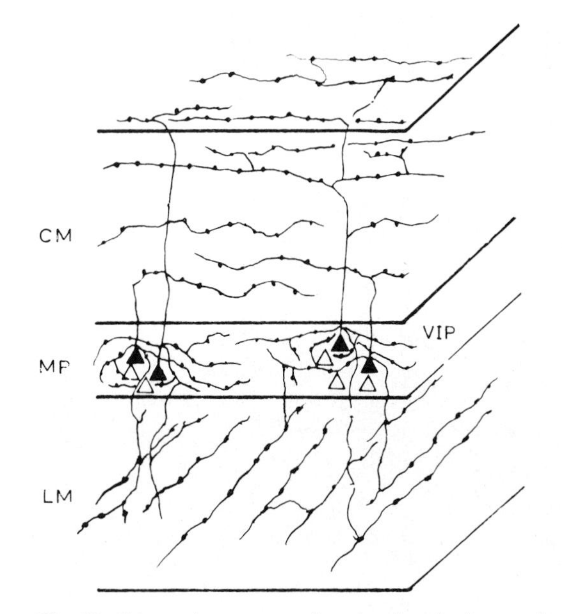

Fig. 30. Schematic representation showing the innervation of VIP-LI in the rat stomach. VIP-LIr fibres in the stomach are supplied by VIP-LIr neurons (▲) located in the myenteric plexus (MP). CM, circular muscle layer; LM, longitudinal muscle layer; △, non-immunoreactive neurons. (From Inoue et al., 1984, with permission.)

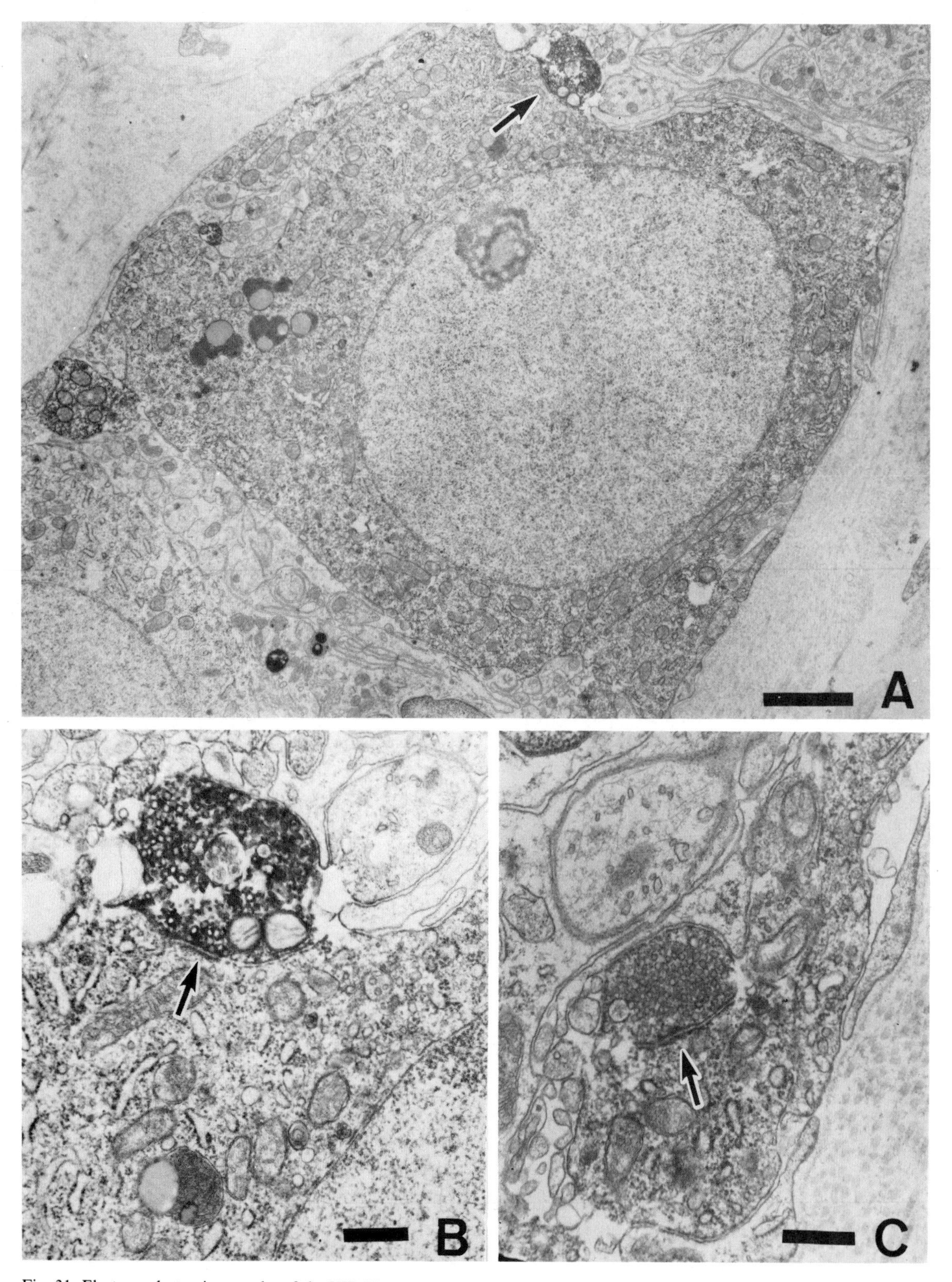

Fig. 31. Electron photomicrographs of the VIP-LIr neuron in the myenteric ganglia of the rat jejunum. It receives synaptic contact from VIP-LIr terminals (arrow) on the soma (A, B) and on the dendrite (C). B is a higher-magnification photomicrograph of a part of A, showing the synaptic contact (arrow). Bars = 2 μm (A) and 0.5 μm (B, C). (Adapted from Maeda et al., 1985, with permission.)

In the rat stomach a very dense VIP-LIr fibre meshwork has been demonstrated in the circular muscle layer, longitudinal muscle layer and myenteric plexus (Inoue et al., 1984). In the muscle layers VIP-LIr fibres run parallel to the muscle. Experimental manipulations using various surgical denervations have shown that the fibres in the stomach originate from the VIP-LIr neurons in the myenteric plexus (Inoue et al., 1984). Fig. 30 schematically shows the innervation of VIP structures in the rat stomach.

Ultrastructural studies have revealed synaptic contacts between VIP-LIr fibres and non-reactive cell somata in the ganglia of the small intestine (Feher and Leranth, 1983), and a direct synaptic contact between VIP-LIr fibres and VIP-LIr neurons in the submucous and myenteric plexuses (Fig. 31) (Maeda et al., 1985).

Somatostatin

SRIF-LIr structures are less numerous than the SP-, ENK-, VIP- and NPY-like immunoreactive structures. The upper parts of the gastrointestinal tract contain only small numbers of immunoreactive structures and some layers seemingly none. In general, more SRIF-LI is observed in the guinea pig than in the rat (Schultzberg et al., 1980).

A moderate number of SRIF-LIr cells are observed in the submucous plexus of the small intestine and colon of the guinea pig but smaller numbers occur in the rest of the gastrointestinal tract of the guinea pig and the rat. No SRIF-LIr cells are seen in the stomach of the rat, in the submucous plexus of rat oesophagus, rat cecum, guinea pig stomach or jejunum, in the myenteric plexus of rat rectum or guinea pig jejunum or cecum.

SRIF-LIr fibres are found in almost all layers of the intestine, but only in a few layers of the oesophagus and stomach, and upper intestine parts. Large numbers of fibres are seen in the submucous and myenteric plexi of the intestine of the guinea pig, while moderate numbers are found in these layers in the rat. No SRIF-LIr fibres are seen around the blood vessels.

Experiments crushing and cutting the nerves have shown that SRIF-LIr neurons are interneurons in the intestine of the guinea pig (Costa et al., 1980b).

Cholecystokinin

CCK-LIr neural structures are found in most parts of the gastrointestinal tract but in small numbers compared to the other peptides described here. In general, the upper tract of guinea pig lacks CCK-LIr fibres. CCK-LIr cell bodies are only seen in small numbers in the submucous plexus and myenteric plexus of the guinea pig duodenum and colon of rat and guinea pig (Larsson et al., 1979; Schultzberg et al., 1980).

In the rat a low density of CCK-LIr fibres is seen throughout all parts of the gastrointestinal tract. In the guinea pig, CCK-LIr fibres are often confined to the myenteric plexus of the intestine and stomach. The other layers mostly have few or single fibres.

Neuropeptide Y

The origins and overall distribution of NPY-LIr structures in the rat stomach were investigated by using whole-mount stomach preparations (Lee et al., 1985b). Numerous NPY-LIr fibres occur in the muscle layers parallel to the muscle, and in the myenteric plexus (Fig. 32A) where groups of NPY-LIr cells are identified. These structures are evenly distributed throughout the entire stomach.

NPY-LIr fibres located in the myenteric plexus (Fig. 32A) contain both TH-LIr and NPY-LIr structures which originate mainly from the coeliac ganglion (see Figs. 11, 33). In addition, a small number of NPY-LIr fibres which are also of intrinsic origin are not TH-immunostained (see Fig. 32E and F).

Colchicine treatment permits the visualization of NPY-LIr cells in the myenteric ganglia, but no immunoreactive cells are identified in the untreated stomach (Fig. 32A). Bilateral transection of the great splanchnic nerve proximal to the junction with the vagus nerve (postganglionic fibres of the coeliac ganglion) results in a marked decrease in the number of NPY-LIr fibres in the myenteric plexus

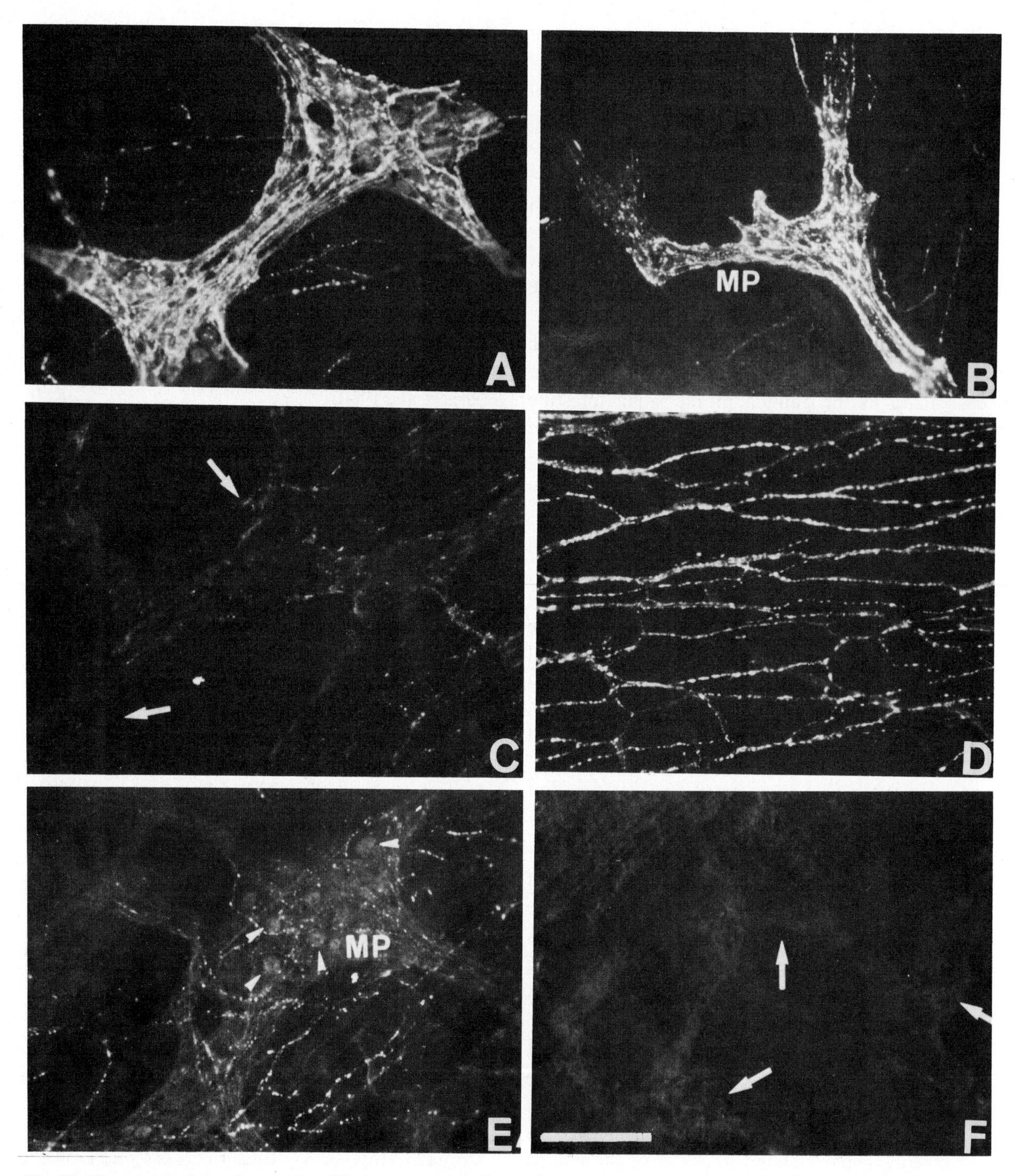

Fig. 32. Fluorescent photomicrographs of flat-mount preparations of the rat stomach showing NPY-LI (A–E) and TH-LI (E, F). Dense NPY-LIr fibre meshwork is found in the myenteric ganglia and internodal strands (A). No alteration in NPY-LIr fibres was detected in the myenteric plexus (MP) after vagotomy (B). By contrast, a marked decrease of NPY-LIr fibres in the myenteric plexus was detected (C) following transection of the great splanchnic nerve proximal to the junction of the vagus (see Fig. 26). On the other hand, this transection did not cause any change in the number of NPY-LIr fibres in the circular and longitudinal muscle layer (D),

(Fig. 32C), but a small number of NPY-LIr fibres still remain intact (Fig. 32C), while no change is seen in the number of NPY-LIr fibres either in the circular or longitudinal muscle layers (Fig. 32D). NPY-LIr fibres along the blood vessels are reduced, however. In these cases NPY-LIr neurons located in the myenteric plexus are easily found in the myenteric plexus of the denervated stomach due to the decrease of NPY-LIr fibres. Bilateral vagotomy has no effect on the number of NPY-LIr fibres in the muscle layers or in the myenteric plexus (Fig. 32B). Furthermore, 6-OHDA injection resulted in the disappearance of TH-LIr fibres throughout the stomach (Fig. 32F), and simultaneously, NPY-LIr fibres located in the myenteric plexus are also decreased markedly in number (Fig. 32E). However, no reduction of NPY-LIr fibres was seen in the circular or longitudinal muscle layers, except for NPY-LIr fibres along blood vessels in the muscle layers which disappeared after 6-OHDA injection. In these cases NPY-LIr neurons located in the myenteric plexus are easily seen due to a decrease in NPY-LIr fibres in this plexus (Fig. 32E). The changes in TH-LIr and NPY-LIr structures seen in the stomach after injection of 6-OHDA are very similar to those seen following bilateral transection of the great splanchnic nerve just proximal to the junction with the vagus (Fig. 32C, see Fig. 26) (Lee et al., 1985b).

These data suggest that NPY-LIr fibres located in the muscle layers, except for those along the blood vessels, do not contain TH immunostaining and are of intrinsic origin. TH-LIr fibres in the muscle layers, excluding those along the blood vessels originate from the coeliac ganglion and lack NPY-LI. NPY fibres located along the blood vessels in the muscle layers originate from the coeliac ganglion and contain TH- and NPY-LI. Thus two types of NPY systems, one coexisting with NA and originating in the coeliac ganglion, and the other afferent from the NA system, are supposed to be present in the gastrointestinal tract (see Fig. 33).

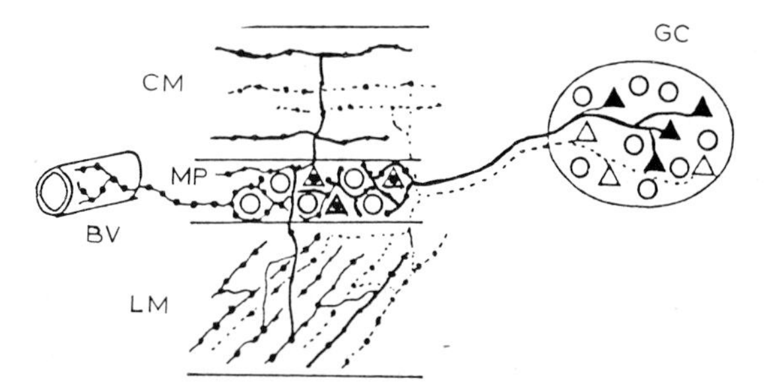

Fig. 33. Schematic representation showing the innervation of NPY-LI and NA in the muscle layer of the rat stomach. NPY/NA neurons (▲) in the coeliac ganglion (GC) innervate the myenteric plexus (MP) and blood vessels (BV). NPY-LIr neurons in the MP innervate the circular (CM) and longitudinal (LM) muscle layers, and partly contribute to intrinsic MP innervation. In the myenteric ganglia, NPY-LIr neurons lack TH-LI. However, in the coeliac ganglia NPY-LIr perikarya with coexistent TH-LI and TH-LIr cells lacking NPY-LI (△) are observed. NA-containing neurons which lack NPY-LI in the coeliac ganglion innervate the circular and longitudinal muscle layers. (From Lee et al., 1985b, with permission.)

Neurotensin

No NT-LIr neuronal cell bodies have so far been identified with certainty.

NT-LIr fibres are observed in the myenteric plexus of the oesophagus, cardia and fundus of the stomach, and duodenum, and in the longitudinal muscle of the rat stomach but not in the guinea pig (Schultzberg et al., 1980).

Bombesin/gastrin-releasing peptide

The presence of bombesin-like immunoreactivity in the rat intestine has been shown by Dockray et al. (1979), using radioimmunoassay and histochemistry. A rich distribution of nerve fibres is found in the mucosa of the rat stomach, but few in

except for NPY-LIr fibres along the blood vessels which decreased following this transection. Chemical sympathectomy (6-OHDA injection) also caused a reduction of NPY-LIr fibres in the MP (E) and along the blood vessels in the muscle layers, but failed to change the number of NPY-LIr fibres in the muscle layers. Note that a significant number of NPY-LIr fibres still remain intact in the MP after 6-OHDA treatment (E). On the other hand, almost total disappearance of TH-LIr fibres (arrows) in the MP was found after 6-OHDA injection. Arrowheads indicate NPY neurons in the MP which could be easily identified because of the decrease of NPY-LIr fibres. Bar = 100 μm. (Adapted from Lee et al., 1985b, with permission.)

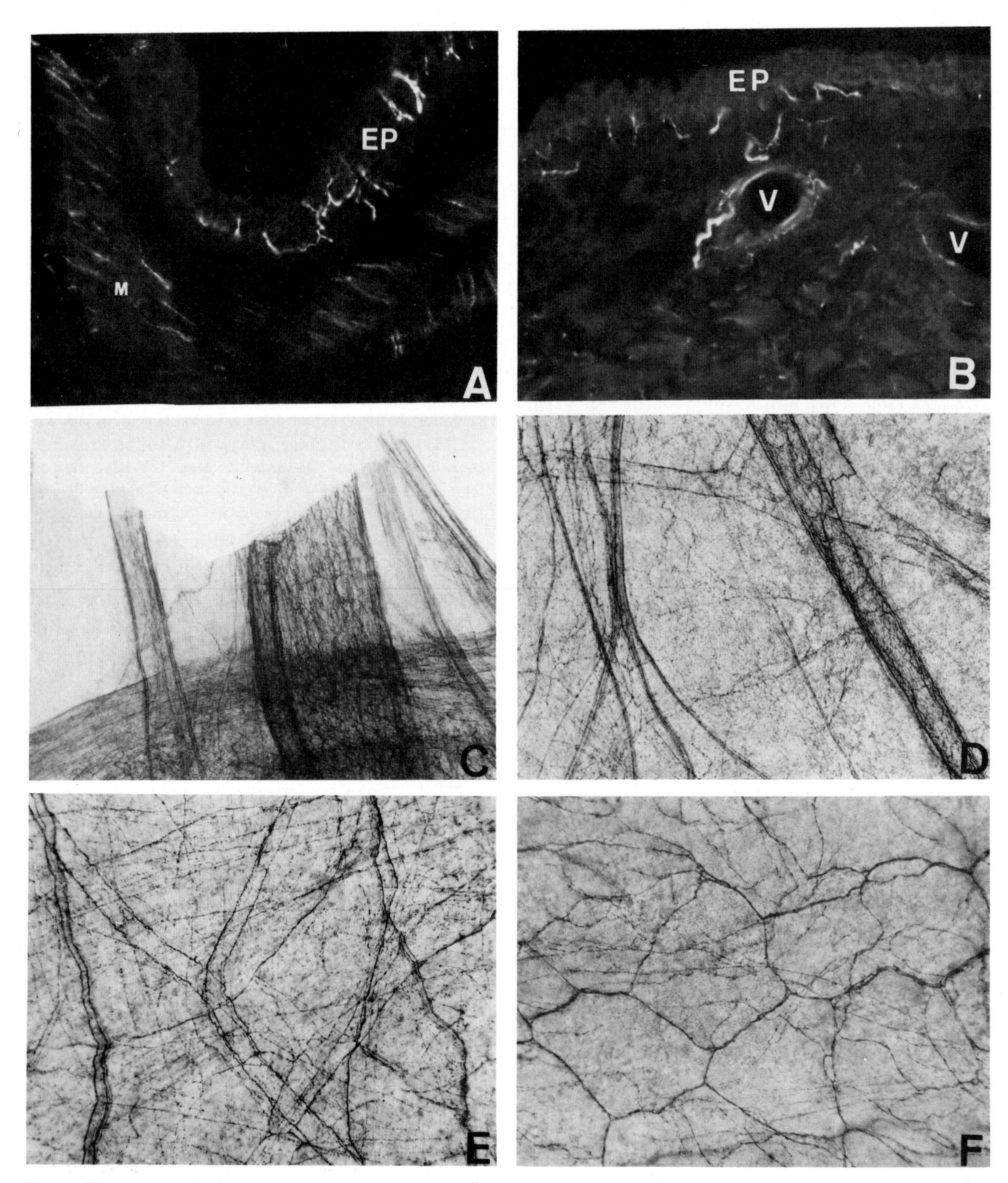

Fig. 34. Fluorescent photomicrographs showing the distribution of SP-LIr fibres in the urinary bladder on the frozen sections (A, B) and bright-field photomicrographs showing the distribution of SP-LIr fibres in the urinary bladder on the whole-mounted preparations. (C–F). A and B show numerous SP-LIr fibres terminating in the epithelium (EP) and just beneath the epithelium, those running parallel to the smooth muscle cells (M) and SP-LIr fibres along the blood vessels (V). C shows SP-LIr fibre bundles that enter the urinary bladder from the neck and those along the blood vessels. D shows SP-LIr fibre bundles which branch into several thinner segments in the neck and immunoreactive fibres along the blood vessels. SP-LIr fibres running parallel to both longitudinal and circular muscle cells are seen (E). F shows a typical meshwork of SP-LIr fibres located just beneath the epithelium. (A, B) × 170; (C, D) × 40; (E, F) × 85. (Adapted from Yokokawa et al., 1985, with permission.)

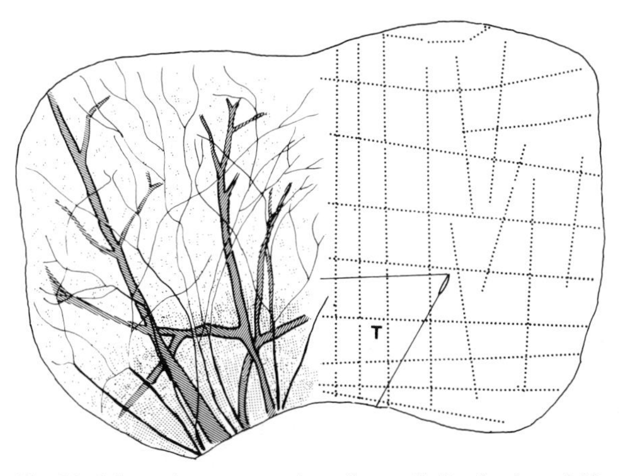

Fig. 35. Schematic representation of overall distribution of SP-LIr fibres in the rat urinary bladder. The SP-LIr fibre bundles (thin curved line) and SP-LIr fibres around the blood vessels (oblique line) enter the urinary bladder from the neck. The large dots (right), and small dots (left) indicate single SP-LIr fibres in the smooth muscle layer and typical SP-LIr meshwork observed just beneath the epithelium, respectively. The SP-LIr nerve endings within the epithelium are not illustrated. T, trigone. (From Yokokawa et al., 1985, with permission.)

the intestinal mucosa. Abundant bombesin-like immunoreactive fibres are seen surrounding nerve cell bodies in the myenteric plexus throughout the gut. No bombesin-like immunoreactive nerve cell bodies are identified.

Serotonin

The presence of 5-HT-LIr nerve cell bodies and fibres has been demonstrated by Furness and Costa (1982). In the guinea pig small intestine, nerve cell bodies are located in the myenteric plexus and varicose fibres are found in the ganglia of the myenteric and submucous plexus in the guinea pig small intestine, stomach and large intestine, and in the intestine of mice, rabbits and rats. They suggest that there are two classes of amine neuron in the guinea pig small intestine, the enteric 5-HT neurons and enteric non 5-HT, amine-handling neurons.

Urinary bladder

Substance P

An early study reported that the SP innervation of the guinea pig bladder was much more conspicuous than that of the rat, in which no or very few SP--LIr structures are observed (Alm et al., 1978; Sharkey et al., 1983). Recent immunohistochemical studies have, however, demonstrated that numerous SP-LIr structures are also present in the rat urinary bladder (Mattiasson et al., 1985; Yokokawa et al., 1985).

Numerous SP-LIr fibres are distributed throughout the smooth muscle, submucosal and epithelial layers of rat urinary bladder (Fig. 34) (Yokokawa et al., 1985). In the muscle layer a number of SP-LIr fibres run parallel to the longitudinal and circular muscle layer. In the submucosal layer SP-LIr fibres are sometimes found around the blood vessels. Two types of fibres are found in the urinary bladder: one forms fibre bundles (non-varicose smooth profiles) and the other has a varicose appearance. Single fibres are found in the ganglionic plexus in the trigonal area, but no SP-LIr cells have been detected so far, even in the colchicine-treated rats (Yokokawa et al., 1985). A higher density of SP-LIr fibres are contained in the bladder neck and trigonal area than the fundus of the bladder (see Fig. 35).

The overall distribution of SP-LI has been studied, using whole-mounted tissues, by Yokokawa et al. (1985), and is schematically shown in Fig. 35. Two types of fibres enter the urinary bladder from the neck: one forms thick fibre bundles and the other extends along the blood vessels (Figs. 34C, 35). These fibre bundles appear to be non-varicose smooth fibres. They extend to the fundus of the bladder where they branch into several thinner segments (Fig. 34D) that often project via collaterals to the smooth muscle and submucosal layers. SP-LIr fibres in the smooth muscle layer are evenly distributed throughout the urinary bladder (Fig. 34E). The density of the SP-LIr fibre meshwork found in the submucosal layer is higher in the neck and trigone than in the fundus of the bladder. Fibre meshwork located in the uppermost part of the submucosal layer sends several fibres that extend to the epithelium (Fig. 34A, B, F). These fibres extend to the superficial part of the epithelium where they terminate and an abundant arborization of fine SP-

LIr fibres is seen (Fig. 34A, B, F). The bladder neck and the trigone contain a higher density of SP-LIr fibres in the epithelium than the fundus of the bladder. Fig. 35 schematically shows the overall distribution of SP-LIr fibres in the rat urinary bladder.

Postganglionic denervation eliminates SP-LIr fibres in the bladder (Mattiasson et al., 1985). Chemical sympathectomy by injection of 6-OHDA does not affect the SP-LIr fibres in the muscle layer of the bladder significantly, while capsaicin treatment of newborn rats causes a loss of SP-LIr fibres in the wall of the bladder, suggesting that these fibres are peripheral branches of the sensory ganglia.

Calcitonin gene-related polypeptide

CGRP-LIr fibres are abundant in the muscle layer, where they run parallel to the muscles, in the submucosa, and in the epithelium (Yokokawa et al., 1986). They form a dense plexus just beneath the epithelium. No CGRP-LIr cells are detected even in the colchicine-injected rats.

CGRP-LIr fibres enter the bladder in various ways, either forming a thick bundle of fibres or extending around the blood vessels. These fibres extend to the fundus of the bladder, branching into several thinner segments.

By using a combined method of tracer and immunohistochemistry as described in this volume by Shiosaka and Tohyama, the origins of CGRP-LIr fibres in the urinary bladder have been investigated (Yokokawa et al., 1986). Injection of the tracer into the bladder results in the demonstration of small- to medium-sized labeled cells in the dorsal root ganglion of L_6 and S_1 that are also CGRP-LIr, and a few are at S_2. No labeled cells are seen in the dorsal root ganglia at the other levels. Therefore, CGRP--LIr fibres in the urinary bladder arise in the CGRP-LIr cell bodies in the dorsal sensory ganglia at the level of L_6 and S_1.

In the spinal sensory ganglia at L_6 and S_1, about 40% of the ganglion cells are CGRP-LIr, which are mostly small to medium in size, 10% are large, and about 20% of the ganglion cells are SP-LIr which are mostly small to medium size. On the same sections, most SP-LIr cells also have CGRP-LI. However, most large-sized CGRP-LIr cells lack SP-LI. As mentioned above, only small- to medium-sized ganglion cells are retrogradely labeled. Thus, it is likely that small- to medium-size CGRP-LIr ganglion cells project to the bladder while large CGRP--LIr cells do not (Yokokawa et al., 1986).

Vasoactive intestinal polypeptide

A moderate number of VIP-LIr fibres is found in the urinary bladder. They are distributed in the smooth muscle and around the blood vessels (Alm et al., 1977; Mattiasson, 1985).

After postganglionic denervation, VIP-LIr fibres disappear from the bladder wall. However, no change in the number of these fibres is detected after chemical sympathectomy using 6-OHDA treatment, nor after injection of capsaicin which causes the total disappearance of SP-LIr fibres in the urinary bladder. These data suggest that most of the VIP-LIr nerve fibres are extrinsic, and separate from the aminergic system, though it cannot be excluded that a part of VIP-LIr fibres may originate in the sympathetic nervous system.

Neuropeptide Y

A very rich distribution of NPY-LI has been found in the detrusor muscle and also around the arteries and arterioles of the bladder wall (Lundberg et al., 1982; Mattiasson, 1985). A sparse distribution of NPY-LIr fibres is seen in the submucosa.

After postganglionic denervation, no NPY-LIr fibres can be visualized in the bladder wall (Mattiasson et al., 1985). In rats given 6-OHDA treatment no definite changes in the density or distribution of NPY-LIr fibres are observed in the muscle layers, whereas the NPY-LIr fibres around the blood vessels disappear and those in the submucosa are reduced in number. On the other hand, almost total disappearance of NA fibres is observed in the bladder of 6-OHDA-injected rats by use of histofluorescence technique, whereas NPY-LIr fibres are numerous in the untreated rat. These data from surgical denervation and chemical sympathectomy using 6-OHDA reveal that a part of the NPY-LIr fibres originate in the cell bodies within the pelvic ganglia or sympathetic ganglia. As found in the gut (Furness, 1983; Sundler et al., 1983), NPY-LI is also distributed in a population of nonadrenergic nerve fibres. No change in the amount or distribution of NPY-LI in the detrusor or pelvic ganglia is found after chemical sympathectomy using 6-OHDA, suggesting that NPY-LI in the detrusor muscle and pelvic ganglia is contained in non-adrenergic nerves.

Other peptides

The occurrence of ENK- (Alm et al., 1978) and SRIF- (Hökfelt et al., 1978) and CCK- (Larsson et al., 1979) -like immunoreactivities has been demon-

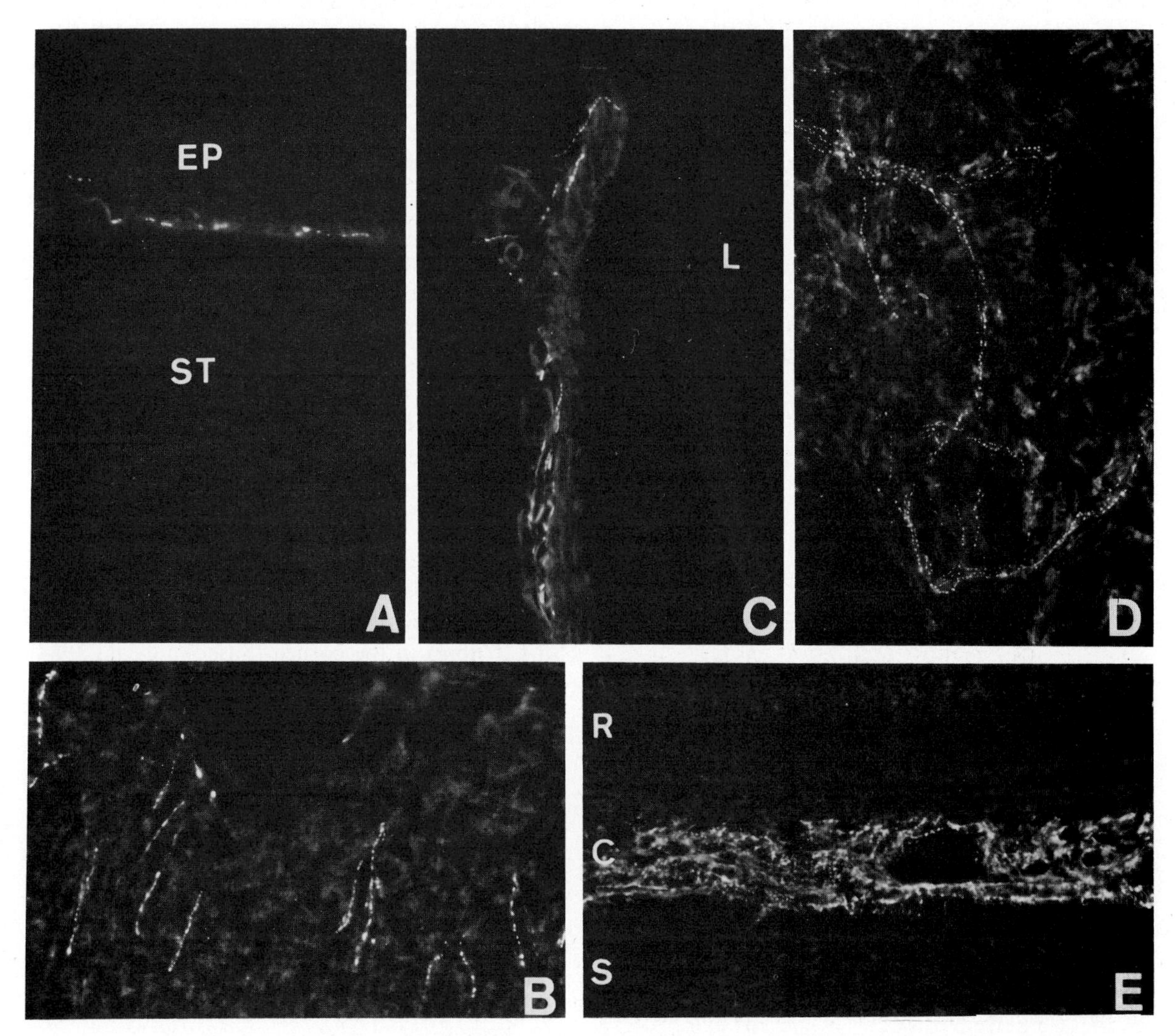

Fig. 36. Fluorescent photomicrographs showing SP-LIr fibres (A–C) and VIP-LIr fibres (D, E) in the anterior eye segments. A shows SP-LIr fibres just beneath the epithelium (EP) of the cornea of the rat. ST, stroma of cornea. ×450. B shows SP-LIr fibres in the ciliary body of the squirrel. Numerous SP-LIr fibres are found in the ciliary muscles ×400. C shows VIP-LIr fibres in the iris of the rat; L, lens. ×110. D shows VIP-LIr fibres in the ciliary body of the cat. ×120. E shows a dense network of VIP-LIr fibres in the choroid of the rat. ×110. R, retina; C, choroid; S, sclera. (From Shimizu, 1982, and Shimizu et al., 1982b, with permission.)

strated in the wall of the urinary bladder of the cat.

With the exception of the urinary bladder, CCK--LIr nerves are not detected in any parts of the male or female genitourinary tract. In the urinary bladder very few CCK-LIr varicose nerves are detected in the muscle wall. Ganglia occurring in the vicinity of the genitourinary organs are devoid of CCK-LIr nerve fibres.

Anterior eye segment

Substance P

Cornea. A large number of SP-LIr fibres in the cornea has been demonstrated by many authors (Miller et al., 1981; Tervo et al., 1981a,b, 1982a,b; Shimizu, 1982; Shimizu et al., 1982b; Stone et al., 1982; Terenghi et al., 1982; Torngvist et al., 1982) (Fig. 36A). In the stroma, thick non-varicose and smooth SP-LIr fibres run progressively more superficially and enter the epithelium. At the upper part of the stroma, SP-LIr fibres have a varicose appearance and form a dense network especially in the uppermost part of the stroma. Some of the varicose fibres diverge to the epithelial layer.

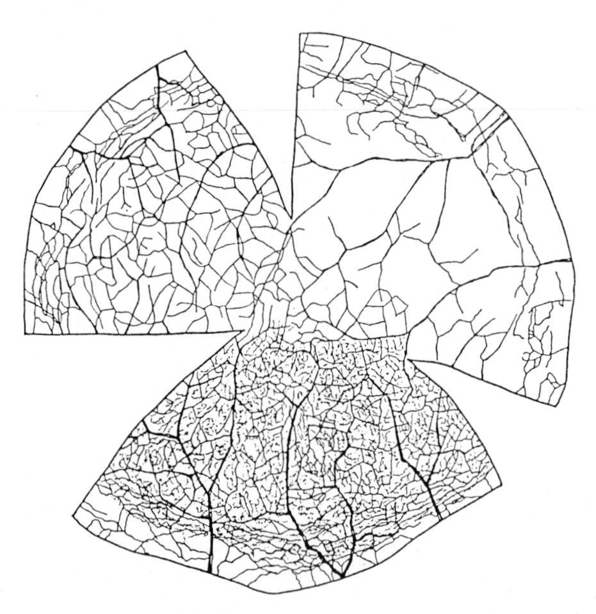

Fig. 37. Schematic representation showing the distribution of SP-LIr fibres in the cornea, reconstructed from the photomontages of the cornea, photographed at three different levels. Upper right shows the SP-LIr fibre trunks that enter the stroma from the sclera. These fibre trunks extend to the central part, subdivide, and approach the epithelium. In the upper left, SP-LIr fibres enter the cornea from the episclera, run circularly at the limbus, often sending some fibres to the central part. In addition, an SP-LIr fibre network is located in the upper most part of the stroma. In the lower half, an abundant arborization of fine SP-LIr terminals is seen within the epithelium. (From Sasaoka et al., 1984, with permission.)

Whole-mount sections show that SP-LIr fibres enter the cornea from two levels: one from the sclera and the other from the episclera (Sasaoka et al., 1984). Fig. 37 schematically shows the overall distribution of SP-LIr fibres in the cornea. From the sclera, a thick fibre trunk, composed of numerous SP-LIr fibres, extends to the central part and subdivides into smaller fibre bundles, which approach the epithelium. The SP-LIr fibre bundles from the episclera are smaller than those from the sclera. However, both fibre bundles form a dense fibre network in the uppermost epithelium part of the stroma. This fibre plexus dissociates into SP-LIr fibres extending to the superficial part of the epithelium where they form an abundant arborization of fine SP-LIr fibres.

Partial or complete transection of the ophthalmic branch of the trigeminal nerve caused the ipsilateral decrease or the almost complete disappearance of SP-LIr fibres in the cornea (Sasaoka et al., 1984). These results suggest that SP-LIr fibres in the cornea originate from SP-LIr neurons in the trigeminal ganglion. Fig. 38 schematically shows the innervation of SP-LIr structures in the anterior eye segment.

Ciliary body. Numerous SP-LIr fibres are seen in the ciliary body (Fig. 36B). The greatest number of SP-LIr fibres are distributed among the ciliary muscle fibres, some of which form many dense fibre bands. In addition, a few SP-LIr fibres are observed around the ciliary cleft. Dense fibre bands situated in the ciliary muscles often give rise to several fibres that enter the ciliary process. These fibres transverse the stroma to reach the area just beneath the pigment epithelial cell layer, giving off several branches along their course (Shimizu, 1982; Shimizu et al., 1982b).

Iris. A moderate number of SP-LIr fibres are found in the rat iris (Fig. 36C) (Shimizu, 1982; Shimizu et al., 1982b). The majority of these fibres are located in the stroma near the endothelial cell layer, although a few SP-LIr fibres are scattered in other parts of the stroma such as around the vessels. A few nerve fibres are directed medially and distributed in the sphincter muscle of the iris. A direct contact has been demonstrated between the SP-LIr nerve terminals and pupillary sphincter muscle of the rat by immunoelectron microscopy (Shimizu et al., 1982b), suggesting the possible contribution of SP to the regulation of smooth muscle tone in addition to any role in sensory transmission. Surgical experiments (Tervo et al., 1982b) suggest that most of the SP-LIr fibres in the anterior eye segment arise in the SP-LIr neurons in the trigeminal ganglion where SP-LIr perikarya occur but not in the ciliary ganglia where no SP-LIr cell bodies are seen (Shimizu et al., 1982b).

Calcitonin gene-related peptide

The occurrence, distribution, and origins of CGRP-LIr fibres in the anterior eye segment of mammals have been demonstrated by recent studies (Terenghi et al., 1985; H. Kiyama et al., in preparation). The distributions of CGRP-LIr fibres and SP-LIr fibres are very similar in the cornea, ciliary body and iris. In the cornea CGRP-LIr fibres are seen in both the stroma and the epithelium. The majority of CGRP-LIr fibres have been shown also to contain SP-LI (H. Kiyama et al., in preparation). A few CGRP-LIr fibres are distributed in the ciliary processes and ciliary muscles, and around the vessels in the pars plicata (corona ciliaris) of the cat ciliary body, and immunoreactive fibres are seen only around the vessels in the pars plana (orbiculus ciliaris). In the cat iris, a large number of CGRP-LIr fibres are found, most of which contain SP-LI. There seems to be a species difference in the distribution of CGRP-LI. For example, a dense plexus of CGRP-LIr fibres is seen in the dilator muscle of the rat, but only a few fibres in this area of the cat (Terenghi et al., 1985; H. Kiyama et al., in preparation). The CGRP-LIr fibres in the anterior eye segment are suggested to arise in the trigeminal sensory CGRP-LIr neurons, most of which also contain SP-LI, and a proportion of CGRP-LIr fibres may also arise in the pterygopalatine parasympathetic ganglion where CGRP-LIr neurons without SP-LI are seen (Lee et al., 1985c; Terenghi et al., 1985; H. Kiyama et al., in preparation).

Enkephalin

A network of ENK-LIr nerve fibres has been demonstrated in rat iris whole-mounts (Björklund et al., 1983). Systemic administration of capsaicin in doses which caused complete disappearance of SP-LIr fibres in the iris did not cause degeneration of ENK-LIr fibres. Thus ENK-LIr fibres in the iris are probably separated from the SP system.

Vasoactive intestinal polypeptide

VIP-LIr fibres are detected in several orbital structures of mammals (Uddman et al., 1980; Shimizu, 1982), but no VIP-LIr fibres are found in the cornea of the rat. Such nerves are numerous in the lacrimal glands and somewhat less numerous in the Harderian glands and tarsal glands. The nerves surround glandular acini and small blood vessels. Intraocularly, VIP-LIr fibres are seen in the ciliary processes and ciliary body (Fig. 36D), in the posterior third of the ciliary muscle, and around or along the blood vessels (Fig. 36D). A dense plexus of VIP-LIr fibres is found in the choroid (Fig. 36E). These fibres outnumber SP-LIr fibres in the choroid. VIP-LIr nerve fibres are absent from vessels in the anterior uvea. This distribution may explain why intracranial stimulation in the exit region of oculomotor fibres dilates the vessels of the choroid but not those of the iris.

A large number of VIP-LIr cell bodies are observed in the pterygopalatine ganglion. Extirpation of this ganglion results in the disappearance of VIP--LIr fibres from the intraocular structures and from the lacrimal and Harderian glands, although re-

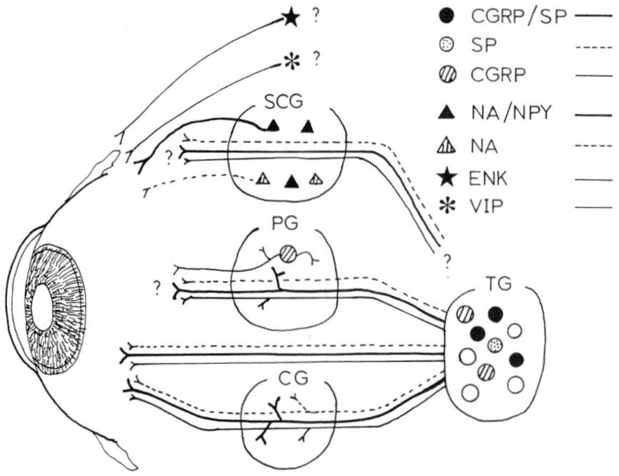

Fig. 38. Schematic drawing of CGRP, SP, NPY, NA, ENK and VIP innervation in the anterior eye segments and related ganglia. TG, trigeminal ganglion; SCG, superior cervical ganglion; PG, pterygopalatine ganglion; CG, ciliary ganglion. (Adapted from H. Kiyama et al., in preparation, with permission.)

moval of the superior cervical ganglion and the ciliary ganglion does not affect the VIP-LIr fibre supply (Uddman et al., 1980). These data suggest that the VIP-LIr fibres in these areas originate in the pterygopalatine ganglion.

Cholecystokinin

The presence of CCK-LIr fibres has been demonstrated in the guinea pig ocular nerves (Stone et al., 1984). The density of CCK-LIr fibres is significantly lower than that of other ocular peptidergic nerve fibres described above. Some of the fibres are CCK-LIr in the ocular nerve, in the choroid and in the posterior ciliary body, in the pectinate ligament near the iris root, rarely within the ciliary process, but not in the cornea. A few fibres are also found near the iris and limbic blood vessels. The uvea contains a small percentage of CCK-LIr fibres in the nerve bundles.

Neuropeptide Y

The presence of APP-LIr fibres has been demonstrated in the guinea pig eye (Stone and Laties, 1983). In all regions of the uvea, APP-LIr fibres were present around large blood vessels and APP-LIr fibres were found throughout the choroid. APP-LIr fibres were also seen in the ciliary processes, iris dilator muscle and, to a lesser degree, in the constrictor muscle. However, convincing data of innervation to the cornea are lacking.

Tongue

Substance P

Three types of papillae contain SP-LIr fibres (Nagy et al., 1982; Nishimoto et al., 1982). The number of fibres varies from papilla to papilla in the rat. The circumvallate papillae contain the greatest numbers of SP-LIr fibres (Fig. 39), followed by the foliate papillae and the fungiform papillae; the filiform papillae lack SP-LIr fibres (Nishimoto et al., 1982). The SP-LIr fibres of the papillae form dense bands in the lamina propria just beneath the epithelium (Fig. 39). Some of the fibres enter the epithelium and the taste buds (Fig. 39). It should be stressed that not every taste bud contains SP-LIr fibres: only 95% of the taste buds in the foliate papillae, 70% of the taste buds in the fungiform papillae, and 40% of the buds in the circumvallate papillae contain detectable fibres (Nishimoto et al., 1982).

Unilateral section of the glossopharyngeal nerve results in a complete disappearance of SP-LIr fibres in the foliate papillae on the operated side, and a slight decrease in the circumvallate papillae on both sides. Bilateral section of the glossopharyngeal nerve results in the complete disappearance of SP-LIr fibres in the foliate and circumvallate papillae. Following unilateral section of the chorda tympani, SP-LIr fibres in the taste buds of the fungiform papillae disappear completely. In addition, unilateral neurotomy of the mandibular nerve results in the complete disappearance of SP-LIr fibres in the epithelium of the fungiform papillae. These data suggest that SP fibres in the foliate and circumvallate papillae are supplied by the glossopharyngeal nerve, SP in the taste buds of the fungiform papillae by the chorda tympani, and SP in the epithelium of the fungiform papillae by the third division of the trigeminal nerve (Nishimoto et al., 1982). Fig. 40

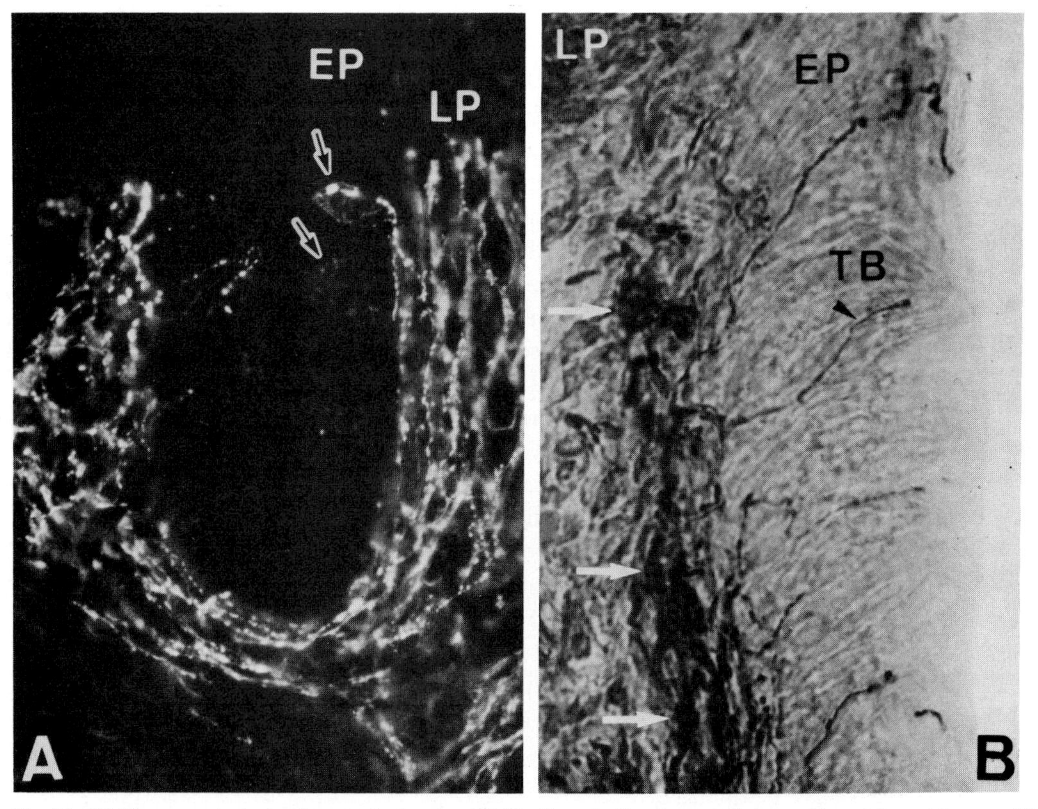

Fig. 39. (A) Fluorescent photomicrograph showing SP-LIr fibres in the circumvallate papillae of the rat tongue. SP-LIr fibres occur in the epithelium (EP) and lamina propria (LP). Arrows indicate taste buds where SP-LIr fibres enter. ×250. (B) Lightfield photomicrograph of SP-LIr fibres (arrows) in the taste buds (TB) and epithelium. Note numerous SP-LIr fibres in the lamina propria. ×570. (From Yamasaki et al., 1984, 1985, with permission.)

schematically shows the SP innervation in the epithelium and lamina propria of the tongue papillae.

The fine structure of SP-LIr fibres in the taste buds of the circumvallate papillae of the rat tongue has been investigated by Yamasaki et al. (1984). Outside of the epithelium, SP-LIr and non SP-LIr fibres are surrounded by the cytoplasm of Schwann cells (Fig. 41). When the SP-LIr fibres enter the epithelium, they immediately lose this cytoplasmic sheath and pass to the taste buds, although no profile suggesting clear synaptic contact between SP-LIr fibres and underlying cells has been identified (Figs. 42, 43). SP-LIr terminals are filled with small synaptic vesicles and contain a few mitochondria. They are surrounded by type I cells (Fig. 42) and type II cells (Fig. 42B) in the taste buds. No SP-LIr structures are found in nerve endings that make synaptic contact with type III cells, the gustatory receptor cells (Fig. 43; schematically shown in Fig. 44) (Yamasaki et al., 1984).

Immunohistochemical ontogenetic development of the SP-LIr structures in the circumvallate papillae of the rat tongue has been investigated by light microscopy (Yamasaki et al., 1985) and electron microscopy (Yamasaki and Tohyama, 1986). SP-LIr fibres in the lamina propria first appear on gestational day 19, and they increase in number up to postnatal day 10. On the other hand, SP-LIr fibres in the extragemmal epithelium are first seen on gestational day 21 and those in the taste buds

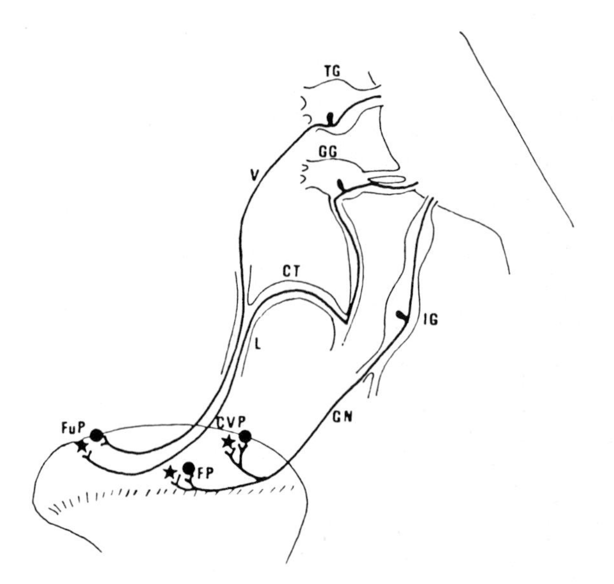

Fig. 40. Schematic representation of the SP-innervation in the rat tongue. Abbreviations: CT, chorda tympani; CVP, circumvallate papillae; FP, foliate papillae; FuP, fungiform papillae; GG, ganglion geniculi; GN, glossopharyngeal nerve; IG, inferior ganglion of the GN; L, ligual nerve; TG, trigeminal ganglion; V, third division of the trigeminal nerve. (●) SP-LIr fibres in the epithelium outside the taste buds; (*) SP-LIr fibres in the taste buds. SP is supplied by GN in the taste buds and epithelium of the CVP and FP, by CT in the taste buds of the FuP, and by V in the epithelium of the FuP. (From Nishimoto et al., 1982.)

on postnatal day 2. These fibres increase remarkably in number after postnatal day 5, reaching their maximum on postnatal day 20, and thereafter decrease slightly. SP-LIr fibres do not form any synaptic contact with underlying cells. In addition, the occurrence of non-immunoreactive fibres in these areas precedes that of the SP-LIr fibres during ontogenetic development. These findings suggest that SP-LIr fibres do not primarily relate to the differentiation of the extragemmal epithelium and taste bud cells, but rather that they are involved in their maintenance or development by being neuromodulator or trophic factors in the epithelium of the tongue.

Calcitonin gene-related peptide

Recently, CGRP-LIr nerve fibres have been demonstrated in the tongue epithelium and striated muscle (Lee et al., 1985c). CGRP-LIr fibres found in the taste buds probably coexist with SP. In the striated muscle of the tongue, CGRP-LIr fibres appear as motor end plates. These fibres found in the muscle lack SP-LI (Lee et al., 1985c).

Comments

Recent advances in immunocytochemical techniques have revealed that the peripheral nervous system contains an abundance of transmitter candidates including neuropeptides, monoamines, and ACH. In order to elucidate the functions of these substances it is necessary to establish the origins of the immunoreactive fibres and projection areas of the axons arising from immunoreactive neurons. Subsequently, detailed analysis of the microcircuitry within a given target organ is required.

Coexistence of neurotransmitters has been found in the peripheral nervous system as in the central nervous system. For example, ENK and ACH coexist in the preganglionic parasympathetic neurons, NA and NPY coexist in the sympathetic ganglia (Fig. 11), ACH and CGRP coexist in the motor neurons innervating striated muscle (Figs. 6–9), and SP and CGRP coexist in the primary sensory ganglionic neurons (Figs. 1–3). Since the neuronal circuitry of the peripheral system is more fully understood than that of the central nervous system, the use of the peripheral nervous system for analysing the significance of the coexistence in single neuron systems may be more suitable than that of the central nervous system. In fact, Takami et al. (1985b) have demonstrated that CGRP enhances the effect

Fig. 41. Electron photomicrographs showing SP-LIr fibres located in the lamina propria just beneath the basal lamina (BL). Numerous SP-LIr fibres are ensheathed by the cytoplasm of the Schwann cell (S) together with non-immunoreactive nerve fibres. (A) × 12 000; (B) × 20 000. (From Yamasaki et al., 1984, with permission.)

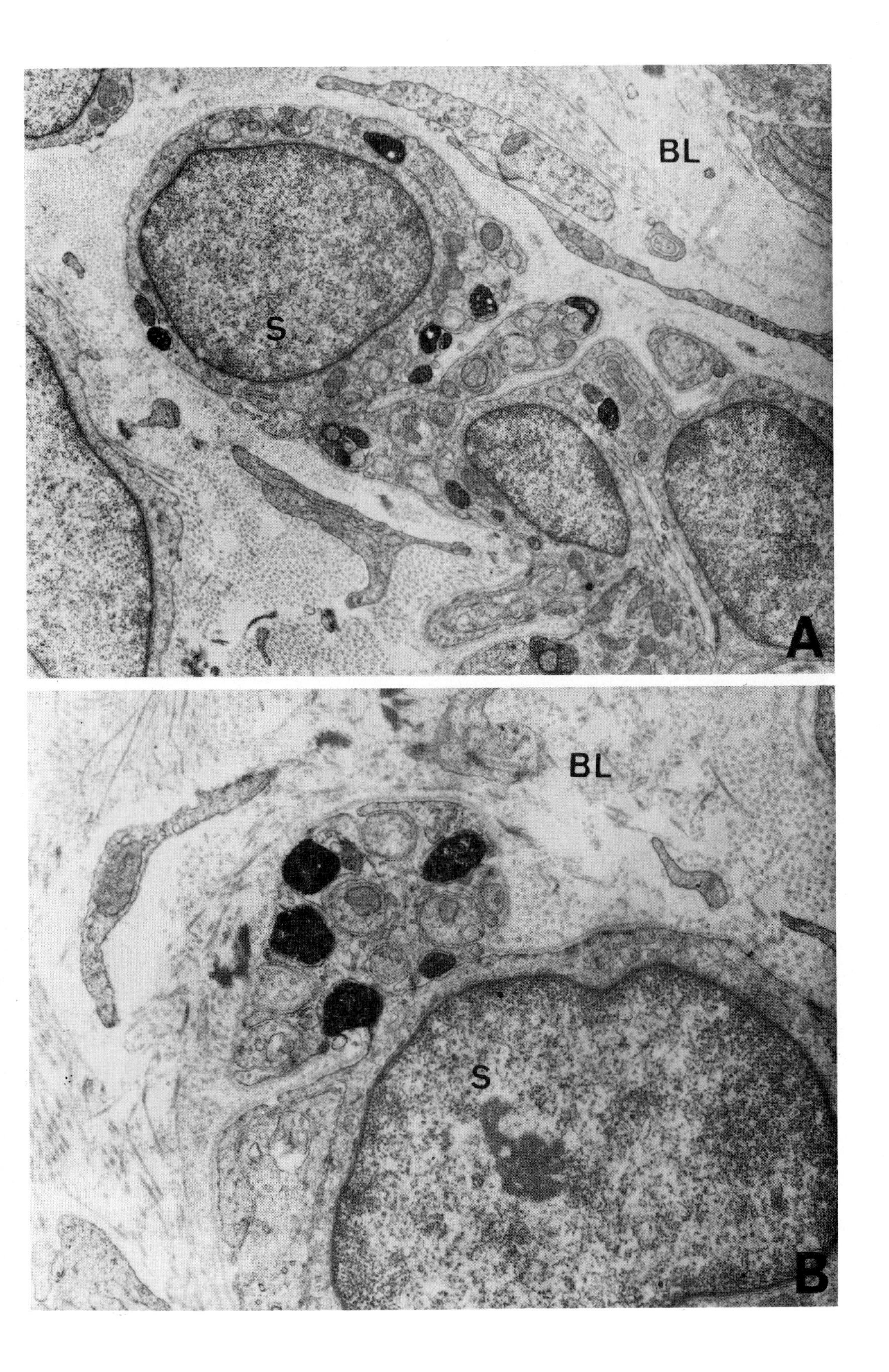
BL
S
A
BL
S
B

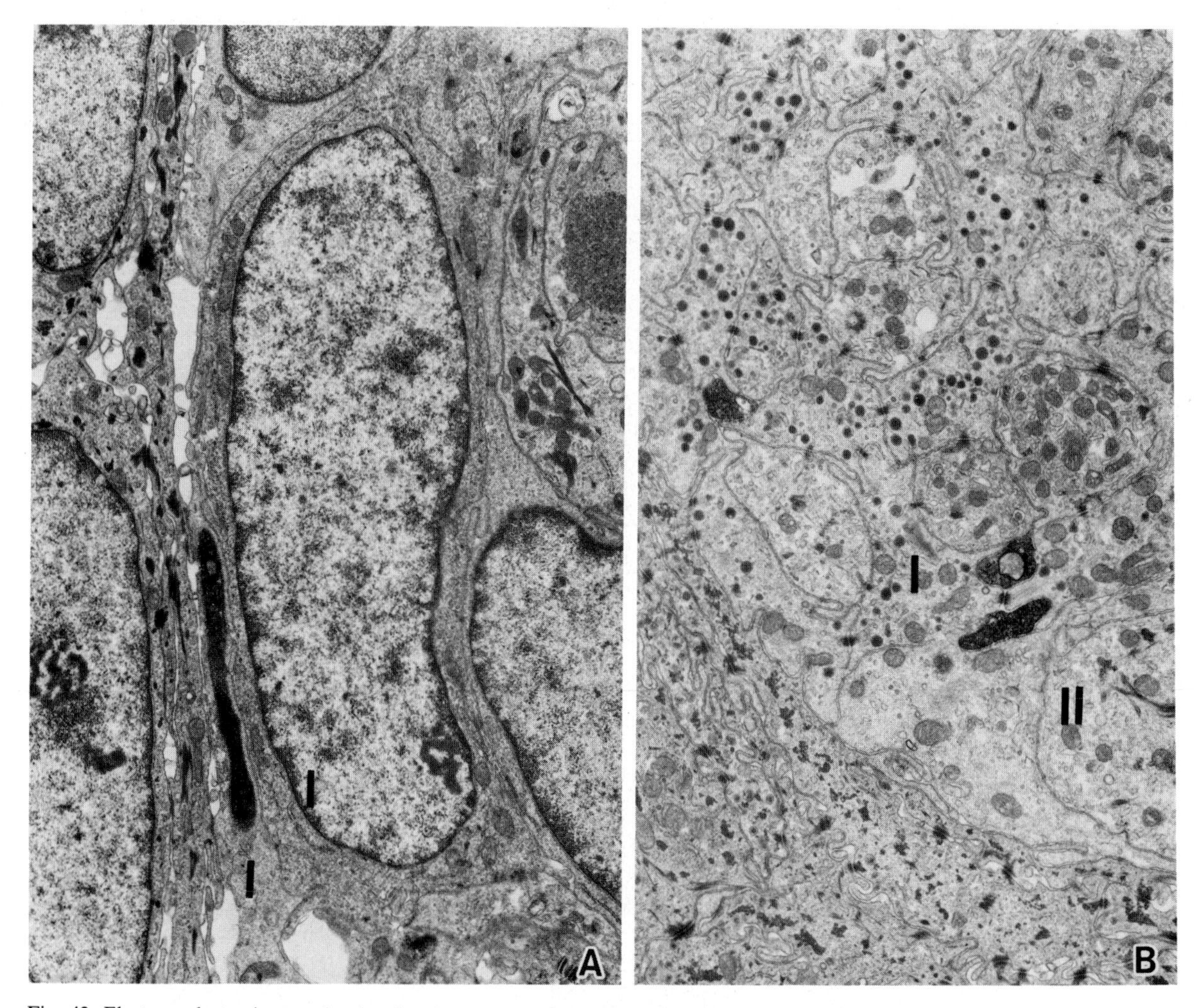

Fig. 42. Electron photomicrographs showing SP-LIr fibres surrounded by the cytoplasm of type I cells (I). No synaptic contacts between SP-LIr fibres and type I cells are identified. (A) Middle portion; (B) area near the outer taste pore. II, type II cell (see Fig. 44). (A) × 10 000; (B) × 11 800. (From Yamasaki et al., 1984, with permission.)

of ACH in motor neurons innervating the striated muscle.

In addition, in the peripheral nervous system, many of the terminals do not form synaptic contacts and show synapses "en passant", particularly in the blood vessels and stomach muscles. In these organs, bioactive substances released from the terminals may directly act on the smooth muscle. Ac-

Fig. 43. (A) Electron photomicrograph showing SP-LIr fibres which show close apposition to type III (III) and type II (II) cells. No special contact is identified between the fibres and type III or type II cells. (B) Electron photomicrograph showing non-immunoreactive nerve ending (N) which makes synaptic contact with type III cells. This nerve ending exhibits a "dendrite-like" profile, while numerous synaptic vesicles are concentrated (axon-like profile) in the SP-LIr fibres shown in A. Since synaptic vesicles in the cytoplasm are located near the contact membrane (arrows) and presynaptic specializations are seen, this non-immunoreactive nerve ending receives gustatory information. (A) × 20 000; (B) × 16 600. (From Yamasaki et al., 1984, with permission.)

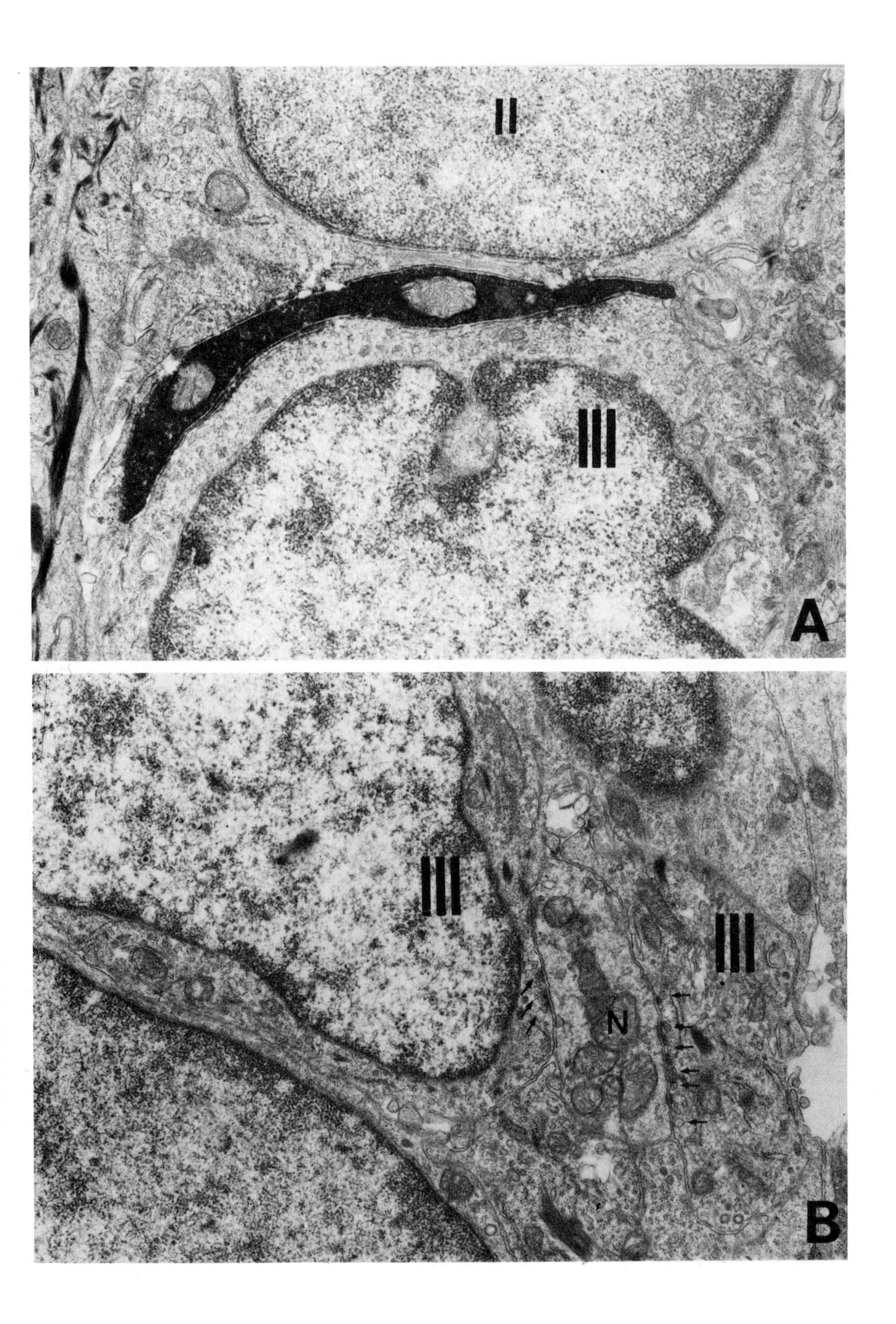
II
III
A
III
III
N
B

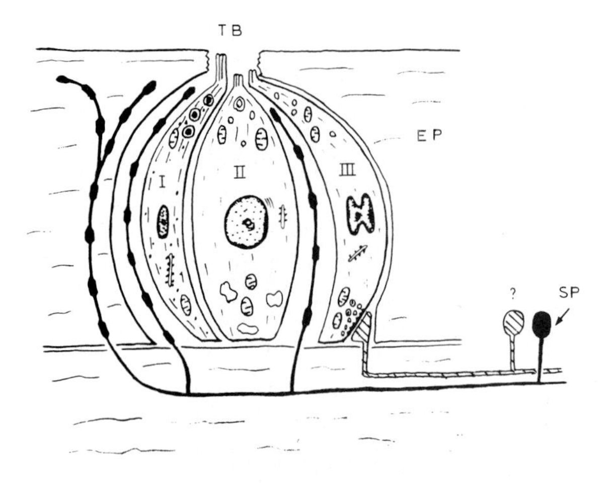

Fig. 44. Schematic representation summarizing the findings. Nerve endings that primarily conduct information and form synaptic contact with a type III cell (III) do not contain SP-LI. SP-LIr fibres that enter the taste buds (TB) or epithelium (EP) outside the taste buds traverse these regions to reach the surface without making synaptic contact with the underlying taste bud cells such as type I (I), type II (II) and type III or epithelial cells. (From Yamasaki et al., 1984, with permission.)

cordingly, in these organs, the localization of the receptor of these transmitter substances must also be explored.

Another interesting problem is that peripheral neurites of the sensory ganglion cells contain SP and CGRP, and probably SRIF. It is well established that many of the SP-LIr fibres in the peripheral organs, such as blood vessels, urinary bladder, myenteric plexus, anterior eye segment and skin epithelium including the taste buds, originate from SP-LIr neurons located in the sensory ganglion. If these SP-LIr fibres primarily convey sensory information from the peripheral organs to the brainstem and spinal cord, these SP-LIr fibres ought to have dendritic profiles, which would be evidenced by a lack of aggregation of synaptic vesicles. In the periphery, the sensory nerves receive information but would not conduct the information to the organs. However, as shown in the blood vessels and tongue epithelium, SP-LIr fibres have axon terminal-like profiles (containing synaptic vesicles), indicating that SP may be located and released at the peripheral terminals in spite of the sensory afferent.

SP-LIr fibres in the blood vessels may also be collaterals of the primary sensory fibres and might have a vasomotor function related to the vascular reflex (Henry, 1977; Lembeck et al., 1977; Hökfelt et al., 1980). Immunoelectron microscopic study by Yamasaki et al. (1984) has shown that SP-LIr terminals in the epithelium of the skin also have axon terminal profiles. In the same study, they show that SP-LIr fibres in the taste buds have axon terminal profiles and do not make synaptic contact with gustatory cells, while nerve terminals which form synaptic contact with gustatory cells have dendritic profiles and lack SP-LI, suggesting that SP-LIr terminals do not primarily conduct sensory information. Apart from the question of whether or not SP--LIr terminals in the blood vessels, skin epithelium and taste buds are peripheral branches of the primary sensory neurons, it could be concluded that nerve terminals which contain SP in the peripheral neurites of the neurons of the sensory ganglion cells are not primarily conductors of sensory information. What then are the roles of the SP in these systems? Further studies are awaited.

Acknowledgements

The authors are grateful to Drs. M. Tohyama, H. Takagi, T. Matsuyama, K. Yokokawa, H. Kiyama, Y. Lee, K. Takami, Y. Kubota and H. Yamasaki for fruitful discussions.

References

Alm, P., Alumets, J., Hakanson, R. and Sundler, F. (1977) Peptidergic (vasoactive intestinal peptide) nerves in the genitourinary tract. *Neuroscience,* 2: 751–754.

Alm, P., Alumets, J., Brodin, E., Hakanson, R., Nilsson, G., Sjoberg, N. O. and Sundler, F. (1978) Peptidergic (Substance P) nerves in the genitourinary tract. *Neuroscience,* 3: 419–425.

Alumets, J., Hakanson, R., Sundler, F. and Chang, K. J. (1978) Leu-enkephalin-like material in nerves and enterochromaffin cells in the gut. *Histochemistry,* 56: 187–196.

Alumets, J., Fahrenkrug, J., Hakanson, R., De Muckadell, O. S., Sundler, F. and Uddman, R. (1979) A rich VIP nerve supply is characteristic of sphincters. *Nature,* 280: 155–156.

Amara, S. G., Jonas, V., Rosenfeld, M. G., Ong, E. S. and Evans, R. M. (1982) Alternative RNA processing in calcitonin gene expression generates mRNA encoding different polypeptide products. *Nature*, 298: 240–244.

Björklund, H., Olson, L. and Seiger, A. (1983) Enkephalin-like immunofluorescence in nerves of the rat iris following systemic capsaicin injection. *Med. Biol.*, 61: 280–282.

Costa, M., Furness, J. B., Buffa, R. and Said, S. I. (1980a) Distribution of enteric nerve cell bodies and axons showing immunoreactivity for vasoactive intestinal polypeptide in the guinea pig ileum. *Neuroscience*, 5: 587–596.

Costa, M., Furness, J. B., Llewellyn-Smith, I. J., Davies, B. and Oliver, J. (1980b) An immunohistochemical study of the projections of somatostatin containing neurons in the guinea-pig intestine. *Neuroscience*, 5: 841–852.

Cuello, A. C., Del Fiacco, M. and Paxinos, G. (1978) The central and peripheral ends of the substance P-containing sensory neurons in the rat trigeminal system. *Brain Res.*, 152: 499–509.

Dalsgaard, C. J., Hökfelt, T., Schultzberg, M., Lundberg, J. M., Terenius, L., Dockray, G. J., Cuello, C. and Goldstein, M. (1983) Origin of peptide-containing fibers in the inferior mesenteric ganglion of the guinea pig: Immunohistochemical studies with antisera to substance P, enkephalin, vasoactive intestinal polypeptide, cholecystokinin/gastrin and bombesin. *Neuroscience*, 9: 191–211.

Dockray, G. J., Valliant, C. and Walsh, J. H. (1979) The neuronal origin of bombesin-like immunoreactivity in the rat gastrointestinal tract. *Neuroscience*, 4: 1561–1568.

Edvinsson, L., Fahrenkrug, J., Hanko, J., Owmann, C., Sundler, F. and Uddman, R. (1980) VIP (vasoactive intestinal polypeptide)-containing nerves of intracranial arteries in mammals. *Cell Tissue Res.*, 208: 135–142.

Edvinsson, L., Degueurce, A., Duverger, D., Mackenzie, E. T. and Scatton, B. (1983a) Central serotonergic nerves project to the pial vessels of the brain. *Nature*, 306: 55–57.

Edvinsson, L., Rosendal-Helgesen, S. and Uddman, R. (1983b) Substance P: Localization, concentration and release in cerebral arteries, choroid plexus and dura mater. *Cell Tissue Res.*, 234: 1–7.

Elde, R., Hökfelt, T., Johansson, O. and Terenius, L. (1976) Immunohistochemical studies using antibodies to leucine-enkephalin. Initial observations on the nervous system of the rat. *Neuroscience*, 1: 349–351.

Feher, E. and Leranth, C. (1983) Light and electron microscopic immunocytochemical localization of vasoactive intestinal polypeptide (VIP)-like activity in the rat small intestine. *Neuroscience*, 10: 97–106.

Fuji, K., Senba, E., Fujii, S., Nomura, I., Wu, J. Y., Ueda, Y. and Tohyama, M. (1985) Distribution, ontogeny and projections of cholecystokinin-8, vasoactive intestinal polypeptide and γ-aminobutyrate-containing neuron systems in the rat spinal cord: an immunohistochemical analysis. *Neuroscience*, 14: 881–894.

Furness, J. B. and Costa, M. (1982) Neurons with 5 hydroxytryptamine-like immunoreactivity in the enteric nervous system: their projections in the guinea-pig small intestine. *Neuroscience*, 7: 341–349.

Furness, J. B., Parka, R. E., Della, M., Costa, M. and Eskay, R. L. (1982) Substance P-like immunoreactivity in nerves associated with the vascular system of guinea-pigs. *Neuroscience*, 7: 447–459.

Furness, J. B., Costa, M., Emson, P. C., Hakanson, R., Moghimzadeh, E., Sundler, F., Taylor, I. L. and Chance, R. E. (1983) Distribution, pathways and reactions to drug treatment of nerves with neuropeptide Y and pancreatic polypeptide-like immunoreactivity in the guinea-pig digestive tract. *Cell Tissue Res.*, 234: 71–92.

Gibson, S. J., Polak, J. M., Bloom, S. R., Sabate, J. M., Mulderry, P. M., Ghatei, M. A., McGregoi, G. P., Morrison, J. F. B., Kelley, J. S., Evans, R. M. and Rosenfeld, M. G. (1984) Calcitonin gene related peptide immunoreactivity in the spinal cord of man and of eight other species. *J. Neurosci.*, 4: 3101–3111.

Glazer, E. J. and Basbaum, A. I. (1978) Leucine enkephalin: localization in and axoplasmic transport by sacral parasympathetic preganglionic neurons. *Science*, 208: 1479–1480.

Griffith, S. G., Lincoln, J. and Burnstock, G. (1982) Serotonin as a neurotransmitter in the cerebral arteries. *Brain Res.*, 247: 388–392.

Hanley, M. R., Benton, H. P., Lightman, S. L., Todd, K., Bone, E. A., Fretten, P., Palmers, S., Kirk, C. J. and Michell, R. H. (1984) A vasopressin-like peptide in the mammalian sympathetic nervous system. *Nature*, 309: 258–261.

Hayashi, H., Ohsumi, K. and Ueda, N. (1982) Effect of spinal gangliotomy on substance P-like immunoreactivity in the gastroduodenal tract of cats. *Brain Res.*, 232: 227–230.

Henry, J. L. (1977) Substance P and pain: A possible relation in afferent transmission. In U. S. von Euler and B. Pernow (Eds.), *Substance P*, Raven Press, New York, pp. 231–239.

Hökfelt, T., Elde, R., Johansson, O., Luft, R. and Arimura, A. (1975a) Immunohistochemical evidence for the presence of somatostatin, a powerful inhibitory peptide, in some primary sensory neurons. *Neurosci. Lett.*, 1: 231–235.

Hökfelt, T., Kellerth, J. O., Nilsson, G. and Pernow, B. (1975c) Experimental immunohistochemical studies on localization and distribution of substance P in cat primary sensory neurons. *Brain Res.*, 100: 235–252.

Hökfelt, T., Elde, R., Johansson, O., Luft, R., Nilsson, G. and Arimura, A. (1976) Immunohistochemical evidence for separate populations of somatostatin-containing and substance P-containing primary afferent neurons in the rat. *Neuroscience*, 1: 131–136.

Hökfelt, T., Elfvin, L. G., Elde, R., Schultzberg, M., Goldstein, M. and Luft, R. (1977a) Occurrence of somatostatin-like immunoreactivity in some peripheral sympathetic noradrenergic neurons. *Proc. Natl. Acad. Sci. U.S.A.*, 74: 3587–3591.

Hökfelt, T., Elfvin, L. G., Schultzberg, M., Goldstein, M. and

Nilsson, G. (1977b) On the occurrence of substance P-containing fibers in sympathetic ganglia: Immunohistochemical evidence. *Brain Res.*, 132: 29–41.

Hökfelt, T., Schultzberg, M., Elde, R., Nilsson, G., Terenius, L., Said, S. and Goldstein, M. (1978) Peptide neurons in the peripheral tissues including the urinary tract: immunohistochemical studies. *Acta Pharmacol. Toxicol.*, Suppl. 11, 43: 79–89.

Hökfelt, T., Johansson, O., Ljungdahl, A., Lundberg, J. M. and Schultzberg, M. (1980) Peptidergic neurons. *Nature*, 284: 515–521.

Honda, C. N., Rethelyi, M. and Petrusz, P. (1983) Preferential immunohistochemical localization of vasoactive intestinal polypeptide (VIP) in the sacral spinal cord of the cat: light and electron microscopic observations. *J. Neurosci.*, 3: 2183–2196.

Hunt, S. P., Emson, P. C., Gilbert, M., Goldstein, M. and Kimmell, J. R. (1981) Presence of avian pancreatic polypeptide-like immunoreactivity in catecholamine and methionine-enkephalin-containing neurons within the central nervous system. *Neurosci. Lett.*, 21: 125–130.

Iijima, T. and Wasano, T. (1980) A histochemical and ultrastructural study of serotonin-containing nerves in cerebral blood vessels of the lamprey. *Anat. Rec.*, 198: 671–680.

Inoue, H., Shiosaka, S., Sasaki, Y., Hayashi, N., Satoh, N., Kamata, T., Tohyama, M. and Shiotani, Y. (1984) Three-dimensional distribution of vasoactive intestinal polypeptide--containing structures in the rat stomach and their origins using whole mount tissue. *J. Neural Trans.*, 59: 195–205.

Jessen, K. R., Polak, J. M., Van Noorden, S., Bloom, S. R. and Burnstock, G. (1980) Peptide-containing neurons connect the two ganglionated plexuses of the enteric nervous system. *Nature*, 283: 391–393.

Kessler, J. A. and Black, I. B. (1981) Nerve growth factor stimulates development of substance P in the embryonic spinal cord. *Brain Res.*, 208: 135–145.

Kiyama, H., Katayama, Y., Hillyard, C. J., Girgis, S., MacIntyre, I., Emson, P. C. and Tohyama, M. Localization of calcitonin gene-related peptide-like structures in the anterior eye segment and its related ganglia of the cat with reference to colocalization of substance P. In preparation.

Konishi, S., Tsunoo, A. and Otsuka, M. (1979) Substance P and noncholinergic excitatory synaptic transmission in guinea pig sympathetic ganglia. *Proc. Jpn. Acad.*, 55: B525–530.

Larsson, L. I. and Rehfeld, J. F. (1979) Localization and molecular heterogeneity of cholecystokinin in the central and peripheral nervous system. *Brain Res.*, 165: 201–218.

Larsson, L. I., Edvinsson, L., Fahrenkrug, J., Håkanson, R., Owmann, C., De Muckadell, S. O. B. and Sundler, F. (1976a) Immunohistochemical localization of vasodilatory polypeptide (VIP) in cerebrovascular nerves. *Brain Res.*, 133: 400–404.

Larsson, L. I., Fahrenkrug, J., De Muckadell, S. O., Sundler, F., Håkanson, R. and Tehfels, J. F. (1976b) Localization of vasoactive intestinal polypeptide (VIP) to central and peripheral neurons. *Proc. Natl. Acad. Sci. U.S.A.*, 73: 3197–3200.

Lee, Y., Kawai, Y., Shiosaka, S., Takami, K., Kiyama, H., Hillyard, C. J., Girgis, S., MacIntyre, I., Emson, P. C. and Tohyama, M. (1985a) Coexistence of calcitonin gene-related peptide (CGRP)- and substance P-like peptide in single cells of the trigeminal ganglion of the rat: Immunohistochemical analysis. *Brain Res.*, 330: 194–196.

Lee, Y., Shiosaka, S., Emson, P. C., Powell, J. F., Smith, A. D. and Tohyama, M. (1985b) Origins and overall distribution of neuropeptide Y-like immunoreactive structures in the rat stomach with special reference to noradrenaline system. *Gastroenterology*, 89: 118–126.

Lee, Y., Takami, K., Kawai, Y., Girgis, S., Hillyard, C. J., MacIntyre, I., Emson, P. C. and Tohyama, M. (1985c) Distribution of calcitonin gene-related peptide in the rat peripheral nervous system with special reference to its coexistence with substance P. *Neuroscience*, 15: 1227–1237.

Lembeck, F., Gamse, R. and Juan, H. (1977) Substance P and sensory nerve endings. In U. S. von Euler and B. Pernow (Eds.), *Substance P*, Raven Press, New York, pp. 169–181.

Linnoila, R. I., Di Augustine, R. P., Miller, R. J., Chang, K. J. and Cuatrecasas, P. (1978) An immunohistochemical and radioimmunological study of the distribution of (met^5)- and (leu^5)-enkephalin in the gastrointestinal tract. *Neuroscience*, 3: 1187–1196.

Lundberg, J. M., Hökfelt, T., Nilsson, G., Terenius, L., Rehfeld, J., Elde, R. and Said, S. (1978) Peptide neurons in the vagus, splanchnic and sciatic nerves. *Acta Physiol. Scand.*, 104: 499–501.

Lundberg, J. M., Hökfelt, T., Angaard, A., Pernow, B. and Emson, P. (1979a) Immunohistochemical evidence for the substance P immunoreactive fibers in the taste buds of the cat. *Acta Physiol. Scand.*, 107: 389–391.

Lundberg, J. M., Hökfelt, T., Schultzberg, M., Uvnäs-Wallensten, K., Köhler, C. and Said, S. (1979b) Occurrence of vasoactive intestinal polypeptide VIP-like immunoreactivity in certain cholinergic neurons of the cat. Evidence from combined immunohistochemistry and acetylcholinesterase staining. *Neuroscience*, 4: 1539–1559.

Lundberg, J. M., Terenius, L., Hökfelt, T., Martling, C. R., Tatemoto, K., Mutt, V., Polak, J., Bloom, S. and Goldstein, M. (1982) Neuropeptide Y-like immunoreactivity in peripheral noradrenergic neurons and effects of NPY on sympathetic function. *Acta Physiol. Scand.*, 116: 477–480.

Maeda, M., Takagi, H., Kubota, Y., Morishima, Y., Akai, F., Hashimoto, S. and Mori, S. (1985) The synaptic relationship between vasoactive intestinal polypeptide (VIP)-like immunoreactive neurons and their axon terminals in the rat small intestine: light and electron microscopic study. *Brain Res.*, 329: 356–359.

Matsuyama, T., Matsumoto, M., Shiosaka, S., Fujisawa, A., Yoneda, S., Kimura, K., Abe, H. and Tohyama, M. (1983a) Overall distribution of substance P and vasoactive intestinal polypeptide in the cerebral arteries: An immunohistochemical study using whole-mount. *J. Cereb. Blood Flow Metab.*, Suppl. 1, 3: S208–209.

Matsuyama, T., Shiosaka, S., Matsumoto, M., Yoneda, S., Kimura, K., Abe, H., Hayakawa, T., Inoue, H. and Tohyama, M. (1983b) Overall distribution of vasoactive intestinal polypeptide-containing nerves on the wall of cerebral arteries: An immunohistochemical study using whole-mounts. *Neuroscience,* 10: 89–96.

Matsuyama, T., Matsumoto, M., Shiosaka, S., Hayakawa, T., Yoneda, S., Kimura, K., Abe, H. and Tohyama, M. (1984a) Dual innervation of substance P-containing neuron system in the wall of the cerebral arteries. *Brain Res.*, 322: 144–147.

Matsuyama, T., Shiosaka, S., Matsumoto, M., Mieno, M., Yoneda, S., Kimura, K., Hayashi, N., Kamata, T., Abe, H., Hayakawa, T., Tsubouchi, H. and Tohyama, M. (1984b) Immunoreactive glucagon in the smooth muscle cells of the rat cerebral artery: an immunohistochemical analysis. *J. Cereb. Blood Flow Metab.*, 4: 305–307.

Matsuyama, T., Shiosaka, S., Wanaka, A., Yoneda, S., Kimura, K., Hayakawa, T., Emson, P. C. and Tohyama, M. (1985) Fine structure of peptidergic and catecholaminergic nerve fibers in the anterior cerebral artery and their interrelationship: An immunoelectron microscopic study. *J. Comp. Neurol.*, 235: 268–276.

Matsuyama, T., Shiosaka, S., Steinbusch, H., Yoneda, S., Kimura, K., Hayakawa, T. and Tohyama, M. Overall distribution and origins of serotonin-like immunoreactive fibers in the cerebral arterial wall of the guinea pig. In preparation.

Mattiasson, A., Ekblad, E., Sundler, F. and Uvelius, B. (1985) Origin and distribution of neuropeptide Y-, vasoactive intestinal polypeptide- and substance P-containing nerve fibers in the urinary bladder of the rat. *Cell Tissue Res.*, 239: 141–146.

Miller, A., Costa, M., Furness, J. B. and Chubb, I. W. (1981) Substance P-immunoreactive sensory nerve supply the rat iris and cornea. *Neurosci. Lett.*, 23: 243–249.

Minagawa, H., Shiosaka, S., Inoue, H., Hayashi, N., Kasahara, A., Kamada, T., Tohyama, M. and Shiotani, Y. (1984) Origins and three dimensional distribution of substance P containing structures on the rat stomach using whole-mount tissue. *Gastroenterology,* 86: 51–59.

Mulderry, P. K., Ghatei, M. A., Rodrigo, J., Allen, J. M., Rosenfeld, M. G., Polak, J. M. and Bloom, S. R. (1985) Calcitonin gene-related peptide in cardiovascular tissues of the rat. *Neuroscience,* 14: 947–950.

Nagy, J. I., Goedert, M., Hunt, S. P. and Bomd, A. (1982) The nature of the substance P-containing nerve fibers in taste papillae of the rat tongue. *Neuroscience,* 7: 3137–3151.

Nishimoto, T., Akai, M., Inagaki, S., Shiosaka, S., Shimizu, Y., Yamamoto, K., Senba, E., Sakanaka, M., Takatsuki, K., Hara, Y., Takagi, H., Matsuzaki, Y., Kawai, Y. and Tohyama, M. (1982) On the distribution and origins of substance P in the papillae of the rat tongue: an experimental and immunohistochemical study. *J. Comp. Neurol.*, 207: 85–95.

Otsuka, M. and Konishi, S. (1976) Substance P an excitatory transmitter of primary sensory neurons. *Cold Spring Harbor Symp. Quant. Biol.*, 15: 135–143.

Panula, P., Hadjiconstantinou, M., Yang, H. Y. T. and Costa, E. (1983) Immunohistochemical localization of bombesin/gastrin releasing peptide and substance P in primary sensory neurons. *J. Neurosci.*, 3: 2021–2029.

Polak, J. M., Bloom, S. R., Sullivan, S. N., Facer, P. and Pearse, A. G. E. (1977) Enkephalin-like immunoreactivity in the human gastrointestinal tract. *Lancet,* 1: 972–974.

Rosenfeld, M. G., Mermod, J. J., Amara, S. G., Swanson, L. W., Sawachenko, P. E., Rivier, J., Vale, W. W. and Evans, R. M. (1983) Production of novel neuropeptide encoded by the calcitonin gene via tissue specific RNA processing. *Nature,* 304: 129–135.

Sasaoka, A., Ishimoto, I., Kuwayama, Y., Sakiyama, T., Manabe, R., Shiosaka, S., Inagaki, S. and Tohyama, M. (1984) Overall distribution of substance P nerves in the rat cornea and their three-dimensional profiles. *Invest. Ophthalmol.,* 25: 351–356.

Schultzberg, M. (1984) Peripheral nervous system. In P. C. Emson (Ed.), *Chemical Neuroanatomy,* Raven Press, New York, pp. 1–51.

Schultzberg, M., Hökfelt, T., Terenius, L., Elfvin, L. G., Lundberg, J. M., Brandt, J., Elde, R. P. and Goldstein, M. (1979) Enkephalin immunoreactive nerve fibers and cell bodies in sympathetic ganglia of the guinea-pig and rat. *Neuroscience,* 4: 249–270.

Schultzberg, M., Hökfelt, T., Nilsson, G., Terenius, L., Rehfeld, J. F., Brown, M., Elde, R., Goldstein, M. and Said, S. I. (1980) Distribution of peptide and catecholamine-containing neurons in the gastrointestinal tract of rat and guinea-pig: Immunohistochemical studies with antisera to substance P, vasoactive intestinal polypeptide, enkephalins, somatostatin, gastrin/cholecystokinin, neurotensin and dopamine β-hydroxylase. *Neuroscience,* 5: 689–744.

Senba, E., Shiosaka, S., Hara, Y., Inagaki, S., Sakanaka, M., Takatsuki, K., Kawai, Y. and Tohyama, M. (1982) Ontogeny of the peptidergic system in the rat spinal cord: Immunohistochemical analysis. *J. Comp. Neurol.,* 208: 54–66.

Sharkey, K. A., Williams, R. G., Schulzberg, M. and Dockray, G. J. (1983) Sensory substance P-innervation of the urinary bladder: possible site of action of capsaicin in causing urine retention in rats. *Neuroscience,* 10: 861–868.

Shimizu, Y. (1982) Localization of neuropeptides in the cornea and uvea of the rat: an immunohistochemical study. *Cell Mol. Biol.,* 28: 103–110.

Shimizu, Y., Ishimoto, I., Shiosaka, S., Kuwayama, Y., Fukuda, M., Inagaki, S., Sakanaka, M. and Tohyama, M. (1982a) A direct contact of substance P-containing nerve fibers with pupillary sphincter muscles of the rat: an immunohistochemical analysis. *Neurosci. Lett.*, 33: 25–28.

Shimizu, Y., Kuwayama, Y., Fukuda, Y., Ishimoto, I., Shiosaka, S., Inagaki, S., Takagi, H., Sakanaka, M., Senba, E., Kawai, Y., Takatsuki, K. and Tohyama, M. (1982b) Localization of substance P-immunoreactivity in the anterior eye segment of squirrels: an immunohistochemical analysis. *Invest. Ophthalmol.*, 22: 259–263.

Stone, R. A. and Laties, A. M. (1983) Pancreatic polypeptide-like immunoreactive eyes in the guinea pig eye. *Invest. Ophthalmol.*, 24: 1620–1623.

Stone, R. A., Laties, A. M. and Brecha, N. C. (1982) Substance P-like immunoreactive nerves in the anterior segment of the rabbit, cat and monkey eye. *Neuroscience*, 7: 2459–2468.

Stone, R. A., Kuwayama, Y., Laties, A. M., McGlinn, A. M. and Schmidt, M. L. (1984) Guinea-pig ocular nerves contain a peptide of the cholecystokinin/gastrin family. *Exp. Eye Res.*, 39: 387–391.

Sundler, F., Moghimzadeh, E., Håkanson, R., Ekelund, M. and Emson, P. (1983) Nerve fibers in the gut and pancreas of the rat displaying neuropeptide-Y immunoreactivity. Intrinsic and extrinsic origin. *Cell Tissue Res.*, 230: 487–493.

Takami, K., Kawai, Y., Shiosaka, S., Lee, Y., Girgis, S., Hillyard, C. J., MacIntyre, I., Emson, P. C. and Tohyama, M. (1985a) Immunohistochemical evidence for the coexistence of calcitonin gene-related peptide- and cholineacetyltransferase-like immunoreactivity in neurons of the rat hypoglossal, facial and ambiguus nuclei. *Brain Res.*, 328: 386–389.

Takami, K., Uchida, S., Kawai, Y., Tohyama, M., Shiotani, Y., Yoshida, H., Emson, P. C., Girgis, S., Hillyard, C. J. and MacIntyre, I. (1985b) Calcitonin gene-related peptide effects in the striated muscle contraction. *Neurosci. Lett.*, 60: 227–231.

Tanaka, J., Shiosaka, S., Tsubouchi, H., Gohda, E., Daikuhara, Y. and Tohyama, M. (1983) Immunoreactive glucagon in the vascular walls of the rat. *Life Sci.*, 33: 1549–1554.

Tatemoto, K., Carlquist, M. and Mutt, V. (1982) Neuropeptide Y – a novel brain peptide with structural similarities to peptide YY and pancreatic polypeptide. *Nature*, 296: 659–660.

Terenghi, G., Polak, J. M., Probert, L., MacGregor, G. P., Ferri, G. L., Blank, M. A., Butler, J. M., Unger, W. G., Zhang, S., Cole, D. F. and Bloom, S. R. (1982) Mapping, quantitative distribution and origin of substance P- and VIP-containing nerves in uvea of guinea pig eye. *Histochemistry*, 75: 399–417.

Terenghi, G., Polak, M. A., Mulderry, P. K., Butler, J. M., Unger, W. G. and Bloom, S. R. (1985) Distribution and origin of calcitonin gene-related peptide (CGRP) immunoreactivity in the sensory innervation of the mammalian eye. *J. Comp. Neurol.*, 233: 506–516.

Tervo, K., Tervo, T., Eränko, L. and Eränko, O. (1981a) Substance P immunoreactive nerves in the rodent cornea. *Neurosci. Lett.*, 25: 95–98.

Tervo, K., Tervo, T., Eränko, L., Eränko, O. and Cuello, A. C. (1981b) Immunoreactivity for substance P in the gasserian ganglion, ophthalmic nerve and anterior segment of the rabbit eye. *Histochem. J.*, 13: 435–441.

Tervo, T., Tervo, K. and Eränko, L. (1982a) Ocular neuropeptides. *Med. Biol.*, 60: 53–60.

Tervo, K., Tervo, T., Eränko, L., Eränko, O., Valtonen, S. and Cuello, A. C. (1982b) Effect of sensory and sympathetic denervation on substance P immunoreactivity in nerve fibers of the rabbit eye. *Exp. Eye Res.*, 34: 577–581.

Torngvist, K., Mandahl, A., Leander, S., Loren, I., Håkanson, R. and Sundler, F. (1982) Substance P-immunoreactive nerve fibers in the anterior segment of the rabbit eye: distribution and possible physiological significance. *Cell Tissue Res.*, 222: 467–477.

Uddman, R., Alumets, J., Ehinger, B., Håkanson, R., Loren, I. and Sunder, F. (1980) Vasoactive intestinal peptide nerve in occular and orbital structures of the cat. *Invest. Ophthalmol.*, 19: 878–885.

Verhofstad, A. A. J., Steinbusch, H. W. M., Penke, B., Varga, J. and Joosten, H. W. J. (1981) Serotonin-immunoreactive cells in the superior cervical ganglion of the rat. Evidence for the existence of separate serotonin- and catecholamine-containing small ganglionic cells. *Brain Res.*, 212: 39–49.

Wanaka, A., Matsuyama, T., Yoneda, S., Kimura, K., Kamada, T., Girgis, S., MacIntyre, I., Emson, P. C. and Tohyama, M. (1986) Origins and distribution of calcitonin gene-related peptide-containing nerves in the walls of the cerebral arteries of the guinea pig with special reference to the coexistence with substance P. *Brain Res.* (in press).

Yamamoto, K., Matsuyama, T., Shiosaka, S., Inagaki, S., Senba, E., Shimizu, Y., Ishimoto, I., Hayakawa, T., Matsumoto, M. and Tohyama, M. (1983) Overall distribution of substance P-containing nerves in the wall of the cerebral arteries of the guinea pig and its origins. *J. Comp. Neurol.*, 215: 421–426.

Yamasaki, H. and Tohyama, M. (1985) Ontogeny of substance P-like immunoreactive fibers in the rat taste buds and their surrounding epithelium of the circumvallate of the rat. II. Electron microscopic analysis. *J. Comp. Neurol.*, 241: 493–502.

Yamasaki, H., Kubata, Y., Takagi, H. and Tohyama, M. (1984) Immuno-electron microscopic study on the fine structures of the substance P-containing fibers in the taste buds of the rat. *J. Comp. Neurol.*, 227: 380–392.

Yamasaki, H., Kubota, Y. and Tohyama, M. (1985) Ontogeny of substance P-containing fibers in the taste buds and the surrounding epithelium. I. Light microscopic analysis. *Dev. Brain Res.*, 18: 301–305.

Yokokawa, K., Sakanaka, M., Shiosaka, S., Tohyama, M., Shiotani, Y. and Tohyama, M. (1985) Three dimensional distribution of substance P-like immunoreactivity in the urinary bladder of rat. *J. Neural Transm.*, 63: 209–222.

Yokokawa, K., Tohyama, M., Shiosaka, S., Shiotani, Y., Sonoda, T., Emson, P. C., Hillyard, C. V., Girgis, S. and MacIntyre, I. (1986) Distribution of calcitonin gene-related peptide-containing fibers in the urinary bladder of the rat and their origin. *Cell Tissue Res.* (in press).

P. C. Emson, M. N. Rossor and M. Tohyama (Eds.),
Progress in Brain Research, Vol. 66.

CHAPTER 13

Peptides in body fluids in pain

S. J. Capper

Pain Relief Foundation, Walton Hospital, Rice Lane, Liverpool L9 1AE, U.K.

Introduction

During recent years the development of the concept of regulatory peptides acting as neurotransmitters and neuromodulators has had a profound influence on pain research due to the large amount of evidence implicating them in the experience of pain. However, we still do not fully understand the role of peptides in the sensory-discriminative perception of pain. Our understanding of their roles in the motivational-affective and cognitive-evaluative aspects of pain is even more limited. Opiates have always been a powerful tool in the treatment of pain and the discovery of the opioid peptides [Met]-enkephalin (MET-ENK), [Leu]enkephalin (LEU-ENK), the endorphins, the dynorphins and their various precursors has been one of the major advances in pain work due to the obvious relevance of their potential analgesic properties. Other regulatory peptides that have been implicated in pain include substance P (SP), somatostatin (SRIF), cholecystokinin (CCK; usually the octapeptide, CCK-8), vasoactive intestinal peptide (VIP) and neurotensin, which are also found in the endocrine cells of the gut. Briefly, the role of peptides can be summarised as a modulation of pain at various loci (Morley, 1982). SP, SRIF, CCK and VIP have been identified in nociceptive primary afferent fibres; enkephalins, neurotensin and SP have been identified in spinal interneurones. Concentrations of several regulatory peptides are high in some brain areas thought to be involved in pain perception and control, such as the periaqueductal grey (PAG) and the nucleus raphe magnus (NRM), and descending fibres containing enkephalin, SP and SP–serotonin (5-HT) fibres have been found. SP has been implicated as a possible neurotransmitter for nociceptive primary afferents (Jessell and Iversen, 1977), and the evidence, although weak, that enkephalins or other opioid peptides modulate pain at the spinal level via presynaptic inhibition of SP release may provide a neurochemical basis for the gate control theory of Melzack and Wall (1965). Furthermore, when administered centrally in standard analgesic tests, opioid peptides, CCK-8, neurotensin and SP produce analgesia or apparent analgesia.

Although much evidence concerning the involvement of peptides in pain has been derived from immunohistochemical and neurophysiological studies these techniques cannot be so easily applied to the study of clinical pain in the human. The endogenous dynamic response of the whole organism to pain does not easily lend itself to these techniques. In measuring peptides in plasma and cerebrospinal fluid (CSF) we may well not be sure of their origin and so there is the further difficulty of relating any changes to the neurophysiological actions of the peptides. Much evidence can be derived from animal studies in which body fluids can be sampled in a way not possible in the patient, although such studies, especially those involving experimentally induced pain, often produce pain that is quite different to that experienced by the patient, particularly those suffering from chronic pain. Pain is an emotional experience and as such it is an erroneous assumption to equate chronic "pain" with the "pain" of a hot-plate test for example.

The methods used to study peptides in body fluids are similar to those used in immunohistochemistry and are prone to the same problems of specificity, but the problem of sensitivity may be particularly exacerbated due to the extremely low circulating concentrations of regulatory peptides that may be encountered.

Peptide measurement

Generally, bioassays and radioreceptor assays are not sufficiently sensitive or specific for the measurement of regulatory peptides in plasma and CSF, although the problem of specificity can be partially overcome by combining the assay with such steps as gel filtration (Terenius and Wahlstrom, 1975) and high pressure liquid chromatography (HPLC) so as to separate the various related peptides that may be present. However, such steps may reduce the sensitivity of the assay, and furthermore, in view of the increasing number of related peptides that may possess similar functional activity in terms of a biological response and/or receptor binding, may not separate them with sufficient resolution, and also may not identify adequately a particular fraction in terms of molecular structure. This latter point is inherent in the nature of bioassays and receptor assays and becomes especially relevant when dealing with families of peptides, such as the opioids where there are many related peptides, some of which may be precursors. Also we would wish to identify the physiologically significant peptide(s) in terms of a specific action. Although such assays can be said to measure functional activity, the complexity of the mediation of the response may undermine their usefulness, and we have to consider carefully the influence of factors such as the route of administration to an animal and the state of the receptor that presumably mediates the action. This point is particularly pertinent in view of the concept of multiple opiate receptors that may be influenced considerably by their molecular environment (Morley, 1983).

Radioimmunoassays (RIA) can achieve greater sensitivity than bio- and receptor-assays, and although specificity is usually greater, it is by no means absolute. Biologically inactive fragments can be detected in RIA and this may be used as a tool to identify precursor molecules of active regulatory peptides. Due to the similarity in amino acid sequence of many of the peptides that one may wish to look at, the development of specific antibodies has been a significant problem. This has been partly circumvented again by the use of RIA in conjunction with techniques such as gel filtration and HPLC. Adequate differentiation is still a problem, particularly with gel filtration and the increasing use of HPLC should resolve this. Other approaches to the problem of specificity include simultaneous assay using two antibodies of differing specificity such as used by Jeffcoate et al. (1978) to differentiate β-endorphin (β-END) and β-lipotropin (β-LPH)-like immunoreactivities (β-END-LI and β-LPH-LI), or by selective alteration of the peptide of interest so only it is recognised by the antibody, e.g. for MET-ENK (Clement-Jones et al., 1980). For a review of endorphin and enkephalin assays see Clement-Jones and Rees (1982).

Sensitivity can be a problem even with RIA due to the very low concentrations of the regulatory peptides found in body fluids when compared with those found in tissue. The use of chromatography as a concentrating step can be used to increase the sensitivity of an RIA and will also serve as an extraction step which is particularly important in the assay of samples from plasma. The assay of regulatory peptides in unextracted samples of body fluids is prone to error due to: (1) the presence of non-specific interfering factors; (2) the presence of specific and non-specific peptide degrading enzymes; and (3) the presence of binding proteins to which the peptide may be bound. The use of an extraction method such as a simple reverse-phase column or HPLC (Bennett et al., 1977) will remove the peptide from these, although this is open to the criticism that extraction may have removed an immunologically reactive compound.

For the assay of regulatory peptides in body fluids the method of choice is a well characterised (in terms of specificity and sensitivity) RIA, preferably used in conjunction with a technique such as HPLC. Most workers have used RIA to analyse peptide changes in pain but have less commonly used it in conjunction with other techniques to improve specificity. We should anticipate that the increased use of HPLC for example will yield much more useful information regarding the nature of the peptides found in plasma and CSF in pain and during the relief of pain.

Peptides and pain

The measurement of the peptides thought to be involved in nociception and antinociception, and during procedures designed to alleviate pain, has enabled us to start to develop concepts of how these peptides modulate the perception of pain in vivo. However, for several peptides a number of in vivo measurements are lacking and we must postulate their mode of action from the results of in vitro, neurophysiological or immunohistochemical work.

Pain

The analgesic action of the opioid peptides obviously provided a stimulus to their study in conditions of chronic pain. In such patients Almay et al. (1978) found that the CSF levels of Fraction I of the opiate-active fraction of Terenius and Wahlstrom (1975) were low. Moreover, Terenius and Wahlstrom (1975) had also found Fraction I to be lower in patients with trigeminal neuralgia than in controls. Fraction I has been identified as being larger than the enkephalins but did not correspond to β-END on chromatography. Another opiate-active fraction identified by this technique, named Fraction II, coelutes with the enkephalins. However, these fractions do not correspond to known opioid peptides and are unidentified. Also, it has been suggested that Fraction I and 5-HT levels are connected in pain perception as there is a correlation between Fraction I and monoamine metabolites in the CSF of chronic pain patients (Almay et al., 1980). Concentrations of β-END-LI in ventricular CSF have been reported as undetectable (Akil et al., 1978a) and as normal (normal range reported as <4–20 pmol/l by Jeffcoate et al., 1978) in lumbar CSF in chronic pain (Clement-Jones et al., 1980); the differences probably being due to differences in sampling site and assay methodology. Similarly, in ventricular CSF LEU-ENK-like immunoreactivity (LEU-ENK-LI) is undetectable in chronic pain using a specific assay (Hosobuchi et al., 1979), while MET-ENK-like immunoreactivity (MET-ENK-LI) is lower than in control patients (Akil et al., 1978b) but is reported as normal using a more specific assay of lumbar CSF (Clement-Jones et al., 1980). Although a depressed opioid peptide concentration appears consistent with chronic pain, the physiological significance of CSF endorphins and enkephalins is, as yet, unclear and the development of more sensitive and specific assays may well be needed before normal basal CSF levels of such peptides can be established in these patients. The presence of peptides in the CSF may well represent non-specific spillover from the CNS and may not provide a measure of pain per se, but may correlate with the amount of analgesia required, e.g. in post-operative pain (Terenius and Tamsen, 1982). Thus, although intracerebroventricular β-END relieves chronic pain (Hosobuchi and Li, 1978; Foley et al., 1979) and intrathecally administered β-END can relieve intractable pain (Oyama et al., 1980), this may well be mediated at a supraspinal level. This may only apply to β-END and the enkephalins may well exert analgesia at spinal levels when administered intrathecally.

Post-operative pain can serve as a model to study clinically relevant pain, and in a series of patients undergoing major surgery there is a significant inverse correlation between the pre-operative endorphin Fraction I CSF level and the post-operative self-administered CSF concentration of pethidine (Terenius and Tamsen, 1982). These authors also report that with such patient-controlled therapy, pethidine decreases CSF SP levels, which may be indicative of a presynaptic inhibition of SP release from primary afferents (Tamsen et al., 1982). These results conform to a situation in which the endogenous endorphin concentration may reflect the extent to which the perception of pain can be modulated, and that this may differ between individuals, so representing an affective dimension of pain. Plasma β-END-LI levels have also been correlated with post-operative morphine usage (Cohen et al., 1982), while post-operative levels of opioid activity, expressed as MET-ENK equivalents in the mouse vas deferens bioassay, were lower than in chronic pain and control groups, with the degree of pain being inversely related to the measured level (Puig et al., 1982). Plasma β-END-LI may be raised by surgical stress (Dubois et al., 1981) but does not appear to induce analgesia in the patient. Tooth pulp stimulation or molar extraction have also been used as a model of clinical pain, although this is probably not as painful or stressful as major surgery. The concentration of MET-ENK-LI in the cisternal CSF of cats rises after tooth pulp stimulation (Cesselin et al., 1982), a stimulus which fires the medial brain-stem reticular formation in the animal and that can be inhibited by γ-aminobutyric acid (GABA), MET-ENK and, for a longer period, by β-END (Lovick and Wolstencroft, 1983). Interestingly, evidence has been presented that epidural MET--ENK is analgesic in post-operative pain (Andersen et al., 1982).

Studies of post-operative dental pain have provided evidence of placebo analgesia, thus illustrating how the responsiveness of individuals to pain can vary (Levine et al., 1978). As assessed using a visual analogue pain rating scale 3–4 hours after surgery, patients fell into two groups of placebo responders and nonresponders; the former had their analgesia reversed by naloxone, thus implicating endorphins in the mediation of placebo analgesia. Partial antagonism of placebo analgesia by naloxone in subjects with ischaemic arm pain may be a result of the drug dose used or could be due to a non-opioid element in the mediation of analgesia (Grevert et al., 1983). In a similar fashion the response of subjects to forearm electrical shock can be divided into higher than average and normal pain threshold groups (Buchsbaum et al., 1977), with only the higher than average group showing naloxone hyperalgesia implicating a tonic endorphinergic or enkephalinergic system. In contrast, El-Sobky et al. (1976) and Grevert and Goldstein (1978) found pain perception following forearm shock to be unaltered by naloxone. Such studies are prone to problems because of (1) the dose of naloxone used, which also only measures opioid activity by inference (a problem in view of the increasing number of opioid peptides); (2) the nature of individual responsiveness to pain, which at present is unclear; (3) how the pain is produced, as experimental pain may not reflect clinical pain; (4) the timing of the experiment; and (5) how pain is assessed quantitatively. With reference to point 4 it is relevant that a diurnal variation in pain sensitivity and opioid levels has been reported so that pain thresholds may only be naloxone reducible when low (Davis et al., 1978; Wesche and Frederickson, 1979). There has even been a circannual variation in CSF endorphins reported that may correlate with a similar variation in pain levels which may need to be taken into account (Von Knorring et al., 1982).

Such considerations of the variations in responsiveness to pain lead us to consider measurements of peptides in disorders such as congenital insensitivity to pain, an interesting group of conditions usually leading to injury and early death. A congenital overabundance of opioid peptides has been hypothesised as mediating the inability to feel pain (Dehen et al., 1978; Yanagida, 1978). However, CSF and plasma β-END concentrations were normal in a chronic schizophrenic who could not feel pain (Matsukura et al., 1981) and also in a child whose reduced pain sensitivity could be reversed by naloxone (Dunger et al., 1980). However, Mohs et al. (1982) have indicated that a high plasma β-END-LI level made no difference to pain sensitivity when naloxone was given. This may not apply in a congenital absence of pain syndrome but may be more pertinent to the naloxone reversible high-threshold group of Buchsbaum et al. (1977). In the monkey, intrathecal administration of β-END raises the nociceptive threshold and is naloxone reversible (Yaksh et al., 1982). The syndrome of chronic absence of pain is an interesting one and merits further study of the physiology and biochemistry, where the measurement of many more of the peptides involved in pain processing may yield some insight into the aetiology of the condition and reveal the most significant peptides involved in the perception of pain.

Pain relieving techniques

The study of the biochemistry of pain transmission and perception in conditions where novel, and as of yet, ill-understood pain relieving techniques are performed provides us with an insight into the physiology of pain and also into the mechanism of the pain relieving procedure. The development of specialist pain clinics has helped such studies, and, with the involvement of a biochemical research element, should encourage further work related to the clinical relief of pain, whether in specialist pain clinics or in the field of anaesthesia. Indeed opioids have been implicated in the mode of action of such well established anaesthetics as nitrous oxide (Williard et al., 1982). General anaesthesia raises plasma β-END-LI levels more than spinal anaesthesia, which may be more related to stress rather than a site of action of the anaesthetic (Finley et al., 1982).

From the first localisation of opiate binding sites it has been felt that areas of the brain possessing a high concentration of receptor sites might be involved in the mediation of analgesia. Stimulation of medial thalamic and periaqueductal sites in the vicinity of the posterior commissure, immediately adjacent to the ventricular wall at the level of the nucleus parafascicularis, was found to be most effective in relieving chronic pain in the human (Richardson and Akil, 1977) without producing the untoward side-effects associated with PAG stimulation alone (Mayer and Liebeskind, 1974). However, this latter effect was noted in the rat. A few minutes of such stimulation can relieve pain for half an hour, and the implantation of electrodes has enabled patients to control their pain due to phantom limb, carcinoma, thalamic syndrome etc., for months or years with little habituation. However, tolerance can develop and cross-tolerance between morphine and brain stimulation is found (Mayer and Hayes, 1975). The analgesia produced by brain stimulation (Richardson and Akil, 1977) was found to be naloxone reversible, which led the authors to infer the release of an opiate-like factor as mediator of the response. Electrical stimulation of periventricular sites resulted in a reduction of pain with a significant rise in ventricular CSF MET-ENK-LI as measured by bioassay and radioreceptor assay (Akil et al., 1978a). The basal level was lower in pain patients than controls. These authors also measured the ventricular CSF concentration of β-END-LI with a specific radioimmunoassay and found a substantial (13–20 fold) rise following stimulation of a medial thalamic site (Akil et al., 1978b). Using an unextracted assay method they could not detect basal levels of β-END-LI in the CSF, and did not specifically identify the change in β-END-LI as their antiserum cross-reacted with β-LPH. Hosobuchi et al. (1979) measured immunoreactive β-END levels in the ventricular fluid of six patients with chronic pain and found that PAG stimulation produced significant increases (of 50–300%) in patients with pain of peripheral origin. Their RIA showed complete cross-reaction with β-LPH and they were unable to characterise β-END by chromatography as well as RIA. In this study they were also unable to detect any LEU-ENK-LI in the CSF of two patients before or after PAG stimulation. Measurement of the peptides in these studies was performed on samples taken within 30 min of stimulation and the rise in opioid peptide concentrations at this time, particularly β-END, has been taken as evidence explaining the long-lasting analgesic effects of stimulation. However, further studies on the time course of the effect of stimulation and a more specific analysis of the peptides released into the CSF are needed. Thalamic relay nucleus stimulation in the human increases lumbar CSF levels of β-END-LI, although not as much as PAG stimulation; however, no correlation was found with the relief of pain by Tsubokawa et al. (1984). These authors indicate that thalamic stimulation may activate descending monoaminergic fibres that mediate the relief observed. Recently small rises in human lumbar CSF SP-LI estimated using a specific assay have been reported following PAG stimulation, which may arise from descending 5-HT–SP neurones or local spinal SP neurones (Brodin and Meyerson, 1983). They report a more marked rise following dorsal column stimulation.

Although the results point to a role for opioid peptides in the mediation of stimulation produced analgesia, we need to integrate this with the neurophysiological mechanisms thought to inhibit pain and to look at animal studies that have shown the possible relationship of neurones involved in pain pathways. The cell bodies of the ventrolateral PAG of the rat have been shown to contain enkephalin-like immunoreactivity (Hökfelt et al., 1977) and this area also contains many fibres projecting from β-END-LIr cells located in the basal tuberal hypothalamus (Bloom et al., 1978). MET-ENK increases the firing rate of PAG neurones in awake rats in the presence of analgesia (Urca et al., 1977). Neurones in this area are thought to project to the nucleus raphe dorsalis and thence to the NRM, which is probably the origin of a descending inhibitory pathway of serotoninergic neurones that synapse in the dorsal horn on primary sensory afferents (Bowsher, 1980). The NRM is excited by PAG stimulation and by opiates (Fields and Anderson, 1978) and lesions of the NRM block opiate analgesia (Proudfit and Anderson, 1975). Furthermore, enkephalin-like immunoreactive cell bodies have been identified in NRM (Hökfelt et al., 1977). It is interesting to note, in view of the implied involvement of a non-opiate element, that stimulation-produced analgesia can be antagonised by *p*-chlorophenylalanine, a 5-HT synthesis inhibitor (Akil and Mayer, 1972). Akil et al. (1978b) found that stimulation-produced analgesia could be blocked by naloxone in 80% of cases, while Le Bars et al. (1976) found that in the spinal cat, dorsal horn responses to Aδ (1st pain) stimulation are only partly suppressed by morphine, an effect that is itself only partially reversed by naloxone. These findings suggest some non-opiate involvement in the mediation of analgesia that could contribute to stimulation-produced analgesia.

Acupuncture is being increasingly used as a pain-relieving technique and in line with this more has been learnt about the biochemical changes which underlie its action. Central mechanisms are thought to be involved in its mode of action, as acupuncture analgesia can be prevented by anaesthetising the needle site or peripheral nerve (Clement-Jones and Rees, 1982). In mice, electroacupuncture (EAP) increased the latency of squeak to noxious heat, an effect that can be blocked by subcutaneous naloxone, which suggests an endorphin involvement (Pomeranz and Chiu, 1976), while in human volunteers receiving tooth stimulation, manual acupuncture increased the pain threshold which was decreased by naloxone (Mayer et al., 1977). Naloxone had no effect on baseline pain threshold in this latter study. The involvement of opioids in acupuncture analgesia has been confirmed by direct measurement of opioid peptide immunoreactivity in the CSF. In 9 patients suffering chronic pain, Sjolund et al. (1977) demonstrated a marked rise in endorphin Fraction I in 4 patients following EAP. There was no change in Fraction II levels, but an interesting feature was that the patients who had a rise in Fraction I had lumbar pain, which may indicate a local or spinal endorphin release near the stimulated site. Using a more specific assay, Clement-Jones et al. (1980) identified a significant rise in β-END-LI in lumbar CSF following low-frequency EAP to relieve pain due to disc pain and cancer secondaries in bone in 10 patients. There was no correlation between the rise in β-END-LI concentration, severity of pain and duration of pain relief. In this study MET-ENK-LI concentrations were measured by a specific RIA and were found not to alter following EAP. These authors characterised the immunoreactive peptides present as β-END, β-LPH and a higher molecular weight precursor which seems to weigh against these peptides originating from retrograde transport from the pituitary. In contrast, in heroin addicts with withdrawal symptoms, high-frequency auricular EAP used as treatment produced a rise in CSF MET-ENK-LI that was significantly associated with relief of symptoms (Clement-Jones et al., 1979). Plasma and CSF β-END-LI levels were high in these patients and did not change with EAP. It has been suggested that an endorphinergic system is involved in mediating analgesia in low-frequency EAP and an enkephalinergic one in high-frequency EAP (Clement-Jones and Rees, 1982). A rise in MET- and LEU-ENK-LI levels as measured by RIA has been found in rabbit CSF following acupuncture (Clement-Jones and Rees, 1982).

Plasma β-END-LI has been found to rise in volunteers after EAP with an increase still being noted 60 min after stimulation (Malizia et al., 1979), a finding confirmed in a study of patients undergoing thoracic surgery who received EAP in addition to anaesthesia (Panerai et al., 1983). These authors also detected no change in plasma MET-ENK-LI concentration following EAP. The assays used by these authors were specific RIAs (Jeffcoate et al., 1978; Clement-Jones et al., 1980) applied to the samples following HPLC. They conclude that pituitary β-END is increased after EAP by a selective change in precursor processing. In contrast, however, Kiser et al. (1983) studying chronic pain patients following EAP found plasma concentrations of MET-ENK-LI significantly raised while plasma β-END-LI levels were unchanged. The reported assay method for MET-ENK was the same as that used by Panerai et al. (1983) and Clement-Jones et al. (1980), while the β-END assay was reported to have 23% cross-reactivity with β-LPH (Feldman et al., 1983). The discrepancy in the results could be due to differences between pain patients and volunteers, although Kiser et al. (1983) report basal levels of both peptides as being within the normal range. There was a significant correlation between the degree of pain relief and the magnitude of the rise in MET-ENK-LI concentration. Kiser et al. (1983) postulate that acupuncture acts to stimulate adrenal processing and release of MET-ENK, as it has been implicated in stress analgesia (Lewis et al., 1982), although they do consider other peripheral sources. Studies showing that MET-ENK may cross the blood brain barrier (Rapoport et al., 1980) indicate that a possible target could be the CNS but no rise in CSF levels has been detected (Clement-Jones et al., 1980). However, a stress-induced release of adrenocorticotropic hormone (ACTH) and β-END is known (Guillemin et al., 1977) which may be found in the stress of surgery found by Panerai et al. (1983) but not in the EAP of Kiser et al. (1983) which may selectively stimulate adrenal enkephalin release. The significance of circulating MET-ENK in EAP must be questioned, however, as it is doubtful that it will act on central receptors from this site.

The mechanism by which acupuncture mediates analgesia is still far from clear but the evidence for a central effect (Clement-Jones and Rees, 1982), and the reduced response of cat spinal cord neurones to noxious stimuli by EAP (Pomeranz et al., 1977) may point to a similar mechanism to that postulated for PAG stimulation. In this context it is noteworthy that a non-opiate system may also be involved in acupuncture, as high-frequency stimulation is not naloxone reversible but partially reversible by inhibition of serotonin synthesis (Cheng and Pomeranz, 1980). Administration of 5-HTP (precursor of serotonin) to patients with chronic central and deafferentation pain relieves the pain and increases CSF and plasma β-END-LI levels (De Benedittis et al., 1983), although recent results show EAP analgesia to be reversed by 5-HTP (Xuan et al., 1982). It could be that high-frequency EAP, which is thought to be enkephalin mediated by Clement-Jones and Rees (1982), may modulate pain inputs in the cord via collaterals from the lateral spinothalamic tract that synapse with the brainstem raphe cells that are activated by intense acupuncture (Le Bars et al., 1979). Descending 5-HT-containing fibres from the raphe cells will mediate the 5-HTP effect, although a non-endorphinergic analgesic system that directly activates the descending inhibitory systems has also been postulated and may be responsible for the naloxone insensitivity described (Hayes et al., 1978). It is likely that a site of action for endorphin-mediated acupuncture analgesia is via brainstem raphe enkephalinergic cells and that together with hypothalamic pituitary endorphins this may well mediate non-segmental acupuncture analgesia (Pomeranz, 1981). The hypothalamic cells may stimulate the PAG via their fibres or by retrograde transport of endorphins in the CSF. The longer half-life of β-END has been cited as evidence for the latter, while the enkephalinase inhibitor phenylalanine potentiates acupuncture analgesia (Hyodo et al., 1983). The presence of the high molecular weight precursor in CSF (Clement-Jones et al., 1980) implies that there might be processing within the CNS, but it is felt this precursor would not be of pituitary origin. The finding of Sjolund et al. (1977) that segmental acupuncture analgesia for lumbar pain raised CSF Fraction I endorphin levels may indicate that this type of analgesia is mediated by dorsal horn enkephalin cells. This is supported by the recent finding that EAP produced a decrease in the rate of glucose utilisation in the spinal cord indicative of nociceptive inhibition (Sjolund and Schouenborg, 1983). The complexity of the potential sites of opioid interaction in the mediation of acupuncture analgesia means that it is necessary to evaluate carefully factors such as the site, intensity and method of stimulation together with the type of pain presented and whether the patient is stressed, anaesthetised, susceptible to analgesia etc., as well as the careful validation of what is actually measured.

Transcutaneous nerve stimulation (TNS) is offered in two different forms to relieve pain; low frequency, high intensity TNS acting in a similar way to acupuncture and being reversed by naloxone, and high-frequency, low intensity TNS being unaffected by naloxone (Sjolund and Erikson, 1979). The high intensity segmental afferent stimulation elevates CSF levels of Fraction I endorphins (Sjolund and Erikson, 1979) while low intensity TNS has been found to lower CSF β-END-LI levels in non-pain subjects (Salar et al., 1981). The evidence has been reviewed by Meyerson (1983) who concludes that, as with acupuncture, the different effects of low and high intensity stimulation need to be considered, as they may act by different mechanisms; low intensity TNS activating only low-threshold skin receptors that presynaptically inhibit primary afferents, while stimulation of the cord and supraspinal structures may occur in high intensity TNS.

Pituitary neuroadenolysis is a treatment that has been successfully used at the Centre for Pain Relief, Walton Hospital, Liverpool for the treatment of cancer pain, and is conveniently accomplished by the transphenoidal injection of alcohol into the pituitary fossa. The mechanism by which this technique relieves pain is still largely unknown but may be via an interference with pain pathways that involve an alteration in hypophyseal peptides. In the monkey, tooth pulp evoked potentials are reduced by pituitary alcohol in a naloxone reversible fashion (Yanagida, 1983) and in the rat this procedure lowers the β-END-LI content of the midbrain, hypothalamus, cerebellum and hindbrain, with an increase in cortical content in the first 5 days post-operatively but a subsequent decline to below basal levels (Yanagida, 1983). This led to the suggestion that the prompt relief of pain afforded by neuroadenolysis may be related to transiently raised cortical β-END-LI levels, but that prolonged relief may involve the interruption of hypophyseal-anterior thalamic fibres. Peptide measurements in CSF and plasma following neuroadenolysis have offered few clues as to its mechanism of action as yet. The major increases in plasma concentrations of ACTH-, β-LPH- and β-END-LI after 2–3 weeks (Deshpande et al., 1983) contrast with the results of Takeda (1983) who found rises in ACTH-LI and α-END-LI in serum and CSF 24 h after injection that declined to normal or below after 6 days. Interestingly, Takeda (1983) found CSF ACTH-LI increases correlated well with pain relief and a much higher CSF level than serum after two months would tend to indicate a non-pituitary origin. An increase in CSF concentrations of α-neoendorphin, β-END, TRH and arginine vasopressin (VP)-like immunoreactivites (VP-LI) was observed 24 h after injection but only TRH-LI and VP-LI remained elevated, although the data points were few. Sites of origin of TRH and VP within the brain are implicated. Preliminary studies from the Pain Relief Foundation have indicated that CSF concentrations of SRIF-LI and SP-LI may rise almost immediately following neuroadenolysis, although this may well represent hypophyseal/hypothalamic damage and does not appear to be well correlated with pain relief. Although more work needs to be done to determine the time course of CSF and blood changes of peptides following the operation, the pain relieving effects of the operation may be produced by profound non-specific changes as yet not understood and unrelated to the present lines of investigation (Miles, 1984).

Stress and analgesia

The finding that rats subjected to inescapable footshock developed a partially naloxone-reversible analgesia (Akil et al., 1976) implicated opioid peptides in its mediation, which was further supported by the report of increased brain opioid levels (Madden et al., 1977). A rise in plasma β-END-LI levels in stress analgesia was subsequently found (Rossier et al., 1977) and a pituitary origin was suggested by Guillemin et al. (1977). A non-opiate system is also implicated (Goldstein et al., 1976; Lewis et al., 1980) that may be selectively activated depending on the region shocked and this is also involved in conditioned analgesia that can arise in environments associated with shock (Watkins et al., 1982) and has been found to be associated with a rise in plasma β-END-LI (Scallet, 1982). Non-pituitary opioids are also implicated in footshock analgesia (Lim et al., 1981). Swim stress can also elicit analgesia associated with a rise in plasma β-END-LI although this is thought not to reflect any opioid pathway activation, whereas changes observed in the intermediate lobe of the pituitary are thought to be indicative of this pathway (Lim and Funder, 1983). LEU-ENK binding to brain is reduced in warm water swim mice (Christie and Chesher, 1983) which may be connected with the reduction in intermediate lobe β-END-LI. No change in immunoreactive dynorphin levels in hypothalamus and cortex of the rat following swim stress has been observed (Morley et al., 1982). In contrast to the results of Lim and Funder (1983) a rise in plasma and neurointermediate lobe β-END-LI and ACTH-LI has been observed, although this is in cold-stressed rats (Giagnoni et al., 1983). Recent evidence has implicated the ventro-medial posterior hypothalamus in a non-endorphinergic mechanism involving SP in the mediation of this type of analgesia (Millan et al., 1983). Possibly related to stress-induced analgesia are the changes in circulating opioid levels found during the stress and pain of labour although β-END may well be acting in a more hormonal role and, this is probably not evidence of a specific neuromodulatory action. Plasma β-END-LI levels rise (Mega et al., 1981) and this can be attenuated by epidural opiate anaesthesia (Abboud et al., 1982). The effects of opioid peptides in stress stimulation are complex and are probably not limited to analgesic actions (μ-receptor mediated) but probably also involve activation of δ and κ receptors as well as possibly acting in a more classical hormonal role to exert complex behavioural effects. Our understanding of the roles of the peptides in this will be enhanced by a better definition of the effects subserved by the various receptor subtypes which will be improved by the development of more specific antagonists.

The role of non-opioid regulatory peptides

In the foregoing sections there has been a great emphasis placed on the measurement of the endorphins and enkephalins and the other regulatory peptides have been somewhat ignored. The great excitement caused by the discovery of these endogenous opioids is partly responsible, and it is only as we become more aware of the presence of other regulatory peptides within the nervous system that attention is turned to their role in pain. The immunohistochemical localisation of many regulatory peptides to parts of the CNS that implicate them in pain processes is well documented (for review see Hökfelt et al., 1983) but the various actions mediated by them are still largely a matter of conjecture and await future developments. Their measurement in body fluids is prone to the usual problems of peptide RIA but further complications that have prevented the assessment of their relevance in pain include a dissociation of a role in pain from a hormonal role, e.g. SRIF, and again low concentrations and short half-life in body fluids, e.g. SP.

SP is thought to act as a transmitter in nociceptive sensory afferents but in addition the presence of SP in many pathways within the brain suggests it has a complex role in the nervous system. The discussion of the differing effects of SP in eliciting analgesia or hyperalgesia (Morley, 1982) certainly highlights this complexity. In addition, the gastrointestinal, endocrine and circulatory effects make identification of a specific neural activity difficult. SP-LI has been identified in normal plasma and CSF (Nobin et al., 1983) but its very low levels and short half-life (a matter of minutes) make it hard to discern changes in these compartments which are also fairly distant from the synaptic site of action. A defined role for somatostatin in nociception has still to be established but its distribution and opiate-like activity (Cox et al., 1975; Terenius, 1976) support this, although there is still discussion as to its excitatory and depressant effects on neurones (Kelly, 1982). More reliable measurements of SRIF by immunoassay have been achieved as extraction methods to remove non-specific interfering factors and possible SRIF binding proteins have been applied, although the presence of multiple molecular weight forms of SRIF, particularly in CSF, where a high molecular weight form has been found (Wass et al., 1982), complicates the interpretation of results. Furthermore, the many gut endocrine functions of SRIF render interpretation of changes in plasma concentration as arising from neural action very uncertain. The same comments can generally be applied to neurotensin both in regard to the evidence for its neurotransmitter/neuromodulator role and to the results and problems of interpreting plasma measurements. The evidence for CCK-8 subserving a neuromodulatory role includes its localisation to primary afferents and PAG, its excitatory action on neurones and the potent analgesia found after injection into the PAG (Morley, 1982). The involvement of opiate receptors is suggested, but a problem with identifying specific nociceptive responses to CCK-8 are its other central and peripheral effects. Interpretation of plasma measurements is complicated by the prospects of cross-reactivity in the assay which result in the wide range of reported values, but it is felt that the circulating concentration is low (Dockray, 1981). However, this again will most likely respond mainly to gut endocrine activity, so providing problems for the study of its neuromodulatory role.

Future directions

The increasing complexity of this field as the number of regulatory peptides discovered increases, means that more and more specificity in what we are measuring will become absolutely essential. For this reason the usefulness of bioassays and receptor assays, as they stand at present, will decline and even radioimmunoassays with their greater sensitivity and specificity will need to be used in conjunction with techniques such as HPLC to identify accurately the peptides involved. The increasing molecular heterogeneity of the identified peptides will make the use of such techniques fundamental as the development of very highly specific antibodies becomes a problem. In conjunction with the use of such chromatographic techniques the development of more highly specific immunoradiometric assays (IRMA), with their potential for greater sensitivity, will become important, and the development of monoclonal antibodies to peptides of interest will facilitate this.

Together with assay development, the discovery of more specific opiate receptor antagonists will help to define more clearly the role of the multiple opiate receptor subtypes in antinociception, to say nothing of the characterisation of SP and CCK-8 receptors involved, about which we know very little. This will help us to define more accurately the particular peptides of relevance in the perception and modulation of pain.

As the field stands there are many areas of research which could be pursued as the number of studies performed is still relatively small. The measurement of peptides such as dynorphin, α-neoendorphin, CCK-8 etc. in pain has hardly been studied and as assays for these become available and are applied we may derive new insights into their role in pain. Several situations such as chronic insensitivity to pain merit further study, and in fairly well documented areas such as stimulation-produced analgesia and acupuncture the varying results (often representing variations in the patient) indicate that a more systematic approach is called for.

Finally, a fundamental problem of all studies of the nervous system in which we measure the response in the plasma and/or CSF is the relevance of such changes, assuming that they exist, and if we can detect them, to the events occurring at the synapse or cell junction which, we hypothesise, is the site of neurotransmission/neuromodulation. If changes in CSF peptide concentrations really do just reflect non-specific spillover of transmitter from the synaptic cleft, how can we specifically identify the site of action, and will any changes we detect only reflect such gross, unphysiological release of transmitter that its relevance is questionable? The solution to these problems may well prove a major step not just in the field of pain research but for neurobiology in general.

Acknowledgement

I wish to thank Peter Duff for the considerable help received during the information search that contributed to this chapter.

References

Abboud, T. K., Sarkis, F., Goebelsmann, T., Hung, T. and Henrikson, E. (1982) Effects of epidural anesthesia during labour on maternal plasma β-endorphin levels. *Anesthesiology*, Suppl. 3, 57: A382.

Akil, H. and Mayer, D. J. (1972) Antagonism of stimulation produced analgesia by p-CPA, a serotonin synthesis inhibitor. *Brain Res.*, 44: 692–697.

Akil, H., Madden, J., Patrick, R. L. and Barchas, J. D. (1976) Stress induced increase in endogenous opiate peptides: Concurrent analgesia and its partial reversal by naloxone. In H. W. Kosterlitz (Ed.), *Opiates and Endogenous Opioid Peptides*, Elsevier/North Holland, Amsterdam, pp. 63–70.

Akil, H., Richardson, D. and Barchas, J. (1978a) Appearance of beta-endorphin-like immunoreactivity in human ventricular cerebrospinal fluid upon analgesic electrical stimulation. *Proc. Natl. Acad. Sci. U.S.A.*, 75: 5170–5172.

Akil, H., Richardson, D. E., Hughes, J. and Barchas, J. D. (1978b) Enkephalin-like material elevated in ventricular cerebrospinal fluid of pain patients after analgetic focal stimulation. *Science*, 201: 463–465.

Almay, B. G. L., Johannson, F., Von Knorring, L., Terenius, L. and Wahlstrom, A. (1978) Endorphins in chronic pain. I. Differences in CSF endorphin levels between organic and psychogenic pain syndromes. *Pain*, 5: 153–162.

Almay, B. G. L., Johansson, F., Von Knorring, L., Sedvall, G. and Terenius, L. (1980) Relationship between CSF levels of endorphins and monoamine metabolites in chronic pain patients. *Psychopharmacology*, 67: 139–142.

Andersen, H. B., Jorgenson, B. C. and Engquist, A. (1982) Epidural methionine-enkephalin (FK 33-824): A dose-effect study. *Acta Anaesthesiol. Scand.*, 26: 69–71.

Bennett, H. P. J., Hudson, A. M., McMartin, C. and Purdon, G. E. (1977) Use of octadecasilyl-silica for the extraction and purification of peptides in biological samples. *Biochem. J.*, 168: 9–13.

Bloom, F., Battenberg, E., Rossier, J., Ling, N. and Guillemin, R. (1978) Neurons containing endorphin in rat brain exist separately from those containing enkephalin: immunocytochemical studies. *Proc. Natl. Acad. Sci. U.S.A.*, 75: 1591–1595.

Bowsher, D. (1980) Peptides and pain. In S. Lipton (Ed.), *Persistent Pain: Modern Methods of Treatment, Vol. 2*, Academic Press, London, pp. 189–202.

Brodin, E. and Meyerson, B. A. (1983) Substance P in human cerebrospinal fluid: effect of central grey or dorsal column stimulation for suppression of chronic pain. In P. Skrabanek and D. Powell (Eds.), *Substance P – Dublin 1983*, Boole Press, Dublin, pp. 190–191.

Buchsbaum, M. S., Davis, G. C. and Bunney, W. E. (1977) Naloxone alters pain perception and somatosensory evoked potentials in normal subjects. *Nature*, 270: 620–622.

Cesselin, F., Oliveras, J. L., Bourgoin, S., Sierralta, F., Michelot, R., Besson, J. M. and Hamon, M. (1982) Increased levels of methionine-enkephalin-like material in the cerebrospinal fluid of anesthetized cats after tooth pulp stimulation. *Brain Res.*, 237: 325–338.

Cheng, R. and Pomeranz, B. (1980) Electroacupuncture analgesia could be mediated by at least two pain-relieving mechanisms: endorphin and non-endorphin systems. *Life Sci.*, 25: 1957–1962.

Christie, M. J. and Chesher, G. B. (1983) Tritium labelled leucine enkephalin binding following chronic swim stress in mice. *Neurosci. Lett.*, 36: 323–328.

Clement-Jones, V. and Rees, L. H. (1982) Neuroendocrine correlates of the endorphins and enkephalins. In G. M. Besser and L. Martini (Eds.), *Clinical Neuroendocrinology, Vol. II*, Academic Press, New York, pp. 139–203.

Clement-Jones, V., McLoughlin, L., Lowry, P. J., Besser, G. M. and Wen, H. L. (1979) Acupuncture in heroin addicts: changes in Met-enkephalin and β-endorphin in blood and cerebrospinal fluid. *Lancet,* ii: 380–382.

Clement-Jones, V., McLoughlin, L., Tomlin, S., Rees, L. H., Besser, G. M. and Wen, H. L. (1980) Increased β-endorphin but not Met-enkephalin levels in human CSF after acupuncture for recurrent pain. *Lancet,* ii: 946–949.

Cohen, M. R., Pickar, D., Dubois, M. and Bunney, W. E., Jr. (1982) Stress-induced plasma β-endorphin immunoreactivity may predict postoperative morphine usage. *Psychiatry Res.,* 6: 7–12.

Cox, B. M., Opheim, K. E., Teschenmacher, H. and Goldstein, A. (1975) A peptide-like substance from pituitary that acts like morphine. *Life Sci.,* 16: 1777–1782.

Davis, G. C., Buchsbaum, M. S. and Bunney, W. E. (1978) Naloxone decreases diurnal variation in pain sensitivity and somatosensory evoked potentials. *Life Sci.,* 23: 1449–1460.

De Benedittis, G., Di Giulio, A. M., Massei, R., Villani, R. and Panerai, A. E. (1983) Effects of 5-hydroxytryptophan on central and deafferentation chronic pain: A preliminary clinical trial. In J. J. Bonica, U. Lindblom and A. Iggo (Eds.), *Advances in Pain Research and Therapy, Vol. 5,* Raven Press, New York, pp. 295–304.

Dehen, H., Willer, J. C., Prier, S., Boureau, F. and Cambier, J. (1978) Congenital insensitivity to pain and the "morphine-like" analgesic system. *Pain,* 5: 351–358.

Deshpande, N., Moricca, G., Saullo, F. and Di Martino, L. (1983) Changes in pituitary function following neuroadenolysis in patients with metastatic cancer. In S. Ischia, S. Lipton and G. F. Maffezzoli (Eds.), *Pain Treatment,* Cortina International, Verona, pp. 51–64.

Dockray, G. J. (1981) Cholecystokinin. In S. R. Bloom (Ed.), *Gut Hormones,* Churchill Livingstone, Edinburgh, pp. 228–239.

Dubois, M., Pickar, D., Cohen, M., Gadde, P., Roth, Y. R., Macnamara, T. E. and Bunney, W. E. (1981) Plasma beta endorphin immunoreactivity is raised by surgical stress, but not anesthetic induction. *Anesthesiology,* Suppl. 3, 55: A244.

Dunger, D. B., Leonard, J. V., Wolff, O. H. and Preece, M. A. (1980) Effect of naloxone in a previously undescribed hypothalamic syndrome. *Lancet,* i: 1277–1281.

El-Sobky, A., Dostrovsky, J. O. and Wall, P. D. (1976) Lack of effect of naloxone on pain perception in humans. *Nature,* 263: 783–784.

Feldman, M., Kiser, R. S., Unger, R. H. and Li, C. H. (1983) Beta-endorphin and the endocrine pancreas. *N. Engl. J. Med.,* 308: 349–353.

Fields, H. L. and Anderson, S. D. (1978) Evidence that raphe-spinal neurons mediate opiate and midbrain stimulation-produced analgesias. *Pain,* 5: 333–349.

Finley, J. H., Cork, R. C., Hameroff, S. R. and Scherer, K. (1982) Comparison of plasma beta-endorphin levels during spinal vs. general anesthesia. *Anesthesiology,* Suppl. 3, 57: A191.

Foley, K. M., Kourides, I. A., Interrisi, C. E., Kaiko, R. F., Zaroulis, C. G., Posner, J. B., Houde, R. W. and Li, C. H. (1979) β-Endorphin: analgesia and hormonal effects in humans. *Proc. Natl. Acad. Sci. U.S.A.,* 76: 5377–5381.

Giagnoni, G., Santagostino, A., Senini, R., Fumagalli, P. and Gori, E. (1983) Cold stress in the rat induces parallel changes in plasma and pituitary levels of endorphin and ACTH. *Pharmacol. Res. Commun.,* 15: 15–21.

Goldstein, A., Pryor, G. T., Otis, L. S. and Larsen, F. (1976) On the role of endogenous opioid peptides: failure of naloxone to influence shock escape threshold in the rat. *Life Sci.,* 18: 599–604.

Grevert, P. and Goldstein, A. (1978) Endorphin: naloxone fails to alter experimental pain or mood in humans. *Science,* 199: 1093–1095.

Grevert, P., Albert, L. H. and Goldstein, A. (1983) Partial antagonism of placebo analgesia by naloxone. *Pain,* 16: 129–143.

Guillemin, R., Vargo, T., Rossier, J., Minick, S., Ling, N., Rivier, C., Vale, W. and Bloom, F. (1977) β-Endorphin and adrenocorticotropin are secreted concomitantly by the pituitary gland. *Science,* 197: 1367–1369.

Hayes, R. L., Bennett, G. J., Newton, P. G. and Mayer, D. J. (1978) Differential effects of spinal cord lesions on narcotic and non-narcotic suppression of nociceptive reflexes: further evidence for the physiological multiplicity of pain modulation. *Brain Res.,* 155: 91–102.

Hökfelt, T., Elde, R., Johansson, O., Terenius, L. and Stein, L. (1977) The distribution of enkephalin-immunoreactive cell bodies in the rat central nervous system. *Neurosci. Lett.,* 5: 25–31.

Hökfelt, T., Skirboll, L., Lundberg, J. M., Dalsgaard, C. J., Johansson, O., Pernow, B. and Jancso, G. (1983) Neuropeptides and pain pathways. In J. J. Bonica, U. Lindblom and A. Iggo (Eds.), *Advances in Pain Research and Therapy, Vol. 5,* Raven Press, New York, pp. 227–246.

Hosobuchi, Y. and Li, C. H. (1978) The analgesic activity of human β-endorphin in man. *Commun. Psychopharmacol.,* 2: 33–37.

Hosobuchi, Y., Rossier, J., Bloom, F. E. and Guillemin, R. (1979) Stimulation of human periaqueductal grey for pain relief increases immunoreactive β-endorphin in ventricular fluid. *Science,* 203: 279–281.

Hyodo, M., Kitode, T. and Hosoka, E. (1983) Study on the enhanced analgesic effect induced by phenylalanine during acupuncture analgesia in humans. In J. J. Bonica, U. Lindblom and A. Iggo (Eds.), *Advances in Pain Research and Therapy, Vol. 5,* Raven Press, New York, pp. 577–582.

Jeffcoate, W. J., Rees, L. H., McLoughlin, L., Ratter, S. J., Hope, J., Lowry, P. J. and Besser, G. M. (1978) Beta-endorphin in human cerebrospinal fluid. *Lancet,* ii: 119–121.

Jessell, T. M. and Iversen, L. L. (1977) Opiate analgesics inhibit substance P release from rat trigeminal nucleus. *Nature,* 268: 549–551.

Kelly, J. S. (1982) Electrophysiology of peptides in the central nervous system. *Br. Med. Bull.*, 38: 283–290.

Kiser, R. S., Gatchel, R. J., Bhatia, K., Khatami, M., Huang, X. and Altshuler, K. Z. (1983) Acupuncture relief of chronic pain syndrome correlates with increased plasma met-enkephalin concentrations. *Lancet*, ii: 1394–1396.

Le Bars, D., Guilbaud, G., Jurna, I. and Besson, J. M. (1976) Differential effects of morphine on responses of dorsal horn lamina V type cells elicited by A and C fibre stimulation in the cat spinal cord. *Brain Res.*, 115: 518–524.

Le Bars, D., Dickenson, A. H. and Besson, J. M. (1979) Diffuse noxious inhibitory controls (DNIC). Lack of effect on non-convergent neurones, supraspinal involvement and theoretical implications. *Pain*, 6: 305–327.

Levine, J. D., Gordon, N. C. and Fields, H. L. (1978) The mechanism of placebo analgesia. *Lancet*, ii: 654–657.

Lewis, R. V., Stern, A. S., Kimura, S., Rossier, J., Stein, S. and Udenfriend, S. (1980) An about 50 000-dalton protein in adrenal medulla: a common precursor of met- and leu-enkephalin. *Science*, 208: 1459–1461.

Lewis, J. W., Tordoff, M. G., Sherman, J. E. and Liebeskind, J. C. (1982) Role of the adrenal medulla in opioid stress analgesia. *Neurosci. Abstr.*, 7: 735.

Lim, A. T. W. and Funder, J. W. (1983) Stress-induced changes in plasma, pituitary and hypothalamic immunoreactive β-endorphin: effects of diurnal variation, adrenalectomy, corticosteroids and opiate agonists and antagonists. *Neuroendocrinology*, 36: 225–234.

Lim, A., Oei, T. and Funder, J. (1981) Stress-induced analgesia: lack of correlation with anterior pituitary immunoreactive β-endorphin. *Neurosci. Lett.*, Suppl. (8): S64.

Lovick, T. A. and Wolstencroft, J. H. (1983) Actions of GABA, glycine, methionine-enkephalin and β-endorphin compared with electrical stimulation of nucleus raphe magnus on responses evoked by tooth pulp stimulation in the medial reticular formation of the cat. *Pain*, 15: 131–144.

Madden, J., Akil, H., Patrick, R. L. and Barchas, J. D. (1977) Stress-induced parallel changes in central opioid levels and pain responsiveness in the rat. *Nature*, 265: 358–360.

Malizia, E., Andreucci, G., Paulucci, D., Crescenzi, F., Fabbri, A. and Fravoli, F. (1979) Electroacupuncture and peripheral β-endorphin and ACTH levels. *Lancet*, ii: 535–536.

Matsukura, S., Yoshinin, H., Sueoka, S., Chihara, K., Fujita, T. and Tanimoto, K. (1981) β-Endorphin in Cotards syndrome. *Lancet*, i: 162–163.

Mayer, D. J. and Hayes, R. L. (1975) Stimulation-produced analgesia: development of tolerance and cross-tolerance to morphine. *Science*, 188: 941–943.

Mayer, D. J. and Liebeskind, J. C. (1974) Pain reduction by focal electrical stimulation of the brain: an anatomical and behavioural analysis. *Brain Res.*, 68: 73–93.

Mayer, D. J., Price, D. D. and Rafii, A. (1977) Antagonism of acupuncture analgesia in man by the narcotic antagonist naloxone. *Brain Res.*, 121: 368–372.

Mega, M., Giorgino, F., Tambuscio, G. and Bonivento, A. (1981) β-endorphins in labour. *Clin. Exp. Obstet. Gynecol.*, 8: 151–155.

Melzack, R. and Wall, P. D. (1965) Pain mechanisms: a new theory. *Science*, 150: 971–979.

Meyerson, B. A. (1983) Electro stimulation procedures: effects, presumed rationale, and possible mechanisms. In J. J. Bonica, U. Lindblom and A. Iggo (Eds.), *Advances in Pain Research and Therapy, Vol. 5*, Raven Press, New York, pp. 495–534.

Miles, J. B. (1984) Pituitary destruction. In P. Wall and R. Melzak (Eds.), *Textbook of Pain*, Churchill Livingstone, Edinburgh, pp. 656–665.

Millan, M. J., Przewlocki, R., Millan, M. H. and Herz, A. (1983) Evidence for a role of the ventro-medial posterior hypothalamus in nociceptive processes in rat. *Pharmacol. Biochem. Behav.*, 18: 901–907.

Mohs, R. C., Davis, B. M., Rosenberg, G. S., Davis, K. L. and Kreiger, D. T. (1982) Naloxone does not affect pain sensitivity, mood or cognition in patients with high levels of β-endorphin in plasma. *Life Sci.*, 30: 1827–1834.

Morley, J. E., Levine, A. S., Elson, M. K. and Shafer, R. B. (1982) Stress and the central nervous system, concentrations of the opioid peptide dynorphin. *Clin. Res.*, 30: 411A.

Morley, J. S. (1982) The potential of centrally acting regulatory peptides as analgesics. In D. Lednicer (Ed.), *Central Analgesics*, Wiley and Sons, New York, pp. 81–135.

Morley, J. S. (1983) A common opiate receptor? *Trends Pharm. Sci.*, 370–371.

Nobin, A., Emson, P., Martensson, H. and Sundler, F. (1983) Substance P in carcinoid patients – tumour content and plasma levels. In P. Skrabanek and D. Powell (Eds.), *Substance P – Dublin 1983*, Boole Press, Dublin, pp. 129–130.

Oyama, T., Jin, T., Yamaka, R., Ling, N. and Guillemin, R. (1980) Profound analgesic effects of β-endorphin in man. *Lancet*, i: 122–124.

Panerai, A. E., Martini, A., Abbate, D., Villani, R. and De Benedittis, G. (1983) β-Endorphin, met-enkephalin, and β-lipotropin in chronic pain and electro acupuncture. In J. J. Bonica, U. Lindblom and A. Iggo (Eds.), *Advances in Pain Research and Therapy, Vol. 5*, Raven Press, New York, pp. 543–547.

Pomeranz, B. (1981) Neural mechanisms of acupuncture analgesia. In S. Lipton and J. Miles (Eds.), *Persistent Pain, Vol. 3*, Academic Press, London, pp. 241–257.

Pomeranz, B. and Chiu, D. (1976) Naloxone blocks acupuncture analgesia and causes hyperalgesia: endorphin is implicated. *Life Sci.*, 19: 1757–1762.

Pomeranz, B., Cheng, R. and Law, P. (1977) Acupuncture reduces electrophysiologica and behavioural responses to noxious stimuli: pituitary is implicated. *Exp. Neurol.*, 54: 172–178.

Proudfit, H. K. and Anderson, E. G. (1975) Morphine analgesia: blockade by raphe magnus lesions. *Brain Res.*, 98: 612–618.

Puig, M. M., Laorden, M. L., Fernando, S., Olaso, M. and Olaso, M. J. (1982) Endorphin levels in cerebrospinal fluid of patients with postoperative and chronic pain. *Anesthesiology,* 57: 1–4.

Rapoport, S. I., Klee, W. A., Pettigrew, K. D. and Ohno, K. (1980) Entry of opioid peptides into the central nervous system. *Science,* 207: 84–86.

Richardson, D. E. and Akil, H. (1977) Pain reduction by electrical brain stimulation in man. *J. Neurosurg.,* 47: 178–194.

Rossier, J., French, E. D., Rivier, G., Ling, N., Guillemin, R. and Bloom, F. E. (1977) Foot-shock induced stress increases β-endorphin levels in blood but not brain. *Nature,* 270: 618–620.

Salar, G., Job, J., Mingrino, S., Bosio, A. and Trabucchi, M. (1981) Effect of transcutaneous electrotherapy on CSF β-endorphin content in patients without pain problems. *Pain,* 10: 169–172.

Scallet, A. (1982) Effects of conditioned fear and environmental novelty on plasma β-endorphin in the rat. *Peptides,* 3: 203–206.

Sjolund, B. H. and Erikson, M. B. E. (1979) The influence of naloxone on analgesia produced by peripheral conditioning stimulation. *Brain Res.,* 173: 295–301.

Sjolund, B. H. and Schouenborg, J. (1983) Site of action antinociceptive acupuncture-like nerve stimulation in the spinal rat as visualized by the C-2-deoxyglucose method. In J. J. Bonica, U. Lindblom and A. Iggo (Eds.), *Advances in Pain Research and Therapy, Vol. 5,* (Proc. World Congr. Pain, 3rd, 1981): 535–541.

Sjolund, B., Terenius, L. and Erikson, M. (1977) Increased cerebrospinal fluid levels of endorphins after electroacupuncture. *Acta Physiol. Scand.,* 100: 382–384.

Takeda, F. (1983) Some considerations on the mechanism of pain relief by means of pituitary neuroadenolysis: a clinical investigation. In S. Ischia, S. Lipton and G. F. Maffezzoli (Eds.), *Pain Treatment,* Cortina International, Verona, pp. 37–50.

Tamsen, A., Sakurada, T., Wahlstrom, A., Terenius, L. and Hartvig, P. (1982) Postoperative demand for analgesics in relation to individual levels of endorphins and substance P in cerebrospinal fluid. *Pain,* 13: 171–184.

Terenius, L. (1976) Somatostatin and ACTH are peptides with partial antagonist-like selectivity for opiate receptors. *Eur. J. Pharmacol.,* 38: 211–213.

Terenius, L. and Tamsen, A. (1982) Endorphins and the modulation of acute pain. *Acta Anaesthesiol. Scand.,* Suppl., (74): 21–24.

Terenius, L. and Wahlstrom, A. (1975) Morphine-like ligand for opiate receptors in human CSF. *Life Sci.,* 16: 1759–1764.

Tsubokawa, T., Yamamoto, T., Katayama, Y., Hirayama, T. and Sibuya, H. (1984) Thalamic relay nucleus stimulation for relief of intractable pain. Clinical results and β-endorphin immunoreactivity in the cerebrospinal fluid. *Pain,* 18: 115–126.

Urca, G., Frenk, H., Liebeskind, J. C. and Taylor, A. N. (1977) Morphine and enkephalin: analgesic and epileptic properties. *Science,* 197: 83–85.

Von Knorring, L., Almay, B. G. L., Johansson, F. and Terenius, L. (1978) Pain perception and endorphin levels in CSF. *Pain,* 5: 359–366.

Von Knorring, L., Almay, B. G. L., Johansson, F., Terenius, L. and Wahlstrom, A. (1982) Circannual variation in concentrations of endorphins in cerebrospinal fluid. *Pain,* 12: 265–272.

Wass, J. A. H. (1982) Somatostatin and its physiology in man in health and disease. In G. M. Besser and L. Martini (Eds.), *Clinical Neuroendocrinology, Vol. II,* Academic Press, New York, pp. 359–395.

Watkins, L. R., Cobelli, D. A., Newsome, H. H. and Mayer, D. J. (1982) Footshock-induced analgesia is dependent neither on pituitary nor sympathetic activation. *Brain Res.,* 245: 81–96.

Wesche, D. L. and Frederickson, R. C. A. (1979) Diurnal differences in opioid peptide levels correlated with nociceptive sensitivity. *Life Sci.,* 24: 1861–1868.

Williard, K. F., Gillmor, S. T., Stanley, T. H., Mueleman, T. R. and Pace, N. L. (1982) The influence of nitrous oxide and nociceptive stimuli on rat plasma and brain endorphin concentrations. *Anesthesiology,* Suppl. 3, 57: A302.

Xuan, Y. T., Zhou, Z. F. and Han, J. S. (1982) Tolerance to electroacupuncture analgesia was reversed by microinjection of 5-hydroxytryptophan into nuclei accumbens in the rabbit. *Int. J. Neurosci.,* 17: 157–161.

Yaksh, T. L., Gross, K. E. and Li, C. H. (1982) The intrathecal effect of β-endorphin in primate. *Brain Res.,* 241: 261–270.

Yanagida, H. (1978) Congenital insensitivity and naloxone. *Lancet,* i: 520–521.

Yanagida, H. (1983) The effect of hypophysectomy on beta-endorphin levels in various brain regions and the effect of naloxone on tooth pulp evoked potentials in the rat. In S. Ischia, S. Lipton and G. F. Maffezzoli (Eds.), *Pain Treatment,* Cortina International, Verona, pp. 21–25.

SECTION V

Therapeutic Perspectives

P. C. Emson, M. N. Rossor and M. Tohyama (Eds.),
Progress in Brain Research, Vol. 66.

CHAPTER 14

The design of antagonists of regulatory peptides, and comments on their specificity

J. S. Morley

Pain Relief Foundation, Walton Hospital, Rice Lane, Liverpool L9 1AE, U.K.

Introduction

The rapid advances in our knowledge of neurotransmission earlier this century owed a great deal to the availability of antagonists that specifically blocked the effect of monoamine transmitters or putative transmitters at the receptor level. Conversely, the present lack of receptor antagonists of neuropeptides, or, when they are available, their generally low specificity, is greatly hindering research on the role of neuropeptides in neurones. This is not for want of effort on the part of neurochemists — indeed the amount of effort expended in attempts to design peptide antagonists has been considerable. Why has the success rate been poor? Are the concepts upon which the work has been based valid? Or do they need modification? This chapter examines these questions, suggests alternative approaches in antagonist design, and describes how a new approach has been applied with some success to the design of specific antagonists of the enkephalin delta opiate receptor.

Past approaches

Two principal models have been proposed to explain the action of regulatory peptides at the receptor level. In one, the participation model (Morley, 1968a; Rudinger et al., 1972), it is envisaged that the peptide agonist occupies binding sites at the receptor, and also supplies a chemical group essential for a chemical event associated with activation of the receptor (proposed, for example, in the case of gastrin/cholecystokinin). In the other, the more generally applicable allosteric model (Monod et al., 1963), it is envisaged that agonist–receptor binding alone can cause activation by inducing a critical displacement of the normal conformation of the receptor. Intrinsic activity (or efficacy) is therefore related to the presence of a specific chemical group in the agonist in the participation model, and to the ability of the agonist to induce favourable conformational changes in the allosteric model.

With few exceptions the search for antagonists of regulatory peptides has been directed towards analogues, i.e. compounds that bear an obvious chemical relationship to the agonist. It is argued that the agonist–receptor interaction involves a number of subsites, and that it is possible to arrive at analogues that still possess affinity for all or some of these subsites, but lack intrinsic activity. Such analogues would be competitive antagonists of the action of the parent agonist. It is further argued that lack of intrinsic activity may arise if the analogue lacks a chemical group essential for the action of the agonist (participation model), if the analogue–receptor binding interaction causes unfavourable conformation changes in the receptor molecule (allosteric model), or if the interaction prevents access of the agonist to at least one of its binding sites (applicable to both models).

Diagrammatic representation

These approaches can be best understood, and new approaches considered, by referring to the mathematical model of De Lean et al. (1979). Agonist–receptor, or more generally ligand–receptor binding is again considered to involve interaction of discrete regions of the ligand molecule with complementary receptor subsites. In the simplified case of a divalent ligand interacting with two subsites (Fig. 1), binding to the two subsites may involve one ligand molecule (represented by the three states P1, P2, and P3) or two (represented by the state P4). Only those ligands able to form the "active state" P3 serve as agonists; this state may be achieved solely from the conformational changes induced by binding of the ligand at both subsites (allosteric model), or by the combination of these changes and the supply, by the ligand, of a chemical group normally absent at the receptor, but critical for its function (participation model).

Past approaches to the design of antagonists may be expressed in terms of this model as follows. If participation pertains, analogues of an agonist which are also capable of binding to both subsites may be antagonists if they are unable to supply (i.e. lack) the missing chemical group (Fig. 2). The most explored case based on participation is that of gastrin, where from structure–function studies the carboxyl group of the penultimate aspartyl residue was

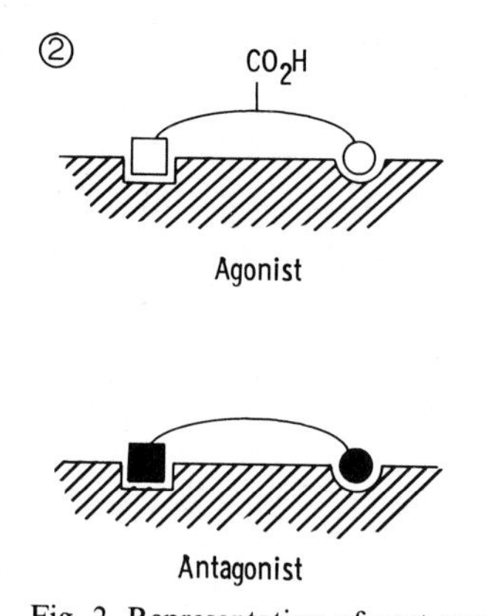

Fig. 2. Representation of past approaches to the design of antagonists based on the participation model. The agonist, in occupying both subsites, supplies a chemical group (carboxyl shown) which is essential for a chemical event associated with activation. If foreign ligands can still occupy both subsites, but lack the activating group, they should act as antagonists.

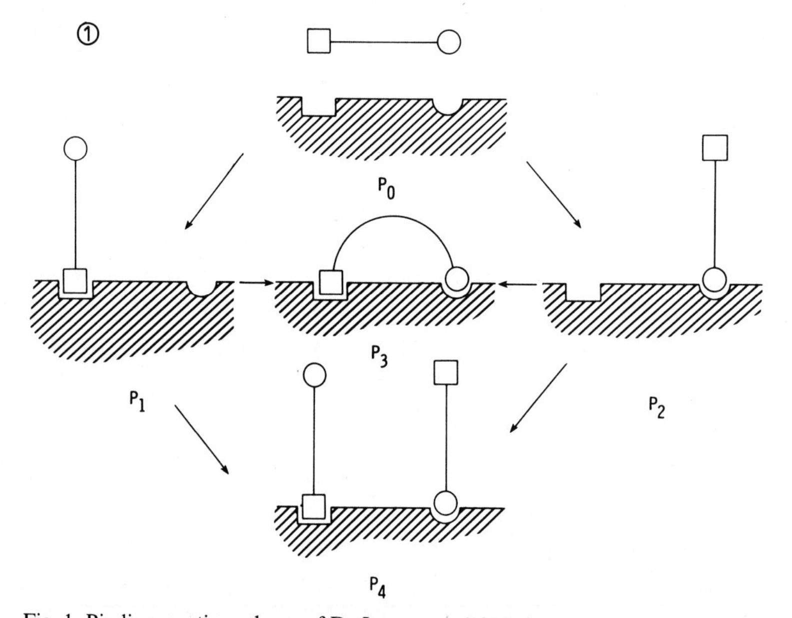

Fig. 1. Binding reaction scheme of De Lean et al. (1979) for a divalent ligand interacting with two subsites of a receptor. In this and subsequent figures the two binding sites of an agonist are shown as ○ or □.

identified as a candidate for such a role (Morley, 1968a). Analogues lacking this carboxyl group were not agonists, but they were also not antagonists (Morley, 1968b). A successful case, held by some to be based on participation, is that of gonadotropin--releasing hormone (GnRH); the imidazole group of the His[2] residue of GnRH may also participate in the same sense as gastrin's carboxyl group, explaining why GnRH analogues lacking a His[2] residue are often antagonists (see Vale et al., 1981). If allosterism pertains, analogues utilising both subsites may be antagonists if they induce conformational changes in the receptor molecule which are only marginally different from those induced by the agonist (Fig. 3). One might hope, for example, that truncated or extended analogues, in which the spacing between the binding sites is less or greater than that in the agonist, would be antagonists. But this approach does not appear to have met with any success.

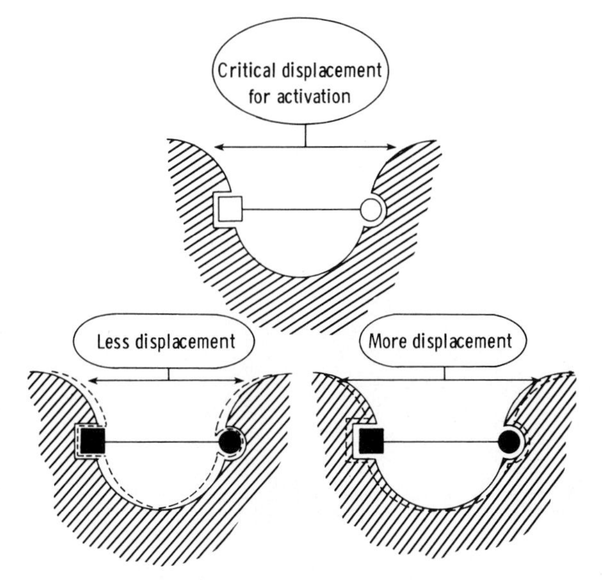

Fig. 3. Representation of approaches to the design of antagonists based on the allosteric model. The agonist, in occupying both subsites, causes a critical displacement of the conformation of the receptor, necessary for activation. Foreign ligands which occupy the same subsites, but which cause less or more displacement, might act as antagonists.

Auto-inhibition — relevance in antagonist design

The interesting phenomenon of auto-inhibition, when low doses of a substance elicit an effect and higher doses cause inhibition of the effect, has frequently been encountered in in vitro and laboratory animal studies of neuropeptides, e.g. substance P (SP) (Hall and Stewart, 1983), gastrin/cholecystokinin (Morley, 1973). The results are reflected in bell-shaped, or biphasic, dose–response curves. The explanation of the phenomenon offered from the De Lean et al. (1979) model is particularly attractive. Thus, if the P4 state (Fig. 1), where each subsite is occupied by separate agonist molecules, is stabilised, then formation of the active P3 state is hindered, and abnormal dose–response curves are predicted. In extreme cases, when the P4 state is favoured to the full exclusion of the P3 state, maximum responses will occur with 50% occupancy of the two subsites, and thereafter increasing concentration of agonist will result in progressively decreased responses; the response becomes nil with 100% occupancy of the subsites. In more likely circumstances, when there is more modest stabilisation of the P4 state, modified bell-shaped dose–response curves will result, dependent on the degree of P4 state stabilisation. Such considerations go a

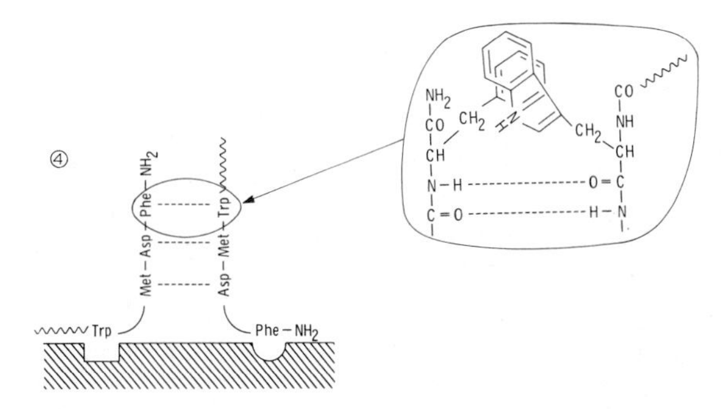

Fig. 4. How stabilisation of the P4 state might be achieved in the case of gastrin/cholecystokinin. It is hypothesised that two gastrin molecules sit on the receptor, one utilising a Trp subsite and the other a Phe subsite; interactions between unbound Phe and Trp residues (pi–pi interaction) are reinforced by interchain hydrogen bonding (exploded representation on the right hand side of the figure).

long way in explaining auto-inhibitory effects of neuropeptides. Stabilisation of the P4 state may arise from various types of inter-molecular bonding, for example pi–pi and hydrogen bonding in the case of gastrin (see Fig. 4).

Let us now consider the model in terms of antagonist design. As with auto-inhibition, we are concerned with circumstances in which ligand binding prevents formation of the active P3 state, but two different ligands are now involved — one the natural agonist, and the other a foreign ligand (the putative antagonist). For antagonism to arise, the foreign ligand needs to have the ability to bind to only one of the two receptor subsites. The state at the receptor recognition sites is again represented by P4, but instead of both subsites being occupied by agonist molecules, one is occupied by agonist, and the other by the foreign ligand (Fig. 5).

If interaction of the two ligands is sufficient to stabilise the P4 state, the foreign ligand will be an antagonist. The interactions concerned must occur at sites other than those involved in receptor binding, and the challenge is how to design ligands with this capability. It is possible that the present breed of SP antagonists act in this way. A common structural feature of these antagonists is an accumulation of C-terminal lipophilic aromatic residues. If we hypothesise that two subsites at the SP receptor involve binding with N- and C-terminal features of the SP molecule, and that SP antagonists utilise only the subsite for N-terminal binding, then antagonism may result from interaction of lipophilic regions of SP and antagonist molecules, and the resulting stabilisation of state P4 (see Fig. 6).

Utilisation of lipid accessory binding sites

It is unlikely that the interactions just discussed are the most efficient means of stabilising the P4 state, bearing in mind the strong forces driving the agonist ligand to displace the foreign ligand with generation of the P3 state. This leads us to consider

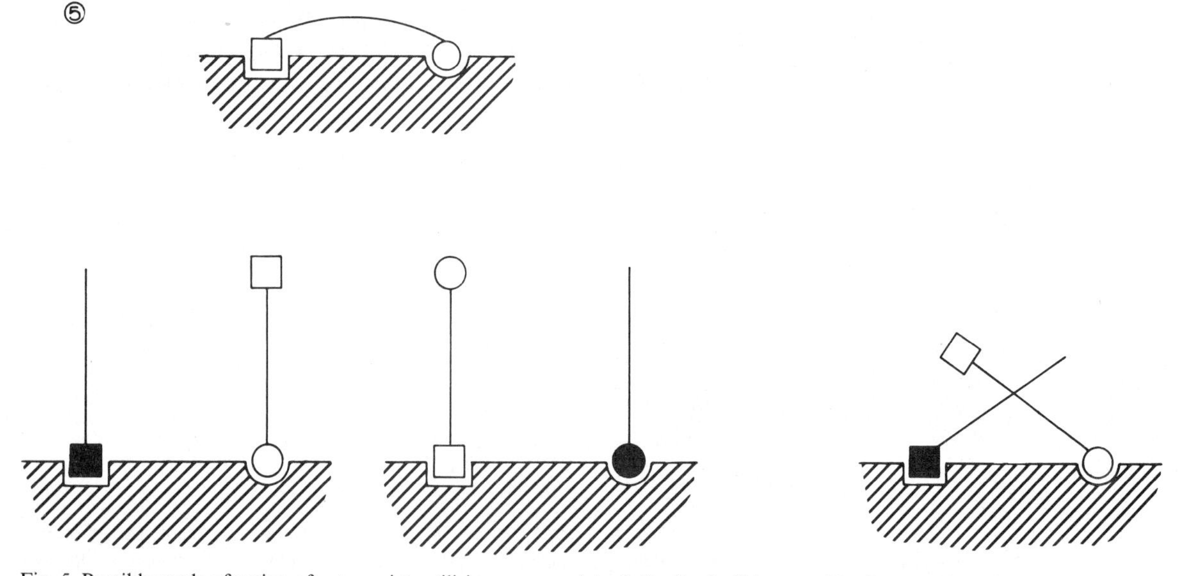

Fig. 5. Possible mode of action of antagonists utilising one agonist subsite. In the P4 state of De Lean et al. (1979) (Fig. 1), one of the subsites is occupied by the agonist (○ and □ represent its two binding sites), and the other by foreign ligand (bottom three diagrams). The analogue is capable of binding to one subsite, but not to the other, therefore it cannot form the active P3 state (top diagram). If the P4 state is stabilised by interaction of agonist and analogue at sites other than those involved in receptor subsite binding (bottom right diagram), then the agonist is prevented or hindered from forming the P3 state, and antagonism results.

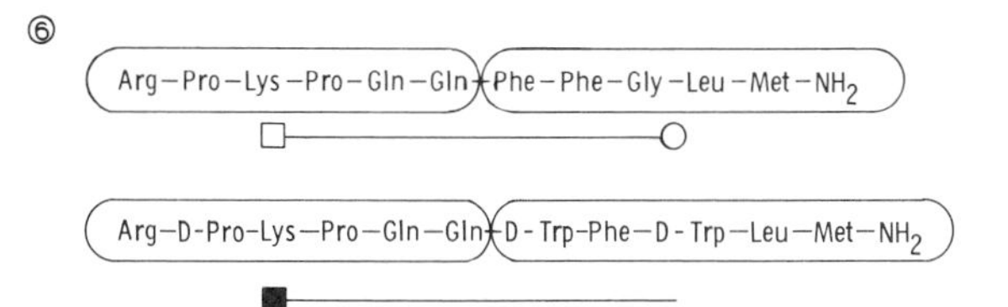

Fig. 6. In the upper diagram, N- and C-terminal features of SP are represented, respectively, as □ or ○; it is hypothesised that these might be involved in binding at two receptor subsites for SP. The SP antagonist, shown in the lower diagram, has the N-terminal feature only (■). The proposed interaction of SP and the antagonist at the receptor can now be represented by the bottom right diagram of Fig. 5. Alternatively, a C-terminal feature of the antagonist may be involved in binding to an accessory subsite, analogous to that represented in Fig. 7.

whether stabilisation might more effectively be achieved by utilisation of binding sites other than those involved in agonist recognition. There is indeed evidence that antagonists of monoamine transmitters act in this way (Ariens et al., 1979). Referring again to the two subsite model, stabilisation would now be achieved by additional binding to a third, "accessory" subsite. This third subsite may be part of the receptor molecule different from that involved in agonist recognition, or part of a different molecule (see Fig. 7). The high lipophilicity of monoamine transmitter antagonists, and their greatly increased affinity, as compared with corresponding monoamine agonists, suggests that the "accessory" sites are located in lipid (or phospholipid) molecules closely associated with membrane receptor proteins. Knowledge of the nature and disposition of such lipids would provide a new basis for antagonist design. The "accessory" subsites could be non-specific, i.e. defined by physical properties as distinct from structure, or highly specific. The latter circumstance seems more probable, since the lipid molecules concerned probably adopt conformations determined by associated membrane proteins.

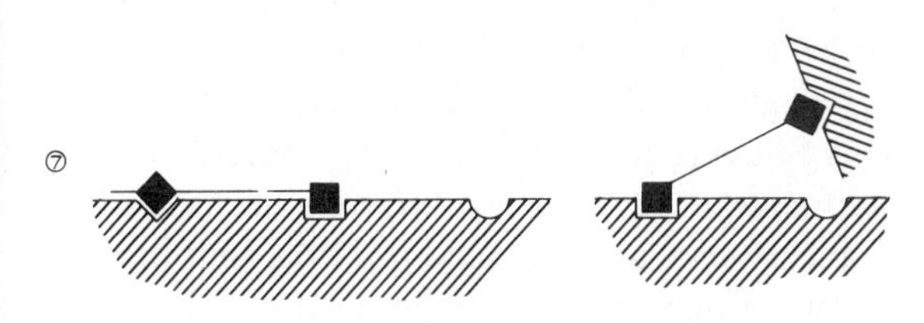

Fig. 7. Stabilisation of the P4 state by utilisation of a third, accessory subsite. Binding of the analogue only is shown. The analogue is capable of binding to only one of the agonist subsites, and also to a third subsite represented as a V cleft. The third subsite may be located in an adjacent part of the receptor (left), or of a different associated molecule (right).

Ligands which do not utilise any of the agonist subsites but only accessory subsites are also capable of acting as antagonists, for example if accessory binding holds the ligand over the receptor in a position that prevents access of the agonist to its subsites. Antagonists acting by this mechanism will clearly bear no structural resemblance with the agonist they antagonise, but may nevertheless show high specificity.

Application in the design of enkephalin delta opiate receptor antagonists

The substitution of certain lipophilic groups, e.g. allyl, cyclopropylmethyl, for the methyl group attached to the nitrogen atom of morphine and related opiates results in dramatic loss of agonist properties and may lead to extremely potent antagonists (e.g. naloxone, naltrexone). We conjectured that antagonism arose from the ability of the lipophilic groups to engage in high affinity binding with lipid/phospholipid molecules in close association with the opiate receptor, the binding being strong enough to cause minor, but highly significant, displacement of the ligand from the opiate receptor (with resultant loss of agonist properties), and strong enough to prevent binding of morphine and other agonists at the receptor. The ligand was envisaged as utilising some, but not all, of the opiate receptor binding sites, the drop in affinity being more than compensated by the extra-receptor binding. This corresponds with the general situation just discussed in which binding is represented by the right hand diagram in Fig. 7. Accordingly, we attempted to generate antagonists of enkephalin by substituting lipophilic groups at the N-terminus of enkephalins. This work resulted in *N,N*-bisallyl-

[Leu]enkephalin, and the metabolically more stable N,N-bisallyl-Tyr-Gly-Gly-$\psi(CH_2S)$-Phe-Leu, where $\psi(CH_2S)$ signifies replacement of the amide CO-NH bond between Gly and Phe by CH_2S. Both compounds were pure antagonists, highly selective for the opiate delta receptor (Gormley et al., 1982; Shaw et al., 1982). Although the potency of these antagonists was modest ($K_e \sim 300$ nM versus leucine enkephalin on the mouse vas deferens preparation), the result lends encouragement to approaches to the design of antagonists based on the utilisation of lipophilic accessory binding sites.

Application in other cases

The emerging proposition is that antagonists of neuropeptides may arise if lipophilic groups are substituted in individual peptides (or active fragments of the peptide) provided the nature and disposition of the lipophilic groups are such as to allow high affinity, extra-receptor binding. The problem in design is how to meet the provisions. The guide lines are as follows:

(a) In the case of the opiate receptor, we were fortunate in being able to draw on experience with non-peptide ligands, which indicated the type of lipophilic group required, and its required disposition in peptide ligands. There is no comparable experience with non-peptide ligands at other receptors, but where there is experience, however fragmentary, this approach should be paramount.

(b) Siting experiments may be carried out in which a restricted range of lipophilic groups are introduced into different, selected positions of the neuropeptide, and the resulting series of substituted neuropeptides is examined for antagonist properties. This is the time-honoured hit-or-miss method used in drug discovery — often criticised, but often successful. If antagonism, however weak, is found in any member of the series, a lead is provided indicating the position in the neuropeptide where substitution is likely to be fruitful. A wider range of lipophilic groups is then substituted at the defined position in the hope of increasing potency.

(c) More rational approaches are possible if study is made of the nature of lipid/phospholipid molecules associated with particular receptors. Evidence that particular lipids/phospholipids predominate at specific receptors may allow more rational choice of lipophilic groups in (b) above. Looking to the future, the spectacular advances in computer graphics may lead to much greater rationality; as more receptors are isolated, the prospect of graphical study of their interaction with lipids/phospholipids becomes real.

(d) A surprising outcome of research on monoamine transmitter antagonists has been the realisation that there is more chemical similarity between different classes of antagonists (anticholinergics, antihistaminics, anti-α-adrenergics, antiserotonins) than between specific pairs of antagonist and agonist (Ariens et al., 1979). Perhaps in synaptic membranes there are lipophilic accessory sites common to different types of receptors, and these are being utilised by different classes of antagonists. This suggests that common, lipophilic features (e.g. bisaryl groups) of monoamine transmitter antagonists may profitably be incorporated in neuropeptides in the search for specific antagonists. For example, bisaryl may be one selected lipophilic group in (b) above.

Specificity of antagonists

It is generally supposed that disparity of pA_2 values for an antagonist against the same agonist on different tissues remains the ultimate proof of the presence of different receptor types on those tissues (Brown and Hill, 1983). Whilst the mathematical basis for the supposition is undisputed, there are difficulties, more than of semantic origin, in deciding what are "different receptor types". Most current thinking revolves round the concept that the recognition site for a neuropeptide lies in a membrane protein, and that binding of the neuropeptide to this recognition site initiates the first stage in the response we associate with that neuropeptide. On this basis we may define a receptor as that specific sequence of a membrane protein that has the ability, after interaction with an agonist, to initiate the sequence of events leading to biological response.

But two main difficulties arise from this definition.

First, the conformation of the sequence will be partly determined by the natural environment of the protein in the cell membrane. If this environment is constant, then the conformation of the sequence is fixed and classical pharmacodynamic calculations based on a single receptor are valid. But does our definition of a receptor now need to include this constant environment? Although seldom stated by the investigators concerned, it is clear that many do indeed embrace the environment in their definition. There seem to be dangers in this practice. Whilst a fixed sequence of a membrane protein is a definite molecular entity, the protein and its environment is a mixture of molecules, whose composition is fixed only if the environment is constant. What, however, is the justification for assuming that the environment will be constant? On the contrary, the more likely situation is that it is subject to constant change. This is the view developed in the multiple environment receptor model (Morley, 1983). Pharmacological observations indicating single or multiple receptors are explained as follows. In any test preparation, response or displacement will be the product of the intrinsic receptor interaction and the ability of the environment to help or hinder approach and/or binding of the agonist/antagonist; there will be an average environment for a receptor leading to average pharmacokinetics or pharmacodynamics. When test preparations are derived from one tissue from one species the chance of arriving at groups of cells with similar average receptor environments will be high. But there will be variation and this might partly explain the variation in ED_{50} or IC_{50} that is found from one day to another. Furthermore, it is not surprising that when the preparation is derived from another species, different average environments result. The multiple receptor model leads us then to postulate different populations of two or more receptors, or a new type of receptor. The multiple environment model says that average environment will lead to average pharmacokinetics depending on the agonist/antagonist/radioligand being used. Even when a common receptor (as defined before) exists in different tissues or species, changed average environment may lead to the conclusion that a different receptor is present. More importantly, the model predicts that the apparent differentiation of receptor types may be diffuse.

The second problem arises from the very arguments used earlier as a basis for antagonist design. If antagonists are capable of utilising extra-receptor, accessory, binding sites, we must assume that some agonists also have this capability. This assumption is explicit, for example, in the model for the beta-endorphin receptor recently proposed by Smith et al. (1983). Pharmacological observations with agonists or antagonists binding in this way may well lead to the conclusion that multiple receptors exist, but consider the danger that arises — different "receptors" could be "proved" for each type of agonist or antagonist that happened to use a different accessory binding site. Whether the multiple environment model is accepted or not, this argues for the receptor definition given in the first paragraph of this section, and stresses the need for caution in using pA_2 value disparity as proof of different receptors.

Finally, the opportunity is taken to answer criticism that has been levelled against the multiple environment model, and to comment further on its practical utility.

Loh et al. (1983) argue that mu and delta opiate receptor types exist in many common brain regions and on the same cell types, and to explain this according to the multiple environment model we must postulate that the same protein exists in two different environments in the same cell membrane. Even if the former part of this argument is correct, why is the postulate unreasonable? Given the wide range of phospholipids present in membranes, there seems no valid reason why the same protein may not exist in two different environments in the same cell membrane. To amplify and be more specific, supposing that one group of phospholipids is dominant in cell membranes in one particular tissue, this will create a receptor conformation and provide a supply of potential accessory binding sites characteristic of one type of binding, say opiate mu type.

In the membranes of a second type of tissue, another group of phospholipids may dominate, creating a changed receptor conformation and a changed supply of potential accessory binding sites characteristic of a second type of binding, say opiate delta type. In the membranes of a third type of tissue, both groups of phospholipid may exist in major amounts, leading to mixed characteristics (mu and delta types). In this third type of tissue, interaction between the two "receptor" types (proposed, but not proved) can be accounted for by assuming that the changed conformation and environment resulting from interaction of a ligand and receptor can affect neighbouring populations allosterically. As to why more multi-forms of the opiate receptor have been evoked than in the case of other receptors, one suspects that the reason relates to the sophisticated state of opioid pharmacology, and the wide range of opioid-type ligands available.

As to the practical utility of the model, the finding of functional specificity in receptors is the significant observation, regardless of the explanatory model. What, however, the multiple environment model predicts, and the multiple receptor model does not, is that it is possible for functionality to change even in one specific tissue.

References

Ariens, E. J., Beld, A. J., Rodrigues De Miranda, J. F. and Simonis, A. M. (1979) The pharmacon–receptor–effector concept. In R. D. O'Brian (Ed.), *The Receptors, Vol. 1,* Plenum, New York, pp. 33–91.

Brown, J. R. and Hill, R. G. (1983) Multiple receptors for substance P. *Trends Pharmacol. Sci.,* 4: 512–514.

De Lean, A., Munson, P. J. and Rodbard, D. (1979) Multisubsite receptors for multivalent ligands. *Mol. Pharmacol.,* 15: 60–70.

Gormley, J. J., Morley, J. S., Priestly, T., Shaw, J. S., Turnbull, M. J. and Wheeler, H. (1982) In vivo evaluation of the opiate delta receptor antagonist ICI 154,129. *Life Sci.,* 31: 1263–1266.

Hall, M. E. and Stewart, J. M. (1983) Substance P and antinociception. *Peptides,* 4: 31–35.

Loh, H. H., Lee, N. M. and Smith, A. P. (1983) Reply. *Trends Pharmacol. Sci.,* 4: 371–372.

Monod, J., Changeux, J.-P. and Jacob, F. (1963) Allosteric proteins and cellular control systems. *J. Mol. Biol.,* 6: 306–329.

Morley, J. S. (1968a) Structure–function relationships in gastrin-like peptides. *Proc. Roy. Soc. B,* 170: 97–111.

Morley, J. S. (1968b) Structure–activity relationships. *Fed. Proc.,* 27: 1314–1317.

Morley, J. S. (1973) Structure–activity relations in GI hormones. In S. Andersson (Ed.), *Frontiers in Gastrointestinal Hormone Research. Nobel Symposium 16,* Almqvist & Wiksell, Stockholm, pp. 143–149.

Morley, J. S. (1983) A common opiate receptor? *Trends Pharmacol. Sci.,* 4: 370–371.

Rudinger, J., Pliska, V. and Krejci, I. (1972) Oxytocin analogues and hormone action. *Recent Progr. Horm. Res.,* 28: 131–172.

Shaw, J. S., Miller, L., Turnbull, M. J., Gormley, J. J. and Morley, J. S. (1982) Selective antagonists at the opiate delta-receptor. *Life Sci.,* 31: 1259–1262.

Smith, A. P., Lee, N. M. and Loh, H. H. (1983) The multiple site beta-endorphin receptor. *Trends Pharmacol. Sci.,* 4: 163–164.

Vale, W. W., Rivier, C., Perrin, M., Smith, M. and Rivier, J. (1981) Pharmacology of gonadotropin releasing hormone: a model regulatory peptide. In J. B. Martin, S. Reichlin and K. L. Bick (Eds.), *Neurosecretion and Brain Peptides, Advances in Biochemical Psychopharmacology, Vol. 28,* Raven, New York, pp. 609–625.

P. C. Emson, M. N. Rossor and M. Tohyama (Eds.),
Progress in Brain Research, Vol. 66.

CHAPTER 15

Studies on peptide comodulator transmission. New perspective on the treatment of disorders of the central nervous system

K. Fuxe[1], L. F. Agnati[2], A. Härfstrand[1], K. Andersson[1], F. Mascagni[1], M. Zoli[2], M. Kalia[3], N. Battistini[2], F. Benfenati[2], T. Hökfelt[1] and M. Goldstein[4]

[1]*Department of Histology, Karolinska Institutet, Stockholm, Sweden;* [2]*Department of Human Physiology, University of Modena, Modena, Italy;* [3]*Department of Neurosurgery, Thomas Jefferson University, Phildelphia, PA, and* [4]*Department of Psychiatry, New York University Medical Center, New York, NY, U.S.A.*

Introduction

Ever since the discovery of central dopamine, noradrenaline, adrenaline and 5-hydroxy-tryptamine (5-HT) neurons in the central nervous system (Dahlström and Fuxe, 1964; Fuxe, 1965a,b; Andén et al., 1966; Fuxe et al., 1970; Goldstein et al., 1973, 1978; Hökfelt et al., 1974, 1980a, 1983c) an increasing number of topological, functional and biochemical heterogeneities have been discovered in the various monoamine pathways (Hökfelt et al., 1978, 1980b, 1983f; Fuxe et al., 1979, 1983a; Agnati et al., 1982b,d, 1983e). In the present paper we will mainly be concerned with the biochemical heterogeneities in various central monoamine systems. Thus, peptide comodulators appear to exist in a large number of central monoamine neuron systems as first indicated in the work of Hökfelt and colleagues (1978, 1980b). During the last few years we have, in a large number of papers, obtained evidence that the peptide comodulators in central monoamine nerve cells, inter alia, represent regulatory transmission lines, which at the postsynaptic level via receptor–receptor interactions can regulate the decoding mechanism at the monoamine (main) transmission line (Fuxe et al., 1981, 1983a,b; Agnati et al., 1983a–e). Thus, the concept has emerged that the synapse is a complex, integrative electro-metabolic unit, which can deliver a large number of messages to the target nerve cell, and which possesses a substantial degree of functional plasticity. In the present study we will analyze the effects of cholecystokinin (CCK) peptides, neurotensin (NT) and neuropeptide Y (NPY) on pre- and post-synaptic mechanisms in central catecholamine neurons. Large numbers of CCK, NT and NPY immunoreactive neuron systems exist in the mammalian brain (Uhl et al., 1977, 1979; Vanderhaeghen et al., 1980; Emson et al., 1982; Jennes et al., 1982; Nemeroff and Prange, 1982; Everitt et al., 1983; Hökfelt et al., 1983). In addition, CCK-8-like immunoreactivity (CCK-LI) exists in a limited subpopulation of mesolimbic dopamine neurons, which is true also for NT-like immunoreactivity (NT-LI) (Hökfelt et al., 1983a). NT also appears to be a comodulator in a certain number of tuberoinfundibular dopamine neurons. Finally, NPY-like immunoreactivity (NPY-LI) (Hökfelt et al., 1983d; Everitt et al., 1983) has been demonstrated in several adrenaline nerve cell groups and within the noradrenaline cell group A1, which gives rise to ascending projections to the hypothalamus and the preoptic area (see Andén et al., 1966). The results obtained in the morphological, neurochemical and functional analysis throw new light on the functional meaning of "coexistence", indicating that coexistence may provide a heterostatic regulation of central monoamine synapses.

Assessment of coexistence of cholecystokinin, neurotensin and neuropeptide Y-like immunoreactivity in central catecholamine nerve cells

By means of the indirect immunofluorescence technique in combination with permanganate elution methodology (Tramu et al., 1978) it has been possible to demonstrate in thin cryostat sections that, for example, some midbrain dopamine nerve cell bodies demonstrated by tyrosine hydroxylase (TH) immunocytochemistry, also contain CCK-LI (Hökfelt et al., 1980b). In these experiments CCK-LI was first demonstrated in a section, which was then photographed in the fluorescence microscope, and restained for TH-like immunoreactivity (TH-LI) after elution of the CCK-LI by the use of potassium permanganate. The section was then rephotographed. It was thus possible, using this method, to demonstrate that some nerve cell bodies contain both TH-LI and CCK-LI (Hökfelt et al., 1980b).

In Figs. 1 and 2 the indirect immunoperoxidase technique has been used to demonstrate the distribution of TH-LIr and CCK-LIr nerve cell bodies in adjacent thick sections of the ventral midbrain of a colchicine-treated male rat. In this procedure free-floating vibratome sections have been used with a section thickness of 50 μm. The CCK-LIr nerve cells are mainly located within the area ventralis tegmenti, and their number is far below that of the TH-LIr nerve cells. At this level (midportion of the interpeduncular nucleus) the CCK-LIr cells present in the substantia nigra are mainly located in the medial and most lateral part (pars lateralis) of the zona compacta. The dopamine nerve cell bodies of the ventral tegmental area project to the subcortical limbic forebrain, to limbic cortical areas and to the anteromedial frontal cortex. In line with these results, a codistribution of certain types of

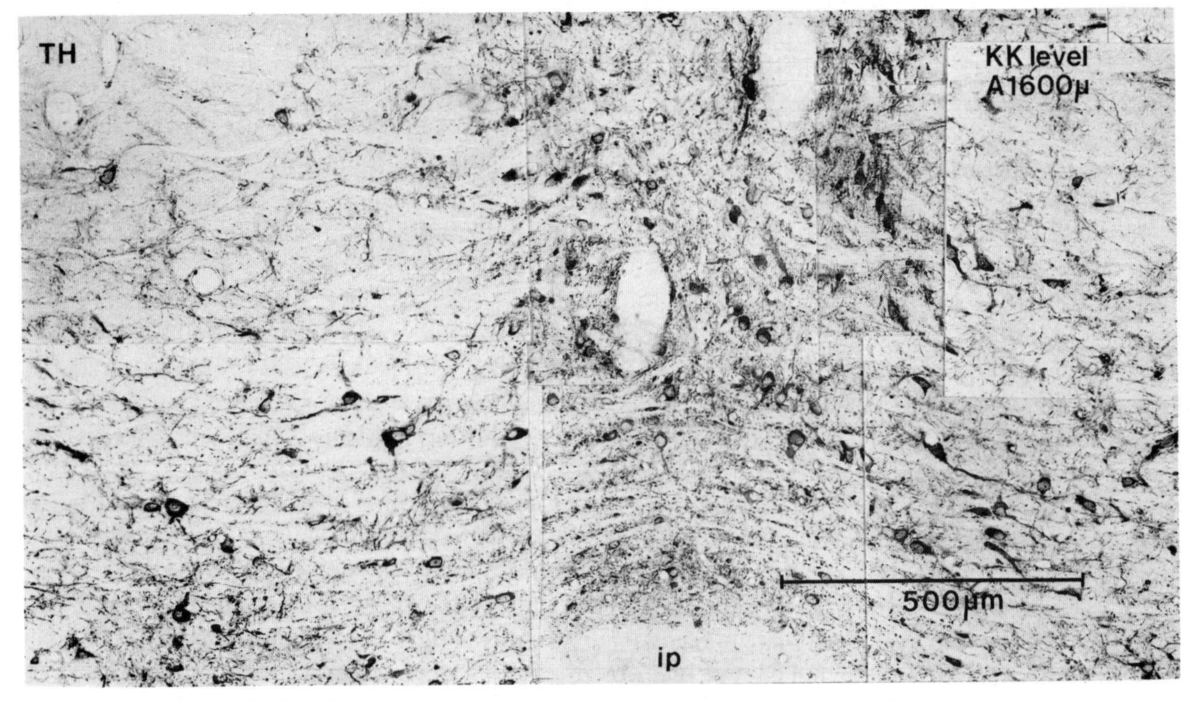

Fig. 1 (A). See page 344 for legend.

dopamine- and CCK-like immunoreactive terminals has been demonstrated within the tuberculum olfactorium and nucleus accumbens, and in the dorsal part of the nucleus interstitialis striae terminalis. These dopamine nerve terminal systems were first described by Fuxe and colleagues (1979) to have a dotted appearance and a lower dopamine turnover than that found in the large diffuse networks of dopamine terminals, innervating the major portion of the nucleus accumbens and tuberculum olfactorium as well as the neostriatum. When analyzing dopamine–CCK peptide interactions, it must also be realized that there exists a CCK innervation of parts of the substantia nigra (Vanderhaeghen et al., 1980).

As demonstrated by several groups (see Nemeroff and Prange, 1982), there exists a highly complex interaction between NT-LIr and dopamine neurons

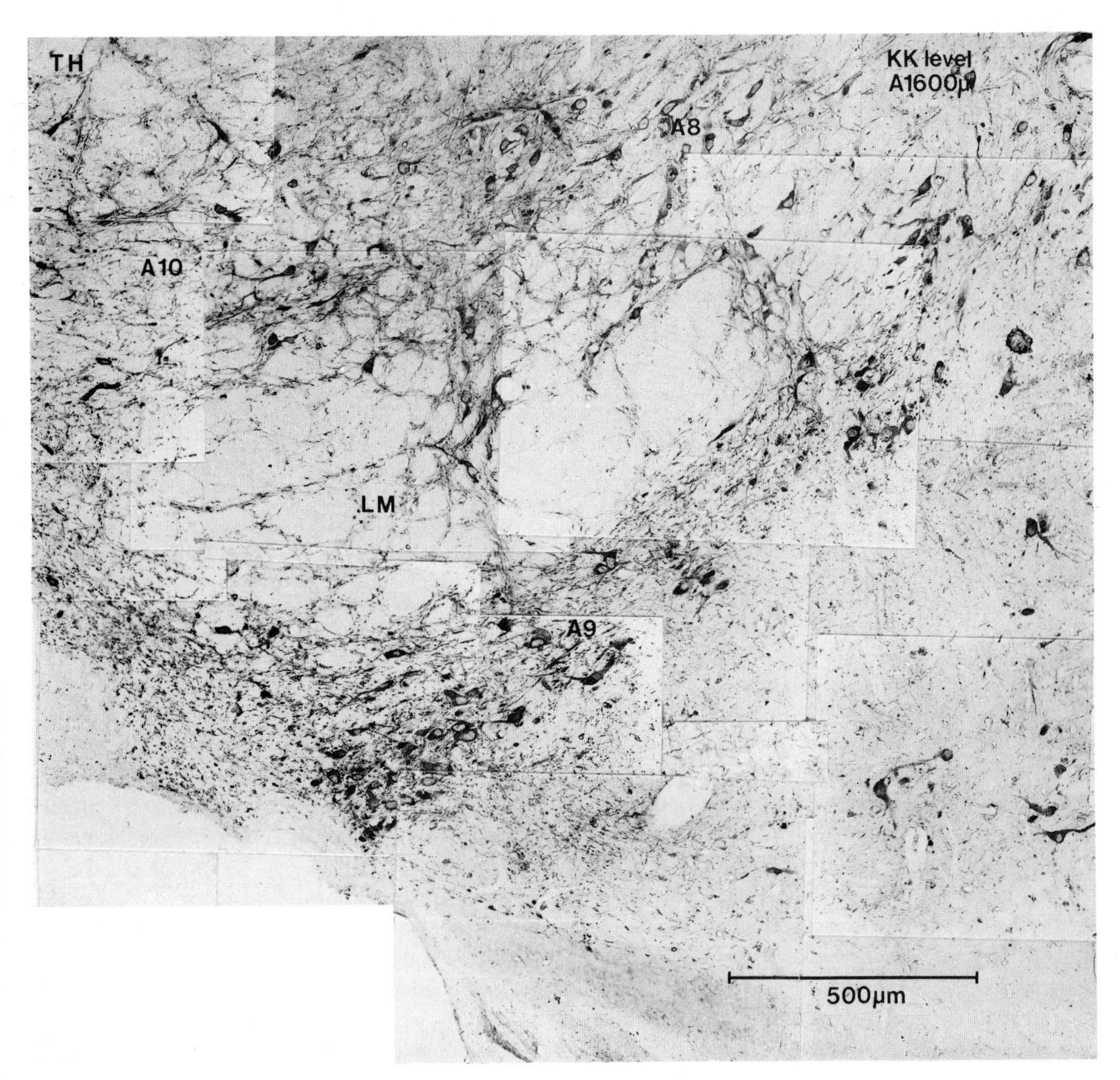

Fig. 1 (B). See page 344 for legend.

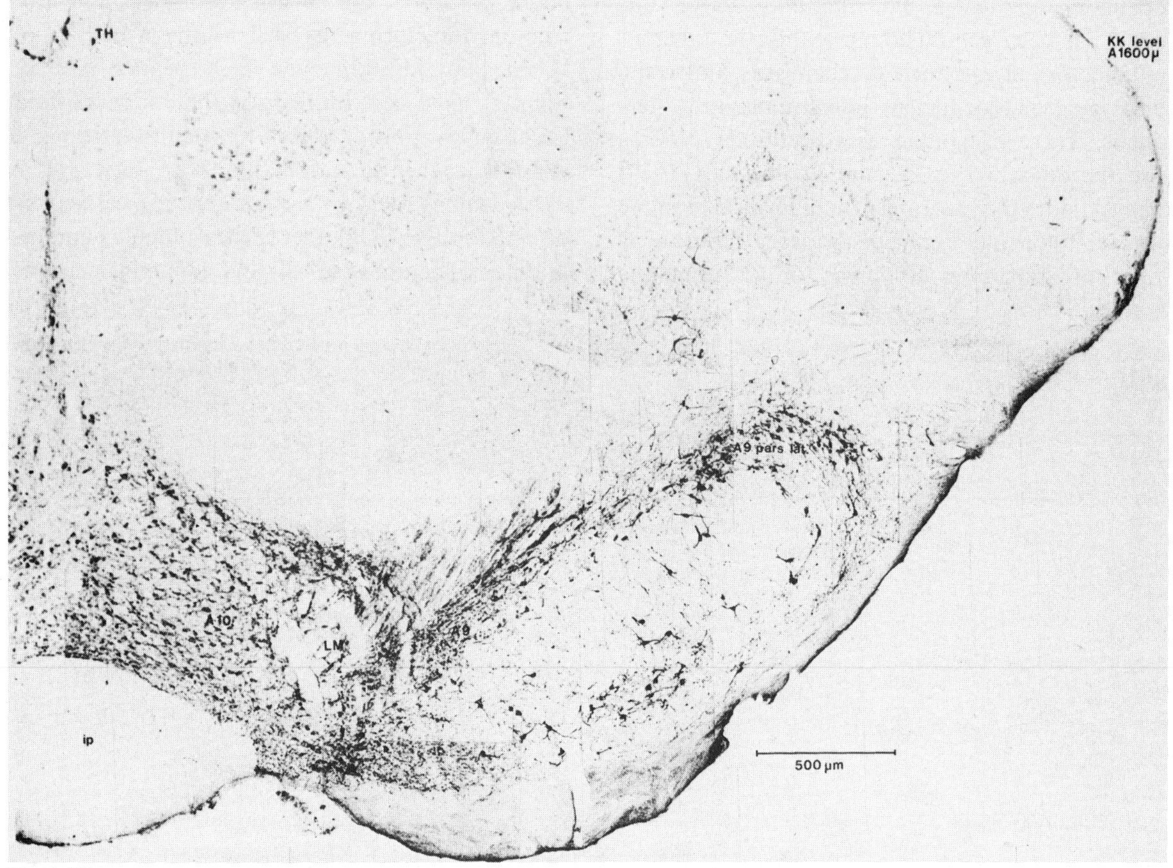

Fig. 1 (C).

Fig. 1. Bright field photomicrographs from a 10 μm thick (A, B) and 50 μm thick (C) coronal section of the ventral midbrain of the male rat. Colchicine treatment (75 μg/30 μl, intraventricularly, 24 h before killing). Indirect immunoperoxidase technique of Sternberger (1979). Tyrosine hydroxylase (TH) antiserum was diluted 1:750. Large numbers of TH-LIr nerve cell bodies are observed in the ventral tegmental area and within the zona compacta of the substantia nigra. LM, lateral lemniscus; ip, interpeduncular nucleus.

within the central nervous system. Thus, within the ventral midbrain there exist both NT-LIr interneurons, which innervate dopamine nerve cell bodies as well as dopamine nerve cell bodies which contain NT-LI and probably use NT as a comodulator (Palacios and Kuhar, 1981; Young and Kuhar, 1981). Furthermore, within the subcortical limbic forebrain, including nucleus accumbens and tuberculum olfactorium, there exist large numbers of NT-LIr cell bodies innervating nucleus accumbens and tuberculum olfactorium as well as dopamine nerve terminal systems which may store NT-LI. However, the major interaction between NT and dopamine in the limbic forebrain seems to be an interaction between two independent nerve terminal systems storing dopamine and NT, respectively, and participating in the same local circuits of the subcortical limbic forebrain (see Nemeroff and Prange, 1982).

Within the dorsal medulla and more precisely in the dorsal strip region of the nucleus tractus solitarius (see Kalia et al., 1983) there exists a well defined adrenaline cell group consisting of small sized nerve cells, some of which may costore NT-LI (Hökfelt et al., 1983a,b). The partial codistribution of phenylethanolamine *N*-methyl transferase

(PNMT) and NT-LIr nerve cell bodies in the dorsal strip area are shown in Figs. 3 and 4. This region receives chemo- and baro-receptor afferents (Kalia and Sullivan, 1982).

In Figs. 5–7 we demonstrate a codistribution of TH-LIr and PNMT-like immunoreactive (PNMT-LIr) nerve cells and of NPY-like immunoreactive (NPY-LIr) nerve cells within group C2 of the dorsal medulla and within the so-called "shell" nucleus of the nucleus tractus solitarius which surrounds the medial part of the nucleus tractus solitarius along most of its rostrocaudal extent, starting 0.5 mm rostral to the obex. In Figs. 3, 8 and 9 a codistribution of PNMT-LIr and NPY-LIr nerve cell bodies is also demonstrated in the dorsal strip region. These results have been obtained in adjacent thick vibratome sections and imply a substantial costorage of NPY-LI and adrenaline in the adrenaline cell groups of the dorsal medulla. Hökfelt and colleagues (1983d) have previously demonstrated that adrenaline nerve cells of group C2 (dorsal medulla) and C1 (ventral medulla) store NPY-LI. The present results, in addition, imply that the adrenaline cell group in the medial part of the nucleus tractus solitarius and the adrenaline cell group in the dorsal strip also contain adrenaline nerve cell bodies which costore NPY-LI. Thus, NPY may be a comodulator or transmitter in all the adrenaline neuron systems of the brain. In view of the existence also of NT-LIr and CCK-LIr neurones in the dorsal strip

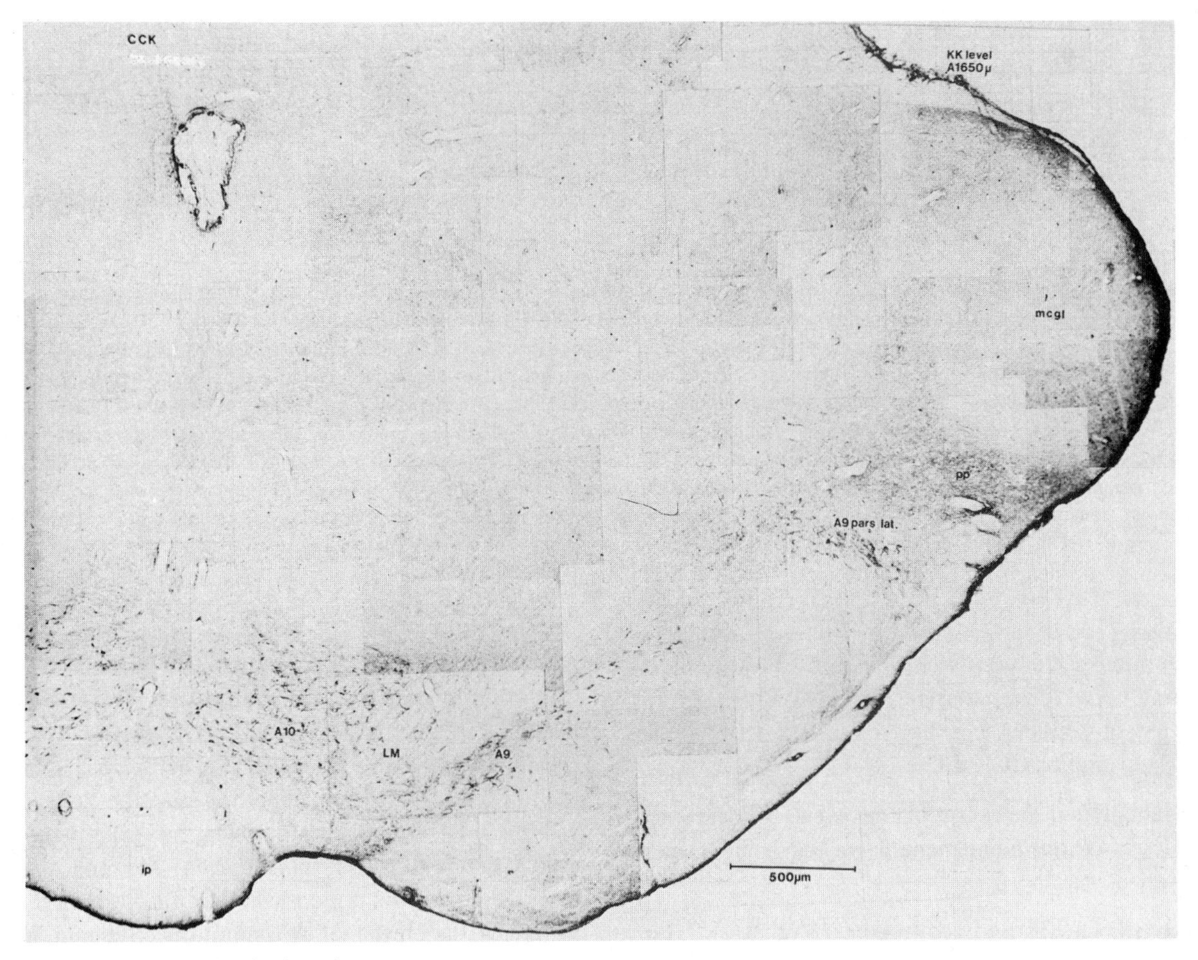

Fig. 2 (A). See page 346 for legend.

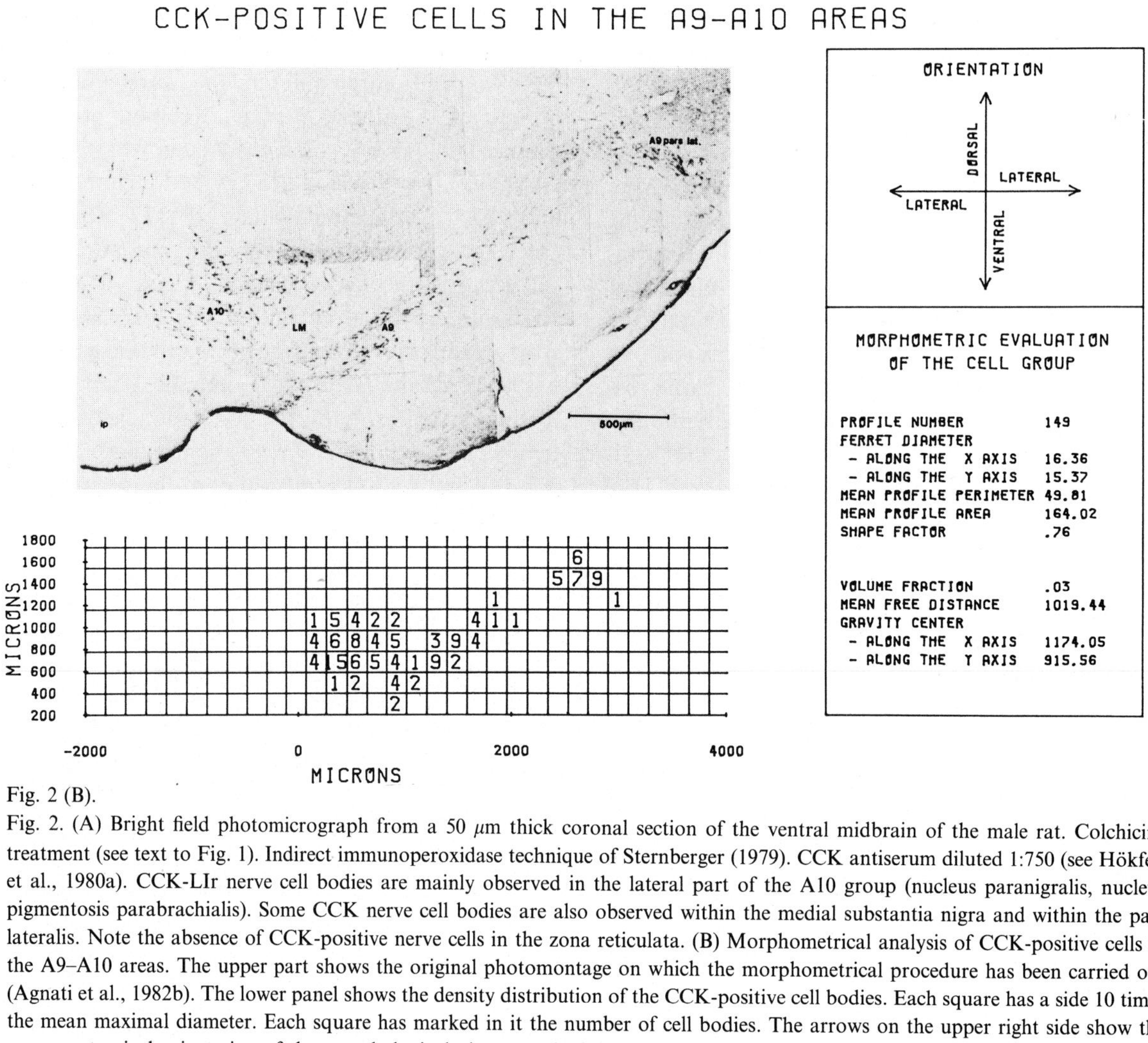
CCK-POSITIVE CELLS IN THE A9-A10 AREAS
A9 pars lat.
A10
LM
A9
ip
500µm
ORIENTATION
DORSAL
LATERAL
LATERAL
VENTRAL
MORPHOMETRIC EVALUATION OF THE CELL GROUP
PROFILE NUMBER 149
FERRET DIAMETER
- ALONG THE X AXIS 16.36
- ALONG THE Y AXIS 15.37
MEAN PROFILE PERIMETER 49.81
MEAN PROFILE AREA 164.02
SHAPE FACTOR .76
VOLUME FRACTION .03
MEAN FREE DISTANCE 1019.44
GRAVITY CENTER
- ALONG THE X AXIS 1174.05
- ALONG THE Y AXIS 915.56
MICRONS
1800
1600
1400
1200
1000
800
600
400
200
-2000
0
2000
4000
MICRONS

Fig. 2 (B).

Fig. 2. (A) Bright field photomicrograph from a 50 μm thick coronal section of the ventral midbrain of the male rat. Colchicine treatment (see text to Fig. 1). Indirect immunoperoxidase technique of Sternberger (1979). CCK antiserum diluted 1:750 (see Hökfelt et al., 1980a). CCK-LIr nerve cell bodies are mainly observed in the lateral part of the A10 group (nucleus paranigralis, nucleus pigmentosis parabrachialis). Some CCK nerve cell bodies are also observed within the medial substantia nigra and within the pars lateralis. Note the absence of CCK-positive nerve cells in the zona reticulata. (B) Morphometrical analysis of CCK-positive cells in the A9–A10 areas. The upper part shows the original photomontage on which the morphometrical procedure has been carried out (Agnati et al., 1982b). The lower panel shows the density distribution of the CCK-positive cell bodies. Each square has a side 10 times the mean maximal diameter. Each square has marked in it the number of cell bodies. The arrows on the upper right side show the neuroanatomical orientation of the morphological picture and of the density distribution. The table gives some cell body and cell group parameters. ip, nuc. interpeduncularis; LM, lemniscus medialis; mcgl, medial geniculate nucleus.

(unpublished data) it seems possible that the adrenaline nerve cells in this region can costore several types of peptides, e.g. not only NPY-LI but also NT-LI and/or CCK-LI.

Evaluation of the extent of coexistence of neuropeptides in central monoamine nerve cell bodies: quantitative aspects

Agnati, Fuxe and colleagues (Agnati et al., 1982a–d; Fuxe et al., 1983a,b) have recently developed three different methods for the quantitative evaluation of coexistence of transmitters and modulators in the same nerve cell body (see Fig. 10). The methods have been called the "gravity centre method", the "occlusion" method and the "overlap" method (Agnati et al., 1982a, 1983a). In the gravity centre method thin adjacent sections are analyzed and it is tested whether the "gravity" centres of the cluster of A immunoreactive and B immunoreactive nerve cell bodies are identical. In

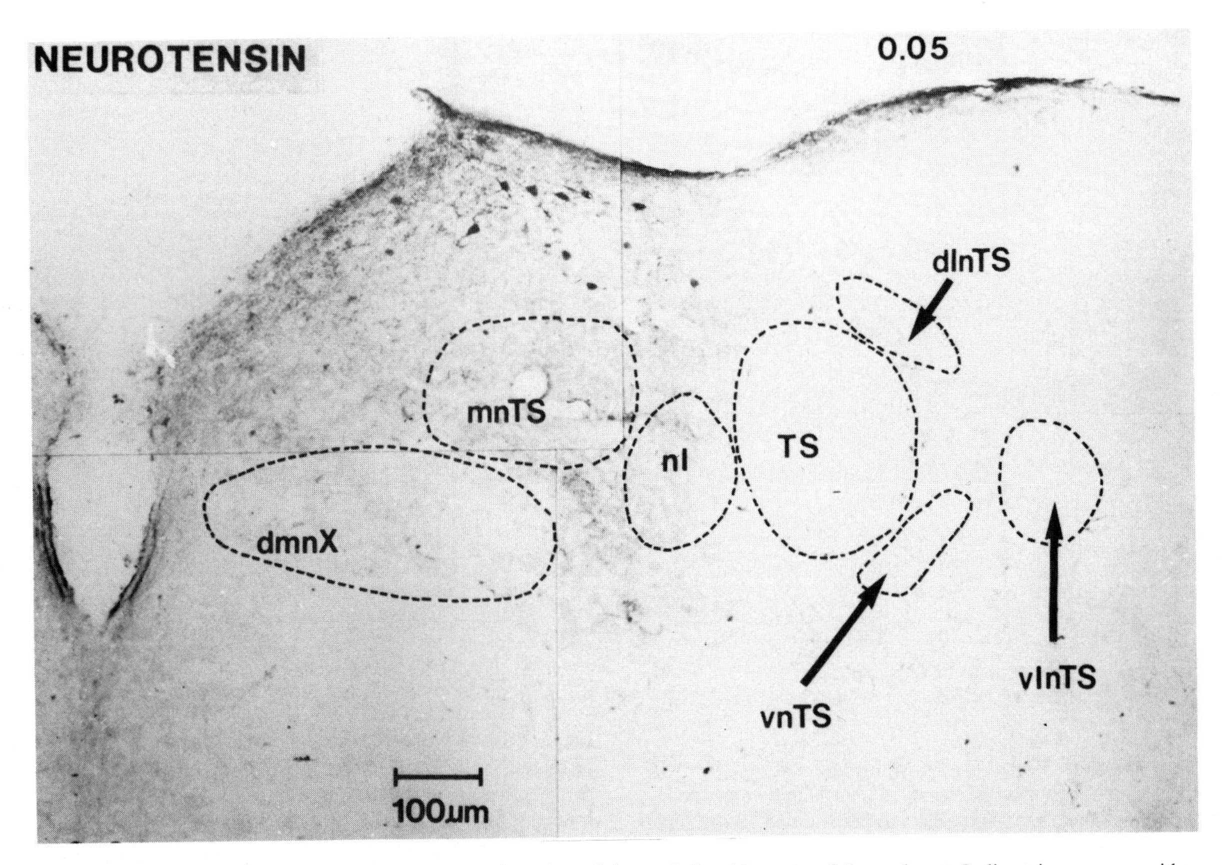

Fig. 3. Bright field photomicrograph from a coronal section of the medulla oblongata of the male rat. Indirect immunoperoxidase technique of Sternberger (1979). NT antiserum diluted 1:600 (MILAB, Copenhagen). NT-LIr nerve cell bodies are observed within the dorsal strip and the periventricular region. NT-LIr nerve terminals appearing as fine dots are observed within the periventricular area and within the medial nucleus of the tractus solitarius (mnTS). Location of section is 0.05 mm rostral to obex. Abbreviations: dmnX, dorsal motor nucleus of the vagus; nI, intermediate nucleus; TS, tractus solitarius; dlnTS, dorsolateral nucleus of the tractus solitarius; vnTS, ventral nucleus of the nucleus tractus solitarius; vlnTS, ventrolateral nucleus of the tractus solitarius.

this procedure, the two photomontages are given an identical x- and y-axis. The x- and y-coordinates can then be calculated for each nerve cell body and subsequently the two gravity centres can be calculated and compared. If a 100% degree of coexistence exists, the two gravity centres should coincide. If the gravity centres, on the other hand, are significantly different from each other, a 100% degree of coexistence does not exist in the cell cluster analyzed. It must be emphasized, however, that the gravity centres of two cell clusters can coincide even if coexistence of the two transmitters does not exist. The identity of gravity centres is a prerequisite for a 100% degree of coexistence but is not a sufficient criterion for the demonstration of coexistence. The occlusion method is illustrated in Fig. 10. This procedure is a statistical approach which allows an overall evaluation of the extent of coexistence in nerve cell bodies of a given area. Three adjacent sections are analyzed and are randomly stained for antigen A, antigen B and antigen A plus antigen B. The extent of coexistence is then obtained by subtracting from the sum of antigen A plus antigen B positive nerve cell bodies, the number of immunoreactive nerve cell bodies obtained when staining for both antigen A and antigen B. The third

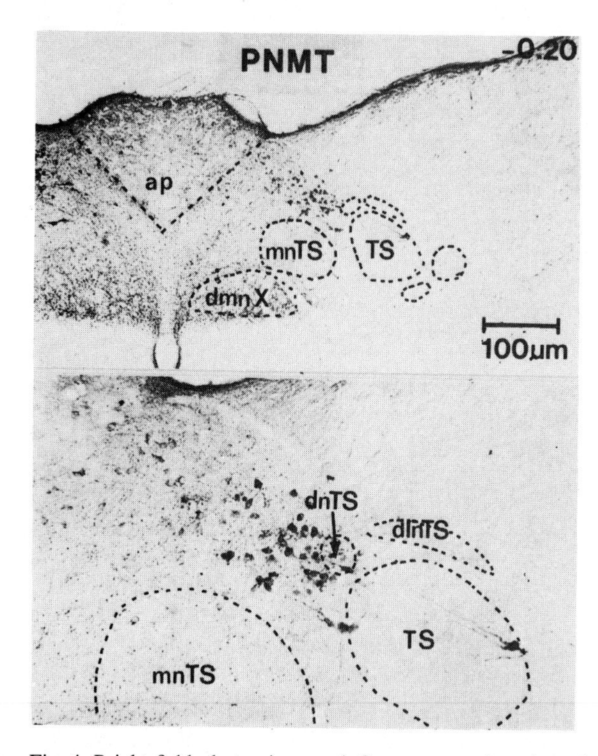

Fig. 4. Bright field photomicrograph from a coronal section of the medulla oblongata of the rat. Indirect immunoperoxidase technique. The PNMT antiserum has been diluted 1:1500. Small PNMT-LIr nerve cell bodies are observed mainly in the dorsal strip and in the dorsal nucleus of tractus solitarius. Large numbers of PNMT-LIr terminals are mainly observed in the dorsal motor nucleus of the vagus and in the commissural nucleus. For abbreviations, see text to previous figure. ap, area postrema. The section is located 0.2 mm caudal to the obex. The bar in the figures refers to the lower figure. Magnification in the upper figure is × 30.

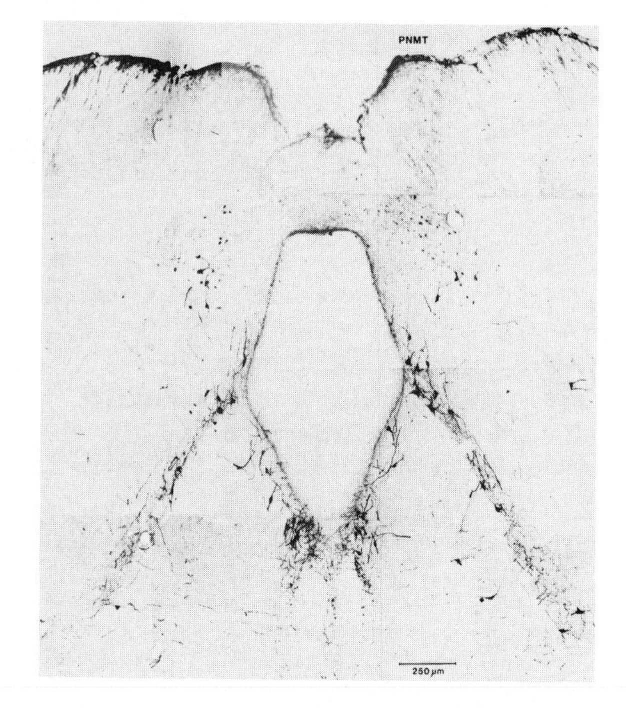

Fig. 5. Bright field photomicrograph from a horizontal section of the dorsal medulla oblongata of the rat. Indirect immunoperoxidase technique. PNMT antiserum has been diluted 1:1500. Caudal is upwards in the figure. The large PNMT-LIr nerve cell bodies with their dendrites are shown to form an M in the section. The inner limbs of the M represent PNMT-LIr nerve cell bodies located within group C2 close to the midline and close to the ventricle. The outer limbs of the M are built up of PNMT nerve cell bodies located in the most medial part of the nucleus tractus solitarius ("Shell nucleus"). At the area postrema level the small PNMT-LIr nerve cell bodies of the dorsal strip area are seen as dots.

method, the so-called "overlap" method, is the only method which allows direct evaluation of coexistence cell by cell. In this analysis a computer assisted morphometrical analysis is performed and each cell body is given an x–y-coordinate and again the x- and y-axes in the two photomontages obtained in the same or adjacent thin sections are identical. The area of each nerve cell body is calculated and when the area of overlap is 30% or more in the two photomontages the nerve cell body in question is considered to show coexistence (see Fig. 10). An example of this procedure is shown in Figs. 11–13.

Effects of cholecystokinin peptides, neurotensin, neuropeptide Y on regional catecholamine levels and turnover

The effects of cholecystokinin peptides

CCK-7, CCK-8 and CCK-39 given intraventricularly in nanomolar amounts selectively reduced dopamine turnover within the diffuse types of dopamine nerve terminal networks of the nucleus accumbens and nucleus caudatus putamen (Fuxe et al., 1980a) (Fig. 14). In contrast, the "dotted" type

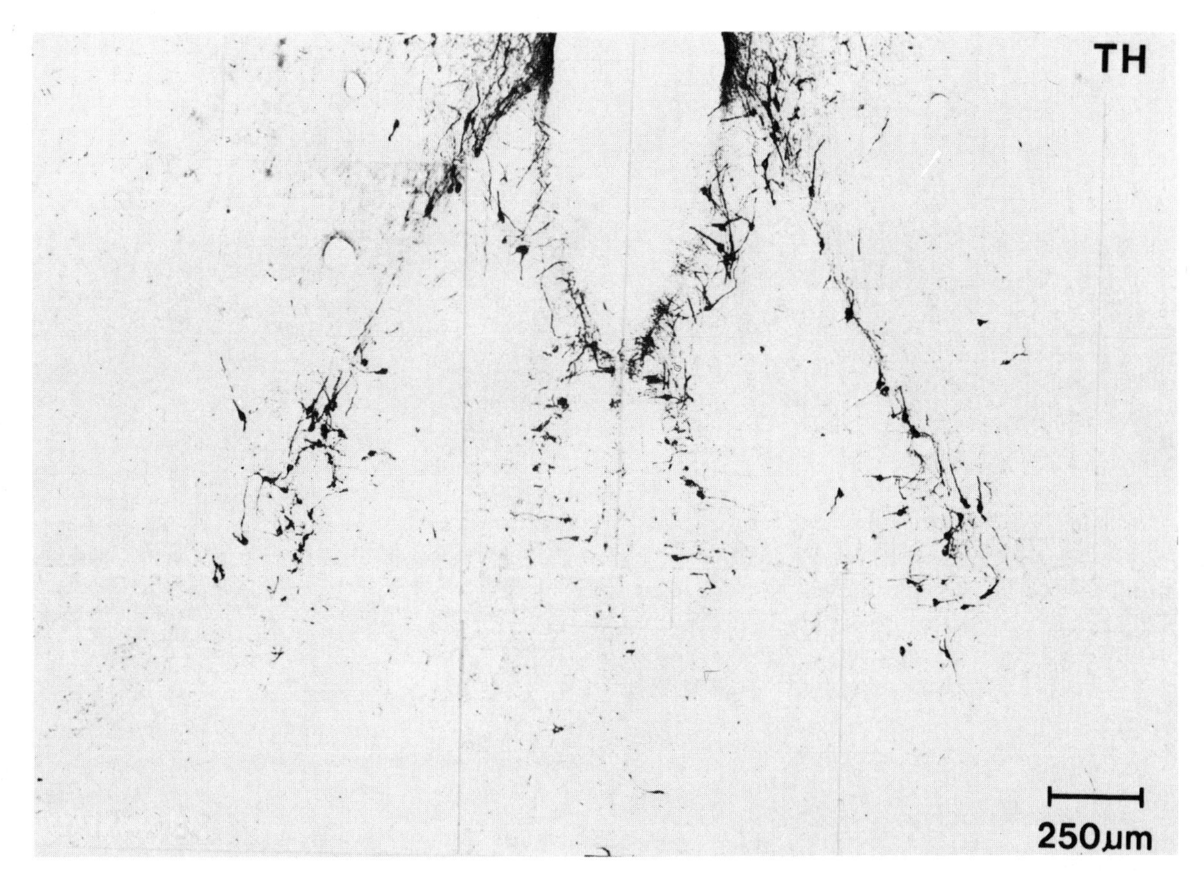

Fig. 6. Bright field photomicrograph of a horizontal section of the dorsal medulla oblongata of the male rat. Indirect immunoperoxidase technique. TH antiserum diluted 1:750 (Markey et al., 1980). The large TH-LIr nerve cell bodies and their dendrites are seen to form an M in the section. The vast majority of the TH-LIr nerve cell bodies represent PNMT nerve cell bodies (see Fig. 5).

of dopamine terminals, which seem to store CCK-LI, do not change their dopamine turnover in response to the intraventricular injections of CCKs (Fig. 14). Thus, it may be that CCK peptides, when present as comodulators in dopamine synapses, do not modulate dopamine turnover and release. Instead, when interacting with dopamine terminals by way of axo-axonic interactions as in the striatum and parts of nucleus accumbens they have the ability to reduce dopamine turnover. This axo-axonic modulation may be direct or indirect in the striatal and accumbens local circuits. Dopamine turnover in the median eminence is not modulated by CCK peptides (Fuxe et al., 1980b, 1982).

Instead, the small fragment CCK-4, when given intraventricularly, produced a selective increase in dopamine turnover within the central part of nucleus caudatus putamen (Fig. 14). This suggests the possibility that small fragments may be formed from large CCK fragments in certain regions (Fig. 15) which can counteract the effects of the large fragments, so forming a feed-back loop to stabilize at a certain level the effects of the larger fragments.

When the CCK fragments are given intravenously in the same amounts as when given intraventricularly, only CCK-39 has the capacity to reduce dopamine turnover in the above-mentioned forebrain regions while CCK-8 (except in the diffuse

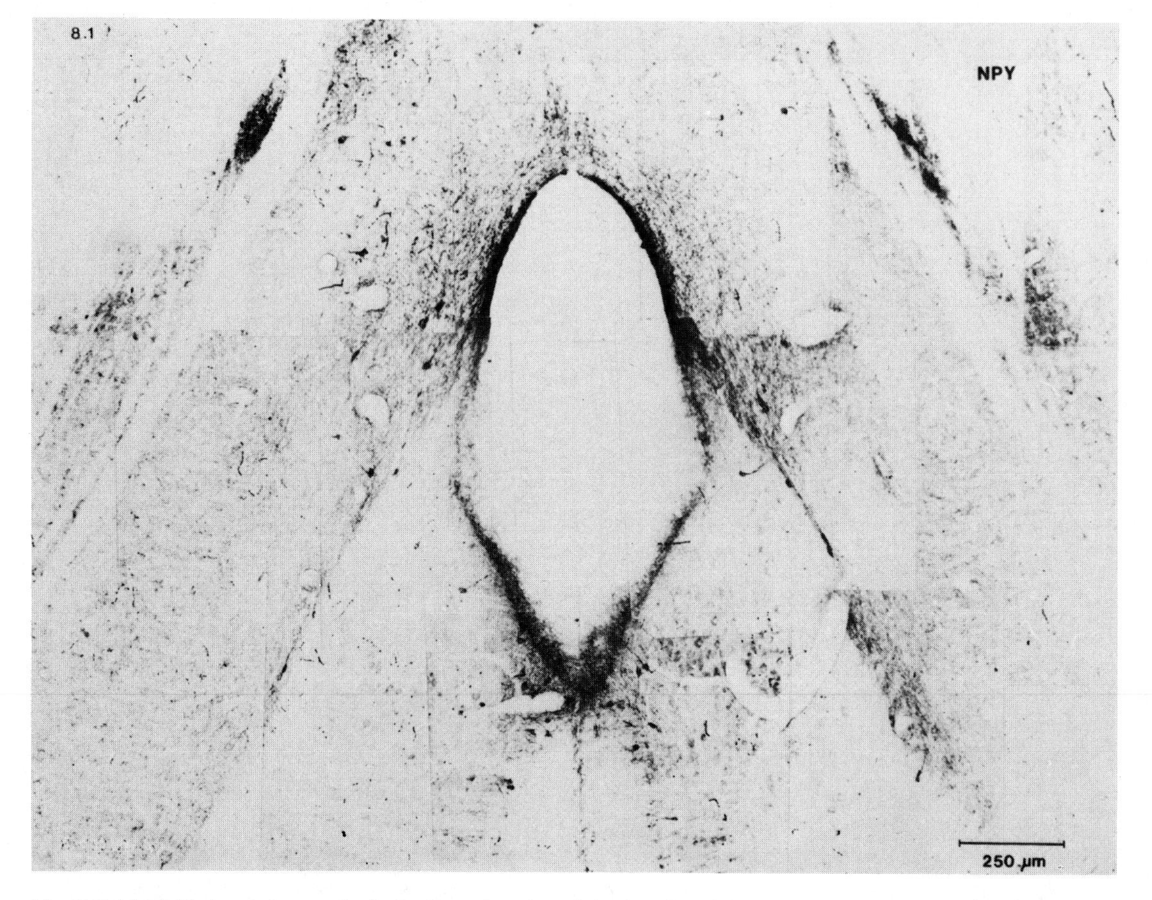

Fig. 7. Bright field photomicrograph of a horizontal section of the dorsal medulla oblongata of the male rat. Indirect immunoperoxidase technique. NPY antiserum diluted 1:1000 (see Hökfelt et al., 1983d). NPY-LIr nerve cell bodies are mainly observed within the group C2 and within the so-called "Shell nucleus" of the nucleus tractus solitarius, where TH-LIr and PNMT-LIr cell bodies have been demonstrated (see previous figures). However, the number of NPY-LIr nerve cell bodies is low compared to the number of TH--LIr and PNMT-LIr nerve cell bodies. Varying densities of NPY-LIr nerve terminals are seen within the nucleus tractus solitarius and within the dorsal motor nucleus of the vagus. A high density of NPY-LIr terminals is shown within the interstitial nucleus of the tractus solitarius on both sides of the brain (see upper part of the figure).

type of accumbens terminalis) and CCK-7 cannot reduce regional dopamine turnover (Fig. 16). In fact, CCK-7 given intravenously now increases the disappearance of dopamine after TH inhibition in the nucleus caudatus putamen (Fig. 16). CCK-4 given intravenously again produces a highly selective increase of dopamine disappearance after TH inhibition in the central part of nucleus caudatus. Different patterns of change in dopamine turnover observed after intravenous or intraventricular injections of CCK peptides probably reflects the existence of a blood brain barrier for CCK peptides, with the possible exception of the small CCK-4 fragment. The fact that CCK-39 can still reduce dopamine turnover may be due to a substantial ability to activate visceral afferents in the vagus nerve over a long period, leading to reflex-induced changes of dopamine turnover in the forebrain. Thus, vagal afferents may be involved in the complex regulation of the ascending mesostriatal and

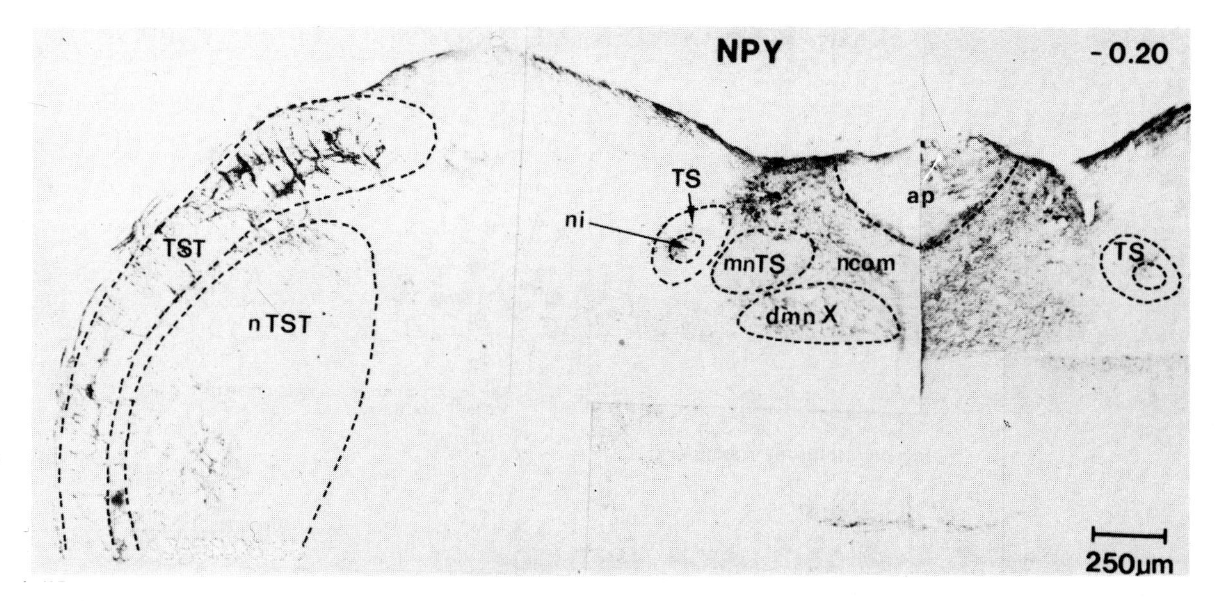

Fig. 8. Bright field photomicrograph of a coronal section of the medulla oblongata of the male rat. Indirect immunoperoxidase technique. NPY antiserum diluted 1:1000. The NPY-LIr nerve cell bodies are located in the dorsal strip area. NPY-LIr terminals are found in high densities in the dorsal motor nucleus of the vagus (dmnX) and in the commissural nucleus (ncom). Some are also found within the interstitial nucleus (ni) and within the spinal tract of the trigeminal nerve (TST). Section located 0.20 mm caudal to the obex. For other abbreviations, see text to previous figures.

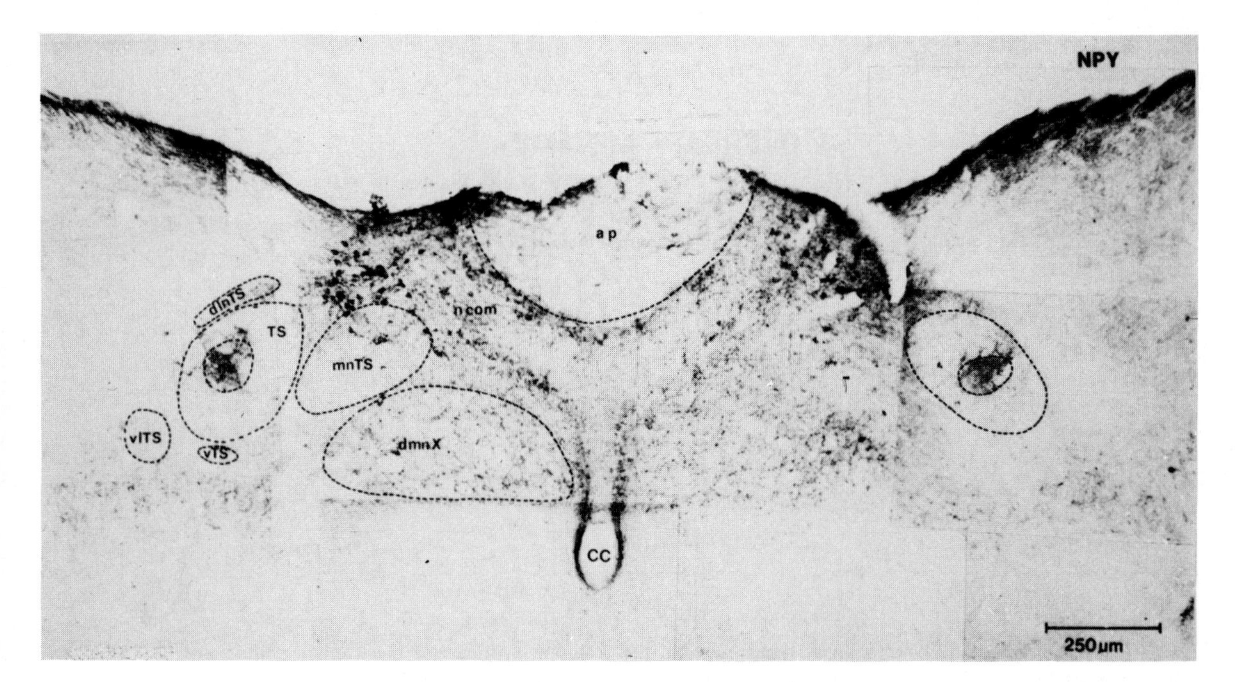

Fig. 9. High magnification of Fig. 8. Large numbers of NPY-LIr nerve cell bodies are found in the dorsal strip area and high densities of NPY-LIr terminals are seen in the commissural nucleus (ncom), within the dorsal motor nucleus of the vagus (dmnX) and within the interstitial nucleus. CC, central canal. For other abbreviations, see text to previous figures.

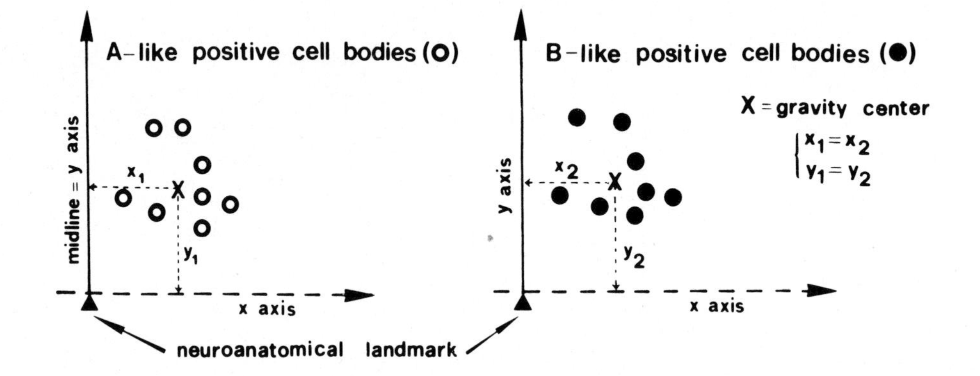

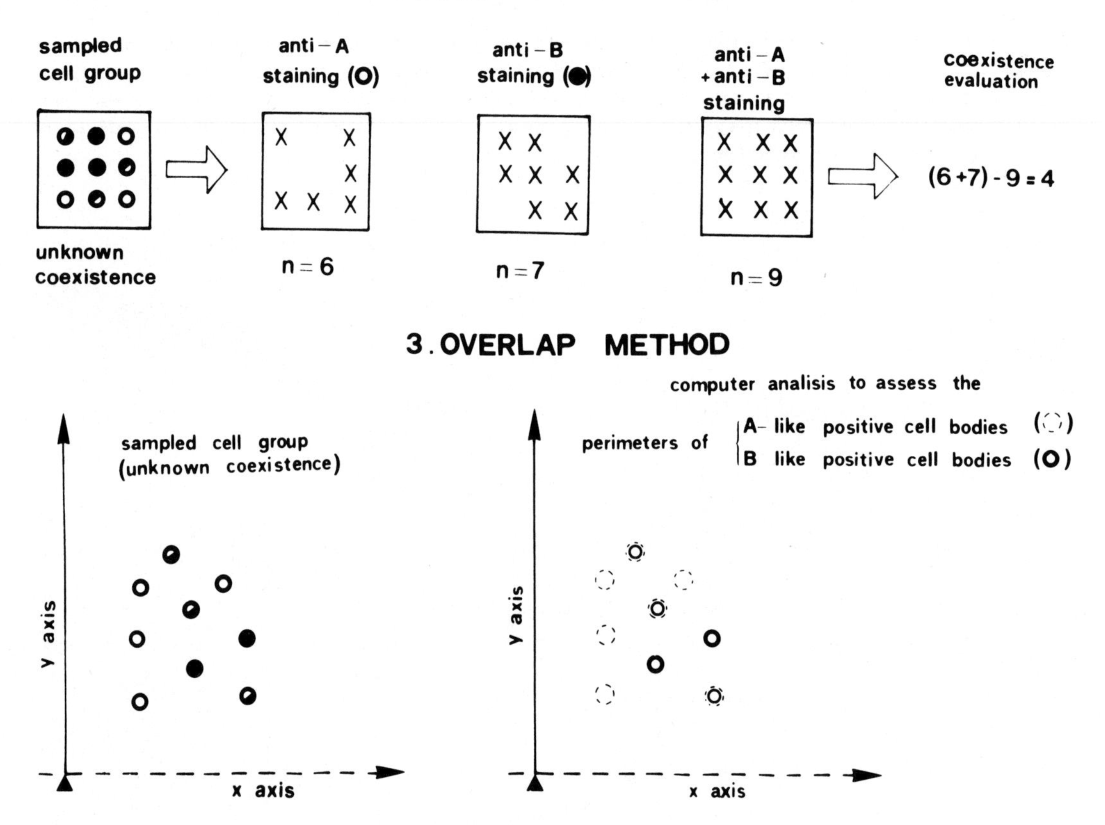

Fig. 10. Schematic illustration of the gravity center method, the occlusion method and the overlap method for the quantitative evaluation of the entity of coexistence of transmitters and modulators in the same cell bodies.

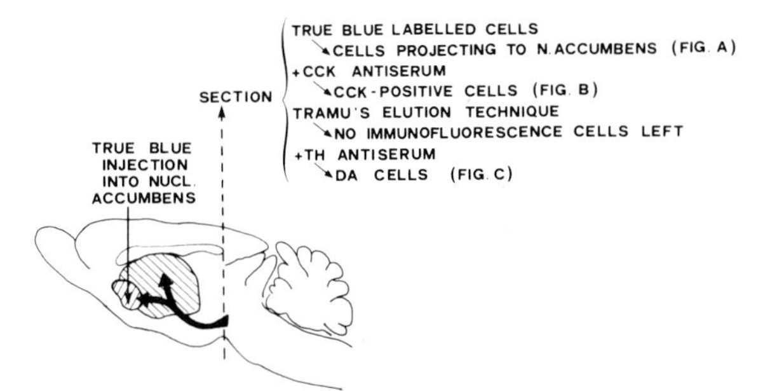

Fig. 11. Schematic illustration of the procedure used by Hökfelt et al. (1980b) to assess the coexistence of CCK-LI and TH-LI in the ventral midbrain. The procedure includes also a mapping of nerve cells in this region, projecting into the nucleus accumbens using a retrograde tracing technique. Thus, true blue was injected into the nucleus accumbens. The overlap method has now been applied on the fluorescence photomicrographs shown by Hökfelt et al. (1980b) (see Figs. 12 and 13).

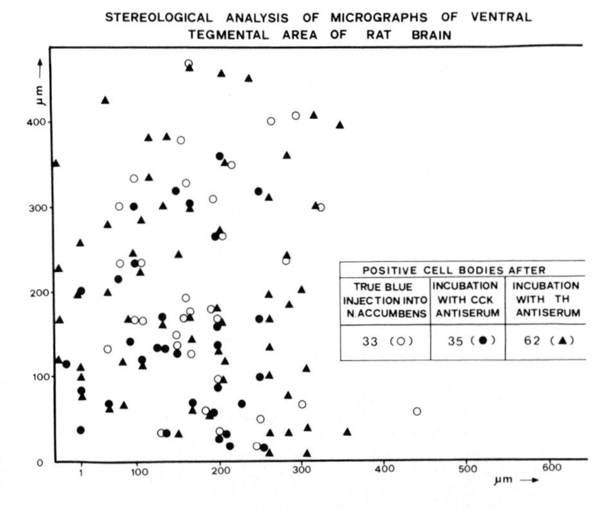

Fig. 12. The overlap method has been applied on photomicrographs taken from an identical region of the ventral tegmental area of the rat brain showing true blue stained cells, CCK-LIr cells and TH-LIr cells. The y-axis is identical to the midline and the x-axis is perpendicular to the midline reaching it at the ventral border of the brain which is the origo of the Cartesian plane. Each nerve cell body has been given x–y coordinates and the computer registers coexistence when there is an overlap of the nerve cell body area by 30% or more. The results of this quantitative analysis of the entity of coexistence are illustrated in Fig. 13.

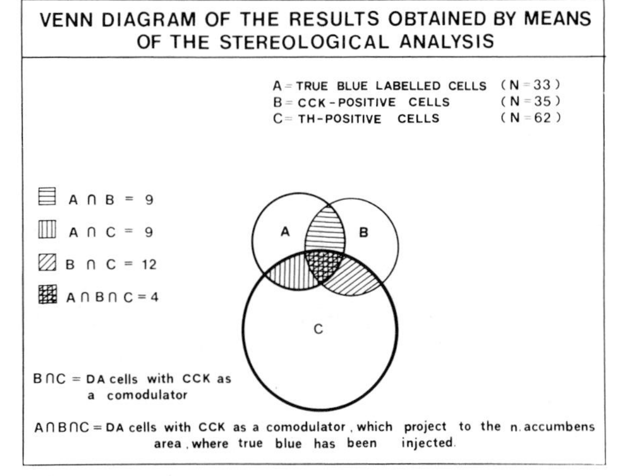

Fig. 13. By means of Venn diagrams the results obtained in the morphometrical analysis on the entity of coexistence are summarized. The three nerve cell body populations are shown by circles A, B and C having a diameter related to the number of cells of the respective nerve cell population. The overlap areas of the circles give an over-all evaluation of the entity of coexistence between the three populations of nerve cell bodies. It is clear that only a minority of the dopamine nerve cell bodies in the analyzed area show CCK-LI. Also the majority of the CCK-LIr nerve cell bodies do not store TH-like material, indicating that most of the CCK-LIr nerve cell bodies represent a separate population of nerve cells different from the dopamine nerve cell population. Furthermore, only a minority of the dopamine nerve cells projecting into the nucleus accumbens store CCK-LI. These results are in agreement with the fact that only the dotted type of the dopamine nerve terminals within the medial and caudal part of the nucleus accumbens store CCK-LI.

mesolimbic cortical dopamine pathways. Furthermore, it should be noted that following intravenous injection, CCK peptides may also influence excitatory CCK mechanisms in the substantia nigra (Skirboll et al., 1981) either via a direct action or via an entry into the third ventricle over the median eminence (Fig. 17).

These results clearly indicate that the effects of CCK peptides on various dopamine neuron systems of the brain are clearly different, depending upon whether the peptide is centrally or peripherally administered. Therefore, when attempting to develop drugs to influence the peptide comodula-

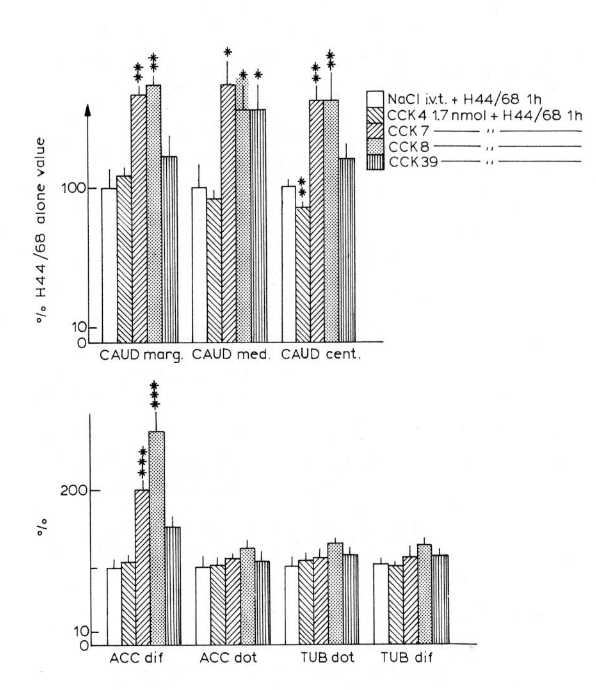

Fig. 14. Effects of intraventricular injections of CCK-4, CCK-7, CCK-8 and CCK-39 (1.7 nmol) on the dopamine fluorescence disappearance in various forebrain areas after TH inhibition (H44/68, α-methyltyrosine methyl ester, i.p., 1 h before killing). Means ± S.E.M. are shown; n = 5–6. Statistical analysis was performed according to Wilcoxon test, treatments vs. controls. * $P < 0.05$; ** $P < 0.01$; *** $P < 0.001$. Abbreviations: CAUD marg., marginal zone of the nucleus caudatus; CAUD med., medial part of the nucleus caudatus; CAUD cent., central part of the nucleus caudatus; ACC dif, diffuse type of dopamine nerve terminals of the anterior nucleus accumbens; ACC dot, dotted type of dopamine nerve terminals of the posterior nucleus accumbens; TUB dot, dotted type of dopamine nerve terminals of the medial-posterior part of the tuberculum olfactorium; TUB dif, diffuse type of dopamine nerve terminals of the lateral-posterior part of the tuberculum olfactorium.

tors in monoamine synapses, multiple sites of action following systemic treatment with peptide comodulators must be taken into account, and the effects induced by changes in visceral afferent activity from the gastrointestinal tract and other visceral organs may be of considerable importance.

Effects of intraventricular injections of neurotensin

As seen in Fig. 18, NT in nanomolar doses (0.06–

Schematic representation of the possible mechanism involved in the CCK modulation of dopamine turnover in caudatus

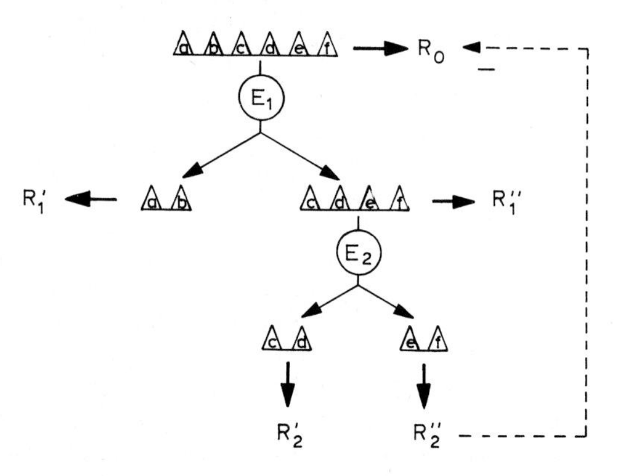

Fig. 15. Schematic illustration of a possible mechanism leading to the formation of small active fragments from large CCK molecules which via a feed-back can stabilize at a certain level the response induced by the large fragment.

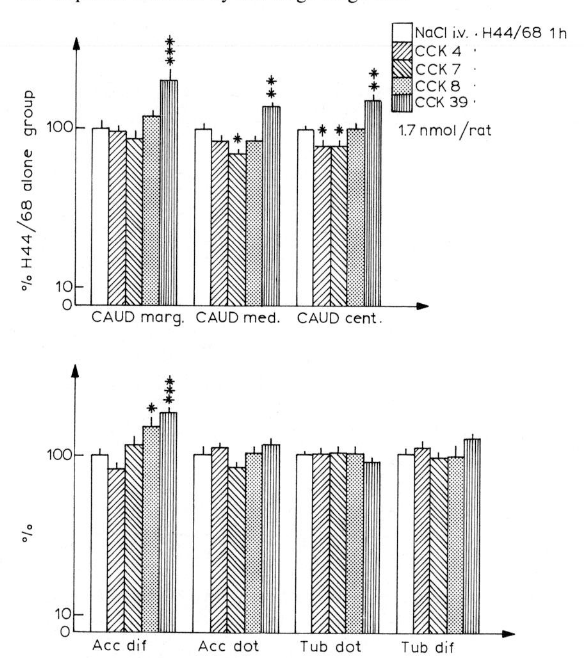

Fig. 16. The effects of intravenous infusions of CCK-4, CCK-7, CCK-8 and CCK-39 (1.7 nmol, 1 h infusion) on the disappearance of the dopamine fluorescence in various forebrain areas following TH inhibition (α-methyltyrosine methyl ester, H44/68, 1 h before killing, i.p., 250 mg/kg). Means ± S.E.M. are shown; n = 5–6. Statistical analysis according to Wilcoxon test, treatment vs. controls. * $P < 0.05$; ** $P < 0.01$; *** $P < 0.001$. For abbreviations, see legend to Fig. 14.

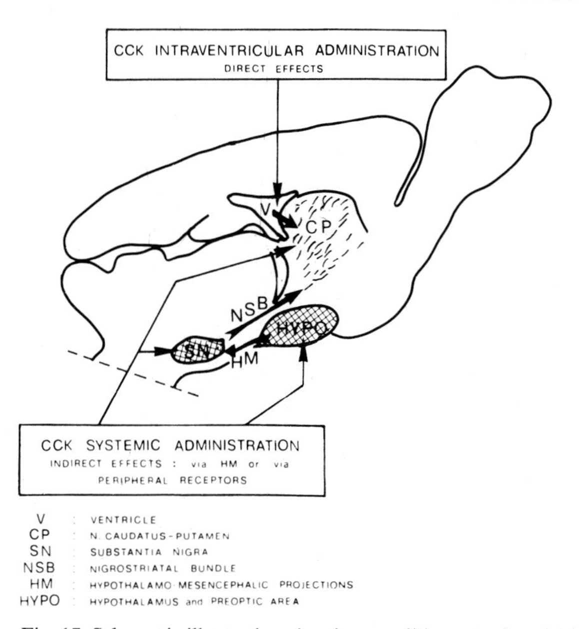

Fig. 17. Schematic illustration showing possible routes by which CCK peptides can modulate dopamine neuronal systems in the forebrain following intravenous or intraventricular injections. Thus, after systemic treatment, activation of peripheral CCK receptors for CCK may lead to reflex-induced changes in regional dopamine turnover but direct effects on substantia nigra or indirect effects via the hypothalamus are also possible.

6 nmol) produces a dose-dependent increase in dopamine turnover in the CCK positive and negative terminals of the nucleus accumbens (see also Widerlöv et al., 1982a). The dopamine levels in the accumbens dopamine nerve terminal systems were unaffected by NT. No significant changes in dopamine turnover could be demonstrated in the nucleus caudatus putamen, in the tuberculum olfactorium nor within the median eminence following NT injection (Figs. 19, 20). This latter observation is of special interest, since NT may be costored with dopamine in some of the tuberoinfundibular dopamine neurons (Hökfelt et al., 1983a). In view of the large number of NT-LIr nerve cell bodies in the limbic forebrain (Jennes et al., 1982), it is likely that the increase in dopamine turnover demonstrated in the nucleus accumbens reflects local circuit interactions between NT and dopamine nerve terminal systems. In the present study it could also be demonstrated that NT induced a dose-dependent increase in noradrenaline turnover in the subependymal layer of the median eminence and in catecholamine turnover in the medial palisade zone of the median eminence. Furthermore, it seems as if NT can selectively increase noradrenaline turnover within the median eminence. This observation is of special interest, since NT-LIr cell bodies and corticotrophin immunoreactive cell bodies overlap in the parvocellular part of the paraventricular hypothalamic nucleus, indicating a coexistence of NT and corticotrophin. Moreover, NT in the present experiments was found to increase adrenocorticotropic hormone (ACTH) secretion (Table I). Previously we have demonstrated that ACTH 1–24 can selectively increase noradrenaline turnover within the median eminence (subependymal layer and medial palisade zone) (Fuxe et al., 1978; Andersson et al., 1980). It therefore seems possible that the present discovery of NT-induced increases of noradrenaline turnover in the median eminence can in part be related to the ability of NT to increase ACTH secretion. ACTH may induce part of its direct inhibitory feed-back action on corticotrophin secretion via activation of an inhibitory noradrenergic mechanism in the median eminence (Andersson et al., 1980). It may be speculated that NT is a facilitatory comodulator in the corticotrophin neurons.

It should be noted that NT in the present experiments (Table I) also induces a lowering of prolactin secretion (see also Rivier et al., 1977; Maeda and Frohman, 1978; Makino et al., 1978; Frohman et al., 1982), which is abolished by TH inhibition, and a lowering of thyroid stimulating hormone (TSH) secretion is also noted. In view of the presence of dopamine nerve terminals in the medial palisade zone releasing dopamine as a prolactin inhibitory factor (Andersson et al., 1981) it should be considered that the increase of catecholamine turnover demonstrated in the medial palisade zone of the median eminence after NT also may reflect an increase of dopamine release in this region. The

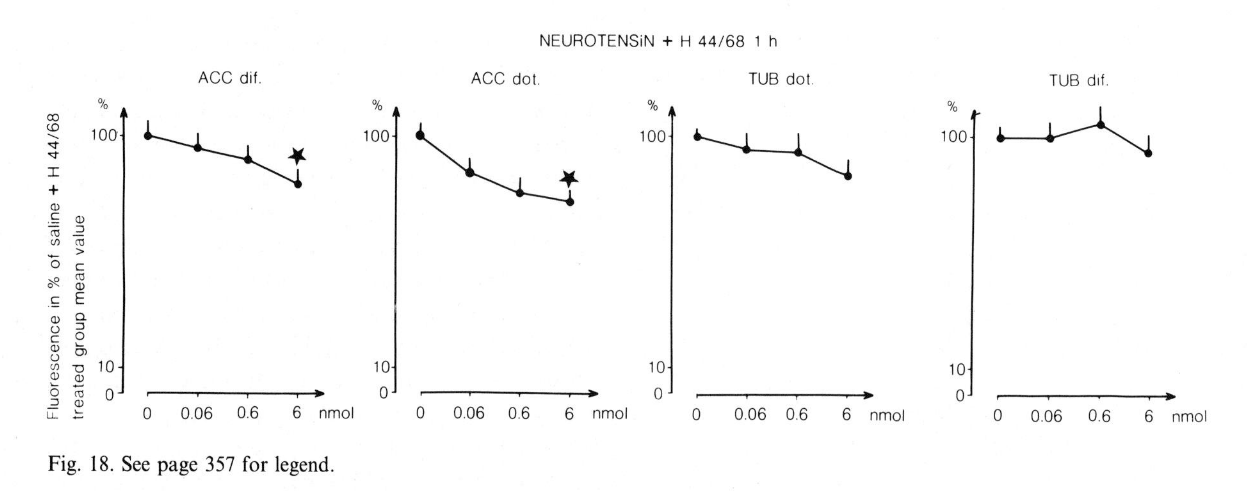

Fig. 18. See page 357 for legend.

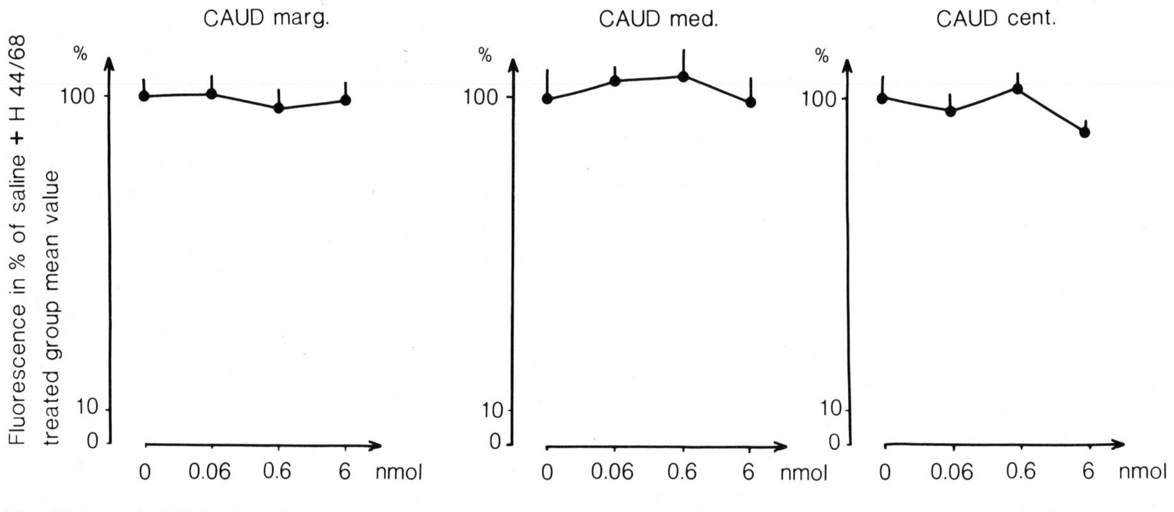

Fig. 19. See page 357 for legend.

NT-induced lowering of prolactin secretion may therefore in part be mediated via a release of dopamine from the median eminence. In view of the inhibitory role of the lateral tuberoinfundibular dopamine neurons innervating the lateral palisade zone of the median eminence (LPZ) in the regulation of luteinizing hormone releasing hormone (LHRH) secretion it is of interest that NT following TH inhibition induced a reduction of luteinizing hormone secretion (Table I). Thus, NT as a comodulator in the tuberoinfundibular dopamine neurons may enhance the inhibitory effects of released dopamine on LHRH release. However, it should be emphasized that NT did not induce any change of dopamine turnover in the LPZ, although it may be a comodulator in some of these terminals.

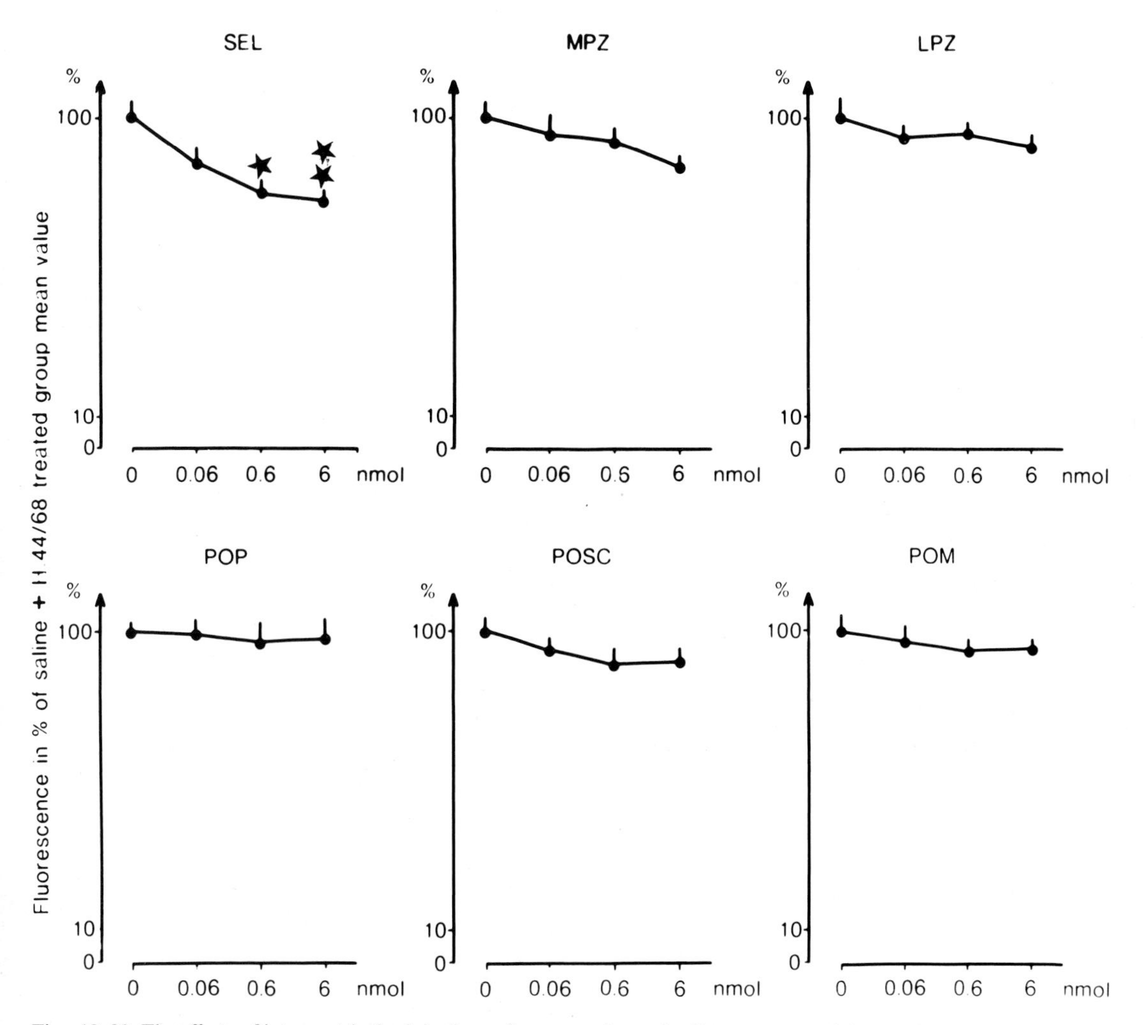

Figs. 18–20. The effects of intraventricular injections of neurotensin on the disappearance of the catecholamine fluorescence in various dopamine and noradrenaline nerve terminal systems of the forebrain and of the hypothalamus in the male unanaesthetized rat. H44/68 was given immediately before the neurotensin injection in a dose of 250 mg/kg (i.p., 1 h before decapitation). Means ± S.E.M. are shown. The statistical analysis was carried out according to the Wilcoxon test, treatment vs. control (* $P < 0.05$; ** $P < 0.01$). In Fig. 18 the effects on dopamine turnover in nucleus accumbens and tuberculum olfactorium are shown. In Fig. 19 the effects on dopamine turnover in nucleus caudatus are shown and in Fig. 20 the effects on dopamine turnover in the median eminence and in the preoptic region are shown. Abbreviations: SEL, subependymal layer of the median eminence; MPZ, medial palisade zone of the median eminence; LPZ, lateral palisade zone of the median eminence; POM, nucleus preopticus medialis; POSC, nucleus preopticus suprachiasmaticus; POP, nucleus preopticus periventricularis.

Effects of neuropeptide Y on noradrenaline and adrenaline levels and turnover in the dorsal medulla

As seen in Fig. 21, NPY given intraventricularly into the male rat in a dose of 1.6 nmol does not change the disappearance rates of the noradrenaline and adrenaline stores in the dorsal medulla following dopamine-β-hydroxylase inhibition using

TABLE I

Effects of intraventricular injections of neurotensin in the presence or absence of tyrosine hydroxylase inhibition on the secretion of adenohypophyseal hormones

Treatment	Dose (nmol)	ACTH (pg/ml)	Prolactin (ng/ml)	TSH (ng/ml)	GH (ng/ml)	LH (ng/ml)
Saline	–	39 ± 9	22 ± 3	318 ± 48	20 ± 11	3.8 ± 0.8
Neurotensin	0.06	72 ± 19	* 5.7 ± 1	199 ± 41	18 ± 10	4.7 ± 0.9
Neurotensin	0.6	209 ± 25 **	<2.5 **	118 ± 8 *	103 ± 80	4.6 ± 0.5
Neurotensin	6	237 ± 26 **	<2.5 **	110 ± 9	2.2 ± 1.0	5.7 ± 0.7
H44/68	–	102 ± 17	113 ± 15	252 ± 38	2.0 ± 0.8	12 ± 4.9
Neurotensin + H44/68	0.06	124 ± 22	104 ± 20	201 ± 46 *	21 ± 13	5.4 ± 1.1 *
Neurotensin + H44/68	0.6	252 ± 36 **	88 ± 6	<100	11 ± 6.6	2.9 ± 0.13
Neurotensin + H44/68	6	180 ± 22	93 ± 10	151 ± 36	12 ± 9.6	6.5 ± 1.5

Neurotensin was given intraventricularly in doses of 0.06–6 nmol/rat to the unrestrained male rat. H44/68 (250 mg/kg, i.p., 1 h before killing) was given immediately before the injection of neurotensin (for technical details, see Fuxe et al., 1980). Means ± S.E.M. (n = 6). Statistical analysis was performed according to Wilcoxon test, treatment vs. controls. * $P < 0.05$; ** $P < 0.01$.

FLA 63 (25 mg/kg, intraperitoneally, 2 h before killing). In this area, NPY-LI exists in the adrenaline neurons and seems to be an adrenaline comodulator. Thus, NPY as a comodulator in adrenaline neurons does not seem capable of modulating adrenaline turnover and release.

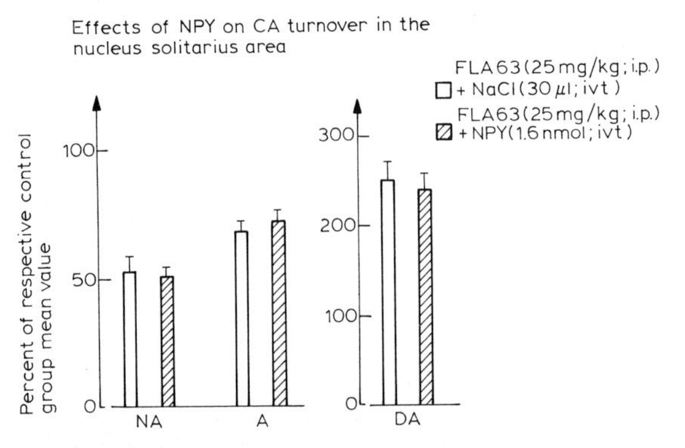

Fig. 21. Effects of intraventricular injections of NPY (1.6 nmol/rat) on catecholamine turnover in the nucleus solitarius area of the rat using the dopamine β-hydroxylase inhibitor FLA 63 (25 mg/kg, i.p., 2 h before killing). NPY was given immediately before FLA 63 injection. The area sampled for the evaluation of catecholamine contents was the medial part of the dorsal caudal medulla oblongata (see Fuxe et al., 1982). Means ± S.E.M. are shown. The evaluation of catecholamine content was carried out by means of HPLC in combination with electrochemical detection. NA, noradrenaline; A, adrenaline; DA, dopamine.

Effects of CCK-8, neurotensin and neuropeptide Y on central catecholamine receptor mechanisms in vitro

Effects of CCK-8 in vitro on [³H]spiperone and N-[³H]propylnorapomorphine binding sites in striatal membranes and in subcortical limbic membranes

CCK-8 (10^{-8} M) did not by itself displace [^{3}H]spiperone or *N*-[^{3}H]propylnorapomorphine (^{3}H-NPA) from their binding sites in striatal membranes and in subcortical limbic membranes. However, following preincubation for 10 min, CCK-8 (10^{-8} M) significantly reduced the affinity and increased the number of ^{3}H-NPA binding sites both in limbic membranes and striatal membranes (Fuxe et al., 1981; Agnati and Fuxe, 1983). The effects of CCK-8 on affinity were most marked within the subcortical limbic membranes (Figs. 22, 23), while a small increase in the number of binding sites was induced in both regions by CCK-8 (10 nM). These results are of substantial interest, since in the subcortical limbic forebrain CCK peptides may act, at least in part, as dopamine comodulators, while in striatum, CCK peptides may exist in neuronal systems independent from the dopamine nerve terminal systems. Thus, in the striatum the CCK re-

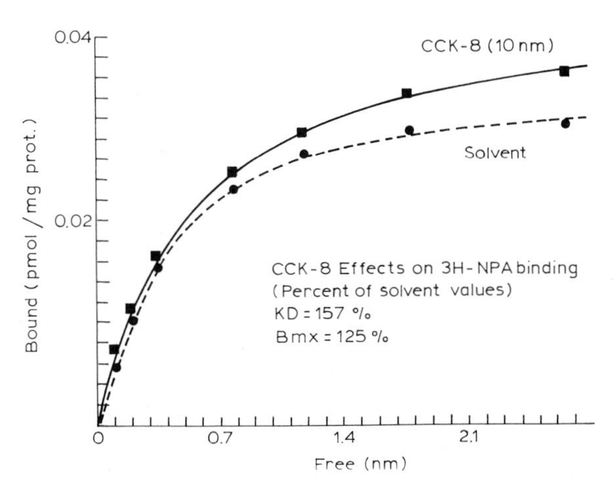

Fig. 22. Effects of CCK-8 (10 nM) on the characteristics of ^{3}H-NPA binding sites in the limbic areas (mainly tuberculum olfactorium and nucleus accumbens together with the medial basal forebrain area). A saturation analysis is shown (non-linear fitting analysis) illustrating the mean effects of CCK-8 on ^{3}H-NPA binding characteristics. Fourteen replications have been made.

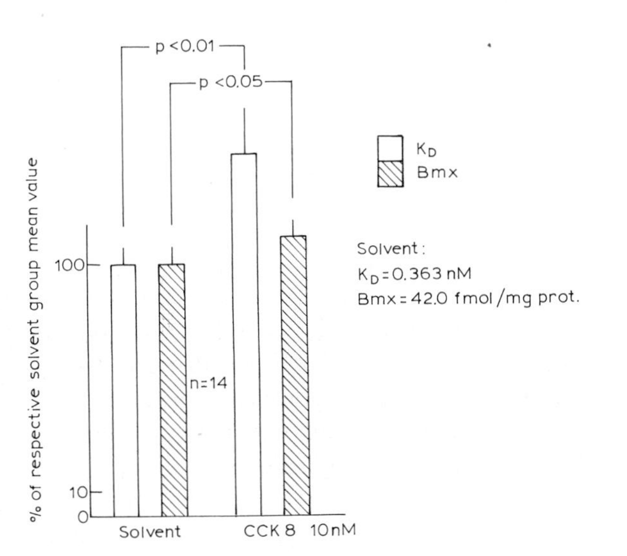

Fig. 23. The effects of CCK-8 (10 nM) on the binding characteristics of ^{3}H-NPA binding sites. The K_D (nM) and B_{max} (fmol/mg prot.) values are given in percent of the respective solvent group mean value. Means ± S.E.M. are shown from 14 experiments. The statistical analysis was carried out by means of Student's paired t-test.

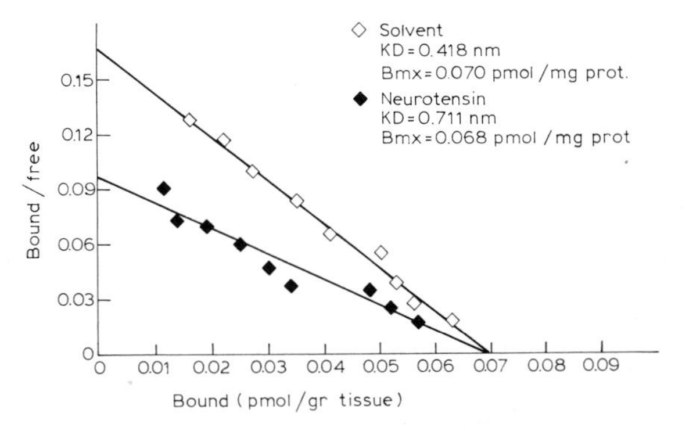

Fig. 24. Effects of neurotensin (10 nM) on the ^{3}H-NPA binding sites in membrane preparations from the subcortical limbic structures of rat brain (mainly nucleus accumbens, tuberculum olfactorium and medial basal forebrain area). ^{3}H-NPA binding assay has been carried out principally according to Fuxe et al. (1983b). The figure shows one out of eight replications. Neurotensin significantly increases K_D values for the ^{3}H-NPA binding sites ($P < 0.01$, Student's paired t-test).

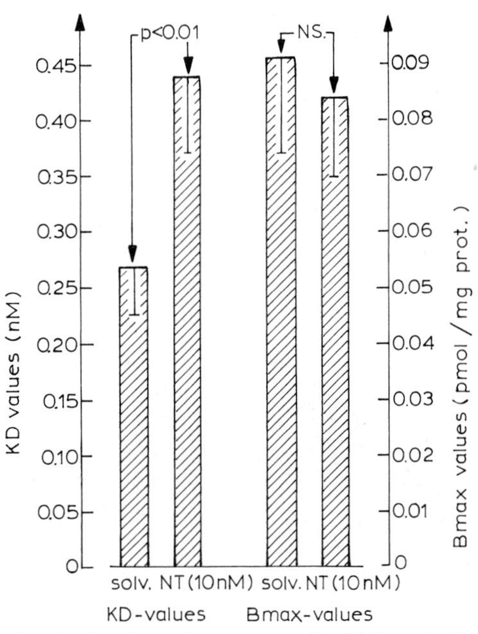

Fig. 25. The effects of neurotensin (10^{-8} M) on the binding characteristics of ^{3}H-NPA binding sites in subcortical limbic membranes. The K_D (nM) and the B_{max} (pmol/mg protein) values are shown on the left and right y-axis, respectively. Means ± S.E.M. from 8 replications were given. The statistical analysis was carried out by means of Student's paired t-test ($P < 0.01$).

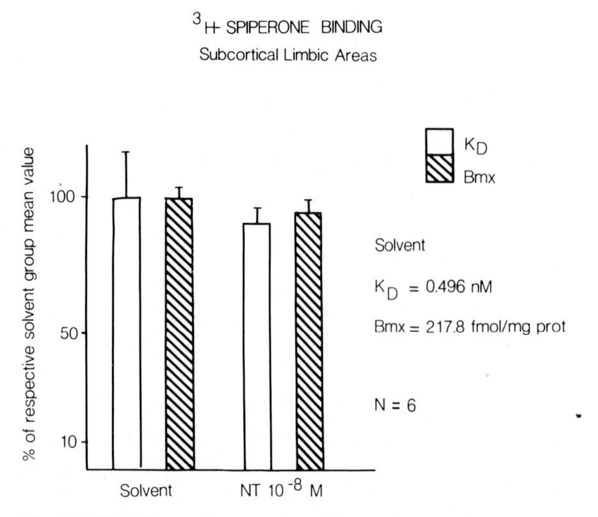

Fig. 26. The effects of neurotensin (10 nM) on the binding characteristics of [^{3}H]spiperone binding sites in subcortical limbic membranes. The 5-HT-2 receptors were blocked by the presence of ketanserin (10^{-7} M). The K_D (nM) and the B_{max} (pmol/mg protein) values are shown. Means ± S.E.M. from 6 replications are given. Statistical analysis was carried out by means of Student's paired t-test. No significant difference was noted.

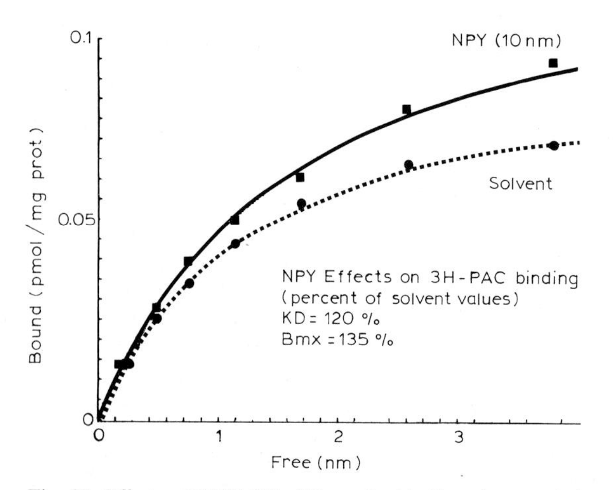

Fig. 27. Effects of NPY (10 nM) on the binding characteristics of ^{3}H-PAC (p-[^{3}H]aminoclonidine) binding sites in membranes of the rat medulla oblongata. For further details, see text to Fig. 22.

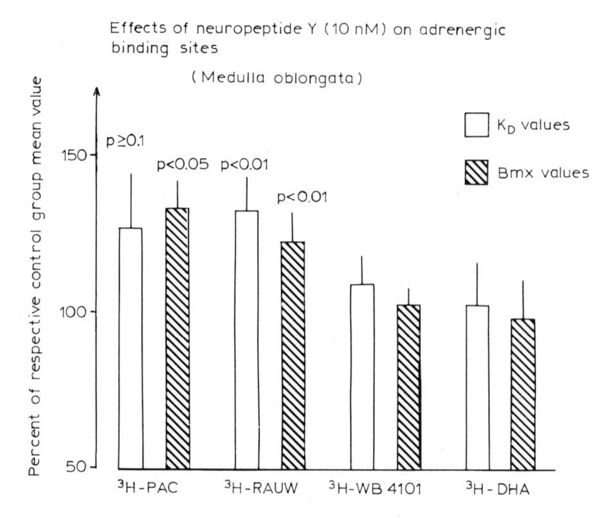

Fig. 28. A summary of the effects of NPY (10 nM) on the p-[^{3}H]aminoclonidine (^{3}H-PAC), [^{3}H]Rauwolscine (^{3}H-RAUW), [^{3}H]2-(2′-6′-dimethoxy-phenoxyethylamine)methyl-benzodioxan (^{3}H-WB4101) and [^{3}H]dihydroalprenolol (^{3}H-DHA) binding sites in membrane preparations of the medulla oblongata. For details on the binding procedures see for example, Agnati et al. (1983d) and Perry and U'Prichard (1981). Means ± S.E.M. are shown in percent of solvent group mean value. The mean values of eight replications observed for the control experiments (no NPY in the incubation medium) were: for ^{3}H-PAC, K_D = 1.28 nM, B_{max} = 0.109 pmol/mg prot.; for ^{3}H-RAUW, K_D = 4.15 nM, B_{max} = 0.93 pmol/mg prot.; for ^{3}H-WB4101, K_D = 0.26 nM, B_{max} = 0.108 pmol/mg prot.; for ^{3}H-DHA, K_D = 2.54 nM, B_{max} = 0.128 pmol/mg prot. The statistical analysis was carried out by means of Student's paired t-test.

ceptors may interact with dopamine receptors at the level of the local circuits of the striatum, while in the subcortical limbic forebrain the CCK- and dopamine receptors may interact with each other at the level of the postsynaptic membrane of the same dopamine synapse. In this case the interaction between the receptors may be more spatially locked, possibly explaining the more clearcut change in affinity seen when analyzing the limbic membranes after preincubation with CCK-8.

The ability of CCK-8 in vitro to modulate [^{3}H]spiperone-labelled dopamine receptors has so far been analyzed only in striatal membranes (Fuxe et al., 1981). In this case, CCK-8 (10^{-8} M) produced a small increase in affinity and a small reduction in the number of [^{3}H]spiperone-labelled dopamine receptors. Thus, CCK-8 can differentially regulate [^{3}H]dopamine antagonist and [^{3}H]dopamine agonist binding sites. Thus, it seems possible that CCK-8 favours the presence of high affinity dopamine agonist binding sites, although with a reduced affinity, while at the same time reducing the number of low affinity dopamine agonist binding sites, since the D2 dopamine receptor antagonist [^{3}H]spiperone labels both high and low affinity types of D2 dopamine receptors (see Seeman, 1980; Creese et al., 1982). Another explanation is that CCK-8 favours the presence of high affinity D1 agonist binding sites, since ^{3}H-NPA labels also these dopamine agonist sites (see Creese et al., 1982). Thus, CCK-8 may regulate a mechanism controlling isoreceptor interconversion (Agnati et al., 1983c).

Effects of neurotensin (10^{-8} M) on the binding characteristics of ^{3}H-NPA and [^{3}H]spiperone binding sites in subcortical limbic membranes

As seen in Figs. 24 and 25, NT can markedly reduce the affinity, but not the density, of the ^{3}H-NPA binding sites without affecting the [^{3}H]spiperone binding sites linked to dopamine receptors (Fig. 26) (see Agnati et al., 1983g). Thus, NT can also differentially regulate the binding characteristics of ^{3}H-NPA and [^{3}H]spiperone binding sites linked to dopamine receptors. In this case a marked and exclusive change was observed in the affinity of the [^{3}H]dopamine agonist binding sites. It should be noted that the change was particularly marked, suggesting that NT receptors in the subcortical limbic forebrain interact with the dopamine receptors at the local circuit level (see above), underlining again the importance of receptor–receptor interactions at the local circuit level. It seems possible that the marked reduction in affinity may be the result of an exclusive steric interaction between the NT receptors and the high affinity D2 and/or D1 dopamine agonist binding sites, since the D2 dopamine antagonist binding sites were not modulated by NT. These results are of special interest in view of the neuroleptic properties of NT (see Nemeroff et al., 1977; Nemeroff and Prange, 1982).

The effects of neuropeptide Y in vitro on the binding characteristics of α_2-adrenergic agonist and antagonist binding sites in membranes from the medulla oblongata

As seen in Figs. 27 and 28, NPY (10^{-8} M) increases the density and reduces the affinity of both [^{3}H]paraaminoclonidine (an α_2-adrenergic agonist) and [^{3}H]Rauwolscine (α_2-antagonist) binding sites in membranes from the rat medulla oblongata (see Agnati et al., 1983d). By contrast, the α_1-adrenergic antagonist binding sites (^{3}H-WB 4101) and the β-adrenergic antagonist binding sites ([^{3}H]dihydroalprenolol) were not modulated by NPY (10^{-8} M) (Agnati et al., 1983d). In agreement with these results, it has been shown that NPY (10^{-8} M) can reduce the ability of clonidine and adrenaline to displace [^{3}H]Rauwolscine from its binding sites in membranes of the medulla oblongata. Thus, an increase in the IC_{50} values for clonidine (4 fold) and adrenaline (2 fold) to displace [^{3}H]Rauwolscine was noted.

These results indicate the existence of a receptor–receptor interaction between NPY and α_2-adrenergic receptors. The interaction seems to involve both the ^{3}H-labelled α_2-agonist and the ^{3}H-labelled α_2-antagonist binding sites. NPY is a comodulator in the adrenaline nerve terminal systems of the medulla oblongata and therefore it seems likely that the NPY receptor–α_2-adrenergic receptor interaction takes place at the comodulator level in the adrenaline synapses of the medulla oblongata.

The concept of heterostatic regulation of monoamine synapses

Our previous (Agnati et al., 1983a,e; Fuxe et al., 1983b) and present results on the effects of neuropeptides on pre- and post-synaptic mechanisms in central catecholamine nerve cells have provided evidence for the existence, not only of a homeostatic, but also of a heterostatic regulation of synapses in the mammalian nervous system (Fig. 29). Thus, as seen from the present studies, CCK peptides, NT and NPY do not modulate amine turnover in those catecholamine neuron systems in which they act as comodulators. Thus, CCK-8 does not produce any change of dopamine turnover in the CCK-8 positive dopamine terminals in the tuberculum olfactorium and nucleus accumbens. NT does not modulate dopamine turnover in the tuberoinfundibular dopamine neurons innervating the lateral palisade zone. Finally, NPY does not change adrenaline turnover in the dorsal medulla. However, in spite of the lack of presynaptic changes in those catecholamine neuron systems, where the above peptides act as comodulators, NPY can still increase the density and reduce the affinity of α_2-adrenergic receptors in membranes of the medulla oblongata and CCK-8 can produce an increase in the number and affinity of [^{3}H]dopamine agonist binding sites in membranes from the subcortical limbic forebrain. Thus, the results indicate that comodulators in CA synapses can change the characteristics of catecholamine receptors (changes in both the affinity and density) without producing any compensatory changes in catecholamine levels and release. Thus, the peptidergic comodulators may effect a heterostatic regulation of monoamine transmission by modulating the decoding of the monoamine transmission line by increasing or decreasing the density of receptors or by increasing or decreasing their affinity. It may be suggested that the peptidergic comodulators have an important function in regulating the gain of monoamine transmission and on this basis we have introduced the concept of "heterostatic regulation" of synaptic transmission (Agnati and Fuxe, 1983; Agnati et al., 1983e). In particular, in the present case, the heterostatic regulation is controlled by peptide comodulators. Thus, the neuropeptide comodulators in monoamine synapses may represent, in part, regulatory transmission lines which via receptor–receptor interactions make possible a heterostatic regulation of the monoamine (main) transmission line at the postsynaptic level, also increasing in this way the functional plasticity of the monoamine synapse.

Possible abnormalities in the peptidergic comodulator transmission line in adrenaline synapses of spontaneous hypertensive rats: physiopathological aspects

NPY can reduce arterial blood pressure and respiratory rate and reduce desynchronized EEG activity in the Sprague–Dawley male rat and mimics in this way the action of the α_2-adrenergic agonist clonidine acting on postsynaptic α_2-adrenergic receptors

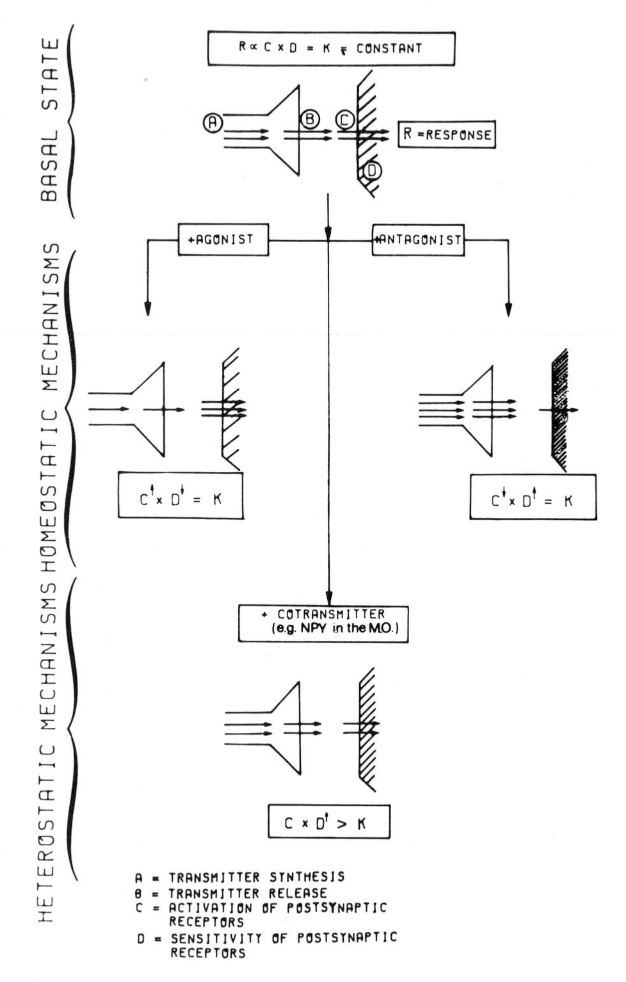

Fig. 29. Schematic representation of the hypothesis on the existence of "heterostatic mechanisms" at synaptic level. It should be stressed that these mechanisms by failing to trigger synaptic feedbacks can provide a further criterion for the distinction between main transmission lines (subserved by classical transmitters) and modulatory transmission lines (subserved, e.g., by some neuropeptides).

on the adrenaline synapses (Fuxe et al., 1974, 1980c, 1983c; Zini et al., 1983). We have suggested that these actions at the network level may be related to the ability of NPY to increase the number of α_2-adrenergic receptors (see above and Agnati et al., 1983d). In agreement, intracisternal injections of the α_2-adrenergic antagonist piperoxane can, in part, counteract the hypotensive effects of NPY (Fig. 30). However, in spontaneous hypertensive rats, the ability of NPY to reduce arterial blood pressure following intracisternal injection is significantly reduced (Härfstrand et al., 1986), and NPY no longer produces desynchronized activity but instead induces an opposite effect, i.e. it increases the amounts of desynchronized EEG activity (see Fig. 31). These abnormal effects induced by NPY at the network levels may be correlated with an inability

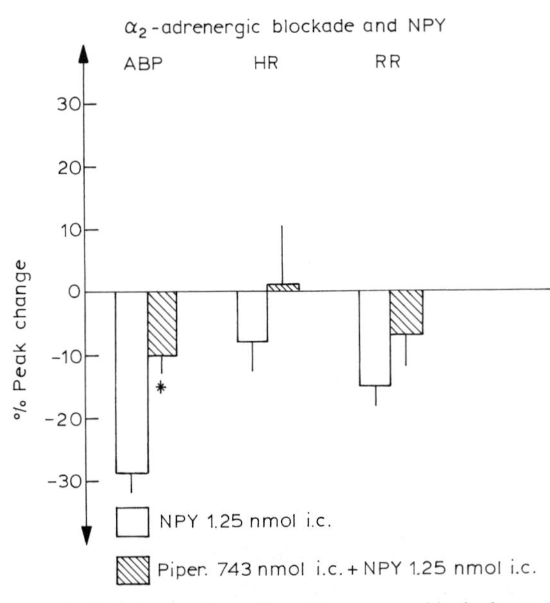

Fig. 30. The effects of α_2-adrenergic receptor blockade on the NPY induced lowering of arterial blood pressure in male rats under α-chloralose anaesthesia. Piperoxane was given intracisternally (740 nmol), 30 min before the intracisternal injection of NPY (1.25 nmol). At this time interval the initial hypertensive effects of piperoxane have disappeared and the basal arterial blood pressure is back to normal value. It is seen that the hypertensive effects of NPY are partly counteracted. The statistical analysis was carried out by means of Student's t-test. * P = 0.05. ABP, arterial blood pressure; HR, heart rate; RR, respiratory rate; piper., piperoxane.

of NPY in vitro to increase α_2-adrenergic receptors. In view of these results it may be speculated that one biochemical error in spontaneous hypertensive rats may be an abnormality in the receptor–receptor interactions at some adrenaline synapses taking place between NPY and α_2-adrenergic receptors, thus leading to a reduced ability of the vasodepressor adrenaline neurons to lower arterial blood pressure.

New openings in the treatment of disorders of the central nervous systems by drug action at the peptidergic comodulator transmission line in monoamine synapses

The synapse must now be considered to contain multiple transmission lines influencing each other via receptor–receptor interactions (Fig. 32). It represents a complex integrative electrometabolic unit allowing the passage of a large number of different messages (Fig. 33).

As outlined in Fig. 34, it seems possible that NPY fragments which have a certain selectivity for central NPY receptors and which can penetrate the blood brain barrier could represent possible new drugs for the treatment of hypertensive disorders in view of their ability to enhance α_2-adrenergic function. Such a fragment of NPY may also be devoid of the sedative effects of clonidine, since in the spontaneous hypertensive rat NPY increases the amounts of desynchronized EEG activity irrespective of the time of day. These results also indicate the possibility that such fragments might be of relevance to the treatment of sleep disorders.

However, it must also be emphasized that peptides acting via receptors which interact with the monoamine receptor at the local circuit level can be important new drugs for the modulation of monoamine synapses and thus for the treatment of disorders related to abnormalities in these systems. An example of a receptor–receptor interaction at the local circuit level is the one occurring between the NT and dopamine receptors within the subcortical limbic forebrain. Thus, NT can markedly reduce the affinity of [^{3}H]dopamine agonist binding sites

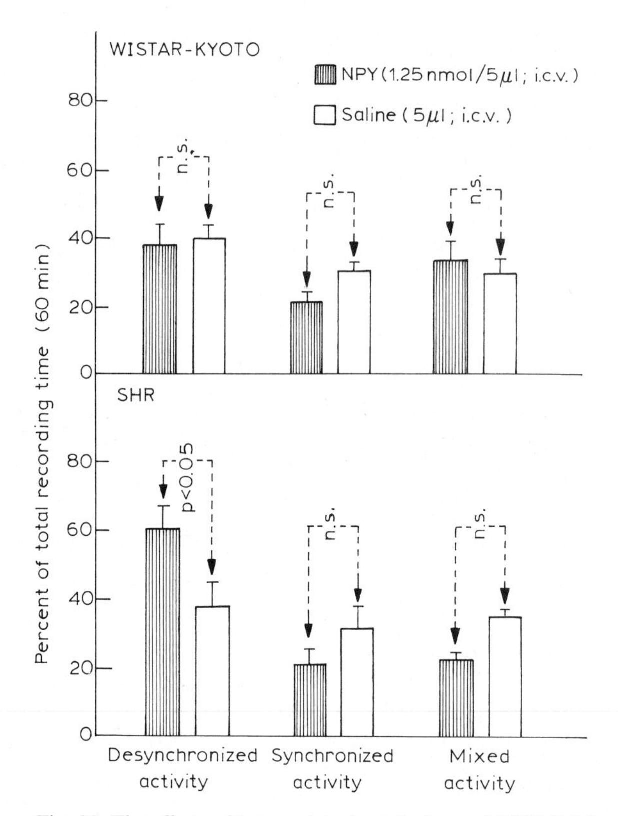

Fig. 31. The effects of intraventricular injections of NPY (1.25 nmol) on EEG activity in Wistar–Kyoto normotensive and spontaneous hypertensive rats. The effects on desynchronized, synchronized and mixed EEG activities are demonstrated. Total recording time was 60 min. Means ± S.E.M. from 8–10 replications are given. The statistical analysis was carried out by means of Student's *t*-test. Note the absence of effects in the Wistar–Kyoto normotensive control rats, while a significant increase in the desynchronized EEG activity is shown in the spontaneous hypertensive rats following central NPY administration.

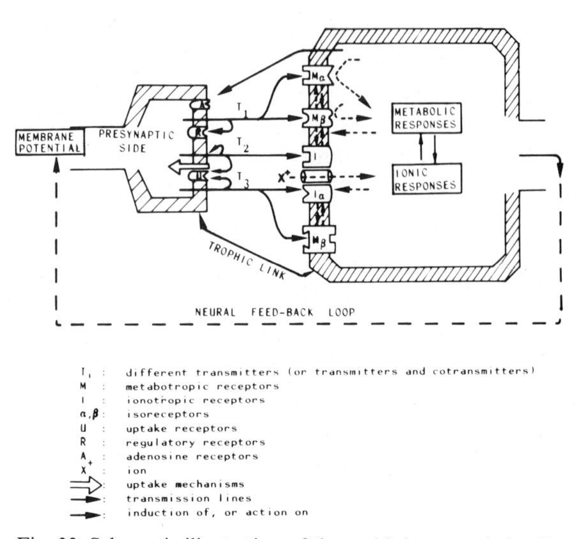

Fig. 32. Schematic illustration of the multiple transmission lines in central synapses. The presence of multiple transmitters and isoreceptors and receptor–receptor interactions are illustrated together with neural feed-back loops and trophic links. Furthermore, the interaction between metabolic responses and ionic responses and their feed-back effects on the receptors in the post-synaptic membrane are indicated.

in the subcortical limbic forebrain, changes which are associated with increases of dopamine turnover in the nucleus accumbens. In view of the reported neuroleptic-like properties of NT and the antipsychotic actions of dopamine receptor blocking drugs (see Cross et al., 1983), these results indicate the possibility that NT fragments, by regulating the affinity of dopamine receptors, could represent novel antipsychotic actions (see also Widerlöv et al., 1982b). The dopamine comodulator CCK-8 possesses similar properties in that it can substantially reduce the affinity of the [^{3}H]dopamine agonist binding sites in the subcortical forebrain. Further testing may lead to the discovery of other powerful peptides involved in the regulation of the decoding mechanisms at the D1 and D2 receptors of the limbic system. Such peptides or their fragments all have the potential of becoming new antipsychotic drugs. Finally, drug action at the peptidergic comodulation line in other types of transmitter-identified synapses may also lead to development of new psychoactive drugs (see Agnati et al., 1983h).

Acknowledgements

This work has been supported by grants (04X-715 and 14X-04246-10B) from the Swedish Medical Research Council, by a grant (MH 25504) from the

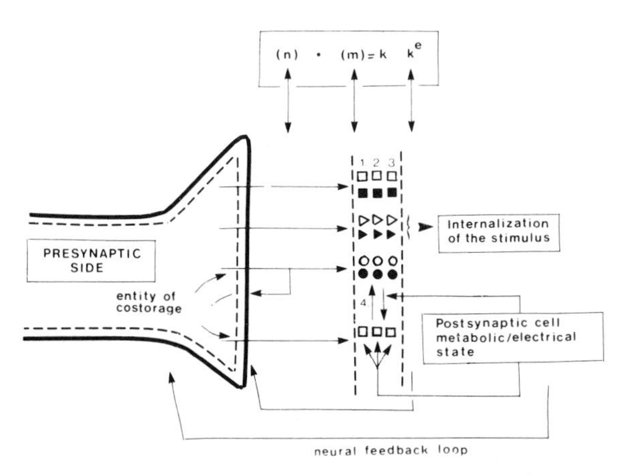

Fig. 33. Schematic illustration of the recognition site, coupling device and the catalytic unit building up the receptor complex (1,2,3) together with the receptor–receptor interactions (4). The number of possible messages reaching the postsynaptic cell is shown to be dependent on the number of transmitters and the number of isoreceptors (m). Thus, the number of messages is $k = n \times m$. The number of these messages (k) can then be further increased and amplified by the presence of receptor–receptor interactions (k^e). It is also indicated that the postsynaptic cell via its metabolic-electrical state can modulate the three components of the receptor complex as well as the receptor–receptor interactions.

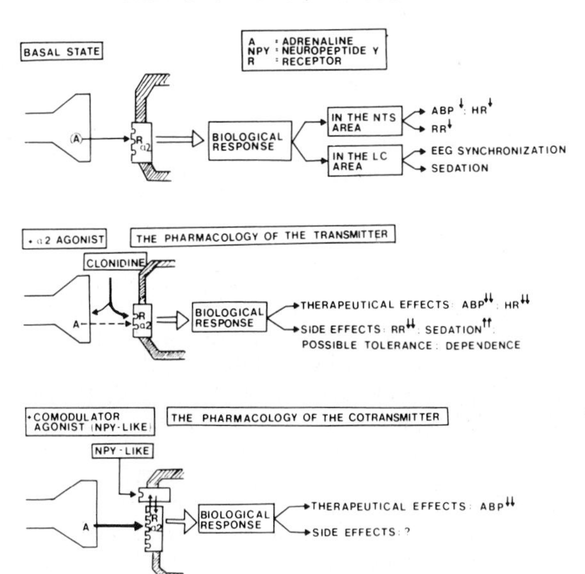

Fig. 34. Schematic illustration of how NPY can modulate the α_2-adrenergic transmission line in the dorsal medulla. It is indicated that NPY-like peptides may produce a therapeutical effect by lowering arterial blood pressure without inducing as many side effects as the α_2-adrenergic agonist clonidine (marked sedation, possible development of tolerance and dependence).

NIH, by a grant from Magnus Bergvall's Stiftelse, by a grant from Knut and Alice Wallenberg's Foundation, and by a CNR international grant. We are grateful to Miss Birgitta Johansson and Mrs. Ulla-Britt Finnman for excellent laboratory assistance. We are also grateful to Mrs. Lena Sunnås for her excellent secretarial assistance.

References

Agnati, L. F. and Fuxe, K. (1983) Subcortical limbic ^{3}H-*N*-propylnorapomorphine binding sites are markedly modulated by cholecystokinin-8 in vitro. *Biosci. Rep.*, 3: 1101–1105.

Agnat, L. F., Fuxe, K., Locatelli, V., Benfenati, F., Zini, I., Panerai, A. E., El Etreby, M. F. and Hökfelt, T. (1982a) Neuroanatomical methods for the quantitative evaluation of coexistence of transmitters in nerve cells. Analysis of the ACTH- and beta-endorphin immunoreactive nerve cell bodies of the mediobasal hypothalamus of the rat. *J. Neurosci. Methods*, 5: 203–214.

Agnati, L. F., Fuxe, K. Zini, I., Benfenati, F., Hökfelt, T. and De Mey, J. (1982b) Principles for the morphological characterization of transmitter identified nerve cell groups. *J. Neurosci. Methods*, 6: 157–167.

Agnati, L. F., Fuxe, K., Calza, L., Hökfelt, T., Johansson, O., Benfenati, F. and Goldstein, M. (1982c) A morphometric analysis of transmitter identified dendrites and nerve terminals. *Brain Res. Bull.*, 9: 53–60.

Agnati, L. F., Fuxe, K., Hökfelt, T., Benfenati, F., Calza, L., Johansson, O. and De Mey, J. (1982d) Morphometric characterization of transmitter-identified nerve cell groups. Analysis of mesencephalic 5-HT nerve cell bodies. *Brain Res. Bull.*, 9: 45–51.

Agnati, L. F., Fuxe, K., Benfenati, F., Calza, L., Battistini, N. and Ögren, S.-O. (1983a) Receptor–receptor interactions: Possible new mechanisms for the action of some antidepressant drugs. In *Frontiers in Neuropsychiatric Research, CINP, Satellite symposium, Corfu, Greece, June 28–30*, MacMillan, New York.

Agnati, L. F., Fuxe, K., Benfenati, F., Celani, M. F., Battistini, N., Mutt, V., Cavicchioli, L., Galli, G. and Hökfelt, T. (1983b) Differential modulation by CCK-8 and CCK-4 of ^{3}H-spiperone binding sites linked to dopamine and 5-hydroxytryptamine receptors in the brain of the rat. *Neurosci. Lett.*, 35: 179–183.

Agnati, L. F., Celani, M. F. and Fuxe, K. (1983c) Cholecystokinin peptides in vitro modulate the characteristics of the striatal ^{3}H-*N*-propylnorapomorphine sites. *Acta Physiol. Scand.*, 118: 79–81.

Agnati, L. F., Fuxe, K., Benfenati, F., Battisini, N., Härfstrand, A., Tatemoto, K. Hökfelt, T. and Mutt, V. (1983d) Neuropeptide Y in vitro selectively increases the number of α_2-adrenergic binding sites in membranes of the medulla oblongata of the rat. *Acta Physiol. Scand.*, 118: 293–295.

Agnati, L. F., Fuxe, K., Benfenati, F., Battistini, N., Zini, I., Camurri, M. and Hökfelt, T. (1983e) Postsynaptic effects of neuropeptide comodulators at central monoamine synapses. In E. Usdiu, A. Carlsson, A. Dahlström and J. Engel (Eds.), *5th International Catecholamine Symposium, June 12–16, Göteborg, Sweden,* Scientific, Medical and Scholarly Publications, Alan R. Liss Inc., New York.

Agnati, L. F., Fuxe, K., Benfenati, F., Battistini, N., Härfstrand, A., Hökfelt, T., Tatemoto, K. and Mutt, V. (1983f) Failure of neuropeptide Y in vitro to increase the number of α_2-adrenergic binding sites in membranes of medulla oblongata of the spontaneous hypertensive rat. *Acta Physiol. Scand.*

Agnati, L. F., Fuxe, K., Benfenati, F. and Battistini, N. (1983g) Neurotensin in vitro markedly reduces the affinity in subcortical limbic ^{3}H-*N*-propylnorapomorphine binding sites. *Acta Physiol. Scand.* (in press).

Agnati, L. F., Fuxe, K., Zini, I., Giardino, L., Zoli, M., Benfenati, F., Calza, L., Toffano, G., Toni, R. and Seyfried, C. A. (1983h) Possible therapeutical approaches for parkinsonian disease: New mechanisms for the modulation of dopamine receptor sensitivity and new strategies for dopaminergic reinnervation of striatum. In A. Agnoli and G. Bertolani (Eds.), *L.I.M.P.E. Conference in Pavia, October 28, 1983.*

Andén, N.-W., Dahlström, A., Fuxe, K., Larsson, K., Olson, L. and Ungerstedt, U. (1966) Ascending monoamine neurons to the telencephalon and diencephalon. *Acta Physiol. Scand.*, 67: 313–326.

Andersson, K., Fuxe, K., Eneroth, P., Agnati, L. F. and Locatelli, V. (1980) Hypothalamic dopamine and noradrenaline nerve terminal systems and their reactivity to changes in pituitary-thyroid and pituitary-adrenal activity and to prolactin. In F. Brambilla, G. Racagni and D. de Wied (Eds.), *Progress in Psychoneuroendocrinology,* Elsevier/North-Holland Biomedical Press, Amsterdam, pp. 395–406.

Andersson, K., Fuxe, K., Eneroth, P., Nyberg, P. and Roos, P. (1981) Rat prolactin and hypothalamic catecholamine nerve terminal systems. Evidence for rapid and discrete increases in dopamine and noradrenaline turnover in the hypophysectomized male rat. *Eur. J. Pharmacol.*, 76: 261–265.

Creese, I., Morrow, A. L., Leff, S. E., Sibley, D. R. and Hamblin, M. W. (1982) Dopamine receptors in the central nervous system. In J. R. Smythies and R. J. Bradley (Eds.), *International Review of Neurobiology, Vol. 23,* Academic Press Inc., New York, pp. 255–301.

Cross, A. J., Crow, T. J., Ferrier, I. N., Johnstone, E. C., McCreadie, R. M., Owen, F., Owens, D. G. C. and Poulter, M. (1983) Dopamine receptor changes in schizophrenia in relation to the disease process and movement disorder. *J. Neural Transm.*, Suppl., 18: 265–272.

Dahlström, A. and Fuxe, K. (1964) Evidence of the existence of monoamine containing neurons in the central nervous system. I. Demonstration of monoamines in the cell bodies of brain stem neurons. *Acta Physiol. Scand.*, Suppl. 232, 62: 1–55.

Emson, P. C., Goedert, M., Horsfield, P., Rioux, F. and St. Pierre, S. (1982) The regional distribution and chromatographic characterisation of neurotensin-like immunoreactivity in the rat central nervous system. *J. Neurochem.*, 38: 992–999.

Everitt, B. J., Hökfelt, T., Tatemoto, K., Mutt, V. and Goldstein, M. (1983) Differential coexistence of neuropeptide Y (NPY)-like immunoreactivity with catecholamines in the central nervous system of the rat. *Neuroscience* (in press).

Frohman, L. A., Maeda, K., Berelowitz, M., Szabo, M. and Thominet, J. (1982) Effects of neurotensin on hypothalamic and pituitary hormone secretion. *Ann. N.Y. Acad. Sci.*, 400: 172–182.

Fuxe, K. (1965a) Evidence for the existence of monoamine neurons in the central nervous system. III. The monoamine nerve terminal. *Z. Zellforsch.*, 65: 573–596.

Fuxe, K. (1965b) Evidence for the existence of monoamine neurons in the central nervous system. IV. Distribution of monoamine nerve terminals in the central nervous system. *Acta Physiol. Scand.*, Suppl. 247, 64: 39–85.

Fuxe, K., Hökfelt, T. and Ungerstedt, U. (1970) Morphological and functional aspects of central monoamine neurons. In J. Smythies and R. Bradley (Eds.), *International Review of Neurobiology, Vol. 13,* Academic Press, New York, pp. 93–126.

Fuxe, K., Lidbrink, P., Hökfelt, T., Bolme, P. and Goldstein, M. (1974) Effects of piperoxane on sleep and waking in the rat. Evidence for increased waking by inhibitory adrenaline receptor on the locus coeruleus. *Acta Physiol. Scand.*, 91: 566–567.

Fuxe, K., Hökfelt, T., Andersson, K., Ferland, L., Johansson, O., Ganten, D., Eneroth, P., Gustafsson, J.-Å., Skett, P., Said, S. I. and Mutt, V. (1978) The transmitters of the hypothalamus. In B. Cox, I. D. Morris and A. H. Weston (Eds.), *Pharmacology of the Hypothalamus,* MacMillan, London, pp. 31–59.

Fuxe, K., Andersson, K., Schwarcz, R., Agnati, L. F., Pérez de la Mora, M., Hökfelt, T., Goldstein, M., Ferland, L., Possani, L. and Tapia, R. (1979) Studies on different types of dopamine nerve terminals in the forebrain and their possible interactions with hormones and with neurons containing GABA, glutamate and opioid peptides. In L. J. Poirier, T. L. Sourkes and P. J. Bédard (Eds.), *Advances in Neurology, Vol. 24,* Raven Press, New York, pp. 199–214.

Fuxe, K., Andersson, K., Locatelli, V., Agnati, L. F., Hökfelt, T., Skirboll, L. and Mutt, V. (1980a) Cholecystokinin peptides produce marked reduction of dopamine turnover in discrete areas in the rat brain following intraventricular injection. *Eur. J. Pharmacol.*, 67: 329–331.

Fuxe, K., Agnati, L. F., Andersson, K., Locatelli, V., Eneroth, P., Hökfelt, T., Mutt, V., McDonald, T., El Etreby, M. F., Zini, I. and Calza, L. (1980b) Concepts in neuroendocrinology with emphasis on neuropeptide–monoamine interactions in neuroendocrine regulation. In F. Brambilla, G. Racagni and D. de Wied (Eds.), *Progress in Psychoneuroendocrinology,* Elsevier/North-Holland Biomedical Press, Amsterdam, pp. 47–61.

Fuxe, K., Bolme, P., Agnati, L. F., Jonsson, G., Andersson, K., Köhler, C. and Hökfelt, T. (1980c) On the role of central adrenaline neurons in central cardiovascular regulation. In K. Fuxe, M. Goldstein, B. Hökfelt and T. Hökfelt (Eds.), *Central Adrenaline Neurons: Basic Aspects and Their Role in Cardiovascular Functions, Wenner-Gren International Symposium Series, Vol. 33,* Pergamon Press, Oxford, pp. 161–182.

Fuxe, K., Agnati, L. F., Benfenati, F., Cimmino, M., Alger, S., Hökfelt, T. and Mutt, V. (1981) Modulation by cholecystokinins of ^{3}H-spiroperidol binding in rat striatum: Evidence for increased affinity and reduction in the number of binding sites. *Acta Physiol. Scand.,* 113: 567–569.

Fuxe, K., Andersson, K., Agnati, L. F., Eneroth, P., Locatelli, V., Cavicchioli, L., Mascagni, F., Tatemoto, K. and Mutt, V. (1982) The influence of cholecystokinin peptides and PYY on the amine turnover in discrete hypothalamic dopamine and noradrenaline nerve terminal systems and possible relationship to neuroendocrine function. *INSERM,* 110: 65–86.

Fuxe, K., Agnati, L. F., Ögren, S.-O., Köhler, C., Calza, L., Benfenati, F., Goldstein, M., Andersson, K. and Eneroth, P. (1983a) The heterogeneity of the dopamine systems in relation to the actions of dopamine agonists. In A. Carlsson and J. L. G. Nilsson (Eds.), *Symposium on Dopamine Receptor Agonists,* Swedish Pharmaceutical Press, Stockholm, pp. 60–79.

Fuxe, K., Agnati, L. F., Benfenati, F., Celani, M. F., Zini, I., Zoli, M. and Mutt, V. (1983b) Evidence for the existence of receptor–receptor interactions in the central nervous system. Studies on the regulation of monoamine receptors by neuroleptics. *J. Neural Transm.,* 18: 165–179.

Fuxe, K., Agnati, L. F., Härfstrand, A., Zini, I., Tatemoto, K., Merlo Pich, E., Hökfelt, T., Mutt, V. and Terenius, L. (1983c) Central administration of neuropeptide Y induces hypotension bradypnea and EEG synchronization in the rat. *Acta Physiol. Scand.,* 118: 189–192.

Goldstein, M., Anagnoste, B., Freedman, L. S., Roffman, M., Ebstein, R. P., Park, D. H., Fuxe, K. and Hökfelt, T. (1973) Characterization, localization and regulation of catecholamine synthesizing enzymes. In E. Usdin and S. H. Snyder (Eds.), *Frontiers in Catecholamine Research,* Pergamon Press Inc., New York, pp. 69–81.

Goldstein, M., Lew, J. Y., Matsumoto, Y., Hökfelt, T. and Fuxe, K. (1978) Localization and function of PNMT in the central nervous system. In M. A. Lipton, A. DiMascio and K. F. Killam (Eds.), *Psychopharmacology,* Raven Press, New York, pp. 261–269.

Härfstrand, A., Fuxe, K., Agnati, L. F., Ganten, D., Tatemoto, K. and Mutt, V. (1986) The effects of central administration of neuropeptide Y on cardiovascular function in α-chloralose anaesthetized, spontaneously hypertensive rats of the Wistar-Kyoto strain. *Acta Physiol. Scand.* (in press).

Hökfelt, T., Fuxe, K., Goldstein, M. and Johansson, O. (1974) Immunohistochemical evidence for the existence of adrenaline neurons in the rat brain. *Brain Res.,* 66: 235–251.

Hökfelt, T., Ljungdahl, A., Steinbusch, H., Verhofstad, A., Nilsson, G., Brodin, E., Pernow, B. and Goldstein, M. (1978) Immunohistochemical evidence of substance P-like immunoreactivity in some 5-hydroxytryptamine-containing neurons in the rat central nervous system. *Neuroscience,* 3: 517–538.

Hökfelt, T., Goldstein, M., Fuxe, K., Johansson, O., Verhofstad, A., Steinbusch, H., Penke, B. and Vargas, J. (1980a) Histochemical identification of adrenaline containing cells with special reference to neurons. In K. Fuxe, M. Goldstein, B. Hökfelt and T. Hökfelt (Eds.), *Central Adrenaline Neurons. Basic Aspects and their Role in Cardiovascular Functions,* Pergamon Press, Oxford, pp. 19–47.

Hökfelt, T., Skirboll, L., Rehfeld, J. F., Goldstein, M., Markey, K. and Dann, O. (1980b) A subpopulation of mesencephalic dopamine neurons projecting to limbic areas contains a cholecystokinin-like peptide: Evidence from immunohistochemistry combined with retrograde tracing. *Neuroscience,* 5: 2093–2124.

Hökfelt, T., Everitt, B. J., Theodorsson-Norheim, E. and Goldstein, M. (1983a) Occurrence of neurotensin-like immunoreactivity in hypothalamic and medullary catecholamine neurons. *J. Comp. Neurol.* (in press).

Hökfelt, T., Everitt, B. J., Theodorsson-Norheim, E., Terenius, L., Tatemoto, K., Mutt, V. and Goldstein, M. (1983b) Neurotensin- and NPY-like immunoreactivities in central catecholamine neurons. In E. Usdin (Ed.), *5th Catecholamine Symposium,* Alan R. Liss Inc., New York (in press).

Hökfelt, T., Johansson, O., Fuxe, K. and Goldstein, M. (1983c) Catecholamine neurons — distribution and cellular localization as revealed by immunohistochemistry. In V. Trendelenburg and N. Weiner (Eds.), *Catecholamines II,* Springer-Verlag, Heidelberg (in press).

Hökfelt, T., Lundberg, J. M., Tatemoto, K., Mutt, V., Terenius, L., Polak, J., Bloom, S., Sasek, C., Elde, R. and Goldstein, M. (1983d) Neuropeptide Y (NPY)- and FMRFamide neuropeptide-like immunoreactivities in catecholamine neurons of the rat medulla oblongata. *Acta Physiol. Scand.,* 117: 315–318.

Hollander, M. and Wolfe, D. A. (1973) *Nonparametric Statistical Methods,* John Wiley, New York, pp. 124–126.

Jennes, L., Stumpf, W. E. and Kalivas, P. W. (1982) Neurotensin: Topographical distribution in rat brain by immunohistochemistry. *J. Comp. Neurol.,* 210: 211–224.

Kalia, M. and Sullivan, M. (1982) Brainstem projections of sensory and motor components of the vagus nerve in the rat. *J. Comp. Neurol.,* 211: 248–264.

Kalia, M., Fuxe, K., Hökfelt, T., Johansson, O., Lang, R. E., Ganten, D., Cuello, C. and Terenius, L. (1983) Distribution of neuropeptide immunoreactive nerve terminals within the subnuclei of the nucleus of the tractus solitarius of the rat. *J. Comp. Neurol.* (in press).

Maeda, K. and Frohman, L. A. (1978) Dissociation of systemic and central effects of neurotensin on the secretion of growth hormone, prolactin and thyrotropin. *Endocrinology,* 103: 1903–1909.

Makino, T., Yokokura, T. and Iisuka, R. (1978) Effect of neurotensin on pituitary gonadotropin release in vivo. *Endocrinol. Jpn.*, 25: 181–183.

Markey, K. A., Kondo, S., Shenkman, I. and Goldstein, M. (1980) Purification and characterization of tyrosine hydroxylase from a clonal phaeochromocytoma cell line. *Mol. Pharmacol.*, 17: 79–85.

Nemeroff, C. B. and Prange, A. J. (1982) Neurotensin, a brain and gastrointestinal peptide. *Ann. N.Y. Acad. Sci.*, 400: 1–430.

Nemeroff, C. B., Bissette, G., Prange, A. J., Loosen, P. T. and Lipton, M. A. (1977) Neurotensin: central nervous system effects of a hypothalamic peptide. *Brain Res.*, 128: 485–496.

Palacios, J. M. and Kuhar, M. J. (1981) Neurotensin receptors are located on dopamine-containing neurons in rat midbrain. *Nature*, 294: 587–589.

Perry, B. and U'Prichard, D. (1981) [^{3}H]Rauwolscine (α-Yohimbine) A specific antagonist radioligand for brain α_2-adrenergic receptors. *Eur. J. Pharmacol.*, 76: 461–464.

Rivier, C., Brown, M. and Vale, W. (1977) Effect of neurotensin, substance P and morphine sulfate on the secretion of prolactin and growth hormone in the rat. *Endocrinology*, 100: 751–754.

Seeman, P. (1980) *Pharmacological Reviews. Brain Dopamine Receptors*, William & Wilkins Co., Baltimore, pp. 229–313.

Skirboll, L., Grace, A. A., Hommer, D. W., Rehfeld, J., Goldstein, M., Hökfelt, T. and Bunney, B. S. (1981) Peptide-monoamine coexistence: studies of the actions of cholecystokinin-like peptide on the electrical activity of midbrain dopamine neurons. *Neuroscience*, 6: 2111–2124.

Sternberger, L. A. (1979) *Immunohistochemistry*, 2nd edn., Wiley and Sons, New York.

Tramu, G., Pillez, A. and Leonardelli, J. (1978) An efficient method of antibody elution for the successive or simultaneous location of two antigens by immunocytochemistry. *J. Histochem. Cytochem.*, 26: 322–324.

Uhl, G. R., Kuhar, M. J. and Snyder, S. H. (1977) Neurotensin: immunohistochemical localization in rat central nervous system. *Proc. Natl. Acad. Sci. U.S.A.*, 74: 4059–4063.

Uhl, G. R., Goodman, R. R. and Snyder, S. H. (1979) Neurotensin-containing cell bodies, fibers and nerve terminals in the brainstem of the rat: Immunohistochemical mapping. *Brain Res.*, 167: 77–91.

Vanderhaeghen, J. J., Lotstra, F., De Mey, J. and Gilles, C. (1980) Immunohistochemical localization of cholecystokinin- and gastrin-like peptides in the brain and hypophysis of the rat. *Proc. Natl. Acad. Sci. U.S.A.*, 77: 1190–1194.

Widerlöv, E., Kilts, C., Mailman, R. B., Nemeroff, C. B., Prange, A. J., Jr. and Breese, G. R. (1982a) Increase in dopamine metabolites in rat brain by neurotensin. *J. Pharmacol. Exp. Ther.*, 221: 1–6.

Widerlöv, E., Lindström, L. H., Besev, G., Manberg, P. J., Nemeroff, C. B., Breese, G. R., Kizer, J. S. and Prange, A. J., Jr. (1982b) Subnormal cerebrospinal fluid levels of neurotensin in a subgroup of schizophrenics: Normalization after neuroleptic treatment. *Am. J. Psychiatry*, 139: 1122–1126.

Young, W. S. III and Kuhar, M. J. (1981) Neurotensin receptor localization by light microscopic autoradiography in rat brain. *Brain Res.*, 206: 273–285.

Zini, I., Merlo Pich, E., Fuxe, K., Lenzi, P. and Agnati, L. F. (1984) Action of centrally administered NPY on EEG activity in different rat strains and in different phases of their circadian cycle. *Acta Physiol. Scand.*, 122: 71–77.

P. C. Emson, M. N. Rossor and M. Tohyama (Eds.),
Progress in Brain Research, Vol. 66.

CHAPTER 16

Clinical relevance of neuropeptide research

C. D. Marsden

University Department of Neurology, Institute of Psychiatry and King's College Hospital Medical School, De Crespigny Park, London SE5 8AF, U.K.

Introduction

The potential of neuropeptide research for the understanding and treatment of neurological and psychiatric disease is vast, but to date it has provided little practical impact on the cause or management of these disorders. This is not surprising, for the field is in its infancy. Three major problems need to be solved before the full clinical impact of the discovery of brain neuropeptides is realised. (a) The large number of peptides discovered so far poses conceptual problems. (b) Most information available on peptides in brain disease concerns steady state levels, which may give misleading information on turnover and functional effects. (c) There are few pharmacological tools available for use in man to manipulate cerebral peptide function. No doubt future research will rectify these gaps, but for the present I must review the data available today.

Each of the many peptides in the brain is in search of a disease, and there are many more neurological diseases of unknown cause than there are peptides! Yet this very abundance causes difficulties. The mathematical possibilities are staggering. Take 30 peptides (and there are more in the brain), which are distributed selectively in 30 different brain areas, and then look at 30 different diseases. A quick calculation indicates that it might be possible to examine 27 000 possible abnormalities of peptide organisation, based upon steady state levels alone. Stretching statistical analysis to make a point, by chance one might expect to pick up 270 possibly irrelevant observations of abnormal peptide levels at the 1% probability level by examination of human necropsy material.

The overwhelming majority of published results of relevance to human neurology (including schizophrenia as a neurological disease), have been descriptive. By this I mean measurement of various peptides in various brain regions, or illustration of morphological differences in peptide neurons, in different diseases. Such data are essential at this stage of development of the field, but one cannot help but reflect upon the potential sources of error in interpretation of such data. Early in the era of brain catecholamine research it became all too apparent that estimates of the steady state levels of these neurotransmitters in the brain often were a poor guide to their functional activity. Obviously there is an urgent need to develop techniques for estimating rates of peptide synthesis and utilisation. However, this is not enough, for the true functional effects of peptide release can only be assessed by reference to the state of the postsynaptic receptors on which they act. Even then, knowing both rates of release and postsynaptic peptide receptor sensitivity may not give an accurate guide to functional action. The latter may only be deduced by measurement of a specific physiological or biochemical effect of target organ activation.

Until such methods are developed we must rely upon the information about steady state levels of peptides in brain necropsy material, and what little is known about their concentrations in human cerebrospinal fluid, to indicate their involvement in specific neurological diseases. Even then there are problems. Attention is drawn to the difficulties of standard dissection of anatomically discrete areas of the human brain, to problems of expression of the results in terms of wet weight, tissue protein or some other marker to allow for shrinkage effects, and to the risks of attributing low levels of a particular peptide identified by an immunoreactive marker to a real loss of specific peptide, when this may only indicate a change in its nature and immunoreactivity. I have laboured these problems, for they must be borne in mind when interpreting published data.

The basal ganglia

The complexity of the organisation of the basal ganglia is now well known. The core unit of the strio-pallidal complex receives inputs to the striatum from the substantia nigra and adjacent ventral tegmental area (dopaminergic), from the cerebral cortex (glutamatergic), from the midline thalamic nuclei (transmitter not established), and to a limited extent from other monoaminergic brainstem nuclei (noradrenergic and serotonergic).

A clear distinction can be drawn between the dorsal striatum, with major inputs from the sensorimotor cortex and substantia nigra, and the ventral striatum, with inputs from the limbic cortex and the ventral tegmental area. The outputs from the dorsal striatum are directed, via the thalamus, to the premotor cortex, while those from the ventral striatum project, also via different regions of the thalamus, to the rest of the frontal cortex. Major outputs from the striatum to the pallidum and substantia nigra, and from the pallidum to the subthalamic nucleus, utilise gamma-aminobutyric acid (GABA).

To this basic scheme must now be added a range of basal ganglia peptidergic pathways. A substance P pathway projects from the striatum to the internal segment of the globus pallidus and to the substantia nigra. A Met-enkephalin and Leu-enkephalin pathway projects from the striatum to the external segment of the globus pallidus and to the substantia nigra. The distribution of these two enkephalin pathways is regionally similar, raising the possibility that they co-exist in the same neurons. An angiotensin-converting enzyme pathway projects from the striatum to both segments of the globus pallidus and to the substantia nigra. Cholecystokinin co-exists in some dopaminergic neurons, particularly those projecting to the ventral striatum. Interneurons in the striatum have been identified as containing not only acetylcholine and GABA, but also somatostatin, neuropeptide Y, vasoactive intestinal polypeptide, and neurotensin. Within the striatum itself there is a complex compartmental organisation of neurotransmitters based upon the weak acetylcholinesterase staining striosomes, which form complex three dimensional labyrinths. (See Tagaki, Emson, this volume.) The organisation of these various peptide systems has been shown to be disrupted in both Parkinson's disease and Huntington's disease.

Parkinson's disease (see Agid, this volume)

Substance P is decreased by 30–40% in the substantia nigra and the internal segment of the globus pallidus, but is normal in the striatum, ventral tegmental area and cerebral cortex. Met-enkephalin is reduced by 70% in the substantia nigra and ventral tegmental area, and by 50% in the putamen and external segment of the pallidum, but is normal in the caudate and cerebral cortex. Leu-enkephalin also is reduced in the putamen and pallidum, but not in the substantia nigra. Cholecystokinin is reduced in the substantia nigra, but not in the ventral tegmental area, striatum or cerebral cortex. Somatostatin is normal in the striatum and substantia nigra, and in the cerebral cortex of patients with intact intellect, but frontal cortical levels are reduced in demented Parkinsonians. Thyroid releasing hormone and vasopressin levels are normal.

What do these changes mean? Many may be explained as secondary to the primary loss of dopaminergic projections to the striatum. Thus, the reduction of nigral substance P and Met-enkephalin may reflect alterations in strio-nigral activity due to lack of normal striatal dopaminergic stimulation. However, it is not clear whether these reduced steady state levels of excitatory peptides indicate trans-synaptic degeneration or reduced activity of strio-nigral pathways, or increased utilisation. On the one hand, such changes could mean a reduced strio-nigral drive to nigral neurons, on the other hand they might signify the reverse. Both substance P and Met-enkephalin synapse with nigral dopamine neurons, and it is tempting to conclude that the changes observed in these peptides represent attempts by strio-nigral feed-back pathways to increase dopaminergic neurotransmission, but this remains to be proved. The preservation of so many striatal peptides, and of acetylcholine and GABA neurons, reinforces the belief that the striatal machinery is not seriously affected in Parkinson's disease. This has implications for understanding the reasons for loss of benefit during long-term levodopa therapy. However, there are no data available to compare peptide levels in those in whom treatment has failed as against those in whom it has continued to work.

The discovery that frontal cortical somatostatin levels are decreased in demented Parkinsonians adds to the understanding of the complication of prolonged Parkinson's disease. Cerebral cortical somatostatin has been found to be reduced in Alzheimer's disease (see below). In both the demented Parkinsonian and in Alzheimer's disease there also is loss of ascending acetylcholine pathways from the substantia innominata to the cerebral cortex. Whether the cortical somatostatin changes observed in these two diseases are secondary to the cholinergic deficit, or due to some other local cortical pathology remains to be established.

Do these peptide changes in Parkinson's disease have implications for treatment? Unfortunately, not yet. Attempts to manipulate brain enkephalin action in Parkinson's disease, using either specific opiate antagonists or synthetic agonists, have had very little effect on the symptoms of the disease.

Huntington's disease (see Emson, this volume)

The massive destruction of striatal neurons that is characteristic of Huntington's disease (maximal in the caudate and putamen, with relative sparing of the ventral striatum) is accompanied by the expected changes in basal ganglia peptides. Thus, there is a massive loss of substance P, enkephalins, and angiotensin converting enzyme, both in the striatum itself and in those areas to which these pathways project. However, concentrations of other striatal peptides, including somatostatin, neuropeptide Y, neurotensin and vasoactive intestinal polypeptide, are preserved or even increased. The increase in striatal somatostatin levels probably is not due to shrinkage, for concentrations in the spared nucleus accumbens are still raised two-fold. Nor are they due to neuroleptic drug treatment, for experimental chronic haloperidol therapy reduced striatal concentrations.

These data lead to some useful conclusions as to the pathogenesis of Huntington's disease. First, whatever destroys most of the intrinsic neurons of the striatum spares the small number (some 2% of striatal neurons) of medium-sized aspiny cells that contain somatostatin. The brunt of the pathological attack on the striatum in Huntington's disease falls on the spiny neurons of the caudate and putamen. Second, the kainic acid model of Huntington's disease is found wanting, for kainic acid injected into the striatum decreases somatostatin concentrations by about 50%. Third, cortical somatostatin levels are not reduced in Huntington's disease, despite the intellectual and personality changes that occur in the illness. The cellular basis of the cortical atrophy of Huntington's disease remains to be established.

Cerebral cortex

Most neurons in the cerebral cortex appear to contain GABA. The identity of the neurotransmitter in large pyramidal output cells is not established for certain, although glutamate and aspartate are candidates. The cortex contains few cholinergic bipolar neurons. Four peptides have been characterised in cortical non-pyramidal intrinsic neurons, namely somatostatin, vasoactive intestinal polypeptide, cholecystokinin and neuropeptide Y. However, these four peptides account for only about 8% of cortical neurons or about 20% of non-pyramidal cells in animal studies. Studies on the rat visual cortex (see Parnavelas, this volume) have identified the localisation of these four cortical peptides, which are confined to specific types of cells and exhibit regional differences in their laminar distribution. Somatostatin and neuropeptide Y may co-exist in the same neurons. The cortex also contains neurotensin, which has been found in a few intrinsic neurons, and substance P, which appears in cortical fibres, but the exact distribution of these and other peptides has not been established.

Alzheimer's disease (see Rossor et al., this volume)

The major identified neurotransmitter abnormalities in Alzheimer's disease have involved subcortical projection systems to the cerebral cortex. Thus, there is considerable loss of cholinergic projections from the substantia innominata and septal nuclei, particularly to the temporal cortex, more than the frontal cortex. In addition, there is loss of noradrenergic pathways from the locus coeruleus to the cortex. Whether there is also loss of dopaminergic mesocortical pathways, or of dopamine striatal systems, is debatable. Some subcortical cholinergic mechanisms, e.g. putamen, appear spared.

As far as peptides are concerned, most subcortical peptide systems appear intact. Thus, levels of vasopressin, oxytocin, neurotensin, thyroid releasing hormone, substance P and enkephalin are normal, although peptides are altered in the amygdala which shows marked histopathological changes. Cortical peptides, however, are affected. There is a considerable loss of cortical somatostatin, although cortical levels of cholecystokinin, vasoactive intestinal polypeptide and neuropeptide Y are normal.

These data bear on the pathogenesis of the disease. The characteristic senile plaque is rich in acetylcholinesterase staining, so may consist of degenerating cholinergic terminals. Indeed, the severity of cortical cholinergic loss correlates with the number of cortical senile plaques. However, there is also a loss of cortical neurons in Alzheimer's disease, and many remaining cells contain neurofibrillary tangles. Some of the cortical neurons lost may be GABAergic, for cortical GABA is reduced by some 20% in the temporal area. Others are likely to contain somatostatin. Thus, there is now unequivocal evidence of selective loss of some cortical neurons in Alzheimer's disease. Whether the additional loss of projection fibres from some subcortical sites is secondary to the cortical pathology, or results from primary damage to their cell bodies, remains to be discovered. It is known that lesions of the nucleus basalis in experimental animals, sufficient to decrease cortical acetylcholine considerably, do not decrease cortical somatostatin; this suggests that the subcortical pathology in Alzheimer's disease is not the cause of the cortical changes, even though the severity of dementia appears to correlate with the degree of cortical choline acetyltransferase loss. The severity of dementia also correlates with the degree of loss of cortical somatostatin. Indeed, the presence of dementia, irrespective of age, seems best related to the loss of cortical acetylcholine and somatostatin. Younger dements (aged 79 years or less) also have added GABA and noradrenaline deficits.

Do these data help to predict likely means of treatment of Alzheimer's disease? So far, attempts to overcome the cholinergic deficiency of the illness, by use of precursers such as choline or lecithin, or by the use of direct acetylcholine agonists, have not met with practical success. An obvious avenue would be to combine such therapy with an agent aimed at restoring cortical somatostatin function, an approach that may also be of benefit in the demented Parkinsonian. However, drugs to manipulate brain somatostatin are not yet available.

Schizophrenia (see Bissette et al., this volume)

The biochemical pharmacology of schizophrenia has been dominated by the "dopamine hypothesis", but it has been difficult to provide concrete support for this notion. Much, if not all of the evidence from biochemical study of the schizophrenic brain has been contaminated by the effects of chronic drug intake in such patients, which may also colour the results of peptide studies in the field. For example, changes have been found in the Met-enkephalin content of the amygdala in early onset cases, and the amygdala has been the focus of recent attention as a specific site for abnormal dopamine metabolism. Decreases of thyrotropin releasing hormone and somatostatin in cerebral cortex have also been described. However, it is obvious that much more work is required before the significance of these and other abnormalities can be assessed.

Spinal cord

There is considerable interest in the arrangement of peptides in the spinal cord because of their possible relevance to pain and spasticity. Unmyelinated afferent C fibres terminate in the upper two laminae of the dorsal horn. They contain substance P, somatostatin, and cholecystokinin. Eighty percent of the content of these peptides in the dorsal horn disappears on sectioning of the dorsal roots, so the remainder must lie in local neurons or in descending fibres. Other peptides in the same region, such as neurotensin, enkephalin and oxytocin are not reduced by dorsal root section, so must also be contained within intrinsic neurons or fibres. Large myelinated fibres do not appear to contain peptides (see elsewhere this volume).

The animal work implicating release of substance P from myelinated C afferents following painful stimuli has provided a valuable basis for the development of novel analgesics aimed at the inhibition of substance P. Similar studies have provided an explanation for the action of opioids at a spinal level via interaction with opiate receptors located pre-synaptically on C fibre afferents.

Human studies of neuropeptide changes in spinal cord disease have been limited and there have been very few examinations of pain states. Peripheral nerve damage in animals results in a depletion of substance P in the dorsal horn and a similar change can be seen following limb amputation in man, with changes persisting for up to two years. Such a depletion could reflect a persistent increase in turnover, an interpretation which could provide a correlate of chronic pain, or a reduction in synthesis. An intriguing speculation is that there may be stable structural alterations and remodelling in response to peripheral injury and that such stable central alterations in peptide circuitry may relate to the chronic pain syndrome that may occur with amputation.

Therapy

If a particular neuropeptide were found to be associated consistently with a given disease and to relate to the functional deficit, what are the possibilities of therapeutic intervention? Neuropeptides possess features pharmacologically which distinguish them from amino acid transmitters, but are similar to catecholamines. The effects of amino acids are very rapid and are involved in point-to-point circuitry. Certain actions of acetylcholine and monoamines are longer and they may not share the precise circuitry of amino acids. For example, some estimates indicate that less than 10% of noradrenergic terminals in cortex make specialised synaptic contact.

Some peptides, such as cholecystokinin, have fast actions but the majority have actions over a much longer time course. A good model for examining peptide action is the sympathetic ganglion of the bullfrog where LHRH is found as a co-existing transmitter (Jan and Jan, 1982). The late excitatory post-synaptic potential which appears due to LHRH extends over 1–2 min and the peptide itself probably diffuses over distances of 10–20 μm from site of release to the site of action.

If a common feature of peptides is to act over long distances, and for long periods of time, to alter the gain in point-to-point circuitry, then replacement therapy (assuming one can bypass the blood brain barrier) may not be an unreasonable expectation.

Reference

Jan, L. Y. and Jan, Y. N. (1982) *J. Physiol.*, 327: 219–246.

Subject Index